Advances in Experimental Medicine and Biology

Volume 851

For further volumes:
http://www.springer.com/series/5584

Advances in Experimental Medicine and Biology

Eugene G. Hrycay • Stelvio M. Bandiera
Editors

Monooxygenase, Peroxidase and Peroxygenase Properties and Mechanisms of Cytochrome P450

Springer

Editors
Eugene G. Hrycay
Faculty of Pharmaceutical Sciences
The University of British Columbia
Vancouver, British Columbia, Canada

Stelvio M. Bandiera
Faculty of Pharmaceutical Sciences
The University of British Columbia
Vancouver, British Columbia, Canada

ISSN 0065-2598 ISSN 2214-8019 (electronic)
Advances in Experimental Medicine and Biology
ISBN 978-3-319-16008-5 ISBN 978-3-319-16009-2 (eBook)
DOI 10.1007/978-3-319-16009-2

Library of Congress Control Number: 2015938603

Springer Cham Heidelberg New York Dordrecht London

Printed on acid-free paper

Springer International Publishing AG Switzerland is part of Springer Science+Business Media (www.springer.com)

Foreword

The cytochrome P450 (CYP) superfamily is one of the most widespread and diverse enzyme systems in nature. CYP enzymes are widely distributed throughout the four biological kingdoms (archaeal, bacterial, eukaryotic and viral) and catalyze a variety of complex oxidative reactions. The most common oxidative reaction performed by CYP enzymes is the monooxygenation of an organic compound by the insertion of an oxygen atom from molecular oxygen. CYP-mediated activation of O_2 by reductive cleavage and concomitant hydroxylation of unactivated substrate C–H bonds is one of the most important processes in nature. Monooxygenation reactions catalyzed by CYP enzymes include aliphatic and aromatic hydroxylation, *N*-hydroxylation, oxygenation of heteroatoms (N, S, P, I), alkene and arene epoxidation, dehalogenation, deamination and *N*-, *O*- and *S*-dealkylation. Understanding the mechanisms of NAD(P)H/O_2-supported monooxygenation reactions and of oxygen atom transfer to substrates by the CYP transitory compound I (Cpd I) species, or by other iron-oxygen intermediates of the CYP catalytic cycle, has been a challenging aspect of CYP biochemistry. This volume focuses on the monooxygenase, peroxidase and peroxygenase properties and mechanisms of CYP enzymes and presents an account of these captivating subjects in 13 stimulating chapters compiled by internationally acclaimed CYP researchers. We hope that this fine interdisciplinary work will become a useful reference book to CYP researchers, professors and students from a variety of scientific disciplines (e.g. pharmacology, toxicology, medical science, microbiology, biochemistry, biochemical engineering, chemistry).

The introductory Chap. 1 discusses the mechanisms by which CYP enzymes catalyze monooxygenase reactions utilizing their peroxidase and peroxygenase functions and the porphyrin π radical ferryl Cpd I species ($Por^{\bullet+}Fe^{IV}{=}O$), or its ferryl radical resonance form ($Fe^{IV}{-}O^{\bullet}$), as the primary oxygenating intermediate. The chapter also describes how the ferric-peroxo anion ($Fe^{III}{-}OO^{-}$) and ferric-hydroperoxo ($Fe^{III}{-}OO^{-}$) intermediates, which form the molecular stations of the CYP catalytic cycle, act as oxygenating species. Also outlined is how CYP enzymes use peroxides, peracids, perborate, percarbonate, periodate, chlorite, *N*-oxides and iodosobenzene as oxygen atom donors to oxygenate substrates via the shunt pathway. Chapter 1 also discusses the invaluable roles played by site-directed mutagenesis and directed evolution in the creation of bacterial, archaeal and mammalian CYP mutant biocatalysts displaying novel or markedly improved monooxygenase and peroxygenase activities, which have found applications in industrial biotechnology, environmental bioremediation, medical science and biosynthesis of pharmaceuticals, human drug metabolites, steroids and other chemicals.

Chapter 2 discusses the mechanistic involvement of CYP enzymes and of the ferryl Cpd I oxygenating intermediate and other multiple oxidizing species of the CYP catalytic cycle in substrate oxygenation reactions, as well as the indispensable role played by the CYP proximal cysteine thiolate ligand. Chapter 3 describes how CYP3A4 functions as the most important human drug-metabolizing enzyme that clears over a half of all administered pharmaceuticals. CYP3A4 has the ability to catalyze diverse oxidative reactions in addition to traditional hydroxylation reactions. An overview is presented of the experimental and theoretical methods used to examine and predict CYP3A4–ligand interactions, which is a defining factor in drug metabolism. Chapter 4 explains how the steroidal CYP17A1 (17α-hydroxylase), CYP19A1 (aromatase) and CYP51A1 (sterol 14α-demethylase) enzymes, when metabolizing substrates that project a carbonyl functionality, utilize the ferric-peroxo anion intermediate formed in the CYP cycle to perform acyl-carbon bond cleavage reactions.

Chapter 5 describes various oxidative reactions catalyzed by a CYP enzyme acting on a single substrate. In the first example, 2,2′,4,4′-tetrabromodiphenyl ether (BDE-47), a halogenated aromatic environmental contaminant, was oxidatively biotransformed by human CYP2B6 to nine different metabolites via monooxygenase reactions that included aromatic hydroxylation, dealkylation and debromination. In the second example, endogenous lithocholic acid serves as a substrate for human CYP3A4 and produces five different metabolites via aliphatic hydroxylation and dehydrogenation reactions. Chapter 6 discusses how members of the CYP superfamily metabolize ω-6 and ω-3 polyunsaturated fatty acids. In mammalian tissues, CYP2C and CYP2J are the major epoxygenases whereas CYP4A and CYP4F function as hydroxylases, using polyunsaturated fatty acids as substrates to produce distinct epoxy and hydroxy metabolites termed eicosanoids. CYP eicosanoids are formed as secondary messengers of hormones, growth factors and cytokines that regulate cardiovascular and renal function and other physiological processes. Imbalances in the formation of CYP eicosanoids are linked to the development of hypertension, myocardial infarction, stroke, maladaptive cardiac hypertrophy, acute kidney injury and inflammatory disorders. The underlying mechanisms can provide novel targets for the prevention and treatment of these disease states.

Chapter 7 describes the use of decoy molecules, whose structures are very similar to natural substrates and can be used to trick the substrate recognition of bacterial CYP enzymes, allowing them to catalyze oxidation reactions of nonnative substrates in the presence of decoy molecules without the substitution of amino acid residues. The hydroxylation of small hydrocarbons such as ethane, propane, butane and benzene can be catalyzed by CYPBM3, a long alkyl-chain hydroxylase, using CYP substrate misrecognition induced by decoy molecules. A number of H_2O_2-dependent bacterial CYP enzymes can catalyze the peroxygenation of a variety of nonnative substrates through a substrate–misrecognition trick, in which catalytic activities and enantioselectivity are dependent on the structure of the decoy molecule.

Chapter 8 describes the ability of CYP enzymes to oxidize unactivated substrate C–H bonds with remarkable chemo-, regio- and stereoselectivity. While bacterial CYP enzymes typically show higher activity, they tend to be highly selective for one or a few substrates. In contrast, mammalian CYP enzymes display astonishing substrate promiscuity. The chapter also discusses the utilization of small molecules for controlling CYP substrate specificity and product selectivity. The first approach involves the use of decoy molecules and provides an effective way for increasing the substrate scope of CYP biocatalysts without the need for protein mutagenesis. The second approach involves the application of substrate engineering by using theobromine as the chemical auxiliary for controlling and predicting the regio- and stereoselectivity of CYP3A4-catalyzed oxidations.

Chapter 9 analyzes the use of electrode/CYP systems for determining the mechanisms involved in CYP-catalyzed reactions. Bioelectrocatalysis-based screening of potential substrates or inhibitors of CYP enzymes, the stoichiometry of the electrocatalytic cycle, redox thermodynamics and the peroxide shunt pathway are examined. Electrochemical techniques for investigating the influence of vitamins, antioxidants and drugs on biocatalysis by CYP enzymes, especially the metabolism of drugs by CYP3A4, are also described, as well as the characteristics, performance and potential applications of CYP electrochemical systems.

Chapter 10 discusses in exquisite detail the structural and functional characteristics of the diverse natural electron transfer redox protein components, as well as the artificial donor constructs, that participate in CYP monooxygenase systems derived from bacterial, archaeal and mammalian organisms. Chapter 11 discusses the biological diversity of CYP redox systems found in mammalian, bacterial and fungal organisms. Also described are the catalytically self-sufficient CYP fusion enzymes, which include the CYPBM3 flavocytochrome and the CYP116B1 and CYP116B2 flavocytochromes. Chapter 12 describes the properties of the bacterial cytochrome P450cin enzyme that catalyzes the enantiospecific hydroxylation of 1,8-cineole to (1*R*)-6β-hydroxycineole when reconstituted with its natural redox partner cindoxin, *E. coli* flavodoxin reductase, and NADPH as an electron source. This enzyme system has become a useful instrument in the study of CYP enzymes, whereby large quantities of P450cin can be prepared and rates of oxidation up to 1,500 min^{-1} achieved. The enzyme also displays a number of unusual characteristics that include an asparagine residue in P450cin that has been found in place of the usual conserved threonine residue observed in most CYP enzymes. Chapter 13, the final chapter, presents an overview on unspecific peroxygenases (UPOs) (e.g. *Agrocybe aegerita* unspecific peroxygenase) and related heme peroxidases such as chloroperoxidase, with focus on their molecular and catalytic properties. The product patterns of UPOs resemble those of human and other mammalian CYP monooxygenases and combine the catalytic cycle of heme peroxidases with the peroxide shunt of CYP enzymes.

Vancouver, Canada — Eugene G. Hrycay
Vancouver, Canada — Stelvio M. Bandiera
Dec, 2014

Research Profile of the Editors

Dr. Eugene G. Hrycay obtained his M.Sc. and Ph.D. degrees (Biochemistry) from Memorial University, Newfoundland, Canada, under the supervision of Dr. Peter J. O'Brien. During his tenure, Dr. Hrycay discovered the crucial peroxidase function for cytochrome P450 (CYP) enzymes. Dr. Hrycay pursued his research interests in the CYP area by carrying out postdoctoral studies with Dr. Ronald Estabrook at the University of Texas Health Science Center, Dallas; with Drs. Anthony Lu and Wayne Levin at Hoffmann-La Roche, Nutley, New Jersey; with Dr. Lars Ernster at the University of Stockholm and Drs. Jan-Åke Gustafsson and Magnus Ingelman-Sundberg at the Karolinska Institutet, Sweden; and with Dr. Colin Jefcoate at the University of Wisconsin, Madison. Dr. Hrycay's postdoctoral research work included the study of the mechanism of oxygen activation by CYP enzymes and involvement of higher CYP ferryl and perferryl heme iron oxidation states in the hydroxylation of drugs and carcinogens using rabbit and rat hepatic microsomes, and purified rat CYP2B1 and CYP1A, as CYP sources; the study of the mechanism of steroid hydroxylation utilizing biological and artificial hydroperoxides, periodate and chlorite as surrogate oxygen atom donors to promote steroid hydroxylations using hepatic and adrenocortical microsomes and mitochondria, and purified rat CYP2B1, as CYP sources; and the study of steroid 11β-hydroxylation by adrenocortical CYP enzymes. During the remaining years until his retirement, Dr. Hrycay held a Senior Research Associate position in the Faculty of Pharmaceutical Sciences, The University of British Columbia, Vancouver, British Columbia, Canada. Dr. Hrycay's most recent focus was the characterization of the hepatic CYP enzymes involved in the metabolism of bile acids and polychlorinated biphenyls (PCBs).

Dr. Stelvio M. Bandiera is a Professor in the Faculty of Pharmaceutical Sciences at the University of British Columbia, Canada. He received a M.Sc. degree (Biochemistry) from Dalhousie University, Canada, and a Ph.D. degree (Biological Chemistry) from the University of Guelph, Canada. He spent a year as a Research Associate with Dr. Stephen Safe at Texas A&M University and completed postdoctoral training at the Roche Institute of Molecular Biology in Nutley, New Jersey, under the supervision of Drs. Allan Conney, Wayne Levin and Paul Thomas. Dr. Bandiera joined the University of British Columbia in 1986. Dr. Bandiera's research interests include hepatic drug metabolism and environmental toxicology. Recent studies in Dr. Bandiera's laboratory have focused on the mechanisms by which hormones regulate expression of CYP enzymes and on CYP-mediated biotransformation of organohalogen compounds such as polychlorinated biphenyls and polybrominated diphenyl ethers.

Contents

Contributors

Muhammad Akhtar School of Biological Sciences, University of the Punjab, Lahore, Pakistan

Centre for Biological Sciences, University of Southampton, Southhampton, UK

Alexander Archakov Institute of Biomedical Chemistry, Moscow, Russia

Karine Auclair Department of Chemistry, McGill University, Montreal, Quebec, Canada

Stelvio M. Bandiera Faculty of Pharmaceutical Sciences, The University of British Columbia, Vancouver, British Columbia, Canada

James Belcher Manchester Institute of Biotechnology, The University of Manchester, Manchester, UK

Tatiana Bulko Laboratory of Bioelectrochemistry, Institute of Biomedical Chemistry, Moscow, Russia

John H. Dawson Department of Chemistry and Biochemistry, University of South Carolina, Columbia, SC, USA

Anand K. Deo Faculty of Pharmaceutical Sciences, The University of British Columbia, Vancouver, British Columbia, Canada

UCB Pharma, Briane l'Alleud, Belgium

James J. De Voss School of Chemistry and Molecular Biosciences, University of Queensland, Brisbane, Australia

Claudio A. Erratico Faculty of Pharmaceutical Sciences, The University of British Columbia, Vancouver, British Columbia, Canada

Toxicology Center, University of Antwerp, Wilrijk, Belgium

Anthony J. Farlow School of Chemistry and Molecular Biosciences, University of Queensland, Brisbane, Australia

David B. Hawkes School of Chemistry and Molecular Biosciences, University of Queensland, Brisbane, Australia

Peter Hlavica Walther-Straub-Institut für Pharmakologie und Toxikologie der LMU, München, Germany

Martin Hofrichter Department of Bio- and Environmental Sciences, International Institute Zittau, Technische Universität Dresden, Zittau, Germany

Eugene G. Hrycay Faculty of Pharmaceutical Sciences, The University of British Columbia, Vancouver, British Columbia, Canada

Harald Kellner Department of Bio- and Environmental Sciences, International Institute Zittau, Technische Universität Dresden, Zittau, Germany

Anne Konkel Max Delbrueck Center for Molecular Medicine, Berlin, Germany

Alexey Kuzikov Laboratory of Bioelectrochemistry, Institute of Biomedical Chemistry, Moscow, Russia

Dominika Luciakova Manchester Institute of Biotechnology, The University of Manchester, Manchester, UK

Anna Makhova The First Moscow State Medical University, Moscow, Russia

Kirsty J. McLean Manchester Institute of Biotechnology, The University of Manchester, Manchester, UK

Anuja R. Modi Department of Chemistry and Biochemistry, University of South Carolina, Columbia, SC, USA

Andrew W. Munro Manchester Institute of Biotechnology, The University of Manchester, Manchester, UK

Marek J. Pecyna Department of Bio- and Environmental Sciences, International Institute Zittau, Technische Universität Dresden, Zittau, Germany

Vanja Polic Department of Chemistry, McGill University, Montreal, Quebec, Canada

Thomas L. Poulos Departments of Molecular Biology and Biochemistry, Chemistry, and Pharmaceutical Sciences, University of California, Irvine, CA, USA

Wolf-Hagen Schunck Max Delbrueck Center for Molecular Medicine, Berlin, Germany

Irina F. Sevrioukova Department of Molecular Biology and Biochemistry, University of California, Irvine, CA, USA

Evgeniya Shich The First Moscow State Medical University, Moscow, Russia

Osami Shoji Bioinorganic Chemistry Laboratory, Department of Chemistry, Nagoya University, Nagoya, Japan

Victoria V. Shumyantseva Laboratory of Bioelectrochemistry, Institute of Biomedical Chemistry, Moscow, Russia

Kate E. Slessor School of Chemistry and Molecular Biosciences, University of Queensland, Brisbane, Australia

Jeanette E. Stok School of Chemistry and Molecular Biosciences, University of Queensland, Brisbane, Australia

Kang Lan Tee Manchester Institute of Biotechnology, The University of Manchester, Manchester, UK

René Ullrich Department of Bio- and Environmental Sciences, International Institute Zittau, Technische Universität Dresden, Zittau, Germany

Yoshihito Watanabe Bioinorganic Chemistry Laboratory, Research Center for Materials Science, Nagoya University, Nagoya, Japan

Christina Westphal Max Delbrueck Center for Molecular Medicine, Berlin, Germany

J. Neville Wright Centre for Biological Sciences, University of Southampton, Southampton, UK

Monooxygenase, Peroxidase and Peroxygenase Properties and Reaction Mechanisms of Cytochrome P450 Enzymes

1

Eugene G. Hrycay and Stelvio M. Bandiera

Abstract

This review examines the monooxygenase, peroxidase and peroxygenase properties and reaction mechanisms of cytochrome P450 (CYP) enzymes in bacterial, archaeal and mammalian systems. CYP enzymes catalyze monooxygenation reactions by inserting one oxygen atom from O_2 into an enormous number and variety of substrates. The catalytic versatility of CYP stems from its ability to functionalize unactivated carbon-hydrogen (C–H) bonds of substrates through monooxygenation. The oxidative prowess of CYP in catalyzing monooxygenation reactions is attributed primarily to a porphyrin π radical ferryl intermediate known as Compound I (CpdI) ($Por{\bullet}^{+}Fe^{IV}{=}O$), or its ferryl radical resonance form ($Fe^{IV}{-}O^{\bullet}$). CYP-mediated hydroxylations occur via a consensus H atom abstraction/oxygen rebound mechanism involving an initial abstraction by CpdI of a H atom from the substrate, generating a highly-reactive protonated Compound II (CpdII) intermediate ($Fe^{IV}{-}OH$) and a carbon-centered alkyl radical that rebounds onto the ferryl hydroxyl moiety to yield the hydroxylated substrate. CYP enzymes utilize hydroperoxides, peracids, perborate, percarbonate, periodate, chlorite, iodosobenzene and *N*-oxides as surrogate oxygen atom donors to oxygenate substrates via the shunt pathway in the absence of NAD(P)H/O_2 and reduction-oxidation (redox) auxiliary proteins. It has been difficult to isolate the historically elusive CpdI intermediate in the native NAD(P)H/O_2-supported monooxygenase pathway and to determine its precise electronic structure and kinetic and physicochemical properties because of its high reactivity, unstable nature ($t_{1/2}$ ~2 ms) and short life cycle, prompting suggestions for participation in monooxygenation reactions of alternative CYP iron-oxygen intermediates such as the ferric-peroxo anion species ($Fe^{III}{-}OO^{-}$), ferric-hydroperoxo species ($Fe^{III}{-}OOH$) and $Fe^{III}{-}(H_2O_2)$ complex.

E.G. Hrycay (✉) • S.M. Bandiera
Faculty of Pharmaceutical Sciences, The University of British Columbia, 2405 Wesbrook Mall, Vancouver, British Columbia V6T1Z3, Canada
e-mail: eugenehrycay@netscape.net

E.G. Hrycay and S.M. Bandiera (eds.), *Monooxygenase, Peroxidase and Peroxygenase Properties and Mechanisms of Cytochrome P450*, Advances in Experimental Medicine and Biology 851,
DOI 10.1007/978-3-319-16009-2_1,

Keywords
Cytochrome P450 • Compound I • Monooxygenase • Peroxygenase • Peroxide shunt • Reaction mechanisms • CYP protein engineering

1.1 Introduction

This review examines the monooxygenase, peroxidase and peroxygenase properties and mechanisms of cytochrome P450 (CYP[1]) enzymes[2] in bacterial, archaeal and mammalian systems and is a more detailed and updated version of our previous review [1].[3] CYP enzymes are widely distributed throughout the four biological kingdoms (archaea, bacteria, eukarya, viruses). As of Aug. 13th, 2013, 1,261 CYP families containing genes that encode CYP proteins have been identified and a total of 21,039 CYP sequences have been named in all groups of organisms [2]. CYP enzymes catalyze a large number of complex oxidative reactions using an enormous number and variety of exogenous and endogenous substrates [3–9]. Remarkably, CYP enzymes are capable of utilizing over one million substrates to catalyze metabolic reactions [10, 11]. The most common oxidative reaction catalyzed by CYP enzymes is the monooxygenation of a substrate, and the equation (eqn) for the reaction adheres to the following formula

$$A + O_2 + NAD(P)H + H^+ + 2e^- \xrightarrow{\text{monooxygenase CYP}} AO + H_2O + NAD(P)^+ \quad (1.1)$$

where A represents the substrate and AO signifies the monooxygenated substrate. Thus, one oxygen atom from atmospheric dioxygen (O_2) is incorporated into the substrate while the other oxygen atom is reduced to H_2O [4, 7]. Two mandatory electrons are supplied to the CYP monooxygenase enzyme by NAD(P)H, through mediation of a reduction-oxidation (redox) flavoprotein and in some cases an iron-sulfur protein.

Early in its evolutionary history, Earth contained an anaerobic reducing atmosphere comprised mainly of hydrogen, nitrogen, methane, ammonia, water vapor, carbon monoxide and carbon dioxide [12, 13]. During that primordial period, it is believed that abiogenic chemicals formed prebiotic amino acids that subsequently self-assembled into autocatalytic peptides, acquiring attributes by which we currently define living organisms [14]. CYP and an electron transfer partner, the iron-sulfur protein (ferredoxin) [12, 15, 16], could have arisen from these prebiotic amino acids in an anaerobic prokaryote some 3.5 billion years ago in a time that predated atmospheric O_2 and the combustion of organic matter [12–17]. CYP enzymes and iron-sulfur proteins display low iron redox potentials [4, 12] and could have participated in anaerobic reductive reactions conducted by ancient prokaryotes in this early anoxic terrestrial environment. Even today, anaerobic reductive reactions play a crucial role in mammalian CYP

[1] *Abbreviations*: *AaeUPO Agrocybe aegerita* unspecific peroxygenase, *CcP* cytochrome *c* peroxidase, *mCPBA m*-chloroperbenzoic acid, *CpdI* Compound I, *CpdII* Compound II, *Cpd0* Compound Zero, *CpdES* Compound ES, *CPO* chloroperoxidase, *CPR* NADPH-cytochrome P450 oxidoreductase, *CYP* cytochrome P450, *HRP* horseradish peroxidase, *KIEs* kinetic isotope effects, *PGG₂* 9α,11-α-epidioxy-15*S*-hydroperoxyprosta-5*Z*,13*E*-dien-1-oic acid, *PGH₂* 9α,11α-epidioxy-15*S*-hydroxyprosta-5*Z*,13*E*-dien-1-oic acid, *TMPD N,N,N′,N′*-tetramethyl-*p*-phenylenediamine. *Amino acid abbreviations*: *Ala* alanine, *Arg* arginine, *Asn* asparagine, *Asp* aspartic acid, *Cys* cysteine, *Glu* glutamic acid, *Gly* glycine, *His* histidine, *Ile* isoleucine, *Leu* leucine, *Lys* lysine, *Met* methionine, *Phe* phenylalanine, *Pro* proline, *Ser* serine, *Thr* threonine, *Trp* tryptophan, *Tyr* tyrosine, *Val* valine.

[2] Members of the CYP superfamily are called enzymes and should not be referred to as isoforms or iso(en)zymes. All CYP members are different enzymes and are products of different genes of the CYP superfamily (Dr. R. Feyereisen, CNRS, Univ. Nice Sophia Antipolis, France, personal communication).

[3] A review of this scope cannot include all references pertaining to the subject matter presented.

enzyme systems [1, 7]. A vital reductive activity of CYP, discovered by Hrycay and coworkers [18–26], is its peroxidase function whereby the hemeprotein reduces and detoxifies biological hydroperoxides such as hydrogen peroxide, lipid hydroperoxides, steroid hydroperoxides and exogenous organic hydroperoxides [1]. The absence of O_2 in the early terrestrial atmosphere precluded prokaryotic CYP from functioning as a monooxygenase. However, the anoxic environment could have been relatively rich in H_2O_2 and peroxygenated organic chemicals [27–29], raising the interesting possibility that the prokaryotic ancestors of contemporary CYP-containing organisms utilized H_2O_2 and other peroxy compounds, instead of the yet unavailable O_2, as oxygen atom donors to conduct the CYP-catalyzed monooxygenation of substrates [1, 29]. Under global anaerobic conditions, ancient CYP enzymes could have functioned as peroxidases and peroxygenases well before their monooxygenase function developed [1]. Support for this proposal is provided by studies showing that archaeal CYP119A1 from *Sulfolobus acidocaldarius* can catalyze the peroxygenation of laurate, *cis*-stilbene and styrene in the presence of H_2O_2 and other peroxy compounds (Table 1.1) [30–33, 58]. The CYP peroxygenase function has been conserved in archaea, bacteria, humans, other mammals (Table 1.2) and additional organisms.

Anaerobic photosynthesizing cyanobacteria are believed to have altered the early terrestrial atmosphere from reducing to oxidizing in nature [13]. Steadily increasing atmospheric O_2 levels, coupled with the reductive metabolism of O_2 by prokaryotic CYP enzymes, generated reactive oxygen species such as the superoxide radical ($O_2^{\bullet -}$), H_2O_2 and the hydroxyl radical ($^{\bullet}OH$) [12, 80] that are toxic to living organisms. Initially, CYP could have removed traces of unwanted O_2 and H_2O_2 in the ancestral prokaryotic cell but after a period of time, CYP activity would not have been adequate. Increasing atmospheric O_2 levels led to the emergence of unicellular eukaryotic species distinct from the original prokaryotic organisms [13, 81]. These evolving eukaryotes had some protection from the damaging effects of reactive oxygen species [12, 13, 80, 82]. More efficient enzymes such as superoxide dismutase, catalase and peroxidase evolved to combat the cytotoxic effects of reactive oxygen species, enabling the catalytic functions of CYP to diversify [12, 13, 80]. The high atmospheric O_2 levels triggered evolution of complex multicellular life-forms containing specific CYP enzymes that could be tailored to recognize distinct structural classes of chemicals [13]. CYP enzymes in the eukaryotic organisms developed critical catalytic functions that were used to oxidize vital endogenous molecules and xenobiotics (foreign compounds) [12, 17, 29]. One major CYP activity to evolve in an oxygen-rich environment was the monooxygenase function that was retained to this day in contemporary organisms because of its importance in anabolic and catabolic processes [1].

The objective of this review is to summarize the earlier literature and highlight recent advances that have been made in elucidating the chemical structures of vital CYP "activated oxygen" intermediates that participate in monooxygenase, peroxidase and peroxygenase reactions in bacterial, archaeal and mammalian systems, focusing particularly on the transitory CYP Compound I (CpdI) oxygenating species. We also examine the characteristics of monooxygenation reactions that proceed via the native NAD(P)H/O_2-supported pathway and the distinctive features of representative CYP monooxygenase reactions that are driven by hydroperoxides and exogenous oxygen atom donors to help us elucidate the mechanisms involved. Finally, we discuss the invaluable roles that site-directed mutagenesis and directed evolution procedures have played in the creation of bacterial, archaeal and mammalian CYP mutant biocatalysts displaying novel or markedly improved monooxygenase and peroxygenase activities that have found applications in industrial biotechnology, environmental bioremediation, medical science and biosynthesis of pharmaceuticals, human drug metabolites, steroids and other chemicals.

Table 1.1 Selected archaeal and bacterial CYP peroxygenase reactions supported by peroxides and peracids

CYP enzyme and source	Peroxide/peracid oxidant	Substrate	Major type of oxidation	Refs.[a]
Archaeal enzyme				
CYP119A1 (*S. acidocaldarius*)	*t*-BuOOH[b], H_2O_2; H_2O_2	Styrene; *cis*-stilbene	Epoxidation	[30, 31]
CYP119A1 (*S. acidocaldarius*)	*m*CPBA, H_2O_2	Laurate	($\omega - 1$)- and ($\omega - 2$)-Hydroxylation	[32–34]
CYP119A2 (*S. tokodaii* strain 7)	H_2O_2	Ethylbenzene; styrene	Hydroxylation; epoxidation	[35]
Bacterial enzyme				
CYP101A1 (*P. putida*)	*m*CPBA, H_2O_2, peracetic acid, PPAA	Camphor	5-*exo*-Hydroxylation	[36–38]
CYP101A1 (*P. putida*)	H_2O_2	Dehydrocamphor; naphthalene, 3-PPA; indole	5-*exo*-Epoxidation; hydroxylation; hydroxylation	[29, 39, 40]
CYP101A1 (*P. putida*)	CuOOH, H_2O_2	7-Isopropoxycoumarin; *N*,*N*-dimethylaniline	*O*-Dealkylation; *N*-Demethylation	[41]
CYP102A1 (*B. megaterium*)	H_2O_2	Caprate, indole, laurate, myristate, 12-*p*NCA; styrene; propranolol	Hydroxylation; epoxidation; hydroxylation	[40, 42–44]
CYP102A1 (*B. megaterium*)	*t*-BuOOH, CuOOH	Indole	3-Hydroxylation	[40]
CYP107A1 (*S. erythraea*)	H_2O_2	7-BQ; testosterone	*O*-Debenzylation; 1-, 11α-, 12-, 16α- hydroxylation	[45]
CYP107AJ1 (*S. peucetius*)	H_2O_2	7-Ethoxycoumarin	*O*-Deethylation	[46]
CYP152A1 (*B. subtilis*)	CuOOH, H_2O_2; H_2O_2; H_2O_2	Myristate; ethylbenzene; styrene	α- and β-Hydroxylation; hydroxylation; epoxidation	[47–49]
CYP152A2 (*C. acetobutylicum*)	H_2O_2	Laurate, myristate, palmitate	α- and β-Hydroxylation	[49, 50]
CYP152B1 (*S. paucimobilis*)	*t*-BuOOH, H_2O_2; H_2O_2	Myristate; styrene	α-Hydroxylation; epoxidation	[51–53]
CYP152OleT$_{JE}$ (*J.* sp. ATCC8456)	H_2O_2	Stearic acid	Decarboxylation	[54]
CYP153A6 (*M.* sp. HXN-1500)	*m*CPBA	Heptane	Hydroxylation	[55]
CYP167A1 (*S. cellulosum*)	H_2O_2	7-EFC	*O*-Deethylation	[56]
CYP175A1 (*T. thermophilus*)	H_2O_2	Palmitoleic acid	Epoxidation	[57]

[a]Data compiled in Table 1.1 were obtained mainly from Ref. [1]

[b]*7-BQ* 7-benzyloxyquinoline, *t-BuOOH t*-butyl hydroperoxide, *mCPBA m*-chloroperbenzoic acid, *CuOOH* cumene hydroperoxide, *7-EFC* 7-ethoxy-4-trifluoromethylcoumarin, *12-pNCA* 12-*p*-nitrophenoxydodecanoic acid, *PPAA* phenylperacetic acid, *3-PPA* 3-phenylpropionate

Table 1.2 Selected mammalian CYP peroxygenase reactions supported by biological hydroperoxides

Biological hydroperoxide	Substrate	Major product(s) or type of oxidation	CYP enzyme or source	Refs.[a]
Arachidonic acid–OOH[c], linolenic acid–OOH	Diethylstilbestrol	Diethylstilbestrol quinone	Rat liver micr[b]	[59]
Linoleic acid–OOH	Androstenedione	6β-, 16α-OH[d]	Rat CYP2B1	[23]
	N,N-Dimethylaniline	*N*-Oxidation	Rabbit liver microsomal CYP2B4[e]	[60]
	Ethanol	Acetaldehyde	Rat liver microsomal CYP2B1[f]	[61]
Citronellol–, geraniol–, myrcene–OOH[g]	Aminopyrine	*N*-Demethylation	Pig liver micr	[62]
H_2O_2	Androstenedione	6β-OH	Rat CYP2B1	[23]
	Aminopyrine, benzphetamine	*N*-Demethylation	Rabbit CYP2B4	[63]
	Aniline	*p*-Aminophenol	Rabbit liver microsomal CYP2B4[e]	[64]
	Benzo[*a*]pyrene; BaP-7,8-diol; DMBA	1,6-, 3,6-, 6,12-Quinones; 9,10-epoxidation; 7-,12-hydroxylation	Human rCYP2S1; rat liver microsomal CYP2B1[f]	[65, 66]
	7-BQ	*O*-Debenzylation	Human rCYP3A4	[67]
	N,N-Dimethylaniline	*N*-Demethylation	Rabbit liver microsomal CYP2B4[e], rabbit rCYP2B4, rat CYP2B1	[60, 68]
	7-EFC	*O*-Deethylation	Rat rCYP2B1	[69]
	IQ; methoxyresorufin	*N*-Oxidation; *O*-demethylation	Human rCYP1A2	[70]
	Lauric acid	10-, 11-, 12-Hydroxylation; 11-hydroxylation	Rat liver micr; rabbit CYP2B4	[71, 72]
	Pinacidil	Pinacidil amide, OH products	Human rCYP3A4	[73]
	Retinoic acid	4-OH- and 4-oxo-Retinoic acid	Human rCYPs 2D6, 2S1	[65, 74]
Urea–H_2O_2	Dextromethorphan; testosterone	Dextrorphan; 6β-OH	Human rCYP2D6; rCYP3A4	[75]
5*S*-HPETE, 13*S*-HPODE	Retinoic acid	4-OH- and 4-oxo-Retinoic acid	Human rCYPs 1A2, 2D6, 3A4	[74]
5*S*-, 12*S*-, 15*S*-HPETE[h]	BaP-7,8-diol	9,10-Epoxidation	Human rCYPs 1A1, 1A2, 1B1, 2S1, 3A4	[76]
5*S*-, 12*S*-, 15*S*-HPETE[i], 9*S*-, 13*S*-HPODE	Diethylstilbestrol	Diethylstilbestrol quinone	Rat liver micr	[59]
Pregnenolone 17α–OOH	Aminopyrine	*N*-Demethylation	Pig liver micr	[62]
	Androstenedione	6β-, 16α-, 17β-OH	Rat liver micr	[24, 25]
	Progesterone	21-OH	Ox adr micr	[26]

[a]Data compiled in Table 1.2 were obtained mainly from Ref. [1]

[b]*Adr* adrenocortical, *BaP-7,8-diol* benzo[*a*]pyrene-7,8-dihydrodiol, *7-BQ* 7-benzyloxyquinoline, *DMBA* 7,12-dimethylbenz[*a*]anthracene, *7-EFC* 7-ethoxy-4-trifluoromethylcoumarin, *5S-HPETE* 5*S*-hydroperoxy-6*E*,8*Z*,11*Z*,14*Z*-eicosatetraenoic acid, *12S-HPETE* 12*S*-hydroperoxy-5*Z*,8*Z*,10*E*,14*Z*-eicosatetraenoic acid [76], *15S-HPETE* 15*S*-hydroperoxy-5*Z*,8*Z*,*10E*,*14Z*-eicosatetraenoic acid [76], *15S-HPETE* 15*S*-hydroperoxy-5*Z*,8*Z*,11*Z*,13*E*-eicosatetraenoic acid [59], *9S-HPODE* 9*S*-hydroperoxy-10*E*,12*Z*-octadecadienoic acid, *13S-HPODE* 13*S*-hydroperoxy-9*Z*,11*E*-octadecadienoic acid, *IQ* 2-amino-3-methylimidazo[4,5-*f*]quinoline, *rCYP* recombinant cytochrome P450, *Retinoic acid*, all-*trans*-Retinoic acid

[c]The OOH moiety signifies the hydroperoxy group of the hydroperoxides

[d]Compounds with the attached OH group signify hydroxy derivatives of substrates

[e]Liver microsomes were prepared from phenobarbital-pretreated rabbits. CYP2B4 is the predominant hepatic CYP enzyme induced by pretreating rabbits with phenobarbital [77, 78]

[f]Liver microsomes were prepared from phenobarbital-pretreated rats. CYP2B1 is the predominant hepatic CYP enzyme induced by pretreating rats with phenobarbital [79]

[g]Hydroperoxides of citronellol, geraniol and myrcene were synthesized from their natural derivative compounds

[h]In this case, 15*S*-HPETE represents 15*S*-hydroperoxy-5*Z*,8*Z*,*10E*,*14Z*-eicosatetraenoic acid [76]

[i]In this case, 15*S*-HPETE represents 15*S*-hydroperoxy-5*Z*,8*Z*,11*Z*,13*E*-eicosatetraenoic acid [59]

1.2 Basic Structural Features of CYP Enzymes

CYP enzymes are widely distributed throughout the archaeal, bacterial, eukaryotic and viral kingdoms [2, 7, 83–88]. Insights into the catalytic mechanisms of CYP enzymes are aided by structural considerations. A common approach to understanding structure-function relationships in CYP enzymes is to compare the three-dimensional structures of family members. Information gathered from analyzing crystal structures, combined with data involving site-directed mutagenesis of amino acids in the CYP active site, as well as biochemical and biophysical data, have been utilized to probe the mechanisms by which CYP enzymes activate molecular oxygen, control heme iron spin state and heme Fe^{III}/Fe^{II} redox potentials, transfer electron equivalents, control proton delivery to dioxygen complexes in the CYP active site, and regulate regio-, chemo- and stereoselective monooxygenation reactions [89–101].

CYP enzymes are ferric heme-thiolate proteins that contain a heme prosthetic group (iron protoporphyrin IX, heme *b*) consisting of a highly-conjugated symmetrical macrocycle coordinated to a ferric iron atom (Fig. 1.1) [93]. The heme moiety is identical to that found in hemoglobin and heme peroxidases such as chloroperoxidase (CPO) and horseradish peroxidase (HRP). The reduced ferrous heme group of CYP has a strong affinity for O_2 but binds CO

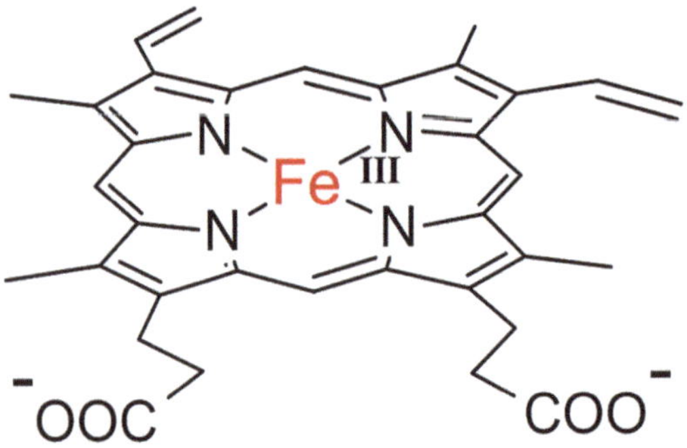

Fig. 1.1 Structure of iron protoporphyrin IX, heme *b*, with the iron atom shown in *red* (Figure 1.1 was adapted from Refs. [93, 102])

with higher affinity, generating a ferrous iron – CO complex that displays a Soret absorption maximum at an unusually long wavelength of ~450 nm. This unique spectral property is a signature of CYP enzymes from which the designation, "P450", was derived [103, 104]. The heme ferric iron is ligated to six ligands, four of which are nitrogen atoms of the planar porphyrin ring. The proximal (fifth) ligand is a thiolate anion side chain of a deprotonated cysteine residue located in the C-terminal region of the polypeptide chain [93, 105–108]. The unique spectral properties of CYP are attributed to the coordination of the heme iron to the thiolate ligand [93, 105]. However, recent genome sequencing revealed that members of the fungal CYP408 family do not contain the thiolate ligand but have sufficient sequence similarity to be classified as CYP enzymes, even though the characteristic CYP spectral signature is absent [109]. The distal (sixth) coordination site is occupied by an easily exchangeable water molecule that is displaced by substrate binding to the active site [110]. The heme resides in a large hydrophobic pocket in the active site. The distal side of the pocket is lined with key hydrophobic amino acid residues that can bind a substrate [93]. Substrate binding involves ionic, H bonding, van der Waals and π–π bond stacking intermolecular forces [102]. The pronounced differences in catalytic activities, nature of the substrates metabolized, types of reactions catalyzed, and reaction mechanisms are accounted for largely by the nature of the CYP apoprotein component and the manner in which it interacts with the heme [111] and will be explored throughout the review.

CYP proteins have variable primary, secondary and tertiary structures essential for accommodating specific substrates and associated redox proteins [96]. CYP proteins also share a common overall three-dimensional protein fold and topology and have a well-conserved core region, despite less than a 20 % sequence similarity among them [94, 97, 99]. The structural core is formed by a four-α-helix bundle comprised of three parallel helices labeled D, L and I and one antiparallel helix E [99]. The heme

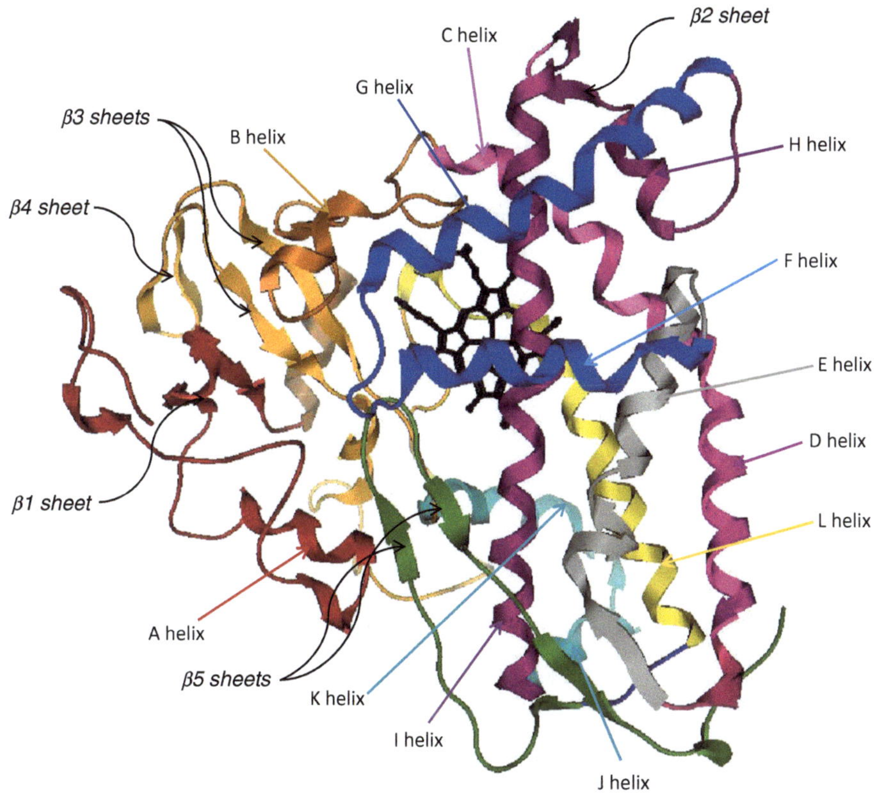

Fig. 1.2 Ribbon diagram of the CYPcam structural core. The diagram illustrates 12 α-helical regions (labeled A to L) and 5 beta (β) sheets (labeled 1–5) defining the two domains. The heme group is shown in *black* near the center of the diagram (Figure 1.2 was adapted from Refs. [91, 112])

is bracketed between the distal I helix and proximal L helix and is bound to the adjacent cysteine – heme-ligand loop. The structural core of bacterial CYP101A1 (CYPcam) comprises approximately 40 % α-helical segments and 10 % antiparallel beta (β) sheets (Fig. 1.2). The general CYP fold consists of six different variable regions known as substrate recognition sites that predetermine substrate specificity [113], enable CYP enzymes to act as highly versatile biocatalysts, and explain the enormous number and variety of substrates metabolized and chemical reactions catalyzed. Our understanding of CYP structure-function relationships has been derived largely from the CYPcam model [89, 110, 114], and the study of CYPcam has served to establish the universal features of CYP enzymes.

1.3 Monooxygenase Cycle of CYP Enzymes

CYP enzymes use molecular oxygen and the formal equivalents of molecular hydrogen ($2H^+ + 2e^-$), donated by NAD(P)H via redox proteins, to catalyze the monooxygenation of a variety of substrates. CYP enzymes function in the reductive activation (splitting) of molecular oxygen [98, 115, 116], a term that describes the ability of CYP to bind O_2, catalytically split it, and utilize one oxygen atom for the oxo-functionalization of substrates. The sequential reaction steps of the CYP monooxygenase cycle (Fig. 1.3) and the systematic generation of CYP intermediates that form the molecular stations of the catalytic cycle were originally

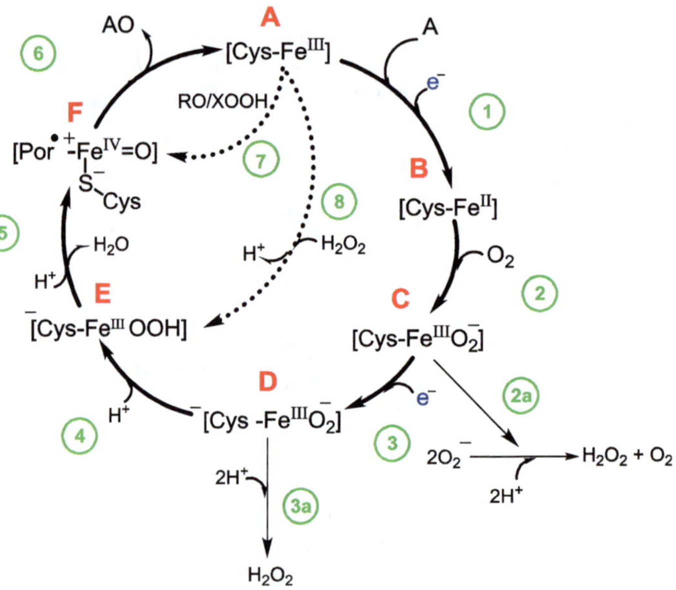

Fig. 1.3 Monooxygenation cycle of CYP enzymes. The sequential steps by which CYP enzyme systems catalyze the NAD(P)H/O_2-mediated oxidation of a substrate (*A*) to a monooxygenated substrate (*AO*) are illustrated numerically in steps 1–6 indicated by encircled numbers shown in *green*. CYP intermediary states are depicted by *bold capital letters* shown in *red*. The substrate (*A*) is bound to states **A** through **F** but is shown only in step 1 and in step 6 as a monooxygenated product (*AO*) to simplify the scheme. States **A** through **F** contain a cysteine thiolate ligand (*Cys*) that displays a delocalized negative charge in states **D** and **E** [117, 118]. Sono et al. [93] and Jin et al. [119] assigned the following net charges to the CYP intermediary states: **A** and **F** are neutral, **B**, **C** and **E** have an overall 1^- charge and **D** has a net 2^- charge. H_2O_2-forming pathways in uncoupled reactions are shown in steps 2A and 3A. Ferric CYP (Fe^{III}) can react with oxotransfer agents (*RO*) and peroxy compounds (XOOH) to generate the Cpd I species [Por•$^+$Fe^{IV}=O] (**F**) via the shunt pathway (step 7). Ferric CYP enzymes such as CYPcam [120], CYP2B4 [63] and CYP3A4 [68] can react with H_2O_2 to form the ferric-hydroperoxo species $^-$[Cys–Fe^{III}–OOH] (**E**) (step 8), which is believed to rapidly dissociate, upon protonation, to produce CpdI (**F**) and H_2O (step 5). Due to spatial considerations or for simplification, the line representing the chemical bond between the iron and oxygen atom (in Fe^{III}–O) is omitted. The + charges on the iron atom are also omitted. All intermediates (**A** through **F**) contain the porphyrin (*Por*) moiety that is shown as a Por•$^+$ radical cation only for CpdI (**F**). See text in Sect. 1.3 for details (Figure 1.3 was adapted from Ref. [1])

documented using the bacterial CYPcam system [110, 121] and the mammalian hepatic microsomal CYP systems [122, 123]. The two electrons required for oxygen activation are delivered to the CYP heme iron in two sequential, one-electron steps [98, 110]. Using CYPcam as a model, the cycle commences with the heme iron in the resting ferric (Fe^{III}) low-spin state (**A**). In the absence of camphor, one water molecule serves as the distal ligand to the heme ferric iron [114]. In step 1, the substrate (A) binds to low-spin, hexacoordinated, water-bound ferric CYP and displaces the water ligand to generate a high-spin, pentacoordinated, camphor-bound ferric CYP complex. The spin shift is accompanied by a concomitant shift in the redox potential of the heme Fe^{III}/Fe^{II} couple, which increases from −340 mV in the camphor-free CYP enzyme to −170 mV in the camphor-bound enzyme [90]. The ferric iron is thermodynamically favored to undergo reduction when the substrate is bound. Then, the high-spin ferric CYPcam-substrate complex is reduced to the ferrous state (**B**) with the first electron donated by NADH and mediated by putidaredoxin reductase and putidaredoxin. Reduction of ferric CYPcam is dependent on the presence of a substrate and is correlated with changes in heme iron spin state and redox potential [90, 110]. However, mammalian CYP enzymes were reduced at equally fast rates in the absence or presence of a substrate, substrate binding was not obligatory for the reduction of the ferric iron of human recombinant (r)CYP1A2 and rCYP2E1, and there was no link between spin state and reduction kinetics [123].

In step 2, ferrous CYP binds O_2 at the distal ligand site and the heme iron returns to the low-spin state. Binding of O_2 produces the ferrous-dioxygen complex [Cys–Fe^{II}–O_2], which is in equilibrium with the nucleophilic, resonance hybrid ferric-superoxo radical anion complex [Cys–Fe^{III}–$OO^{\bullet -}$] (**C**) that contains an unpaired electron on the distal (furthest from ferric iron) oxygen atom. In step 3, transfer of a second electron generates a supernucleophilic dinegatively-charged ferric-peroxo intermediate $^-$[Cys–Fe^{III}–OO^-] (**D**) containing one negative charge on the distal oxygen atom and a second negative charge delocalized over the cysteine (Cys) thiolate ligand [117]. The Cys designation is often not shown in CYP complexes throughout the text for simplification. In step 4, addition of a proton to the distal oxygen atom of $^-$[Cys–Fe^{III}–OO^-] (**D**) generates the ferric-hydroperoxo intermediate $^-$[Cys–Fe^{III}–OOH] (**E**), also known as Compound Zero (Cpd0) [124], containing one negative charge delocalized over the thiolate ligand. In step 5, delivery of a second proton to the distal oxygen atom of $^-$[Cys–Fe^{III}–OOH] generates an unstable iron oxo–H_2O adduct precursor $^-$[Cys–Fe^{III}–O–OH_2] [98] (not shown in Fig. 1.3), which rapidly dissociates through heterolytic cleavage of the O–O bond to produce a H_2O molecule and a simultaneously-created porphyrin (Por) π radical ferryl intermediate [$Por\bullet^+Fe^{IV}$=O] (**F**) identified provisionally as CpdI [11, 34, 98, 125]. In step 6, CpdI transfers its ferryl oxygen atom to the substrate (A) to generate ferric CYP and the monooxygenated substrate (AO), which is released from the active site. Coordination of a water molecule restores the resting, hexacoordinated, ferric CYP state. In the second protonation step (step 5), the formal oxidation state of the heme iron increases from Fe(III) to Fe(V) with the formation of CpdI. The formal Fe(V) oxidation state in CpdI is the combination of $Por\bullet^+$ and Fe^{IV}=O.

The majority of monooxygenation reactions catalyzed by CYP are thought to involve CpdI as the primary "activated oxygen" form of CYP [1, 11, 34, 98, 125–128]. The intricate nature, mechanism of formation and fundamental properties of this remarkable CpdI species are examined in greater detail in Sects. 1.6 and 1.7. Other transitory CYP iron-oxygen species, such as the ferryl radical anion resonance form of CpdI, ferric oxenoid complex, perferryl entity, CpdII species, protonated CpdII complex, CpdES species, ferric-peroxo anion species, ferric-hydroperoxo intermediate and ferric-(H_2O_2) complex, have also been postulated to act as oxygenating species (see Sect. 1.7) [1, 25, 55, 72, 118, 119, 127, 129–136].

CYP-mediated oxygenative metabolism of substrates is often highly inefficient. If the transfer of an oxygen atom to a substrate is not tightly coupled to NAD(P)H utilization, a portion of the electron equivalents derived from NAD(P)H is not utilized to oxygenate substrates but instead is unproductively deployed in the reduction of CYP-oxygen complexes formed in the CYP catalytic cycle. The complexes then dissociate and are liberated as reactive oxygen species (O_2^- and H_2O_2) in a process known as uncoupling [98, 99]. A likely reason for uncoupling in CYP systems is the substrate-dependent access of water to the CYP active site [92, 99, 137], which destabilizes the ferric-superoxo intermediate [Cys–Fe^{III}–O_2^-] (**C**) formed in the CYP cycle and results in dissociation of the complex to liberate ferric CYP and O_2^-. Subsequent dismutation of O_2^- produces H_2O_2 (see Fig. 1.3, step 2A). A second mode of uncoupling involves reduction of the ferric-peroxo intermediate $^-$[Cys–Fe^{III}–O_2^-] (**D**) and subsequent dissociation to produce H_2O_2 (see Fig. 1.3, step 3A). A third mode of uncoupling involves dissociation of the hydroperoxide anion from ferric iron (**E**) to form H_2O_2. A fourth mode of uncoupling is the reduction of CpdI (**F**) by two additional electrons and protons via the NAD(P)H oxidase pathway, which generates water instead of monooxygenated substrate [98, 99]. Remarkably, the CYPcam enzyme system catalyzes the regio- and stereospecific 5-*exo*-hydroxylation of the natural substrate, (1*R*)-camphor, with ~100 % efficiency with respect to coupling of NADH utilization to hydroxylated product formation [36].

1.4 Monooxygenase Properties of Mammalian CYP Enzymes

The monooxygenase properties of human and other mammalian CYP enzymes have been reviewed extensively in previous articles and will not be discussed in detail in this chapter. Readers are referred to the following articles and comprehensive reviews dealing with CYP monooxygenase mechanisms [1, 3, 4, 11, 97–99, 120, 138–145], oxygen activation and reactivity [11, 98, 115, 116, 145], human [7–9, 146], rat [3, 4] and mouse [8, 87, 147] CYP enzymes, steroidogenic CYP enzymes [83, 148], diversity and complexity of CYP reactions [7, 8, 144], CYP chemistry and oxidative reactions [6, 120, 149] and unusual CYP enzymes and reactions [5, 150]. Mammalian CYP enzymes catalyze a variety of monooxygenation reactions (shown in Table 1.3) including aliphatic and aromatic hydroxylations, *N*-hydroxylations, alcohol and amine oxidations, oxygenations of heteroatoms (N, S, P and I), alkene and arene epoxidations, dehalogenations, deaminations and *N*-, *O*- and *S*-dealkylations

Table 1.3 Selected monooxygenase reactions catalyzed by mammalian CYP enzymes[a]

Types of monooxygenase reactions	Examples of substrates	Product(s) formed
Alicyclic carbon hydroxylation	Acetohexamide	*trans*-4-Hydoxyacetohexamide
	Cyclophosphamide	4-Hydroxycyclophosphamide
Aliphatic amine hydroxylation	Amantadine, benzylamphetamine, nicotine, phenmetrazine, phentermine	*N*-Hydroxy derivatives
Aliphatic carbon hydroxylation	Alcohols, *n*-alkanes, bile acids, cyclohexane, drugs, fatty acids, prostaglandins, steroids	Hydroxylation at different sites
Alkene epoxidation	Aflatoxin B_1	Aflatoxin B_1 2,3-oxide
	Styrene	Styrene oxide
Aromatic carbon epoxidation	Benzene, benzo[*a*]pyrene, naphthalene	Epoxide derivatives
Aromatic carbon hydroxylation	Amphetamine, benzo[*a*]pyrene, coumarin, phenobarbital, propranolol, BDE-47[b], BDE-99, PCBs 45, 47, 52, 77, 84, 91, 95, 136	Hydroxylation at different sites
N-Dealkylation	Aminopyrine	Monomethyl-4-aminoantipyrine
	Caffeine	1,7-Dimethylxanthine
O-Dealkylation	Codeine	Morphine
	Phenacetin	Acetaminophen
S-Dealkylation	Methitural	Desmethylmethitural
	Methylmercaptan	Mercaptan
Dehalogenation	Chloramphenicol	Oxamic acid derivative
	Chloroform	Phosgene
N-Hydroxylation	Acetaminophen, phenacetin, phenytoin	*N*-Hydroxy derivatives
Methyl hydroxylation	Dacarbazine, ibuprofen, midazolam, naproxen	Hydroxymethyl derivatives
Sulfoxidation	Chlorpromazine, cimetidine, dimethyl sulfide	Sulfoxide derivatives

[a]Examples are not meant to be comprehensive but are meant to illustrate the diversity of substrates and reactions. Data were compiled from Refs. [7, 8, 151–163]

[b]*BDE-47* 2,2′,4,4′-tetrabromodiphenyl ether, *BDE-99* 2,2′,4,4′,5-pentabromodiphenyl ether, *PCB 45* 2,2′,3,6-tetrachlorobiphenyl, *PCB 47* 2,2′,4,4′-tetrachlorobiphenyl, *PCB 52* 2,2′,5,5′-tetrachlorobiphenyl, *PCB 77* 3,3′,4,4′-tetrachlorobiphenyl, *PCB 84* 2,2′,3,3′,6-pentachlorobiphenyl, *PCB 91* 2,2′,3,4′,6-pentachlorobiphenyl, *PCB 95* 2,2′,3,5′,6-pentachlorobiphenyl, *PCB 136* 2,2′,3,3′,6,6′-hexachlorobiphenyl

Table 1.4 Human CYP enzymes arranged according to their major substrate class[a]

Major substrate class/CYP enzyme	Major characteristic activities
Sterols	
7A1[b], 7B1, 8B1, 27A1, 39A1	Bile acid synthesis or metabolism
11A1, 11B1, 11B2, 17A1, 21A2	Steroid hormone synthesis
19A1	Estrogen hormone synthesis
46A1	Cholesterol metabolism
51A1	Cholesterol synthesis
Fatty acids	
2J2	Arachidonic acid epoxidation, linoleic acid metabolism
2U1, 4A11, 4B1, 4F3B, 4F12	Fatty acid hydroxylation
Eicosanoids	
2S1	Metabolism of prostaglandin G_2, prostaglandin H_2, 5*S*-, 12*S*- and 15*S*-HPETE[c]
4F2, 4F3A, 4F11	Leukotriene metabolism
4F8	19*R*-Hydroxyprostaglandin synthesis
5A1	Thromboxane A_2 synthesis
8A1	Prostaglandin I_2 synthesis
Vitamins	
2R1	Vitamin D_3 25-hydroxylation
26A1, 26B1, 26C1	Retinoic acid metabolism
24A1	1,25-Dihydroxyvitamin D_3 24-hydroxylation
27B1	25-Hydroxyvitamin D_3 1α-hydroxylation
Xenobiotics	
1A1, 1A2, 1B1, 2A6, 2A13, 2B6, 2C8, 2C9, 2C18, 2C19, 2D6, 2E1, 2F1, 3A4, 3A5, 3A7	*Oxygenation of xenobiotics*: drugs, carcinogens, PAHs (benz[*a*]anthracene, benzo [*a*]pyrene, naphthalene, nitropyrene, phenanthrene), alkanes, alcohols, anesthetics, organic solvents, pesticides, plant products (alkaloids, flavonoids, terpenes), BDE-47, BDE-99, PCBs 45, 91, 136, 153
Unknown	
2A7, 2W1, 3A43, 4A22, 4F22, 4V2, 4X1, 4Z1, 20A1, 27C1	

[a]Substrate classification for the human CYP enzymes is somewhat arbitrary as several CYP enzymes can metabolize substrates from more than one class [8, 9]. Data were compiled from Refs. [7–9, 146, 151–153, 156, 159, 160, 164–167]
[b]Due to spatial considerations, the CYP designation is not shown for the CYP enzymes
[c]*BDE-47* 2,2′,4,4′-tetrabromodiphenyl ether, *BDE-99* 2,2′,4,4′,5-pentabromodiphenyl ether, *PAHs* polycyclic aromatic hydrocarbons, *PCB 45* 2,2′,3,6-tetrachlorobiphenyl, *PCB 91* 2,2′,3,4′,6-pentachlorobiphenyl, *PCB 136* 2,2′,3,3′,6,6′-hexachlorobiphenyl, *PCB 153* 2,2′,4,4′,5,5′-hexachlorobiphenyl, *5S-HPETE* 5*S*-hydroperoxy-6*E*,8*Z*,11*Z*,14*Z*-eicosatetraenoic acid, *12S-HPETE* 12*S*-hydroperoxy-5*Z*,8*Z*,10*E*,14*Z*-eicosatetraenoic acid, *15S-HPETE* 15*S*-hydroperoxy-5*Z*,8*Z*,*10E*,*14Z*-eicosatetraenoic acid, *Retinoic acid* all-*trans*-Retinoic acid

[4–9, 93]. Human CYP enzymes and their major characteristic activities are listed in Table 1.4. Mammalian CYP systems function in the deactivation of xenobiotics as well as their metabolic bioactivation. CYP enzymes are the human body's first line of defence in the modification and detoxification of drugs and other xenobiotics [9], with hepatic CYP enzymes being of foremost importance. CYP enzymes catalyze the oxidative biotransformation of countless exogenous compounds including drugs, carcinogens, alkanes, alcohols, organic solvents, anesthetics, plant products (alkaloids, flavonoids, terpenes) and environmental chemical contaminants such as pesticides, polycyclic aromatic hydrocarbons (PAHs), polybrominated diphenyl ethers (PBDEs) and polychlorinated biphenyls (PCBs) [3–9, 146, 149–158]. CYP enzymes also function in the biosynthesis or metabolism of endogenous or natural bioactive substances such as alkaloids, amino acids, bile acids, bilirubin, cholesterol, eicosanoids (eicosaenoic acids, leukotrienes, prostaglandins), fatty acids, indole, melatonin, steroid hormones, terpenes and vitamins [3–9, 146, 147, 159–162, 164–167]. The reaction mechanisms by which CYP enzymes catalyze monooxygenation reactions are examined in Sect. 1.7.

1.5 Peroxidase Properties of CYP and Other Heme Enzymes

1.5.1 Heme Peroxidase Models for CYP

Heme peroxidases such as HRP, CPO and *Agrocybe aegerita* unspecific peroxygenase (*Aae*UPO) use H_2O_2 and other hydroperoxides to catalyze one-electron oxidation of substrates (peroxidase activity) and can also incorporate an oxygen atom from O_2 into organic substrates (monooxygenase activity) [78, 93, 111, 118, 168, 169]. CYP monooxygenases can also function as peroxidases [18–22]. HRP, CPO and *Aae*UPO have structural and functional similarities to CYP enzymes and have been used as models to study CYP reaction mechanisms. The catalytic functions of HRP, CPO and *Aae*UPO helped characterize the reactive oxygen intermediates that participate in the CYP monooxygenase cycle, an area that produced numerous articles and reviews [1, 34, 78, 93, 98, 117, 118, 120, 124, 125, 141–143, 168–198]. Characteristic structural and optical features of selected CYP enzymes, heme peroxidases and their CpdI derivatives are listed in Table 1.5.

1.5.1.1 Horseradish Peroxidase (HRP)

HRP has long been the preferential paradigm for biochemical investigations of heme peroxidases. Early studies revealed that spectrally-observable transitory ferryl CpdI and CpdII complexes were formed upon reacting HRP with H_2O_2 [118, 168,

Table 1.5 Proximal ligands and compound I features in selected CYP enzymes and heme peroxidases[a]

Hemeprotein	Source	Proximal ligand[b]	CpdI[c] (or CpdES) structure	CpdI (or CpdES) spectral peaks (nm)	Refs.
rCYP2B4	Rabbit	Thiolate	$Por\bullet^{+}Fe^{IV}{=}O$	~417	[199]
CYP101A1	*Pseudomonas putida*	Thiolate	$Por\bullet^{+}Fe^{IV}{=}O$	367, ~694	[108, 139, 176, 180]
CYP101A1	*Pseudomonas putida*	Thiolate	$Tyr\bullet Fe^{IV}{=}O$[d]	406[d]	[180]
CYP102A1 (Phe87Gly)[e]	*Bacillus megaterium*	Thiolate	$Por\bullet^{+}Fe^{IV}{=}O$	370, 422	[136]
rCYP119A1	*Sulfolobus acidocaldarius*	Thiolate	$Por\bullet^{+}Fe^{IV}{=}O$	370, ~416, 610, 690	[32, 34]
CPO	*Caldariomyces fumago*	Thiolate	$Por\bullet^{+}Fe^{IV}{=}O$	~367, 610, ~690	[118, 139, 175]
*Aae*UPO	*Agrocybe aegerita*	Thiolate	$Por\bullet^{+}Fe^{IV}{=}O$	361, 694	[200]
HRP	Horseradish root	Imidazole	$Por\bullet^{+}Fe^{IV}{=}O$	400, 577, ~620, 651	[118, 169, 194]
CcP	Yeast	Imidazole	$Trp\bullet Fe^{IV}{=}O$[f]	421, 530, 561	[118, 178]
Catalase	Ox liver	Tyrosinate	$Por\bullet^{+}Fe^{IV}{=}O$	405, 660	[194]

[a]Data compiled in Table 1.5 were obtained mainly from ref. [1]

[b]A cysteine contributes the proximal thiolate ligand for CYP enzymes, CPO and *Aae*UPO, a histidine supplies the imidazole ligand for HRP and CcP, while a tyrosine provides the tyrosinate ligand for catalase

[c]*AaeUPO Agrocybe aegerita* unspecific peroxygenase, *mCPBA* *m*-chloroperbenzoic acid, *CPO* chloroperoxidase, *CpdI* Compound I, *CpdES* Compound ES, *CcP* cytochrome *c* peroxidase, *HRP* horseradish peroxidase, *rCYP* recombinant cytochrome P450

[d]Spolitak et al. [180] reacted ferric CYP101A1 with *m*CPBA to generate the tyrosine radical-ferryl intermediate ($Tyr\bullet Fe^{IV}{=}O$) known as CpdES, which elicited an absorption maximum at 406 nm

[e]Raner et al. [136] reacted the Phe87Gly mutant of CYP102A1 with *m*CPBA to generate the CpdI species

[f]The tryptophan (Trp191) radical-ferryl complex ($Trp\bullet Fe^{IV}{=}O$) of CcP is known as CpdES [177]

169, 177]. Histidine imidazole-ligated HRP typically catalyzes single-electron oxidations of substrates (see reactions below) and does not readily perform the demanding two-electron oxygen atom transfer reactions catalyzed by cysteine thiolate-ligated CPO and CYP enzymes.

$$Fe^{III} + H_2O_2 \xrightarrow{-H^+} Fe^{III}{-}O{-}OH \xrightarrow{+H^+} Fe^{III}{-}O{-}OH_2 \rightarrow Por\bullet^{+}Fe^{IV}{=}O\ (CpdI) + H_2O \quad (1.2)$$

$$Por\bullet^{+}Fe^{IV}{=}O + AH \rightarrow Fe^{IV}{=}O\ (CpdII) + A^{\bullet} + H^+ \quad (1.3)$$

$$Fe^{IV}{=}O + AH + H^+ \rightarrow Fe^{III} + A^{\bullet} + H_2O \quad (1.4)$$

Thus, the peroxidase mechanism commences with the binding of H_2O_2 to ferric HRP (Fe^{III}). A histidine residue located in the distal binding pocket of HRP then withdraws a proton from the proximal (closest to ferric iron) oxygen atom of the ligated peroxide and donates the same proton to the distal (furthest from ferric iron) oxygen atom of the $Fe^{III}{-}O{-}OH$ species, generating the iron-oxo–H_2O adduct precursor ($Fe^{III}{-}O{-}OH_2$) and the $Por\bullet^{+}Fe^{IV}{=}O$ species (CpdI) (Eq. 1.2) [98, 118]. The next step involves the one-electron reduction of the porphyrin radical ($Por\bullet^{+}$) of CpdI by the first molecule of substrate (AH) to form the $Fe^{IV}{=}O$ species (CpdII), and the concomitant one-electron oxidation of AH to its free radical product ($A^{\bullet}$) (Eq. 1.3). The final step is the one-electron reduction of CpdII by a second molecule of AH, regenerating the ferric enzyme (Fe^{III}) and releasing the ferryl oxygen atom as H_2O (Eq. 1.4). The rate of substrate oxidation by CpdI is normally 10–100-times faster than the rate of oxidation by CpdII [118].

The heme iron oxidation states of HRP have been investigated thoroughly using spectroscopic techniques [118, 173, 177, 194]. The UV/visible absorption spectrum of CpdI is characterized by a broad Soret band at ~400 nm and a weak absorption band at 651 nm [118]. HRP contains histidine and arginine residues located distal to the heme that are believed to promote O–O bond cleavage in $Fe^{III}{-}O{-}OH_2$ and expedite CpdI formation (Eq. 1.2) [118]. Once CpdI has been generated, the heme environment plays a crucial role in controlling the subsequent reactivity of the catalytic system and ultimately determining if HRP will function as a peroxidase or an oxotransferase. Hydroperoxide-dependent HRP-catalyzed oxidations of substrates are generally one-electron oxidations that do not involve oxo transfer from the CpdI ferryl moiety to the electron donor substrate [78, 117, 170]. Substrates interact exclusively with the heme edge near the -*meso*-carbon of HRP, and substrate oxidation involves electron transfer to the heme periphery between the -*meso*-carbon and 8-methyl substituent [78, 170]. In addition, the peroxygenase (oxotransferase) activity of HRP is suppressed because of the steric and polar constraints of the protein-imposed barrier, which limit access of organic substrates to the ferryl oxygen atom of CpdI and prevent substrates from being oxygenated [170]. In contrast, in CYP-catalyzed monooxygenation reactions, CYP functions as an oxotransferase by transferring the ferryl oxygen atom of CpdI to the substrate, demonstrating that oxidizable substrates are readily accessible to the CYP ferryl moiety [34, 133].

1.5.1.2 Chloroperoxidase (CPO)

CPO has been the most intensely studied heme haloperoxidase [185] and was originally isolated from the ascomycetous fungus, *Caldariomyces fumago* [197]. Directed evolution of CPO created catalytic mutants with improved epoxidation and chlorination activities [198]. The signature function of CPO is the H_2O_2-dependent halogenation of organic substrates in the presence of a halide anion (bromide, chloride, iodide) [107, 118, 173, 185, 201], as outlined in the reactions shown below.

$$Fe^{III} + H_2O_2 \rightarrow Fe^{III}{-}O{-}OH + H^+ \rightarrow Por\bullet^{+}Fe^{IV}{=}O\ (CpdI) + H_2O \quad (1.5)$$

$$Por\bullet^{+}Fe^{IV}{=}O + X^- \rightarrow Fe^{IV}{-}OX(CpdX) \quad (1.6)$$

$$Fe^{IV}{-}OX + H^+ \rightarrow XOH + Fe^{III} \quad (1.7)$$

$$XOH + AH \rightarrow AX + H_2O \quad (1.8)$$

The halogenation cycle begins with the binding of H_2O_2 to ferric CPO (Fe^{III}), which is followed by heterolytic cleavage of the peroxy O–O bond to form CpdI (Eq. 1.5). CpdI then reacts with a halide anion (X^-) to generate the CpdX halogenating species (Fe^{IV}–OX), which is chemically unstable and dissociates to produce a hypohalous acid (XOH) and ferric enzyme (Eq. 1.7). The final step is the halogenation of the substrate (AH) in the presence of XOH and the release of the halogenated product (AX) and H_2O (Eq. 1.8) [185, 186]. CPO can use chlorite (ClO_2^-) as an oxidant (replacing H_2O_2) and a chloride donor (replacing the chloride anion) in the oxidative chlorination of monochlorodimedone [173]. Representative halogenation activities catalyzed by CPO include the chlorination and bromination of β-diketones [198], halogenation of alkenes, alkynes and cycloalkanes [202] and chlorination of PAHs [203].

Although CPO is a poor peroxidase, it mimics CYP enzymes and readily performs heteroatom dealkylations, allylic and benzylic hydroxylations, alcohol oxidations and oxygen atom transfer to alkenes, alkynes, alkyl- and arylamines, and sulfides [78, 174, 175, 204, 205]. Ferric CPO can generate CpdI upon reacting with H_2O_2, organic hydroperoxides and peracids [118, 185]. As in CYP-catalyzed hydroxylations, CPO-mediated hydroperoxide-driven hydroxylations proceed via the H atom abstraction/oxygen rebound pathway involving CpdI [127, 206]. The spectroscopic features of CpdI complexes of CPO have been investigated extensively (see Table 1.5) [34, 118, 175, 179, 206] and the CPO CpdI species represents the best characterized reactivity model for CYP CpdI [34, 118, 206]. The cysteine thiolate axial ligand of CPO has long been thought to be essential for chlorination, epoxidation and other monooxygenation activities. In addition, polar amino acid residues located distal to the heme group of CPO are vital in maintaining its diverse catalytic activities [118, 132].

1.5.1.3 *Agrocybe aegerita* Unspecific Peroxygenase (*Aae*UPO)

In addition to CPO, a second heme-thiolate haloperoxidase named *Aae*UPO was discovered in the agaric basidiomycetous fungus, *Agrocybe aegerita*, a popular edible mushroom found in Mediterranean countries [184–186]. Hofrichter and coworkers carried out extensive studies on the spectroscopic and catalytic properties of *Aae*UPO and showed its close relationship to CPO and CYP enzymes [184–190, 200]. The Soret absorption band of the reduced iron-CO complex of *Aae*UPO displays an absorbance maximum at 445 nm [184] indicative of thiolate ligation to the heme iron [4]. Recently, a highly reactive CpdI species of *Aae*UPO was characterized by determining its UV/visible spectral features using rapid-mixing, stopped-flow spectroscopy [200]. Mixing of ferric *Aae*UPO with *m*-chloroperbenzoic acid (*m*CPBA) generated a short-lived intermediate that elicited an absorption band at 361 nm, which indicated the presence of a porphyrin π radical CpdI species. Kinetic measurements of the reactivity of *Aae*UPO CpdI using a panel of substrates revealed rapid hydroxylation rates similar to those found for CpdI of CYP119A1 [32, 34] and for the most reactive synthetic iron porphyrin model compounds [207].

*Aae*UPO uses its peroxidase function to catalyze the metabolism of H_2O_2 and the H_2O_2-driven halogenation of monochlorodimedone and phenol [185, 186]. Extracellular *Aae*UPO catalyzes H_2O_2-dependent aromatic and aliphatic hydroxylation and epoxidation reactions usually accredited to intracellular CYP enzymes [185, 186, 188]. *Aae*UPO regioselectively catalyzes the hydroxylation of aromatic substrates (benzene, naphthalene, toluene) and aliphatic compounds (alkanes, fatty acids) [184–186, 189], the benzylic hydroxylation of *n*-alkyl- and cycloalkylbenzenes and the epoxidation of styrenes [190]. *Aae*UPO also catalyzes the peroxygenation of drug substrates (e.g. diclofenac, ibuprofen, metoprolol, naproxen, phenacetin, propranolol, sildenafil, tolbutamide) to produce relevant human drug metabolites [187, 188]. The H_2O_2-mediated monooxygenase cycle of *Aae*UPO consists of intermediates and reactions that are similar to those that occur in CPO and CYP monooxygenase cycles [185, 186, 200].

1.5.2 Peroxidase Nature of CYP Enzymes

In early studies, Hrycay and coworkers established that CYP enzymes of rat liver microsomes and bovine adrenocortical microsomes functioned as peroxidases in the metabolism of lipid hydroperoxides, steroid hydroperoxides and alkyl hydroperoxides [18–22], while purified rat liver CYP2B1 catalyzed the reduction of H_2O_2 [23]. The transitory CYP species postulated to participate in the peroxidase mechanism was the perferryl entity ($Fe^{V}{=}O$) displaying a formal Fe(V) oxidation state [1, 18, 19] and therefore being isomerically equivalent to the ferryl CpdI species ($Por\bullet^{+}Fe^{IV}{=}O$). The combination of $Por\bullet^{+}$ and $Fe^{IV}{=}O$ in CpdI represents a formal Fe(V) oxidation state. Electron donor substrates in peroxidase reactions, such as *N,N,N′,N′*-tetramethyl-*p*-phenylenediamine (TMPD) [18, 19, 208], NAD(P)H [21, 22], alcohols, phenols, diaminobenzidine, dimethylphenylenediamine, sulfite [130], ascorbate and guaiacol [180], were able to reduce and discharge the high-valent iron-oxo intermediates [1, 172]. To measure the peroxidase activity of CYP enzymes, Hrycay and O'Brien [18, 19] developed a sensitive and convenient colorimetric assay whereby metabolism of the hydroperoxide cosubstrate by CYP enzymes was coupled to the one-electron oxidation of TMPD and formation of its Wurster's blue free radical metabolite, as shown below.

$$Fe^{III} + XOOH \rightarrow Fe^{III}{-}O{-}OH + X^{+} \rightarrow Por\bullet^{+}Fe^{IV}{=}O\ (CpdI) + XOH \tag{1.9}$$

$$Por\bullet^{+}Fe^{IV}{=}O + A \rightarrow Fe^{IV}{=}O\ (CpdII) + A^{\bullet} \tag{1.10}$$

$$Fe^{IV}{=}O + A + 2H^{+} \rightarrow Fe^{III} + A^{\bullet} + H_2O \tag{1.11}$$

Thus, the CYP peroxidase mechanism commences with the reaction of ferric CYP (Fe^{III}) and a hydroperoxide (XOOH) (X is an organic substituent or H atom), which generates the CpdI species and the reduced hydroperoxide (XOH) or H_2O. The porphyrin π radical ($Por\bullet^{+}$) of CpdI is subsequently reduced via a one-electron reduction by the first molecule of substrate A (TMPD) to produce CpdII ($Fe^{IV}{=}O$). A concomitant one-electron oxidation of substrate A (TMPD) generates its free radical oxidation product $A^{\bullet}$ (Wurster's blue). The final step is a one-electron reduction of CpdII by a second molecule of substrate A (TMPD), regenerating ferric CYP (Fe^{III}) and releasing the ferryl oxygen atom as H_2O (Eq. 1.11) [1].

Mammalian hepatic microsomal CYP enzymes can function as dioxygenases by using lipid substrates as hydrogen donors to catalyze the O_2-dependent formation of lipid hydroperoxides [209]. CYP enzymes subsequently use their peroxidase function to decompose the lipid hydroperoxides [18]. Alternatively, CYP enzymes can use lipid hydroperoxides as oxygen atom donors to oxygenate a variety of xenobiotic and endobiotic substrates. Mammalian CYP enzymes metabolize biological and exogenous hydroperoxides by a reductive process, generating the corresponding alcohols as major products and sometimes other metabolites [18–20, 59, 63, 82, 209–216]. Biological hydroperoxides are generated in vivo at various mammalian cellular sites and are subsequently metabolized by CYP enzymes. Biological hydroperoxide substrates include H_2O_2 [18, 23, 82], lipid or fatty acid hydroperoxides (arachidonic acid hydroperoxides [59, 211, 213], linoleic acid hydroperoxide [18, 19], linolenic acid hydroperoxide [59]), steroid hydroperoxides (6α- and 6β-hydroperoxyandrostenedione [214], pregnenolone 17α- and progesterone 17α-hydroperoxide [19, 20], cholesterol hydroperoxides [19]), prostaglandin endoperoxides (PGG_2, PGH_2) [164, 167, 215, 216] and hydroperoxides of other eicosanoids [167, 210, 212, 217]. Because mechanistic interest in reactions of CYP with hydroperoxides resides in its vital peroxygenase rather than peroxidase function, the peroxygenase properties of CYP are examined in Sect. 1.6.

1.6 Peroxygenase Properties of CYP Enzymes

1.6.1 Peroxygenase Nature of CYP Enzymes

In addition to their monooxygenase and peroxidase functions, CYP enzymes function as peroxygenases by utilizing hydroperoxides, peracids and other peroxy compounds as oxygen atom donors to catalyze the peroxygenation of substrates by a reaction not involving the reduction of ferric CYP and not requiring NAD(P)H/O_2, the CYP oxidoreductase and other redox proteins. The oxygen atom incorporated into the substrates is derived from the peroxy compound rather than from the medium [38, 63, 169, 218]. The peroxygenase activity of CYP enzymes is illustrated in simplified form in Eq. 1.12, as shown below.

$$A + XOOH \xrightarrow{\text{peroxygenase CYP}} AO + XOH \quad (1.12)$$

Thus, one oxygen atom and two oxidizing equivalents are extracted directly by ferric CYP from the peroxy compound (XOOH) (X is an organic substituent or H atom) and delivered to the electron donor substrate (A), generating the oxygenated substrate (AO) and reduced peroxy compound (XOH) or H_2O. Peroxygenase reactions have yielded valuable insights into the catalytic mechanisms by which CYP enzymes oxygenate substrates. Peroxygenase reactions catalyzed by CYP enzymes of mammalian, bacterial and archaeal systems include aliphatic, aromatic, fatty acid and steroid hydroxylations, *N*- and *O*-dealkylations, alkene and arene epoxidations and alcohol oxidations. Ability of peroxygenase reactions to mimic NAD(P)H/O_2-mediated monooxygenase reactions has been shown with a variety of substrates (e.g. alcohols, carcinogens, coumarins, drugs, fatty acids, PAHs, resorufins, Retinoic acids, steroids). CYP substrates and peroxygenase reactions driven by hydroperoxides and peracids are listed in Tables 1.1, 1.2, and 1.6. Monooxygenase reactions driven by peroxides and other peroxy compounds often generate structurally similar products to those formed in NAD(P)H/O_2-mediated reactions. It has therefore been assumed that a common mechanism operates in the two systems.

In the peroxygenase pathway, ferric CYP reacts with peroxy compounds to form a CpdI intermediate [1, 34, 63, 68] believed to be identical to the provisional CpdI intermediate formed in the NAD(P)H/O_2-mediated monooxygenase pathway. In both pathways, substrates undergoing monooxygenation act as electron donors in a peroxidase-type mechanism [1, 172]. The peroxygenase pathway involving substitution of peroxy compounds for NAD(P)H/O_2 and redox proteins is commonly referred to as the peroxide shunt or simply the shunt [1, 11, 23–26, 149] (see Fig. 1.3, steps 7 and 8). The shunt pathway leading to Cpd I formation is equivalent to the first steps of the heme peroxidase pathways involving HRP, CPO or *Aae*UPO (see Sect. 1.5.1). The shunt pathway can be used to oxygenate diverse substrates and serves a vital function in humans and other mammals in view of the many biological peroxides that are generated in vivo [1, 18, 19, 23, 25, 26, 59, 62, 65, 74, 76, 82, 209, 214, 216, 217]. Representative examples of peroxygenase reactions catalyzed by mammalian CYP enzymes and supported by hydroperoxides are listed in Table 1.6. Some of these reactions warrant a brief discussion.

1.6.1.1 Mammalian Monooxygenase Reactions Supported by Organic Hydroperoxides or NADPH/O_2

Despite the prevalence of common features in organic hydroperoxide – and NADPH/O_2-supported monooxygenation reactions, the reactions driven by hydroperoxides sometimes generate structurally different products from those formed in NADPH/O_2-sustained reactions, which have been postulated to occur via the involvement of different CYP enzymes [25, 59, 130, 227, 228] or different reaction mechanisms [5, 111, 222–224, 229, 235–238]. The catalytic mechanisms underlying these differences are explored in Sects. 1.6.1.1 and 1.6.1.2. The majority of mammalian organic hydroperoxide-driven

Table 1.6 Selected mammalian CYP peroxygenase reactions supported by exogenous organic hydroperoxides

Exogenous hydroperoxide	Substrate	Major product(s) or type of oxidation	CYP enzyme or source	Refs.[a]
t-BuOOH[b]	Aminopyrine	*N*-Demethylation	Rat liver micr	[219]
	Benzphetamine	*N*-Demethylation	Rabbit CYP2B4	[63]
	Dextromethorphan; testosterone	Dextrorphan; 6β-OH[c]	Human rCYP2D6, rCYP3A4	[75]
	Ethanol	Acetaldehyde	Rat liver micr	[61]
	Androstenedione	6β-, 7α-, 16α-OH[c]; 6β-, 11β-, 19-OH	Rat CYP2B1; ox adr mito	[23, 26]
	17α-Hydroxyprogesterone, progesterone	21-OH	Ox adr micr	[26]
CuOOH	Alcohols (ethanol, propanol, butanol, pentanol)	Corresponding aldehydes	Rat liver microsomal CYP2B1[d]	[61]
	Aminopyrine	*N*-Demethylation	Rabbit CYP2B4, rat liver micr	[63, 219]
	Amitryptiline	*N*-Demethylation	Human rCYP2D6	[220]
	Aniline; biphenyl; coumarin	*p*-Hydroxyaniline; hydroxylation; 7-OH	Rabbit liver micr[e]; rat liver micr; rabbit liver microsomal CYP2B4[f]	[221]
	Benzo[*a*]pyrene	1,6-, 3,6-, 6,12-Quinones	Human rCYPs 1A1, 2S1, rat liver micr[g]	[65, 222, 223]
	Benzphetamine	*N*-Demethylation	Pig and rat liver micr, rabbit CYP2B4	[62, 63]
	4-Chloroaniline	1-Chloro-4-nitro-benzene	Rabbit CYP2B4	[224]
	Cyclohexane; cyclohexene	Cyclohexanol; epoxide, enol	Rabbit CYP2B4	[63, 138]
	Dextromethorphan	*O*-Demethylation	Human rCYP2D6	[75, 220]
	N,N-Dimethylaniline	*N*-Demethylation	Pig and rat liver micr, rabbit CYP2B4	[62, 63]
	Ethanol	Acetaldehyde	Rat liver micr, rat CYP2B1, human rCYP2E1	[225]
	7-Ethoxycoumarin	*O*-Deethylation	Hamster liver micr, olfactory micr; rabbit liver microsomal CYP2B4[f]	[208, 226]
	Lauric acid; 5β-cholestane-3α,7α,12α-triol	11-Hydroxylation; 25-OH	Rat liver micr; kidney micr	[227, 228]
	p-Methyltoluene	*p*-Methylbenzyl alcohol	Rabbit CYP2B4	[229]
	Phenacetin	*O*-Deethylation	Rabbit liver microsomal CYP2B4[f], human rCYP1A2	[226, 230]
	Phenanthrene	Phenanthrene 9,10-oxide	Rat liver micr	[218]
	Propranolol	Hydroxylation	Human liver micr	[231]
	Retinoic acid	4-OH- and 4-oxo-Retinoic acid	Human rCYPs 1A1, 1A2, 2B6, 2D6, 2E1, 2S1, 3A4	[65, 74]

(continued)

Table 1.6 (continued)

Exogenous hydroperoxide	Substrate	Major product(s) or type of oxidation	CYP enzyme or source	Refs.[a]
	Androstenedione	6β-, 7α-, 16α-OH; 6β-, 7α-, 15-, 16α-, 17β-OH; 6β-, 11β-, 16β-, 19-OH	Rat CYP2B1; rat liver micr; ox adr micr, mito	[23–26]
	17β-Estradiol	6α-, 6β-, 16α-OH; 2-, 4-OH	Rat liver micr; human placental micr, human rCYP1B1	[25, 232, 233]
	Progesterone	2α-, 6β-, 15β-, 16α-OH; 6β-, 7β-, 21-OH	Rat liver micr; ox adr micr	[25, 26]
	Testosterone	6β-, 7α-, 16α-OH; 6β-OH	Rat liver micr; human rCYP3A4	[25, 75]
p-MenthylOOH	Aminopyrine	*N*-Demethylation	Pig liver micr	[62]
	Benzphetamine	*N*-Demethylation	Rabbit CYP2B4	[63]
	Ethanol	Acetaldehyde	Rat liver micr	[61]

[a]Data compiled in Table 1.6 were obtained mainly from ref. [1]

[b]*Adr* adrenocortical, *t-BuOOH t*-butyl hydroperoxide, *CuOOH* cumene hydroperoxide, *7-EFC* 7-ethoxy-4-trifluoromethylcoumarin, *p-menthylOOH p*-menthyl hydroperoxide, *micr* microsomes, *mito* mitochondria, *rCYP* recombinant cytochrome P450, *Retinoic acid* all-*trans*-Retinoic acid

[c]Compounds with the attached OH group signify hydroxy derivatives of substrates

[d]Liver microsomes were prepared from phenobarbital-pretreated rats. CYP2B1 is the predominant hepatic CYP enzyme induced by pretreating rats with phenobarbital [79]

[e]Liver microsomes were prepared from 3-methylcholanthrene-pretreated rabbits. CYP1A2 is the predominant hepatic CYP enzyme induced by pretreating rabbits with 3-methylcholanthrene [169]

[f]Liver microsomes were prepared from phenobarbital-pretreated rabbits. CYP2B4 is the predominant hepatic CYP enzyme induced by pretreating rabbits with phenobarbital [77, 169]

[g]Liver microsomes were prepared from phenobarbital- or 3-methylcholanthrene-pretreated rats [222, 223]. CYP2B1 and CYP1A1 are the predominant hepatic CYP enzymes induced by pretreating rats with phenobarbital and 3-methylcholanthrene, respectively [79, 234]

Table 1.7 Selected mammalian and bacterial CYP monooxygenase reactions supported by exogenous oxidants

Exogenous oxidant	Substrate	Major product(s) or type of oxidation	CYP enzyme or source	Refs.
N,*N*-Dimethyl-aniline *N*-oxide	*N*,*N*-Dimethylaniline *N*-oxide	*N*-Demethylation	Rabbit CYP2B4, rat rCYP2E1, CYP101A1	[242, 244]
	N-Cyclopropyl-*N*-methylaniline; *N*,*N*-dimethylaniline	*N*-Demethylation	CYP101A1	[245]
PFDMAO[a]	*N*-Cyclopropyl-*N*-methylaniline; *N*,*N*-dimethylaniline	*N*-Demethylation	CYP101A1	[245]
Pyridine *N*-oxide	Camphor	5-*exo*-Hydroxylation	CYP101A1	[281]
Iodosobenzene	Androstenedione	6β-, 7α-, 16α-OH[b]	Rat liver micr	[282]
	Camphor	5-*exo*-Hydroxylation	CYP101A1	[36]
	N,*N*-Dimethylaniline	*N*-Demethylation, *N*-oxide	Rabbit rCYP2B4, rat CYP2B1	[68, 283]
	7-Ethoxycoumarin	*O*-Deethylation	Rat liver microsomal CYP2B1[c]	[284]
	Laurate	11-Hydroxylation	Rat liver micr	[285]
	Progesterone	15α-, 15β-OH	CYP106A2	[286]
	Testosterone	6β-OH	Human rCYP3A4	[287]
	U46619	Hydroxylation	Human platelet microsomal CYP5A1, rat liver microsomal CYP2B1[c]	[215]
Sodium chlorite	Androstenedione	6β-, 16α-OH; 6β-, 7α-, 15-, 16α-, 17β-OH; 6β-, 11β-, 16β-, 19-OH	Rat CYP2B1; rat liver micr; ox adr micr, mito	[23–26]
	Benzphetamine	*N*-Demethylation	Rabbit CYP2B4	[63]
	17α-Hydroxyprogesterone	21-OH	Ox adr micr	[26]
	Laurate, palmitate, stearate	11-Hydroxylation	Rat liver micr	[285]
	Progesterone	6β-, 7β-, 21-OH; 15β-OH; 15α-, 15β-OH	Ox adr micr; adr mito; CYP106A2	[26, 286]
Sodium periodate	Androstenedione	6β-, 7α-, 16α-OH; 6β-, 7α-, 15-, 16α-, 17β-OH; 6β-, 11β-, 16β-, 19-OH	Rat CYP2B1; rat liver micr; ox adr micr, mito	[23–26]
	5β-Cholestane-3α,7α,12α-triol; taurochenodeoxycholic acid	25-OH; 6β-OH	Rat liver micr	[228]
	17β-Estradiol	6α-, 6β-, 16α-OH	Rat liver micr	[25]
	17α-Hydroxyprogesterone	21-OH	Ox adr micr	[26]
	Laurate, palmitate, stearate	11-Hydroxylation	Rat liver micr	[285]
	Progesterone	2α-, 6β-, 7α-, 15α-, 15β-, 16α-OH; 6β-, 7β-, 21-OH; 1β-, 6β-, 15β-OH; 15α-, 15β-OH	Rat liver micr; ox adr micr; adr mito; CYP106A2	[25, 26, 286]
	Testosterone	6β-, 7α-, 16α-OH	Rat liver micr	[25]

(continued)

Table 1.7 (continued)

Exogenous oxidant	Substrate	Major product(s) or type of oxidation	CYP enzyme or source	Refs.
Sodium perborate	Dextromethorphan; testosterone	Dextrorphan; 6β-OH	Human rCYP2D6; rCYP3A4	[75]
Sodium percarbonate	Dextromethorphan; testosterone	Dextrorphan; 6β-OH	Human rCYP2D6; rCYP3A4	[75]
*m*CPBA	Benzphetamine	*N*-Demethylation	Rabbit CYP2B4	[63]
p-Nitroperbenzoic acid	Benzphetamine	*N*-Demethylation	Rabbit CYP2B4	[63]
Phenylperacetic acid	Cumene; cyclohexane; ethylbenzene	2-Phenyl-2-propanol; cyclohexanol; α-methylbenzyl alcohol	Rabbit CYPs 1A2, 2B4; CYP2B4	[38, 288ok]
	Cyclohexene; *p*-chlorotoluene	Hydroxylation; benzylic hydroxylation	Rabbit CYP2B4	[237]

[a] *Adr* adrenocortical, *mCPBA m*-chloroperbenzoic acid, *micr* microsomes, *mito* mitochondria, *PFDMAO* 2,3,4,5,6-pentafluoro-*N*,*N*-dimethylaniline *N*-oxide, *rCYP* recombinant cytochrome P450, *U46619* 15*S*-hydroxy-11α,9α-epoxymethano-5*Z*,13*E*-prostadienoic acid

[b] Compounds with the attached OH group signify hydroxy derivatives of substrates

[c] Liver microsomes were prepared from phenobarbital-pretreated rats. CYP2B1 is the predominant hepatic CYP enzyme induced by pretreating rats with phenobarbital [79]

peroxygenase reactions have been performed using the model compound, cumene hydroperoxide (α,α-dimethylbenzyl hydroperoxide), as the organic oxidant (see Table 1.6).

Capdevila et al. [222] and Cavalieri et al. [223] examined the reaction mechanisms involved in cumene hydroperoxide- versus $NADPH/O_2$-supported benzo[*a*]pyrene oxidation by CYP enzymes of hepatic microsomes prepared from phenobarbital- or 3-methylcholanthrene-pretreated rats. In the hydroperoxide-driven reaction, the 1,6-, 3,6- and 6,12-dione quinone isomers were generated from benzo[*a*]-pyrene while in the $NADPH/O_2$-sustained reaction, the metabolic profile shifted to the production of mainly phenols (3-, 6- and 9-hydroxybenzo[*a*]pyrene) and dihydrodiols [222, 223]. Previous investigators demonstrated that quinones are generated by one-electron oxidations involving oxobenzo[*a*]pyrene radicals [235, 236]. Radical cationic chemical experiments using fluoro substitution as a probe for one-electron oxidation in aromatic substrates such as benzo[*a*]pyrene enabled Cavalieri et al. [223] to demonstrate that the CYP-catalyzed formation of quinones in hydroperoxide- and $NADPH/O_2$-mediated reactions are derived only from their intermediate free radical cations. A mechanism was proposed whereby the ferryl oxygen atom of the CpdI intermediate, formed in the peroxide shunt pathway or in the native $NADPH/O_2$ route, initiates a nucleophilic attack on the C-6 position in benzo[*a*]pyrene to produce a transient free radical cation, with the + charge initially localized at C-6 and the radical delocalized to various positions. Subsequent oxidation steps that involve insertion of an oxygen atom at C-6 and additional one-electron oxidations and oxygen atom insertions at C-1, C-3 and C-12 generate the quinone metabolites.

1.6.1.2 Heterolytic Versus Homolytic Peroxy O-O Bond Scission

When peroxides or peracids serve as terminal oxygenating agents in peroxygenase reactions, the requisite oxidizing equivalents for monooxygenation are derived from scission of the peroxy O–O bond. CYP enzymes catalyze the reductive cleavage of peroxy compounds using either the heterolytic (two-electron) or homolytic (one-electron) mechanism, the mode of cleavage depending on the structure of the peroxy compound and nature of the substrate [38, 111, 229, 237, 238]. Hydroperoxide cleavage via the heterolytic pathway generates the corresponding reduced hydroperoxide and a catalytically-viable CpdI intermediate [1, 11, 68]. When the peroxygenase activities involve hydrocarbon hydroxylation, reactions then proceed via the two-step H atom abstraction/oxygen rebound mechanism first formulated by Groves and coworkers (Eqs. 1.14 and 1.15) [126, 127], and based on the large intrinsic intramolecular kinetic isotope effects (KIEs) and net retention of stereochemistry at the oxidized carbon center of aliphatic substrates.

$$Fe^{III} + XOOH \rightarrow Por\bullet^{+}Fe^{IV}{=}O\ (CpdI) + XOH \tag{1.13}$$

$$Por\bullet^{+}Fe^{IV}{=}O + AH \xrightarrow{\text{H abstraction}} Fe^{IV}{-}OH + A^{\bullet} \xrightarrow{\text{oxygen rebound}} A^{\bullet}{-}Fe^{IV}{-}OH \rightarrow \tag{1.14}$$

$$AOH{-}Fe^{III} \rightarrow AOH + Fe^{III} \tag{1.15}$$

Thus, the mechanistic process commences with the reaction of ferric CYP (Fe^{III}) with the hydroperoxide (XOOH) (X is an organic substituent or H atom), which generates the primary CpdI intermediate and the reduced hydroperoxide (XOH) or H_2O (Eq. 1.13). CpdI subsequently abstracts a H atom from the hydrocarbon substrate (AH) and produces the ultimate protonated CpdII hydroxylating intermediate ($Fe^{IV}{-}OH$) and a carbon-centered substrate radical ($A^{\bullet}$), which then rebounds in the oxygen rebound step onto the nascent ferryl hydroxyl moiety to produce the substrate radical-bound protonated CpdII complex ($A^{\bullet}{-}Fe^{IV}{-}OH$) (Eq. 1.14) [239, 240]. The bound substrate radical is subsequently hydroxylated by the ferryl hydroxyl moiety to generate the ferric CYP-alcohol product complex ($AOH{-}Fe^{III}$), which then dissociates and releases the alcohol (AOH) and regenerated ferric CYP in the final step (Eq. 1.15). A similar mechanistic process operates in the classical

NAD(P)H/O_2-mediated hydroxylation pathway. The ferric-hydroperoxo intermediate (Fe^{III}–OOH) formed in the CYP cycle (see Fig. 1.3, step 4) accepts a proton from the CYP proton relay system (see Sect. 1.7.7.2), generating an iron-oxo–H_2O adduct precursor (Fe^{III}–O–OH_2) that rapidly dissociates through heterolytic cleavage of the O–O bond [98, 241] to produce the primary CpdI intermediate. CpdI then reacts via the H atom abstraction/oxygen rebound pathway, producing the ultimate protonated CpdII species that performs substrate hydroxylations [127, 239, 240]. The CpdI and protonated CpdII intermediates thus play critical roles in both the CYP-mediated NAD(P)H/O_2-supported, and the hydroperoxide-driven, hydroxylation pathway.

Cleavage of hydroperoxides and the subsequent peroxygenation of hydrocarbon substrates also occurs via the homolytic pathway [5, 38, 138, 222, 224, 229]. As an example, Blake and Coon [229] investigated the mechanism of hydroxylation of toluene and its *m*- and *p*-substituted derivatives using cumene hydroperoxide as the oxygenating agent and purified rabbit CYP2B4 as the biocatalyst. To explain structure-activity relationships for the hydroxylation of toluene and its substituted derivatives, a homolytic mechanism for hydroperoxide cleavage was invoked, as outlined in the following reactions.

$$Fe^{III} + XO{-}OH \rightarrow Fe^{IV}{-}OH + XO^{\bullet} \quad (1.16)$$

$$XO^{\bullet} + AH \rightarrow XOH + A^{\bullet} \quad (1.17)$$

$$A^{\bullet} + Fe^{IV}{-}OH \rightarrow A^{\bullet} - Fe^{IV}{-}OH \rightarrow AOH{-}Fe^{III} \rightarrow AOH + Fe^{III} \quad (1.18)$$

Thus, the mechanistic process begins with the reaction of ferric CYP2B4 (Fe^{III}) with cumene hydroperoxide (XO–OH) (X signifies cumene), which generates the ultimate protonated CpdII hydroxylating intermediate (Fe^{IV}–OH) and a cumyloxy radical ($XO^{\bullet}$) formed by homolytic cleavage of the O–O bond (Eq. 1.16). The cumyloxy radical then reacts with the substrate (AH) (*p*-chlorotoluene used as an example) and generates the reduced hydroperoxide (XOH) (cumenol) and a carbon-centered substrate benzylic radical ($A^{\bullet}$), which reacts with Fe^{IV}–OH to form the benzylic radical-bound protonated CpdII complex ($A^{\bullet}$–Fe^{IV}–OH). The bound benzylic radical is then hydroxylated by the ferryl hydroxyl moiety, forming the ferric CYP2B4-benzylic alcohol product complex (AOH–Fe^{III}) that rapidly dissociates and releases the benzylic alcohol product (AOH) (*p*-chlorohydroxytoluene) and regenerated ferric CYP2B4 in the final steps (Eq. 1.18) [111, 229]. In peroxygenase reactions that involve heterolytic or homolytic cleavage of peroxy compounds, both mechanistic processes are proposed to utilize the common protonated CpdII intermediate as the ultimate hydroxylating species [111, 240]. Although heterolytic peroxy O–O bond cleavage to produce the primary CpdI intermediate and the ultimate protonated CpdII hydroxylating species is considered to be the likely mechanistic process operating in most hydroxylation reactions driven by hydroperoxides and other peroxy compounds [1, 11, 63, 68, 111, 125, 239–245], cumulative evidence indicates that CYP enzymes can also sometimes perform homolytic O–O bond scission, producing the protonated CpdII intermediate that acts as the hydroxylating species in a proposed mechanistic process not involving CpdI (see Eqs. 1.16 to 1.18) [5, 38, 111, 138, 222, 224, 229, 238].

Hlavica et al. [224] examined the mechanism of NADPH/O_2- and cumene hydroperoxide-supported CYP2B4-catalyzed *N*-oxidation of 4-chloroaniline by comparing metabolic profiles in the two systems. In the NADPH/O_2-mediated pathway, 4-chloroaniline was hydroxylated to the *N*-(4-chlorophenyl)hydroxylamine product in a reconstituted system containing CYP2B4, CPR, NADPH/O_2 and substrate. Formation of the hydroxylated product likely proceeded by the traditional two-electron hydroxylation pathway involving CpdI. Substituting cumene hydroperoxide for NADPH/O_2 and CPR shifted the metabolic profile from the preponderant production of *N*-(4-chlorophenyl)hydroxylamine to the formation of 1-chloro-4-nitrobenzene.

Endogenous lipid hydroperoxides, prepared from isolated rabbit liver microsomal phospholipids, were also able to support the CYP2B4-catalyzed *N*-oxidation of 4-chloroaniline [246]. A sequential one-electron oxidation of the substrate, triggered by homolytic cleavage of the O–O bond of the hydroperoxide, was proposed as the operating mechanism for 1-chloro-4-nitrobenzene formation.

1.7 CYP Iron-Oxygen Intermediates as Possible Oxygenating Species

The following CYP iron-oxygen intermediates, which include the primary CpdI species, have been postulated to function as oxygenating species in monooxygenation reactions: (1) singly-bonded ferryl radical resonance form ($Fe^{IV+}–O^{\bullet-}$ or simply $Fe^{IV}–O^{\bullet}$) of CpdI; (2) ferric oxenoid complex ($Fe^{III}=O$); (3) perferryl entity ($Fe^{V}=O$); (4) CpdII species ($Fe^{IV}=O$); (5) protonated CpdII complex ($Fe^{IV}–OH$); (6) CpdES intermediates ($Tyr{\bullet}Fe^{IV}=O$ or $Trp{\bullet}Fe^{IV}=O$); (7) Cpd I species ($Por{\bullet}^{+}Fe^{IV}=O$); (8) ferric-peroxo anion species ($Fe^{III}–OO^{-}$); (9) ferric-hydroperoxo intermediate ($Fe^{III}–OOH$); and (10) $Fe^{III}–(H_2O_2)$ complex [1, 11, 25, 34, 55, 111, 118, 119, 125–136, 179–181, 206, 239, 240, 247–250]. Experimental and theoretical evidence supporting the proposed major oxygenating species is examined in Sects. 1.7.1, 1.7.2, 1.7.3, 1.7.4, 1.7.5, 1.7.6, 1.7.7, 1.7.8, 1.7.9, and 1.7.10.

1.7.1 Singly-Bonded Ferryl Radical Anion Species

In 1974, Rahimtula and O'Brien [221] showed that CYP enzymes of rat and rabbit hepatic microsomes catalyzed the cumene hydroperoxide-driven hydroxylation of aromatic compounds (e.g. aniline, biphenyl, benzo[*a*]-pyrene, coumarin) and suggested that the CYP singly-bonded ferryl radical anion resonance form ($Fe^{IV+}–O^{\bullet-}$ or simply $Fe^{IV}–O^{\bullet}$) of CpdI was the transitory hydroxylating species. In 1975, Hrycay, Gustafsson and coworkers conducted an extensive investigation of steroid hydroxylations in varying positions of the steroid nucleus using $NADPH/O_2$, organic hydroperoxides, H_2O_2, periodate and chlorite as oxygenating agents, purified rat liver CYP2B1 [23], liver microsomes from untreated or phenobarbital-pretreated rats [24, 25] and bovine adrenocortical microsomes [26] as CYP sources, and androstenedione and other steroids as substrates. Of the oxygenating agents tested, periodate was by far the most effective, followed by cumene hydroperoxide, $NADPH/O_2$ and other agents (i.e. chlorite, *t*-butyl hydroperoxide, pregnenolone 17α-hydroperoxide, linoleic acid hydroperoxide), in promoting androstenedione 6β-, 7α-, 15- and 16α-hydroxylations by liver microsomal CYP enzymes. H_2O_2 was ineffective due to the presence of contaminating catalase in the microsomal preparations. However, H_2O_2, periodate and other agents promoted androstenedione hydroxylation using purified rat liver CYP2B1 as the catalyst [23]. Also, periodate was the most effective oxygenating agent, followed by cumene hydroperoxide and other agents, in promoting progesterone 21-hydroxylation by bovine adrenocortical microsomal CYP enzymes. A mechanism for steroid hydroxylation was proposed whereby the ferryl radical resonance hybrid of CpdI functions as the common primary oxidant in the $NADPH/O_2$-mediated, and shunt, pathway.

There are three possible resonance forms of CpdI where the radical can be localized [11]: on the (1) conjugated porphyrin ring ($Por{\bullet}^{+}Fe^{IV}=O$); (2) sulfur atom of the thiolate ligand; and (3) oxygen atom of the singly-bonded ferryl radical species ($Fe^{IV}–O^{\bullet}$). CpdI species are represented collectively as porphyrin-centered π radical ferryl functionalities ($Por{\bullet}^{+}Fe^{IV}=O$). However, theoretical investigations of the CYP intermediary ferryl states [117, 249, 250] and supporting experimental data [131–133] implicated a mesomeric singly-bonded ferryl oxygen-centered radical entity ($Fe^{IV}–O^{\bullet}$) as the active oxidant. Theoretical calculations performed by Loew et al. [249]

Table 1.8 Kinetic isotope effects and ^{18}O incorporations in the *O*-demethylation of *p*-dimethoxybenzene by iron porphyrin oxidant systems and CYP-NADPH/O_2 systems[a]

Iron porphyrin system	Axial ligand	Oxidant	KIEs[b] (k_H/k_D)	^{18}O incorp. (%)
Swan-Resting porphyrin[c]	Thiolate	PPAA	11.7	<1 %
Fe(TPP)Cl[c]	Cl^-	PPAA	1.0	90
Fe(TMP)Cl[c]	Cl^-	PPAA	1.2	51
Fe(TPFPP)Cl[c]	Cl^-	PPAA	1.1	89
Fe(TPP)Cl[d]	Cl^-	PhIO	1.9	nd
Fe(TMP)(1-MeIm)$_2$ClO$_4$[d]	Imidazole	PPAA	1.2	nd
Rat hepatic microsomes[e]	Thiolate	O_2	11.9	1
Human rCYP1A2[e]	Thiolate	O_2	11.6	nd

[a]KIEs and ^{18}O incorporations were determined by GC/SIM from the M^+ peak area ratio of the products. Hepatic microsomes were prepared from phenobarbital-pretreated rats [252] and reactions were performed in the presence of NADPH/O_2. CYP2B1 is the predominant hepatic CYP enzyme induced by pretreating rats with phenobarbital [79]. Human rCYP1A2 was obtained from GENTEST Corp. and combined with CPR and NADPH/O_2 for activity measurements (Table 1.8 was adapted from ref. [133])

[b]*CPR* NADPH-cytochrome P450 oxidoreductase, *PhIO* iodosobenzene, *KIEs* kinetic isotope effects, *1-MeIm* 1-methylimidazole, *nd* not determined, *PPAA* phenylperacetic acid, *rCYP* recombinant cytochrome P450, *TPFPP meso*-tetrakis-(pentafluorophenyl)porphyrin, *TMP meso*-tetramesitylporphyrin, *TPP meso*-tetraphenylporphyrin

[c,d,e]Solvent used in reactions with these agents was benzene[c], dichloromethane[d] and water[e], respectively

indicated a longer Fe–O bond distance in CpdI of CYPcam compared to CpdI of HRP, which is believed to stabilize the CYPcam ferryl radical state. The CYP thiolate ligand was proposed to pull the Fe atom toward it, lengthening the Fe–O bond and stabilizing the ferryl radical species. Based on resonance Raman spectroscopy, Champion [132] suggested that the CYPcam ferryl radical entity, rather than the porphyrin π radical species, is favored as an oxidant because of the strong π electron-donating effects of the thiolate ligand [11]. CpdI of CYP is thought to have an overall neutral charge [118, 119] and the ferryl radical configuration is favored over the higher-energy porphyrin π radical ferryl state. CpdI of histidine-ligated hemeproteins such as HRP has a net charge of +1 that favors the porphyrin π radical configuration over the ferryl radical state [118, 132]. Mössbauer and electron paramagnetic resonance (EPR) spectroscopic data indicate that in thiolate-ligated CPO, the porphyrin π radical configuration is favored due to the presence of polar amino acid residues located on the distal side of the heme that aid in stabilizing the porphyrin π radical ferryl state [118, 132].

Higuchi and coworkers examined the effect of the thiolate ligand on the reactivity of CYP oxidizing intermediates and synthetic thiolate-ligated iron porphyrin systems. Mechanisms involving the CYP-mediated *O*-demethylation of *p*-dimethoxybenzene [131, 133, 251] and hydroxylation of cyclooctane [134] were used as probes to differentiate the nature of the oxidizing intermediates involved in chloride anion-, imidazole- and thiolate-ligated iron porphyrin systems, using five synthetic *meso*-tetraaryl iron porphyrin probes (Table 1.8). An alkylthiolate-ligated iron porphyrin designated the Swan-Resting (SR) complex, named in this manner because the shape topologically resembles a swan that is resting [131, 133], was designed to introduce bulky pivaloyl groups onto the axial thiolate coordination surface of the iron porphyrin, which stabilizes and protects the thiolate ligation from oxidation (Fig. 1.4) [131]. Phenylperacetic acid (PPAA) was used as a probe to differentiate between the modes of peroxy O–O bond cleavage in oxygenative reactions mediated by iron porphyrins. In the *O*-demethylation mechanism, the synthetic iron porphyrin SR-PPAA complex and NADPH/O_2-mediated rat liver microsomal CYP2B1 system [251] and human rCYP1A2-CPR systems were utilized as thiolate-ligated porphyrin complexes,

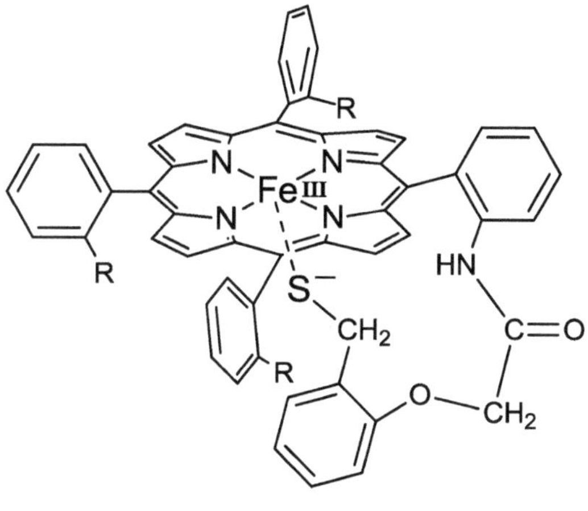

Fig. 1.4 The Swan-Resting (*SR*) complex. The SR complex is a unique alkylthiolate-ligated iron porphyrin complex that retains its axial thiolate coordination during catalytic oxidative reactions (Figure 1.4 was adapted from Ref. [133])

three synthetic iron porphyrin-PPAA complexes and one iron porphyrin-iodosobenzene complex were used as chloride anion-ligated systems, and one porphyrin-PPAA complex was used as an imidazole-ligated system [133]. In the cyclooctane hydroxylation mechanism, NADPH/O_2 was used in the rat liver microsomal thiolate-ligated CYP2B1 system and human rCYP3A4-CPR systems, while *m*CPBA was used as the oxidant in SR-thiolate-, chloride anion- and imidazole-ligated iron porphyrin systems [134].

Oxidative *O*-dealkylation of alkyl-aryl ethers such as *p*-dimethoxybenzene is one of the major reactions catalyzed by CYP. Two generally accepted mechanisms prevail, as shown in the reactions below.

$$\begin{aligned}&H_3CO\text{–phenyl–}OCH_3 \xrightarrow{\text{CYP or SR porphyrin+oxidant}} H_3CO\text{–phenyl–}OCH_2^{\bullet} \rightarrow\\&H_3CO\text{–phenyl–}OCH_2OH \xrightarrow{-H_2C=O} H_3CO\text{–phenyl–}OH\end{aligned} \tag{1.19}$$

$$\begin{aligned}&H_3CO\text{–phenyl–}OCH_3 \xrightarrow{\text{iron porphyrin+oxidant}} H_3CO\text{–phenyl–}(OH)OCH_3 \xrightarrow{-CH_3OH}\\&H_3CO\text{–phenyl–}O^{\bullet} \rightarrow H_3CO\text{–phenyl–}OH\end{aligned} \tag{1.20}$$

Thus, in the CYP-mediated oxidative reactions and iron porphyrin-iodosobenzene oxidative systems, the H atom abstraction mechanism predominates [133] whereas in the hydroxyl radical-mediated reaction, the *ipso*-substitution mechanism prevails [133, 252]. In the H atom abstraction mechanism (Eq. 1.19), a H atom is abstracted from a methoxy group of *p*-dimethoxybenzene to generate a carbon-centered radical species, H_3CO–phenyl–$OCH_2^{\bullet}$, which is oxygenated to release formaldehyde and the demethylated *p*-methoxyhydroxybenzene product (H_3CO–phenyl–OH). In the *ipso*-substitution mechanism (Eq. 1.20), *p*-substituted dimethoxybenzene undergoes oxidative attack at what is termed the *ipso* position to generate the hydroxylated H_3CO–phenyl–(OH)OCH_3 intermediate. This is followed by release of the CH_3OH moiety and *p*-substitution by a hydroxyl group to generate the *p*-methoxyhydroxybenzene product.

Pronounced differences are observed between the two reaction mechanisms in the KIEs and origin of the oxygen atom of the phenolic hydroxy group in phenylacetic acid (PAA). In thiolate-ligated SR and CYP iron porphyrin systems, *p*-dimethoxybenzene was *O*-demethylated with high KIE values (>10) and low ^{18}O incorporation (~1 %), indicating that the reactions proceeded by H atom abstraction. In chloride anion- and imidazole-ligated porphyrin systems, low KIE values (~1.0) and high ^{18}O incorporation (51–90 %) were observed (see Table 1.8), demonstrating that the reactions proceeded in the *ipso*-substitution manner. In addition, *p*-dimethoxybenzene was *O*-demethylated at a much higher rate in the thiolate-ligated systems than in the chloride anion- and imidazole-ligated systems. These combinatorial results provided compelling evidence that the iron-oxo intermediate of thiolate-ligated iron porphyrin systems has a different reactivity to that of the chloride anion- and imidazole-ligated systems. The differences in reactivity between thiolate-ligated CYP enzymes and imidazole-ligated peroxidases such as HRP likely originate from the isoelectronic states of the oxidizing intermediates as suggested by Champion [132]. The studies indicated that the high-valent iron-oxo intermediate of thiolate-ligated SR and CYP iron porphyrin systems has a stronger H atom abstraction activity than the synthetic chloride anion- and imidazole-ligated iron porphyrins, suggesting that the preferred chemical structure of the oxidizing intermediate of thiolate-ligated iron porphyrins contains an iron(IV) oxygen-centered radical such as in $Fe^{IV}{-}O^{\bullet}$, rather than a porphyrin-centered π radical such as in $Por^{\bullet+}Fe^{IV}{=}O$ [133]. The large cyclooctane hydroxylation activities of thiolate-ligated CYP complexes [134] were ascribed to their CYP reactive oxygen intermediates, which were projected to display an iron(IV) oxygen radical configuration identical to the ferryl radical resonance form of CpdI. It was further proposed that proximal ligands with strong electron-donating properties, such as the thiolate ligand [11], increase the radical character of the iron-oxo group of CpdI to generate its reactive $Fe^{IV}{-}O^{\bullet}$ resonance hybrid that participates in aliphatic C–H hydroxylation reactions via the H atom abstraction/oxygen rebound pathway [1, 127, 131–134].

1.7.2 Ferric Oxenoid Complex

The CYP ferric oxenoid species ($Fe^{III}{=}O$) contains a six-valence electron oxene atom coordinated to a heme iron(III) center. Hamilton [129] used the term "oxene" to describe a free oxygen atom encircled by six electrons after recognizing that oxene is isoelectronic with carbene and nitrene that also have six electrons in their outer valence shells and function in insertion reactions analogous to those catalyzed by CYP monooxygenases. Hamilton [129] postulated that the transitory ferric oxenoid species generated in the CYP monooxygenation cycle [23, 25] was able to transfer its oxygen atom to the substrate by an oxene transfer mechanism, a proposal that was also later considered by other investigators [23, 25, 145]. However, an iron oxenoid species has not been observed in any CYP-mediated oxidative reaction and, therefore, its independent existence must be regarded as hypothetical and too transitory a species to be detectable in a CYP oxygenation system [145]. Furthermore, a $Fe^{III}{=}O$ electronic structure has not been detected upon reacting iron(III) porphyrin compounds with oxygen atom donors and in the cases studied, the formal iron oxidation state of the heme compounds was found to be at least IV [117, 120, 149].

1.7.3 Perferryl Species

CpdI is sometimes depicted as a perferryl entity ($Fe^{V}{=}O$) with two extra oxidizing equivalents localized on the heme iron(V) center. Based on studies with CPO, HRP and biomimetic iron porphyrin compounds, CpdI was presumed to exist in a $Por^{\bullet+}Fe^{IV}{=}O$ configuration rather than an $Fe^{V}{=}O$ state [120, 177]. Newcomb challenged this assumption and used porphyrin model compounds to explore involvement of $Fe^{V}{=}O$ in hydroxylation reactions. The greater reactivity displayed by model $Fe^{V}{=}O$ perferryl complexes compared to $Por^{\bullet+}Fe^{IV}{=}O$ led Newcomb and coworkers [135, 253] to postulate that the first oxygenating species formed in hydroxylation reactions could be the $Fe^{V}{=}O$ entity that hydroxylates unactivated substrate C–H bonds more quickly than it transforms to the more stable, isomeric CpdI configuration.

Isobe et al. [254] performed DFT calculations on the reactivities of low-lying doublet and quartet ferryl ($Fe^{IV}{=}O$) oxidants and a doublet perferryl ($Fe^{V}{=}O$) oxidant as key intermediary oxygenating species for CYP enzymes, addressing several aspects of the mechanism of H atom abstraction using propane as a substrate. DFT calculations indicated that the perferryl species can abstract a H atom from propane with a low activation energy barrier of 0.6–2.5 kcal mol^{-1}, which is considerably smaller than that of the ferryl species (13.4–17.8 kcal mol^{-1}).

$Fe^{V}=O$ contributes to the reactivity of CpdI due to the presence of the highly reactive π atomic radical character of the oxo ligand. Meanwhile, the interplay between the accessible ferryl and perferryl states of CpdI displaying different reactivities was suggested to be a reason for the "elusiveness" of CpdI, a characteristic acquired from its high reactivity and short life cycle, in NAD(P)H/O_2-mediated monooxygenation reactions [32, 34, 128, 240].

1.7.4 CpdII Species

Spolitak et al. [247] created Tyr75Phe, Tyr96Phe and Tyr96Phe/Tyr75Phe variants of CYPcam in which tyrosine at positions 75 and 96 was replaced by phenylalanine. The variants contained more hydrophobic active sites than wild-type CYPcam and after reacting with peracids or cumene hydroperoxide, the variants produced changes in H-bonding patterns that favored extensive formation of CpdII species ($Fe^{IV}=O$). Very few CpdI or CpdES species were detected when the variants reacted with peracids or cumene hydroperoxide. Although CpdII species were the predominant entities formed, they were ineffective at performing camphor hydroxylation [55, 247]. Recently, Spolitak et al. [55] conducted kinetic studies using bacterial CYP153A6 and *m*CPBA as the artificial oxidant and compared the results to those obtained with CYPcam and CYP_{BM3}. The more hydrophobic active site of CYP153A6 allowed accumulation of the transitory CpdII species in the reaction with peracid. However, CpdI, CpdII and CpdES species could all be observed at pH 8.0 after reacting CYP153A6 with *m*CPBA, due to the unique nature of the CYP153A6 active site compared to that of CYPcam and CYP_{BM3}. Theoretical analyses suggested that the CpdII species, which can be generated by the one-electron reduction of CpdI [11], were far less reactive than CpdI species in performing hydroxylation reactions [255], in agreement with experimental model iron porphyrin studies. Although CpdII species of porphyrin iron(IV)-oxo complexes can perform a few hydroxylation and epoxidation reactions [256, 257], other reports with model iron porphyrin systems imply that CpdII species are sluggish oxidants compared to CpdI species [255, 258].

1.7.5 Protonated CpdII Complex

The CpdI ($Por^{\bullet +}Fe^{IV}=O$) and protonated CpdII ($Fe^{IV}-OH$) species of CYP enzymes have been proposed to function as highly-reactive intermediates in NAD(P)H/O_2-supported and hydroperoxide-driven hydroxylation reactions, which proceed via the H atom abstraction/oxygen rebound pathway [127, 240]. Early studies have shown that during the course of productive CYP-mediated substrate C−H bond activation in hydroxylation reactions, the CpdI species abstracts a H atom from the substrate to generate a substrate radical and a protonated CpdII intermediate that acts as the ultimate hydroxylating species [127, 240] (see Sect. 1.6.1.2). Green and coworkers [206, 239, 259] have argued that the CYP axial cysteine thiolate ligand promotes C−H bond activation through the generation of basic protonated CpdII ($Fe^{IV}-OH$) species. In addition, the oxidation of tyrosine residues contained within the CYP protein framework to generate CpdES ($Tyr^{\bullet}Fe^{IV}=O$) species (see Sect. 1.7.6) [179, 181, 260], a reaction known to dominate nonproductive decay in CYP enzyme systems, represents an energetically favorable "short circuit" for the oxidizing equivalents of CpdI formed during hydroxylation reactions.

Based on the free energies of the productive (involving formation of CpdI) and nonproductive (involving formation of $Tyr^{\bullet}Fe^{IV}=O$ species) pathways, it has been proposed that a critical role of cysteine thiolate ligation in CYP enzymes is to alter the free energy landscape, shifting the relative free energy for the productive and nonproductive pathways where the rate constant for C−H bond activation dominates that for nonproductive decay [259]. Recently, Green and coworkers [259] demonstrated that cysteine thiolate coordination to ferric iron in bacterial CYP158A2 increased the pK_a (acid dissociation constant) of the protonated CpdII species, correspondingly lowering the one-electron reduction potential of CpdI, the catalytically-active oxidizing intermediate, and decreasing the driving force for deleterious auto-oxidation of tyrosine and tryptophan residues located in the enzyme's framework. A pK_a value of 11.9 was determined for the protonated CpdII species. Using insights gained from site-directed mutagenesis studies, Green and coworkers [259] also determined the CpdII pK_a in an additional CYP enzyme. Thus, incorporation of a tyrosine residue at the position corresponding to that of Tyr-352 in CYP158A2 produced a Leu316Tyr variant of CYP119A1 that generated CpdII in high yield, giving a pK_a value of 12.2. These elevated pK_a values resulted in a >10,000-fold reduction in the rate constant for oxidations of tyrosine (and tryptophan) residues found in the CYP protein framework, making these oxidative processes noncompetitive with CpdI-mediated substrate oxidation.

1.7.6 CpdES Species

Schünemann, Jung and coworkers [179, 181, 260] reacted ferric CYPcam and CYP102A1$_{BMP}$ (CYP$_{BMP}$) with peracetic acid and characterized the trapped species using rapid-freeze-quench EPR and Mössbauer spectroscopy. A species identified as a protein tyrosine radical (spin $S' = 1/2$) bound to a ferryl ($Fe^{IV}{=}O$) moiety (d^4, spin $S = 1$) and designated CpdES (Tyr•$Fe^{IV}{=}O$) was detected by EPR spectroscopy within 8 ms after mixing each CYP enzyme with peracetic acid. Tyr96 of wild-type CYPcam provided the aromatic radical observed by EPR spectroscopy [261]. A tryptophan radical CpdES species (Trp•$Fe^{IV}{=}O$) was also detected in EPR spectra of CYP$_{BMP}$ after mixing the enzyme with peracid [181]. Trp•$Fe^{IV}{=}O$ of CYP$_{BMP}$ is similar to CpdES of cytochrome *c* peroxidase (CcP) that also contains a ferryl center coupled with a tryptophan radical derived from Trp191 [178]. The CpdI intermediates were proposed to be the first species formed upon reacting CYPcam and CYP$_{BMP}$ with peracetic acid. The CpdI species were generated on a very rapid time scale but were quickly reduced within 8 ms by an intramolecular one-electron transfer from nearby tyrosine or tryptophan residues, generating the corresponding Tyr•$Fe^{IV}{=}O$ or Trp•$Fe^{IV}{=}O$ CpdES intermediates and quenching the initial rapidly-formed porphyrin π radicals [181].

Spolitak et al. [180] reacted ferric CYPcam with *m*CPBA and detected an intermediate that displayed the characteristic CpdI absorption band at ~367 nm. CpdI then converted within 0.038 s to a species that displayed an absorbance maximum at ~406 nm, and was consequently assigned a tyrosine radical-ferryl structure (Tyr•$Fe^{IV}{=}O$) and named CpdES. Although both the CpdES and CpdII species contain a ferryl ($Fe^{IV}{=}O$) chromophore, CpdES (Soret band at ~406 nm) was readily detected at lower pH (pH 6.2), while CpdII (Soret band at ~420 nm) was observed at higher pH (pH 7.5). It was proposed that the species displaying the Soret band at ~420 nm at higher pH was the unprotonated CpdII species ($Fe^{IV}{=}O$), whereas the entity absorbing maximally at ~406 nm at lower pH was the tyrosine radical-protonated ferryl CpdES species (Tyr•$Fe^{IV}{-}OH$) [55, 247, 261]. It was also suggested that the Tyr•$Fe^{IV}{=}O$ and Tyr•$Fe^{IV}{-}OH$ species of CYPcam were unlikely to perform hydroxylation reactions using hydrocarbon substrates [180, 261]. Raner et al. [136] replaced the CYP$_{BM3}$ Phe87 residue with glycine and reacted the mutant with surrogate oxidants (e.g. *m*CPBA, peracetic acid, periodate, iodosobenzene) to generate CpdI and CpdES intermediates. The CYP$_{BM3}$ mutant has a larger more accommodating active-site pocket that is well-suited for binding small aromatic oxygen atom donors such as *m*CPBA and iodosobenzene. Rapid mixing of the mutant with *m*CPBA resulted in the sequential formation of two transitory ferryl species. The first species was spectrally similar to CpdI of CYPcam [139, 180] and CYP119A1 [32, 34]. CpdI of the CYP$_{BM3}$ mutant converted to a more stable intermediate characterized by an intense Soret peak at 406 nm, similar to that observed with CYPcam. EPR spectroscopic studies suggested that this stable intermediate was a CpdES species containing a protein-based radical localized on a nearby tyrosine or tryptophan residue.

1.7.7 CpdI Species

The transitory CYP species that function in the CYP catalytic cycle have been studied by high resolution cryocrystallography, radiolytic reduction methodology, and EPR, UV/visible, magnetic circular dichroism, Mössbauer, resonance Raman, electron nuclear double resonance (ENDOR), extended X-ray absorption fine structure spectroscopies, and theoretical computational methods [106, 124, 125, 262–266]. Most monooxygenation reactions supported by NAD(P)H/O_2 can be rationalized in the context of the porphyrin π radical CpdI oxygenating species (Por•$^+$Fe^{IV}=O) believed to function in the CYP catalytic cycle. CpdI of CYP enzymes is one of the most highly sought intermediates in biochemistry. In 1976, Dawson et al. [106] proposed that the highly polarizable CYP cysteine ligand provides a strong proximal "push" of electron density via the heme and onto the O–O bond in Fe^{III}–O–OH, thus promoting heterolytic O–O bond scission (after protonation) and expediting CpdI formation [11, 108, 267]. Once the O–O bond has been cleaved, the electron-donating character of the cysteine ligand helps to stabilize the high-valent nature of the CpdI species [11]. Despite countless efforts to capture and characterize the CpdI species in the NAD(P)H/O_2-mediated CYP cycle, it has been difficult to isolate it and determine its precise electronic structure and kinetic and physicochemical properties because of its high reactivity, unstable nature ($t_{½}$ ~2 ms) and short life cycle [32, 128, 240].

The CpdI species, readily formed by reacting H_2O_2 with heme peroxidases, were recognized for many years as transitory reactants in the catalytic cycles of HRP, CPO and *Aae*UPO [118, 168, 175, 177, 185, 200, 268]. Because of the similar coordination sphere of the heme iron in thiolate-ligated CPO and CYP, CPO CpdI is electronically equivalent to CYP CpdI. The electronic structure of CYP CpdI is best described as a Fe^{IV}=O unit (spin $S = 1$) coupled with a ligand-based radical (spin $S = 1/2$) delocalized over the thiolate ligand and porphyrin ring [34]. As reasoned by Shaik et al. [250] using bonding principles, the Fe^{IV}=O moiety of CpdI is a triplet state and has two triplet coupled electrons. An odd electron is present formally on the porphyrin, making the heme macrocycle a π radical species [269]. The three electrons are, in turn, coupled to two closely-lying spin states consisting of a ferromagnetic high-spin quartet $^4A_{2u}$ state with all three electrons having the same spin, and an antiferromagnetic low-spin doublet $^2A_{2u}$ state where the spin of the electron in the porphyrin ring is opposite to the two spins on the Fe^{IV}=O moiety [269]. The CpdI species is, therefore, a triradicaloid characterized by three singly-occupied π^*_{xz}, π^*_{yz} and a_{2u} orbitals. The low-spin and high-spin states of CpdI are believed to play critical roles in its biochemical reactivity in monooxygenation reactions (see Sects. 1.7.9.2 and 1.7.9.3).

1.7.7.1 CpdI Formation from the CYP Ferric-Superoxo Anion Intermediate

CYPcam residues that are extremely important for oxygen activation in oxygenation reactions are the proximal cysteine ligand (Cys357) and the acid-alcohol pair (Asp251–Thr252), located in the active site of the enzyme [11, 98, 270]. Schlichting et al. [270] used freeze-trapping and X-ray cryocrystallography to determine the crystal structures of the iron-oxo intermediates that participate in the CYPcam hydroxylation cycle. Three X-ray data sets were collected from the CYPcam crystals (frozen in liquid nitrogen) during various stages of the catalytic cycle. One-electron reduction of the CYPcam ferric heme-camphor crystal complex represented the first step of oxygen activation and the requisite first electron was provided by sodium dithionite. Forced diffusion of O_2 into the reduced CYPcam crystal lattice produced the ferric-superoxo anion intermediate (**C**) (see Fig. 1.3) at which point the first data set was collected. The second electron was transferred to Fe^{III}–O_2^- at cryogenic temperatures by the solvent matrix that released free electrons after exposure to monochromatic synchrotron long wavelength X-rays, generating the ferric-peroxo anion intermediate (**D**) that, after protonation, formed the ferric-hydroperoxo intermediate (**E**), at which point the second data set was collected. After thawing and refreezing this latter crystal, the third data set was collected. The ferric-hydroperoxo complex (**E**) decayed to an intermediate assigned a $Por•^{+}Fe^{IV}=O$ (CpdI) structure (**F**) (see Fig. 1.3) because the latter species displayed an iron-to-oxygen bond distance of ~1.65 Å (suggestive of a Fe=O double bond), which was too short to be assigned a Fe−O single bond distance of 1.8 Å, as was present in hydroxide-ligated ferric heme [270].

Indications that the reactive oxidant of CYPcam was CpdI originated from studies with CYPcam mutants lacking a highly conserved threonine residue (Thr252), which forms part of the distal O_2-binding site of CYPcam [271, 272]. When the Thr252 residue was replaced by alanine, the Thr252Ala mutant was fully uncoupled, resulting almost entirely in H_2O_2 production rather than 5-*exo*-hydroxycamphor product formation [98]. CYPcam was only capable of efficiently hydroxylating its substrate when an amino acid containing a hydroxyl group (e.g. threonine, serine) was present at position 252 [271]. In addition, the threonine to alanine mutation produced a dramatic reduction in H_2O_2-driven camphor hydroxylation activity, indicating that the decreased activity was caused by a disruption of a common mechanism of peroxygenase and monooxygenase catalysis [272]. Insights into the role of CYP intermediates in steps 3–6 of the classical CYP cycle (see Fig. 1.3) were provided by Davydov et al. [263], who used ^{60}Co γ-irradiation at cryogenic temperatures (77 K) to reduce the ferric-superoxo-substrate complexes of wild-type and mutant CYPcam. EPR spectroscopic signals were detected for the primary reduction product, the ferric-peroxo anion intermediate ($Fe^{III}-OO^{-}$) (g ~2.25) and its immediate protonated conversion product, the ferric-hydroperoxo species ($Fe^{III}-OOH$) (g ~2.3). Stepwise annealing at high temperatures converted the wild-type CYPcam ferric-hydroperoxo-camphor complex to a product state whereby the natural product of camphor hydroxylation, 5-*exo*-hydroxycamphor, was coordinated to ferric heme. Further annealing produced low-spin regenerated ferric CYPcam and dissociated 5-*exo*-hydroxycamphor. However, the ferric-hydroperoxo-camphor complex of a Thr252Ala CYPcam mutant reverted to the ferric state without generating hydroxylated substrate, and no spectroscopic evidence was found for the accumulation of the CpdI intermediate expected from O−O bond cleavage, presumably because of its high reactivity. Furthermore, ENDOR spectroscopic studies in normal and deuterated buffers revealed that the oxygen atom of the hydroxyl group in the 5-*exo*-hydroxylated substrate, rather than a H_2O molecule, was coordinated to heme iron and that the hydroxyl proton originated from the C−5 position of camphor. This result was construed as evidence that camphor hydroxylation occurred via the ferryl CpdI-mediated H atom abstraction/oxygen rebound mechanism and not through the $Fe^{III}-OOH$ functionality, because hydroxylation by $Fe^{III}-OOH$ would be expected to produce a water-coordinated ferric heme and a "scrambled" hydroxyl proton derived from the surrounding solvent [193, 263].

More recently, Davydov et al. [266] utilized EPR and ^{1}H ENDOR spectroscopy to analyze the intermediary CYP states produced during the hydroxylation of (1*R*)-camphor (H_2-camphor) and (1*R*)-5,5-dideuterocamphor (D_2-camphor), which were formed by cryoreduction at 77 K and annealing of the ferrous CYPcam−O_2-camphor precursor complex. Hydroxylation of H_2-camphor generated a primary product state in which 5-*exo*-hydroxycamphor was coordinated to ferric CYPcam. ENDOR spectra contained signals derived from two protons [Fe(III)-bound C5-OH_{exo} and C5-OH_{endo}] of camphor. When D_2-camphor was hydroxylated in H_2O or D_2O buffer, both ENDOR spectroscopic H_{exo} and H_{endo} signals were absent. For D_2-camphor in H_2O buffer, H/D exchange caused the C5-OH_{exo} signal to reappear while for H_2-camphor in D_2O, the C5-OH_{exo} signal decreased in magnitude via H/D exchange.

These results indicated the involvement of CpdI as the primary oxygenating species and ruled out the participation of the ferric-peroxo and ferric-hydroperoxo intermediate as the reactive oxidant.

1.7.7.2 Proton Delivery Pathways for CYPcam

A vital catalytic function of CYP is the sequential transfer of protons to the distal oxygen atom of the ferric-peroxo and ferric-hydroperoxo intermediates generated in the CYP catalytic cycle. CYP proton delivery pathways have been investigated thoroughly using CYPcam and its mutants as primary sources for experimental investigations [98, 241, 273–276]. The proton relay system involves backbone and side-chain moieties of the CYPcam protein structure and localized protein-bound water molecules, which are appropriately stabilized in the active center of the enzyme through specific H-bond interactions [98, 273, 276]. Examining the role of Thr252 in the proton delivery pathway, Raag et al. [274] created a Thr252Ala mutant of CYPcam whereby camphor hydroxylation was uncoupled from electron transfer. H_2O_2 and excess water were generated in the mutant instead of the major product, 5-*exo*-hydroxycamphor. Solvent protons were more accessible to O_2 in the mutant, a factor that promoted H_2O_2 and water formation instead of substrate hydroxylation. Based on crystallographic and mutagenesis data, proton delivery pathways involving Lys178/Arg186, Asp251 and Thr252 residues were proposed for CYPcam [274].

Davydov et al. [241] further investigated the proton relay system of CYPcam and its Asp251Asn and Thr252Ala mutants, utilizing EPR and ENDOR spectroscopic examination of the intermediary states of CYPcam. Their investigations indicated that the Asp251Asn mutation impeded delivery of the first proton to the distal oxygen atom of the ferric-peroxo anion species $^{-}$[Cys$-$Fe^{III}$-$OO^{-}] (**D**) (see Fig. 1.3, step 4), while the Thr252Ala mutation interfered with delivery of the second proton to the distal oxygen atom of the ferric-hydroperoxo species $^{-}$[Cys$-$Fe^{III}$-$OOH] (**E**) (see Fig. 1.3, step 5). This second protonation was likely assisted by the negative charge accumulated on the thiolate ligand of the $^{-}$[Cys$-$Fe^{III}$-$OOH] intermediate [11, 99]. Alternatively, if the protonation of $^{-}$[Cys$-$Fe^{III}$-$OOH] occurs on the proximal (adjacent to ferric iron) oxygen atom, then iron-complexed hydrogen peroxide is formed and H_2O_2 is released as a product in an unproductive uncoupled reaction [277]. In a subsequent complementary study, Schlichting et al. [270] utilized cryocrystallography to characterize the oxy-CYPcam complex, and demonstrated the existence on the distal side of the heme of two H bond-stabilized water molecules (Wat901 and Wat902) that were interpreted to be part of the proton delivery pathway involving Asp251 and Thr252. Asp251 indirectly stabilizes Wat901 whereas the side chain of Thr252 rotates to interact with both the dioxygen molecule and the Wat901 and Wat902 molecules. The two water molecules likely act as sources of protons required for the formation of the catalytic CpdI species [98, 270, 273]. Thus, a critical role for CYPcam is its ability to control delivery of vital protons to the developing negative charge of the ferric-peroxo and ferric-hydroperoxo intermediates as electrons are introduced into the heme iron center, with the acid-alcohol pair, Asp251$-$Thr252, working in concert at the proper place and time to deliver the required protons [98, 270]. The CYPcam redox partner proteins, putidaredoxin reductase and putidaredoxin, also play crucial roles in oxygen activation by binding to the active site of CYPcam and providing the active-site rearrangement required for proper proton-coupled electron transfer [278, 279].

1.7.7.3 CpdI Formation Using Hydroperoxides and Peracids

The transitory CpdI species are readily formed by reacting hydroperoxides or peracids with CYP enzymes [32, 34, 139, 180, 240] and heme peroxidases [175, 177, 200, 206]. Because the acid in peracids is an excellent leaving group, heterolytic cleavage of the O$-$O bond to generate CpdI species with hemeproteins is considerably more facile than the analogous reaction with H_2O_2 in which OH^- (or H_2O after protonation) is the leaving group [180]. Hydroperoxides and peracids both contain an O−O bond and are therefore better in mimicking native O_2-supported monooxygenation reactions than are single oxygen atom-containing surrogate oxidants such as iodosobenzene and *N*-oxides. In 1974, Rahimtula et al. [172] examined the influence of cumene hydroperoxide on the EPR spectral properties of hepatic microsomal CYP2B4. After adding cumene hydroperoxide to hepatic microsomes prepared from phenobarbital-pretreated rabbits and containing high levels of ferric CYP2B4, an EPR spectroscopic radical signal (g ~2.0) was detected, which was comparable to the signal observed when cumene hydroperoxide or H_2O_2 reacted with metmyoglobin or CcP [280]. A similar radical signal was detected when *m*CPBA reacted with ferric CYP119A1 [34] and when H_2O_2 was added to ferric HRP [177]. The radical signal observed after reacting liver microsomal CYP2B4 with hydroperoxide was proposed to be due to the formation of CpdI [172]. Egawa et al. [139] examined the reaction of ferric low-spin CYPcam with *m*CPBA using rapid-scan stopped-flow spectroscopy. Rapid scanning indicated the formation, within 6 ms, of a primary CYPcam intermediate displaying a Soret absorption maximum at 367 nm and a visible maximum at 694 nm, which were nearly identical to the absorption maxima elicited by CpdI of CPO. Based on spectral similarities, the primary spectral intermediate of CYPcam was suggested to be the CpdI species.

Kellner et al. [32] used rapid-scan stopped-flow spectroscopy to examine the reaction of *m*CPBA with CYP119A1 derived from archaeal *Sulfolobus acidocaldarius*. After reacting ferric rCYP119A1 with *m*CPBA, an intermediate displaying an optical signature characteristic of CpdI was identified by absorption maxima at 370, 610 and 690 nm. A rate constant for CpdI formation of $3.2 \times 10^5\ M^{-1}\ s^{-1}$ at pH 7.0 and 4 °C was reported. However, only a small percentage of the intermediate was trapped in this form, likely due to the presence at the active CYP center of an adventitious reducing equivalent such as a fatty acid substrate [32, 240] that quickly reduced the CpdI species. Green and coworkers [34, 240] isolated the long-sought CpdI intermediate of CYP119A1 at unprecedented yields (~75 %) after reacting highly purified ferric rCYP119A1 with *m*CPBA. Capture and characterization of CpdI was achieved using low-temperature stopped-flow UV/visible, Mössbauer and EPR spectroscopy. The newly-formed spectral species displayed similar UV/visible and Mössbauer spectra to those elicited by CPO CpdI and was assigned a CpdI structure [34]. The rates at which CpdI of CYP119A1 hydroxylated hydrocarbon substrates (e.g. hexanoic acid, lauric acid, camphor) were determined using stopped-flow techniques. Gas chromatography/mass spectrometry product analysis following addition of *m*CPBA to a CYP119A1-lauric acid mixture revealed an equal distribution of 11- and 10-hydroxylauric acid. An apparent second-order rate constant for lauric acid hydroxylation of $k_{app} = 1.1 \times 10^7\ M^{-1}\ s^{-1}$ was reported. Large intrinsic KIEs ($\geq$12.5) were observed for the direct oxidation of substrates by CpdI, supporting the H atom abstraction/oxygen rebound mechanism for substrate hydroxylation [127]. Previous preparations of CYP119A1 yielded low levels of CpdI (<5 %) due to contaminating substrates such as fatty acids that bound to the enzyme's active site, hindering CpdI formation and expediting its decay [240]. The contaminating molecules were removed by additional ion exchange chromatography to produce highly purified CYP119A1, which allowed the preparation of CpdI in high yields (~75 %) [34]. Krest et al. [240] found an additional archaeal CYP enzyme, CYP119A2 from the thermoacidophilic crenarchaeon *Sulfolobus tokodaii* strain 7, which after extensive purification and reaction with *m*CPBA generated the CpdI species in ~65 % yield. More recently, Hayakawa et al. [35] demonstrated that archaeal CYP119A2 catalyzed the hydroxylation of ethylbenzene and the epoxidation of styrene in the presence of H_2O_2. The general structure of CpdI of CYP enzymes is shown in Fig. 1.5.

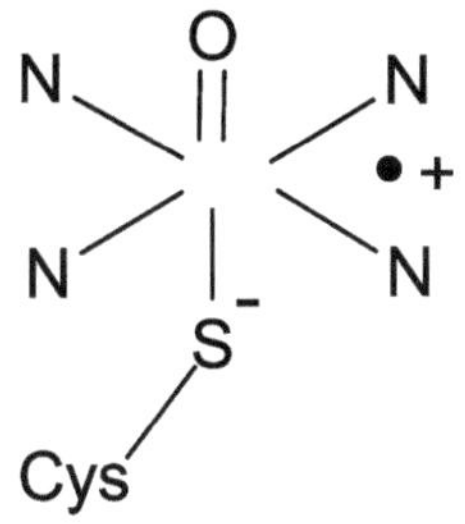

Fig. 1.5 The general structure of the CYP CpdI species. The diagram illustrates the four nitrogen ligands of the porphyrin macrocycle, the radical cation of the porphyrin, and the cysteine thiolate ligand. The central Fe atom shown in *red* (Figure 1.5 was adapted from Ref. [1])

1.7.7.4 CpdI Formation Using Periodate, Chlorite, Perborate, Percarbonate, Iodosobenzene and *N*-Oxides

In addition to hydroperoxides and peracids, exogenous oxygen atom donors such as periodate, chlorite, perborate, percarbonate, iodosobenzene and *N*-oxides have been used to drive monooxygenation reactions catalyzed by CYP enzymes in mammalian and bacterial systems (Table 1.7) [23–26, 75, 281–286]. In 1975, Hrycay, Gustafsson and coworkers demonstrated that periodate, chlorite and hydroperoxides promoted the hydroxylation of androstenedione and other steroid substrates using purified rat liver CYP2B1 [23], rat liver microsomes [24, 25] and bovine adrenocortical microsomes and mitochondria [26] as CYP sources. Periodate was by far the most effective oxygenating agent, compared to chlorite and the tested hydroperoxides, in the hydroxylation reactions studied. A mechanism for steroid hydroxylation was proposed whereby the exogenous oxygen atom donor transfers an oxygen atom to ferric CYP via the shunt pathway, generating the ferryl radical anion resonance hybrid of CpdI that functions as the transitory oxidant [23–26]. More recently, Auclair and coworkers [75] demonstrated that perborate and percarbonate also acted as oxotransfer agents in the *O*-demethylation and 6β-hydroxylation of dextromethorphan and testosterone, respectively, catalyzed by human rCYP2D6 and rCYP3A4.

Iodosobenzene (iodosylbenzene), a lipophilic two-electron oxotransfer agent that contains a single oxygen atom, has frequently been used as an oxygen atom donor in monooxygenation reactions. Its use developed out of earlier studies in which periodate was utilized as an oxygen atom donor [142] in steroid hydroxylation reactions [23–26, 228]. Lichtenberger et al. [284] showed that iodosobenzene supported the *O*-deethylation of 7-ethoxycoumarin by rat liver microsomal CYP and proposed an oxotransfer mechanism whereby iodosobenzene relinquishes its single oxygen atom to CYP, generating the ferryl radical anion species that functions as the transitory oxidant. Gustafsson and coworkers showed that iodosobenzene promoted the CYP-catalyzed 11-hydroxylation of laurate [285] and the hydroxylation of androstenedione [282] in rat liver microsomes. CYP106A2 isolated from *Bacillus megaterium* catalyzed the 15α- and 15β-hydroxylation of progesterone using periodate, chlorite and iodosobenzene as oxotransfer agents [286]. Hecker et al. [215] discovered that human platelet CYP5A1 catalyzed the transfer of an oxygen atom from iodosobenzene to the C–H bond of a prostaglandin H_2 analog substrate, 15*S*-hydroxy-11α,9α-epoxymethano-5*Z*,13*E*-prostadienoic acid (U46619). Sligar and coworkers showed that CYPcam isolated from *Pseudomonas putida* catalyzed the 5-*exo*-hydroxylation of camphor using iodosobenzene [36] and pyridine *N*-oxide [281] as oxygen atom donors. Blake and Coon [199] used stopped-flow spectroscopy to examine the spectral intermediates formed upon reacting ferric CYP2B4 with iodosobenzene compounds. A transitory invariant CYP2B4 spectral species was generated that elicited an absorption maximum at ~393 nm in the absolute spectrum. The CYP2B4 species was proposed to contain one atom of oxygen, derived from the iodosobenzene compound, bound to heme Fe(IV) iron.

In addition to CpdI, the ferric-hydroperoxo intermediate (Cpd0) has also been implicated as a possible oxidant in monooxygenation reactions [119, 220, 277]. Dowers et al. [244] attempted to establish if CpdI or the ferric-hydroperoxo intermediate functions as the primary transitory oxidant in *N*-demethylation reactions. The premise was used that donation of an oxygen atom from the *N*-oxide of *N,N*-dimethylaniline oxidants to the heme-ferric iron of CYPcam or CYP2E1 generates CpdI, simultaneously producing the corresponding *N,N*-dimethylaniline analog to serve as a substrate and eliminating the possibility for the formation of the Cpd0 intermediate. KIE profiles of *N*-oxide *N,N*-dimethylaniline substrates and CYPcam or CYP2E1 were compared to profiles of *N,N*-dimethylaniline substrates and NAD(P)H/O_2/CYP-oxidoreductase systems. The results revealed that the KIE profiles obtained with the *N*-oxide substrates and CYPcam or CYP2E1 closely matched those of the *N,N*-dimethylaniline substrates and NAD(P)H/O_2/CYP-oxidoreductase systems. Because the *N*-oxide–CYP systems directly generated the CpdI species and bypassed the preceding intermediates of the CYP cycle, the results provided compelling evidence for CpdI serving as the primary transitory oxidant in *N*-demethylation reactions. In monooxygenation reactions supported by exogenous oxidants, the mechanistic process likely involves a two-electron transfer of a single oxygen atom to ferric CYP, generating CpdI that serves as the primary oxidant [1, 25, 34, 39, 63, 240, 244, 281, 289, 290]. The use of the single oxygen atom-containing oxidants, iodosobenzene and *N*-oxides, instead of hydroperoxides, to promote CYP-mediated monooxygenation reactions greatly simplifies the task of assigning the correct structure to the oxidizing species because it circumvents the necessity of considering Fe^{III}–O–OH intermediates as oxidizing species [1] and eliminates the need to examine reactions associated with O–O bond scission.

The inability of CYP researchers to capture and characterize the historically elusive CpdI intermediate generated in the classical NAD(P)H/O_2-mediated CYP monooxygenation cycle prompted suggestions for participation of alternative CYP iron-oxygen intermediates such as the ferric-peroxo anion intermediate (Fe^{III}–OO^-), ferric-hydroperoxo species (Fe^{III}–OOH) and Fe^{III}–(H_2O_2) complex of CYP. The nature and viability of these alternative oxidants are examined in Sects. 1.7.8, 1.7.9, and 1.7.10.

1.7.8 Ferric-Peroxo Anion Intermediate

1.7.8.1 Steroid Aromatization and Aldehyde Decarbonylation

A challenge to the prevailing assumption of CpdI being the singular oxygenating species involved in monooxygenation reactions originated from the mechanistic studies of Akhtar et al. [291–294] with steroid CYP19A1 (aromatase). Aromatase catalyzes the aromatization and 19-demethylation of androgen substrates to form aromatic estrogens by three sequential oxidative steps. The first and second steps are conventional NADPH/O_2-mediated steroid C19 hydroxylations postulated to involve the CYP ferryl radical intermediate (Fe^{IV}–$O^{\bullet}$) [292, 293]. The first hydroxylation occurs at the C19 methyl group of an androgen substrate to generate the 19-hydroxylated androgen, which is then oxidized in a second hydroxylation step to produce the *gem*-diol intermediate. The latter species undergoes dehydration and forms the C19 aldehyde. The third (terminal) step initially proceeds by an obligatory stereospecific abstraction of H atoms from the 1β and 2β positions of the steroid A ring to produce an aromatic ring. A unique mechanism for aromatization was originally proposed by Akhtar et al. [291, 292] whereby a CYP ferric-peroxo anion intermediate (Fe^{III}–O–O^-) produced in the CYP cycle initiates a nucleophilic attack on the C19-aldehyde of the androgen and "adds on" to generate a peroxohemiacetal adduct species, which rearranges and fragments to liberate the aromatic estrogen product and formic acid (Fig. 1.6). This aromatase mechanism received endorsement from a recent study involving EPR spectroscopic analysis of the aromatase reaction catalyzed by human aromatase [295]. Related steroidal CYP enzymes such as 17α-hydroxylase-17,20-lyase (CYP17A1) and sterol 14α-demethylase (CYP51A1) also catalyze C–C bond cleavage and aldehyde

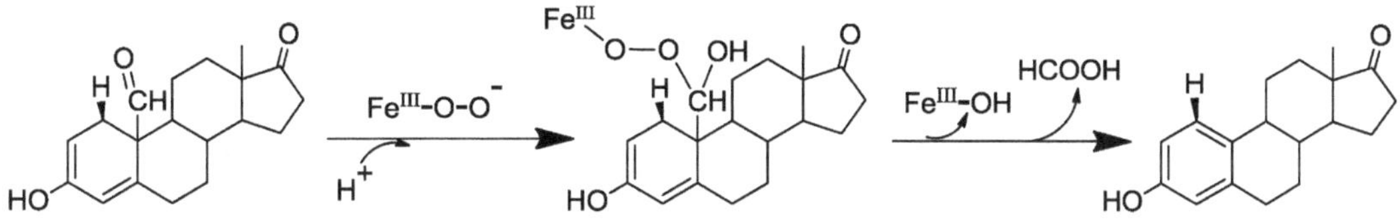

Fig. 1.6 Aromatase mechanism of CYP enzymes (Figure 1.6 was adapted from Refs. [291, 292])

decarbonylation reactions using a similar mechanism [293, 294].

Hackett et al. [296] utilized DFT calculations and constructed models of the peroxohemiacetal intermediate of CYP to assess feasibility of its involvement in aromatization. H atom abstraction from the 1β position of the steroid A ring by the ferriperoxo-carbonyl adduct, a prerequisite step for aromatization, required excessive energy and did not initiate fragmentation to the experimentally observed products. An alternative pathway for aromatization was delineated whereby the widely-accepted CpdI species, as opposed to the ferric-peroxo anion intermediate, initiates H atom abstraction to generate a steroid radical cation that undergoes deprotonation and cleavage to liberate aromatized estrogen and formic acid.

1.7.8.2 Aldehyde Deformylation and Olefin Formation

Vaz and coworkers [297, 298] demonstrated that purified CYPs 1A2, 2B4, 2C3, 2E1 and 3A6 obtained from rabbit catalyzed the oxidative deformylation of cyclohexane carboxaldehyde in the presence of NADPH/O_2 and CPR. A mechanism was proposed, similar to that formulated for the aromatase reaction, whereby an O_2-derived nucleophilic ferric-peroxo anion intermediate formed in the CYP cycle reacts with the electrophilic aldehyde carbonyl group and “adds on” to generate a CYP-bound peroxohemiacetal adduct species that rearranges and fragments to produce formic acid and cyclohexene. Of the well-known surrogate oxidants (i.e. *m*CPBA, cumene hydroperoxide, H_2O_2, iodosobenzene), only H_2O_2 was able to effectively promote deformylation of the aldehyde, likely by directly generating the ferric-peroxo anion species to serve as the oxidant. Further support for this mechanism was provided by a site-directed mutagenesis study in which the highly conserved Thr302 residue of CYP2B4 was replaced by alanine and the mutant tested for deformylation activity [299]. The mutant demonstrated increased deformylation activity with cyclohexane carboxaldehyde (tenfold) but displayed decreased hydroxylation activity with the benzphetamine (ninefold), cyclohexane (fourfold) and phenylethanol (twofold) substrates. The threonine to alanine mutation was speculated to produce a perturbation in the active site of CYP2B4 that interfered with proton delivery, leading to an elevation in the level of the ferric-peroxo anion species and an increase in deformylation activity. Aldehyde deformylation was also shown to occur with other cyclic and acyclic aldehydes in reconstituted systems containing CYP2B4 or other purified CYP enzymes [298].

1.7.8.3 Ferric-Peroxo Anion Intermediate of Selenocysteine-Substituted CYP125A1

CYP125A1 from *Mycobacterium tuberculosis* is a steroid C26-monooxygenase that catalyzes the sequential oxidation of the cholesterol side-chain terminal methyl group to the alcohol, aldehyde and final carboxylic acid product [262]. The enzyme also catalyzed the formation of five other products, which arose from deformylation of the sterol side chain. The aldehyde intermediate was the precursor of the conventional carboxylic acid metabolite and the five deformylation products. Formation of the carboxylic acid involved initial protonation of the ferric-peroxo anion intermediate ($Fe^{III}-OO^-$) and subsequent formation of the CpdI intermediate ($Por^{\bullet+}Fe^{IV}=O$), followed by H atom abstraction and oxygen atom transfer. The deformylation products arose by addition of the ferric-peroxo anion to the aldehyde intermediate, forming a peroxyhemiacetal adduct complex that fragmented through C–C bond cleavage to liberate the deformylation products. Replacing the proximal thiolate ligand of CYP125A1 by the stronger electron-donating selenocysteine allowed investigation of the role of excess electron donation by the proximal ligand on partitioning of the products between the two reaction pathways that lead to formation of the carboxylic acid product (heterolytic CpdI pathway) versus the C–C bond cleavage products (homolytic deformylation pathway). Substitution of the thiolate ligand by a selenocysteine produced UV/visible, EPR and resonance Raman spectral changes indicative of an increased electron donation from the selenolate ligand to the heme iron. Product distribution analysis in the reaction of the selenocysteine-substituted enzyme revealed a gain in the formation of the carboxylic acid product (via the CpdI pathway) at the expense of the deformylation products. The results provided direct evidence that increased proximal electron donation to the heme iron increases protonation of the ferric-peroxo anion intermediate, probably by increasing its pK_{α}, and therefore promotes formation of the CpdI species.

1.7.9 Ferric-Hydroperoxo Intermediate

To postulate involvement of intermediary CYP species other than CpdI to function as "electrophilic" oxidants in CYP-catalyzed monooxygenation reactions must be considered carefully, in view of the fact that the three preceding transitory species in the CYP cycle, i.e. the ferric-hydroperoxo (**E**), ferric-peroxo (**D**) and ferric-superoxo (**C**) species (see Fig. 1.3), carry negative charges that makes them good nucleophiles but not good electrophiles [269]. The possible involvement of the ferric-hydroperoxo species in CYP-catalyzed monooxygenation reactions is examined in Sects. 1.7.9.1, 1.7.9.2, and 1.7.9.3.

1.7.9.1 Olefin Epoxidation

The involvement of a negatively-charged ferric-hydroperoxo intermediate $^{-}$[Cys–Fe^{III}–OOH] as an "electrophile" in the oxidation of CYP substrates containing electron-rich double bonds [119, 300] and hydrocarbon bonds [301–306] has been postulated, but the evidence marshalled in support of this putative oxidizing species has been mostly circumstantial and unconvincing (see [6, 141, 143, 257, 307, 308] for critical comments). In principle, electrophilic reactivity could be feasible if the ferric-hydroperoxo species is sufficiently long-lived. A ferric-hydroperoxo intermediate is believed to function as an electrophilic oxidant in the conversion of heme to biliverdin by heme oxygenase, a heme enzyme distinct from CYP [141]. To determine the participation of the ferric-hydroperoxo species in olefin epoxidation reactions, Vaz et al. [300] created threonine to alanine mutants of CYP2B4 and CYP2E1 and conducted oxidative experiments involving olefin epoxidation and allylic hydroxylation using styrene, cyclohexene and 2-butene as model substrates. Styrene epoxidation, cyclohexene epoxidation and hydroxylation, and butene hydroxylation activities decreased significantly using the Thr302Ala mutant of CYP2B4 as a catalyst. By comparison, epoxidation activities of the three olefins increased whereas allylic hydroxylation activities of cyclohexene and butene decreased using the Thr303Ala mutant of CYP2E1. The threonine to alanine mutation was speculated to cause a partial block in proton delivery to the CYP2E1 active site and shift the protonation equilibrium in favor of the ferric-hydroperoxo intermediate, thus increasing its steady-state concentration and epoxidation capability at the expense of CpdI. Two electrophilic oxygenating species, the ferric-hydroperoxo intermediate that predominantly epoxidizes substrates and the CpdI species that primarily hydroxylates (but also epoxidizes) substrates, were proposed to function in oxidative reactions catalyzed by CYP2E1. In addition to being implicated as an electrophilic oxidant, the ferric-hydroperoxo species was also proposed to function as a nucleophilic oxidant, as shown by the large increase in aldehyde deformylation activity observed with the Thr303Ala mutant of CYP2E1 [300].

In a theoretical treatise, Watanabe [308] examined if involvement of the ferric-hydroperoxo species as a second electrophilic oxidant was sufficiently plausible to account for increased olefin epoxidation in the Thr303Ala mutant of CYP2E1. Instead of increasing the lifetime or population of the ferric-hydroperoxo species, Watanabe proposed that the threonine to alanine mutation could have disrupted the proton regulation machinery and caused release of the hydroperoxo group from ferric heme through uncoupling, as was shown to occur with the Thr252Ala mutant of CYPcam [271]. Alternatively, the threonine to alanine mutation in CYP2E1 could have affected epoxide and allylic oxidation selectivity of the mutant or the rate-limiting steps for olefin epoxidation and allylic hydroxylation. Then again, the mutation could have disturbed a well-organized network of water molecules surrounding the CpdI species of CYP2E1, altering its H-bonding interactions and reactivity. Meunier et al. [269] disputed the suggestion of Vaz et al. [300] that the ferric-hydroperoxo species can function as both a nucleophilic oxidant in aldehyde deformylation activities and as an electrophilic oxidant in olefin epoxidation reactions. Using DFT calculations, Shaik and coworkers proposed that the negatively-charged ferric-hydroperoxo intermediate is a strongly basic species displaying poor electron-accepting capability that does not favor electrophilic reactivity. Furthermore, the relatively large energy barriers (37–53 kcal mol^{-1}) computed for ethylene epoxidation by Fe^{III}–OOH [258] compared to the small energy barriers (14 kcal mol^{-1}) found for ethylene epoxidation by CpdI support the conclusion that the ferric-hydroperoxo intermediate is a poor electrophile and a very sluggish epoxidizing species that cannot compete in the presence of CpdI [257, 258, 289, 309].

1.7.9.2 Sulfoxidation and *N*-Demethylation

To account for the multiple pathways that function in epoxidation and hydroxylation reactions, Newcomb and coworkers [277, 301–306] formulated the two-oxidant hypothesis that involves CpdI as the principal oxygenating species and the ferric-hydroperoxo intermediate as a second oxidant. As an alternative proposal, Shaik and coworkers [250, 257, 258, 289] formulated the two-state reactivity theory that implicates high-spin and low-spin states of CpdI as leading reactants in epoxidation and hydroxylation reactions. Volz et al. [307] conducted site-directed mutagenesis and deuterium KIE experiments using bacterial CYP_{BM3} and its Thr268Ala mutant to examine the involvement of the ferric-hydroperoxo intermediate and CpdI species in the sulfoxidation and *N*-demethylation of dimethyl-(4-methylsulfanylphenyl)amine. The threonine to alanine mutation, believed to increase intrinsic levels of the ferric-hydroperoxo intermediate, produced a fourfold increase in the rate of sulfoxidation relative to *N*-demethylation. Substitution of H atoms by deuterium on the *N*-methyl groups of the substrate led to an intrinsic KIE but did not alter the ratio of sulfoxidation to *N*-demethylation. These patterns were consistent with sulfoxidation and *N*-demethylation activities arising from two separate enzyme-substrate complexes and two different oxidizing species. Volz first entertained the idea that sulfoxidation could be performed by the ferric-hydroperoxo species whereas *N*-demethylation is mediated exclusively by CpdI [68, 243, 244]. An alternative hypothesis patterned after Shaik's two-state reactivity theory was advanced. The threonine to alanine mutation in CYP_{BM3} was proposed to affect the reactivity of the Thr268Ala mutant by producing alterations in the CYP active-site structure and not in the oxygenating species [307].

Volz and coworkers studied the mechanism of sulfoxidation and *N*-demethylation further, utilizing the Phe87Ala mutant of CYP_{BM3} whereby conversion of Fe^{III}–OOH to CpdI in the mutant's catalytic cycle proceeds normally. A decrease in the rate of sulfoxidation relative to *N*-demethylation was observed, indicating that the change in regioselectivity reflected a change in the CYP_{BM3} active site and not in the oxygenating species. To rationalize their deuterium KIE data, the high-spin and low-spin states of CpdI were suggested to generate noninterchangeable enzyme-substrate complexes [307] whereby sulfoxidation was mediated by a spin-state complex computed to be high-spin CpdI [163], while *N*-demethylation was mediated by low-spin CpdI [310], the two spin states of CpdI thereby behaving as different oxidants. To rationalize the increased sulfoxidation activity in the Thr268Ala mutant, the threonine to alanine mutation, rather than increasing the amount of Fe^{III}–OOH, was proposed to change the organized network of water molecules surrounding CpdI, thereby altering its H-bonding interactions [308] that changed the ratio of high-spin to low-spin CpdI and altered the reactivity of the CYP_{BM3} mutant [307].

Sharma and coworkers [163] found the two-oxidant reactivity theory puzzling because the CpdI species by itself performs both sulfoxidation [311] and *N*-demethylation reactions [243, 244]. Consequently, a DFT study was conceived to further examine the mechanism of sulfoxidation by CpdI vis-à-vis Fe^{III}–OOH using dimethyl sulfide as a substrate [163]. The computational data revealed relatively large energy barriers (49.3 kcal mol^{-1}) for sulfoxidation by Fe^{III}–OOH and smaller energy barriers (16.8 and 19.2 kcal mol^{-1}) for sulfoxidation by the high-spin and low-spin states of CpdI, respectively, indicating that Fe^{III}–OOH is a sluggish oxidant and cannot compete in the presence of CpdI [163, 257, 289]. In contrast to alkane C–H hydroxylation (the intermediary step in *N*-demethylation) [310] whereby the H atom abstraction step has a lower energy barrier on the low-spin surface of CpdI, sulfoxidation displays a significantly lower energy barrier on the high-spin surface of CpdI [163], suggesting that low-spin CpdI mediates *N*-demethylation while high-spin CpdI mediates sulfoxidation.

1.7.9.3 Hydroxylation

Three major mechanisms were formulated for CYP-catalyzed hydroxylation reactions: (1) a radical stepwise non-concerted mechanism known as H atom abstraction/oxygen rebound that involves the primary CpdI species or its ferryl radical resonance hybrid [127, 133]; (2) a non-radical concerted mechanism based on utilization of ultrafast substrate probes and involving the insertion of a hydroxonium cation (OH^+) from the ferric-hydroperoxo intermediate into C–H bonds of substrates [302–304]; and (3) a two-state reactivity mechanism involving energetically-accessible low-spin and high-spin states of CpdI [250, 257, 309]. Early conceptions of the hydroxylation mechanism focused on an oxygen atom insertion pathway involving a CYP species equivalent to CpdI of heme peroxidases [25, 63]. This mechanism was refined and supplanted by the H atom abstraction/oxygen rebound mechanism whereby highly reactive CpdI, or its ferryl radical resonance hybrid [133], abstracts a H atom from the carbon center of an aliphatic substrate to produce an alkyl radical that is partitioned between two competing processes [127]. The alkyl radical can either instantly recombine by rebound with the ferryl-bound hydroxyl moiety to generate an unrearranged alcohol product, keeping the original stereochemical integrity of the substrate, or the alkyl radical can undergo rearrangement and then rebound to form a rearranged alcohol product [149, 309]. The rebound mechanism convincingly explained experimental observations of CYP hydroxylation reactions including (1) large intrinsic intramolecular KIEs (k_H/k_D >10) observed when the H atom of the scissible C–H bond is replaced by a deuterium atom, during C–H activation reactions with hydroxylatable substrates such as *N,N*-dimethylaniline [283], norbornane [127], norcarane [140] and testosterone [312]; (2) partial loss of regio- or stereochemistry in the hydroxylation of camphor [36] and norbornane [127]; and (3) allylic rearrangements in the hydroxylation of unsaturated hydrocarbons such as linoleic acid and cyclohexenes [313–315]. The large degree of stereoselectivity (>90 %) often found in the hydroxylation of substrates by CYP enzymes indicated that the carbon radicals were short-lived [142].

The advent of ultrafast radical clock substrates allowed rates of the rebound step and lifetimes of the substrate radical intermediates in hydroxylation reactions to be probed [1, 141].

Radical clock substrates contain a cyclopropyl ring bonded to the adjacent carbon atom, which becomes a radical center during the catalytic reaction. The strain intrinsic to the three-membered cyclopropyl ring causes cyclopropyl-carbinyl radicals to irreversibly rearrange to the corresponding homoallylic radical. The ratio of products derived from the cyclopropyl-carbinyl radical versus the homoallylic radical is obtained from the relative magnitude of the rate constants for radical quenching (oxygen rebound) and rearrangement. Rat liver CYP2B1 catalyzed the oxidation of a highly strained bicyclo[2.1.0]pentane radical clock substrate to a radical that rearranged to a monocyclic radical at a rate (k_r) of 2.4×10^9 s^{-1}, generating a mixture of ring-opened and unrearranged alcohol products from which a moderate rebound rate (k_{ox}) of 1.4×10^{10} s^{-1} and a radical lifetime of $\tau = 50$ ps were calculated [141, 316, 317]. With norcarane as a substrate probe, Auclair et al. [140] observed the formation of products indicative of a radical intermediate with lifetimes ranging from 16 to 52 ps and a radical rearrangement rate (k_r) of 2×10^8 s^{-1} during catalysis by recombinant CYP_{BM3}, CYPcam, rat CYP2B1 and human CYP2E1 enzymes.

Newcomb initiated investigations with radical clock and carbocation clock substrate probes to assess the validity of the radical rebound hydroxylation mechanism. The mechanistic picture began to cloud when ultrafast radical clock substrates yielded abnormally short radical lifetimes and extremely large rate constants for oxygen rebound, ranging from 2×10^{12} s^{-1} to 1.4×10^{13} s^{-1} [141–143, 302, 318]. Amounts of rearranged products from probes did not correlate with radical rearrangement rate constants, indicating that the rearranged products originated from cationic rather than radical intermediates. Sophisticated sensitive probes were then constructed that distinguished between radical and carbocation intermediates based on the identity of the rearranged products [304, 305]. CYP-mediated oxidation of the newly-designed probes produced carbocation-derived rearrangement products, disproving the assumption that the rearrangements originated from radical intermediates. Ultrashort radical lifetimes ranging from 70 to 200 fs were obtained that could not be attributed to true radical intermediates but instead corresponded to vibrational lifetimes or lifetimes of transition states [302, 305]. The short-lived radicals could not be the origin of the carbocation intermediates, casting doubt on the radical rebound mechanism and prompting Newcomb to invoke a non-radical mechanistic pathway for their formation. Aliphatic hydroxylation was proposed to proceed by competing mechanisms nascent from the two transitory CpdI and Fe^{III}–OOH species. In this model, CpdI reacts via a non-radical process by inserting an oxygen atom into the C–H bond of the substrate represented as H–CH–R (R is an organic substituent), forming the unrearranged alcohol product (HO–CH–R) (Eq. 1.21). Fe^{III}–O–OH reacts by inserting a hydroxonium cation (OH^+) into the C–H bond to generate a protonated alcohol (H_2O^+–CH–R) (Eq. 1.22), which in a bifurcated reaction process is either deprotonated to the unrearranged alcohol (HO–CH–R) (Eq. 1.23) or converted to carbocation rearrangement products via solvolysis (Eq. 1.24) [304, 305]. The cationic pathway requires heterolytic O–O bond scission in Fe^{III}–O–OH to release OH^+ and Fe^{III}=O (Eq. 1.22), as shown below.

$$\mathrm{H{-}CH{-}R + Por{\bullet}^{+}Fe^{IV}{=}O \rightarrow HO{-}CH{-}R + Fe^{III}} \quad (1.21)$$

$$\mathrm{H{-}CH{-}R + Fe^{III}{-}O{-}OH \rightarrow [H_2O^{+}{-}CH{-}R] + Fe^{III} = O} \quad (1.22)$$

$$\mathrm{[H_2O^{+}{-}CH{-}R] \xrightarrow{-H^{+}} HO{-}CH{-}R} \quad (1.23)$$

$$\mathrm{[H_2O^{+}{-}CH{-}R] \xrightarrow{-H_2O}} \text{carbocation rearrangement products} \quad (1.24)$$

Theoretical computational approaches contributed to the clarification of the mechanism of substrate hydroxylation and a proposed solution for discrepancies in the radical clock data. To account for the different pathways that participate in hydroxylation and epoxidation reactions, Shaik and coworkers [124, 250, 257, 258, 309] formulated the two-state reactivity theory whereby low-spin and high-spin states of CpdI function as the major reactants. The low-spin and high-spin states arise from two different coupling modes between the spin of the porphyrin radical (spin $S = 1/2$) and the $Fe^{IV} = O$ triplet electrons (spin $S = 1$) in CpdI [11]. The two-state reactivity theory stipulates that in hydroxylation reactions, H atom abstraction from the substrate by CpdI generates two intermediary CpdI states consisting of a low-spin ($S = 1/2$) (doublet) state and a high-spin ($S = 3/2$) (quartet) state. The low-spin component, having essentially a zero radical lifetime, abstracts a H atom from the C–H bond of the radical clock substrate to generate a carbon-centered unpaired electron, which rapidly recombines with the electron in the iron-hydroxyl orbital of the transitory ferryl hydroxyl species (Fe^{IV}–OH) formed from CpdI after H atom abstraction, to generate the unrearranged hydroxylated substrate. In the low-spin state, the unpaired electron residing on the substrate after H atom abstraction has an opposite spin to the electron in the iron-hydroxyl orbital [142]. The low-spin transition state collapses in a barrierless manner and the substrate is thus not able to undergo radical rearrangement. Alternatively, the high-spin component, characterized by normal radical lifetimes, either abstracts a H atom from the substrate to yield a carbon-centered unpaired electron that undergoes radical rebound recombination by the normal route to form the unrearranged hydroxylated substrate, or the high-spin species having to overcome an energy barrier becomes susceptible to substrate ring opening and rearrangement reactions prior to substrate radical rebound recombination with the ferryl hydroxyl species to yield the rearranged hydroxylated substrate. In the high-spin state, the substrate-based radical has the same spin orientation as that of the ferryl hydroxyl species and a spin inversion barrier must be overcome to generate the hydroxylated substrate [142].

The KIE on product distribution was used by Newcomb to formulate the two-oxidant hypothesis involving CpdI and Fe^{III}–OOH in the hydroxylation of the *trans*-2-methylphenylcyclopropane substrate probe [306]. Because differences in product isotope effect data can be interpreted differently, Kumar et al. [319] used theoretical calculations of the product isotope effect involving Newcomb's substrate probe to assess which of the two mechanisms prevail. The two-state reactivity theory, which predicts a product isotope effect in perfect accord with experiment, provided the best model for explaining the experimental observations. Newcomb's two-oxidant model and Shaik's two-state reactivity theory both accommodate iron-oxo insertion reactions [277, 309] but the difference rests in the identity of the "other" oxidant. Is the other transitory oxidant a ferric-hydroperoxo intermediate, a second spin state of the ferryl CpdI species, or a yet undefined entity of CYP? A recent theoretical study of de Visser et al. [320] provided an overview of the two-oxidant theory versus the two-state reactivity hypothesis. The authors discussed how early simplistic models involving a single reactive CpdI species have been invalidated by experimental studies that indicate involvement of two reactive oxidizing species in monooxygenation reactions. Thus, a two-oxidant theory was advanced whereby CpdI and its precursor in the CYP catalytic cycle, the ferric-hydroperoxo intermediate (Fe^{III}–OOH), are competitive oxidants. However, DFT studies suggest an alternative hypothesis consisting of a two-state reactivity scenario involving energetically-accessible low-spin and high-spin states of CpdI. The authors discussed how the two spin states of CpdI react differently as two distinct oxidants with substrates via aliphatic and aromatic C–H hydroxylation, C–C epoxidation, and sulfoxidation reactions, and explain experimentally observed product distributions and KIEs.

1.7.10 Fe^{III}–(H_2O_2) Complex of CYP as an Oxygenating Species

In an early study, Pratt et al. [72] examined the mechanism of lauric acid hydroxylation using purified CYP2B4 as the biocatalyst and H_2O_2 or iodosobenzene as the surrogate oxidant to determine the involvement of either the $Fe^{III}-(H_2O_2)$ complex, or the CpdI intermediate, of CYP2B4 as the primary transitory oxidant. CpdI formation was monitored spectroscopically as described by Blake and Coon [199]. The spectral data indicated that the CpdI complex was formed after CYP2B4 reacted with iodosobenzene, but this preformed CpdI intermediate was incapable of hydroxylating lauric acid. In contrast, when CYP2B4 reacted with H_2O_2, the spectral absorption bands in the 380–390 nm region indicative of CpdI formation were absent, but CYP2B4 was capable of catalyzing H_2O_2-mediated lauric acid 11-hydroxylation. These results prompted Pratt and coworkers to postulate that the intermediary $Fe^{III}-(H_2O_2)$ complex of CYP2B4, rather than the CpdI intermediate, acted as the primary transitory oxidant in lauric acid hydroxylation. More recently, Wang et al. [321] used theoretical computational analysis to demonstrate that the $Fe^{III}-(H_2O_2)$ complex of CYP is a highly efficient oxidant in sulfoxidation reactions. Theoretical calculations showed that $Fe^{III}-(H_2O_2)$ undergoes a low-barrier nucleophilic attack by sulfur on the distal oxygen atom, resulting in heterolytic O–O bond cleavage coupled to proton transfer. $Fe^{III}-(H_2O_2)$ also performed sulfoxidation much faster than could CpdI and will therefore bypass CpdI in the presence of a thioether. Theoretical calculations also revealed that $Fe^{III}-(H_2O_2)$ was an efficient sulfoxidation catalyst for synthetic iron porphyrin and iron corrolazine compounds, while the ferric-hydroperoxo intermediate was inactive in performing sulfoxidation reactions. The theoretical study highlights a new mechanistic pathway for sulfoxidation reactions.

1.8 Engineered CYP Monooxygenase and Peroxygenase Biocatalysts

Interest in CYP enzymes stems not only from their pharmacological and medical importance but also from the relentless drive to harness their biosynthetic potential. CYP enzymes display attractive properties that can be readily exploited in biotechnological, bioremedial, biomedical and biosynthetic applications [86, 322–329]. CYP proteins can also be engineered through rational design or directed evolution into mutant biocatalysts displaying novel, altered or improved oxidative enzymatic activities [86, 288, 322–334]. Rational design involves site-directed mutagenesis of amino acids in the CYP active site. Directed evolution entails creating CYP variants using error-prone PCR random/semi-random mutagenesis, saturation mutagenesis and recombination of CYP gene fragments, followed by screening and selection of variants for desired enzymatic characteristics [334–340]. In the next subsections, we will examine the utilization of site-directed mutagenesis and directed evolution procedures to create modified CYP mutant biocatalysts displaying novel, altered or markedly improved monooxygenase and peroxygenase activities.

1.8.1 Engineered CYP Monooxygenases

1.8.1.1 Engineered Bacterial CYP Monooxygenases

The high catalytic activity and stability of CYPcam and CYP_{BM3} have made these biocatalysts ideal candidates for protein engineering. Using site-directed mutagenesis, Wong and coworkers substituted the Tyr96 residue of CYPcam by alanine or phenylalanine and created mutants that dramatically increased the oxidation of alkanes, naphthalene, pyrene and styrene. The ability of the Tyr96Phe mutant to catalyze naphthalene and styrene oxidation was improved 142-fold [340] and 25-fold [341], respectively, relative to wild-type CYPcam. Table 1.9 lists the increases in NAD(P)H/O_2-supported monooxygenation activities (or catalytic efficiencies) displayed by bacterial and mammalian CYP mutants. Mutations of the active-site Phe87 and Tyr96 residues of CYPcam greatly enhanced its oxidative activity toward PAH substrates [335]. Notably, the activity of the Phe87Ala/Tyr96Phe double mutant toward

Table 1.9 Selected NAD(P)H/O_2-supported monooxygenase reactions catalyzed by bacterial, archaeal and mammalian CYP mutants

CYP mutant enzyme[a]	Substrate	Major activity	Increase over parent[b]	Refs.
Bacterial/archaeal mutant				
CYP101A1 (Tyr96Phe[c])	Naphthalene	Hydroxylase	142[d]	[340, 341]
CYP101A1 (Phe87Ala/Tyr96Phe)	Styrene	Epoxidase Hydroxylase	25	[335]
	Phenanthrene		500	
	Pyrene		420	
CYP101A1 (Phe87Try/Tyr96Phe/Thr101Leu/ Val247Leu	Butane	Hydroxylase	2,000	[342]
CYP101A1 (EB/Leu294Met/Thr185Met/ Leu1358Pro/Gly248Ala)	Ethane	Hydroxylase	8[e]	[343]
CYP102A1 (Phe87Val)	Benzyloxyphenol	Hydroxylase	963	[344]
	Benzothiophene		220	
	2,6-Dichlorophenol		99	
	Indan		66	
CYP102A1 (Arg47Leu/Tyr51Phe/Ala264Gly)	Pyrene	Hydroxylase	~1,700	[345]
	Phenanthrene		182	
	Fluoranthene		175	
CYP102A1 (Ala74Gly/Phe87Val/Leu188Glu)	Acenaphthene	Hydroxylase	~40,000	[346]
	9-Methylanthracene		11,500	
	Acenaphthylene		2,400	
	Fluorene		380	
	Naphthalene		210	
CYP102A1 (Asp168His/3X, Glu435Thr/3X)	Indole	3-Hydroxylase	6[f]	[347]
CYP102A1 (CYP_{PMO})	Propane	Hydoxylase	~9000[f, g]	[332]
	Ethane		10[h]	
CYP119A1 (Asp77Arg/Thr214Val)	Lauric acid	$(\omega - 1)$-Hydroxylase	~24	[348]
Mammalian mutant				
Human CYP1A2 (Glu163Lys/Val193Met/ Lys170Glu)	7-Methoxyresorufin	*O*-Demethylase	5	[349]
Human CYP2A6 (Phe209Thr)	Coumarin	7-Hydroxylase	13	[350]
Rat CYP2B1 (Phe202Leu/-Leu209Ala/Ser334Pro)	7-EFC	*O*-Deethylase	6	[69]
Rabbit CYP2B4 (Ile363Ala)	Testosterone	16α-Hydroxylase	55[f,i]	[101]

[a]Mutants of the wild-type enzymes are shown in parentheses
[b]Numbers shown are the fold increases in catalytic activity (k_{cat}) of the mutants over the parent wild-type enzymes, except where specifically noted
[c]*7-EFC* 7-ethoxy-4-trifluoromethylcoumarin. For amino acid abbreviations, see footnote "1" in the actual text
[d]The Tyr96Phe mutant of CYP101A1 was the parent enzyme
[e]The EB/Leu294Met/Thr185Met/Leu1358Pro mutant of CYP101A1 was the parent enzyme
[f]Fold increases shown are for catalytic efficiency (k_{cat}/K_m)
[g]The 139-3 mutant of CYP102A1 was the parent enzyme
[h]The 35E11 mutant of CYP102A1 was the parent enzyme
[i]The 2B4dH/H226Y mutant of CYP2B4 was the parent enzyme

phenanthrene and pyrene was increased 500- and 420-fold, respectively. The Phe87Trp/Tyr96Phe/Thr101Leu/Val247Leu quadruple variant catalyzed the hydroxylation of butane at a rate of 750 min^{-1}, which was 2,000 times greater than that of native CYPcam [342]. Site-directed mutagenesis and directed evolution techniques were used to improve the catalytic efficiency of CYP_{BM3} or to alter its substrate specificity and selectivity. Notably, a Phe87Val variant was highly active toward the oxidation of 2-(benzyloxy)phenol, benzothiophene, dichlorophenol and indan, producing reaction rates that were 963-, 220-, 99- and 66-fold greater than those of native CYP_{BM3}, respectively [344]. Variants of CYP_{BM3} were also created that catalyzed the biodegradation of PAHs. The Arg47Leu/Tyr51Phe/Ala264Gly triple mutant displayed a ~1,700-, 182- and 175-fold increase in oxidative activity toward pyrene, phenanthrene and fluoranthene, respectively [345]. The Ala74Gly/Phe87Val/Leu188Glu triple mutant was highly active in oxidizing acenaphthene, 9-methylanthracene, acenaphthylene, fluorene and naphthalene, producing reaction rates that were ~40,000-, 11,500-, 2,400-, 380- and 210-times higher than those of CYP_{BM3}, respectively [346].

Arnold and coworkers [330–332, 351–353] utilized directed evolution procedures to design libraries of CYP_{BM3} variants that displayed altered substrate specificities and novel or improved oxidative activities toward natural and unnatural substrates. Notably, CYP_{BM3} was engineered from a long-chain fatty acid hydroxylase into a short-chain alkane hydroxylase using a directed evolution strategy in which mutations were accumulated over multiple generations of random mutagenesis, site saturation and site-directed mutagenesis, recombination and screening [330]. A 53-5H mutant with 15 heme-domain substitutions was created that catalyzed the regioselective hydroxylation of octane to produce 2-octanol (89 %) at a rate of 660 min^{-1}, favoring the *S* enantiomer (65 % *ee*) and displaying a total turnover number of 8,000 with a coupling efficiency of 80 %. Wild-type CYP_{BM3} produced mainly 3- and 4-octanol. Multiple rounds of directed evolution combined with site-directed mutagenesis were used to engineer CYP_{BM3} from a fatty acid hydroxylase with no measurable propane oxygenase activity into a highly proficient propane monooxygenase ($CYP_{PMO}R2$), which displayed a hydroxylation rate of 370 min^{-1}, a coupling efficiency of 98.2 % and a high total turnover number on propane of ~46,000 [331]. Further directed evolution generated a CYP_{PMO} variant that displayed a total turnover number for ethane hydroxylation of 2,450, which was a tenfold improvement over the previously-created 35E11 parent mutant [332]. Although the catalytic rate and total turnover number for ethane hydroxylation were too low for practical application, continual improvements of CYP_{PMO} by directed evolution could eventually produce an ethane hydroxylase, or even a methane hydroxylase, with similar productivity as obtained with propane [331, 332], thus creating new opportunities for "green" conversion of gaseous hydrocarbons into liquid fuels and chemicals. CYP_{BM3} variants have also been designed for the production of bioactive drugs such as artemisinin and ibuprofen [352, 354] and for the biosynthesis of human drug metabolites from parent drugs such as acetaminophen, buspirone, clozapine, dextromethorphan, diclofenac, phenacetin, propranolol, verapamil, lovastatin and simvastatin [44, 333, 337, 355–359]. Human drug metabolites are required for toxicity tests as part of the development process for safe and effective pharmaceuticals [44].

1.8.1.2 Engineered Mammalian CYP Monooxygenases

Of the diverse CYP monooxygenases found in nature, microsomal CYP enzymes from mammalian liver display the widest substrate specificities. However, mammalian CYP enzymes also have disadvantages that include low catalytic activities, poor stability, short lifetimes, inadequate recombinant expression in *E. coli*, dependence on CPR and often cytochrome b_5 as auxiliary electron transfer proteins and poor NADPH coupling efficiencies [86, 323]. Most of these disadvantages have recently been resolved [86]. Moreover, the requirement for NADPH, CPR and cytochrome b_5 has been overcome by substitution of peroxides (see Sect. 1.8.2.2). Mammalian CYP monooxygenases with their catalytic versatility and broad substrate specificity offer possibilities for widespread applications in industrial biotechnology [86], environmental bioremediation [86], medical science (e.g. gene-directed enzyme prodrug therapy) [86, 325–327], biosensor design [86, 360], dye production [350, 361] and biosynthesis of novel drug substrates [86, 325, 352], human drug metabolites [86] and steroids [67, 69, 86]. Mammalian CYP enzymes can also be engineered into newly-designed mutant biocatalysts displaying novel, altered or markedly improved monooxygenation activities that can be readily exploited in biotechnological, bioremedial, biomedical and biosynthetic applications [69, 86, 104, 326, 334–338, 349].

Guengerich and coworkers constructed mutants of human CYP1A2 and CYP2A6 that displayed enhanced oxidative activities toward alkylresorufins, heterocyclic amines, phenacetin, coumarins and indole [165, 230, 323, 324, 349, 350, 362, 363]. Bicistronic expression systems, where two proteins are expressed from a single plasmid, were developed for coexpression of the CYP mutant enzymes with their CPR redox partners. After three rounds of mutagenesis and screening, a Glu163Lys/Val193Met/Lys170Glu triple mutant of CYP1A2 was constructed that showed a fivefold greater activity for 7-methoxyresorufin demethylation compared to the native enzyme [362]. CYP1A2 variants were also created that displayed a 3- to 12-fold greater catalytic efficiency in the oxidation of 2-amino-3,5-dimethylimidazo[4,5-*f*]-quinolone, phenacetin and ethoxyresorufin [230, 349, 362]. Similarly, a Phe209Thr mutant of CYP2A6 showed a 13-fold greater activity for coumarin 7-hydroxylation [350]. Several CYP2A6 variants displayed increased activities for indole 3-hydroxylation and catalyzed the oxidation of substituted indoles to new mixtures of dyes including those that contained indigo and indirubin [339, 350, 361]. Because indigo and indirubin are high affinity ligands for the mammalian aryl hydrocarbon receptor [363] and are inhibitors of kinases [323], the dyes have the potential for being used as therapeutic drugs.

Kumar, Halpert and coworkers [69, 86, 101, 158, 324, 364] applied rational and directed evolution approaches to engineer mammalian CYP2B enzymes for enhanced catalytic activity and altered substrate specificity and regio- and stereoselectivity. A fluorescence-based activity screening assay was developed to measure enhanced oxidative activity of rat CYP2B1 using 7-ethoxy-4-trifluoromethylcoumarin (7-EFC) as a fluorogenic substrate [69]. The Phe202Leu/Leu209Ala/Ser334Pro triple mutant of CYP2B1 demonstrated a ~6-fold increase in NADPH/O_2-mediated oxidative activity toward 7-EFC, compared to native CYP2B1, and also displayed increased activity toward benzphetamine, benzyloxyresorufin and testosterone [69]. The triple mutant catalyzed testosterone 16-α-hydroxylation at a catalytic rate of 150 min^{-1}, which is among the highest rates reported for any mammalian CYP enzyme in an NADPH/O_2-supported reaction [69]. An Ile363Ala mutant of rabbit CYP2B4 showed an 8- and 12-fold increase in testosterone 16α- and 16β-hydroxylase activity, respectively, compared to the 2B4dH/H226Y parent enzyme [101]. The mutant also displayed a >5-fold lower K_m for testosterone than the parent enzyme while catalyzing 16α-hydroxylase activity, leading to a 55-fold improvement in catalytic efficiency (k_{cat}/K_m).

1.8.2 Engineered CYP Peroxygenases

1.8.2.1 Engineered Bacterial and Archaeal CYP Peroxygenases

Arnold and coworkers [29] engineered CYPcam by directed evolution to function as an effective peroxygenase in the hydroxylation of naphthalene, using H_2O_2 as the surrogate oxygen atom donor in place of O_2, NADH and redox proteins. Approximately 200,000 random CYPcam mutants produced by mutagenic PCR and coexpressed with HRP1A6 in *E. coli* were screened for peroxygenase activity by

Table 1.10 Selected peroxide-supported peroxygenase reactions catalyzed by bacterial, archaeal and mammalian CYP mutants

CYP mutant enzyme[a]	Peroxide oxidant	Substrate	Major activity	Increase over parent[b]	Refs.
Bacterial/archaeal mutant					
CYP101A1 (S3-20, S3-27)	H_2O_2	Naphthalene	Hydroxylase	20	[29]
CYP102A1 (Phe87Ala[c])	H_2O_2	12-*p*NCA	Hydroxylase	320	[44]
CYP102A1 (21B3)	H_2O_2	Styrene	Epoxidase	24[d]	[42]
CYP102A1 (2C11)	H_2O_2	Propranolol	4′-Hydroxylase	4[e]	[43]
CYP102A1 (Phe87Val)	H_2O_2	Indole	3-Hydroxylase	92	[40]
	t-BuOOH			100	
CYP102A1 (SH-44)	H_2O_2	1-Hexene	Epoxidase	55	[351]
CYP107A1 (Gly91Ala/Ala245Thr)	H_2O_2	7-BQ	*O*-Debenzylase	4[f]	[45]
CYP119A1 (Thr214Ala/Thr214Val)	H_2O_2	Styrene	Epoxidase	3	[30]
CYP119A2 (Phe310Ala/Ala320Gln)	H_2O_2	Ethylbenzene	Hydroxylase	~20	[35]
		styrene	epoxidase	~20	
CYP152B1 (Leu78Phe)	H_2O_2	Myristate	α-Hydroxylase	~4	[365]
Mammalian mutant					
Rat CYP2B1 (Val183Leu/Phe202Leu/-Leu209Ala/Ser334Pro)	H_2O_2	7-EFC	*O*-Deethylase	6[g]	[69]
Human CYP2D6 (Thr309Val)	CuOOH	Bufuralol	4-Hydroxylase	74	[366]
		Dextromethorphan	*O*-Demethylase	~3	
		7-MAMC	*O*-Demethylase	~3	
Human CYP3A4 (Leu216Try)	H_2O_2	7-BQ	*O*-Debenzylase	~3[h]	[67]
Human CYP3A4 (Phe228Ile/Thr309Ala)	CuOOH	7-BQ	*O*-Debenzylase	11[h]	[67]
Human CYP3A4 (Thr309Val)	CuOOH	7-BFC	*O*-Debenzylase	~3	[67]

[a]Mutants of the wild-type enzymes are shown in parentheses
[b]Numbers shown are the fold increases in catalytic activity (k_{cat}) of the mutants over the parent wild-type enzymes, except where specifically noted
[c]*7-BQ* 7-benzyloxyquinoline, *7-BFC* 7-benzyloxy-4-trifluoromethylcoumarin; *t-BuOOH* *t*-butyl hydroperoxide, *CuOOH* cumene hydroperoxide, *7-EFC* 7-ethoxy-4-trifluoromethylcoumarin, *7-MAMC* 7-methoxy-4-aminomethyl-coumarin, *12-pNCA* 12-*p*-nitrophenoxydodecanoic acid. For amino acid abbreviations, see footnote "1" in the actual text
[d]The HPhe87Ala mutant of CYP102A1 was the parent enzyme
[e]The 9C1 mutant of CYP102A1was the parent enzyme
[f]The Ala245Thr mutant of CYP107A1 was the parent enzyme
[g]The Leu209Ala mutant of CYP2B1 was the parent enzyme
[h]The increases shown are for catalytic efficiency (k_{cat}/K_m)

fluorescence digital imaging in the presence of naphthalene and H_2O_2. A second-generation CYPcam library prepared by staggered extension process in vitro recombination of five improved variants yielded two variants (S3–20 and S3–27) that displayed a 20-fold improvement in H_2O_2-driven naphthalene hydroxylase activity over wild-type CYPcam [29]. Khan et al. [47] performed site-directed mutagenesis of five active-site residues in CYP107A1 and created a Gly91Ala/Ala245Thr double mutant that showed a fourfold higher activity for H_2O_2-driven 7-benzyloxyquinoline (7-BQ) debenzylation compared to the Ala245Thr mutant parent enzyme. Table 1.10 lists the increases in peroxide-supported peroxygenase activities and catalytic efficiencies displayed by bacterial, archaeal and mammalian CYP mutants.

CYP_{BM3} was engineered to function as an effective peroxygenase in the oxygenation of fatty acids, alkenes and chemotherapeutic drugs [43–45]. Notably, the Phe87Ala mutant of CYP_{BM3} catalyzed the H_2O_2-driven oxygenation of 12-*p*NCA at a catalytic rate that was

320 times greater than that of native CYP_{BM3} [45]. CYP_{BM3} was engineered into a heme-domain 21B3 variant that displayed a H_2O_2-sustained peroxygenase activity toward lauric acid (50 min^{-1}) and styrene (54 min^{-1}) [43]. A 9C1 variant catalyzed the H_2O_2-driven hydroxylation of propranolol with a total turnover number of 180, producing 70 mg of propranolol metabolites per liter of *E. coli* culture [44]. A unique approach was used to construct an artificial family of ~3,000 chimeric CYP proteins by recombining fragments of the genes encoding the heme domains of three bacterial CYP proteins, CYP102A1 (CYP_{BM3}), CYP102A2 and CYP102A3, which share ~65 % sequence similarity [336]. After screening the chimeric CYP proteins, 73 % were found to be active peroxygenases and catalyzed the H_2O_2-driven oxygenation of 2-phenoxyethanol. In a subsequent complementary study involving the three CYP102A proteins, several newly-constructed chimeric CYP peroxygenase enzymes were created that, in the presence of H_2O_2, catalyzed the production of metabolites from the parent drugs astemizole and verapamil [337].

Recently, Hayakawa et al. [35] investigated the pH dependence of the peroxygenase reactions catalyzed by the archaeal wild-type CYP119A2 enzyme and its Phe310Ala/Ala320Gln double mutant. At pH 5, the mutant was approximately 20 times more active than the wild-type enzyme in catalyzing the hydroxylation of ethylbenzene and the epoxidation of styrene in the presence of H_2O_2 (see Table 1.10).

1.8.2.2 Engineered Mammalian CYP Peroxygenases

Mammalian CYP enzymes have been engineered to act as effective peroxygenases using H_2O_2 and cumene hydroperoxide as oxidants. A Val183Leu/Phe202Leu/Leu209Ala/Ser334Pro quadruple mutant of rat CYP2B1, created through random and site-directed mutagenesis and expressed in *E. coli*, showed a sixfold improvement in catalytic activity over the Leu209Ala mutant parent for H_2O_2-sustained deethylation of 7-EFC [71]. A Thr309Val mutant of CYP2D6 engineered by site-directed mutagenesis demonstrated a 74-fold increase in activity, compared to native CYP2D6, for the cumene hydroperoxide-supported 4-hydroxylation of bufuralol [366]. The mutant also showed a 2.4- to 3.4-fold enhanced activity for the cumene hydroperoxide-driven *O*-demethylation of dextromethorphan and 7-methoxy-4-aminomethylcoumarin, and for the *O*-demethylenation of 3,4-methylenedioxymethylamphetamine [366]. Kumar and coworkers [67] developed a fluorescence-based activity screening assay system using 7-BQ as the substrate and H_2O_2 or cumene hydroperoxide as oxidants to screen CYP3A4 variants for 7-BQ debenzylase activity. After an initial screening of several thousand random clones derived from the directed evolution of CYP3A4, a Leu216Trp variant was created that showed a ~3-fold increase in catalytic efficiency using H_2O_2 as the oxidant [67]. In addition, a Phe228Ile/Thr309Ala double mutant demonstrated an 11-fold improvement in catalytic efficiency for cumene hydroperoxide-supported 7-BQ debenzylation. A Thr443Ser mutant catalyzed an 80 % conversion of testosterone to 6β-hydroxytestosterone with a total turnover of 180 min^{-1} in a cumene hydroperoxide-sustained reaction [67, 86].

1.9 Conclusions

This review examined the monooxygenase, peroxidase and peroxygenase properties and reaction mechanisms of CYP enzymes in bacterial, archaeal and mammalian systems. We reported that CYP enzymes typically catalyze the monooxygenation of a vast number and variety of exogenous and endogenous substrates by utilizing molecular oxygen and two electrons donated by NAD(P)H through the mediation of redox proteins. We conducted a detailed investigation of the reaction mechanisms by which CYP enzymes catalyze monooxygenation reactions by examining the invaluable roles played by the primary CYP CpdI oxidizing intermediate and other reactive intermediary iron-oxygen species that form the molecular stations of the CYP catalytic cycle. We then reviewed the peroxygenase properties of CYP enzymes by which peroxides, peracids, periodate, perborate, percarbonate, chlorite, iodosobenzene and *N*-oxides function as surrogate oxygen atom donors to drive monooxygenation reactions via the shunt pathway. We discussed the ability of surrogate oxidant-driven monooxygenase reactions to mimic native NAD(P)H/O_2-supported monooxygenase reactions and examined the initial role of CpdI, or its ferryl radical resonance hybrid, as the common primary oxidizing intermediate in the two pathways.

CYP enzymes catalyze the oxidation of a vast number and variety of exogenous and endogenous compounds, most commonly by inserting an oxygen atom from O_2 into unactivated aliphatic sp^3-hybridized, and aryl sp^2-hybridized, C–H bonds of substrates. Other prominent monooxygenation reactions catalyzed by CYP enzymes include the epoxidation of C=C double bonds, *N*-hydroxylations, dehalogenations, deaminations, oxygenations of heteroatoms (N, S, P and I) and *N*-, *O*- and *S*-dealkylations, thus making CYP enzymes attractive biocatalysts from a biosynthetic standpoint [93, 95, 329]. Widespread interest has accumulated in harnessing the invaluable monooxygenase activity of CYP enzymes, particularly from bacterial CYPcam and CYP_{BM3} and from mammalian CYP enzymes. In addition, site-specific mutagenesis and directed evolution techniques have been utilized to create CYP mutants displaying novel, altered or markedly improved monooxygenase and peroxygenase activities that have found applications in industrial biotechnology, medical science, environmental bioremediation and biosynthesis of pharmaceuticals, human drug metabolites, steroids and other chemicals.

Inspired by Charles Darwin's and Captain George Nares's nineteenth century voyages of scientific discovery, Dr. J. Craig Venter and his team embarked on a 2.5 year circumnavigation sea odyssey aboard the sailing yacht, Sorcerer II, to collect and analyze mainly surface seawater samples that would provide a comprehensive genomic survey of microbial (bacterial, archaeal, viral) life in the world's oceans [367–369]. This Global Ocean Sampling (GOS) expedition began in Halifax, Canada, in August 2003, traversed several oceans and ended in Florida in January 2006. Venter's sailing vessel circumnavigated the world and seawater samples containing mostly bacteria were collected from 150 open-ocean and coastal sites. During the first leg of the expedition (Halifax to the Galapagos Islands), 41 samples of marine planktonic microbiota were collected from seawater mainly at the ocean's surface (about one foot deep in Ecuador) and at more than 4,500 m deep off of Mexico's Yucatan Peninsula. Samples were collected every 200 nautical miles across a several-thousand km transect from the North Atlantic Ocean through the Panama Canal and ending in the South Pacific Ocean. Filtered samples were subjected to genome shotgun sequencing, yielding 7.7 million GOS sequences encoding 6.12 million new proteins [367–369]. Venter also sorted these protein sequences by protein family and found 3,305 new marine CYP sequences [370], mostly from the bacterial kingdom. It will be interesting to see what novel enzymatic activities and unexpected properties are displayed by the bacterial (and archaeal) CYP enzymes.

We conclude by proposing that under the global anaerobic conditions of early Earth, CYP enzymes of ancient prokaryotic organisms functioned as peroxidases and peroxygenases well before their monooxygenase function developed [1]. Support for this proposal is provided by studies showing that archaeal CYP119A1 from *Sulfolobus acidocaldarius* can catalyze the peroxygenation of laurate, *cis*-stilbene and styrene in the presence of H_2O_2 and other peroxy compounds [30–32, 58]. In addition, a recent study has demonstrated that archaeal CYP119A2 from the *Sulfolobus tokodaii* strain 7 can catalyze the peroxygenation of ethylbenzene and styrene in the presence of H_2O_2 [35]. We speculate that CYP119A1 and CYP119A2 emerged from an ancient anaerobic prokaryote [12] in an era that resembled the iron-sulfur world envisioned by Wächtershäuser [15] and in a time that predated atmospheric O_2 and the combustion of organic matter [12, 17]. Early primordial CYP enzymes could have emerged under global anaerobic conditions to provide the organism with peroxidative and peroxygenative abilities to metabolize critical endogenous compounds, as well as to biodegrade environmental chemicals utilized for energy [1, 12, 17, 29]. There were 39 named CYP members[4] identified from the archaeal kingdom as of April 21st, 2011 [371]. The inherent catalytic properties of contemporary archaeal CYP119A1 and CYP119A2 [30, 32, 34, 35, 240] that evolved from an ancient prokaryote and can function as efficient peroxygenases allow us to recreate the primordial functions of the ancient ancestral CYP enzymes.

Acknowledgments The authors acknowledge current and previous support from the Natural Sciences and Engineering Research Council of Canada and the Canadian Institutes of Health Research.

References

1. Hrycay EG, Bandiera SM (2012) The monooxygenase, peroxidase, and peroxygenase properties of cytochrome P450. Arch Biochem Biophys 522:71–89
2. http://drnelson.uthsc.edu/P450.statsfile.html
3. Lewis DFV (1996) Cytochromes P450: structure, function and mechanism. Taylor & Francis, London
4. Lewis DFV (2001) Guide to cytochromes P450: structure and function. Taylor & Francis, London
5. Guengerich FP (2001) Common and uncommon cytochrome P450 reactions related to metabolism and chemical toxicity. Chem Res Toxicol 14:611–650
6. Ortiz de Montellano PR, De Voss JJ (2005) Substrate oxidation by cytochrome P450 enzymes. In: Ortiz de Montellano PR (ed) Cytochrome P450: structure, mechanism, and biochemistry, 3rd edn. Kluwer Academic/Plenum Publishers, New York, pp 183–245
7. Hrycay EG, Bandiera SM (2008) Cytochrome P450 enzymes. In: Gad SC (ed) Preclinical development handbook: ADME and biopharmaceutical properties. Wiley, Hoboken, pp 627–696
8. (a) Hrycay EG, Bandiera SM (2009) Expression, function and regulation of mouse cytochrome P450 enzymes: comparison with human cytochrome P450 enzymes. Curr Drug Metab 10:1151–1183. (b) Hrycay EG, Bandiera SM (2010) Addendum to ref. 8(a). Curr Drug Metab 11:560
9. Guengerich FP (2005) Human cytochrome P450 enzymes. In: Ortiz de Montellano PR (ed) Cytochrome P450: structure, mechanism, and biochemistry, 3rd edn. Kluwer Academic/Plenum Publishers, New York, pp 377–530
10. Shumyantseva VV, Bulko TV, Archakov AI (2005) Electrochemical reduction of cytochrome P450 as an approach to the construction of biosensors and bioreactors. J Inorg Biochem 99:1051–1063
11. Perera R, Jin S, Sono M, Dawson JH (2007) Cytochrome P450-catalyzed hydroxylations and epoxidations. Met Ions Life Sci 3:319–359
12. Wickramasinghe RH, Villee CA (1975) Early role during chemical evolution for cytochrome P450 in oxygen detoxification. Nature 256:509–511
13. Lewis DFV, Sheridan G (2001) Cytochromes P450, oxygen, and evolution. Sci World 1:151–167
14. Kauffman SA (1986) Autocatalytic sets of proteins. J Theor Biol 119:1–24
15. Wächtershäuser G (1992) Groundworks for an evolutionary biochemistry: the iron–sulphur world. Prog Biophys Mol Biol 58:85–201
16. Davis BK (2002) Molecular evolution before the origin of species. Prog Biophys Mol Biol 79:77–133
17. Nelson DR, Kamataki T, Waxman DJ, Guengerich FP, Estabrook RW, Feyereisen R, Gonzalez FJ, Coon MJ, Gunsalus IC, Gotoh O, Okuda K, Nebert DW (1993) The P450 superfamily: update on new sequences, gene mapping, accession numbers, early trivial names of enzymes, and nomenclature. DNA Cell Biol 12:1–51
18. Hrycay EG, O'Brien PJ (1971) Cytochrome P-450 as a microsomal peroxidase utilizing a lipid peroxide substrate. Arch Biochem Biophys 147:14–27
19. Hrycay EG, O'Brien PJ (1972) Cytochrome P-450 as a microsomal peroxidase in steroid hydroperoxide reduction. Arch Biochem Biophys 153:480–494
20. Hrycay EG, O'Brien PJ, Van Lier JE, Kan G (1972) Pregnene 17α-hydroperoxides as possible precursors of the adrenosteroid hormones. Arch Biochem Biophys 153:495–501
21. Hrycay EG, O'Brien PJ (1973) Microsomal electron transport: I. Reduced nicotinamide adenine dinucleotide phosphate–cytochrome c reductase and cytochrome P-450 as electron carriers in microsomal NADPH–peroxidase activity. Arch Biochem Biophys 157:7–22
22. Hrycay EG, Prough RA (1974) Reduced nicotinamide adenine dinucleotide-cytochrome b_5 reductase and cytochrome b_5 as electron carriers in NADH-supported cytochrome P-450-dependent enzyme activities in liver microsomes. Arch Biochem Biophys 165:331–339
23. Hrycay EG, Gustafsson JÅ, Ingelman-Sundberg M, Ernster L (1975) Sodium periodate, sodium chlorite,

[4] The number of discovered CYP members in the archaeal kingdom as of April 21st, 2011 was 39, and was obtained through a BLAST search of the reference protein sequence database of the NCBI [371].

organic hydroperoxides and H_2O_2 as hydroxylating agents in steroid hydroxylation reactions catalyzed by partially purified cytochrome P−450. Biochem Biophys Res Commun 66:209–216
24. Hrycay EG, Gustafsson JÅ, Ingelman-Sundberg M, Ernster L (1975) Sodium periodate, sodium chlorite, and organic hydroperoxides as hydroxylating agents in hepatic microsomal steroid hydroxylation reactions catalyzed by cytochrome *P*-450. FEBS Lett 56:161–165
25. Hrycay EG, Gustafsson JÅ, Ingelman-Sundberg M, Ernster L (1976) The involvement of cytochrome *P*-450 in hepatic microsomal steroid hydroxylation reactions supported by sodium periodate, sodium chlorite, and organic hydroperoxides. Eur J Biochem 61:43–52
26. Gustafsson JÅ, Hrycay EG, Ernster L (1976) Sodium periodate, sodium chlorite, and organic hydroperoxides as hydroxylating agents in steroid hydroxylation reactions catalyzed by adrenocortical microsomal and mitochondrial cytochrome P_{450}. Arch Biochem Biophys 174:440–453
27. McKay CP, Hartman H (1991) Hydrogen peroxide and the evolution of oxygenic photosynthesis. Orig Life Evol Biosph 21:157–163
28. Samuilov VD (1997) Photosynthetic oxygen: the role of H_2O_2. A review. Biochem Mosc 62:451–454
29. Joo H, Lin Z, Arnold FH (1999) Laboratory evolution of peroxide-mediated cytochrome P450 hydroxylation. Nature 399:670–673
30. Koo LS, Tschirret-Guth RA, Straub WE, Moënne-Loccoz P, Loehr TM, Ortiz de Montellano PR (2000) The active site of the thermophilic CYP119 from *Sulfolobus solfataricus*. J Biol Chem 275:14112–14123
31. Rabe KS, Kiko K, Niemeyer CM (2008) Characterization of the peroxidase activity of CYP119, a thermostable P450 from *Sulfolobus acidocaldarius*. ChemBioChem 9:420–425
32. Kellner DG, Hung SC, Weiss KE, Sligar SG (2002) Kinetic characterization of compound I formation in the thermostable cytochrome P450 CYP119. J Biol Chem 277:9641–9644
33. Jiang Y, Sivaramakrishnan S, Hayashi T, Cohen S, Moënne-Loccoz P, Shaik S, Ortiz de Montellano PR (2009) Calculated and experimental spin state of seleno cytochrome P450. Angew Chem Int Ed 48:7193–7195
34. Rittle J, Green MT (2010) Cytochrome P450 compound I: capture, characterization, and C-H bond activation kinetics. Science 330:933–937
35. Hayakawa S, Matsumura H, Nakamura N, Yohda M, Ohno H (2014) Identification of the rate-limiting step of the peroxygenase reactions catalyzed by the thermophilic cytochrome P450 from *Sulfolobus tokodaii* strain 7. FEBS J 281:1409–1416
36. Gelb MH, Heimbrook DC, Mälkönen P, Sligar SG (1982) Stereochemistry and deuterium isotope effects in camphor hydroxylation by the cytochrome P450cam monooxygenase system. Biochemistry 21:370–377
37. Sligar SG, Shastry BS, Gunsalus IC (1977) Oxygen reactions of the P450 heme protein. In: Ullrich V, Roots I, Hildebrandt A, Estabrook RW, Conney AH (eds) Microsomes and drug oxidations. Pergamon Press, Oxford, pp 202–209
38. White RE, Sligar SG, Coon MJ (1980) Evidence for a homolytic mechanism of peroxide oxygen−oxygen bond cleavage during substrate hydroxylation by cytochrome P-450. J Biol Chem 255:11108–11111
39. Sligar SG, Gelb MH, Heimbrook DC (1984) Bio-organic chemistry and cytochrome P-450-dependent catalysis. Xenobiotica 14:63–86
40. Li QS, Ogawa J, Schmid RD, Shimizu S (2005) Indole hydroxylation by bacterial cytochrome P450 BM-3 and modulation of activity by cumene hydroperoxide. Biosci Biotechnol Biochem 69:293–300
41. Shinohara A, Kamataki T, Iizuka T, Ishimura Y, Ogoshi H, Okuda K, Kato R (1987) Drug oxidation activities of horseradish peroxidase, myoglobin and cytochrome $P\text{-}450_{CAM}$ reconstituted with synthetic hemes. Jpn J Pharmacol 45:107–114
42. Cirino PC, Arnold FH (2003) A self-sufficient peroxide-driven hydroxylation biocatalyst. Angew Chem Int Ed 42:3299–3301
43. Otey CR, Bandara G, Lalonde J, Takahashi K, Arnold FH (2006) Preparation of human metabolites of propranolol using laboratory-evolved bacterial cytochromes P450. Biotechnol Bioeng 93:494–499
44. Li QS, Ogawa J, Shimizu S (2001) Critical role of the residue size at position 87 in H_2O_2-dependent substrate hydroxylation activity and H_2O_2 inactivation of cytochrome P450BM-3. Biochem Biophys Res Commun 280:1258–1261
45. Khan KK, He YA, He YQ, Halpert JR (2002) Site-directed mutagenesis of cytochrome P450eryF: implications for substrate oxidation, cooperativity, and topology of the active site. Chem Res Toxicol 15:843–853
46. Niraula NP, Kanth BK, Sohng JK, Oh TJ (2011) Hydrogen peroxide-mediated dealkylation of 7-ethoxycoumarin by cytochrome P450 (CYP107AJ1) from *Streptomyces peucetius* ATCC27952. Enzyme Microb Technol 48:181–186
47. Matsunaga I, Ueda A, Sumimoto T, Ichihara K, Ayata M, Ogura H (2001) Site-directed mutagenesis of the putative distal helix of peroxygenase cytochrome P450. Arch Biochem Biophys 394:45–53
48. Shoji O, Fujishiro T, Nakajima H, Kim M, Nagano S, Shiro Y, Watanabe Y (2007) Hydrogen peroxide dependent monooxygenations by tricking the substrate recognition of cytochrome $P450_{BS\beta}$. Angew Chem Int Ed 46:3656–3659
49. Girhard M, Kunigk E, Tihovsky S, Shumyantseva V, Urlacher V (2013) Light-driven biocatalysis with cytochrome P450 peroxygenases. Biotechnol Appl Biochem 60:111–118

50. Girhard M, Schuster S, Dietrich M, Dürre P, Urlacher VB (2007) Cytochrome P450 monooxygenase from *Clostridium acetobutylicum*: a new α-fatty acid hydroxylase. Biochem Biophys Res Commun 362:114–119
51. Matsunaga I, Yamada M, Kusunose E, Miki T, Ichihara K (1998) Further characterization of hydrogen peroxide-dependent fatty acid α-hydroxylase from *Sphingomonas paucimobilis*. J Biochem 124:105–110
52. Shoji O, Watanabe Y (2014) Peroxygenase reactions catalyzed by cytochromes P450. J Biol Inorg Chem 19:529–539
53. Fujishiro T, Shoji O, Kawakami N, Watanabe T, Sugimoto H, Shiro Y, Watanabe Y (2012) Chiral-substrate-assisted stereoselective epoxidation catalyzed by H_2O_2-dependent cytochrome $P450_{SP\alpha}$. Chem Asian J 7:2286–2293
54. Rude MA, Baron TS, Brubaker S, Alibhai M, Del Cardayra SB, Schirmer A (2011) Terminal olefin (1-alkane) biosynthesis by a novel P450 fatty acid decarboxylase from *Jeotgalicoccus* species. Appl Environ Microbiol 77:1718–1727
55. Spolitak T, Funhoff EG, Ballou DP (2010) Spectroscopic studies of the oxidation of ferric CYP153A6 by peracids: insights into P450 higher oxidation states. Arch Biochem Biophys 493:184–191
56. Ogura H, Nishida CR, Hoch UR, Perera R, Dawson JH, Ortiz de Montellano PR (2004) EpoK, a cytochrome P450 involved in biosynthesis of the anticancer agents epothilones A and B. Substrate-mediated rescue of a P450 enzyme. Biochemistry 43:14712–14721
57. Goyal S, Banerjee S, Mazumdar S (2012) Oxygenation of monoenoic fatty acids by CYP175A1, an orphan cytochrome P450 from *Thermus thermophilis* HB27. Biochemistry 51:7880–7890
58. Sheng X, Horner JH, Newcomb M (2008) Spectra and kinetic studies of the compound I derivative of cytochrome P450 119. J Am Chem Soc 130:13310–13320
59. Wang MY, Liehr JG (1994) Identification of fatty acid hydroperoxide cofactors in the cytochrome P450-mediated oxidation of estrogens to quinone metabolites. J Biol Chem 269:284–291
60. Hlavica P, Künzel-Mulas U (1993) Metabolic N-oxide formation by rabbit-liver microsomal cytochrome P-4502B4: involvement of superoxide in the NADPH-dependent N-oxygenation of N,N-dimethylaniline. Biochim Biophys Acta 1158:83–90
61. Rahimtula AD, O'Brien PJ (1977) The role of cytochrome P-450 in the hydroperoxide-catalyzed oxidation of alcohols by rat-liver microsomes. Eur J Biochem 77:201–208
62. Kadlubar FF, Morton KC, Ziegler DM (1973) Microsomal-catalyzed hydroperoxide-dependent C-oxidation of amines. Biochem Biophys Res Commun 54:1255–1261
63. Nordblom GD, White RE, Coon MJ (1976) Studies on hydroperoxide-dependent substrate hydroxylation by purified liver microsomal cytochrome P-450. Arch Biochem Biophys 175:524–533
64. Renneberg R, Scheller F, Ruckpaul K, Pirrwitz J, Mohr P (1978) NADPH and H_2O_2-dependent reactions of cytochrome $P\text{-}450_{LM}$ compared with peroxidase catalysis. FEBS Lett 96:349–353
65. Bui PH, Hankinson O (2009) Functional characterization of human cytochrome P450 2S1 using a synthetic gene-expressed protein in *Escherichia coli*. Mol Pharmacol 76:1031–1043
66. Renneberg R, Capdevila J, Chacos N, Estabrook RW, Prough RA (1981) Hydrogen peroxide-supported oxidation of benzo[*a*]pyrene by rat liver microsomal fractions. Biochem Pharmacol 30:843–848
67. Kumar S, Liu H, Halpert JR (2006) Engineering of cytochrome P450 3A4 for enhanced peroxide-mediated substrate oxidation using directed evolution and site-directed mutagenesis. Drug Metab Dispos 34:1958–1965
68. Guengerich FP, Vaz ADN, Raner GN, Pernecky SJ, Coon MJ (1997) Evidence for a role of a perferryl-oxygen complex, FeO^{3+}, in the *N*-oxidation of amines by cytochrome P450 enzymes. Mol Pharmacol 51:147–151
69. Kumar S, Chen CS, Waxman DJ, Halpert JR (2005) Directed evolution of mammalian cytochrome P450 2B1: mutations outside of the active site enhance the metabolism of several substrates, including the anticancer prodrugs cyclophosphamide and ifosfamide. J Biol Chem 280:19569–19575
70. Anari MR, Josephy PD, Henry T, O'Brien PJ (1997) Hydrogen peroxide supports human and rat cytochrome P450 1A2-catalyzed 2-amino-3-methylimidazo[4,5-*f*]quinoline bioactivation to mutagenic metabolites: significance of cytochrome P450 peroxygenase. Chem Res Toxicol 10:582–588
71. Romano MC, Straub KM, Yodis LA, Eckardt RD, Newton JF (1988) Determination of microsomal lauric acid hydroxylase activity by HPLC with flow-through radiochemical quantitation. Anal Biochem 170:83–93
72. Pratt JM, Ridd TI, King LJ (1995) Activation of H_2O_2 by P450: evidence that the hydroxylating intermediate is iron(III)-coordinated H_2O_2 and not the ferryl FeO^{3+} complex. J Chem Soc Chem Commun 22:2297–2298
73. Zhang Z, Li Y, Stearns RA, Ortiz de Montellano PR, Baillie TA, Tang W (2002) Cytochrome P450 3A4-mediated oxidative conversion of a cyano to an amide group in the metabolism of pinacidil. Biochemistry 41:2712–2718
74. Muindi JF, Young CW (1993) Lipid hydroperoxides greatly increase the rate of oxidative catabolism of all-*trans*-retinoic acid by human cell culture

microsomes genetically enriched in specified cytochrome P-450 isoforms. Cancer Res 53:1226–1229
75. Chefson A, Zhao J, Auclair K (2006) Replacement of natural cofactors by selected hydrogen peroxide donors or organic peroxides results in improved activity for CYP3A4 and CYP2D6. ChemBioChem 7:916–919
76. Bui PH, Hsu EL, Hankinson O (2009) Fatty acid hydroperoxides support cytochrome P450 2S1-mediated bioactivation of benzo[*a*]pyrene-7,8-dihydrodiol. Mol Pharmacol 76:1044–1052
77. Sheng X, Zhang H, Im SC, Horner JH, Waskell L, Hollenberg PF, Newcomb M (2009) Kinetics of oxidation of benzphetamine by compounds I of cytochrome P450 2B4 and its mutants. J Am Chem Soc 131:2971–2976
78. Ortiz de Montellano PR, Choe YS, DePillis G, Catalano CE (1987) Structure-mechanism relationships in hemoproteins: oxygenations catalyzed by chloroperoxidase and horseradish peroxidase. J Biol Chem 262:11641–11646
79. Kania-Korwel I, Hrycay EG, Bandiera SM, Lehmler HJ (2008) 2,2′,3,3′,6,6′-Hexachlorobiphenyl (PCB 136) atropisomers interact enantioselectively with hepatic microsomal cytochrome P450 enzymes. Chem Res Toxicol 21:1295–1303
80. Lewis DFV (2002) Oxidative stress: the role of cytochromes P450 in oxygen activation. J Chem Technol Biotechnol 77:1095–1100
81. Doolittle RF, Feng DF, Tsang S, Cho G, Little E (1996) Determining divergence times of the major kingdoms of living organisms with a protein clock. Science 271:470–477
82. Chance B, Sies H, Boveris A (1979) Hydroperoxide metabolism in mammalian organs. Physiol Rev 59:527–605
83. Payne AH, Hales DB (2004) Overview of steroidogenic enzymes in the pathway from cholesterol to active steroid hormones. Endocr Rev 25:947–970
84. Kelly SL, Kelly DE, Jackson CJ, Warrilow AGS, Lamb DC (2005) The diversity and importance of microbial cytochromes P450. In: Ortiz de Montellano PR (ed) Cytochrome P450: structure, mechanism, and biochemistry, 3rd edn. Kluwer Academic/Plenum Publishers, New York, pp 585–617
85. Lamb DC, Lei L, Warrilow AGS, Lepesheva GI, Mullins JGL, Waterman MR, Kelly SL (2009) The first virally encoded cytochrome P450. J Virol 83:8266–8269
86. Kumar S (2010) Engineering cytochrome P450 biocatalysts for biotechnology, medicine and bioremediation. Expert Opin Drug Metab Toxicol 6:115–131
87. Renaud HJ, Cui JY, Khan M, Klaassen CD (2011) Tissue distribution and gender-divergent expression of 78 cytochrome P450 mRNAs in mice. Toxicol Sci 124:261–277
88. Norlin M, Wikvall K (2007) Enzymes in the conversion of cholesterol into bile acids. Curr Mol Med 7:199–218
89. Hasemann CA, Kurumbail RG, Boddupalli SS, Peterson JA, Deisenhofer J (1995) Structure and function of cytochromes P450: a comparative analysis of three crystal structures. Structure 2:41–62
90. Gunsalus IC, Pederson TC, Sligar SG (1975) Oxygenase-catalyzed biological hydroxylations. Annu Rev Biochem 44:377–407
91. Poulos TL, Finzel BC, Howard AJ (1987) High-resolution crystal structure of cytochrome P450cam. J Mol Biol 195:687–700
92. Poulos TL, Raag R (1992) Cytochrome P450cam: crystallography, oxygen activation, and electron transfer. FASEB J 6:674–679
93. Sono M, Roach MP, Coulter ED, Dawson JH (1996) Heme-containing oxygenases. Chem Rev 96:2841–2887
94. Peterson JA, Graham SE (1998) A close family resemblance: the importance of structure in understanding cytochromes P450. Structure 6:1079–1085
95. Larsen AT, May EM, Auclair K (2011) Predictable stereoselective and chemoselective hydroxylations and epoxidations with P450 3A4. J Am Chem Soc 133:7853–7858
96. Scott EE, He YA, Wester MR, White MA, Chin CC, Halpert JR, Johnson EF, Stout CD (2003) An open conformation of mammalian cytochrome P450 2B4 at 1.6-Å resolution. Proc Natl Acad Sci U S A 100:13196–13201
97. Poulos TL, Johnson EF (2005) Structures of cytochrome P450 enzymes. In: Ortiz de Montellano PR (ed) Cytochrome P450: structure, mechanism, and biochemistry, 3rd edn. Kluwer Academic/Plenum Publishers, New York, pp 87–114
98. Makris TM, Denisov I, Schlichting I, Sligar SG (2005) Activation of molecular oxygen by cytochrome P450. In: Ortiz de Montellano PR (ed) Cytochrome P450: structure, mechanism, and biochemistry, 3rd edn. Kluwer Academic/Plenum Publishers, New York, pp 149–182
99. Denisov IG, Makris TM, Sligar SG, Schlichting I (2005) Structure and chemistry of cytochrome P450. Chem Rev 105:2253–2277
100. Vatsis KP, Peng HM, Coon MJ (2005) Abolition of oxygenase function, retention of NADPH oxidase activity, and emergence of peroxidase activity upon replacement of the axial cysteine-436 ligand by histidine in cytochrome P450 2B4. Arch Biochem Biophys 434:128–138
101. Hernandez CE, Kumar S, Liu H, Halpert JR (2006) Investigation of the role of cytochrome P450 2B4 active site residues in substrate metabolism based on crystal structures of the ligand-bound enzyme. Arch Biochem Biophys 455:61–67
102. Baj-Rossi C, De Micheli G, Carrara S (2011) P450-based nano-bio-sensors for personalized medicine. In: Serra PA (ed) Biosensors–emerging materials and applications, InTech, Rijeka, Croatia, pp 447–482, ISBN: 978-953-307-328-6
103. Omura T, Sato R (1964) The carbon monoxide-binding pigment of liver microsomes: I. Evidence

for its hemoprotein nature. J Biol Chem 239:2370–2378
104. Estabrook RW (2003) A passion for P450s (remembrances of the early history of research on cytochrome P450). Drug Metab Dispos 31:1461–1473
105. Mason HS, North JC, Vanneste M (1965) Microsomal mixed-function oxidations: the metabolism of xenobiotics. Fed Proc 24:1172–1180
106. Dawson JH, Holm RH, Trudell JR, Barth G, Linder RE, Bunnenberg E, Djerassi C, Tang SC (1976) Oxidized cytochrome P-450: magnetic circular dichroism evidence for thiolate ligation in the substrate-bound form. Implications for the catalytic mechanism. J Am Chem Soc 98:3707–3709
107. Dawson JH, Sono M (1987) Cytochrome P-450 and chloroperoxidase: thiolate-ligated heme enzymes. Spectroscopic determination of their active site structures and mechanistic implications of thiolate ligation. Chem Rev 87:1255–1276
108. Auclair K, Moënne-Loccoz P, Ortiz de Montellano PR (2001) Roles of the proximal heme thiolate ligand in cytochrome P450cam. J Am Chem Soc 123:4877–4885
109. Nelson DR (2009) The cytochrome P450 homepage. Hum Genomics 4:59–65
110. Mueller EJ, Loida PJ, Sligar SG (1995) Twenty-five years of P450cam research: mechanistic insights into oxygenase catalysis. In: Ortiz de Montellano PR (ed) Cytochrome P450: structure, mechanism, and biochemistry, 2nd edn. Plenum Press, New York, pp 83–124
111. Testa B (1995) Reactions catalyzed by peroxidases. In: Testa B (ed) The metabolism of drugs and other xenobiotics: biochemistry of redox reactions. Academic Press, London, pp 346–388
112. Porro CS (2011) Quantum mechanical/molecular mechanics studies of cytochrome P450BM3, PhD thesis, The University of Manchester, Manchester
113. Gotoh O (1992) Substrate recognition sites in cytochrome P450 family 2 (CYP2) proteins inferred from comparative analysis of amino acid and coding nucleotide sequences. J Biol Chem 267:83–90
114. Poulos TL, Finzel BC, Howard AJ (1986) Crystal structure of substrate-free *Pseudomonas putida* cytochrome P450. Biochemistry 25:5314–5322
115. Ortiz de Montellano PR (1986) Oxygen activation and transfer. In: Ortiz de Montellano PR (ed) Cytochrome P-450: structure, mechanism, and biochemistry, 1st edn. Plenum Press, New York, pp 217–271
116. Ortiz de Montellano PR (1995) Oxygen activation and reactivity. In: Ortiz de Montellano PR (ed) Cytochrome P450: structure, mechanism, and biochemistry, 2nd edn. Plenum Press, New York, pp 245–304
117. Meunier B, Bernadou J (2000) Active iron-oxo and iron-peroxo species in cytochromes P450 and peroxidases: oxo-hydroxo tautomerism with water-soluble metalloporphyrins. Struct Bond 97:1–35
118. Isaac IS, Dawson JH (1999) Haem iron-containing peroxidases. Essays Biochem 34:51–69
119. Jin S, Makris TM, Bryson TA, Sligar SG, Dawson JH (2003) Epoxidation of olefins by hydroperoxo–ferric cytochrome P450. J Am Chem Soc 125:3406–3407
120. Groves JT (2005) Models and mechanisms of cytochrome P450 action. In: Ortiz de Montellano PR (ed) Cytochrome P450: structure, mechanism, and biochemistry, 3rd edn. Kluwer Academic/Plenum Publishers, New York, pp 1–43
121. Katagari M, Ganguli BN, Gunsalus IC (1968) A soluble cytochrome P-450 functional in methylene hydroxylation. J Biol Chem 243:3543–3546
122. Estabrook RW, Martinez-Zedillo G, Young S, Peterson JA, McCarthy J (1975) The interaction of steroids with liver microsomal cytochrome P-450– a general hypothesis. J Steroid Biochem 6:419–425
123. Guengerich FP, Johnson WW (1997) Kinetics of ferric cytochrome P450 reduction by NADPH–cytochrome P450 reductase: rapid reduction in the absence of substrate and variations among cytochrome P450 systems. Biochemistry 36:14741–14750
124. Shaik S, Cohen S, Wang Y, Chen H, Kumar D, Thiel W (2010) P450 enzymes: their structure, reactivity, and selectivity–modeled by QM/MM calculations. Chem Rev 110:949–1017
125. Jung C (2011) The mystery of cytochrome P450 compound I: a mini-review dedicated to Klaus Ruckpaul. Biochim Biophys Acta 1814:46–57
126. Groves JT, McClusky GA (1976) Aliphatic hydroxylation via oxygen rebound. Oxygen transfer catalyzed by iron. J Am Chem Soc 98:859–861
127. Groves JT, McClusky GA, White RE, Coon MJ (1978) Aliphatic hydroxylation by highly purified liver microsomal cytochrome P-450. Evidence for a carbon radical intermediate. Biochem Biophys Res Commun 81:154–160
128. Jung C, de Vries S, Schünemann V (2011) Spectroscopic characterization of cytochrome P450 compound I. Arch Biochem Biophys 507:44–55
129. Hamilton GA (1969) Mechanisms of two- and four-electron oxidations catalyzed by some metalloenzymes. Adv Enzymol Relat Areas Mol Biol 32:55–96
130. O'Brien PJ (1978) Hydroperoxides and superoxides in microsomal oxidations. Pharmacol Ther 2:517–536
131. Higuchi T, Urano Y, Hirobe M, Nagano T (1998) Pronounced effects of axial thiolate ligation on oxygen activation by iron porphyrin. In: Ishimura Y, Shimada H, Suematsu M (eds) Oxygen homeostasis and its dynamics: Keio University symposia for life science and medicine, vol 1. Springer, Tokyo, pp 181–188
132. Champion PM (1989) Elementary electronic excitations and the mechanism of cytochrome P450. J Am Chem Soc 111:3434–3436

133. Urano Y, Higuchi T, Hirobe M, Nagano T (1997) Pronounced axial thiolate ligand effect on the reactivity of high-valent oxo–iron porphyrin intermediate. J Am Chem Soc 119:12008–12009
134. Ohno T, Suzuki N, Dokoh T, Urano Y, Kikuchi K, Hirobe M, Higuchi T, Nagano T (2000) Remarkable axial thiolate ligand effect on the oxidation of hydrocarbons by active intermediate of iron porphyrin and cytochrome P450. J Inorg Biochem 82:123–125
135. Harischandra DN, Zhang R, Newcomb M (2005) Photochemical generation of a highly reactive iron–oxo intermediate: a true iron(V)–oxo species? J Am Chem Soc 127:13776–13777
136. Raner GM, Thompson JI, Haddy A, Tangham V, Bynum N, Reddy GR, Ballou DP, Dawson JH (2006) Spectroscopic investigations of intermediates in the reaction of cytochrome $P450_{BM3}$–F87G with surrogate oxygen atom donors. J Inorg Biochem 100:2045–2053
137. Loida PJ, Sligar SG (1993) Molecular recognition in cytochrome P-450: mechanism for the control of uncoupling reactions. Biochemistry 32:11530–11538
138. McCarthy MB, White RE (1983) Functional differences between peroxidase compound I and the cytochrome P-450 reactive oxygen intermediate. J Biol Chem 258:9153–9158
139. Egawa T, Shimada H, Ishimura Y (1994) Evidence for compound I formation in the reaction of cytochrome P450cam with *m*-chloroperbenzoic acid. Biochem Biophys Res Commun 201:1464–1469
140. Auclair K, Hu Z, Little DM, Ortiz de Montellano PR, Groves JT (2002) Revisiting the mechanism of P450 enzymes with the radical clocks norcarane and spiro [2,5]octane. J Am Chem Soc 124:6020–6027
141. Ortiz de Montellano PR, De Voss JJ (2002) Oxidizing species in the mechanism of cytochrome P450. Nat Prod Rep 19:477–493
142. Ortiz de Montellano PR (2010) Hydrocarbon hydroxylation by cytochrome P450 enzymes. Chem Rev 110:932–948
143. Hlavica P (2004) Models and mechanisms of O-O bond activation by cytochrome P450: a critical assessment of the potential role of multiple active intermediates in oxidative catalysis. Eur J Biochem 271:4335–4360
144. Schuler MA, Sligar SG (2007) Diversities and similarities in P450 systems: an introduction. Met Ions Life Sci 3:1–26
145. Lewis DFV, Pratt JM (1998) The P450 catalytic cycle and oxygenation mechanism. Drug Metab Rev 30:739–786
146. Rendic S (2002) Summary of information on human CYP enzymes: human P450 metabolism data. Drug Metab Rev 34:83–448
147. Hrycay E, Forrest D, Liu L, Wang R, Tai J, Deo A, Ling V, Bandiera S (2014) Hepatic bile acid metabolism and expression of cytochrome P450 and related enzymes are altered in *Bsep*$^{-/-}$ mice. Mol Cell Biochem 389:119–132
148. Hanukoglu I (1992) Steroidogenic enzymes: structure, function, and role in regulation of steroid hormone biosynthesis. J Steroid Biochem Mol Biol 43:779–804
149. Groves JT, Han YZ (1995) Models and mechanisms of cytochrome P450 action. In: Ortiz de Montellano PR (ed) Cytochrome P450: structure, mechanism, and biochemistry, 2nd edn. Plenum Press, New York, pp 3–48
150. Guengerich FP, Munro AW (2013) Unusual cytochrome P450 enzymes and reactions. J Biol Chem 288:17065–17073
151. Erratico CA, Szeitz A, Bandiera SM (2012) Oxidative metabolism of BDE-99 by human liver microsomes: predominant role of CYP2B6. Toxicol Sci 129:280–292
152. Erratico C, Szeitz A, Bandiera SM (2013) Biotransformation of 2,2′,4,4′-tetrabromodiphenyl ether (BDE-47) by human liver microsomes: identification of cytochrome P450 2B6 as the major enzyme involved. Chem Res Toxicol 26:721–731
153. Schnellmann RG, Putman CW, Sipes IG (1983) Metabolism of 2,2′,3,3′,6,6′-hexachlorobiphenyl and 2,2′,4,4′,5,5′-hexachlorobiphenyl by human hepatic microsomes. Biochem Pharmacol 32:3233–3239
154. Bandiera SM (2001) Cytochrome P450 enzymes as biomarkers of PCB exposure and modulators of toxicity. In: Robertson LW, Hansen LG (eds) PCBs: recent advances in environmental toxicology and health effects. University Press of Kentucky, Lexington, pp 185–192
155. Wu X, Pramanik A, Duffel MW, Hrycay EG, Bandiera SM, Lehmler HJ, Kania-Korwel I (2011) 2,2′,3,3′,6,6′–hexachlorobiphenyl (PCB 136) is enantioselectively oxidized to hydroxylated metabolites by rat liver microsomes. Chem Res Toxicol 24:2249–2257
156. Lehmler HJ, Harrad SJ, Hühnerfuss H, Kania-Korwel I, Lee CM, Lu Z, Wong CS (2010) Chiral polychlorinated biphenyl transport, metabolism, and distribution: a review. Environ Sci Technol 44:2757–2766
157. Ishida C, Koga N, Hanioka N, Saeki HK, Yoshimura H (1991) Metabolism in vitro of 3,4,3′,4′- and 2,5,2′,5′-tetrachlorobiphenyl by rat liver microsomes and highly purified cytochrome P-450. J Pharmacobio Dyn 14:276–284
158. Waller SC, He YA, Harlow GR, He YQ, Mash EA, Halpert JR (1999) 2,2′,3,3′,6,6′-Hexachlorobiphenyl hydroxylation by active site mutants of cytochrome P450 2B1 and 2B11. Chem Res Toxicol 12:690–699
159. Deo AK, Bandiera SM (2008) Identification of human hepatic cytochrome P450 enzymes involved in the biotransformation of cholic and chenodeoxycholic acid. Drug Metab Dispos 36:1983–1991
160. Deo AK, Bandiera SM (2009) 3-Ketocholanoic acid is the major in vitro human hepatic microsomal metabolite of lithocholic acid. Drug Metab Dispos 37:1938–1947

161. Capdevila JH, Holla VR, Falck JR (2005) Cytochrome P450 and the metabolism and bioactivation of arachidonic acid and eicosanoids. In: Ortiz de Montellano PR (ed) Cytochrome P450: structure, mechanism, and biochemistry, 3rd edn. Kluwer Academic/Plenum Publishers, New York, pp 531–551
162. Konkel A, Schunck WH (2011) Role of cytochrome P450 enzymes in the bioactivation of polyunsaturated fatty acids. Biochim Biophys Acta 1814:210–222
163. Sharma PK, de Visser SP, Shaik S (2003) Can a single oxidant with two spin states masquerade as two different oxidants? A study of the sulfoxidation mechanism by cytochrome P450. J Am Chem Soc 125:8698–8699
164. Hecker M, Ullrich V (1989) On the mechanism of prostacyclin and thromboxane A_2 biosynthesis. J Biol Chem 264:141–150
165. Gillam EMJ, Notley LM, Cai H, De Voss JJ, Guengerich FP (2000) Oxidation of indole by cytochrome P450 enzymes. Biochemistry 39:13817–13824
166. Ma X, Idle JR, Krausz KW, Gonzalez FJ (2005) Metabolism of melatonin by human cytochromes P450. Drug Metab Dispos 33:489–494
167. Bui P, Imaizumi S, Beedanagari SR, Reddy ST, Hankinson O (2011) Human CYP2S1 metabolizes cyclooxygenase- and lipoxygenase-derived eicosanoids. Drug Metab Dispos 39:180–190
168. George P (1953) The chemical nature of the second hydrogen peroxide compound formed by cytochrome *c* peroxidase and horseradish peroxidase. 1. Titration with reducing agents. Biochem J 54:267–276
169. Hollenberg PF (1992) Mechanisms of cytochrome P450 and peroxidase-catalyzed xenobiotic metabolism. FASEB J 6:686–694
170. Ator MA, Ortiz de Montellano PR (1987) Protein control of prosthetic heme reactivity: reaction of substrates with the heme edge of horseradish peroxidase. J Biol Chem 262:1542–1551
171. Newmyer SL, Sun J, Loehr TM, Ortiz de Montellano PR (1996) Rescue of the horseradish peroxidase His-170→Ala mutant activity by imidazole: importance of proximal ligand tethering. Biochemistry 35:12788–12795
172. Rahimtula AD, O'Brien PJ, Hrycay EG, Peterson JA, Estabrook RW (1974) Possible higher valence states of cytochrome P-450 during oxidative reactions. Biochem Biophys Res Commun 60:695–702
173. Hollenberg PF, Rand-Meir T, Hager LP (1974) The reaction of chlorite with horseradish peroxidase and chloroperoxidase: enzymatic chlorination and spectral intermediates. J Biol Chem 249:5816–5825
174. Kedderis GL, Koop DR, Hollenberg PF (1980) *N*-demethylation reactions catalyzed by chloroperoxidase. J Biol Chem 255:10174–10182
175. Palcic MM, Rutter R, Araiso T, Hager LP, Dunford HB (1980) Spectrum of chloroperoxidase compound I. Biochem Biophys Res Commun 94:1123–1127
176. Wagner GC, Palcic MM, Dunford HB (1983) Absorption spectra of cytochrome P450cam in the reaction with peroxy acids. FEBS Lett 156:244–248
177. Rutter R, Valentine M, Hendrich MP, Hager LP, Debrunner PG (1983) Chemical nature of the porphyrin π cation radical in horseradish peroxidase. Biochemistry 22:4769–4774
178. Sivaraja M, Goodin DB, Smith M, Hoffman BM (1989) Identification by ENDOR of Trp191 as the free-radical site in cytochrome c peroxidase compound ES. Science 245:738–740
179. Schünemann V, Jung C, Terner J, Trautwein AX, Weiss R (2002) Spectroscopic studies of peroxyacetic acid reaction intermediates of cytochrome P450cam and chloroperoxidase. J Inorg Biochem 91:586–596
180. Spolitak T, Dawson JH, Ballou DP (2005) Reaction of ferric cytochrome P450cam with peracids: kinetic characterization of intermediates on the reaction pathway. J Biol Chem 280:20300–20309
181. Jung C, Schünemann V, Lendzian F, Trautwein AX, Contzen J, Galander M, Böttger LH, Richter M, Barra AL (2005) Spectroscopic characterization of the iron-oxo intermediate in cytochrome P450. Biol Chem 386:1043–1053
182. Jung C, Schünemann V, Lendzian F (2005) Freeze-quenched iron-oxo intermediates in cytochromes P450. Biochem Biophys Res Commun 338:355–364
183. Stone KL, Behan RK, Green MT (2005) X-ray absorption spectroscopy of chloroperoxidase compound I: insight into the reactive intermediate of P450 chemistry. Proc Natl Acad Sci U S A 102:16563–16565
184. Ullrich R, Hofrichter M (2005) The haloperoxidase of the agaric fungus *Agrocybe aegerita* hydroxylates toluene and naphthalene. FEBS Lett 579:6247–6250
185. Hofrichter M, Ullrich R (2006) Heme-thiolate haloperoxidases: versatile biocatalysts with biotechnological and environmental significance. Appl Microbiol Biotechnol 71:276–288
186. Ullrich R, Hofrichter M (2007) Enzymatic hydroxylation of aromatic compounds. Cell Mol Life Sci 64:271–293
187. Kinne M, Poraj-Kobielska M, Aranda E, Ullrich R, Hammel KE, Scheibner K, Hofrichter M (2009) Regioselective preparation of 5-hydroxypropranolol and 4′-hydroxydiclofenac with a fungal peroxygenase. Bioorg Med Chem Lett 19:3085–3087
188. Poraj-Kobielska M, Kinne M, Ullrich R, Scheibner K, Kayser G, Hammel KE, Hofrichter M (2011) Preparation of human drug metabolites using fungal peroxygenases. Biochem Pharmacol 82:789–796

189. Gutirrez A, Babot ED, Ullrich R, Hofrichter M, Martínez AT, del Río JC (2011) Regioselective oxygenation of fatty acids, fatty alcohols and other aliphatic compounds by a basidiomycete heme-thiolate peroxidase. Arch Biochem Biophys 514:33–43
190. Kluge M, Ullrich R, Scheibner K, Hofrichter M (2012) Stereoselective benzylic hydroxylation of alkylbenzenes and epoxidation of styrene derivatives catalyzed by the peroxygenase of *Agrocybe aegerita*. Green Chem 14:440–446
191. Denisov IG, Dawson JH, Hager LP, Sligar SG (2007) The ferric-hydroperoxo complex of chloroperoxidase. Biochem Biophys Res Commun 363:954–958
192. Meharenna YT, Doukov T, Li H, Soltis SM, Poulos TL (2010) Crystallographic and single-crystal spectral analysis of the peroxidase ferryl intermediate. Biochemistry 49:2984–2986
193. Rittle J, Younker JM, Green MT (2010) Cytochrome P450: the active oxidant and its spectrum. Inorg Chem 49:3610–3617
194. Dolphin D, Forman A, Borg DC, Fajer J, Felton RH (1971) Compounds I of catalase and horseradish peroxidase: π-cation radicals. Proc Natl Acad Sci U S A 68:614–618
195. Yamada H, Yamazaki I (1974) Proton balance in conversions between five oxidation–reduction states of horseradish peroxidase. Arch Biochem Biophys 165:728–738
196. Loew GH, Harris DL (2000) Role of the heme active site and protein environment in structure, spectra, and function of the cytochrome P450s. Chem Rev 100:407–419
197. Morris DR, Hager LP (1966) Chloroperoxidase. I. Isolation and properties of the crystalline glycoprotein. J Biol Chem 241:1763–1768
198. Rai GP, Sakai S, Florez AM, Mogollon L, Hager LP (2001) Directed evolution of chloroperoxidase for improved epoxidation and chlorination catalysis. Adv Synth Catal 343:638–645
199. Blake RC II, Coon MJ (1989) On the mechanism of action of cytochrome P-450: spectral intermediates in the reaction with iodosobenzene and its derivatives. J Biol Chem 264:3694–3701
200. Wang X, Peter S, Kinne M, Hofrichter M, Groves JT (2012) Detection and kinetic characterization of a highly reactive heme–thiolate peroxygenase compound I. J Am Chem Soc 134:12897–12900
201. Wannstedt C, Rotella D, Siuda JF (1990) Chloroperoxidase mediated halogenation of phenols. Bull Environ Contam Toxicol 44:282–287
202. Geigert J, Neidleman SJ, Dalietos DJ (1983) Novel haloperoxidase substrates. Alkanes and cyclopropanes. J Biol Chem 258:2273–2277
203. Niu J, Yu G (2004) Molecular structural characteristics governing biocatalytic chlorination of PAHs by chloroperoxidase from *Caldariomyces fumago*. Environ Res 15:159–167
204. Doerge DR, Corbett MD (1991) Peroxygenation mechanism for chloroperoxidase-catalyzed *N*-oxidation of arylamines. Chem Res Toxicol 4:556–560
205. Colonna S, Gaggero N, Manfredi A, Casella L, Gullotti M, Carrea G, Pasta P (1990) Enantioselective oxidations of sulfides catalyzed by chloroperoxidases. Biochemistry 29:10465–10468
206. Green MT, Dawson JH, Gray HB (2004) Oxoiron (IV) in chloroperoxidase compound I is basic: implications for P450 chemistry. Science 304:1653–1656
207. Bell SR, Groves JT (2009) A highly reactive P450 model compound I. J Am Chem Soc 131:9640–9641
208. Reed CJ, De Matteis F (1989) Cumene hydroperoxide-dependent oxidation of *N N N′ N′*-tetramethyl-*p*-phenylenediamine and 7-ethoxycoumarin by cytochrome P-450. Biochem J 261:793–800
209. O'Brien PJ, Rahimtula A (1975) Involvement of cytochrome P-450 in the intracellular formation of lipid peroxides. J Agr Food Chem 23:154–158
210. Lindstrom TD, Aust SD (1984) Studies on cytochrome P-450-dependent lipid hydroperoxide reduction. Arch Biochem Biophys 233:80–87
211. Weiss RH, Arnold JL, Estabrook RW (1987) Transformation of an arachidonic acid hydroperoxide into epoxyhydroxy and trihydroxy fatty acids by liver microsomal cytochrome P-450. Arch Biochem Biophys 252:334–338
212. Vaz ADN, Coon MJ (1987) Hydrocarbon formation in the reductive cleavage of hydroperoxides by cytochrome P-450. Proc Natl Acad Sci U S A 84:1172–1176
213. Vaz ADN, Roberts ES, Coon MJ (1990) Reductive β-scission of the hydroperoxides of fatty acids and xenobiotics: role of alcohol-inducible cytochrome P-450. Proc Natl Acad Sci U S A 87:5499–5503
214. Tan L, Hrycay EG, Matsumoto K (1983) Synthesis and properties of the epimeric 6-hydroperoxyandrostenediones, new substrates/inhibitors of human placental aromatase. J Steroid Biochem 19:1329–1338
215. Hecker M, Baader WJ, Weber P, Ullrich V (1987) Thromboxane synthase catalyses hydroxylations of prostaglandin H_2 analogs in the presence of iodosylbenzene. Eur J Biochem 169:563–569
216. Plastaras JP, Guengerich FP, Nebert DW, Marnett LJ (2000) Xenobiotic-metabolizing cytochromes P450 convert prostaglandin endoperoxide to hydroxyheptadecatrienoic acid and the mutagen, malondialdehyde. J Biol Chem 275:11784–11790
217. Chang MS, Boeglin WE, Guengerich FP, Brash AR (1996) Cytochrome P450-dependent transformations of 15*R*- and 15*S*-hydroperoxyeicosatetraenoic acids: stereoselective formation of epoxy alcohol products. Biochemistry 35:464–471

218. Rahimtula AD, O'Brien PJ, Seifried HE, Jerina DM (1978) The mechanism of action of cytochrome *P*-450. Occurrence of the 'NIH shift' during hydroperoxide-dependent aromatic hydroxylation. Eur J Biochem 89:133–141
219. Cederbaum AI (1983) Organic hydroperoxide-dependent oxidation of ethanol by microsomes: lack of a role for free hydroxyl radicals. Arch Biochem Biophys 227:329–338
220. Hutzler JM, Powers FJ, Wynalda MA, Wienkers LC (2003) Effect of carbonate anion on cytochrome P450 2D6–mediated metabolism in vitro: the potential role of multiple oxygenating species. Arch Biochem Biophys 417:165–175
221. Rahimtula AD, O'Brien PJ (1974) Hydroperoxide catalyzed liver microsomal aromatic hydroxylation reactions involving cytochrome P-450. Biochem Biophys Res Commun 60:440–447
222. Capdevila J, Estabrook RW, Prough RA (1980) Differences in the mechanism of NADPH- and cumene hydroperoxide-supported reactions of cytochrome P-450. Arch Biochem Biophys 200:186–195
223. Cavalieri EL, Rogan EG, Cremonesi P, Devanesan PD (1988) Radical cations as precursors in the metabolic formation of quinones from benzo[*a*]pyrene and 6-fluorobenzo[*a*]pyrene: fluoro substitution as a probe for one-electron oxidation in aromatic substrates. Biochem Pharmacol 37:2173–2182
224. Hlavica P, Golly I, Mietaschk J (1983) Comparative studies on the cumene hydroperoxide- and NADPH-supported *N*-oxidation of 4-chloroaniline by cytochrome P-450. Biochem J 212:539–547
225. Miwa GT, Levin W, Thomas PE, Lu AYH (1978) The direct oxidation of ethanol by a catalase- and alcohol dehydrogenase-free reconstituted system containing cytochrome P-450. Arch Biochem Biophys 187:464–475
226. Rahimtula AD, O'Brien PJ (1975) Hydroperoxide dependent O-dealkylation reactions catalyzed by liver microsomal cytochrome P450. Biochem Biophys Res Commun 62:268–275
227. Ellin Å, Orrenius S (1975) Hydroperoxide-supported cytochrome P-450-linked fatty acid hydroxylation in liver microsomes. FEBS Lett 50:378–381
228. Danielsson H, Wikvall K (1976) On the ability of cumene hydroperoxide and $NaIO_4$ to support microsomal hydroxylations in biosynthesis and metabolism of bile acids. FEBS Lett 66:299–302
229. Blake RC II, Coon MJ (1981) On the mechanism of action of cytochrome P-450: evaluation of homolytic and heterolytic mechanisms of oxygen-oxygen bond cleavage during substrate hydroxylation by peroxides. J Biol Chem 256:12127–12133
230. Yun CH, Miller GP, Guengerich FP (2000) Rate-determining steps in phenacetin oxidations by human cytochrome P450 1A2 and selected mutants. Biochemistry 39:11319–11329
231. Otton SV, Gillam EM, Lennard MS, Tucker GT, Woods HF (1990) Propranolol oxidation by human liver microsomes–the use of cumene hydroperoxide to probe isoenzyme specificity and regio- and stereoselectivity. Br J Clin Pharmacol 30:751–760
232. Bui Q, Weisz J (1988) Identification of microsomal, organic hydroperoxide-dependent catechol estrogen formation: comparison with NADPH-dependent mechanism. Pharmacology 36:356–364
233. Spink DC, Spink BC, Zhuo X, Hussain MM, Gierthy JF, Ding X (2000) NADPH- and hydroperoxide-supported 17β-estradiol hydroxylation catalyzed by a variant form (432L, 453S) of human cytochrome P450 1B1. J Steroid Biochem Mol Biol 74:11–18
234. Edwards PR, Hrycay EG, Bandiera SM (2007) Differential inhibition of hepatic microsomal alkoxyresorufin *O*-dealkylation activities by tetrachlorobiphenyls. Chem Biol Interact 169:42–52
235. Nagata C, Tagashira Y, Kodama M (1974) Metabolic activation of BaP: significance of the free radical. In: Ts'o POP, DiPaolo JA (eds) Chemical carcinogenesis, part A. Marcel Dekker, New York, pp 87–111
236. Lesko S, Caspary W, Lorentzen R, Ts'o POP (1975) Enzymic formation of 6-oxobenzo[*a*]pyrene radical in rat liver homogenates from carcinogenic benzo[*a*]pyrene. Biochemistry 14:3978–3984
237. McCarthy MB, White RE (1983) Competing modes of peroxyacid flux through cytochrome P-450. J Biol Chem 258:11610–11616
238. Lee WA, Bruice TC (1985) Homolytic and heterolytic oxygen-oxygen bond scissions accompanying oxygen transfer to iron (III) porphyrins by percarboxylic acids and hydroperoxides: a mechanistic criterion for peroxidase and cytochrome P-450. J Am Chem Soc 107:513–514
239. Green MT (2009) C–H bond activation in heme proteins: the role of thiolate ligation in cytochrome P450. Curr Opin Chem Biol 13:84–88
240. Krest CM, Onderko EL, Yosca TH, Calixto JC, Karp RF, Livada J, Rittle J, Green MT (2013) Reactive intermediates in cytochrome P450 catalysis. J Biol Chem 288:17074–17081
241. Davydov R, Macdonald IDG, Makris TM, Sligar SG, Hoffman BM (1999) EPR and ENDOR of catalytic intermediates in cryoreduced native and mutant oxy-cytochromes P450cam: mutation-induced changes in the proton delivery system. J Am Chem Soc 121:10654–10655
242. Heimbrook DC, Murray RI, Egeberg KD, Sligar SG, Nee MW, Bruice TC (1984) Demethylation of N, N-dimethylaniline and p-cyano-N, N-dimethylaniline and their N-oxides by cytochromes P450LM2 and P450CAM. J Am Chem Soc 106:1514–1515
243. Shaffer CL, Harriman S, Koen YM, Hanzlik RP (2002) Formation of cyclopropanone during cytochrome P450-catalyzed N-dealkylation of a cyclopropylamine. J Am Chem Soc 124:8268–8274
244. Dowers TS, Rock DA, Rock DA, Jones JP (2004) Kinetic isotope effects implicate the iron–oxene as

the sole oxidant in P450-catalyzed N-dealkylation. J Am Chem Soc 126:8868–8869
245. Roberts KM, Jones JP (2010) Anilinic *N*-oxides support cytochrome P450-mediated *N*-dealkylation through hydrogen-atom transfer. Chem Eur J 16:8096–8107
246. Golly I, Hlavica P, Wolf J (1984) The role of lipid peroxidation in the *N*-oxidation of 4-chloroaniline. Biochem J 224:415–421
247. Spolitak T, Dawson JH, Ballou DP (2008) Replacement of tyrosine residues by phenylalanine in cytochrome P450cam alters the formation of Cpd II-like species in reactions with artificial oxidants. J Biol Inorg Chem 13:599–611
248. Green MT (1999) Evidence for sulfur-based radicals in thiolate compound I intermediates. J Am Chem Soc 121:7939–7940
249. Loew GH, Collins J, Luke B, Waleh A, Padzianowski A (1986) Theoretical studies of cytochrome P-450: characterization of stable and transient active states, reaction mechanisms and substrate-enzyme interactions. Enzyme 36:54–78
250. Shaik S, Filatov M, Schröder D, Schwarz H (1998) Electronic structure makes a difference: cytochrome P-450 mediated hydroxylations of hydrocarbons as a two-state reactivity paradigm. Chem Eur J 4:193–199
251. Urano Y, Higuchi T, Hirobe M (1996) Substrate-dependent changes of the oxidative *O*-dealkylation mechanism of several chemical and biological oxidizing systems. J Chem Soc Perkin Trans 2:1169–1173
252. Lindsay-Smith JR, Sleath PR (1983) Model systems for cytochrome P450 dependent mono-oxygenases. Part 2. Kinetic isotope effects for the oxidative demethylation of anisole and [Me-H_3]anisole by cytochrome P450 dependent mono-oxygenases and model systems. J Chem Soc Perkin Trans 2:621–628
253. Pan Z, Wang RQ, Sheng X, Horner JH, Newcomb M (2009) Highly reactive porphyrin-iron-oxo derivatives produced by photolysis of metastable porphyrin-iron(IV) diperchlorates. J Am Chem Soc 131:2621–2628
254. Isobe H, Yamaguchi K, Okumura M, Shimada J (2012) Role of perferryl-oxo oxidant in alkane hydroxylation catalyzed by cytochrome P450: a hybrid density functional study. J Phys Chem B 116:4713–4730
255. Altun A, Shaik S, Thiel W (2007) What is the active species of cytochrome P450 during camphor hydroxylation? QM/MM studies of different electronic states of compound I and of reduced and oxidized iron-oxo intermediates. J Am Chem Soc 129:8978–8987
256. Nam W, Park SE, Lim IK, Lim MH, Hong J, Kim J (2003) First direct evidence for stereospecific olefin epoxidation and alkane hydroxylation by an oxoiron (IV) porphyrin complex. J Am Chem Soc 125:14674–14675
257. Shaik S, Kumar D, de Visser SP, Altun A, Thiel W (2005) Theoretical perspective on the structure and mechanism of cytochrome P450 enzymes. Chem Rev 105:2279–2328
258. Ogliaro F, de Visser SP, Cohen S, Sharma PK, Shaik S (2002) Searching for the second oxidant in the catalytic cycle of cytochrome P450: a theoretical investigation of the iron(III)-hydroperoxo species and its epoxidation pathways. J Am Chem Soc 124:2806–2817
259. Yosca TH, Rittle J, Krest CM, Onderko EL, Silakov A, Calixto JC, Behan RK, Green MT (2013) Iron(IV)hydroxide pK_a and the role of thiolate ligation in C–H bond activation by cytochrome P450. Science 342:825–829
260. Schünemann V, Lendzian F, Jung C, Contzen J, Barra AL, Sligar SG, Trautwein AX (2004) Tyrosine radical formation in the reaction of wild type and mutant cytochrome P450cam with peroxy acids: a multifrequency EPR study of intermediates on the millisecond time scale. J Biol Chem 279:10919–10930
261. Spolitak T, Dawson JH, Ballou DP (2006) Rapid kinetics investigations of peracid oxidation of ferric cytochrome P450cam: nature and possible function of compound ES. J Inorg Biochem 100:2034–2044
262. Sivaramakrishnan S, Ouellet H, Matsumura H, Guan S, Moënne-Loccoz P, Burlingame AL, Ortiz de Montellano PR (2012) Proximal ligand electron donation and reactivity of the cytochrome P450 ferric-peroxo anion. J Am Chem Soc 134:6673–6684
263. Davydov R, Makris TM, Kofman V, Werst DE, Sligar SC, Hoffman BM (2001) Hydroxylation of camphor by reduced oxy-cytochrome P450cam: mechanistic implications of EPR and ENDOR studies of catalytic intermediates in native and mutant enzymes. J Am Chem Soc 123:1403–1415
264. Davydov R, Hoffman BM (2011) Active intermediates in heme monooxygenase reactions as revealed by cryoreduction/annealing, EPR/ENDOR studies. Arch Biochem Biophys 507:36–43
265. Luthra A, Denisov IG, Sligar SG (2011) Spectroscopic features of cytochrome P450 reaction intermediates. Arch Biochem Biophys 507:26–35
266. Davydov R, Dawson JH, Perera R, Hoffman BM (2013) The use of deuterated camphor as a substrate in ^{1}H ENDOR studies of hydroxylation by cryoreduced oxy P450cam provides new evidence of the involvement of compound I. Biochemistry 52:667–671
267. Yoshioka S, Takahashi S, Ishimori K, Morishima I (2000) Roles of the axial push effect in cytochrome P450cam studied with the site-directed mutagenesis at the heme proximal site. J Inorg Biochem 81:141–151
268. Rutter R, Hager LP, Dhonau H, Hendrich MP, Hager LP, Debrunner PG (1984) Chloroperoxidase compound I: electron paramagnetic resonance and Mössbauer studies. Biochemistry 22:6808–6816

269. Meunier B, de Visser SP, Shaik S (2004) Mechanism of oxidation reactions catalyzed by cytochrome P450 enzymes. Chem Rev 104:3947–3980
270. Schlichting I, Berendzen J, Chu K, Stock AM, Maves SA, Benson DE, Sweet RM, Ringe D, Petsko GA, Sligar SG (2000) The catalytic pathway of cytochrome P450cam at atomic resolution. Science 287:1615–1622
271. Imai M, Shimada H, Watanabe Y, Matsushima-Hibiya Y, Makino R, Koga H, Horiuchi T, Ishimura Y (1989) Uncoupling of the cytochrome P-450cam monooxygenase reaction by a single mutation, threonine-252 to alanine or valine: a possible role of the hydroxy amino acid in oxygen activation. Proc Natl Acad Sci U S A 86:7823–7827
272. Martinis SA, Atkins WM, Stayton PS, Sligar SG (1989) A conserved residue of cytochrome P-450 is involved in heme-oxygen stability and activation. J Am Chem Soc 111:9252–9253
273. Vidakovic M, Sligar SG, Li H, Poulos TL (1998) Understanding the role of the essential Asp251 in cytochrome P450cam using site-directed mutagenesis, crystallography, and kinetic solvent isotope effect. Biochemistry 37:9211–9219
274. Raag R, Martinis SA, Sligar SG, Poulos TL (1991) Crystal structure of the cytochrome P-450cam active site mutant Thr252Ala. Biochemistry 30:11420–11429
275. Gerber NC, Sligar SG (1992) Catalytic mechanism of cytochrome P-450: evidence for a distal charge relay. J Am Chem Soc 114:8742–8743
276. Gerber NC, Sligar SG (1994) A role for Asp-251 in cytochrome P-450cam oxygen activation. J Biol Chem 269:4260–4266
277. Chandrasena REP, Vatsis KP, Coon MJ, Hollenberg PF, Newcomb M (2004) Hydroxylation by the hydroperoxy-iron species in cytochrome P450 enzymes. J Am Chem Soc 126:115–126
278. Sevrioukova IF, Poulos TL (2011) Structural biology of redox partner interactions in P450cam monooxygenase: a fresh look at an old system. Arch Biochem Biophys 507:66–74
279. Poulos TL, Madrona Y (2013) Oxygen activation and redox partner binding in cytochromes P450. Biotechnol Appl Biochem 60:128–133
280. Yonetani T, Schleyer H (1967) Studies on cytochrome *c* peroxidase: IX. The reaction of ferrimyoglobin with hydroperoxides and a comparison of peroxide-induced compounds of ferrimyoglobin and cytochrome *c* peroxidase. J Biol Chem 242:1974–1979
281. Sligar SG, Kennedy KA, Pearson DC (1982) The chemical basis of mixed function oxidation. In: King TE, Mason HS, Morrison M (eds) Oxidases and related redox systems. Permagon Press, Oxford, pp 837–856
282. Gustafsson JÅ, Rondahl L, Bergman J (1979) Iodosylbenzene derivatives as oxygen donors in cytochrome P-450 catalyzed steroid hydroxylations. Biochemistry 18:865–870
283. Macdonald TL, Gutheim WG, Martin RB, Guengerich FP (1989) Oxidation of substituted N, N-dimethylanilines by cytochrome P-450: estimation of the effective oxidation-reduction potential of cytochrome P-450. Biochemistry 28:2071–2077
284. Lichtenberger F, Nastainczyk W, Ullrich V (1976) Cytochrome P450 as an oxene transferase. Biochem Biophys Res Commun 70:939–946
285. Gustafsson JÅ, Bergman J (1976) Iodine- and chlorine-containing oxidation agents as hydroxylating catalysts in cytochrome P-450-dependent fatty acid hydroxylation reactions in rat liver microsomes. FEBS Lett 70:276–280
286. Berg A, Ingelman-Sundberg M, Gustafsson JÅ (1979) Purification and characterization of cytochrome $P-450_{meg}$. J Biol Chem 254:5264–5271
287. Yamazaki H, Ueng YF, Shimada T, Guengerich FP (1995) Roles of divalent metal ions in oxidations catalyzed by recombinant cytochrome P450 3A4 and replacement of NADPH−cytochrome P450 reductase with other flavoproteins, ferredoxin, and oxygen surrogates. Biochemistry 34:8380–8389
288. Glieder A, Farinas ET, Arnold FH (2002) Laboratory evolution of a soluble, self-sufficient, highly active alkane hydroxylase. Nat Biotechnol 20:1135–1139
289. Shaik S, de Visser SP, Kumar D (2004) One oxidant, many pathways: a theoretical perspective of monooxygenation mechanisms by cytochrome P450 enzymes. J Biol Inorg Chem 9:661–668
290. Cho KB, Moreau Y, Kumar D, Rock DA, Jones JP, Shaik S (2007) Formation of the active species of cytochrome P450 by using iodosylbenzene: a case for spin-selective reactivity. Chem Eur J 13:4103–4115
291. Akhtar M, Corina DL, Pratt J, Smith T (1976) Studies on the removal of C-19 in oestrogen biosynthesis using $^{18}O_2$. J Chem Soc Chem Commun 21:854−856
292. Akhtar M, Calder MR, Corina DL, Wright JN (1982) Mechanistic studies on C-19 demethylation in oestrogen biosynthesis. Biochem J 201:569–580
293. Akhtar M, Njar VCO, Wright JN (1993) Mechanistic studies on aromatase and related C−C bond cleaving P-450 enzymes. J Steroid Biochem Mol Biol 44:375–387
294. Akhtar M, Wright JN, Lee-Robichaud P (2011) A review of mechanistic studies on aromatase (CYP19) and 17α-hydroxylase-17,20-lyase (CYP17). J Steroid Biochem Mol Biol 125:2–12
295. Gantt SI, Denisov IG, Grinkova YV, Sligar SG (2009) The critical iron−oxygen intermediate in human aromatase. Biochem Biophys Res Commun 387:169–173
296. Hackett JC, Brueggemeier RW, Hadad CM (2005) The final catalytic step of cytochrome P450 aromatase: a density functional theory study. J Am Chem Soc 127:5224–5237
297. Vaz ADN, Roberts ES, Coon MJ (1991) Olefin formation in the oxidative deformylation of aldehydes by cytochrome P-450: mechanistic implications for catalysis by oxygen-derived peroxide. J Am Chem Soc 113:5886–5887

298. Roberts ES, Vaz ADN, Coon MJ (1991) Catalysis by cytochrome P-450 of an oxidative reaction in xenobiotic aldehyde metabolism: deformylation with olefin formation. Proc Natl Acad Sci U S A 88:8963–8966
299. Vaz ADN, Pernecky SJ, Raner GM, Coon MJ (1996) Peroxo-iron and oxenoid-iron species as alternative oxygenating agents in cytochrome P450-catalyzed reactions: switching by threonine-302 to alanine mutagenesis of cytochrome P450 2B4. Proc Natl Acad Sci U S A 93:4644–4648
300. Vaz ADN, McGinnity DF, Coon MJ (1998) Epoxidation of olefins by cytochrome P450: evidence from site-specific mutagenesis for hydroperoxo-iron as an electrophilic oxidant. Proc Natl Acad Sci U S A 95:3555–3560
301. Newcomb M, Le Tadic MH, Putt DA, Hollenberg PF (1995) An incredibly fast apparent oxygen rebound rate constant for hydrocarbon hydroxylation by cytochrome P-450 enzymes. J Am Chem Soc 117:3312–3313
302. Newcomb M, Le Tadic-Biadatti MH, Chestney DL, Roberts ES, Hollenberg PF (1995) A nonsynchronous concerted mechanism for cytochrome P-450 catalyzed hydroxylation. J Am Chem Soc 117:12085–12091
303. Toy PH, Newcomb M, Coon MJ, Vaz ADN (1998) Two distinct electrophilic oxidants effect hydroxylation in cytochrome P-450-catalyzed reactions. J Am Chem Soc 120:9718–9719
304. Newcomb M, Toy PH (2000) Hypersensitive radical probes and the mechanisms of cytochrome P450-catalyzed hydroxylation reactions. Acc Chem Res 33:449–455
305. Newcomb M, Shen R, Choi SY, Toy PH, Hollenberg PF, Vaz ADN, Coon MJ (2000) Cytochrome P450-catalyzed hydroxylation of mechanistic probes that distinguish between radicals and cations: evidence for cationic but not for radical intermediates. J Am Chem Soc 122:2677–2686
306. Newcomb M, Aebisher D, Shen R, Chandrasena REP, Hollenberg PF, Coon MJ (2003) Kinetic isotope effects implicate two electrophilic oxidants in cytochrome P450-catalyzed hydroxylations. J Am Chem Soc 125:6064–6065
307. Volz TJ, Rock DA, Jones JP (2002) Evidence for two different active oxygen species in cytochrome P450 BM3 mediated sulfoxidation and N-dealkylation reactions. J Am Chem Soc 124:9724–9725
308. Watanabe Y (2001) Alternatives to the oxoferryl porphyrin cation radical as the proposed reactive intermediate of cytochrome P450: two-electron oxidized Fe(III) porphyrin derivatives. J Biol Inorg Chem 6:846–856
309. Shaik S, Cohen S, de Visser SP, Sharma PK, Kumar D, Kozuch S, Ogliaro F, Danovich D (2004) The "rebound controversy": an overview and theoretical modeling of the rebound step in C–H hydroxylation by cytochrome P450. Eur J Inorg Chem 2004:207–226
310. de Visser SP, Ogliaro F, Sharma PK, Shaik S (2002) What factors affect the regioselectivity of oxidation by cytochrome P450? A DFT study of allylic hydroxylation and double bond epoxidation in a model reaction. J Am Chem Soc 124:11809–11826
311. Goto Y, Matsui T, Ozaki S, Watanabe Y, Fukuzumi S (1999) Mechanisms of sulfoxidation catalyzed by high-valent intermediates of heme enzymes: electron-transfer vs oxygen-transfer mechanism. J Am Chem Soc 121:9497–9502
312. Krauser JA, Guengerich FP (2005) Cytochrome P450 3A4-catalyzed testosterone 6β-hydroxylation stereochemistry, kinetic deuterium isotope effects, and rate-limiting steps. J Biol Chem 280:19496–19506
313. Tanaka K, Kurihara N, Nakajima M (1979) Oxidative metabolism of tetrachlorocyclohexenes, pentachlorocyclohexenes, and hexachlorocyclohexenes with microsomes from rat liver and house fly abdomen. Pestic Biochem Physiol 10:79–95
314. Groves JT, Subramanian DV (1984) Hydroxylation by cytochrome P-450 and metalloporphyrin models: evidence for allylic rearrangement. J Am Chem Soc 106:2177–2181
315. Oliw EH, Brodowsky ID, Hörnsten L, Hamberg M (1993) Bis-Allylic hydroxylation of polyunsaturated fatty acids by hepatic monooxygenases and its relation to the enzymatic and nonenzymatic formation of conjugated hydroxy fatty acids. Arch Biochem Biophys 300:434–439
316. Ortiz de Montellano PR, Stearns RA (1987) Timing of the radical recombination step in cytochrome P-450 catalysis with ring-strained probes. J Am Chem Soc 109:3415–3420
317. Bowry VW, Ingold KU (1991) A radical clock investigation of microsomal cytochrome P-450 hydroxylation of hydrocarbons: rate of oxygen rebound. J Am Chem Soc 113:5699–5707
318. Atkinson JK, Ingold KU (1993) Cytochrome P450 hydroxylation of hydrocarbons: variation in the rate of oxygen rebound using cyclopropyl radical clocks including two new ultrafast probes. Biochemistry 32:9209–9214
319. Kumar D, de Visser SP, Shaik S (2003) How does product isotope effect prove the operation of a two-state "rebound" mechanism in C–H hydroxylation by cytochrome P450? J Am Chem Soc 125:13024–13025
320. de Visser SP, Porro CS, Quesne MG, Sainna MA, Munro AW (2014) Overview on theoretical studies discriminating the two-oxidant versus two-state-reactivity models for substrate monooxygenation by cytochrome P450 enzymes. Curr Topics Med Chem 13:2218–2232
321. Wang B, Li C, Cho K, Nam W, Shaik S (2013) The $Fe^{III}(H_2O_2)$ complex as a highly efficient

oxidant in sulfoxidation reactions: revival of an underrated oxidant in cytochrome P450. J Chem Theory Comput 9:2519–2525
322. Gillam EMJ, Guengerich FP (2001) Exploiting the versatility of human cytochrome P450 enzymes: the promise of blue roses from biotechnology. IUBMB Life 52:271–277
323. Guengerich FP (2002) Cytochrome P450 enzymes in the generation of commercial products. Nat Rev Drug Discov 1:359–366
324. Kumar S, Halpert JR (2005) Use of directed evolution of mammalian cytochromes P450 for investigating the molecular basis of enzyme function and generating novel biocatalysts. Biochem Biophys Res Commun 338:456–464
325. Roy P, Waxman DJ (2006) Activation of oxazaphosphorines by cytochrome P450: application to gene-directed enzyme prodrug therapy for cancer. Toxicol In Vitro 20:176–186
326. Purnapatre K, Khattar SK, Saini KS (2008) Cytochrome P450s in the development of target-based anticancer drugs. Cancer Lett 259:1–15
327. Hlavica P (2009) Assembly of non-natural electron transfer conduits in the cytochrome P450 system: a critical assessment and update of artificial redox constructs amenable to exploitation in biotechnological areas. Biotechnol Adv 27:103–121
328. Di Nardo G, Fantuzzi A, Sideri A, Panicco P, Sassone C, Giunta C, Gilardi G (2007) Wild-type CYP102A1 as a biocatalyst: turnover of drugs usually metabolised by human liver enzymes. J Biol Inorg Chem 12:313–323
329. Fasan R (2012) Tuning P450 enzymes as oxidation catalysts. ACS Catal 2:647–666
330. Meinhold P, Peters MW, Chen MMY, Takahashi K, Arnold FH (2005) Direct conversion of ethane to ethanol by engineered cytochrome P450 BM3. ChemBioChem 6:1765–1768
331. Fasan R, Chen MM, Crook NC, Arnold FH (2007) Engineered alkane-hydroxylating cytochrome $P450_{BM3}$ exhibiting native-like catalytic properties. Angew Chem Int Ed 46:8414–8418
332. Fasan R, Meharenna YT, Snow CD, Poulos TL, Arnold FH (2008) Evolutionary history of a specialized P450 propane monooxygenase. J Mol Biol 383:1069–1080
333. Sawayama AM, Chen MMY, Kulanthaivel P, Kuo MS, Hemmerle H, Arnold FH (2009) A panel of cytochrome P450 BM3 variants to produce drug metabolites and diversify lead compounds. Chem Eur J 15:11723–11729
334. Jung ST, Lauchli R, Arnold FH (2011) Cytochrome P450: taming a wild type enzyme. Curr Opin Biotechnol 22:1–9
335. Harford-Cross CF, Carmichael AB, Allan FK, England PA, Rouch DA, Wong LL (2000) Protein engineering of cytochrome $P450_{CAM}$ (CYP101) for the oxidation of polycyclic aromatic hydrocarbons. Protein Eng 13:121–128
336. Otey CR, Landwehr M, Endelman JB, Hiraga K, Bloom JD, Arnold FH (2006) Structure-guided recombination creates an artificial family of cytochromes P450. PLoS Biol 4:789–798
337. Li Y, Drummond DA, Sawayama AM, Snow CD, Bloom JD, Arnold FH (2007) A diverse family of thermostable cytochrome P450s created by recombination of stabilizing fragments. Nat Biotechnol 25:1051–1056
338. Huang W, Johnston WA, Hayes MA, De Voss JJ, Gillam EMJ (2007) A shuffled CYP2C library with a high degree of structural integrity and functional diversity. Arch Biochem Biophys 467:193–205
339. Wu ZL, Podust LM, Guengerich FP (2005) Expansion of substrate specificity of cytochrome P450 2A6 by random and site-directed mutagenesis. J Biol Chem 280:41090–41100
340. England PA, Harford-Cross CF, Stevenson JA, Rouch DA, Wong LL (1998) The oxidation of naphthalene and pyrene by cytochrome P450cam. FEBS Lett 424:271–274
341. Nickerson DP, Harford-Cross CF, Fulcher SR, Wong LL (1997) The catalytic activity of cytochrome P450cam towards styrene oxidation is increased by site-specific mutagenesis. FEBS Lett 405:153–156
342. Bell SG, Stevenson JA, Boyd HD, Campbell S, Riddle AD, Orton EL, Wong LL (2002) Butane and propane oxidation by engineered cytochrome P450cam. Chem Commun 5:490–491
343. Xu F, Bell SG, Lednik J, Insley A, Rao Z, Wong LL (2005) The heme monooxygenase cytochrome $P450_{cam}$ can be engineered to oxidize ethane to ethanol. Angew Chem Int Ed 44:4029–4032
344. Sulistyaningdyah WT, Ogawa J, Li QS, Maeda C, Yano Y, Schmid RD, Shimizu S (2005) Hydroxylation activity of P450 BM-3 mutant F87V towards aromatic compounds and its application to the synthesis of hydroquinone derivatives from phenolic compounds. Appl Microbiol Biotechnol 67:556–562
345. Carmichael AB, Wong LL (2001) Protein engineering of *Bacillus megaterium* CYP102. The oxidation of polycyclic aromatic hydrocarbons. Eur J Biochem 268:3117–3125
346. Li QS, Ogawa J, Schmid RD, Shimizu S (2001) Engineering cytochrome P450 BM-3 for oxidation of polycyclic aromatic hydrocarbons. Appl Environ Microbiol 67:5735–5739
347. Li HM, Mei LE, Urlacher VB, Schmid RD (2008) Cytochrome P450 BM-3 evolved by random and saturation mutagenesis as an effective indole-hydroxylating catalyst. Appl Biochem Biotechnol 144:27–36
348. Koo LS, Immoos CE, Cohen MS, Farmer PJ, Ortiz de Montellano PR (2000) Enhanced electron transfer and lauric acid hydroxylation by site-directed mutagenesis of CYP119. J Am Chem Soc 124:5684–5691
349. Kim D, Guengerich FP (2004) Enhancement of 7-methoxyresorufin O-demethylation activity of

human cytochrome P450 1A2 by molecular breeding. Arch Biochem Biophys 432:102–108
350. Nakamura K, Martin MV, Guengerich FP (2001) Random mutagenesis of human cytochrome P450 2A6 and screening with indole oxidation products. Arch Biochem Biophys 395:25–31
351. Kubo T, Peters MW, Meinhold P, Arnold FH (2006) Enantioselective epoxidation of terminal alkenes to (*R*)- and (*S*)-epoxides by engineered cytochromes P450 BM-3. Chem Eur J 12:1216–1220
352. Rentmeister A, Arnold FH, Fasan R (2009) Chemo-enzymatic fluorination of unactivated organic compounds. Nat Chem Biol 5:26–28
353. Lewis JC, Mantovani SM, Fu Y, Snow CD, Komor RS, Wong CH, Arnold FH (2010) Combinatorial alanine substitution enables rapid optimization of cytochrome $P450_{BM3}$ for selective hydroxylation of large substrates. ChemBioChem 11:2502–2505
354. Dietrich JA, Yoshikuni Y, Fisher KJ, Woolard FX, Ockey D, McPhee DJ, Renninger NS, Chang MC, Baker D, Keasling JD (2009) A novel semi-biosynthetic route for artemisinin production using engineered substrate-promiscuous P450(BM3). ACS Chem Biol 4:261–267
355. Landwehr M, Hochrein L, Otey CR, Kasrayan A, Bäckvall JE, Arnold FH (2006) Enantioselective α-hydroxylation of 2-arylacetic acid derivatives and buspirone catalyzed by engineered cytochrome P450 BM-3. J Am Chem Soc 128:6058–6059
356. van Vugt-Lussenburg BMA, Stjernschantz E, Lastdrager J, Oostenbrink C, Vermeulen NPE, Commandeur JNM (2007) Identification of critical residues in novel drug metabolizing mutants of cytochrome P450 BM3 using random mutagenesis. J Med Chem 50:455–461
357. Damsten MC, van Vugt-Lussenburg BMA, Zeldenthuis T, de Vlieger JSB, Commandeur JNM, Vermeulen NPE (2008) Application of drug metabolising mutants of cytochrome P450 BM3 (CYP102A1) as biocatalysts for the generation of reactive metabolites. Chem Biol Interact 171:96–107
358. Kim DH, Kim KH, Kim D, Jung HC, Pan JG, Chi YT, Ahn T, Yun CH (2010) Oxidation of human cytochrome P450 1A2 substrates by *Bacillus megaterium* cytochrome P450 BM3. J Mol Catal B Enzym 63:179–187
359. Kim KH, Kang JY, Kim DH, Park SH, Park SH, Kim D, Park KD, Lee YJ, Jung HC, Pan JG, Ahn T, Yun CH (2011) Generation of human chiral metabolites of simvastatin and lovastatin by bacterial CYP102A1 mutants. Drug Metab Dispos 39:140–150
360. Bistolas N, Wollenberger U, Jung C, Scheller FW (2005) Cytochrome P450 biosensors – a review. Biosens Bioelectron 20:2408–2423
361. Gillam EMJ, Aguinaldo AMA, Notley LM, Kim D, Mundkowski RG, Volkov AA, Arnold FH, Souček P, De Voss JJ, Guengerich FP (1999) Formation of indigo by recombinant mammalian cytochrome P450. Biochem Biophys Res Commun 265:469–472
362. Kim D, Guengerich FP (2004) Selection of human cytochrome P450 1A2 mutants with enhanced catalytic activity for heterocyclic amine N-hydroxylation. Biochemistry 43:981–988
363. Adachi J, Mori Y, Matsui S, Takigami H, Fujino J, Kitagawa H, Miller CA III, Kato T, Saeki K, Matsuda T (2001) Indirubin and indigo are potent aryl hydrocarbon receptor ligands present in human urine. J Biol Chem 276:31475–31478
364. Kumar S, Scott EE, Liu H, Halpert JR (2003) A rational approach to re-engineer cytochrome P450 2B1 regioselectivity based on the crystal structure of cytochrome P450 2C5. J Biol Chem 278:17178–17184
365. Fujishiro T, Shoji O, Nagano S, Sugimoto H, Shiro Y, Watanabe Y (2011) Crystal structure of H_2O_2-dependent cytochrome $P450_{SP\alpha}$ with its bound fatty acid substrate: insight into the regioselective hydroxylation of fatty acids at the α position. J Biol Chem 286:29941–29950
366. Keizers PHJ, Schraven LHM, de Graaf C, Hidestrand M, Ingelman-Sundberg M, van Dijk BR, Vermeulen NPE, Commandeur JNM (2005) Role of the conserved threonine 309 in mechanism of oxidation by cytochrome P450 2D6. Biochem Biophys Res Commun 338:1065–1074
367. Yooseph S, Sutton G, Rusch DB, Halpern AL, Williamson SK, Remington K, Eisen JA, Heidelberg KB, Manning G, Li W, [---], Venter JC (2007) The *sorcerer II* global ocean sampling expedition: expanding the universe of protein families. PLoS Biol 5:432–466. e16. doi: 10.1371/journal.pbio.0050016
368. Rusch DB, Halpern AL, Sutton G, Heidelberg KB, Williamson S, Yooseph S, Wu D, Eisen JA, Hoffman JM, Remington K, [---], Venter JC (2007) The *sorcerer II* global ocean sampling expedition: Northwest Atlantic through Eastern Tropical Pacific. PLoS Biol 5:398–431. e77. doi: 10.1371/journal.pbio.0050077
369. Kannan N, Taylor SS, Zhai Y, Venter JC, Manning G (2007) Structural and functional diversity of the microbial kinome. PLoS Biol 5(467–478):e17. doi:10.1371/journal. pbio.0050017
370. Nelson DR (2007) The evolution and genomics of cytochrome P450, drnelson.uthsc.edu/talks/UTSAshorttalk.pdf
371. Müller WJ (2012) Cytochrome P450 monooxygenases from extremophiles, PhD thesis, University of the Free State, Bloemfontein, Republic of South Africa

Oxidizing Intermediates in P450 Catalysis: A Case for Multiple Oxidants

2

Anuja R. Modi and John H. Dawson

Abstract

Cytochrome P450 (P450 or CYP) catalysis involves the oxygenation of organic compounds via a series of catalytic intermediates, namely, the ferric-peroxo, ferric-hydroperoxo, Compound I (Cpd I) and $Fe^{III}-(H_2O_2)$ intermediates. Now that the structures of P450 enzymes have been well established, a major focus of current research in the P450 area has been unraveling the intimate details and activities of these reactive intermediates. The general consensus is that the Cpd I intermediate is the most reactive species in the reaction cycle, especially when the reaction involves hydrocarbon hydroxylation. Cpd I has recently been characterized experimentally. Other than Cpd I, there is a multitude of evidence, both experimental as well as theoretical, supporting the involvement of other intermediates in various types of oxidation reactions. The involvement of these multiple oxidants has been experimentally demonstrated using P450 active-site mutants in epoxidation, heteroatom oxidation and dealkylation reactions. In this chapter, we will review the P450 reaction cycle and each of the reactive intermediates to discuss their role in oxidation reactions.

Keywords

Cytochrome P450 • Reaction cycle • Compound I • Ferric-peroxo • Ferric-hydroperoxo • Reactive intermediates • Multiple oxidants

Abbreviations: *Cpd I or Cpd II* compound I or II of a heme enzyme, an $Fe^{IV}=O$ radical cation species, *CPO* chloroperoxidase, *CYP or P450* cytochrome P450, *CYP119A1* orphan P450 from *Sulfolobus acidocaldarius*, *CYP2B4* phenobarbital-inducible rabbit liver microsomal P450 enzyme (P450LM2), *CYP2E1* alcohol-inducible rabbit liver microsomal P450 enzyme (P450LM3), *CYP51A1* lanosterol 14α-demethylase, *ENDOR* electron-nuclear double resonance, *EPR* electron paramagnetic resonance, *ES* enzyme-substrate complex, *FAD* flavin adenine dinucleotide, *FMN* flavin mononucleotide, heme iron protoporphyrin IX (heme-b), *KIE* kinetic isotope effect, *NOS* nitric oxide synthase enzyme, *P450BM3* fatty acid hydroxylating P450 enzyme from *Bacillus megaterium*, *P450CAM* camphor-hydroxylating P450 enzyme from *Pseudomonas putida*, *RH* substrate, *ROH* oxidized substrate.

A.R. Modi • J.H. Dawson (✉)
Department of Chemistry and Biochemistry, University of South Carolina, Columbia, SC 29208, USA
e-mail: jdawson@mailbox.sc.edu

E.G. Hrycay and S.M. Bandiera (eds.), *Monooxygenase, Peroxidase and Peroxygenase Properties and Mechanisms of Cytochrome P450*, Advances in Experimental Medicine and Biology 851,
DOI 10.1007/978-3-319-16009-2_2,

2.1 Introduction

Cytochrome P450 (P450 or CYP) enzymes are heme-thiolate ligated monooxygenases that are ubiquitous in the biological kingdom and catalyze a variety of oxidation reactions covering a wide range of substrates [1, 2]. Hemeproteins are classified as P450s when their $Fe^{II}-CO$ complex has a maximum Soret absorbance at 450 nm [3]. P450s were discovered five decades ago because of their important role in xenobiotic clearance from the human body, but the interesting nature of their chemistry has attracted attention from chemists, biochemists, biophysicists, structural biologists and now even biotechnologists. Oxygen activation is central to life as spin forbiddance makes ground-state triplet molecular oxygen by itself inert toward organic molecules [4]. Living beings therefore use enzyme systems for oxygen activation to perform biologically important reduction-oxidation (redox) reactions. P450s are one of the metal-containing oxygenases that utilize molecular oxygen to stereo- and regioselectively oxygenate substrates under physiological conditions. While P450s are capable of diverse reactions, they are in fact mostly known for their ability to catalyze the oxidation of inert substrate C−H bonds under physiological conditions. To put into perspective, the bond strength of a typical secondary C−H bond is about 101 kcal mol^{-1} [5]. Present understanding of the P450 catalytic mechanism has been developed over the course of the last four decades by advances in genomics, molecular biology and spectroscopy [6]. Comparison with analogous heme oxygen activation systems has also greatly contributed to our current understanding of its mechanism. Knowledge of the P450 intermediates is now being used for development of efficient inorganic catalysts for laboratory and commercial use [7]. Alternatively, in biotechnological setups, P450s are being modified to catalyze stereoselective oxidation reactions [8]. Oxidation of substrates by P450 enzymes can be summarized by the following equation:

$$R-H+NAD(P)H+O_2+H^+ \\ \rightarrow R-OH+NAD(P)^+ +H_2O$$

The catalytic mechanism of P450s occurs in a cyclic fashion involving systematic generation of intermediates, some of which are transient [9]. Electrons for this oxidation reaction are provided by NAD(P)H and are shuttled to the P450 active site with the aid of reductase enzymes. Water molecules in the active site donate protons. Based on associated redox systems, P450s can be classified as type I or type II as shown in Fig. 2.1 [10]. Type I P450s are mainly the mammalian mitochondrial and bacterial P450s, which utilize a flavoprotein to transfer electrons to P450s via an intervening iron-sulfur cluster protein (Fe_2S_2). Type II P450s are the mammalian xenobiotic-metabolizing enzymes that receive electrons via FAD- and FMN-containing reductases. There are certain exceptions such as P450BM3 in which the heme and FMN/FAD-containing reductase domains are part of a single polypeptide that functions as a self-sufficient unit [11, 12].

2.2 The P450 Catalytic Cycle

P450CAM (CYP101A1) is a bacterial P450 enzyme from *Pseudomonas putida* that converts 1R-(+)-camphor to 5-*exo*-hydroxycamphor. This enzyme is soluble, thus easy to purify and was the first P450 whose crystal structure was solved [13]. Since then, its structure has been extensively studied and it has served as a prototype for structure-function studies of the entire P450

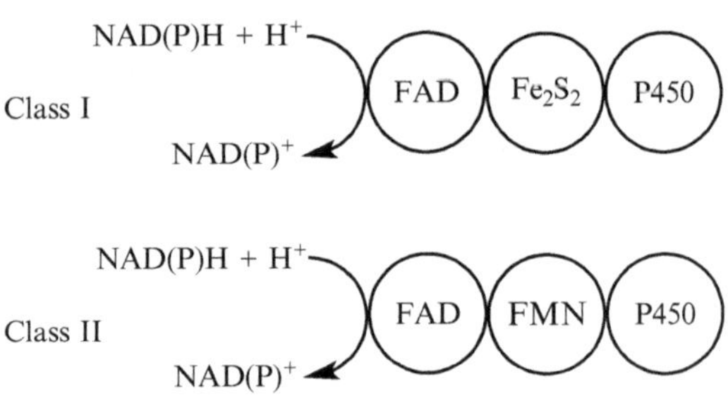

Fig. 2.1 Electron transfer chains in class I and class II of P450 enzymes (See Ref. [10])

family. The reaction cycle for P450CAM also holds true for the entire P450 family. The putative catalytic mechanism of P450s in which the substrate RH is oxidized to R-OH in a series of steps is shown in Fig. 2.2. The catalytic cycle begins with the reversible substrate binding to the water-coordinated low-spin ($S = 1/2$) resting state of the ferric enzyme (Fig. 2.2, **1**). Substrate binding causes displacement of water as the sixth ligand to the heme with formation of the high-spin ($S = 5/2$) pentacoordinate enzyme-substrate adduct (Fig. 2.2, **2**), resulting in the shift of the midpoint redox potential of the heme to a more positive value (from −330 to −173 mV) [14]. This sharp shift in the reduction potential enables electrons to flow from NAD(P)H to the P450 enzyme via an associated reductase. The first electron generates the reduced ferrous-substrate adduct (Fig. 2.2, **3**). Subsequent binding of dioxygen generates the oxyferrous complex or a resonance-stabilized ferric-superoxide complex $Fe^{+3}{-}OO\bullet^{-}$, a η^1 superoxide radical anion coordinated to the ferric heme center with an unpaired electron on the terminal oxygen atom (Fig. 2.2, **4**) [15]. The second electron from NAD(P)H then reduces the oxyferrous complex resulting in the $Fe^{+3}{-}OO^{-}$ (Fig. 2.2, **5**) ferric-peroxo intermediate. This is also the rate-limiting step. Protonation of this intermediate leads to the $Fe^{+3}{-}OOH$ (Fig. 2.2, **6**) ferric-hydroperoxo intermediate, also known as Compound 0 (Cpd 0). A second protonation of this intermediate leads to O−O bond heterolysis forming the transient and highly reactive porphyrin π radical cation ferryl complex (Fig. 2.2, **7**) known as Cpd I. Cpd I derives its name from the analogous high-valent Cpd I species of heme peroxidases [16, 17]. According to the now well-accepted mechanism for hydrocarbon hydroxylation, Cpd I abstracts a H atom from the substrate resulting in a ferryl hydroxyl intermediate (Fig. 2.2, **8**) known as protonated Compound II (Cpd II) and a substrate radical. In what is known as the oxygen rebound, the hydroxyl moiety on the iron combines with the substrate radical to give the hydroxylated product, while the enzyme returns to its resting ferric state.

In addition to the normal catalytic pathway, there are three uncoupling reactions within the cycle that lead back to the enzyme-substrate adduct without any product formation. The first is the auto-oxidation of the oxyferrous enzyme with simultaneous generation of a superoxide anion (Fig. 2.2, **I**). In the second shunt pathway, the hydroperoxo anion dissociates from the ferric-hydroperoxo intermediate (Fig. 2.2, **II**). Heterolytic cleavage of the O−O bond is critical for Cpd I formation. Incorrect protonation at the distal oxygen generates the $Fe^{+3}{-}(H_2O_2)$ intermediate (Fig. 2.2, **9**) followed by dissociation of hydrogen peroxide without substrate turnover. This pathway is often seen in the active-site alcohol-alanine mutant [18]. In the oxidase shunt (Fig. 2.2, **III**), the ferryl intermediate is reduced to water in lieu of substrate oxidation. In an alternative pathway to the normal reaction cycle, the enzyme can be turned over without the nucleotide-reducing equivalents via the peroxide shunt (Fig. 2.2, **IV**). Cpd I can be generated from this pathway using oxygen atom donors such as peracids, peroxides and iodosobenzene [19–21].

2.3 Nature of the P450 Active Site

Currently, there are over 20,000 known CYP genes, a summary of which can be found here, (http://drnelson.uthsc.edu/CytochromeP450.html). Notably, P450 enzymes share their catalytic capabilities with certain heme-containing enzymes such as catalases, peroxidases, oxygenases, etc. but all these enzymes have remarkably different structures. The architecture of the active site in P450s plays a crucial role in the sequential generation of intermediates in the P450 catalytic cycle. For a complete understanding of P450 monooxygenation chemistry, the majority of research has focused on the factors influencing electron delivery and dioxygen binding to the heme iron, proton addition to the bound dioxygen in the distal pocket of the heme and cleavage of the O−O bond. Residues most important for oxygen activation are the heme proximal cysteine ligand (Cys357 in P450CAM) and the acid-alcohol pair

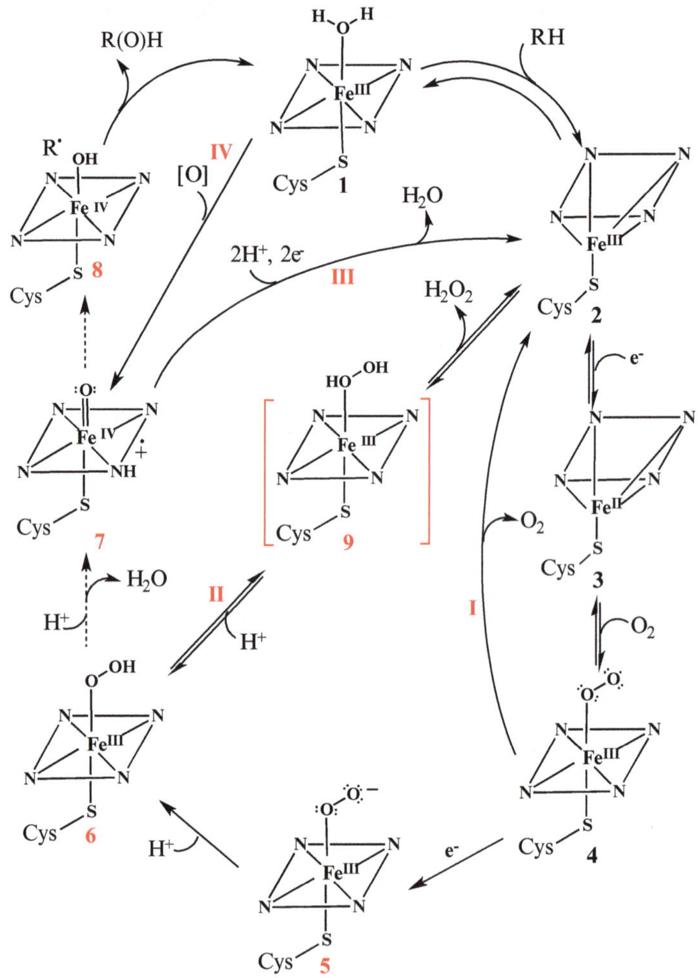

Fig. 2.2 Reactive intermediates and shunt pathways in the catalytic cycle of P450 enzymes

in the distal pocket of the heme (Asp251-Thr252 in P450CAM). These amino acids are conserved in the active site of almost all P450s. Since the active site is hydrophobic, a chain of water molecules provides access to bulk solvent outside and is held in place by H-bonding to each other and to the alcohol residue in the active site (Thr252 in P450CAM) [22]. The active site of camphor-bound oxy-P450CAM is shown in Fig. 2.3.

In this landmark work, Schlichting and coworkers structurally characterized the oxyferrous complex of P450CAM using cryocrystallography [9]. Important residues involved in oxygen activation were Cys357, Asp251 and Thr252 as well as the bound camphor substrate and are shown in Fig. 2.3. The hydroxyl group of Tyr96 H-bonds to the keto group of camphor to orient the site of oxygenation above the heme. The water molecules Wat901 and Wat902 seen in the active site are implicated as the source of protons required for formation of the active $Fe^{IV}{=}O$ species. Accordingly, the Thr252 H-bonds to Wat901, which serves as the H-bond donor to the distal oxygen atom of the heme-bound dioxygen in P450CAM.

2.3.1 Role of Cysteine as Proximal Heme Ligand

The heme in P450 enzymes is of the heme-b type, where the iron protoporphyrin-IX is covalently linked to the protein backbone via a Fe-S

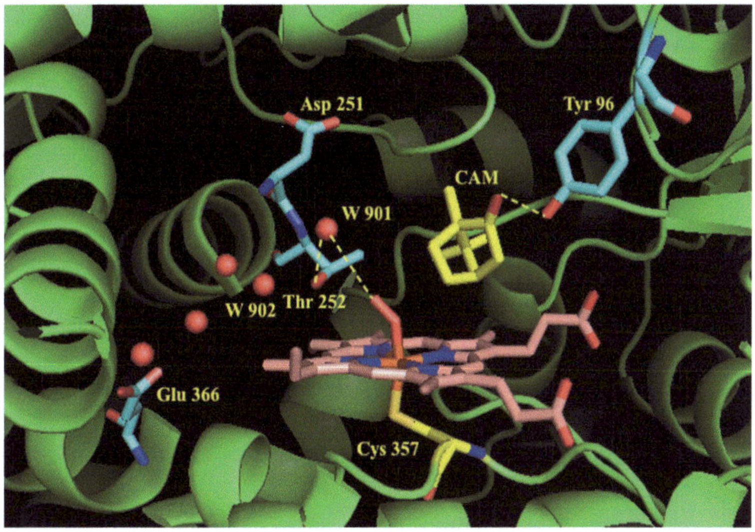

Fig. 2.3 Active site of camphor-bound oxy-P450CAM constructed using PDB file 1DZ8 [9]

bond to cysteine. In the case of P450CAM, Cys357 serves as the cysteine residue as seen in Fig. 2.3. The proximal cysteine thiolate ligand is indispensible for P450 catalytic activity and mutation of the cysteine residue leads to loss of activity [23]. In the P450 catalytic cycle, one electron reduction of the oxyferrous state followed by protonation of the distal oxygen leads to the ferric-hydroperoxo intermediate. A second protonation of the ferric-hydroperoxo intermediate followed by heterolytic cleavage of the O–O bond leads to formation of Cpd I, which is the primary oxidant in the cycle. Maintaining the cysteine as a thiolate anion on the proximal side of the heme at the same time as the iron in the ferrous state is crucial for Cpd I generation [24]. The thiolate anion is stabilized by H bonds from the protons of the adjacent residues, Leu358 (3.5 Å), Gly359 (3.3 Å) and Gln360 (3.3 Å). Mutation of these residues led to distortion in H-bonding and an increase in the uncoupling of the ferric-hydroperoxo intermediate [25, 26].

Dawson and coworkers suggested that the polarizable nature of the cysteine thiolate anion ligand provides a strong 'push' of electron density via the heme onto the O–O bond of the ferric-hydroperoxo intermediate, thus promoting heterolytic O–O bond cleavage [27, 28]. Furthermore, the electron-donating nature of the thiolate ligand also helps to stabilize the resulting Cpd I intermediate. This result is similar to the effect seen in cytochrome *c* peroxidase that contains a partially deprotonated proximal histidine ligand, wherein the imidazolate 'push' in concert with a 'pull' from the conserved distal His-Arg amino acids lead to heterolytic cleavage of the O–O bond to generate Cpd I [29].

2.3.2 Role of the Acid-Alcohol Pair in Oxygen Activation

An acid-alcohol pair that is highly conserved in almost all P450 enzymes aids oxygen activation in the distal heme pocket. The alcohol in most cases is threonine or serine and the acid can be aspartate or glutamate. In the case of P450CAM, these residues are Asp251-Thr252. Given their highly conserved nature and proximity to the heme-dioxygen binding site, the role of this acid-alcohol pair in catalysis has been examined in several mutagenesis studies. Specifically, the role of Thr was investigated by changing the residue to Ala. In P450CAM, the Thr252Ala mutant was almost completely uncoupled, leading to normal NADH and O_2 consumption but essentially no product formation [18, 30]. Based on this result in P450CAM and other P450s as well [31, 32], the alcohol residue is thought to stabilize water molecules in the active site by H-bonding during substrate oxygenation [18, 30, 33]. Ishimura and coworkers demonstrated that the uncoupling reaction is promoted when the Thr in P450CAM is mutated to a Ser or Asn, thereby ascertaining the role of Thr in stabilizing the H-bonding network in the distal pocket and controlling proton delivery to the distal oxygen of bound dioxygen [22]. The Thr252Ser and Thr252Asn mutant enzymes retained more than half of the hydroxylating capability of the enzyme. The Thr252 residue also participated in H-bonding with the distal oxygen of the oxyferrous-P450 complex in P450CAM [9]. The high resolution crystal structure of the P450CAM Thr252Ala mutant showed a clearly perturbed H-bonding network and excess water molecules in the active site [33]. It is thought that this perturbation leads to uncoupling due to incorrect delivery of the second proton to the proximal oxygen [34]. Just as in P450CAM, the Thr268 in P450BM3 has been shown to play an important role in sustaining the proton delivery pathway from the bulk solvent to the dioxy-bound heme. Mutation of Thr268 to Ala also leads to uncoupling followed by reduced substrate oxidation [31, 35].

Unlike the alcohol residue, the acid residue has an important role in electron transfer following oxyferrous intermediate formation. In P450CAM, the mutagenesis of Asp251 to Asn leads to decreased turnover in the mutant enzyme rather than uncoupling [36, 37]. The Asp251Asn mutant displays an increased kinetic solvent isotope effect compared to the wild-type enzyme and a directly linear correlation to NADH consumption on bulk proton concentration, indicating that the proton delivery pathway has been modified in the Asp251Asn mutant [36]. Structural analysis of the Asp251Asn mutant reveals significant changes in the active site. The Asn251 and Lys178 side chains rotate away from the active site and the Asn251 H-bonds to Asp182, causing open access to the heme [36]. The flexibility of the Asp251 side chain stabilized by electrostatic bonding plays an important role in dioxygen scission in P450CAM and suggests a similar role for the conserved acid functionality in other P450 enzymes.

2.4 Multiple Oxidants in P450 Catalysis

Although the P450s are a single family of enzymes, the wide variety of substrates oxidized by P450 is quite astounding. The catalytic cycle of P450s has been well established based on P450CAM as the prototype. The key catalytic intermediates have been detected and well characterized. The ferryl Cpd I intermediate, initially thought to be too short lived to detect, has now been well characterized [38]. Despite the fact that this intermediate has not been detected in the normal P450 catalytic cycle, it has been observed in the peroxide shunt pathway and there is little doubt about its involvement in substrate oxidation. While the ferryl Cpd I intermediate is thought to be the oxidant of choice in most oxidation reactions, the nature of certain catalytic intermediates and comparison with analogous reactions catalyzed in other enzyme systems make it difficult to deny the existence of multiple oxidizing species in the catalytic cycle. As seen in Fig. 2.4, several reaction intermediates other than Cpd I are thought to be capable of catalyzing some oxidation reactions depending on the type of substrate.

2.4.1 The P450 Dioxygen Complex

In P450 enzymes, binding of dioxygen to the ferrous heme traps O_2 for substrate oxidation. This generates the ferrous-dioxygen complex (Fe^{+2}–OO), which is in resonance with the ferric-superoxide complex (Fe^{+3}–OO•$^-$) (Fig. 2.2, **4**). The binding constant of dioxygen to P450CAM is $1.7 \times 10^6\ M^{-1}\ s^{-1}$ at 4 °C [39]. The oxyferrous complex of P450s is not as

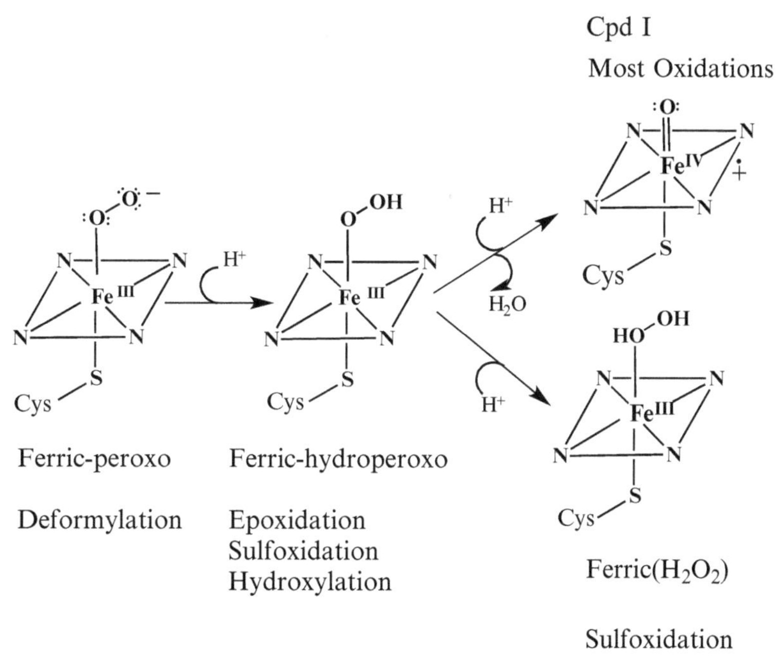

Fig. 2.4 Multiple P450 oxidants and types of reactions catalyzed

stable as that of oxygen carrier proteins. In P450CAM, it is moderately stable in the presence of camphor and auto-oxidizes back to the ferric state at the rate of 0.01 s^{-1} at room temperature [39, 40]. The oxyferrous-P450 complex is similar to that generated in many analogous hemeproteins such as myoglobin, hemoglobin, CPO, NOS, etc. [41–44]. The oxyferrous stretching band of oxyferrous P450CAM as determined by resonance Raman spectroscopy is 1,141 cm^{-1}, which is typical for superoxide complexes [45]. Using cryocrystallization, the oxyferrous complex of P450CAM was determined at atomic resolution. A representative figure is shown in Fig. 2.5 [9].

The oxygen is coordinated to the heme iron in a slightly bent fashion with the Fe-O−O angle being 142°. The oxyferrous complex is stabilized with the aid of a H bond between the distal oxygen and hydroxyl of the nearby Thr252 residue. While the oxyferrous complex by itself is not known to catalyze any oxidation reaction, its formation is necessary to generate the subsequent catalytic intermediates in the reaction cycle.

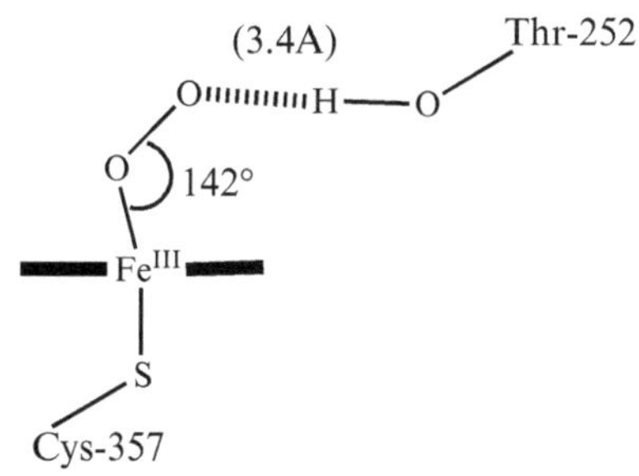

Fig. 2.5 Oxyferrous complex of P450-CAM determined using PDB file 1DZ8 [9]

2.4.2 Ferric-Peroxo Intermediate as a Nucleophilic Oxidant

Akhtar and coworkers first proposed a role for the ferric-peroxo intermediate in the final step of oxidative deformylation catalyzed by lanosterol 14α-demethylase [46, 47]. The enzyme catalyzes the oxidative deformylation of lanosterol, concomitantly forming olefin in three oxidative steps, and each step utilizing a single equivalent of NADPH and O_2 as seen in Fig. 2.6. The final

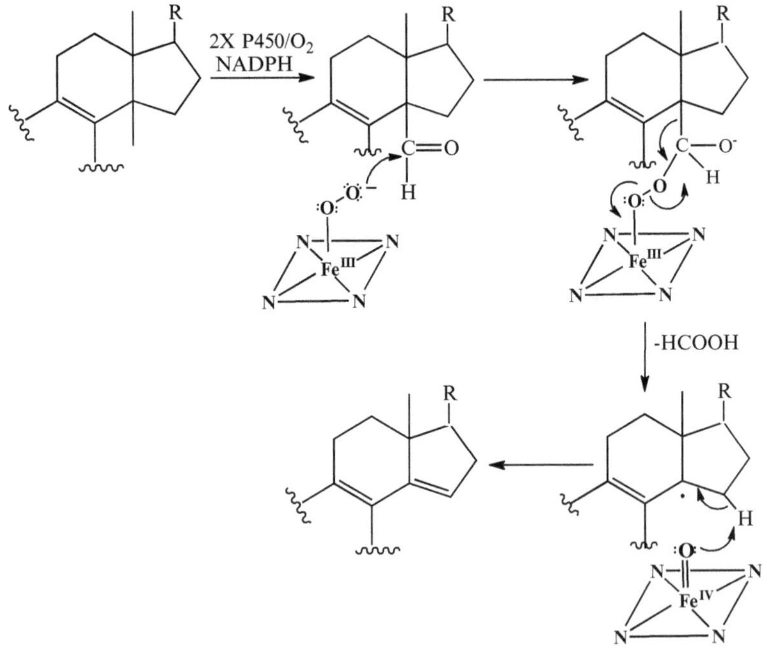

Fig. 2.6 Mechanism of oxidative deformylation catalyzed by lanosterol 14α-demethylase (See Ref. [47])

step results in cleavage of the C14–C32 bond with stereoselective removal of 15α-H, resulting in the formation of a 14,15 double bond and release of formic acid. The proposed mechanism involves homolytic cleavage of the O–O bond in a peroxy-aldehyde adduct to give an alkoxy free radical that decays to the olefin as a result of H abstraction by the simultaneously-created ferryl species.

Similar mechanisms have also been proposed for demethylation in estrogen formation by aromatase (CYP19A1) and in the CYP17A1-catalyzed C−C bond scission of 17-α-hydroxyprogesterone [47, 48]. The formic acid formed in these P450-catalyzed oxidative deformylations has been shown to retain the original carbonyl oxygen and hydrogen as well as an atom from molecular oxygen, clearly pointing to the involvement of the ferric-peroxo intermediate in the mechanism. In the CYP17A1 (17α-hydroxylase-17,20-lyase)−catalyzed reaction, oxygen labeling experiments also point to homolytic scission of the O−O bond in the peroxo-substrate adduct [49]. Vaz and coworkers analyzed the elimination reaction of the aliphatic aldehyde in the rabbit drug-metabolizing CYP2B4 enzyme. These reactions also seem to corroborate the involvement of peroxy anion-supported homolytic scission, followed by fragmentation of the adduct into a carbon radical and a formyl species that yields olefin products [50]. Further evidence supporting this mechanism is found when the carbon radical formed during the reaction inactivates the heme in P450 [51, 52].

The electrophilic nature of aldehydes makes them easily susceptible to attack from the nucleophilic peroxy anion. Such an example of nucleophilic attack is also seen in nitric acid synthase (NOS) (Fig. 2.7). NOS is a heme-containing enzyme that catalyzes the conversion of arginine to *N*-hydroxyarginine and then to citrulline and nitric oxide. The second step of this reaction has been proposed to involve nucleophilic addition of the ferric-peroxo species to the –C=NOH bond of the substrate [53]. Inorganic metalloporphyrins mimicking the ferric-peroxo intermediate have also been shown to catalyze the deformylation, as well as epoxidation, of α,--unsaturated carbonyl groups [54, 55].

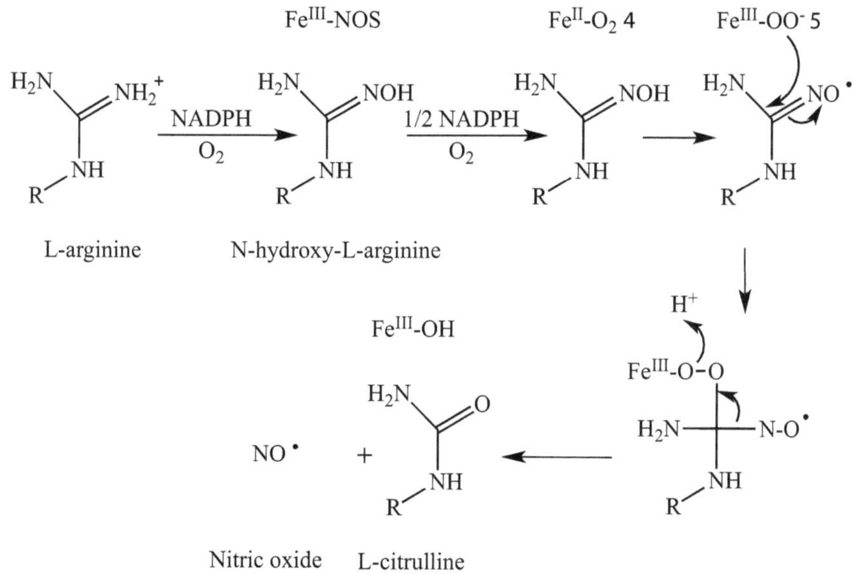

Fig. 2.7 Mechanism of oxidative deformylation catalyzed by NOS (See Ref. [53])

2.4.3 The Ferric-Hydroperoxo Intermediate as an Electrophilic Oxidant

The ferric-hydroperoxo intermediate has been proposed as an oxidant in catalysis involving nucleophilic substrates. However, unlike in the case of the ferric-peroxo intermediate, the hypothetical involvement of the ferric-hydroperoxo species in oxidative catalysis is not supported by solid evidence. Given the electrophilic nature of most substrates oxidized by P450s, Cpd I is the clear favorite oxidant in these cases due to its high reactivity. The most compelling evidence for involvement of the ferric-hydroperoxo intermediate was demonstrated via substrate oxidation by active-site mutants in P450s. In P450CAM, the conserved Thr252 alcohol side chain was mutated to Ala and the resulting Thr252Ala mutant was unable to catalyze the hydroxylation of camphor. Instead, the mutant was highly capable of accepting electrons from the nucleotide cofactor to convert dioxygen to hydrogen peroxide, due to improper protonation to the proximal oxygen of the ferric-peroxo intermediate, thus leading to uncoupling (Fig. 2.2, **II**) [18, 30]. This mutant is, therefore, unable to form Cpd I but generates both the ferric-peroxo and ferric-hydroperoxo intermediates. ENDOR spectroscopic analysis of the cryoreduced Thr252Ala mutant shows a buildup of the ferric-hydroperoxo intermediate at 77 K, annealing at high temperatures yield the ferric enzyme but no hydroxylated product [56, 57]. As such, the Thr mutant of several P450s has been used in the study of a number of electrophilic oxidation reactions.

Vaz, Coon and coworkers were the first to study the effects of the active-site Thr to Ala mutation in rabbit drug-metabolizing CYP2B4 and CYP2E1 enzymes using various alkene substrates (Fig. 2.8) [58]. The researchers observed a decrease in allyllic oxidation of the alkenes. In contrast, the Thr303Ala mutation in CYP2E1 significantly increased the rates of epoxidation compared to the wild-type enzyme.

On the other hand, the corresponding Thr302Ala mutant of CYP2B4 demonstrated reduced rates of both allylic hydroxylation and epoxidation. Increased epoxidation versus allylic hydroxylation was observed in the CYP2E1 Thr303Ala mutant and was construed as evidence that epoxidation could be catalyzed by the ferric-hydroperoxo intermediate, while decreased Cpd I formation led to decreased

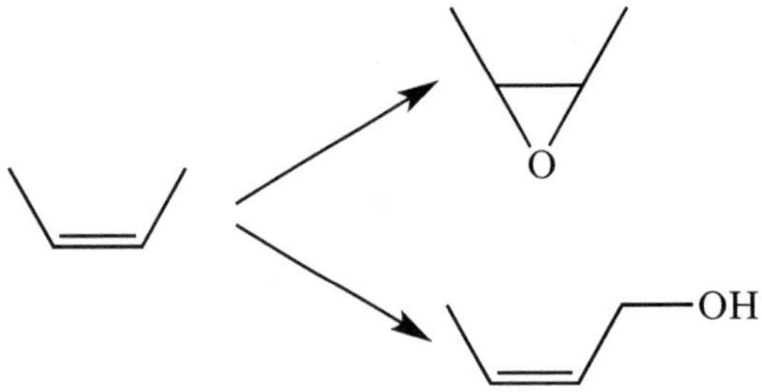

Fig. 2.8 Possible epoxidation reaction products in P450 enzymes

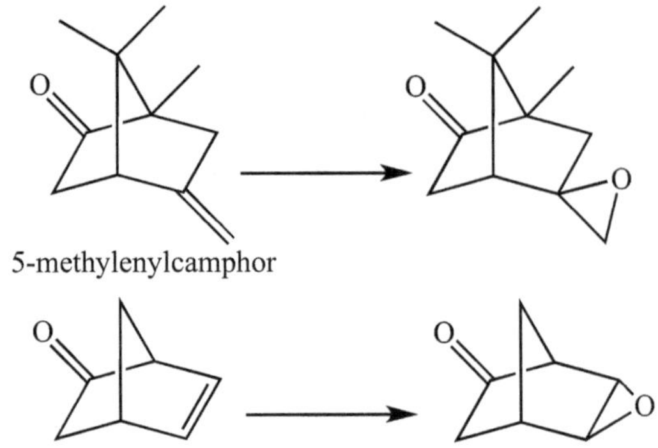

Fig. 2.9 Olefin epoxidation by P450 CAM (See Ref. [59])

hydroxylation. However, failure to observe similar results with CYP2B4 along with high quantities of hydroxylated product made the data somewhat less reliable, apparently because Cpd I was still being generated to a significant extent.

Dawson and coworkers, in collaboration with the Sligar laboratory, studied the reactivity of the ferric-hydroperoxo intermediate in the Thr252Ala mutant of P450CAM using the alkene epoxidation reaction. Unlike the CYP2B4 and CYP2E1 enzymes, the Thr252Ala P450CAM mutant catalyzed the formation of less than 1 % of the hydroxylated product, thus providing a robust system to analyze the presence of a second oxidant, in this case the ferric-hydroperoxo intermediate. Both substrates were easily oxidized to epoxides (Fig. 2.9) at a rate that was ~15–20 % compared to that of wild-type P450CAM [59]. These results substantiated the work of Vaz and coworkers regarding the involvement of a second electrophilic oxidant.

Shaik and coworkers examined alkene epoxidation in the context of the two-state reactivity theory involving Cpd I [60, 61] and proposed that the ferric-hydroperoxo species is a sluggish oxidant compared to the highly reactive Cpd I species. The researchers concluded that the ferric-hydroperoxo species has a large energy barrier to overcome, whereas ferric-hydroperoxo conversion to Cpd I is barrierless [62]. P450 reactivity also appears to be influenced by H-bonding to the proximal thiolate ligand and polarity changes in the vicinity [62–64].

P450 reactivity is also proposed to be influenced by changes in the relative amounts of high-spin and low-spin Cpd I, rather than amounts of the ferric-hydroperoxo and Cpd I species [62]. However, the 'two-state' reactivity theory cannot clearly explain why the Thr252Ala mutant does not hydroxylate camphor. If the Cpd I oxidizing species only displays variation in the amounts of high-spin and low-spin states, the mutant enzyme should also have displayed significant hydroxylation activity.

The ferric-hydroperoxo intermediate has also been investigated as a potential oxidant in heteroatom oxidations. Jones and coworkers have looked particularly at sulfoxidation and *N*-dealkylation reactions utilizing P450BM3. Using clever substrate design and the Thr268Ala mutant, the researchers sought to test whether the two products originated from the same oxidant species (Fig. 2.10) [65]. Thus, substrate **10a** showed four times increased sulfoxidation activity compared to *N*-dealkylation activity. Next, the investigators used an isotopically sensitive *N*-dealkylation substrate, **10b**, to test the premise that due to a large kinetic isotope effect (KIE), sulfoxidation activity would be higher than *N*-dealkylation activity if both products arose from a single oxidant.

However, a negligible KIE was observed leading to several possible conclusions: (1) both products arose from different oxidants, (2) binding of substrate to the P450BM3 enzyme caused an interchange in the position of substituents, thereby changing their position in the catalytic

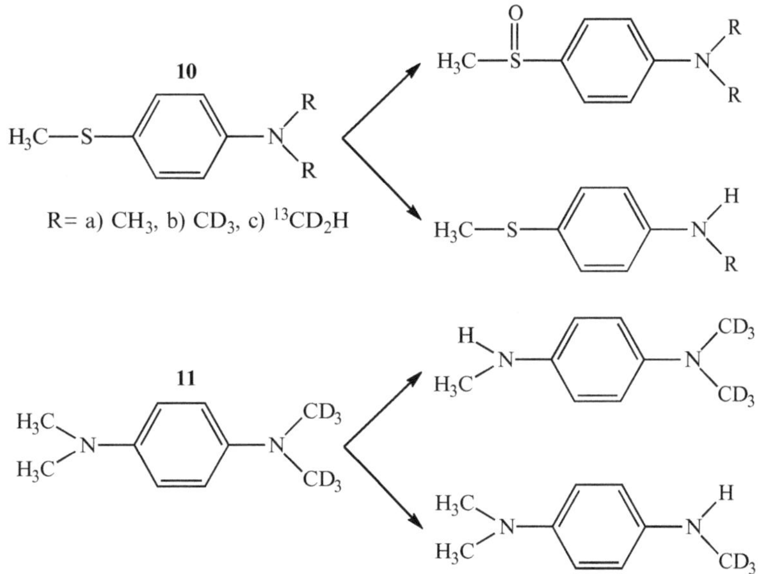

Fig. 2.10 Heteroatom oxidation by P450BM3 (See Ref. [65])

site, and (3) the inherent KIE of *N*-dealkylation was very small. Using substrate **11**, the researchers were able to demonstrate the rapid interchange of substituents at the end of the molecule. This result combined with an intramolecular KIE for substrate **10c** eliminated the last two possibilities, leading to the proposal that both the *N*-dealkylation and sulfoxidation products arose from two different oxidants. The authors suggested that *N*-dealkylation was a product of Cpd I-mediated oxidation while sulfoxidation likely involved the ferric-hydroperoxo intermediate, without eliminating the possibility that the data could result from different forms of the same active oxygen species, i.e. Cpd I. In accordance with the 'two-state' reactivity theory, where the low-spin and high-spin states form different enzyme-substrate (ES) complexes, the *N*-dealkylation and sulfoxidation reaction products can result from two non-interchangeable ES complexes [63, 65]. Watanabe has also proposed that the modified reactivity in the active-site threonine to alanine mutants may be a result of the altered water molecule network in the active site, which affects the H bonding of the Cpd I-ES complex [66]. This alteration can skew the ratios of the low-spin and high-spin state of Cpd I, thereby affecting the mutant reactivity.

The ferric-hydroperoxo species has also been implicated as an oxidant in hydrocarbon hydroxylation reactions. Catalysis by the ferric-hydroperoxo species was proposed to involve a cationic protonated alcohol intermediate [67] (Fig. 2.11) as opposed to a radical intermediate formed in the radical rebound pathway (Fig. 2.2).

Newcomb and coworkers used 'radical clock' experiments to provide evidence for involvement of a cationic intermediate [67, 69–72]. In the first of such studies, the oxidation of *trans*-1-methyl-2-(4-trifluoromethyl)-phenylcyclopropane was examined. The substrate could be oxygenated either on the methyl group yielding methyl alcohol as an unrearranged product, or on the phenyl ring giving a ring-opened alcohol as a rearranged product. Using CYP2B1 as a biocatalyst, the substrate **12b** was shown to generate a ring-opened product, characteristic of a cationic rearrangement pathway (Fig. 2.12) [73].

Reaction of substrate **12a** with the CYP2B4 Thr302Ala mutant enzyme showed a mixture of products, both unrearranged and rearranged methyl oxidation as well as phenyl ring oxidation products. There was little difference in the ratio

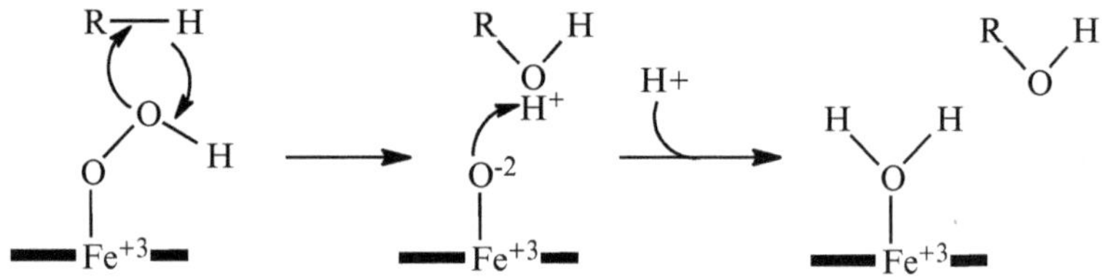

Fig. 2.11 Proposed hydrocarbon hydroxylation mechanism by the ferric-hydroperoxo species (See Refs. [67, 68])

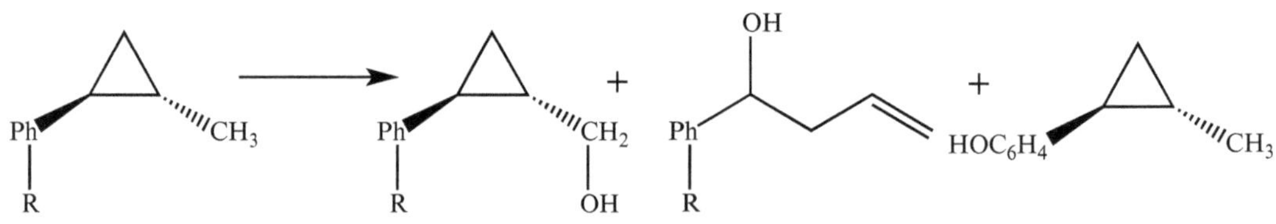

12a) R = H
12b) R= CF_3

Fig. 2.12 Radical clock probes for mechanistic elucidation in CYP2B4 (See Ref. [73])

of rearranged to unrearranged product between the wild-type and mutant enzyme. However, a higher ratio of phenyl oxidation was seen in the mutant enzyme. The authors suggested that this result clearly indicated an alternative oxidant at play in the mutant enzyme that preferred the easier phenyl ring oxidation. To suppress phenyl ring oxidation in substrate **12b**, the phenyl ring was replaced with an electron withdrawing–CF_3 group, which produced an altered ratio between the ring-opened and ring-closed products. This intimated a change in the oxidant in the hydroxylation reaction for that substrate. While these results satisfyingly conveyed involvement of the ferric-hydroperoxo species in the hydroxylation of certain substrates, the species is indeed a sluggish oxidant whereas Cpd I appears to be the oxidant of choice in hydrocarbon hydroxylations [61]. While the ferric-hydroperoxo species appears in almost all heme-based oxygen activation enzymes, there are a few examples where it plays a primary role in substrate oxidation. For example, heme oxygenase catalyzes the oxidation of heme to biliverdin [74] and the first step of this oxidation involves an α-*meso*-hydroxylation of the heme group that is thought to be catalyzed by an electrophilic oxidant, most likely, the ferric-hydroperoxo intermediate [75–77].

2.4.4 Cpd I as the Most Powerful Oxidant

The mechanism of oxidation in P450s has been established by comparison with other heme-based oxygen activating enzymes as well as spectroscopic characterization of the reaction intermediates. P450s are similar to other metalloenzymes such as NOS and chloroperoxidase (CPO) in that they all have heme coordinated to a cysteine thiolate ligand. P450s and NOS are oxidoreductases that activate molecular oxygen [78, 79]. The P450s have long been presumed to oxidize substrates via a reactive porphyrin radical cation ferryl species known as Cpd I. Additional evidence for reactive intermediates was also collected by direct observation through a combination of various spectroscopic techniques [80]. Based on the observed activation of P450s by hydrogen peroxide, alkyl hydroperoxides, periodate and iodosobenzene, oxygen activation was assumed to occur by a two-electron reduction of dioxygen to the level of H_2O_2 followed by formation of the ferryl intermediate as seen in heme-containing peroxidases [81]. Synthetic metalloporphyrins could form a porphyrin radical cation ferryl species at low temperature on reaction with peroxy acids and this intermediate had the ability to

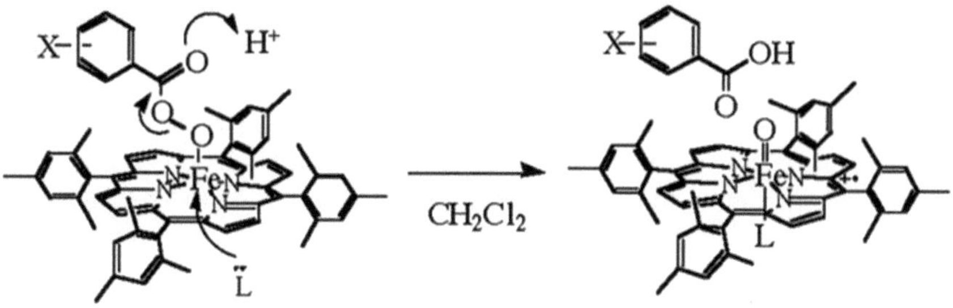

Fig. 2.13 Cpd I formation in synthetic metalloporphyrins by peroxy acids (Reproduced with permission from Groves [81], Copyright 2003 National Academy of Sciences, USA)

insert an oxygen atom into hydrocarbon substrates (Fig. 2.13) [82].

When the transfer of an oxygen atom from the peroxy acid to produce the ferryl intermediate occurs, then the substrate is referred to as an 'oxygen-rebound' substrate [83]. Cpd I has been well characterized in CPO [84] and was thought to be elusive in P450 until recently. In 2010, Green and coworkers were successfully able to directly observe Cpd I in CYP119A1 for the first time [38]. Cpd I was formed in about 75 % yield by the reaction of ferric CYP119A1 with *m*-chloroperbenzoic acid. The resulting Cpd I species could then hydroxylate C–H bonds in lauric acid with an apparent rate constant of $k_{app} = 1.1 \times 10^7\ M^{-1}\ s^{-1}$. The Mossbauer spectrum of this Cpd I species was similar to that seen using Cpd I of CPO. The mechanism of oxygen atom transfer from Cpd I to form the hydroxylated product has been a hotly debated topic. The initially proposed concerted mechanism of oxygen insertion [85] fell aside in favor of the two-step H atom abstraction/oxygen rebound mechanism [83]. As explained in Sect. 2.4.3, the ferric-hydroperoxo intermediate has also been implicated as an oxidant in a few hydrocarbon hydroxylations. ENDOR spectroscopic studies with cryoreduced wild-type P450CAM and its active-site mutants provided compelling evidence in favor of H atom abstraction/hydroxyl rebound [57]. Active oxidant species of P450CAM were prepared by cryoreduction at 77 K of the oxyferrous intermediate in the P450CAM-camphor complex. The ferric-peroxo and ferric-hydroperoxo intermediates were observed upon slowly warming to 119 K and were subsequently characterized by EPR and ENDOR spectroscopy. Around 200 K, the ferric-hydroperoxo species was quantitatively converted to 5-*exo*-hydroxycamphor, the natural product of camphor hydroxylation. While the ferryl intermediate was not observed directly, this oxidation was assumed to proceed through the hydroxyl intermediate due to the following observations in the experiment. After formation of the ferric-hydroperoxo species upon slowly warming the sample, the first species observed had the hydroxyl group bound to the heme iron, as was expected for the H atom abstraction in the ferryl-mediated mechanism. Had the ferric-hydroperoxo species been involved in the oxidation, it would have initially formed hydroxycamphor via hydroxy insertion of the distal oxygen atom of the ferric-hydroperoxo species. Hydroxycamphor would be required to displace the hydroxyl/water that was bound to heme, but this displacement reaction was implausible to occur at 200 K. Furthermore, ENDOR spectroscopy showed that the hydrogen attached to the hydroxyl oxygen in the hydroxycamphor product originated from the C-5 position of camphor, further supporting the ferryl mechanism. Involvement of the ferric-hydroperoxo species would have required this H atom to originate from the surrounding solvent.

Shaik and coworkers examined the mechanism of hydrocarbon hydroxylation using theoretical calculations and proposed a two-state reactivity instead of two-oxidant reactivity [86–89]. The researchers proposed that the porphyrin radical cation ferryl species exists in two spin states, a quartet spin state and a doublet spin state that are close in energy. Both species initiate the reaction by nearly identical H atom abstraction transition states. The species in the doublet state can quickly collapse to the product in a barrierless reaction with no formation of an intermediate. This almost-concerted mechanism is aided by the increased interaction of the cysteine thiolate ligand with the heme iron, the push effect. On the other hand, the quartet state must overcome a significant energy barrier to form the product, thus allowing formation and rearrangements of radicals if any. The two-state reactivity model has provided a good explanation for the stereochemical scrambling and structural rearrangement resulting from the radical clock experiments. Direct observation and characterization of Cpd I has cemented the H atom abstraction/hydroxyl rebound mechanism of P450 enzymes [90].

2.4.5 Fe^{III}–(H_2O_2) as an Oxidant in Sulfoxidation Reactions

As described earlier, Jones and coworkers, based on the oxidation of a substrate with both amine and thioether functional groups in P450BM3, proposed that two different oxidants can be responsible for the oxidation of the two classes of substrates. The ferric-hydroperoxo species was proposed to be the oxygenating species responsible for sulfur oxidation [65]. In the same vein, De Voss and coworkers analyzed the oxidation of thia fatty acids using P450BM3 as the biocatalyst, O_2 as the oxidant and NADPH as the cofactor (Fig. 2.14) [91].

Analysis of the products indicated that presence of the thioether functionality dramatically shifted the regiochemistry of the reaction. With substrates **13** and **14**, the oxidation was distributed across the last three methylene groups. However, replacement of the second methylene group with sulfur resulted in the oxidation in **15** and **16** occurring exclusively at the sulfur. Interestingly, the sulfoxides were S enantiomers whereas the alcohols were R enantiomers. While it was speculated that thioethers undergo unusual binding to yield S sulfoxides, it has been shown in the past that modified fatty acids undergo R oxidation exclusively [91–93]. Substrates were also reacted with the P450BM3 Thr268Ala active-site mutant, based on previous studies showing that the mutant enzyme formed very little Cpd I and was able to accumulate the ferric-hydroperoxo species [18, 30, 31]. Despite the low turnover in the mutant for substrate **14**, product distribution and enantioselectivity remained unchanged between the wild-type and mutant enzyme. This demonstrated that there occurred reduced Cpd I formation in the mutant. For the thia fatty acid substrates, negligible change was observed in the turnover, product distribution and

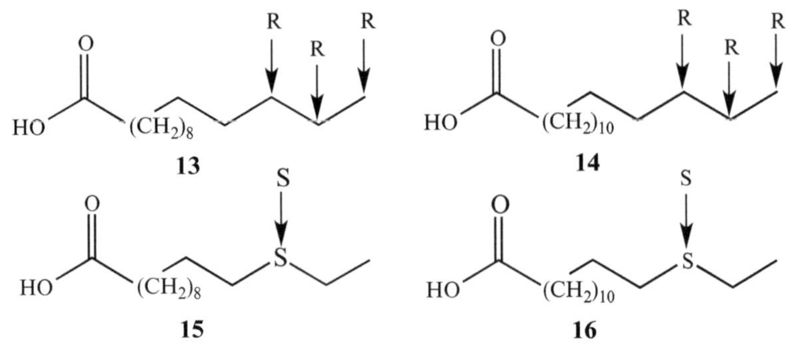

Fig. 2.14 Fatty acid oxidation by P450BM3, *arrows* indicate sites of oxidation and stereochemistry of products formed (See Ref. [91])

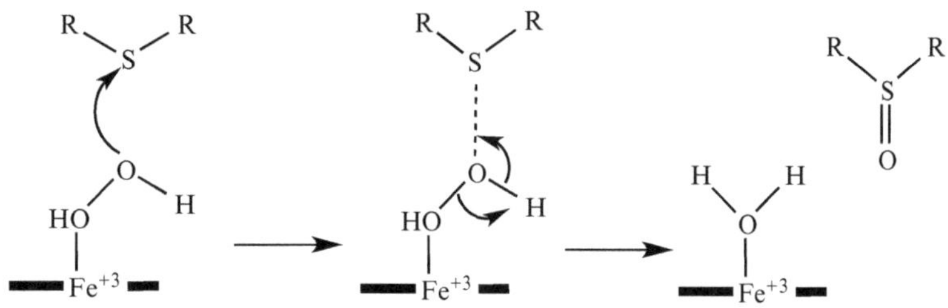

Fig. 2.15 Proposed mechanism of sulfoxidation by the $Fe^{III}-(H_2O_2)$ intermediate (See Ref. [94])

enantioselectivity of the products between the products of the wild-type and mutant enzyme. The authors proposed that sulfur oxidation must be easily catalyzed by the ferric-hydroperoxo species, thereby enabling the mutant enzyme to form comparable amounts of product compared to that of the wild-type enzyme.

However, Shaik and coworkers have recently used theoretical calculations to show that the $Fe^{III}-(H_2O_2)$ complex (Fig. 2.2, **9**) is a very efficient oxidant for sulfoxidation reactions in P450s and iron corrolazine compounds [94]. The $Fe^{III}-(H_2O_2)$ complex was shown to undergo a nucleophilic attack from the distal oxygen atom of the peroxo complex, resulting in heterolytic O–O bond scission that is coupled to proton transfer (Fig. 2.15). The $Fe^{III}-(H_2O_2)$ complex could also catalyze the oxidation on sulfur much faster than could Cpd I. The ferric-hydroperoxo intermediate, in contrast, had a high barrier via the homolysis pathway of oxygen insertion [94]. This finding offers a new paradigm for sulfoxidation reactions in P450s and their synthetic monologues.

2.5 Conclusions

The mechanistic complexity of P450 enzymes has been intensely debated for the last few decades. The recent direct observation of P450 Cpd I and resulting studies of its reactivity have provided strong support for the validity of the H atom abstraction/radical rebound mechanism in hydrocarbon hydroxylation reactions. The role of the ferric-peroxo intermediate as a nucleophilic oxidant is also well established with experimental evidence. The proposed role of the ferric-hydroperoxo intermediate as an electrophilic oxidant remains to be established. Its role as an oxidant has been proposed mainly based on turnover studies in P450 mutants with impaired ability to form Cpd I. A recent addition to this oxidant puzzle is the $Fe^{III}-(H_2O_2)$ intermediate, which has been proposed to be more active than Cpd I in thio-ether oxidation reactions. An alternative explanation for the multiple oxidant hypothesis is provided by the theoretical two-state reactivity hypothesis involving Cpd I, which has advanced some explanations for the disparate experimental data. However, additional experimental and theoretical data are still needed to provide further insights into the mechanisms of P450 catalysis.

Acknowledgment The NIH (GM-26730) has supported cytochrome P450 research in the Dawson laboratory. We would like to thank Dr. Masanori Sono for pertinent advice.

References

1. Poulos TL, Johnson EF (2005) Structures of cytochrome P450 enzymes. In: Ortiz de Montellano PR (ed) Cytochrome P450: structure, mechanism, and biochemistry, 3rd edn. Kluwer Academic/Plenum Publishers, New York, pp 87–114
2. Sono M, Roach MP, Coulter ED, Dawson JH (1996) Heme-containing oxygenases. Chem Rev 96:2841–2888
3. Omura T, Sato R (1964) The carbon monoxide-binding pigment of liver microsomes: I. Evidence for its hemoprotein nature. J Biol Chem 239:2370–2378
4. Filatov M, Reckien W, Peyerimhoff SD, Shaik S (2000) What are the reasons for the kinetic stability of a mixture of H_2 and O_2? J Phys Chem A 104:12014–12020
5. Blanksby SJ, Ellison GB (2003) Bond dissociation energies of organic molecules. Acc Chem Res 36:255–263

6. Sligar SG, Makris TM, Denisov IG (2005) Thirty years of microbial P450 monooxygenase research: peroxo-heme intermediates—the central bus station in heme oxygenase catalysis. Biochem Biophys Res Commun 338:346–354
7. Meunier B (1992) Metalloporphyrins as versatile catalysts for oxidation reactions and oxidative DNA cleavage. Chem Rev 92:1411–1456
8. Glieder A, Farinas ET, Arnold FH (2002) Laboratory evolution of a soluble, self-sufficient, highly active alkane hydroxylase. Nat Biotechnol 20:1135–1139
9. Schlichting I, Berendzen J, Chu K, Stock AM, Maves SA, Benson DE, Sweet RM, Ringe D, Petsko GA, Sligar SG (2000) The catalytic pathway of cytochrome P450cam at atomic resolution. Science 287:1615–1622
10. Nebert DW, Gonzalez FJ (1987) P450 genes: structure, evolution, and regulation. Annu Rev Biochem 56:945–993
11. Narhi LO, Fulco AJ (1986) Characterization of a catalytically self-sufficient 119,000-dalton cytochrome P-450 monooxygenase induced by barbiturates in *Bacillus megaterium*. J Biol Chem 261:7160–7169
12. Ruettinger RT, Wen LP, Fulco AJ (1989) Coding nucleotide, 5′ regulatory, and deduced amino acid sequences of P-450BM-3, a single peptide cytochrome P-450:NADPH-P-450 reductase from *Bacillus megaterium*. J Biol Chem 264:10987–10995
13. Poulos TL, Finzel BC, Gunsalus IC, Wagner GC, Kraut J (1985) The 2.6-Å crystal structure of *Pseudomonas putida* cytochrome P-450. J Biol Chem 260:16122–16130
14. Gunsalus IC, Pederson TC, Sligar SG (1975) Oxygenase-catalyzed biological hydroxylations. Annu Rev Biochem 44:377–407
15. Meunier B, de Visser SP, Shaik S (2004) Mechanism of oxidation reactions catalyzed by cytochrome P450 enzymes. Chem Rev 104:3947–3980
16. Winfield ME (1965) Mechanisms of oxygen uptake: the autoxidation of myoglobin and of reduced cyanocobaltates and their significance to oxidase reactions. In: King TE, Mason HS, Morrison M (eds) Oxidases and related redox systems, vol 1. Wiley, New York, pp 115–130
17. Dunford HB, Stillman JS (1976) On the function and mechanism of action of peroxidases. Coord Chem Rev 19:187–251
18. Imai M, Shimada H, Watanabe Y, Matsushima-Hibiya Y, Makino R, Koga H, Horiuchi T, Ishimura Y (1989) Uncoupling of the cytochrome P-450cam monooxygenase reaction by a single mutation, threonine-252 to alanine or valine: possible role of the hydroxy amino acid in oxygen activation. Proc Natl Acad Sci U S A 86:7823–7827
19. Egawa T, Shimada H, Ishimura Y (1994) Evidence for compound I formation in the reaction of cytochrome P450cam with m-chloroperbenzoic acid. Biochem Biophys Res Commun 201:1464–1469
20. Schünemann V, Lendzian F, Jung C, Contzen J, Barra AL, Sligar SG, Trautwein AX (2004) Tyrosine radical formation in the reaction of wild type and mutant cytochrome P450cam with peroxy acids: a multifrequency EPR study of intermediates on the millisecond time scale. J Biol Chem 279:10919–10930
21. Spolitak T, Dawson JH, Ballou DP (2005) Reaction of ferric cytochrome P450cam with peracids: kinetic characterization of intermediates on the reaction pathway. J Biol Chem 280:20300–20309
22. Shimada H, Watanabe Y, Imai M, Makino R, Koga H, Horiuchi T, Ishimura Y (1991) The role of threonine 252 in the oxygen activation by cytochrome P-450 cam: mechanistic studies by site-directed mutagenesis. In: Simandi LI (ed) Dioxygen activation and homogeneous catalytic oxidation. Elsevier, Amsterdam, pp 3136–3319
23. Auclair K, Moënne-Loccoz P, Ortiz de Montellano PR (2001) Roles of the proximal heme thiolate ligand in cytochrome P450cam. J Am Chem Soc 123:4877–4885
24. Perera R, Sono M, Sigman JA, Pfister TD, Lu Y, Dawson JH (2003) Neutral thiol as a proximal ligand to ferrous heme iron: implications for heme proteins that lose cysteine thiolate ligation on reduction. Proc Natl Acad Sci U S A 100:3641–3646
25. Yoshioka S, Takahashi S, Ishimori K, Morishima I (2000) Roles of the axial push effect in cytochrome P450cam studied with the site-directed mutagenesis at the heme proximal site. J Inorg Biochem 81:141–151
26. Yoshioka S, Tosha T, Takahashi S, Ishimori K, Hori H, Morishima I (2002) Roles of the proximal hydrogen bonding network in cytochrome P450cam-catalyzed oxygenation. J Am Chem Soc 124:14571–14579
27. Dawson JH (1988) Probing structure-function relations in heme-containing oxygenases and peroxidases. Science 240:433–439
28. Dawson JH, Holm RH, Trudell JR, Barth G, Linder RE, Bunnenberg E, Djerassi C, Tang SC (1976) Oxidized cytochrome P-450. Magnetic circular dichroism evidence for thiolate ligation in the substrate-bound form. Implications for the catalytic mechanism. J Am Chem Soc 98:3707–3709
29. Poulos TL, Kraut J (1980) The stereochemistry of peroxidase catalysis. J Biol Chem 255:8199–8205
30. Martinis SA, Atkins WM, Stayton PS, Sligar SG (1989) A conserved residue of cytochrome P-450 is involved in heme-oxygen stability and activation. J Am Chem Soc 111:9252–9253
31. Yeom H, Sligar SG, Li H, Poulos TL, Fulco AJ (1995) The role of Thr268 in oxygen activation of cytochrome P450BM-3. Biochemistry 34:14733–14740
32. Imai Y, Nakamura M (1988) The importance of threonine-301 from cytochromes P-450 (laurate (ω-1)-hydroxylase and testosterone 16α-hydroxylase) in substrate binding as demonstrated by site-directed mutagenesis. FEBS Lett 234:313–315
33. Raag R, Martinis SA, Sligar SG, Poulos TL (1991) Crystal structure of the cytochrome P-450CAM

active site mutant Thr252Ala. Biochemistry 30:11420–11429
34. Harris DL, Loew GH (1996) Investigation of the proton-assisted pathway to formation of the catalytically active, ferryl species of P450s by molecular dynamics studies of P450eryF. J Am Chem Soc 118:6377–6387
35. Truan G, Peterson JA (1998) Thr268 in substrate binding and catalysis in P450BM-3. Arch Biochem Biophys 349:53–64
36. Vidakovic M, Sligar SG, Li H, Poulos TL (1998) Understanding the role of the essential Asp251 in cytochrome P450cam using site-directed mutagenesis: crystallography, and kinetic solvent isotope effect. Biochemistry 37:9211–9219
37. Gerber NC, Sligar SG (1994) A role for Asp-251 in cytochrome P-450cam oxygen activation. J Biol Chem 269:4260–4266
38. Rittle J, Green MT (2010) Cytochrome P450 compound I: capture, characterization, and C-H bond activation kinetics. Science 330:933–937
39. Loida PJ, Sligar SG (1993) Molecular recognition in cytochrome P-450: mechanism for the control of uncoupling reactions. Biochemistry 32:11530–11538
40. Eisenstein L, Debey P, Douzou P (1977) P450cam: oxygenated complexes stabilized at low temperature. Biochem Biophys Res Commun 77:1377–1383
41. Antonini E, Brunori M (1971) Hemoglobin and myoglobin in their reactions with ligands. North-Holland, Amsterdam
42. Couture M, Stuehr DJ, Rousseau DL (2000) The ferrous dioxygen complex of the oxygenase domain of neuronal nitric-oxide synthase. J Biol Chem 275:3201–3205
43. Macdonald IDG, Sligar SG, Christian JF, Unno M, Champion PM (1998) Identification of the Fe–O–O bending mode in oxycytochrome P450cam by resonance Raman spectroscopy. J Am Chem Soc 121:376–380
44. Sono M, Eble KS, Dawson JH, Hager LP (1985) Preparation and properties of ferrous chloroperoxidase complexes with dioxygen, nitric oxide, and an alkyl isocyanide. Spectroscopic dissimilarities between the oxygenated forms of chloroperoxidase and cytochrome P-450. J Biol Chem 260:15530–15535
45. Bangcharoenpaurpong O, Rizos AK, Champion PM, Jollie D, Sligar SG (1986) Resonance Raman detection of bound dioxygen in cytochrome P-450cam. J Biol Chem 261:8089–8092
46. Fischer RT, Trzaskos JM, Magolda RL, Ko SS, Brosz CS, Larsen B (1991) Lanosterol 14α-methyl demethylase. Isolation and characterization of the third metabolically generated oxidative demethylation intermediate. J Biol Chem 266:6124–6132
47. Akhtar M, Corina D, Miller S, Shyadehi AZ, Wright JN (1994) Mechanism of the acyl-carbon cleavage and related reactions catalyzed by multifunctional P-450s: studies on cytochrome P-$450_{17\alpha}$. Biochemistry 33:4410–4418
48. Akhtar M, Alexander K, Boar RB, McGhie JF, Barton DH (1978) Chemical and enzymic studies on the characterization of intermediates during the removal of the 14α-methyl group in cholesterol biosynthesis. The use of 32-functionalized lanostane derivatives. Biochem J 169:449–463
49. Corina DL, Miller SL, Wright JN, Akhtar M (1991) The mechanism of cytochrome P-450 dependent C-C bond cleavage: studies on 17α-hydroxylase-17,20-lyase. J Chem Soc Chem Commun 782–783
50. Roberts ES, Vaz AD, Coon MJ (1991) Catalysis by cytochrome P-450 of an oxidative reaction in xenobiotic aldehyde metabolism: deformylation with olefin formation. Proc Natl Acad Sci U S A 88:8963–8966
51. Kuo CL, Raner GM, Vaz ADN, Coon MJ (1999) Discrete species of activated oxygen yield different cytochrome P450 heme adducts from aldehydes. Biochemistry 38:10511–10518
52. Bestervelt LL, Vaz AD, Coon MJ (1995) Inactivation of ethanol-inducible cytochrome P450 and other microsomal P450 isozymes by trans-4-hydroxy-2-nonenal, a major product of membrane lipid peroxidation. Proc Natl Acad Sci U S A 92:3764–3768
53. Korth HG, Sustmann R, Thater C, Butler AR, Ingold KU (1994) On the mechanism of the nitric oxide synthase-catalyzed conversion of N^{ω}-hydroxy-L-arginine to citrulline and nitric oxide. J Biol Chem 269:17776–17779
54. Wertz DL, Sisemore MF, Selke M, Driscoll J, Valentine JS (1998) Mimicking cytochrome P-450 2B4 and aromatase: aromatization of a substrate analogue by a peroxo Fe(III) porphyrin complex. J Am Chem Soc 120:5331–5332
55. Sisemore MF, Burstyn JN, Valentine JS (1996) Epoxidation of electron-deficient olefins by a nucleophilic iron(III) peroxo porphyrinato complex, peroxo (tetramesitylporphyrinato)ferrate(1–). Angew Chem Int Ed 35:206–208
56. Davydov R, Macdonald IDG, Makris TM, Sligar SG, Hoffman BM (1999) EPR and ENDOR of catalytic intermediates in cryoreduced native and mutant oxy-cytochromes P450cam: mutation-induced changes in the proton delivery system. J Am Chem Soc 121:10654–10655
57. Davydov R, Makris TM, Kofman V, Werst DE, Sligar SG, Hoffman BM (2001) Hydroxylation of camphor by reduced oxy-cytochrome P450cam: mechanistic implications of EPR and ENDOR studies of catalytic intermediates in native and mutant enzymes. J Am Chem Soc 123:1403–1415
58. Vaz ADN, McGinnity DF, Coon MJ (1998) Epoxidation of olefins by cytochrome P450: evidence from site-specific mutagenesis for hydroperoxo-iron as an electrophilic oxidant. Proc Natl Acad Sci U S A 95:3555–3560
59. Jin S, Makris TM, Bryson TA, Sligar SG, Dawson JH (2003) Epoxidation of olefins by hydroperoxo–ferric cytochrome P450. J Am Chem Soc 125:3406–3407
60. de Visser SP, Ogliaro F, Harris N, Shaik S (2001) Multi-state epoxidation of ethene by cytochrome

P450: a quantum chemical study. J Am Chem Soc 123:3037–3047
61. de Visser SP, Ogliaro F, Sharma PK, Shaik S (2002) What factors affect the regioselectivity of oxidation by cytochrome P450? A DFT study of allylic hydroxylation and double bond epoxidation in a model reaction. J Am Chem Soc 124:11809–11826
62. Ogliaro F, Cohen S, de Visser SP, Shaik S (2000) Medium polarization and hydrogen bonding effects on compound I of cytochrome P450: what kind of a radical is it really? J Am Chem Soc 122:12892–12893
63. Ogliaro F, de Visser SP, Cohen S, Sharma PK, Shaik S (2002) Searching for the second oxidant in the catalytic cycle of cytochrome P450: a theoretical investigation of the iron(III)-hydroperoxo species and its epoxidation pathways. J Am Chem Soc 124:2806–2817
64. de Visser SP, Ogliaro F, Sharma PK, Shaik S (2002) Hydrogen bonding modulates the selectivity of enzymatic oxidation by P450: chameleon oxidant behavior by compound I. Angew Chem Int Ed 41:1947–1951
65. Volz TJ, Rock DA, Jones JP (2002) Evidence for two different active oxygen species in cytochrome P450 BM3 mediated sulfoxidation and *N*-dealkylation reactions. J Am Chem Soc 124:9724–9725
66. Watanabe Y (2001) Alternatives to the oxoferryl porphyrin cation radical as the proposed reactive intermediate of cytochrome P450: two-electron oxidized Fe(III) porphyrin derivatives. J Biol Inorg Chem 6:846–856
67. Newcomb M, Shen R, Choi SY, Toy PH, Hollenberg PF, Vaz ADN, Coon MJ (2000) Cytochrome P450-catalyzed hydroxylation of mechanistic probes that distinguish between radicals and cations. Evidence for cationic but not for radical intermediates. J Am Chem Soc 122:2677–2686
68. Jin S, Bryson T, Dawson J (2004) Hydroperoxoferric heme intermediate as a second electrophilic oxidant in cytochrome P450-catalyzed reactions. J Biol Inorg Chem 9:644–653
69. Toy PH, Newcomb M, Coon MJ, Vaz ADN (1998) Two distinct electrophilic oxidants effect hydroxylation in cytochrome P-450-catalyzed reactions. J Am Chem Soc 120:9718–9719
70. Newcomb M, Le Tadic-Biadatti MH, Chestney DL, Roberts ES, Hollenberg PF (1995) A nonsynchronous concerted mechanism for cytochrome P-450 catalyzed hydroxylation. J Am Chem Soc 117:12085–12091
71. Newcomb M, Aebisher D, Shen R, Chandrasena REP, Hollenberg PF, Coon MJ (2003) Kinetic isotope effects implicate two electrophilic oxidants in cytochrome P450-catalyzed hydroxylations. J Am Chem Soc 125:6064–6065
72. Chandrasena REP, Vatsis KP, Coon MJ, Hollenberg PF, Newcomb M (2003) Hydroxylation by the hydroperoxy-iron species in cytochrome P450 enzymes. J Am Chem Soc 126:115–126
73. Toy PH, Dhanabalasingam B, Newcomb M, Hanna IH, Hollenberg PF (1997) A substituted hypersensitive radical probe for enzyme-catalyzed hydroxylations: synthesis of racemic and enantiomerically enriched forms and application in a cytochrome P450-catalyzed oxidation. J Org Chem 62:9114–9122
74. Ortiz de Montellano PR, Wilks A (2000) Advances in inorganic chemistry. In: Sykes G, Mauk AG (eds) Iron porphyrins, vol 51. Academic, San Diego, pp 359–407
75. Tenhunen R, Marver H, Pinstone NR, Trager WF, Cooper DY, Schmid R (1972) Enzymic degradation of heme. Oxygenative cleavage requiring cytochrome P-450. Biochemistry 11:1716–1720
76. Wilks A, Ortiz de Montellano PR (1993) Rat liver heme oxygenase. High level expression of a truncated soluble form and nature of the meso-hydroxylating species. J Biol Chem 268:22357–22362
77. Wilks A, Torpey J, Ortiz de Montellano PR (1994) Heme oxygenase (HO-1). Evidence for electrophilic oxygen addition to the porphyrin ring in the formation of α-meso-hydroxyheme. J Biol Chem 269:29553–29556
78. Sundaramoorthy M, Terner J, Poulos TL (1995) The crystal structure of chloroperoxidase: a heme peroxidase–cytochrome P450 functional hybrid. Structure 3:1367–1378
79. Groves JT, Wang CCY (2000) Nitric oxide synthase: models and mechanisms. Curr Opin Chem Biol 4:687–695
80. Makris TM, Schlichting I, Sligar SG (2005) Activation of molecular oxygen by cytochrome P450. In: Ortiz de Montellano PR (ed) Cytochrome P450: structure, mechanism, and biochemistry, 2nd edn. Plenum Press, New York, pp 149–182
81. Groves JT (2003) The bioinorganic chemistry of iron in oxygenases and supramolecular assemblies. Proc Natl Acad Sci U S A 100:3569–3574
82. Groves JT, Haushalter RC, Nakamura M, Nemo TE, Evans BJ (1981) High-valent iron-porphyrin complexes related to peroxidase and cytochrome P-450. J Am Chem Soc 103:2884–2886
83. Groves JT, McClusky GA (1976) Aliphatic hydroxylation via oxygen rebound. Oxygen transfer catalyzed by iron. J Am Chem Soc 98:859–861
84. Groves JT, Watanabe Y (1988) Reactive iron porphyrin derivatives related to the catalytic cycles of cytochrome P-450 and peroxidase. Studies of the mechanism of oxygen activation. J Am Chem Soc 110:8443–8452
85. Shapiro S, Piper JU, Caspi E (1982) Steric course of hydroxylation at primary carbon atoms. Biosynthesis of 1-octanol from (1R)- and (1S)-[1-3H, 2H, 1H; 1-14C]octane by rat liver microsomes. J Am Chem Soc 104:2301–2305
86. Shaik S, Filatov M, Schröder D, Schwarz H (1998) Electronic structure makes a difference: cytochrome P-450 mediated hydroxylations of hydrocarbons as a two-state reactivity paradigm. Chem Eur J 4:193–199
87. Harris N, Cohen S, Filatov M, Ogliaro F, Shaik S (2000) Two-state reactivity in the rebound step of alkane hydroxylation by cytochrome P-450: origins of free radicals with finite lifetimes. Angew Chem Int Ed 39:2003–2007
88. Ogliaro F, de Visser SP, Groves JT, Shaik S (2001) Chameleon states: high-valent metal-oxo species of

cytochrome P450 and its ruthenium analogue. Angew Chem Int Ed 40:2874–2878
89. Ogliaro F, Harris N, Cohen S, Filatov M, de Visser SP, Shaik S (2000) A model "rebound" mechanism of hydroxylation by cytochrome P450: stepwise and effectively concerted pathways, and their reactivity patterns. J Am Chem Soc 122:8977–8989
90. Krest CM, Onderko EL, Yosca TH, Calixto JC, Karp RF, Livada J, Rittle J, Green MT (2013) Reactive intermediates in cytochrome P450 catalysis. J Biol Chem 288:17074–17081
91. Cryle MJ, De Voss JJ (2006) Is the ferric hydroperoxy species responsible for sulfur oxidation in cytochrome P450s? Angew Chem Int Ed 45:8221–8223
92. Truan G, Komandla MR, Falck JR, Peterson JA (1999) P450BM-3: absolute configuration of the primary metabolites of palmitic acid. Arch Biochem Biophys 366:192–198
93. Capdevila JH, Wei S, Helvig C, Falck JR, Belosludtsev Y, Truan G, Graham-Lorence SE, Peterson JA (1996) The highly stereoselective oxidation of polyunsaturated fatty acids by cytochrome P450BM-3. J Biol Chem 271:22663–22671
94. Wang B, Li C, Cho KB, Nam W, Shaik S (2013) The $Fe^{III}(H_2O_2)$ complex as a highly efficient oxidant in sulfoxidation reactions: revival of an underrated oxidant in cytochrome P450. J Chem Theory Comput 9:2519–2525

Current Approaches for Investigating and Predicting Cytochrome P450 3A4-Ligand Interactions

3

Irina F. Sevrioukova and Thomas L. Poulos

Abstract

Cytochrome P450 3A4 (CYP3A4) is the major and most important drug-metabolizing enzyme in humans that oxidizes and clears over a half of all administered pharmaceuticals. This is possible because CYP3A4 is promiscuous with respect to substrate binding and has the ability to catalyze diverse oxidative chemistries in addition to traditional hydroxylation reactions. Furthermore, CYP3A4 binds and oxidizes a number of substrates in a cooperative manner and can be both induced and inactivated by drugs. In vivo, CYP3A4 inhibition could lead to undesired drug-drug interactions and drug toxicity, a major reason for late-stage clinical failures and withdrawal of marketed pharmaceuticals. Owing to its central role in drug metabolism, many aspects of CYP3A4 catalysis have been extensively studied by various techniques. Here, we give an overview of experimental and theoretical methods currently used for investigation and prediction of CYP3A4-ligand interactions, a defining factor in drug metabolism, with an emphasis on the problems addressed and conclusions derived from the studies.

Keywords

Cytochrome P450 • CYP3A4 • Ligand binding • Enzyme inhibition • Drug metabolism • Drug-drug interactions

3.1 Introduction

Cytochrome P450 3A4 (CYP3A4[1]) is one of many human cytochrome P450 (P450) enzymes

I.F. Sevrioukova (✉)
Department of Molecular Biology and Biochemistry, University of California, Irvine, CA 92697, USA
e-mail: sevrioui@uci.edu

T.L. Poulos
Departments of Molecular Biology and Biochemistry, Chemistry, and Pharmaceutical Sciences, University of California, Irvine, CA 92697, USA

[1] *Abbreviations*: *ANF* α-naphthoflavone, *BEC* bromoergocryptine, *BFC* 7-benzyloxy-4-trifluoromethylcoumarin, *BQ* 7-benzyloxyquinoline, b_5 cytochrome b_5, *CPR* NADPH-cytochrome P450 oxidoreductase, *CYP or P450* cytochrome P450.

E.G. Hrycay and S.M. Bandiera (eds.), *Monooxygenase, Peroxidase and Peroxygenase Properties and Mechanisms of Cytochrome P450*, Advances in Experimental Medicine and Biology 851, DOI 10.1007/978-3-319-16009-2_3, that plays a central role in drug metabolism.[2] While some P450s specialize in catalyzing very specific reactions (e.g. synthesis of cholesterol or fatty acid oxidation), CYP3A4 biotransforms a wide range of endogenous and exogenous compounds including drugs, toxins and pollutants. This is possible due to a large and malleable active site of CYP3A4 (Fig. 3.1), capable of accommodating substrates varying in size and chemical nature, as well as its ability to catalyze diverse chemical reactions such as alkyl carbon and aromatic ring hydroxylation, *O*- and *N*-dealkylation, and epoxidation [1].

CYP3A4 is also known for atypical (sigmoidal, non-Michaelis-Menten) enzyme kinetics and complex ligand-binding behavior resulting from homotropic and heterotropic cooperativity of some of its substrates. Different categories of atypical kinetic profiles and underlying mechanisms were discussed in detail previously [2–7]. In short, cooperative effects in CYP3A4 are thought to occur when the second (or third) substrate molecule of the same or different nature binds remotely or within the active-site pocket and increases turnover by promoting the productive orientation of the first substrate, which can occur with or without a conformational change in CYP3A4.

Despite substrate promiscuity and a highly plastic active site cavity, CYP3A4 with many substrates displays considerable regio- and stereoselectivity in product formation, indicating that structural features of substrates and/or the active site result in selective substrate binding modes. Another important aspect of CYP3A4-drug interactions is that some drugs as well as natural compounds consumed with food can act as CYP3A4 inhibitors. In vivo, this may lead to drug-drug interactions, perturbed pharmacokinetics and toxicity. There has been a continuous effort to unravel and better understand the CYP3A4 inhibitory mechanisms, the full knowledge of which could help medicinal chemists to develop safer drugs.

[2] A review of this scope cannot include all the references pertaining to the subject matter.

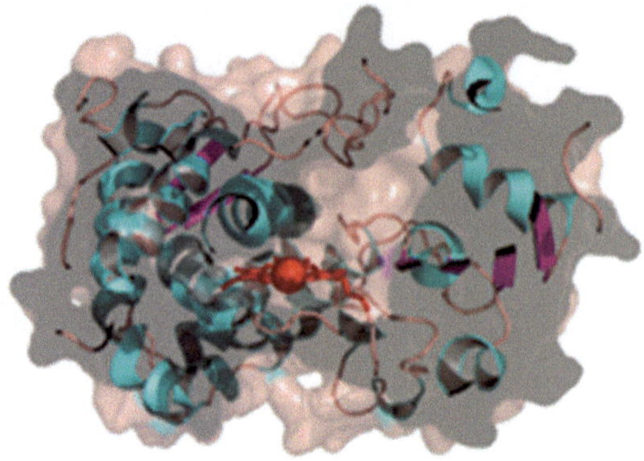

Fig. 3.1 Crystal structure of ligand-free CYP3A4 (Protein Data Base ID 1TQN). A cross-section of solvent-accessible surface is shown to display the active-site cavity. Helices, strands and loops are in *cyan*, *magenta* and *beige*, respectively, and the heme is *red*

Here, we give an overview of experimental and theoretical approaches that have been used for investigation and prediction of CYP3A4-ligand interactions, including absorbance, fluorescence, nuclear magnetic and electron paramagnetic resonance (EPR) spectroscopy, X-ray crystallography, various computational and other techniques. Both soluble and membrane-bound forms of CYP3A4 (liver and insect microsomes, proteoliposomes, lipid bilayer nanodiscs, etc.) have been investigated. Preparation of the protein forms is not discussed in this review but the type of model system used in the experimental work will be specified when necessary.

3.2 Experimental Approaches

3.2.1 Absorbance Spectroscopy

All P450s contain the heme cofactor, whose absorption wavelength (λ_{max}) and amplitude depend on the heme iron oxidation-reduction (redox) and coordination state [8, 9]. The oxidized and reduced ligand-free forms absorb at 415–418 nm and 407–409 nm, respectively. Upon binding in the active site, substrates displace a coordinated water ligand and shift the Soret band to 385–395 nm (type I spectral changes; low- to high-spin shift). Depending on the substrate affinity and spatial fit, the spectral change can be partial or complete. In contrast, molecules that contain unhindered nitrogen atoms can ligate to the heme iron directly or via the axial water molecule, shifting the Soret band to 420–425 nm (type II spectral changes). These compounds usually act as inhibitors but, in some cases, can be metabolized by P450. Finally, the ferrous ligand-free and ligand-bound species can react with carbon monoxide and form a long-lived CO-adduct absorbing at ~450 nm. Such spectral properties enable researchers to monitor formation and measure affinity of the P450-substrate/inhibitor complexes using conventional and stopped-flow spectrophotometers.

3.2.1.1 Equilibrium Titrations

One parameter reflecting ligand affinity, a spectral dissociation constant (K_s), can be determined from a plot of absorbance changes observed during equilibrium titrations of P450 with a substrate or inhibitor vs. ligand concentration. A hyperbolic fitting is usually satisfactory for weaker ligands, whereas quadratic nonlinear regression is used for strong binders. Two such examples, binding of bromoergocryptine (BEC) and ritonavir to CYP3A4 (K_s of 0.3 μM and 50 nM, respectively), are shown in Fig. 3.2.

Equilibrium titrations could also provide evidence for multiple ligand binding to CYP3A4. If several molecules enter the active site and affect the iron spin equilibrium, then the best fit to the absorbance change vs. [ligand] plot will be a two-site binding (or higher order) hyperbolic equation. The sigmoidal shape of a titration curve, in turn, would be indicative of two cooperative ligand interaction sites. Testosterone is one of the substrates shown to cooperatively bind to two sites in CYP3A4 with the Hill coefficient (n_H) of 1.3 [10, 11]. Positive cooperativity was also detected in the binding of aflatoxin B1 (n_H of 2.3) [12], α-naphthoflavone (ANF; n_H of 1.2–1.7) [11, 13], and Nile Red [NR] (n_H of 1.6) [14]. Acetaminophen and midazolam bind to CYP3A4 with a negative cooperativity [15, 16], whereas progesterone and 7-benzyloxyquinoline (BQ) display both positive and negative cooperativity [10, 17]. Structures of these substrates are shown in Fig. 3.3.

A variation of equilibrium titrations is mixed titrations, when two substrates are added simultaneously at fixed molar ratios. These types of experiments were conducted with testosterone, ANF and nanodisc-incorporated CYP3A4, where the spin shifts caused by the substrates added separately or in mixtures were analyzed [18]. Based on the properties of the two-dimensional spin shift surfaces, it was possible to separate specific heterotropic cooperative interactions from the additive affinities of the two substrates and conclude that the apparent positive heterotropic effect of ANF on testosterone binding is due to an additive spin shift caused by ANF rather than specific favorable ANF-testosterone interactions.

No cooperativity in testosterone and ANF binding was detected using global analysis of equilibrium substrate binding, steady-state NADPH consumption and product formation by nanodisc-incorporated CYP3A4 [19, 20]. Considering the individual testosterone and ANF dissociation constants and fractional contributions of the binding intermediates to the overall enzyme behavior, it was suggested that (1) up to three molecules of each substrate could simultaneously bind to CYP3A4 with little or no cooperativity; (2) the first binding event does not lead to a notable spin-state transition and product formation but accelerates NADPH consumption due to uncoupling; (3) spin shift, product formation and NADPH consumption rates reach maximum when the second substrate molecule binds; (4) association of the third substrate improves coupling efficiency but does not affect turnover rate; and (5) functional cooperativity between substrate molecules underlies cooperativity in testosterone and ANF metabolism observed during steady-state kinetics.

If ligands strongly influence heme absorption, normalized instead of simple absorbance difference spectroscopy can be used [13, 21]. In this case, the absorbance amplitude of the low- or high-spin forms of P450 is normalized before changes in other absorption bands are measured. A value related to the equilibrium constant between the high and low spin (K_{spin}) is then determined by subtracting the normalized absorbance of ligand-bound P450 from that of the ligand-free form.

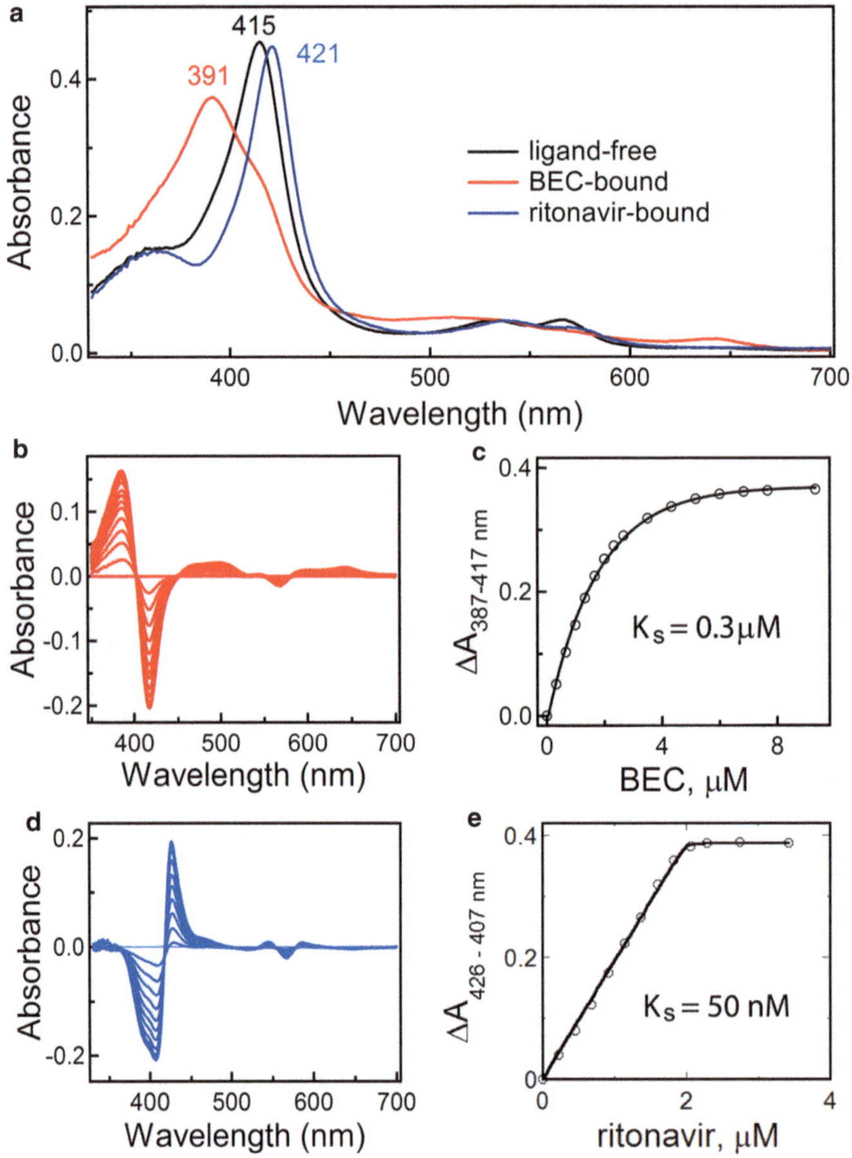

Fig. 3.2 Spectral changes induced by BEC and ritonavir in CYP3A4. (**a**) Upon binding, BEC causes a *blue* shift in the Soret band, whereas ritonavir induces a *red* shift (type I and type II spectral changes, respectively). (**b, c**) Difference absorbance spectra recorded during BEC binding and a plot of absorbance changes vs. ligand concentration, respectively. (**d, e**) Difference absorbance spectra recorded during ritonavir binding and a plot of absorbance changes vs. ligand concentration, respectively. Spectral dissociation constants (K_s) calculated from titration plots are indicated

3.2.1.2 Job's Titration and Titration by Dilution

Intensity of light during titration experiments may change with an increase in ligand concentrations (internal filter effect). Job's method of continuous variations and a titration by dilution approach allow one to overcome this problem. Job's method involves absorbance measurements in a series of solutions with a constant total molarity but different protein-ligand ratios. In the titration by dilution experiments, the light path of the sample

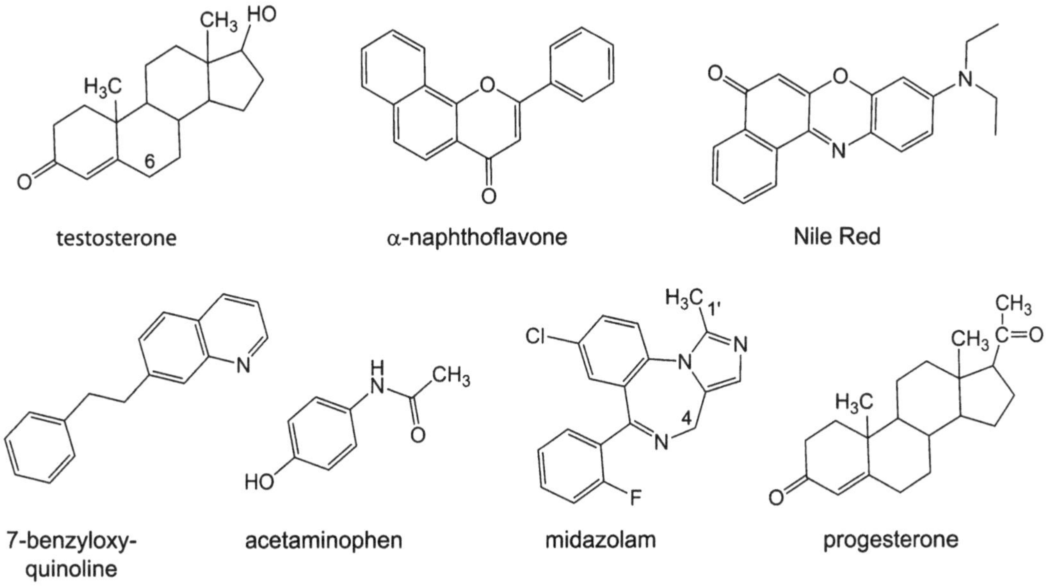

Fig. 3.3 CYP3A4 substrates that display binding cooperativity. Sites of metabolism in testosterone (C6) and midazolam (C4 and C1′) are indicated

increases simultaneously with dilution. Both methods were utilized for determination of binding affinity for various CYP3A4 substrates [22–24].

3.2.1.3 Quantitative Spectral Analysis

The principal component analysis technique, also known as bilinear factor analysis and singular value decomposition analysis, was applied for monitoring spin equilibrium in microsomal CYP3A4 at different temperatures and BEC concentrations [25]. Thermodynamic parameters for BEC binding (dissociation constant, ΔH, ΔS and ΔG) and spin transitions in CYP3A4 were evaluated using spectral standards for the high-spin, low-spin and cytochrome P420 (P420) states, leading to a conclusion that the substrate causes profound changes in the protein-heme interactions that favor dissociation of the fifth heme ligand and a low- to high-spin transition.

3.2.1.4 Ligand Binding Kinetics

Individual steps in the ligand-binding reaction can be resolved by stopped-flow spectrophotometry where multi- or single-wavelength kinetic data is collected after rapid mixing of protein and ligand solutions. The CYP3A4-ligand association is usually a complex process that proceeds in several steps [26–31]. The kinetic dissociation constant (K_d) for the CYP3A4-ligand complex can be estimated from a plot of the observed rate constant for the ligand binding reaction (k_{obs} or k_{on}) vs. ligand concentration. In most cases, the K_s and K_d values are close. However, for the CYP3A4-ritonavir complex, the K_d was found to be 17-fold higher than the K_s [28]. To better understand why there was such a big difference between the two related parameters, kinetics of ritonavir binding was reexamined using a wider range of ligand concentrations to include both supra- and sub-equimolar protein:ligand ratios. Under these conditions, the k_{obs} vs. [ritonavir] plot was V-shaped, with a minimum at a protein:ligand ratio of ~1.0 [30], which could arise from ritonavir docking to a peripheral site prior to moving into the active-site cavity.

3.2.1.5 Heme Reduction Kinetics

The P450 heme iron can accept electrons delivered by chemical compounds (e.g. sodium dithionite) or by NADPH via an associated redox partner, NADPH-cytochrome P450 oxidoreductase (CPR). When CO is present in the reaction mixture, heme reduction can be conveniently monitored by following heme-CO adduct formation. Using this approach, it was shown that the electron transfer rate from CPR to purified CYP3A4 is enhanced in the presence of testosterone, Mg^{2+} and cytochrome b_5 (b_5), but neither of these additives was required for optimal electron transfer in the presence of ethylmorphine (90 % low-spin state) [32]. Differences in the CYP3A4 reduction kinetics measured in human liver microsomes, baculovirus membranes (CPR:3A4 $\approx$ 8), *E. coli* membranes (CPR:CYP3A4 $\approx$ 1), a reconstituted system with phospholipids (CPR:3A4 $\approx$ 2) and a CYP3A4-CPR fusion protein suggested that one regulatory factor may be CYP3A4 clustering/aggregation, as it could create protein pools differing in the electron accepting ability due to spatial or conformational effects [33].

That two types of CYP3A4 oligomers, substrate-sensitive and substrate-insensitive, co-exist under equilibrium was demonstrated by measuring the dithionite-driven reduction of low- and high-spin forms of soluble, nanodisc- and liposome-incorporated CYP3A4 [34]. Another evidence for functional heterogeneity in CYP3A4 was obtained when the flavin domain of cytochrome P450 BM3 was utilized as a redox partner. Only partial heme reduction was observed in soluble CYP3A4 aggregates [35] but complete heme reduction occurred in nanodisc- or liposome-incorporated CYP3A4 monomers [36].

3.2.1.6 CO Rebinding Kinetics

Flash photolysis allows measuring CO recombination kinetics after the CO-Fe bond is disrupted by a laser flash. The CO rebinding kinetics reflects the rate of CO diffusion through the protein matrix and depends on the protein conformation, heme environment and a substrate binding mode. The higher the protein flexibility and the wider the ligand access channel, the higher the CO binding rate. With this methodical approach, it was shown that different CYP3A4 conformers exist [37], and that substrates can accelerate or reduce the CO binding rate by modulating the protein conformation and dynamics [38]. Flavonoids such as ANF, for instance, are thought to enhance the CYP3A4 activity by binding and activating a subpopulation that otherwise is metabolically inactive [39].

3.2.1.7 Heme Depletion Kinetics

Conformation-dependent changes in accessibility of the active site can be analyzed using a heme depletion assay. P450 heme absorbance starts decaying upon addition of an excess of H_2O_2 and can be followed spectroscopically over time. The heme depletion in CYP3A4 is multiphasic, which may result from conformational heterogeneity [40]. Since the rate of the Soret band bleaching depends on the presence of b_5, the assay can also be utilized for investigating CYP3A4-redox partner interactions.

3.2.1.8 High Pressure Spectroscopy

High-pressure spectroscopy can be applied for studying conformational heterogeneity and allosteric mechanisms in CYP3A4 as well [41, 42]. High hydrostatic pressure induces a low-spin shift and a P450-to-P420 conversion in a manner that depends on protein-protein and protein-ligand interactions. Both isolated and microsomal CYP3A4 display barotropic heterogeneity, with no interconversion between the distinct conformers [41]. However, there is no pressure-induced spin shift in microsomal CYP3A4, possibly due to a stabilizing effect of the membrane environment and/or CPR and b_5 that could limit the water access to the active site. In a separate study, a notable difference in the pressure-induced transitions was observed in the presence of allosteric (testosterone and 1-pyrenebutanol) and non-allosteric (BEC) substrates regardless of whether CYP3A4 was in solution or incorporated into a nanodisc. The high- to low-spin shift was complete in the BEC-bound form and partial in the testosterone- and 1-pyrenebutanol-bound P450 [42]. The allosteric substrates were suggested to induce conformational changes that decrease the water flux into the heme pocket and stabilize the high-spin state.

3.2.2 Isothermal Titration Calorimetry (ITC)

ITC titrations, allowing determination of binding stoichiometry based on heat changes, were carried out to estimate how many molecules of BEC and the inhibitor clotrimazol bind to CYP3A4 [26, 27]. Both binding reactions were found to be exothermic, with saturation at an equimolar ligand:CYP3A4 ratio. During ITC titrations, the ligand is usually added to the protein solution. However, owing to limited solubility of BEC and clotrimazol, these compounds were placed into the ITC cell and titrated with CYP3A4.

3.2.3 Equilibrium Dialysis

Ligand binding to CYP3A4 can also be examined by equilibrium dialysis. To determine the stoichiometry of CYP3A4-BEC binding [26], the ligand solution was placed into two cells separated by a dialysis membrane, and CYP3A4 was added to one of the cells. After equilibration, the protein was precipitated and the ligand concentration in both cells was estimated fluorimetrically. The [BEC_{free}] vs. [BEC_{bound}] plot reached a plateau at equal concentrations of BEC and CYP3A4, implying a 1:1 binding stoichiometry.

3.2.4 Inhibitor-Induced Cooperativity

Substrate stoichiometry in CYP3A4 can be estimated based on steady-state kinetics measured in the presence of large inhibitors. Large molecules compete with a substrate which, in turn, affects the Hill coefficient. The n_H value for the BQ debenzylation reaction increased from 1.74 to 2.1–3.7 when bulky troleandomycin, erythromycin, ketoconazole, cyclosporine A or BEC, but not smaller midazolam and testosterone, were present in the reaction mixture, meaning that up to four molecules of BQ can be simultaneously bound to CYP3A4 [43]. The obtained n_H values are still thought to reflect the lower limits of the number of substrate molecules that can enter the CYP3A4 active site.

3.2.5 Fluorescence Spectroscopy

Fluorescent properties of tryptophan residues, substrates, inhibitors and other fluorescent compounds were exploited to investigate protein-ligand interactions, multiple ligand binding sites, and cooperativity in CYP3A4 [14, 26, 27, 44, 45].

3.2.5.1 Fluorescent Substrates

α-Naphthoflavone (ANF) (Fig. 3.3) fluoresces with an excitation wavelength (λ_{ex}) of 320 nm and emission wavelength (λ_{em}) of 440 nm. The fluorescence yield decreases upon ANF ligation to CYP3A4 and, therefore, it was possible to fluorimetrically determine the binding affinity and contribution of ANF to CYP3A4 conformational heterogeneity and substrate cooperativity [23, 26, 46].

The fluorescent properties of bromoergocryptine (BEC) ($\lambda_{ex} = 320$ nm; $\lambda_{em} = 440$ nm; Fig. 3.4) were utilized to resolve a multi-step

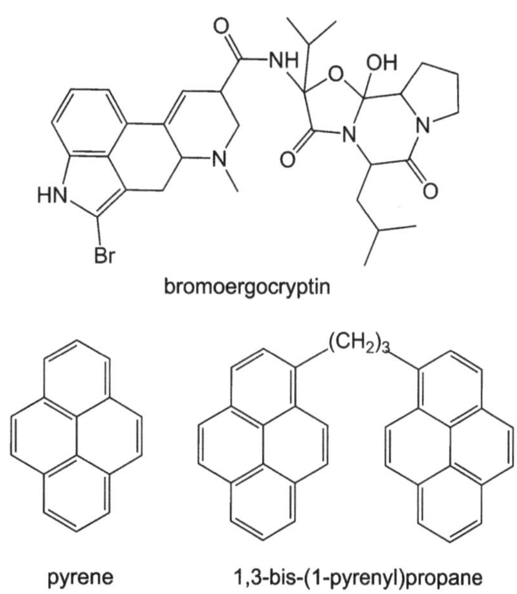

Fig. 3.4 Fluorescent substrates of CYP3A4

ligand binding process. Comparison of BEC-dependent fluorescence and absorbance changes observed during interaction with CYP3A4 enabled the detection of an 'absorbance-silent' step, attributed to substrate association to a peripheral site prior to translocation into the active-site cavity [26].

The fluorescent properties of NR (λ_{ex} = 550 nm; λ_{em} = 620 nm; Fig. 3.3) helped identify two-site binding to CYP3A4 [45] and probe allostery and sequential metabolism [14, 47]. Similar to testosterone and ANF [13, 21], NR has a minor effect on the heme spin-state equilibrium when it binds to a high-affinity site (K_d of 0.3 μM), whereas its association to a low-affinity site (K_d of 2.2 μM) leads to the majority of the ligand-induced spin shift [45].

Based on differences in the fluorescence emission of monomers and dimers of pyrene (λ_{ex} = 295 nm; $\lambda_{em}^{monomer}$ = 370–390 nm; λ_{em}^{dimer} = 480 nm; Fig. 3.4), it was shown that two pyrene molecules bind simultaneously in the CYP3A4 active site and that a π-π stacked pyrene-pyrene complex rather than a single monomer undergoes oxidation [44].

3.2.5.2 Fluorescent Products

7-Hydroxy-4-trifluoromethylcoumarin (λ_{ex} = 410 nm; λ_{em} = 538 nm) is formed upon CYP3A4-dependent debenzylation of 7-benzyloxy-4-trifluoromethylcoumarin (BFC). BFC is widely used for rapid fluorimetric measurements of CYP3A4 activity and detection of drug-drug interactions in conventional and high-throughput screening assays.

7-Hydroxyquinoline (λ_{ex} = 410 nm; λ_{em} = 538 nm) is the product formed upon BQ debenzylation. In the inhibitory assays, BQ was shown to be a less sensitive fluorimetric probe than BFC [48].

3.2.5.3 Fluorescent Inhibitors

Several fluorescent inhibitors specific for CYP3A4 have been synthesized by attaching a dansyl, deazaflavin or pyrene group to the C6 atom of testosterone [49]. Fluorescence of steroid derivatives is quenched upon reaction with the heme but can be restored when the active-site-bound fluorophore is displaced by another compound, which makes possible fluorimetric determination of relative affinities of various compounds and monitoring of drug-drug interactions.

3.2.5.4 Fluorescent Probes

2-*p*-Toluidinylnaphthalene-6-sulfonic acid (TNS) (λ_{ex} = 320 nm; λ_{em} = 440 nm) is non-fluorescent in aqueous solutions but emits light in a hydrophobic environment, such as the protein interior. Upon binding to CYP3A4, TNS induces type II spectral changes and fluoresces with a high quantum yield [50]. Based on changes in steady-state and time-resolved TNS fluorescence, it was suggested that a remote, high-affinity binding site for TNS exists, occupation of which could affect the active-site environment.

6-(Bromoacetyl)-2-(dimethylamino)naphthalene (BADAN; λ_{ex} = 387 nm, λ_{em} = 520 nm), 7-(diethylamino)-3-(4′-maleimidylphenyl)-4-methylcoumarin (CPM; λ_{ex} = 405 nm, λ_{em} = 530 nm) and monobromobimane (mBBr; λ_{ex} = 395 nm, λ_{em} = 490 nm) are environment-sensitive thiol-reactive fluorescent probes used for studying the interaction between cysteine-depleted CYP3A4 (only Cys58 and/or Cys64 left) and several substrates that do or do not display binding cooperativity [51]. Analysis of the substrate- and probe-dependent fluorescence changes and H_2O_2-induced heme destruction kinetics provided evidence for a distinct low-affinity ANF binding site, association to which is not accompanied by a heme spin shift. BADAN-labeled CYP3A4 was also used for investigating the effects of reduced glutathione (GSH) on BFC and BQ oxidation [52]. GSH eliminates homotropic cooperativity of BFC and BQ but amplifies the activating effect of ANF on BFC oxidation, and is suggested to have two binding modes, one of which is direct coordination to the heme iron via the SH-group.

3.2.5.5 Single Molecule Fluorescence Spectroscopy

Single molecule fluorescence studies were conducted on nanodisc-incorporated CYP3A4 where the evanescent NR excitation, generated using internal reflection fluorescent microscopy, was followed by measurements of residence times in the low-occupancy NR-bound complexes [47]. The observed biphasic dwell-time distribution is thought to reflect two phases of the NR dissociation reaction: a fast off-rate from the nanodisc lipid bilayer and a slow off-rate from the protein (30 and 1.5 s^{-1}, respectively). Based on a fivefold ANF-induced decrease in the slow phase of NR dissociation, it was concluded that the CYP3A4 effectors could modulate the substrate off-rates by altering structure/dynamics of monomeric CYP3A4.

3.2.6 Fluorescence Resonance Energy Transfer (FRET)

FRET is a mechanism through which energy is transferred between two closely positioned fluorophores. If the distance is short enough, an excited donor can transfer energy to an acceptor through nonradiative dipole-dipole coupling. FRET efficiency, therefore, can serve as a measure of a distance between two fluorophores. FRET between BEC, 1-pyrenemethylamine ($\lambda_{ex} = 340$ nm; $\lambda_{em} = 380$ nm) or PB ($\lambda_{ex} = 331$ nm; $\lambda_{em} = 380$ nm) and the CYP3A4 heme was utilized in combination with absorbance spectroscopy to monitor individual ligand-binding events [22–24]. Unlike BEC and 1-pyrenemethylamine that bind to a single site in CYP3A4, association of 1-pyrenebutanol was consistent with a two-step sequential model, where the spin transition takes place upon 1-pyrenebutanol binding to a low-affinity site [22]. In the F213W, F304W and L211F/D214E mutants, however, cooperativity in 1-pyrenebutanol binding was altered and a partial spin shift was observed in the binary rather than ternary enzyme-substrate complex [24]. In another study, substrate-dependent conformational transitions in CYP3A4 were analyzed by measuring FRET from the BADAN and CPM labels to the heme, and from tryptophan residues to BADAN [51]. Finally, a FRET-based assay employing *N*-(4,4-difluoro-5,7-dimethyl-4-bora-3α,4α-diaza-*s*-indacene-3-yl)methyliodacetamide (BODIPY-FL iodoacetamide) was utilized for testing how the surface density of liposome-bound CYP3A4 affects spin equilibrium and heme reduction kinetics [36].

3.2.7 Luminescence Resonance Energy Transfer (LRET)

LRET uses the long-lived triplet state of a phosphorescent probe as an energy donor. Its advantage over FRET is the longer distance at which the energy can be transferred and the longer donor life-time, which makes LRET measurements more accurate and orientation independent. Using an LRET-based method, the relation between the concentration of membrane-bound CYP3A4 and its oligomeric state was examined [46]. Cysteine-depleted CYP3A4 was labeled with either erythrosine 5′-iodacetamide (ERIA) or DY-731 maleimide (DYM), serving as a LRET donor and acceptor, respectively. Addition of CYP3A4-DYM to CYP3A4-ERIA-containing liposomes led to a drastic decrease in the donor emission and delayed fluorescence of the acceptor. The sigmoidal dependence of the LREP amplitude on the surface density of liposomal CYP3A4 became hyperbolic in the presence of ANF. This finding and a correlation between the oligomerization state of CYP3A4 and its susceptibility to activation by ANF support the hypothesis that protein-protein interactions can play a role in the allosteric mechanism.

3.2.8 Photoaffinity Labeling

Protein labeling with photoaffinity probes can provide information on the substrate recognition sites without carrying out the enzymatic reaction. The method utilizes a labeling reagent that upon photolysis converts to an extremely reactive intermediate and covalently binds to the active-site residues. Chromene-like molecules act as photoaffinity ligands for CYP3A4 [53, 54]. Upon UV light irradiation, these compounds rearrange into conjugated tricyclic structures with different life-times. Plant benzochromene, lapachenole, serves as a substrate and a

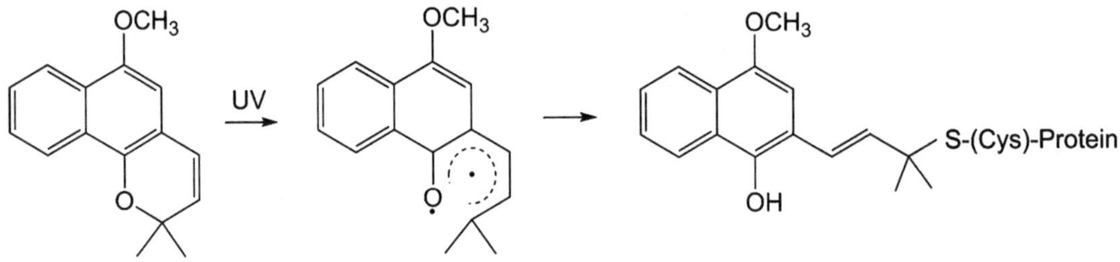

Fig. 3.5 CYP3A4 labeling by photoactivated lapachenole [54]

competitive inhibitor of CYP3A4 but when photoactivated, it becomes an irreversible inactivator and fluorescently modifies thiol groups of cysteine residues (Fig. 3.5) [53]. Analysis of the lapachenole-linked protein identified Cys98 and Cys468 as the primary modification sites [54]. The functional role of the surface Cys468 remains unclear, whereas Cys98 is part of the substrate access channel and, hence, could affect substrate recognition and catalysis.

3.2.9 Circular Dichroism (CD) Spectroscopy

CD spectroscopy allows rapid determination of the secondary structure and folding properties of proteins and, in conjunction with other methods, was used for investigation of CYP3A4 inhibition by Cu^{2+} and Zn^{2+} ions [55, 56]. One of the inhibitory effects of Cu^{2+} on testosterone hydroxylation was induction of a conformational change in CYP3A4 (7 % decrease in the α-helix content) [55]. Zn^{2+}, on the other hand, had not only a more pronounced effect on CYP3A4 secondary structure (11 % decrease in α-helix content) but also prevented the stimulatory action of b_5 on testosterone metabolism [56]. Thus, the cytosol metal ion balance could be one of the factors regulating CYP3A4 activity in vivo.

3.2.10 Linear Dichroism (LD) Spectroscopy

LD is observed when plane-polarized light is absorbed by samples that are oriented intrinsically or by external forces. With this spectroscopic technique, information on the orientation of a chromophore or structures within molecules can be obtained. LD measurements were performed to characterize the heme tilt angle in nanodisc-incorporated CYP3A4 [57]. A low deviation between experimental values (average of 59.7 ± 4.1°) indicated that CYP3A4 is specifically orientated relative to the lipid bilayer.

3.2.11 Surface Plasmon Resonance (SPR)

SPR on metallic surfaces is a powerful optic sensing method for monitoring label-free bimolecular interactions. SPR analysis was utilized to investigate CYP3A4 binding to antifungal azoles, itraconazole and ketoconazole [58]. Based on the binding kinetics, absorbance spectroscopy and catalytic studies, two orientation modes for both drugs were identified: a catalytically productive mode and a slowly dissociating inhibitory mode. In combination with other methods, the SPR technique was used to determine the off-rates for structurally related quinolone carboxamide compounds acting as type I and type II ligands of CYP3A4, and helped to clarify the kinetic mechanism for their metabolism [59].

A related method, SPR in nanometer-sized structures or localized SPR (LSPR) that also depends on the refractive index of the surrounding media, was developed for detection of CYP3A4-drug interactions [60]. Nanodisc-bound CYP3A4 was covalently immobilized on the surfaces of silver nanoparticles and the drug binding was monitored by measuring the resonant coupling between the nanoparticles and the CYP3A4 heme. Ligand-induced changes in position and amplitude of the LSPR spectrum maxima correlated well with the spectral changes observed in solution.

3.2.12 Nuclear Magnetic Resonance (NMR)

NMR T_1 paramagnetic relaxation studies provided the first physicochemical evidence for the allosteric substrate binding in the CYP3A4 active site [15, 61, 62]. T_1 relaxation experiments are suited for analyzing CYP3A4 allosterism because they allow determination of distances from a paramagnetic center (the heme iron) to the protons of multiple substrates and, hence, can detect changes in their relative orientation. One NMR study investigated heterotropic cooperativity between midazolam and two effectors, testosterone and ANF [61]. Midazolam is hydroxylated at C1′ and C4′ positions, and the ratio between metabolite formation rates depends on midazolam concentration and the presence of testosterone or ANF. Owing to negative homotropic cooperativity, the 1′-hydroxyproduct is preferably formed at low midazolam concentrations, whereas the 4′-hydroxylation rate increases at higher substrate concentrations [16]. The midazolam protons-heme distances measured in the absence and presence of the effectors suggest that midazolam can rotate within the active site or slide parallel to the heme plane [61]. The NMR data also indicate that ANF and testosterone exert their allosteric effects through direct binding in the vicinity of the heme and by reorienting midazolam, to bring the C1′ or C4′ atoms closer to the heme, which explains the kinetics of activation and preferable formation of 1′-hydroxy- and 4′-hydroxymidazolam in the presence of ANF and testosterone, respectively.

In combination with the molecular docking technique, T_1 longitudinal NMR relaxation was applied to probe the cooperativity of midazolam metabolism with carbamazepine serving as a heterotropic effector [62]. Similar to testosterone, carbamazepine inhibits formation of 1′-hydroxymidazolam and, as this study revealed, assumes a stacked configuration with midazolam and brings its C4′ atom closer to the heme. Stacking of two midazolam molecules gives the same result, whereas a single midazolam docks with the C1′ atom closest to the heme. Since many CYP3A4 substrates have a planar aromatic structure, ligand cooperativity through direct stacking interactions was proposed to be one of the possible allosteric mechanisms.

Using NMR T_1 paramagnetic relaxation, positioning of two other substrates, acetaminophen and caffeine, in the CYP3A4 active site was investigated [15]. CYP3A4-dependent acetaminophen oxidation exhibits negative homotropic cooperativity ($n_H = 0.7$) but follows Michaelis-Menten kinetics in the presence of caffeine. The calculated distances were consistent with acetaminophen coordination to the heme through the amide group. Upon entering the active site, caffeine is thought to disrupt the weak Fe-N coordination, thereby promoting acetaminophen oxidation. However, it remains unclear whether caffeine, more remote from the heme than acetaminophen, directly interacts with acetaminophen and precludes its coordination or affects the active-site conformation leading to acetaminophen reorientation.

A magic-angle spinning solid-state NMR (MAS SSNMR) spectroscopic study on ^{13}C, ^{15}N-enriched nanodisc-incorporated CYP3A4 demonstrated its structural integrity and proper folding [63]. Analysis of the BEC binding reaction showed that CYP3A4 remains fully active after precipitation with polyethylene glycol, required for SSNMR measurements. Despite good quality, 2D MAS SSNMR spectra were not sufficient for determining site-specific assignments.

3.2.13 Electron Paramagnetic Resonance (EPR) Spectroscopy

Testosterone-dependent spin state equilibrium in CYP3A4 was compared by EPR and optical spectroscopic titrations [21]. To quantify and characterize the single and double occupancy testosterone binding sites, the protein concentration in two sets of experiments either exceeded or was below the K_d for testosterone. Using this combined approach, it was possible to construct a free energy landscape for multiple ligand binding, which suggested that the first testosterone binds with a higher affinity to a 'non-productive' site, whereas the second testosterone displaces the bound water and drives the heme to the high-spin state more efficiently than the first.

EPR and UV-vis spectroscopy were also used to examine heterotropic cooperativity and individual binding events for ANF and testosterone [13]. Two types of binding sites for both substrates were identified: high-affinity spin-state insensitive (peripheral) and lower-affinity spin-state sensitive (proximal, near the heme). Based on the thermodynamic analysis of the testosterone- and ANF-induced spin shifts, testosterone was proposed to bind to CYP3A4 sequentially and occupy the proximal site after the peripheral site is saturated. It was speculated also that testosterone affinity for the proximal site increases upon a conformational change caused by ANF association to the peripheral site.

A combination of conventional continuous-wave and pulsed EPR (hyperfine sublevel correlation spectroscopy (HYSCORE)) techniques helped to investigate and compare the binding mechanism of 1,2,3-triazole and 17α-(2H-2,3,4-triazolyl)estradiol [64]. Although both compounds are type II ligands of CYP3A4, they induce different perturbations in the EPR *g* values and, as the HYSCORE analysis showed, the 1,2,3-triazole moiety does not displace the axial water molecule when incorporated into the 17α-estradiol scaffold. Instead, 17α-(2H-2,3,4-triazolyl)estradiol hydrogen-bonds to the coordinated water ligand and, by altering its field strength, causes the spin-state change. This study raises concerns on interpreting the P450-ligand structure based on optical spectra, as some ligands that produce type II spectral changes are not always directly ligated to the heme.

3.2.14 Resonance Raman (RR) Spectroscopy

RR spectroscopy was applied to identify substrate- and redox-dependent structural changes in nanodisc-incorporated CYP3A4 [65]. Analysis of the high and low frequency RR spectra of ligand-free and BEC-, testosterone- or erythromycin-bound CYP3A4 led to a conclusion that the size and number of substrate molecules bound have no significant effect on the ferric heme structure but greatly influence the conformation of gas ligands (CO and O_2) and the ferrous heme structure. In particular, BEC and testosterone induce changes in the low frequency RR spectrum associated with the heme peripheral group dispositions or out-of-plane macrocycle distortion, which could have an impact on the reactivity of intermediates and CYP3A4 function.

3.2.15 Isotope Fractionation (Kinetic Isotope Effect)

Deuterium-containing substrate analogs were used for examining the mechanism of CYP3A4-dependent ezlopitant dehydrogenation and testosterone hydroxylation [66, 67]. Ezlopitant is metabolized to a benzyl alcohol and a benzyl alkene, where the latter is formed directly rather than via sequential dehydration of the benzyl alcohol. When deuterium was incorporated into the benzylic position, a low isotope effect (ratio between the reaction rate constant of the light and heavy isotope) was observed for both products, indicating that benzylic hydrogen abstraction is obligatory in the formation of both metabolites and there is no metabolic switching (change in regional specificity of ezlopitant metabolism). However, placement of deuterium atoms at adjacent positions led to a small inverse isotope effect on benzyl alcohol but not alkene formation. This suggested that formation of the alkene requires benzylic hydrogen abstraction and that the benzylic radical partitions between the formation of the two metabolites.

Analysis of metabolites of ^{2}H- and ^{3}H-labeled forms of testosterone showed that the testosterone 6β-hydroxylation reaction is stereoselective, as CYP3A4 abstracts hydrogen and rebounds oxygen only at the β-face [67]. A high intrinsic isotope effect (^{D}k of 15) for the labeled testosterone 6β-hydroxylation was consistent with the initial hydrogen atom abstraction. In non-competitive reactions, some metabolic switching occurred and the ^{D}k value was attenuated (<3). Because considerable attenuation in ^{D}k was also observed for BQ *O*-debenzylation, the C-H bond breaking reaction is unlikely to be rate-limiting.

3.2.16 Isotope Dilution Analysis

Sequential NR metabolism and heterotropic allosteric activation by ANF were investigated by isotope dilution analysis to quantitatively measure the relative flux of a reactive cycle intermediate [68]. NR is metabolized by CYP3A4 to monodesethyl-NR (M1), which is sequentially oxidized to didesethyl-NR (M2) (Fig. 3.6). Comparison of metabolites produced by effector-free and ANF-bound CYP3A4 upon incubation with a mixture of deuterated NR and unlabeled M1 suggested that ANF increases the velocity of M1 and M2 production, and modulates the branching k_{cat}/k_{off} ratio in favor of M2 by affecting the substrate off-rate.

3.2.17 Electrochemistry

Replacement of the natural electron delivery system, a CPR/NADPH redox pair, by an electromotive force is one of the approaches for developing biosensors for rapid monitoring of CYP3A4-mediated drug metabolism. Purified CYP3A4 was immobilized on gold electrodes coated with 3-mercapto-1-propenesulfonic acid [69], carbon nanofibers [70] and glassy carbon electrodes modified with poly(diallyldimethylammonium chloride [71] or Nafio-cobalt (III) sepulchrate [72], whereas CYP3A4-containing didodecyldimethylammonium bromide vesicles were attached to a platinum disc electrode [73]. Thiolate-coated gold electrodes, in turn, were used for immobilization of hepatic microsomes [74]. Using these systems, it was possible to monitor CYP3A4-drug interactions, drug metabolism and inhibition reactions.

Cyclic voltammograms of immobilized CYP3A4 have two peaks attributed to the Fe^{+3}/Fe^{2+} redox couple. When a substrate and oxygen are present, heme reduction is coupled to substrate oxidation, manifested as an increase in the cathodic peak current. The surface concentration of an electro-active enzyme (n) is estimated based on the Faraday's Law: $Q = nF$, where F is the Faraday's constant and Q is the total charge transferred upon reduction of CYP3A4. Q is calculated from the integration of the reduction peak recorded under anaerobic conditions, whereas the enzyme turnover is estimated based on n and the substrate-dependent catalytic current measured by chronoamperometry.

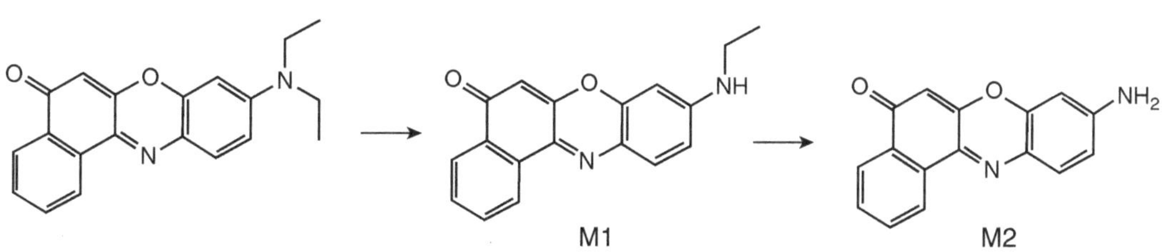

Fig. 3.6 Sequential demethylation of Nile Red catalyzed by CYP3A4 [68]

Electrode-immobilized CYP3A4 metabolizes drugs similar to the microsomal protein, with comparable product formation rates and small contribution of H_2O_2 to the catalytic cycle [69]. The electrochemically-driven CYP3A4 reactions are sensitive to the substrate concentration and can be inhibited by ketoconazole, cimetidine and diclofenac [69, 71], which enables estimation of the k_{cat} and K_m values for the substrate turnover and IC_{50} for the inactivators (a concentration of an inhibitor that reduces substrate metabolism by 50 %). Some biosensors were reported to have a rapid response time and ability to detect very low concentrations of drugs and pollutants and, hence, could be utilized for electrochemical detection in biological samples [72, 73].

3.2.18 Chemical Auxiliary Approach

A chemical auxiliary was utilized to control the selectivity of CYP3A4 reactions [75]. By linking substrates to inexpensive, achiral, cell-permeable theobromine, it was possible to achieve predictable stereo- and chemoselective hydroxylation and epoxidation at the fourth carbon from the auxiliary (Fig. 3.7). The method is limited to substrates that are not larger than theobromine, but the advantage is that it does not yield overoxidation products and is tolerant to various functional groups.

3.2.19 X-ray Crystallography

X-ray crystallography provides direct insights into the protein structure and protein-ligand interactions. Recombinant mammalian P450s can be crystallized upon deletion of the membrane-binding fragment and, in some cases, modification of the N-terminus [76]. Multiple CYP3A4 crystal structures have been solved recently but mostly with type II inhibitors bound [28–31, 77–80]. Obtaining co-crystals of CYP3A4 with substrates is challenging because they dissociate from the active site during crystallization. Thus far, only BEC- and erythromycin-bound structures are available, where only BEC is bound in a productive mode [29, 79].

Inhibitors co-crystallized with CYP3A4 include metyrapone, ketoconazole, ritonavir, desthiazolylmethyloxycarbonyl ritonavir and seven desoxyritonavir analogs (GS2-GS8) that widely vary in K_d (22 nM–4 μM) [28–31, 77, 79, 80]. Interestingly, the ketoconazole-, GS4- and GS5-bound structures contain two inhibitor molecules bound to the active site: ketoconazole1 and ketoconazole2 associate in an antiparallel tandem fashion, GS5-1 and GS5-1 in an intertwined parallel mode, and GS4-1 and GS4-2 in a perpendicular mode (Fig. 3.8). Thus, X-ray data proves that multiple molecules differing in size and chemical nature can simultaneously bind to CYP3A4. Comparison of the

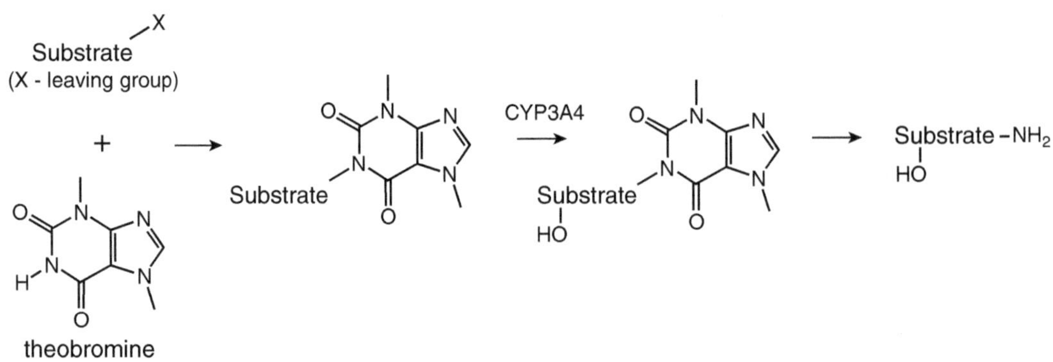

Fig. 3.7 Theobromine auxiliary for controllable oxidations by CYP3A4 [75]

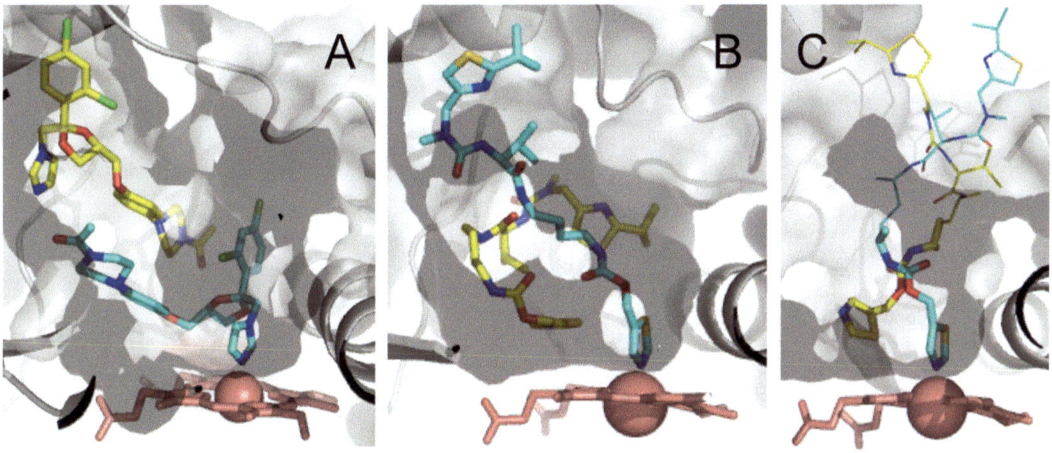

Fig. 3.8 Crystal structures of CYP3A4 with two inhibitors bound in the active site. (**a**) Ketoconazole molecules bind in a parallel tandem fashion (2VOM structure) [79]. (**b**, **c**) Ritonavir analogs GS4 and GS5 associate in a perpendicular and intertwined parallel mode, respectively (PDB ID 4K9T and 4K9U) [80]. Disordered parts of GS5 that are not seen in the X-ray structure are shown as *thin lines*. Heme is in *pink* and the heme-bound ligands are in *cyan*

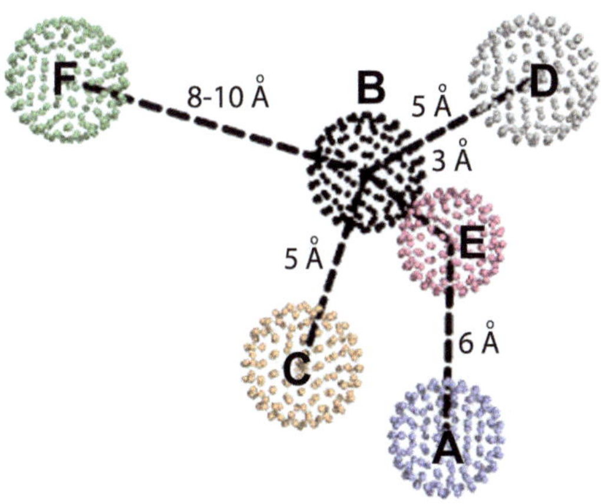

Fig. 3.9 Pharmacophore for a potent CYP3A4 inhibitor. Pharmacophoric features were derived based on studies with ritonavir analogs [28, 30, 31, 80] and include: (**a**) strong heme-ligating nitrogen donor; (**b**) flexible backbone; (**c**) aromatic group; (**d**) hydrophobic group; (**e**) hydrogen donor/acceptor, and (**f**) polyfunctional end-group

binding affinity, IC_{50} and orientation modes of ritonavir-like compounds (reviewed elsewhere [81]) helped to better understand the inhibitory mechanism and derive a pharmacophore for a CYP3A4-specific inactivator that could guide structure-based inhibitor design (Fig. 3.9).

Another interesting finding was a peripheral sterol binding site in the CYP3A4-progesterone structure [77]. Instead of associating to the active site, progesterone was found to dock 17 Å away from the heme in a surface hydrophobic pocket comprised by the F′-G′-loop residues. Since progesterone displays both negative and positive binding cooperativity [10, 17], it was hypothesized that the progesterone-binding pocket represents a peripheral site where hydrophobic substrates associate before moving into the active-site cavity.

When all available X-ray models of CYP3A4 are superimposed, it becomes evident that very little structural change is needed to accommodate bulky and structurally diverse compounds, even two at a time. Conformational changes take place primarily in the F-G- and C-terminal loop regions, the I-helix adjacent to the heme, and the 369–371 peptide (Fig. 3.10). Such local and minor rearrangements argue against the conformational-heterogeneity-driven allostery in CYP3A4 and rather support the multiple substrate binding mechanism without major conformational changes, as proposed for P450eryF [82].

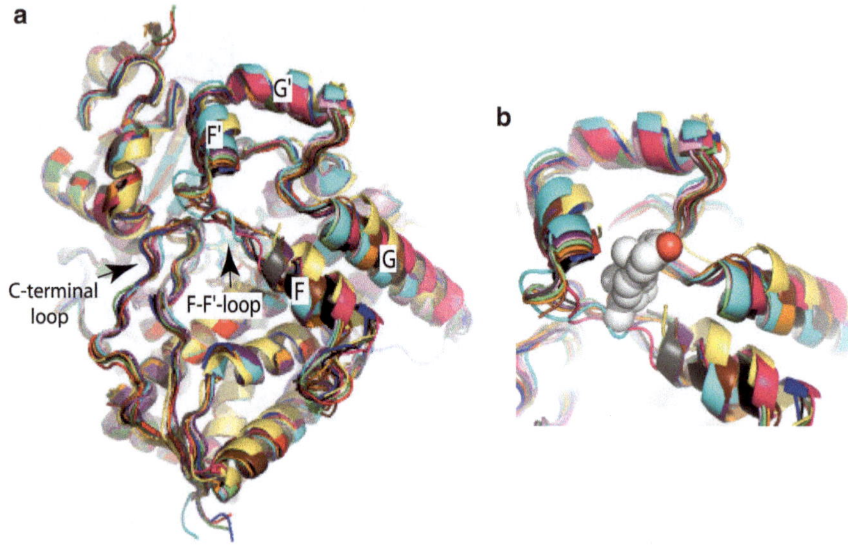

Fig. 3.10 Superposition of all available structures of CYP3A4. (**a**) A general view showing that ligand-induced rearrangement occurs mainly in helices F and G, and the F-F′- and C-terminal loops. The superimposed structures are: 1TQN, 1WOE, 1WOF, 1WOG, 2JOD, 2VOM, 3NXU, 3TJS, 3UA1, 4I3Q, 4I4G, 4I4H, 4K9T, 4K9U, 4K9V, 4K9W and 4K9X. To simplify viewing, all ligands were excluded from the active site. (**b**) A closer view at the F-F′-G′-G-helical region that serves as a binding site for progesterone (shown in cpk representation). The hydrophobic pocket is thought to represent a peripheral docking site involved in effector/substrate recognition and could play a role in modulating cooperativity [77]

3.3 Computational Approaches

Because X-ray crystallography has its limitations and co-crystallization of CYP3A4 with the compound(s) of interest is not always possible, theoretical studies are more frequently used these days to investigate the dynamics of ligand binding and catalytic mechanism, predict ligand association modes, and identify substrate and solvent channels. Moreover, computer modeling techniques are indispensible for prediction of CYP3A4-mediated drug metabolism and drug-drug interactions, some of which will be briefly described in this section.

3.3.1 Molecular Dynamics (MD) Simulations

One of the principal tools to theoretically study biological molecules is MD simulations that derive time-dependent dynamic behavior and give a view of the atomic motions. In combination with other approaches, MD simulations were conducted on CYP3A4 in a number of studies, which (1) suggested that the F-F′-loop (residues 211–218) defines the promiscuity and broad substrate selectivity of CYP3A4 [83], and Ser119, Phe205, Arg212, Phe213 and Phe304 are key residues that help orient substrates in the active site [47, 83–86]; (2) identified preferred substrate access/egress and solvent channels [87–90]; (3) clarified the mechanism of cooperative binding of diazepam and ketoconazole [84, 91] and the role of CPR in activating water channels [92]; and (4) characterized the membrane-bound state of CYP3A4 [57, 93, 94].

In steered MD (SMD), a harmonic restraining potential is applied to a protein or ligand in order to manipulate the ligand by pulling it along desired degrees of freedom. SMD with adaptive direction adjustments is an improved method that could accelerate MD simulations, find optimal pathways for ligand dissociation, and bypass a barrier along the failed direction. This method was used to investigate dissociation of the CYP3A4-bound inhibitor metyrapone and led to the discovery of a pathway with a lower energy barrier, shorter dissociation time and shorter motion trajectory compared to those predicted by conventional SMD [95, 96].

3.3.2 Quantum Mechanics/Molecular Mechanics (QM/MM) and Density Functional Theory (DFT) Calculations

A hybrid QM/MM approach is a molecular simulation method for studying chemical processes in solution and in proteins, whereas DFT is a quantum mechanical modeling method for electronic structure calculation by using functionals of the spatially dependent electron density. Theoretical calculations were applied to CYP3A4 to estimate the activation energy of the intermediate formation and predict the binding modes for indapamide and 4-aminopiperidine [85, 86], investigate interaction between nevirapine, carbamazepine and endogenous steroids [97], explore features of the catalytic oxyferryl species (compound I) [98], model the regioselectivity preference of testosterone hydroxylation [99], and test the reactivity of various sites on flunitrazepam and progesterone [100].

Reaction dynamic calculations for the testosterone 6β-hydrogen/deuterium abstraction were performed at the level of canonical variational transition state theory with a multiconfigurational MM technique, allowing the construction of a semiglobal full-dimensional potential energy surface, to gain deeper insights into the quantum tunneling in testosterone hydroxylation by CYP3A4 [101]. In agreement with the experimental results [67], the calculated multidimensional tunneling coefficients indicated substantial contributions by quantum tunneling which, however, only modestly contributed to the kinetic isotope effects. The use of a gas-phase model without considering the protein-solvent interactions was suggested to be one of the reasons for the discrepancy between the theoretical and experimental results.

Gas phase DFT calculations were also conducted to examine the binding of unsubstituted imidazole, 1,2,4- and 1,2,3-triazole to a model Fe^{3+} heme in an attempt to understand underrepresentation of the latter functional group among P450 inhibitors [64]. It was found that 1,2,3-triazole interacts with the heme weaker than other azoles, ligates to heme iron with either N1 or N2 atoms, and forms a longer Fe^{3+}-N bond due to lower basicity.

The binding free energies of 16 structurally diverse CYP3A4 inhibitors to the iron porphyrin model, calculated using DFT and the implicit solvation methods in water, were shown to be a good descriptor in interpreting the CYP3A4-inhibitor interaction [102]. The relative free energies in the gas phase were mainly responsible for the total binding free energies in water, although desolvation could affect the inhibitor affinity.

3.3.3 In Silico Drug Metabolism Prediction

Computational techniques are indispensable and increasingly used in early drug design along with in vitro methods to predict substrate affinity, liability and metabolic pathways. Before the X-ray structure of CYP3A4 became available, homology modeling, quantitative structure-activity relationship (QSAR) modeling and building a 3D-pharmacophore were utilized for drug metabolism prediction [103]. Pharmacophore modeling requires a large and rigid template to identify structural determinants (conformation, shape and electronic properties) of substrates, inhibitors or metabolites critical for catalytic specificity, which could indirectly provide information on the protein active site. QSAR modeling, in turn, is applied to large datasets of molecules and molecular descriptors to derive 3D-features of substrates and inhibitors that can interact with a specific enzyme. Both approaches generated useful information about CYP3A4. In particular, they helped develop pharmacophores for substrates and inhibitors [104–109] and suggested that hydrophobicity and hydrogen bonding are dominant factors guiding ligand binding [104, 107, 110]. Semi-empirical AM1 molecular orbital calculations of the energy of hydrogen radical abstraction is another modeling approach that can predict likely sites of CYP3A4-mediated metabolism by relying solely on the electronics and intermolecular sterics of drug-like molecules [111].

QSAR modeling is still used in the pharmaceutical industry to analyze very large high-throughput screening sets and to predict CYP3A4 inhibition and substrate potential. Owing to recent structural and computer modeling advances, QSAR methodology was utilized in combination with the multiple pharmacophore hypothesis approach [112], advanced docking techniques (MetSite, GLUE, AutoDock and other) [113–116], GALAS (Global Adjusted Locally According to Similarity) method [117, 118], structure-based comparative molecular field analysis [119], and NMR spectroscopy data [120]. Gaussian kernel weighted *k*-nearest neighbor models were also used for in silico prediction of CYP3A4 inhibitors [121]. Accuracy, sensitivity and specificity of some of the modeling methods were recently compared [118, 122, 123]. The modern and most powerful approach for predicting drug metabolism and drug-drug interactions is a combination of structure-based docking, MD and quantum chemical calculations, which is reviewed elsewhere [124].

3.4 Conclusions

Owing to a central role of CYP3A4 in drug metabolism, it is crucial to fully understand the mechanism of CYP3A4-ligand interactions. In this review, we summarized methodical approaches currently used in the CYP3A4 research, highlighting the problems addressed and conclusions made in relevant studies. Although utilization of a wide array of biochemical, biophysical, structural and computational techniques led to breakthroughs in our understanding of how CYP3A4 functions, some of the processes, such as substrate cooperativity and the mechanism of drug-drug interactions, still need to be clarified. Future development and employment of new methodologies to the CYP3A4 research could lead to new discoveries that may help resolve these issues.

Acknowledgments Financial support from the National Institute of General Medical Sciences (Grant GM57353) and the California Center for Antiviral Drug Discovery is gratefully appreciated.

References

1. Rendic S, Di Carlo FJ (1997) Human cytochrome P450 enzymes: a status report summarizing their reactions, substrates, inducers, and inhibitors. Drug Metab Rev 29:413–580
2. Atkins WM, Wang RW, Lu AY (2001) Allosteric behavior in cytochrome P450-dependent in vitro drug-drug interactions: a prospective based on conformational dynamics. Chem Res Toxicol 14:338–347
3. Hutzler JM, Tracy TS (2002) Atypical kinetic profiles in drug metabolism reactions. Drug Metab Dispos 30:355–362
4. Atkins WM (2005) Non-Michaelis-Menten kinetics in cytochrome P450-catalyzed reactions. Annu Rev Pharmacol Toxicol 45:291–310
5. Sligar SG, Denisov IG (2007) Understanding cooperativity in human P450 mediated drug-drug interactions. Drug Metab Rev 39:567–579
6. Davydov DR, Halpert JR (2008) Allosteric P450 mechanisms: multiple binding sites, multiple conformers or both? Expert Opin Drug Metab Toxicol 4:1523–1535
7. Denisov IG, Sligar SG (2012) A novel type of allosteric regulation: functional cooperativity in monomeric proteins. Arch Biochem Biophys 519:91–102
8. Jefcoate CR (1978) Measurement of substrate and inhibitor binding to microsomal cytochrome P-450 by optical-difference spectroscopy. Methods Enzymol 52:258–279
9. Schenkman JB, Sligar SG, Cinti DL (1981) Substrate interaction with cytochrome P-450. Pharmacol Ther 12:43–71
10. Harlow GR, Halpert JR (1998) Analysis of human cytochrome P450 3A4 cooperativity: construction and characterization of a site-directed mutant that displays hyperbolic steroid hydroxylation kinetics. Proc Natl Acad Sci U S A 95:6636–6641
11. Hosea NA, Miller GP, Guengerich FP (2000) Elucidation of distinct ligand binding sites for cytochrome P450 3A4. Biochemistry 39:5929–5939
12. Ueng YF, Kuwabara T, Chun YJ, Guengerich FP (1997) Cooperativity in oxidations catalyzed by cytochrome P450 3A4. Biochemistry 36:370–381

13. Roberts AG, Atkins WM (2007) Energetics of heterotropic cooperativity between α-naphthoflavone and testosterone binding to CYP3A4. Arch Biochem Biophys 463:89–101
14. Lampe JN, Fernandez C, Nath A, Atkins WM (2008) Nile Red is a fluorescent allosteric substrate of cytochrome P450 3A4. Biochemistry 47:509–516
15. Cameron MD, Wen B, Roberts AG, Atkins WM, Campbell AP, Nelson SD (2007) Cooperative binding of acetaminophen and caffeine within the P450 3A4 active site. Chem Res Toxicol 20:1434–1441
16. Maekawa K, Yoshimura T, Saito Y, Fujimura Y, Aohara F, Emoto C, Iwasaki K, Hanioka N, Narimatsu S, Niwa T, Sawada J (2009) Functional characterization of CYP3A4. 16: catalytic activities toward midazolam and carbamazepine. Xenobiotica 39:140–147
17. Domanski TL, He YA, Khan KK, Roussel F, Wang Q, Halpert JR (2001) Phenylalanine and tryptophan scanning mutagenesis of CYP3A4 substrate recognition site residues and effect on substrate oxidation and cooperativity. Biochemistry 40:10150–10160
18. Frank DJ, Denisov IG, Sligar SG (2009) Mixing apples and oranges: analysis of heterotropic cooperativity in cytochrome P450 3A4. Arch Biochem Biophys 488:146–152
19. Denisov IG, Baas BJ, Grinkova YV, Sligar SG (2007) Cooperativity in cytochrome P450 3A4: linkages in substrate binding, spin state, uncoupling, and product formation. J Biol Chem 282:7066–7076
20. Frank DJ, Denisov IG, Sligar SG (2011) Analysis of heterotropic cooperativity in cytochrome P450 3A4 using α-naphthoflavone and testosterone. J Biol Chem 286:5540–5545
21. Roberts AG, Campbell AP, Atkins WM (2005) The thermodynamic landscape of testosterone binding to cytochrome P450 3A4: ligand binding and spin state equilibria. Biochemistry 44:1353–1366
22. Fernando H, Halpert JR, Davydov DR (2006) Resolution of multiple substrate binding sites in cytochrome P450 3A4: the stoichiometry of the enzyme-substrate complexes probed by FRET and Job's titration. Biochemistry 45:4199–4209
23. Fernando H, Davydov DR, Chin CC, Halpert JR (2007) Role of subunit interactions in P450 oligomers in the loss of homotropic cooperativity in the cytochrome P450 3A4 mutant L211F/D214E/F304W. Arch Biochem Biophys 460:129–140
24. Fernando H, Rumfeldt JA, Davydova NY, Halpert JR, Davydov DR (2011) Multiple substrate-binding sites are retained in cytochrome P450 3A4 mutants with decreased cooperativity. Xenobiotica 41:281–289
25. Renaud JP, Davydov DR, Heirwegh KP, Mansuy D, Hui Bon Hoa GH (1996) Thermodynamic studies of substrate binding and spin transitions in human cytochrome P-450 3A4 expressed in yeast microsomes. Biochem J 319(Pt 3):675–681
26. Isin EM, Guengerich FP (2006) Kinetics and thermodynamics of ligand binding by cytochrome P450 3A4. J Biol Chem 281:9127–9136
27. Isin EM, Guengerich FP (2007) Multiple sequential steps involved in the binding of inhibitors to cytochrome P450 3A4. J Biol Chem 282:6863–6874
28. Sevrioukova IF, Poulos TL (2010) Structure and mechanism of the complex between cytochrome P4503A4 and ritonavir. Proc Natl Acad Sci U S A 107:18422–18427
29. Sevrioukova IF, Poulos TL (2012) Structural and mechanistic insights into the interaction of cytochrome P4503A4 with bromoergocryptine, a type I ligand. J Biol Chem 287:3510–3517
30. Sevrioukova IF, Poulos TL (2012) Interaction of human cytochrome P4503A4 with ritonavir analogs. Arch Biochem Biophys 520:108–116
31. Sevrioukova IF, Poulos TL (2013) Pyridine-substituted desoxyritonavir is a more potent cytochrome P450 3A4 inhibitor than ritonavir. J Med Chem 56:3733–3741
32. Yamazaki H, Ueng YF, Shimada T, Guengerich FP (1995) Roles of divalent metal ions in oxidations catalyzed by recombinant cytochrome P450 3A4 and replacement of NADPH-cytochrome P450 reductase with other flavoproteins, ferredoxin, and oxygen surrogates. Biochemistry 34:8380–8389
33. Guengerich FP, Johnson WW (1997) Kinetics of ferric cytochrome P450 reduction by NADPH-cytochrome P450 reductase: rapid reduction in the absence of substrate and variations among cytochrome P450 systems. Biochemistry 36:14741–14750
34. Davydov DR, Fernando H, Baas BJ, Sligar SG, Halpert JR (2005) Kinetics of dithionite-dependent reduction of cytochrome P450 3A4: heterogeneity of the enzyme caused by its oligomerization. Biochemistry 44:13902–13913
35. Fernando H, Halpert JR, Davydov DR (2008) Kinetics of electron transfer in the complex of cytochrome P450 3A4 with the flavin domain of cytochrome P450BM-3 as evidence of functional heterogeneity of the heme protein. Arch Biochem Biophys 471:20–31
36. Davydov DR, Sineva EV, Sistla S, Davydova NY, Frank DJ, Sligar SG, Halpert JR (2010) Electron transfer in the complex of membrane-bound human cytochrome P450 3A4 with the flavin domain of P450BM-3: the effect of oligomerization of the heme protein and intermittent modulation of the spin equilibrium. Biochim Biophys Acta 1797:378–390
37. Koley AP, Buters JT, Robinson RC, Markowitz A, Friedman FK (1995) CO binding kinetics of human cytochrome P450 3A4. Specific interaction of substrates with kinetically distinguishable conformers. J Biol Chem 270:5014–5018
38. Koley AP, Robinson RC, Friedman FK (1996) Cytochrome P450 conformation and substrate

interactions as probed by CO binding kinetics. Biochimie 78:706–713
39. Koley AP, Buters JT, Robinson RC, Markowitz A, Friedman FK (1997) Differential mechanisms of cytochrome P450 inhibition and activation by α-naphthoflavone. J Biol Chem 272:3149–3152
40. Kumar S, Davydov DR, Halpert JR (2005) Role of cytochrome b_5 in modulating peroxide-supported CYP3A4 activity: evidence for a conformational transition and cytochrome P450 heterogeneity. Drug Metab Dispos 33:1131–1136
41. Davydov DR, Halpert JR, Renaud JP, Hui Bon Hoa G (2003) Conformational heterogeneity of cytochrome P450 3A4 revealed by high pressure spectroscopy. Biochem Biophys Res Commun 312:121–130
42. Davydov DR, Baas BJ, Sligar SG, Halpert JR (2007) Allosteric mechanisms in cytochrome P450 3A4 studied by high-pressure spectroscopy: pivotal role of substrate-induced changes in the accessibility and degree of hydration of the heme pocket. Biochemistry 46:7852–7864
43. Kapelyukh Y, Paine MJ, Marechal JD, Sutcliffe MJ, Wolf CR, Roberts GC (2008) Multiple substrate binding by cytochrome P450 3A4: estimation of the number of bound substrate molecules. Drug Metab Dispos 36:2136–2144
44. Dabrowski MJ, Schrag ML, Wienkers LC, Atkins WM (2002) Pyrene-pyrene complexes at the active site of cytochrome P450 3A4: evidence for a multiple substrate binding site. J Am Chem Soc 124:11866–11867
45. Nath A, Fernandez C, Lampe JN, Atkins WM (2008) Spectral resolution of a second binding site for Nile Red on cytochrome P4503A4. Arch Biochem Biophys 474:198–204
46. Davydov DR, Davydova NY, Sineva EV, Kufareva I, Halpert JR (2013) Pivotal role of P450-P450 interactions in CYP3A4 allostery: the case of α-naphthoflavone. Biochem J 453:219–230
47. Nath A, Koo PK, Rhoades E, Atkins WM (2008) Allosteric effects on substrate dissociation from cytochrome P450 3A4 in nanodiscs observed by ensemble and single-molecule fluorescence spectroscopy. J Am Chem Soc 130:15746–15747
48. Stresser DM, Blanchard AP, Turner SD, Erve JC, Dandeneau AA, Miller VP, Crespi CL (2000) Substrate-dependent modulation of CYP3A4 catalytic activity: analysis of 27 test compounds with four fluorometric substrates. Drug Metab Dispos 28:1440–1448
49. Chougnet A, Grinkova Y, Ricard D, Sligar S, Woggon WD (2007) Fluorescent probes for rapid screening of potential drug-drug interactions at the CYP3A4 level. ChemMedChem 2:717–724
50. Lampe JN, Atkins WM (2006) Time-resolved fluorescence studies of heterotropic ligand binding to cytochrome P450 3A4. Biochemistry 45:12204–12215
51. Tsalkova TN, Davydova NY, Halpert JR, Davydov DR (2007) Mechanism of interactions of α-naphthoflavone with cytochrome P450 3A4 explored with an engineered enzyme bearing a fluorescent probe. Biochemistry 46:106–119
52. Davydov DR, Davydova NY, Tsalkova TN, Halpert JR (2008) Effect of glutathione on homo- and heterotropic cooperativity in cytochrome P450 3A4. Arch Biochem Biophys 471:134–145
53. Gartner CA, Wen B, Wan J, Becker RS, Jones G 2nd, Gygi SP, Nelson SD (2005) Photochromic agents as tools for protein structure study: lapachenole is a photoaffinity ligand of cytochrome P450 3A4. Biochemistry 44:1846–1855
54. Wen B, Doneanu CE, Gartner CA, Roberts AG, Atkins WM, Nelson SD (2005) Fluorescent photoaffinity labeling of cytochrome P450 3A4 by lapachenole: identification of modification sites by mass spectrometry. Biochemistry 44:1833–1845
55. Kim JS, Ahn T, Yim SK, Yun CH (2002) Differential effect of copper (II) on the cytochrome P450 enzymes and NADPH-cytochrome P450 reductase: inhibition of cytochrome P450-catalyzed reactions by copper (II) ion. Biochemistry 41:9438–9447
56. Kim JS, Yun CH (2005) Inhibition of human cytochrome P450 3A4 activity by zinc(II) ion. Toxicol Lett 156:341–350
57. Baylon JL, Lenov IL, Sligar SG, Tajkhorshid E (2013) Characterizing the membrane-bound state of cytochrome P450 3A4: structure, depth of insertion, and orientation. J Am Chem Soc 135:8542–8851
58. Pearson JT, Hill JJ, Swank J, Isoherranen N, Kunze KL, Atkins WM (2006) Surface plasmon resonance analysis of antifungal azoles binding to CYP3A4 with kinetic resolution of multiple binding orientations. Biochemistry 45:6341–6353
59. Pearson J, Dahal UP, Rock D, Peng CC, Schenk JO, Joswig-Jones C, Jones JP (2011) The kinetic mechanism for cytochrome P450 metabolism of type II binding compounds: evidence supporting direct reduction. Arch Biochem Biophys 511:69–79
60. Das A, Zhao J, Schatz GC, Sligar SG, Van Duyne RP (2009) Screening of type I and II drug binding to human cytochrome P450-3A4 in nanodiscs by localized surface plasmon resonance spectroscopy. Anal Chem 81:3754–3759
61. Cameron MD, Wen B, Allen KE, Roberts AG, Schuman JT, Campbell AP, Kunze KL, Nelson SD (2005) Cooperative binding of midazolam with testosterone and α-naphthoflavone within the CYP3A4 active site: a NMR T1 paramagnetic relaxation study. Biochemistry 44:14143–14151
62. Roberts AG, Yang J, Halpert JR, Nelson SD, Thummel KT, Atkins WM (2011) The structural basis for homotropic and heterotropic cooperativity of midazolam metabolism by human cytochrome P450 3A4. Biochemistry 50:10804–10818
63. Kijac AZ, Li Y, Sligar SG, Rienstra CM (2007) Magic-angle spinning solid-state NMR spectroscopy

of nanodisc-embedded human CYP3A4. Biochemistry 46:13696–13703
64. Conner KP, Vennam P, Woods CM, Krzyaniak MD, Bowman MK, Atkins WM (2012) 1,2,3-Triazole-heme interactions in cytochrome P450: functionally competent triazole-water-heme complexes. Biochemistry 51:6441–6457
65. Mak PJ, Denisov IG, Grinkova YV, Sligar SG, Kincaid JR (2011) Defining CYP3A4 structural responses to substrate binding. Raman spectroscopic studies of a nanodisc-incorporated mammalian cytochrome P450. J Am Chem Soc 133:1357–1366
66. Obach RS (2001) Mechanism of cytochrome P4503A4- and 2D6-catalyzed dehydrogenation of ezlopitant as probed with isotope effects using five deuterated analogs. Drug Metab Dispos 29:1599–1607
67. Krauser JA, Guengerich FP (2005) Cytochrome P450 3A4-catalyzed testosterone 6β-hydroxylation stereochemistry, kinetic deuterium isotope effects, and rate-limiting steps. J Biol Chem 280:19496–19506
68. Woods CM, Fernandez C, Kunze KL, Atkins WM (2011) Allosteric activation of cytochrome P450 3A4 by α-naphthoflavone: branch point regulation revealed by isotope dilution analysis. Biochemistry 50:10041–10051
69. Joseph S, Rusling JF, Lvov YM, Friedberg T, Fuhr U (2003) An amperometric biosensor with human CYP3A4 as a novel drug screening tool. Biochem Pharmacol 65:1817–1826
70. Xue Q, Kato D, Kamata T, Guo Q, You T, Niwa O (2013) Human cytochrome P450 3A4 and a carbon nanofiber modified film electrode as a platform for the simple evaluation of drug metabolism and inhibition reactions. Analyst 138:6463–6468
71. Sadeghi SJ, Ferrero S, Di Nardo G, Gilardi G (2012) Drug-drug interactions and cooperative effects detected in electrochemically driven human cytochrome P450 3A4. Bioelectrochemistry 86:87–91
72. Hendricks NR, Waryo TT, Arotiba O, Jahed N, Baker PGL, Iwuoha EI (2009) Microsomal cytochrome P450-3A4 (CYP3A4) nanobiosensor for the determination of 2,4-dichlorophenol – an endocrine disruptor compound. Electrochim Acta 54:1925–1931
73. Ignaszak A, Hendricks N, Waryo T, Songa E, Jahed N, Ngece R, Al-Ahmed A, Kgarebe B, Baker P, Iwuoha EI (2009) Novel therapeutic biosensor for indinavir – a protease inhibitor antiretroviral drug. J Pharm Biomed Anal 49:498–501
74. Mie Y, Suzuki M, Komatsu Y (2009) Electrochemically driven drug metabolism by membranes containing human cytochrome P450. J Am Chem Soc 131:6646–6647
75. Larsen AT, May EM, Auclair K (2011) Predictable stereoselective and chemoselective hydroxylations and epoxidations with P450 3A4. J Am Chem Soc 133:7853–7858
76. Johnson EF, Stout CD (2005) Structural diversity of human xenobiotic-metabolizing cytochrome P450 monooxygenases. Biochem Biophys Res Commun 338:331–336
77. Williams PA, Cosme J, Vinkovic DM, Ward A, Angove HC, Day PJ, Vonrhein C, Tickle IJ, Jhoti H (2004) Crystal structures of human cytochrome P450 3A4 bound to metyrapone and progesterone. Science 305:683–686
78. Yano JK, Wester MR, Schoch GA, Griffin KJ, Stout CD, Johnson EF (2004) The structure of human microsomal cytochrome P450 3A4 determined by X-ray crystallography to 2.05-Å resolution. J Biol Chem 279:38091–38094
79. Ekroos M, Sjogren T (2006) Structural basis for ligand promiscuity in cytochrome P450 3A4. Proc Natl Acad Sci U S A 103:13682–13687
80. Sevrioukova IF, Poulos TL (2013) Dissecting cytochrome P450 3A4-ligand interactions using ritonavir analogues. Biochemistry 52:4474–4481
81. Sevrioukova IF, Poulos TL (2014) Ritonavir analogues as a probe for deciphering the cytochrome P450 3A4 inhibitory mechanism. Curr Topics Med Chem 14:1348–1355
82. Cupp-Vickery J, Anderson R, Hatziris Z (2000) Crystal structures of ligand complexes of P450eryF exhibiting homotropic cooperativity. Proc Natl Acad Sci U S A 97:3050–3055
83. Park H, Lee S, Suh J (2005) Structural and dynamical basis of broad substrate specificity, catalytic mechanism, and inhibition of cytochrome P450 3A4. J Am Chem Soc 127:13634–13642
84. Fishelovitch D, Hazan C, Shaik S, Wolfson HJ, Nussinov R (2007) Structural dynamics of the cooperative binding of organic molecules in the human cytochrome P450 3A4. J Am Chem Soc 129:1602–1611
85. Sun H, Moore C, Dansette PM, Kumar S, Halpert JR, Yost GS (2009) Dehydrogenation of the indoline-containing drug 4-chloro-N-(2-methyl-1-indolinyl)-3-sulfamoylbenzamide (indapamide) by CYP3A4: correlation with in silico predictions. Drug Metab Dispos 37:672–684
86. Sun H, Scott DO (2011) Metabolism of 4-aminopiperidine drugs by cytochrome P450s: molecular and quantum mechanical insights into drug design. ACS Med Chem Lett 2:638–643
87. Li W, Liu H, Luo X, Zhu W, Tang Y, Halpert JR, Jiang H (2007) Possible pathway(s) of metyrapone egress from the active site of cytochrome P450 3A4: a molecular dynamics simulation. Drug Metab Dispos 35:689–696
88. Fishelovitch D, Shaik S, Wolfson HJ, Nussinov R (2009) Theoretical characterization of substrate access/exit channels in the human cytochrome P450 3A4 enzyme: involvement of phenylalanine residues in the gating mechanism. J Phys Chem 113:13018–13025

89. Krishnamoorthy N, Gajendrarao P, Thangapandian S, Lee Y, Lee KW (2009) Probing possible egress channels for multiple ligands in human CYP3A4: a molecular modeling study. J Mol Model 16:607–614
90. Shahrokh K, Cheatham TE 3rd, Yost GS (2012) Conformational dynamics of CYP3A4 demonstrate the important role of Arg212 coupled with the opening of ingress, egress and solvent channels to dehydrogenation of 4-hydroxytamoxifen. Biochim Biophys Acta 1820:1605–1617
91. Bren U, Oostenbrink C (2012) Cytochrome P450 3A4 inhibition by ketoconazole: tackling the problem of ligand cooperativity using molecular dynamics simulations and free- energy calculations. J Chem Inf Model 52:1573–1582
92. Fishelovitch D, Shaik S, Wolfson HJ, Nussinov R (2010) How does the reductase help to regulate the catalytic cycle of cytochrome P450 3A4 using the conserved water channel? J Phys Chem 114:5964–5970
93. Denisov IG, Shih AY, Sligar SG (2012) Structural differences between soluble and membrane bound cytochrome P450s. J Inorg Biochem 108:150–158
94. Berka K, Paloncyova M, Anzenbacher P, Otyepka M (2013) Behavior of human cytochromes P450 on lipid membranes. J Phys Chem 117:11556–11564
95. Liu X, Wang X, Jiang H (2008) A steered molecular dynamics method with direction optimization and its applications on ligand molecule dissociation. J Biochem Biophys Methods 70:857–864
96. Yang K, Liu X, Wang X, Jiang H (2009) A steered molecular dynamics method with adaptive direction adjustments. Biochem Biophys Res Commun 379:494–498
97. Torimoto N, Ishii I, Hata M, Nakamura H, Imada H, Ariyoshi N, Ohmori S, Igarashi T, Kitada M (2003) Direct interaction between substrates and endogenous steroids in the active site may change the activity of cytochrome P450 3A4. Biochemistry 42:15068–15077
98. Fishelovitch D, Hazan C, Hirao H, Wolfson HJ, Nussinov R, Shaik S (2007) QM/MM study of the active species of the human cytochrome P450 3A4, and the influence thereof of the multiple substrate binding. J Phys Chem 111:13822–13832
99. Zhang Y, Morisetti P, Kim J, Smith L, Lin H (2008) Regioselectivity preference of testosterone hydroxylation by cytochrome P450 3A4. Theor Chem Acc 121:313–319
100. Kongsted J, Ryde U (2009) An improved method to predict the entropy term with the MM/PBSA approach. J Comput Aided Mol Des 23:63–71
101. Zhang Y, Lin H (2009) Quantum tunneling in testosterone 6β-hydroxylation by cytochrome P450: reaction dynamics calculations employing multiconfiguration molecular-mechanical potential energy surfaces. J Phys Chem 113:11501–11508
102. Lee JY, Kang NS, Kang YK (2012) Binding free energies of inhibitors to iron porphyrin complex as a model for cytochrome P450. Biopolymers 97:219–228
103. de Groot MJ, Ekins S (2002) Pharmacophore modeling of cytochromes P450. Adv Drug Deliv Rev 54:367–383
104. Wang Y, Han KL, Sheng-Li Y, Yang L (2004) Structural determinants of steroids for cytochrome P450 3A4-mediated metabolism. J Mol Struct 710:215–221
105. Ekins S, Bravi G, Wikel JH, Wrighton SA (1999) Three-dimensional-quantitative structure activity relationship analysis of cytochrome P-450 3A4 substrates. J Pharmacol Exp Ther 291:424–433
106. Ekins S, Bravi G, Binkley S, Gillespie JS, Ring BJ, Wikel JH, Wrighton SA (1999) Three- and four-dimensional quantitative structure activity relationship analyses of cytochrome P-450 3A4 inhibitors. J Pharmacol Exp Ther 290:429–438
107. Riley RJ, Parker AJ, Trigg S, Manners CN (2001) Development of a generalized, quantitative physicochemical model of CYP3A4 inhibition for use in early drug discovery. Pharm Res 18:652–655
108. Ekins S, Stresser DM, Williams JA (2003) *In vitro* and pharmacophore insights into CYP3A enzymes. Trends Pharmacol Sci 24:161–166
109. Ekins S, Berbaum J, Harrison RK (2003) Generation and validation of rapid computational filters for CYP2D6 and CYP3A4. Drug Metab Dispos 31:1077–1080
110. Regev-Shoshani G, Shoseyov O, Kerem Z (2004) Influence of lipophilicity on the interactions of hydroxy stilbenes with cytochrome P450 3A4. Biochem Biophys Res Commun 323:668–673
111. Singh SB, Shen LQ, Walker MJ, Sheridan RP (2003) A model for predicting likely sites of CYP3A4-mediated metabolism on drug-like molecules. J Med Chem 46:1330–1336
112. Mao B, Gozalbes R, Barbosa F, Migeon J, Merrick S, Kamm K, Wong E, Costales C, Shi W, Wu C, Froloff N (2006) QSAR modeling of in vitro inhibition of cytochrome P450 3A4. J Chem Inf Model 46:2125–2134
113. Lill MA, Dobler M, Vedani A (2006) Prediction of small-molecule binding to cytochrome P450 3A4: flexible docking combined with multidimensional QSAR. Chem Med Chem 1:73–81
114. Sheridan RP, Korzekwa KR, Torres RA, Walker MJ (2007) Empirical regioselectivity models for human cytochromes P450 3A4, 2D6, and 2C9. J Med Chem 50:3173–3184
115. Kjellander B, Masimirembwa CM, Zamora I (2007) Exploration of enzyme-ligand interactions in CYP2D6 and 3A4 homology models and crystal structures using a novel computational approach. J Chem Inf Model 47:1234–1247
116. Jayakanthan M, Chandrasekar S, Muthukumaran J, Mathur PP (2010) Analysis of CYP3A4-HIV-1

protease drugs interactions by computational methods for highly active antiretroviral therapy in HIV/AIDS. J Mol Graph Model 28:455–463

117. Dapkunas J, Sazonovas A, Japertas P (2009) Probabilistic prediction of the human CYP3A4 and CYP2D6 metabolism sites. Chem Biodivers 6:2101–2106
118. Didziapetris R, Dapkunas J, Sazonovas A, Japertas P (2010) Trainable structure-activity relationship model for virtual screening of CYP3A4 inhibition. J Comput Aided Mol Des 24:891–906
119. Handa K, Nakagome I, Yamaotsu N, Gouda H, Hirono S (2013) Three-dimensional quantitative structure-activity relationship analysis of inhibitors of human and rat cytochrome P4503A enzymes. Drug Metab Pharmacokinet 28:345–355
120. Tie Y, McPhail B, Hong H, Pearce BA, Schnackenberg LK, Ge W, Buzatu DA, Wilkes JG, Fuscoe JC, Tong W, Fowler BA, Beger RD, Demchuk E (2013) Modeling chemical interaction profiles: II. Molecular docking, spectral data-activity relationship, and structure-activity relationship models for potent and weak inhibitors of cytochrome P450 CYP3A4 isozyme. Molecules 17:3407–3460
121. Jensen BF, Vind C, Padkjaer SB, Brockhoff PB, Refsgaard HH (2007) In silico prediction of cytochrome P450 2D6 and 3A4 inhibition using Gaussian kernel weighted *k*-nearest neighbor and extended connectivity fingerprints, including structural fragment analysis of inhibitors versus noninhibitors. J Med Chem 50:501–511
122. Zhou D, Afzelius L, Grimm SW, Andersson TB, Zauhar RJ, Zamora I (2006) Comparison of methods for the prediction of the metabolic sites for CYP3A4-mediated metabolic reactions. Drug Metab Dispos 34:976–983
123. Vedani A, Dobler M, Lill MA (2006) The challenge of predicting drug toxicity *in silico*. Basic Clin Pharmacol Toxicol 99:195–208
124. Sun H, Scott DO (2010) Structure-based drug metabolism predictions for drug design. Chem Biol Drug Des 75:3–17

Acyl-Carbon Bond Cleaving Cytochrome P450 Enzymes: CYP17A1, CYP19A1 and CYP51A1

4

Muhammad Akhtar and J. Neville Wright

Abstract

Cytochrome P450 (P450 or CYP) enzymes in their resting state contain the heme-iron in a high-spin Fe^{III} state. Binding of a substrate to a P450 enzyme allows transfer of the first electron, producing a Fe^{II} species that reacts with oxygen to generate a low-spin iron superoxide intermediate (Fe^{III}–O–O$^{\bullet}$) ready to accept the second electron to produce an iron peroxy anion intermediate (**a**, Fe^{III}–O–O^{-}). In classical monooxygenation reactions, the peroxy anion upon protonation fragments to form the reactive Compound I intermediate (Por$\bullet^{+}Fe^{IV}$=O), or its ferryl radical resonance form (Fe^{IV}–O$^{\bullet}$). However, when the substrate projects a carbonyl functionality, of the type **b**, at the active site as is the case for reactions catalyzed by CYP17A1, CYP19A1 and CYP51A1, the peroxy anion (Fe^{III}–O–O^{-}) is trapped, yielding a tetrahedral intermediate (**c**) that fragments to an acyl-carbon cleavage product (**d** plus an acid). Analogous acyl-carbon cleavage reactions are also catalyzed by certain hepatic P450s and CYP125A1 from *Mycobacterium tuberculosis*. A further improvisation on the theme is provided by aldehyde deformylases that convert long-chain aliphatic aldehydes to hydrocarbons. CYP17A1 is involved in the biosynthesis of corticoids as well as androgens. The flux toward these two classes of hormones seems to be regulated by cytochrome b_5, at the level of the acyl-carbon cleavage reaction. It is this regulation of CYP17A1 that provides a safety mechanism, ensuring that during corticoid biosynthesis, which requires 17α-hydroxylation by CYP17A1, androgen formation is avoided (Fig. 4.1).

Keywords

Aromatase • Sterol 14α-demethylase • 17α-hydroxylase-17,20-lyase • P450 2B4 • CYP125A1 • Aldehyde deformylase • Cytochrome b_5 in CYP17A1 catalysis

M. Akhtar (✉)
School of Biological Sciences, University of the Punjab, New Campus, Lahore 54590, Pakistan

Centre for Biological Sciences, University of Southampton, Southampton SO17 1BJ, UK
e-mail: ma3@soton.ac.uk

J.N. Wright
Centre for Biological Sciences, University of Southampton, Southampton SO17 1BJ, UK

E.G. Hrycay and S.M. Bandiera (eds.), *Monooxygenase, Peroxidase and Peroxygenase Properties and Mechanisms of Cytochrome P450*, Advances in Experimental Medicine and Biology 851,
DOI 10.1007/978-3-319-16009-2_4,

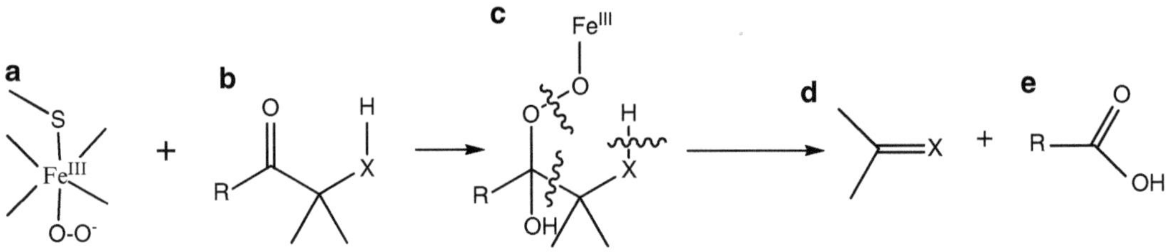

Fig. 4.1 Intermediary P450 complexes formed in reactions catalyzed by steroidal P450 enzymes

4.1 Introduction and Historical Background

Classically, the P450 class of enzymes is known to catalyze monooxygenations according to the reactions of equations (eqns) 1 and 2, in which an oxygen atom is inserted either into a C–H bond or added to an olefinic linkage [1, 2]. Then, our studies on the enzymes involved in certain oxidative reactions in steroid hormone and sterol biosynthesis pathways highlighted transformations in which P450s catalyze three different generic reactions at a single active site [3–6]. Two of these pathways follow the basic chemistry of eqn 1. However, a third pathway represents a novel acyl-carbon cleavage process shown in eqn 3 (Scheme 4.1).

We argued that in the normal catalytic cycle of P450, two key reactive oxygen intermediates are involved [7–9], and these are recruited differentially depending on the nature of the substrate functional group resident on the active site [3–6]. We reviewed the evidence then available and viewed the individual reactions of the cycle as shown in Scheme 4.2 [6].

There is substantial evidence that P450s in the resting state contain Fe^{III} in a high-spin state, the active-site cysteine thiolate ligand and a protein-bound water molecule (**1**). Binding of the substrate to the resting state of P450s is regarded as the first step of the catalytic cycle, which prepares the P450-substrate complex for further manipulations at the heme-iron atom.

Delivery of an electron from NADPH via NADPH–P450 oxidoreductase (CPR) converts Fe^{III} to Fe^{II} ready to react with oxygen to generate a species (**2**) that may be regarded to mimic the chemistry of oxyhemoglobin. However, we reasoned that, based on the original proposal of Weiss [10] but opposed by Pauling [11] and supported by subsequent biophysical studies (see citations in [6]), oxyhemoglobin should be formulated as a low-spin adduct of Fe^{III} and a superoxide anion. By analogy, the oxygen adduct in the P450 cycle has a low-spin structure as in **2**.

Delivery of the second electron converts the superoxide moiety to a peroxy anion (**3**), which upon protonation yields the ferrihydroperoxo intermediate (Fe^{III}–O–OH) (**4**). We regard the overall conversion **2**–**4** as a two-step process in which the peroxy anion (**3**) is a discrete intermediate. Its subsequent fate depends on the nature of the target C atom projected into the active site. In the case of the C-atom requiring hydroxylation, the peroxy anion is protonated, then cleaved to the intermediate **5**, while with substrates requiring an acyl-carbon cleavage it forms a tetrahedral adduct (**7**).

The O–O bond in a peroxide of the type **4** undergoes a rapid cleavage, yielding an iron-oxygen species with one of the canonical structures shown in **5**, which is reminiscent of Compound I normally formed during the catalytic cycle of heme peroxidases. Compound I of P450 is currently represented as $Por\bullet^{+}Fe^{IV}{=}O$ where Por signifies the porphyrin group. Here, it suffices to include the steps involved in the hydroxylation reaction (Scheme 4.3). These include the initial abstraction of a hydrogen atom from the substrate to produce a carbon radical **11**, which is quenched by a hydroxyl radical equivalent donated by the Fe^{IV}–OH species, forming the hydroxylated product and regenerating the resting state of P450.

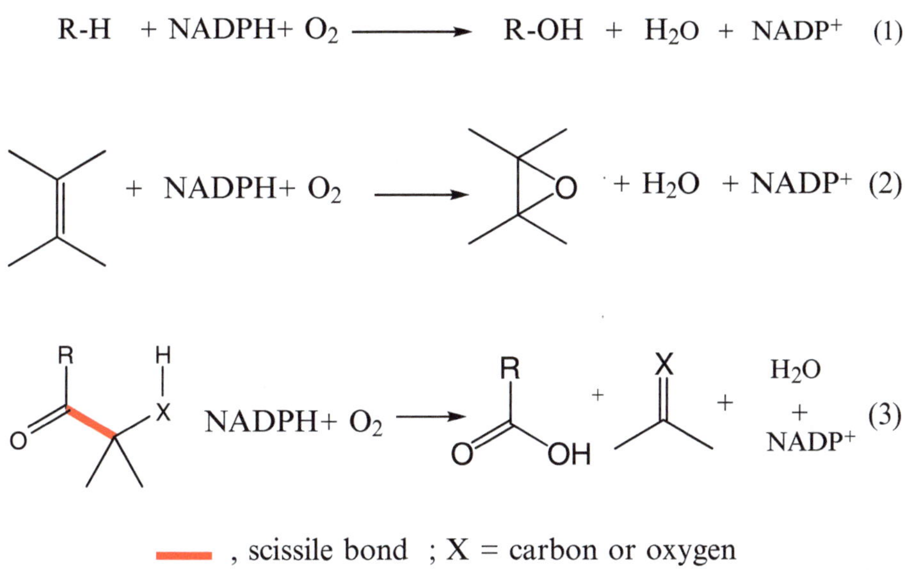

Scheme 4.1 Different generic reactions catalyzed by P450 enzymes

4.2 Aromatase (CYP19A1)

The first clue for the existence of multifunctional P450s resulted from our studies on aromatase (CYP19A1). The enzyme catalyzes the conversion of androgens to estrogens, which involves the loss of the C-19 methyl group of the substrate.

Chemical considerations had suggested that removal of a methyl group must somehow involve its functionalization, preferably by an oxygen substituent. Indeed, the earliest studies carried out in the late 1950s had shown that 19-hydroxyandrostenedione could be converted to estrone (Scheme 4.4) by a placental microsomal preparation [12] then introduced by Ryan [13]. Much of the subsequent work in the aromatase field has been greatly facilitated by the discovery of Ryan and the human placental microsomal system exploited for the purification of aromatase. Using the placental enzyme, we made two observations that proved to be important for understanding the nature of aromatase catalysis. The first observation was that 19-hydroxyandrostenedione is converted into a 19-aldehyde intermediate by a reaction involving NADPH and oxygen [14], a process whereby one of the C-19 *pro*-chiral hydrogen atoms of **14**, H_{Re}, is stereospecifically removed [15–17]. The second observation was that in the aromatization process, the 19 carbon atom is released as formic acid [15]. Thus, the latter species was the target of our future studies.

We set out to determine the origin of the two oxygen atoms in formic acid and the single carbon-bound hydrogen. For this purpose, a variety of isotopically-labeled androstendione derivatives were synthesized and used as mechanistic probes [3–6]. Noteworthy among these were 19-hydroxyandrostenedione and 19-oxoandrostenedione, containing ^{18}O at C-19 (structures of the type **14** and **15b**, Scheme 4.5). Their conversion into estrone was studied with placental aromatase, using these precursors containing ^{16}O under an atmosphere of $^{18}O_2$ or those labeled with ^{18}O under $^{16}O_2$. Formic acid released in the process was isolated, derivatized and its isotopic composition determined by gas-liquid chromatography linked to mass

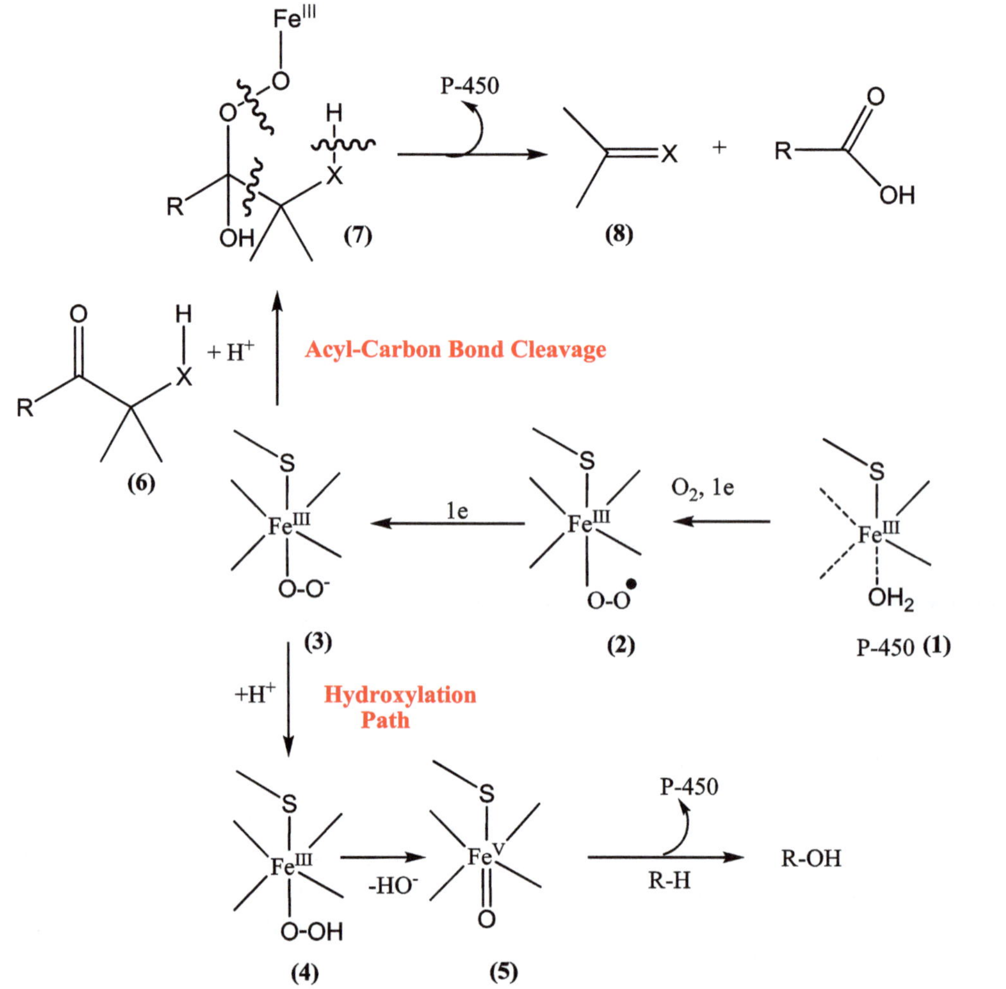

Scheme 4.2 Catalytic cycle of P450 enzymes. The resting state of P450 shown by structure (**1**) is converted to the peroxy anion (**3**), which either produces the oxo-derivative (**5**, known as Compound I) involved in monooxygenation reactions or is trapped by a substrate carbonyl group to produce **7**, which undergoes an acyl-carbon cleavage

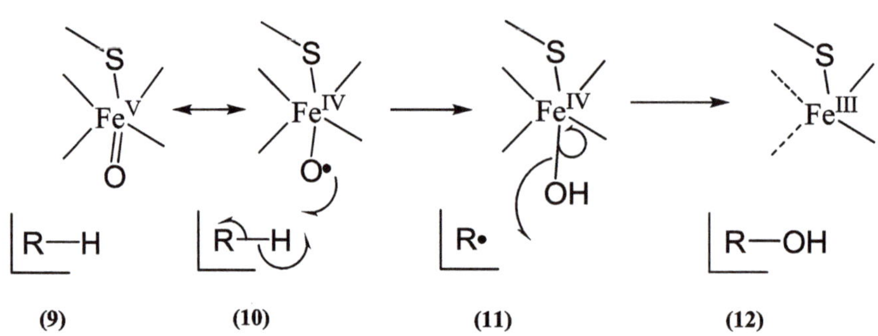

Scheme 4.3 Mechanism of the hydroxylation reaction

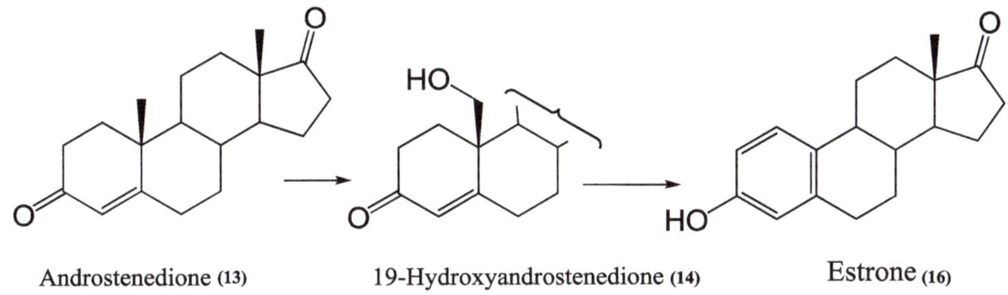

Scheme 4.4 The role of 19-hydroxyandrostenedione as an intermediate in estrone biosynthesis

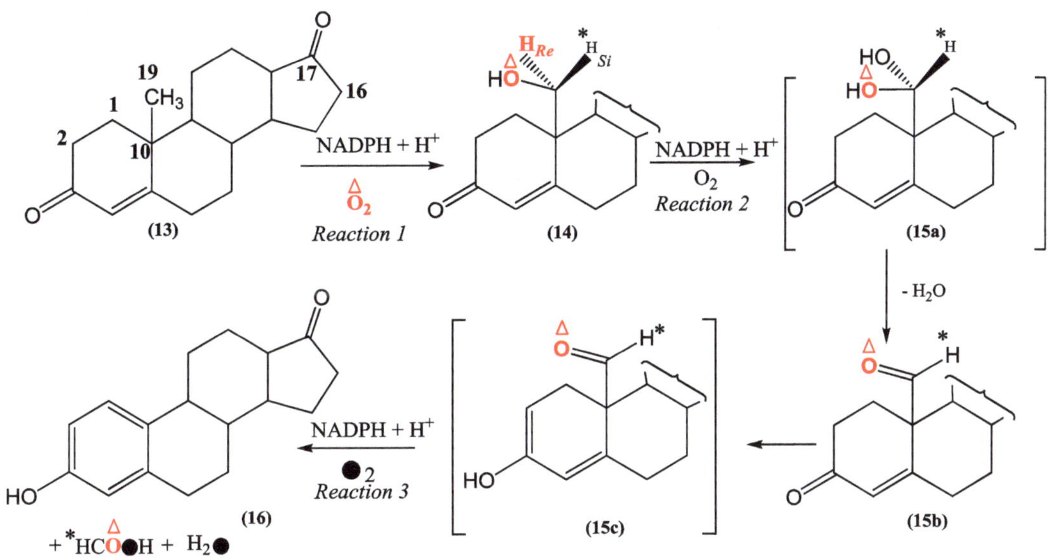

Scheme 4.5 Sequence of reactions catalyzed by aromatase. Atoms of oxygen from O_2 incorporated in *Reactions 1 and 3* are found in formic acid

spectrometry. The salient features of the finding are illustrated in Scheme 4.5 and are summarized as follows:

As expected for a hydroxylation reaction, in the presence of NADPH, an oxygen atom from molecular oxygen is incorporated at C-19 producing 19-hydroxyandrostenedione (**14**). The next stage also requires NADPH and oxygen to give **15b**, whereby the oxygen atom introduced in the first step is retained. What happens to the oxygen used in the second step is not known but there are two possible scenarios. The first is that in this step, another hydroxylation reaction occurs, generating a *gem*-diol intermediate (**15a**) that decomposes with the elimination, in a stereospecific fashion, of the hydroxyl group introduced in the second step. The elimination reaction must be under a strict steric control because the use of ^{18}O-labeled 19-oxoandrostenedione as a substrate leads to a quantitative retention of the ^{18}O label in formic acid. The dehydration reaction, **15a** to **15b**, is expected to be reversible and in the event of a less stringent steric control over this process, incorporation of oxygen from water into the eliminated formic acid will be expected. However, this was not observed. An alternative possibility is that following the abstraction of a hydrogen atom from the substrate by the conventional P450 cycle, the substrate carbon radical is quenched not by oxygen-rebound but by the removal of an alternate hydrogen from the –OH group. In such an event, the conversion **14** to **15b** occurs without involvement of a true covalent intermediate. Irrespective of which alternative is followed, the initial hydrogen abstraction is common to both. The stereochemistry of the removal of a *pro*-chiral hydrogen from C-19 of **14** has been elucidated and found to be H_{Re}, as shown for *Reaction 2* (Scheme 4.5) [16, 17].

Reaction 3 of the aromatase sequence also requires NADPH and O_2 and results in the cleavage of the C-10—C-19 bond by a process whereby the aldehydic hydrogen, as well as oxygen atom of the substrate (**15b**), are retained and an oxygen atom from molecular oxygen is incorporated into released formic acid. Thus, the overall aromatization process requires three molecules of NADPH and of O_2 [18]. The focus of all subsequent efforts to explain the mechanism of cleavage of the C-10—C-19 bond has been to accommodate the isotopic inventory of *Reaction 3*. We will defer any discussion on this facet until equivalent data on two additional related enzyme systems, sterol 14α-demethylase (CYP51A1) and 17α-hydroxylase-17,20-lyase (CYP17A1), have been reviewed.

4.3 Sterol Demethylation Reactions

4.3.1 Demethylations in Cholesterol Biosynthesis

From the very early days of interest in the understanding of chemical mechanisms underpinning reactions involved in sterol biosynthesis, it was assumed that the removal of the three methyl groups of lanosterol (**17**, Scheme 4.6), during its multistep conversion to cholesterol (**18**, Scheme 4.6), must involve their prior functionalization. It seemed attractive that the two C-4 methyl groups are removed following their oxidation to the corresponding carboxylic acid, which is decarboxylated using the electron withdrawing property of a neighboring group. Because the C-3 hydrogen atom was removed in the conversion of lanosterol to cholesterol, this suggested that the electron withdrawing substituent could be a 3-ketone formed in a reversible dehydrogenation reaction, to serve as a trigger for decarboxylation as shown in Scheme 4.7 [19–21]. Subsequent studies on a variety of sterol 4-demethylases demonstrated that such a view is by and large correct and also applicable to plant and fungal demethylases [22].

4.3.2 Sterol 14α-Demethylation (CYP51A1)

The situation regarding 14α-demethylation was less straight forward. Here, the release of the methyl carbon as CO_2 had been claimed and the assumption was that decarboxylation may be aided by the 8,9-double bond of lanosterol, which following decarboxylation occupies the 8,14-position. The latter then somehow reverts to the 8,9-position for further conversion [23].

The initial pointer to the possibility that 14-demethylation can occur through a novel process originated from the observation that, in the

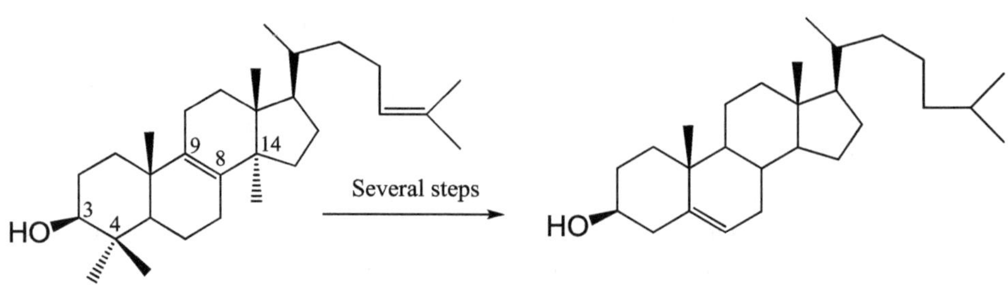

Lanosterol **(17)** Cholesterol **(18)**

Scheme 4.6 In the conversion of lanosterol to cholesterol, three methyl groups of lanosterol, two from C-4 and one from C-14, are removed

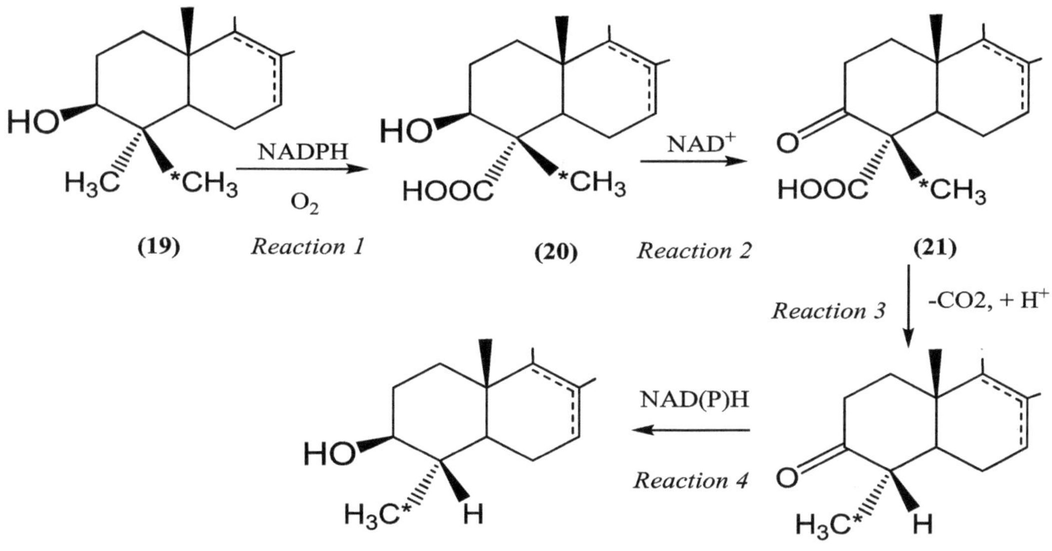

Scheme 4.7 Sequence of reactions for lanosterol demethylation at C-4. It is to be noted that following removal of the 4α-methyl group, the 4β-methyl group occupies the α-position to undergo another round of *Reactions 1–4*

conversion of lanosterol to cholesterol, the 15-α-hydrogen of the former is removed (Scheme 4.8) [24–26]. This finding gave rise to various speculations, invoking that the labilization of the hydrogen may somehow be linked to the 14α-demethylation process. We then showed that during the conversion of the sterols of the type **27** and **28** into cholesterol, a 7,14-diene (**29**) (Scheme 4.9) played an intermediary role and that its further conversion involved a reductase, introducing a medium-derived hydrogen atom at C-15 [27].

The latter results prompted us to extend the approaches used in the study of aromatase to 14α-demethylation. As in the study of the mechanism of aromatase, underpinning the work on 14α-demethylation was our ability to prepare potential intermediates (**27** and **28**) of the process, functionalized at the target C-atom with isotopic hydrogen (^{2}H or ^{3}H) together with ^{16}O or ^{18}O. These precursors were then subjected to demethylation using CYP51A1 ($P450_{14DM}$) from rat liver microsomes [28, 29] or *Candida albicans* expressed in *Saccharomyces cerevisiae* [30]. The key features of the results are as follows.

As background information, it should be borne in mind that in mammalian systems, 14-α-demethylation occurs with sterols containing a 8,9-double bond such as lanosterol. However, our extensive studies demonstrated that corresponding sterols containing a 7,8-double bond are also demethylated [28–30]. The terminal products of the demethylation of 8,9- and 7,8-sterols are the 8,14- and 7,14-dienes, respectively. The synthetic methodology for the introduction of an oxygen function at C-32, leads directly to the creation of a double bond in the 7,8-position. Therefore, most of the labeling experiments with isotopes of oxygen are performed with sterols containing a 7,8-double bond (**27** and **28**) [30].

The first step in the demethylation must be a hydroxylation reaction, deduced primarily from precedence, and the fact that the metabolic conversion of the precursor sterols requires NADPH and oxygen, which is a well-recognized indicator of a hydroxylation reaction. The use of isotopically labeled **27** showed it to be converted first to **28**. Then, cleavage of the C-14—C-32 bond occurs to produce the diene (**29**) with elimination of formic acid [28–30]. When the conversion is

Scheme 4.8 Loss of the 15α-hydrogen in the overall conversion of lanosterol to cholesterol. The 14α-demethylation occurs with sterols containing a double bond in the 8,9- as well as 7,8-positions

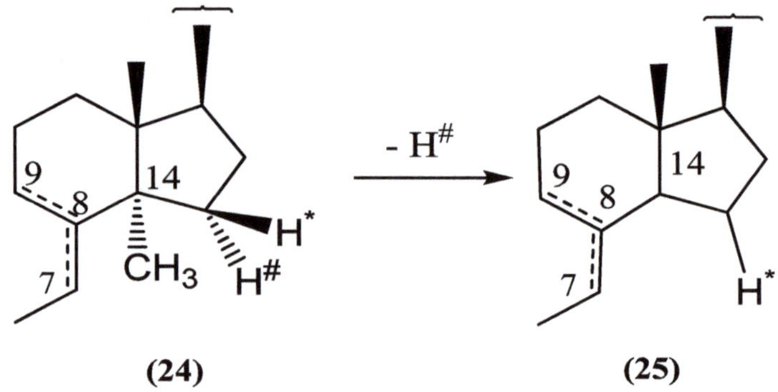

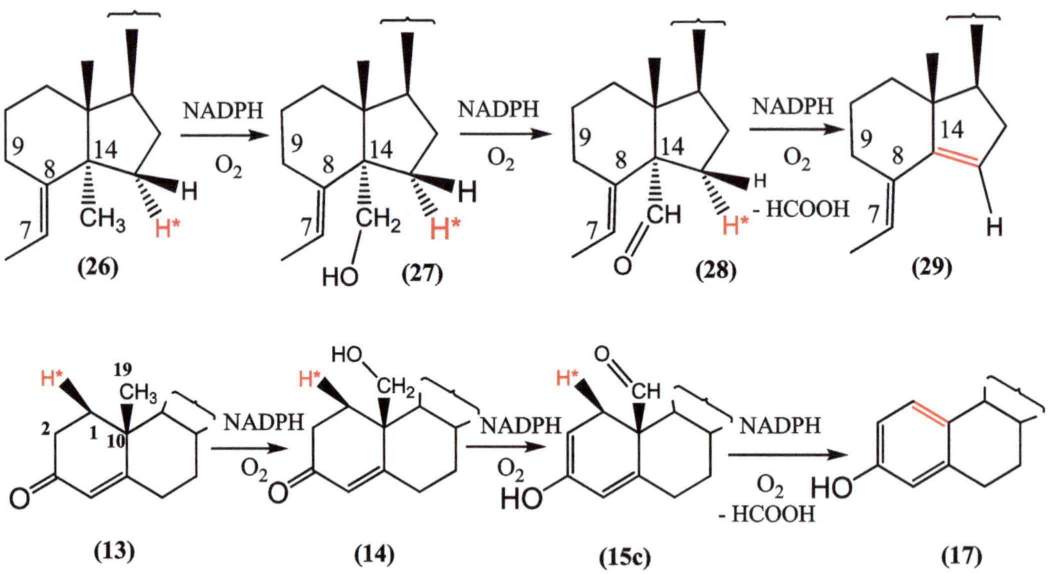

Scheme 4.9 Superimposable sequence of reactions involved in the aromatase (CYP19A1) and 14α-demethylase (CYP51A1) reactions

performed with ^{18}O labeled **27**, the hydroxyl oxygen atom is found in the released formic acid. The incubation of unlabeled precursors **27** and **28** under $^{18}O_2$ led to the incorporation of one atom of ^{18}O into formic acid. Cumulatively, these experiments show that in the oxidation of the hydroxyl group of **27** to **28**, the original oxygen atom at the target C-atom is retained and one oxygen atom from molecular oxygen is incorporated into formic acid during the final cleavage process. All these features are reminiscent of the reaction catalyzed by aromatase. Thus, an essentially superimposable sequence of chemical events is seen for the removal of the 19-methyl group by aromatase and the 14-α-methyl group by the demethylase (Scheme 4.9). The similarity necessitates that in the case of aromatase, at the point of cleavage of the C-10–C-19 bond, the 3-keto group can exist in an enolic form (**15c**) (Scheme 4.5).

4.4 17α-Hydroxylase-17,20-Lyase (CYP17A1)

17α-Hydroxylase-17,20-lyase (CYP17A1) acts on the pregnene nucleus, catalyzing a hydroxylation reaction necessary for corticoid as well as androgen biosynthesis, and a main side-chain cleavage reaction (*Reaction 2*, Scheme 4.10) required only for formation of biologically-active androgens [31, 32]. Depending on the source of CYP17A1 and presence of the cytochrome b_5 protein, two other side-chain cleavage products, 16-ene (**34**) and 17α-hydroxyandrogen (**35**), are produced [32–34]. When viewed superficially, the three side-chain cleavage products seem not to conform to the pattern of acyl-carbon cleavage reactions catalyzed by aromatase and 14α-demethylase. However, our mechanistic studies have shown that these reactions represent minor variations [33, 35–38] on the theme underpinning the acyl-carbon cleavages, seen for the aforementioned two enzyme systems.

Apart from our ability to synthesize suitably labeled pregnene derivatives, the success of the work depended on the availability of methods for purification of CYP17A1 from neonatal pig testes [31] and subsequently of the recombinant human enzyme expressed in *E. coli* [39]. Studies using ^3H, ^2H and ^{18}O labelings were performed with pregnenolone containing a Δ^5-3β-hydroxy system, rather than progesterone, because of ease of labeling with the former steroid that contains a single carbonyl group required for insertion of isotopes [35–38]. Incubation of [20-^{18}O; 21-$^2H_3$] pregnenolone (**30**) with pig CYP17A1 resulted in the release of acetic acid containing all three D atoms at C-21 of the precursor as well as the C-20 carbonyl ^{18}O. A complementary experiment with unlabeled **30** but under an atmosphere of $^{18}O_2$ led to the incorporation of one atom of ^{18}O into acetic acid [35–38]. These results indicated that the incorporation of the oxygen label had occurred during the second step of the reaction involving cleavage of the side chain. Experiments with 17α-hydroxypregnenolone (**32**) confirmed this conclusion, when its 20-carbonyl oxygen was retained in acetic acid and the second oxygen atom was derived from molecular oxygen.

When pregnenolone is the substrate, with pig CYP17A1, the cleavage of the side chain occurs either after hydroxylation at C-17, at the level of 17α-hydroxypregnenolone (**32**), or directly with pregnenolone to give **34** and **35**. Evidence has been presented to show that during all the three cleavage courses, the three hydrogen atoms at C-21 of the precursor as well as the C-20 carbonyl oxygen are retained in the released acetic acid, while its second oxygen atom originates from molecular oxygen [36–38].

Now focusing on the steroid nucleus [36–38], examination of isotopic composition showed that the C-17 carbonyl oxygen atom of **33**, as expected, originates from molecular oxygen incorporated during 17α-hydroxylation. However, the 17α-hydroxyl group of 17-α-hydroxyandrogen is also derived from oxygen. The status of the two C-16 hydrogen atoms of pregnenolone has also been investigated to show that these are retained in **32** and **35**, while in the formation of the Δ^{16} bond of the diene (**34**) it is the 16α-hydrogen atom of the precursor that is eliminated. Both steroids (**34** and **35**) retain the 17α-hydrogen atom of the pregnenolone precursor. We will use the isotopic information on the formation of **34** and **35** as key pointers to understand the mechanism of acyl-carbon cleavage by P450s (vide infra).

In light of the observations above, we consider the mechanism of the acyl-carbon bond cleavage reaction, dealing with aromatase first. Incorporation of an oxygen atom from O_2 and the retention of the original aldehydic oxygen, as well as hydrogen, in the released formic acid [3–6], were suggested to occur as follows. It was argued that when the carbonyl group of the substrate is projected into the active site of aromatase, the normal P450 cycle in which the peroxy anion is destined to be protonated for conversion into Compound I is interrupted. Instead, the peroxy anion is trapped (Scheme 4.2) by attack on the carbonyl group, producing a tetrahedral adduct (for example, **37**). In principle, the latter by a Baeyer-Villiger rearrangement could produce a formyl ester derivative that releases formic acid by an elimination reaction. The isotopic data will be consistent with this mechanism. However, we have unambiguously shown that the predicted formate derivative was neither used by aromatase as a substrate nor did it trap any radioactivity during the conversion of the C-19 labelled aldehyde (**15b**) into estrone [4]. Hence, it was proposed

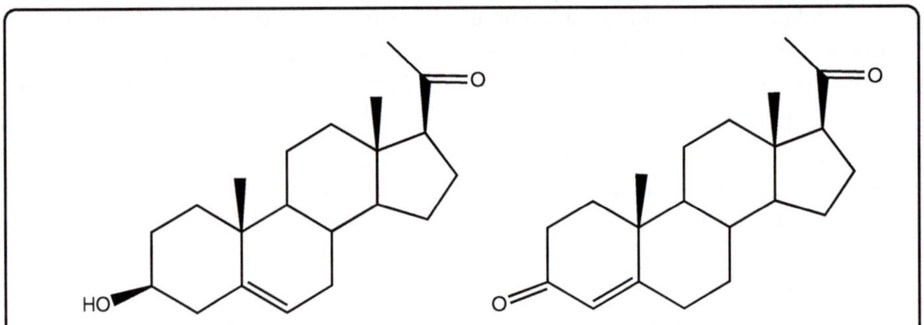

Pregnenolone (30) Progesterone (31)

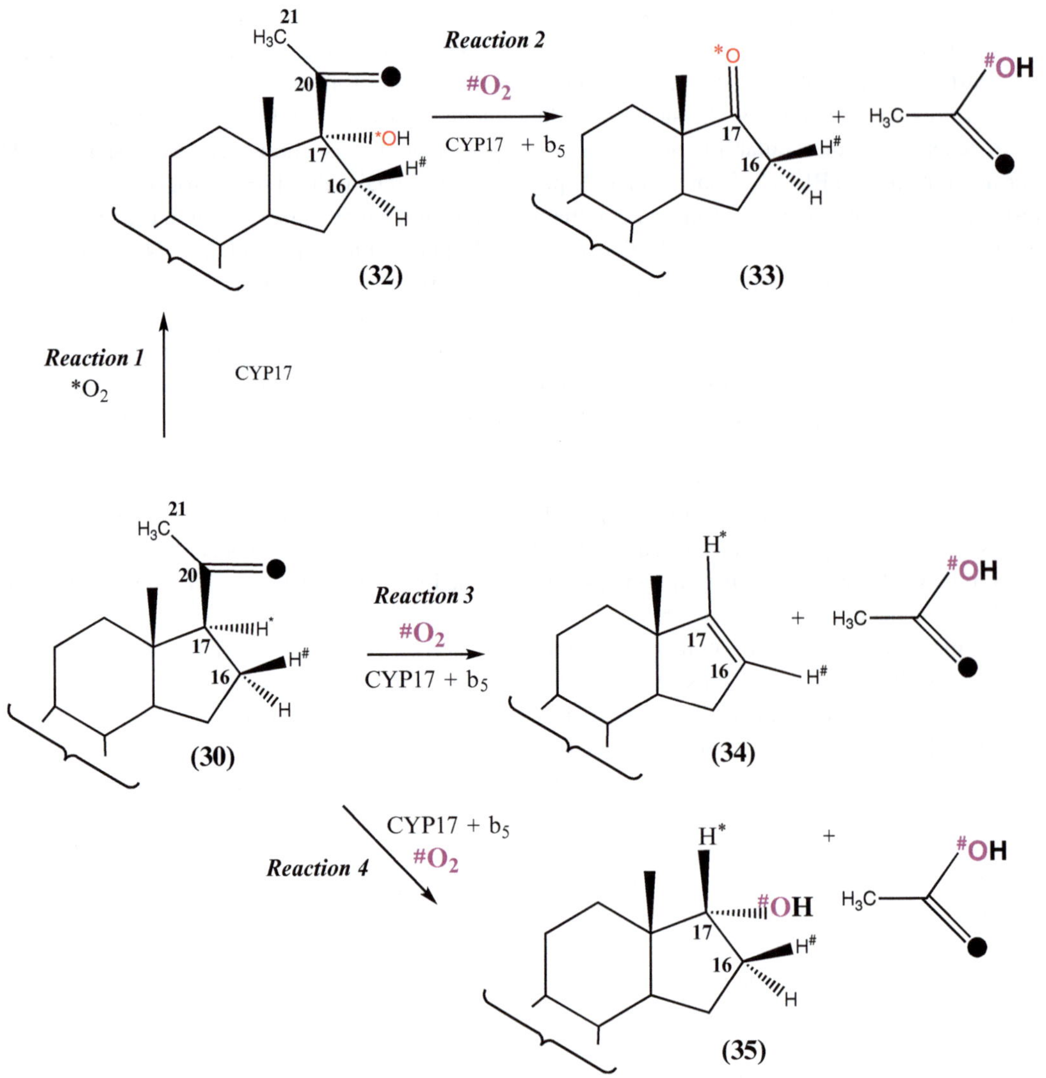

Scheme 4.10 Reactions catalyzed by CYP17A1. The box contains the two key progestogens, pregnenolone and progesterone. The hydroxylation *Reaction 1* and cleavage by *Reaction 2* occur with both substrates. However, with the human CYP17A1 enzyme, the cleavage *Reactions 3* and *4* are observed only with pregnenolone

that the peroxy – adduct fragments to create a new double bond between C-1 and C-10, releasing formic acid with the desired status of its various atoms. This basic concept has stood the test of time since it was proposed in 1976 [40], also see [4–6]. How does the peroxy – adduct fragment? Although a concerted process for the fragmentation is possible, we favored a step-wise process involving a radical species as shown in Scheme 4.11. Here, the key event is the cleavage of the O–O bond of the peroxy – adduct (**38**), presumably triggered by the same factors that lead to formation of the oxo derivative in the classical P450 cycle. Precedents from the behavior of alkoxy radicals would predict that the radical of the type (**39**) would decompose to furnish the carbon radical (**40**). The driving force for this process is predominantly provided by the creation of a carbonyl double bond of the released formic acid. In the case of aromatase, the radical formed by fragmentation is also energetically favorable by virtue of being on a tertiary carbon atom and further stabilized by conjugation to a Δ^4-3-one system. The latter features may make additional contributions to the energetics of the fragmentation reaction. However, as we will see later, when the mechanism is extended to other acyl-carbon cleaving P450s, the stabilization of the carbon radical is not available. Returning to the mechanistic sequence, a hydrogen abstraction from C-1 of the steroid nucleus by Fe^{III}–O$^{\bullet}$ creates the C-1–C-10 double bond of estrone. There is room for debate whether the aforementioned hydrogen abstraction step occurs with the conjugated ketone system of ring A (**40a**) or following an enolization process (via **40b**) (Scheme 4.11). We have no view on this issue.

The role of a peroxy anion in acyl-carbon cleavage is indirectly highlighted by resonance Raman spectroscopy studies on the enzyme-substrate complex of CYP17A1. The data have been interpreted to suggest that with its preferred substrate (17α-hydroxypregnenolone) for acyl-carbon cleavage, the terminal oxygen of the peroxy anion is directed differently from that in the complex involving a steroid that is not cleaved (17α-hydroxyprogesterone) [41]. Support for the involvement a peroxy anion in acyl-carbon cleavage is also provided by molecular dynamic simulations [42]. However, with the latter approach, a revised view has been presented for the steps following formation of the tetrahedral adduct (**37**). It is argued that the adduct fragments, not by a homolytic mechanism as in Scheme 4.11 but by a heterolytic process in which the oxyanion in (**37**, Scheme 4.12), using the electron withdrawing properties of the Δ^4-3-one system of the substrate, aids in cleavage of the C-19–C-10 bond. We will later see that in other related enzymes, such a sink is not available, nor is this so in dihydrotestosterone, which recently has been shown to be a substrate for aromatase [43]. According to the proposal, the tri-oxy species (upper structure in **41**), which was produced following the cleavage step, undergoes O–O bond cleavage homolytically, producing a formyl radical and the Fe^{III}–O$^{\bullet}$ species [42]. The conversion of the formyl radical into the formate anion is then achieved by a roundabout process involving the transfer of an electron from Fe^{III}, which in turn recovers the electron from the enolate ion in ring A of **42**. The subsequent steps in the mechanism (Scheme 4.12) are essentially the same as in Scheme 4.11.

Attempts were made to explain the incorporation of the second oxygen atom from O_2 into formic acid by invoking a hydroxylation at the 2β-position that, via the intermediacy of a hemiacetal derivative, fragments by a heterolytic process using the electron withdrawing properties of the Δ^4-3-one system to produce the aromatic ring system of estrone [44]. Thus, this proposal retains the property of P450s to catalyze only the classical hydroxylation reaction during all three steps of the aromatase reaction. However, there are several pieces of evidence against this proposal [45–47]. Noteworthy among them is the finding of aromatization of the diene (**44**) where the possibility of hydroxylation at C-2 does not exist [46] (Scheme 4.13).

Yet another variant, in which all three stages in the aromatization of androstenedione use the oxo derivative, is the variant involving the compulsory role of the *gem*-diol (**15a**) in the cleavage process. In the advocacy of such a mechanism, it was asserted that the *gem*-diol formed in

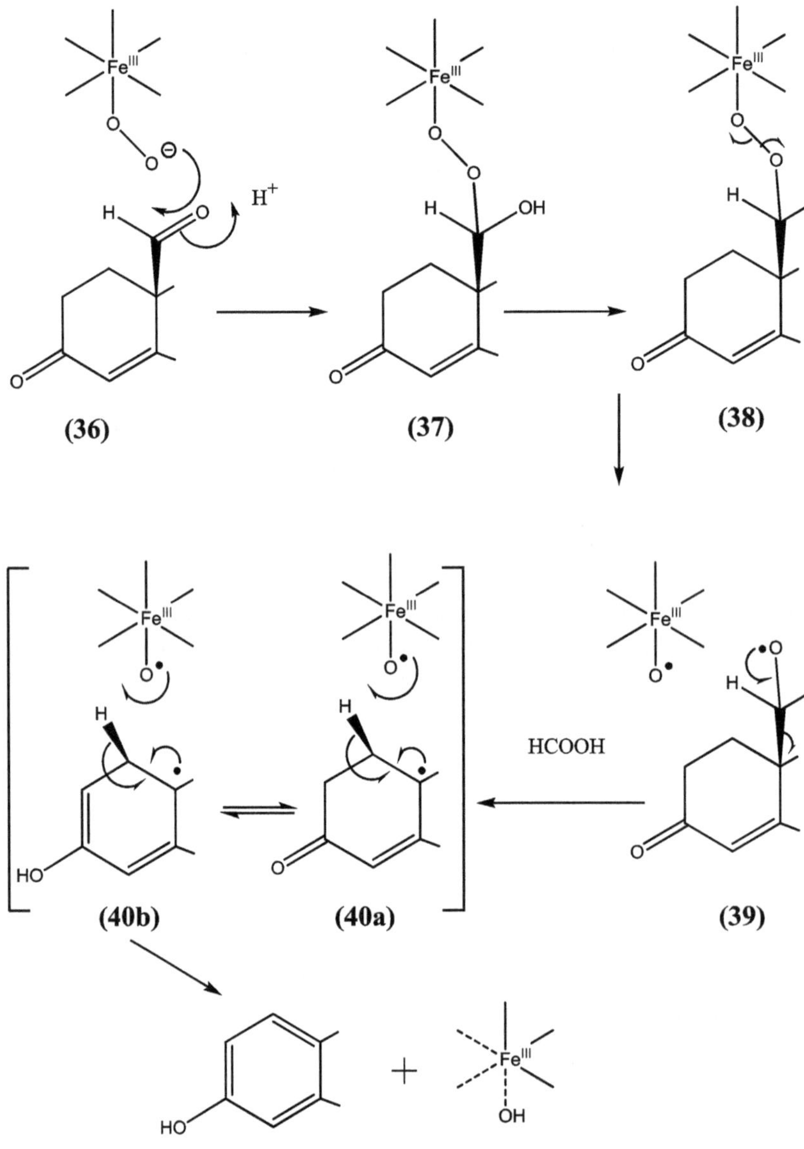

Scheme 4.11 Mechanism of the *Reaction 3* (Scheme 4.5) catalyzed by aromatase

Reaction 2 of the sequence in Scheme 4.5 is faithfully taken further without conversion to the aldehyde (**15b**) [48]. This assertion overlooked the clear demonstration that in the aromatization of the 19-hydroxysteroid, the 19-oxocompound (**15b**) accumulates in the medium before its utilization in *Reaction 3* (see Fig. 2 in [4]). Had the *gem*-diol been on the main path to acyl-carbon cleavage, aromatization of the 19-oxocompound itself would have required its hydration using water from the medium and then, retention of the medium oxygen in the released formate, which clearly was not the case [4]. Although this view [48] has been revised [42] by the original proposers, the mechanism continues to be cited by other workers [43].

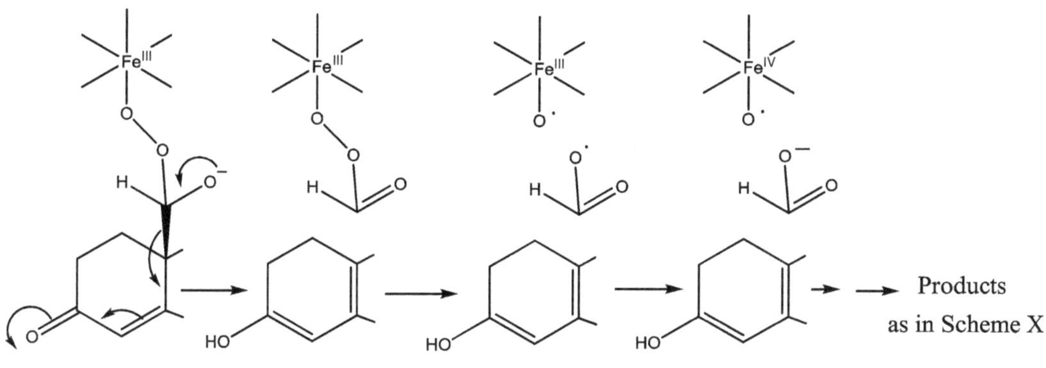

Scheme 4.12 An alternative mechanism for cleavage of the C-10–C-19 bond in the reaction catalyzed by aromatase as suggested in [42]

Scheme 4.13 Aromatization of a 3-deoxy precursor by aromatase

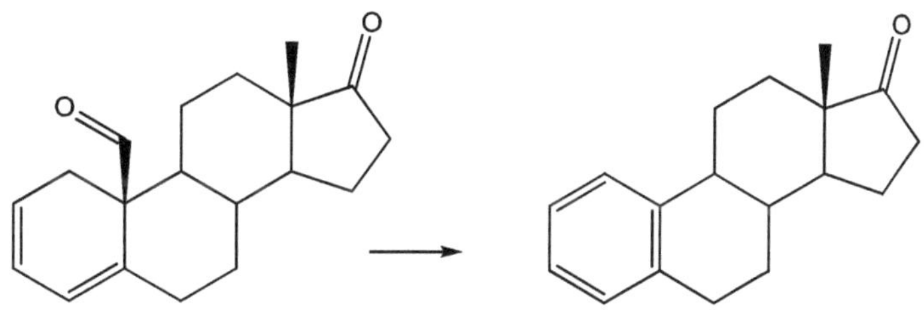

In 2009, a 2.9 Å resolution X-ray crystal structure of aromatase was published [49]. This was remarkable because the structure comprised the full length native protein purified from human placenta and showing the enzyme complexed with its natural androstenedione substrate (**13**). Examination of the positioning of the substrate in the active site enabled the authors to assign individual amino acids to catalytic roles predicted by chemical studies on the mechanism (and supported by site-specific mutagenesis studies) [50].

The structure shows that Asp309 is positioned so that, in its protonated form, it can hydrogen-bond to the 3-keto oxygen of androstenedione. This would not only be to aid substrate binding but would also promote the possible enolization of the ketone to form the 2,3-double bond as discussed above. To direct the enolization, it would be desirable to have groups aiding in the removal of the 2β hydrogen. It appears that the carbonyl of Ala306, with participation of the hydroxyl of Thr310 and possibly a catalytic water, is positioned to carry out this role. In the X-ray crystal structures of soluble bacterial P450 hydroxylases, the equivalent of these two residues has been proposed to take part in the proton relay network that provides the two protons required to break the O–O bond of the iron-bound peroxy species [51, 52]. This forms the iron monooxygen species required for hydroxylation. Thus, in aromatase, it is proposed that in the first two oxidative steps, Ala306 and Thr310 are involved in the formation of the oxo derivative and in the final step, they participate in the removal of the 2β hydrogen.

The A ring of 19-oxoandrostenedione, in a dienol form (**15c**), needs to be aromatized with a stereospecific removal of the 1β hydrogen (Scheme 4.5). It is interesting that no amino acid residue could be identified as participating in this process. This is consistent with our view of the mechanism, whereby removal of the 1β hydrogen occurs concomitantly with the breaking of the C10–C19 bond. In the radical fragmentation mechanism, the 1β hydrogen is removed as a hydrogen radical, quenching the oxygen radical remaining on the heme iron. The participation of other groups is not required [4].

The mechanism for aromatase considered in Scheme 4.11 has been directly applied to 14-α-demethylation as elaborated in Scheme 4.14. The key feature of the mechanism again is the crucial role of the peroxy anion that is trapped by the aldehydic carbonyl group to yield the tetrahedral intermediate (**47**). Fragmentation of the latter by a homolytic process generates the C-14 radical **48**, which through removal of the hydrogen atom by Fe^{III}–O$^\bullet$ produces the diene (**29**), regenerating the resting state of CYP51A1 [30]. Theoretical studies support all the elements of the mechanism of Scheme 4.14 but persist with the alternative possibility (thin arrows in Scheme 4.14) of the fragmentation of **50** by a heterolytic process though in this case, the active site will be needed to stabilize an otherwise unstable allylic carbanion intermediate (**51**) [53].

4.5 A General Mechanism of the Acyl-Carbon Bond Cleavage Step Applicable to CYP17A1

The three side-chain cleavage reactions catalyzed by CYP17A1 provide interesting diversity to test the basic tenets of the fragmentation process leading to acyl-carbon cleavage. The chemical mechanism for the cleavage step of CYP19A1 and CYP51A1 can be directly extended to the conversion of a 17α-hydroxyprogestogen to androgen (**33**), Scheme 4.15 by substituting an O–H for a C–H bond at the β-position with respect to the scissile acyl-carbon bond [36]. Now, an atom of oxygen from O_2, via the intermediacy of a peroxy-hemiketal intermediate (**51**), is incorporated into the expelled acetic acid through its homolytic cleavage, and the C-17 carbon radical thus generated is neutralized by a hydrogen transfer from the –OH group to Fe^{III}–O$^\bullet$ (**53** to **33**).

We now address the isotopic profile involved in the formation of the other two side-chain cleavage products, **34** and **35** (Scheme 4.15). The conversion of pregnenolone into **34** was shown to involve the *trans* elimination of two substituents, 17β-side chain and a 16α-hydrogen atom [38]. This stereochemical outcome makes it mandatory that the overall reaction is the consequence of a stepwise and not concerted process. The mode of formation of the 17α-androgen (**35**) is even more informative. Here, the 17β-oriented acetyl side chain in the substrate is replaced by a molecular oxygen-derived 17α-hydroxyl group, with the retention of the original 17α-hydrogen [38]. In other words, the C-17 hydrogen of the substrate has undergone an inversion of stereochemistry.

It is interesting to point out that the two cleavage products are also produced when the aldehyde analogue (**30**, 21-methyl group replaced by hydrogen) is the substrate [38, 54]. That the process with the analogue is ten times faster than with the natural substrate is consistent with the stronger electrophilicity of an aldehydic carbonyl compared to the ketonic carbonyl of the natural substrate, thus supporting the view that the conversion may involve a nucleophilic attack on the carbonyl carbon, as is embodied in our mechanism using a peroxy anion. In this particular case, the fragmentation of the peroxy-hemiketal (**54**) then produces the C-17 carbon radical, which can either yield a hydroxylated product by an oxygen-rebound process or the 16-ene by the abstraction of a hydrogen atom, as is the case in the last step of the reaction catalyzed by aromatase and 14α-demethylase [30].

The mechanism of the formation of **34** and **35** from a common intermediate establishes a kinship between the hydroxylation and desaturation process. In the case under consideration, that the ultimate event in the two processes occurs from the α-face of the steroid molecule minimizes the constraints at the active site of CYP17A1 that are required for the operation of two courses.

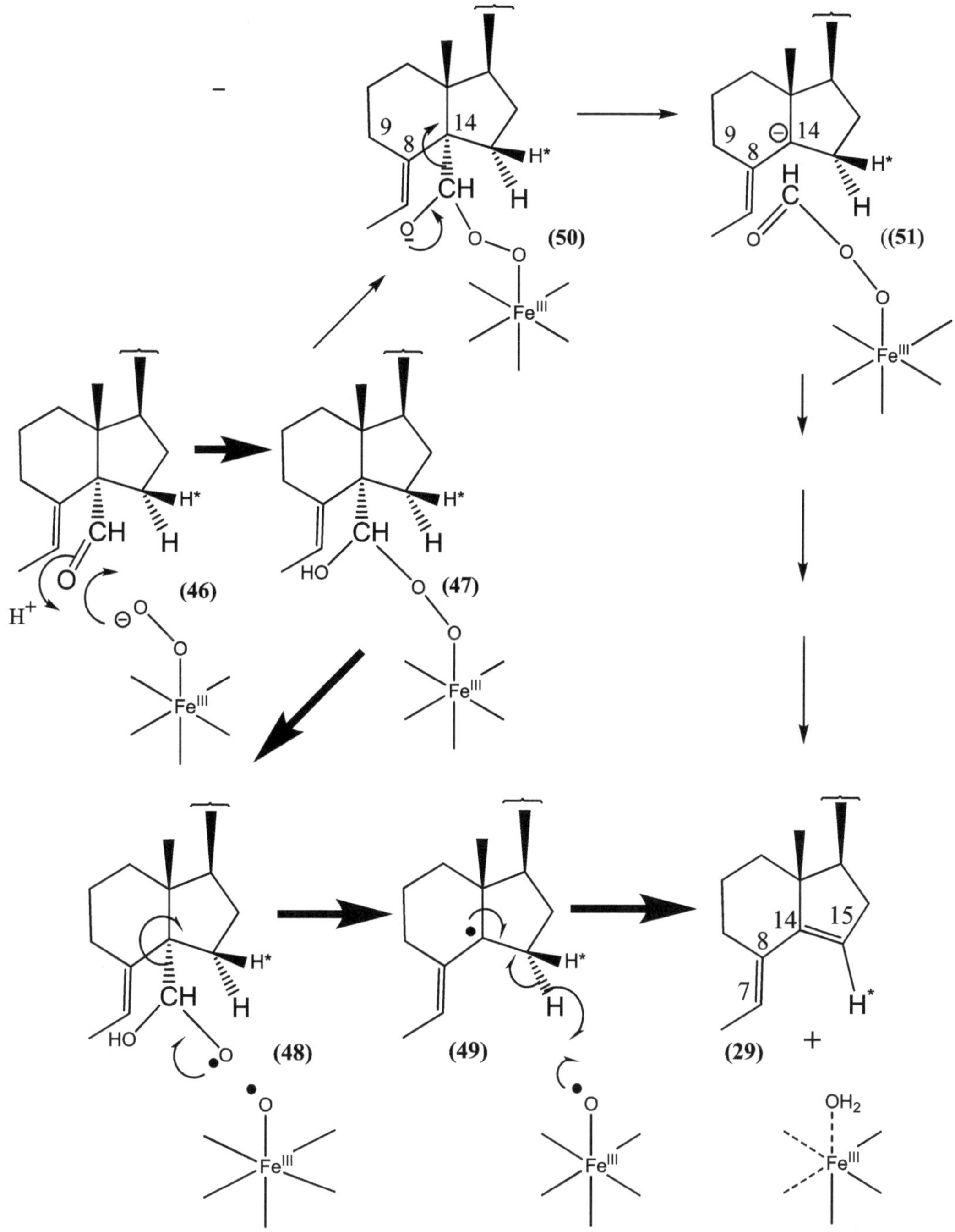

Scheme 4.14 The mechanism of the acyl-carbon bond cleavage by 14α-demethylase (CYP51A1) (*heavy arrows*). Suggestion in [53] (*thin arrows*)

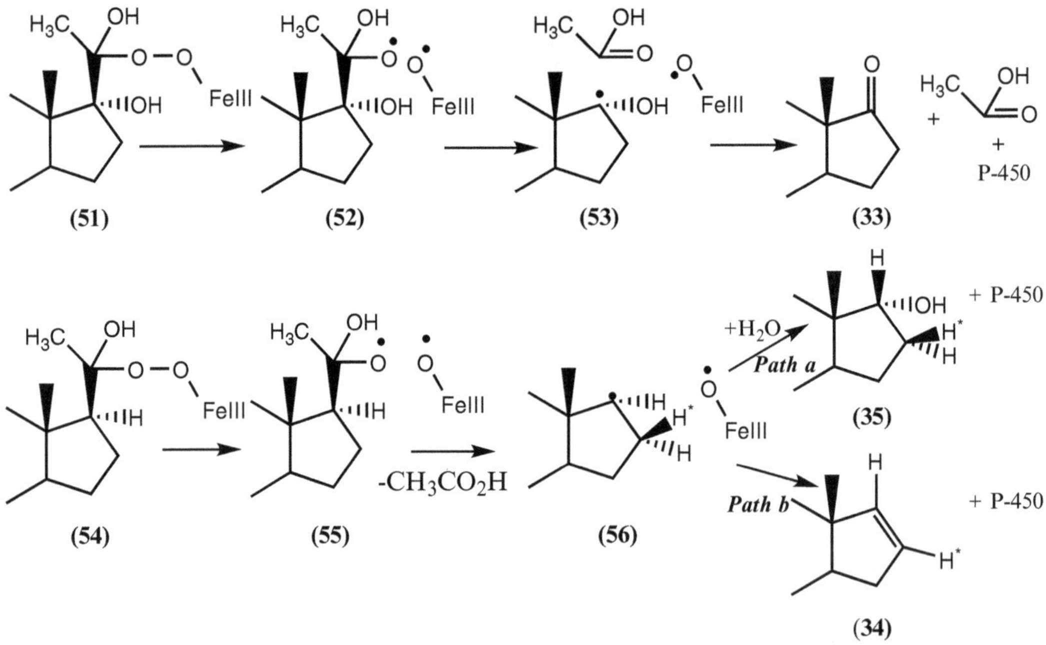

Scheme 4.15 Pathway for the formation of three cleavage products (**33**, **34** and **35**) from the peroxy adducts (**51** and **54**) in the reaction catalyzed by CYP17A1

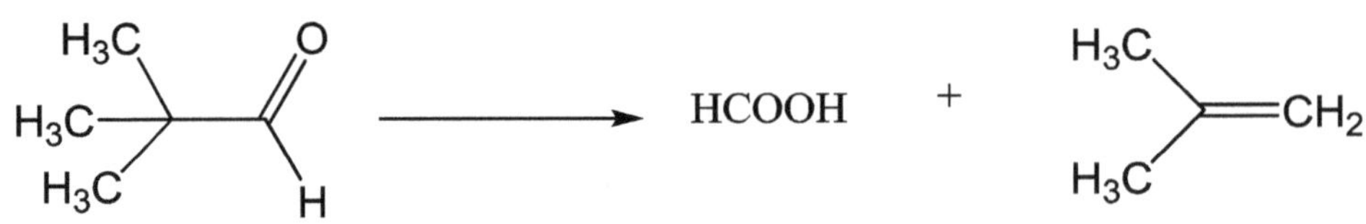

Scheme 4.16 Deformylation of an aliphatic aldehyde

4.6 Oxygen Dependent Acyl-Carbon Cleavage by Other Enzyme Systems

The earliest example of P450s, other than those involved in the steroid transformations, was the observation that P450 2B4 is able to use aliphatic aldehydes as substrates to generate products expected from an acyl-carbon type of cleavage [55]. The reaction was interpreted to follow the mechanism involved in the third reaction of aromatase (Scheme 4.16).

Mycobacterium tuberculosis degrades cholesterol through the oxidation of one of the *gem*-methyl groups in its side chain by CYP125A1. The overall sequence involving three reactions bears remarkable similarity to the reactions catalyzed by aromatase and 14α-demethylase. Thus, the reaction sequence includes two monooxygenation steps to generate an aldehyde that is deformylated by an acyl-carbon cleavage to generate an olefin or alternative products (Scheme 4.17) formed by the quenching of the key radical species (of the type **40**, Scheme 4.11) [56]. CYP125A1 differs from CYP19A1 and CYP51A1 in the sense that in this case, a substantial amount of the aldehyde is converted further into a carboxylic acid derivative by a monooxygenation reaction. A mutant form of CYP125A1 containing a selenocysteine axial ligand makes a small but significant contribution to enhance formation of the carboxylic acid relative to that of the deformylation products. Thus,

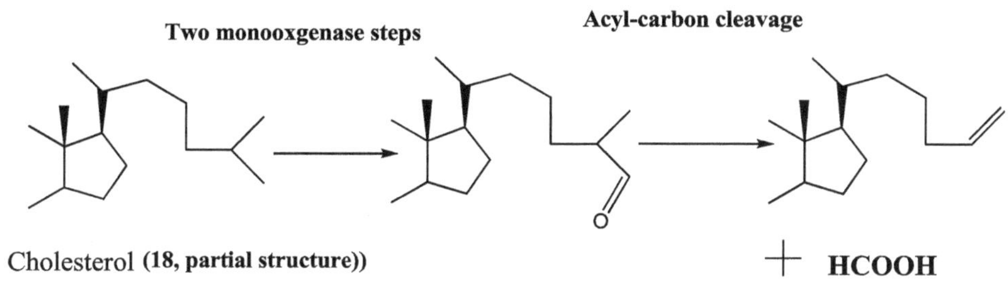

Scheme 4.17 The reactions catalyzed by *Mycobacterium tuberculosis* CYP125A1

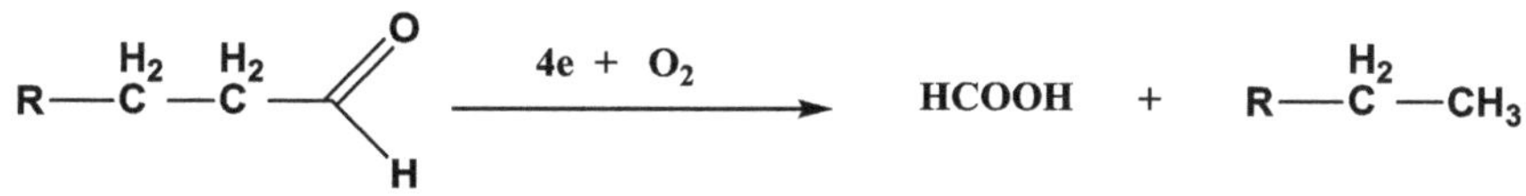

Scheme 4.18 Deformylation leading to hydrocarbon biosynthesis

it appears that in the selenium-containing CYP125A1, the balance of the branched pathway of Scheme 4.2 is shifted toward the formation of Compound I, presumably as a consequence of increasing the pK_{α} of the anion, thus favoring the formation of the protonated species, hence of Compound I, involved in monooxygenation.

Another recent development in the field has been the discovery of an enzyme system involved in the deformylation of long-chain aliphatic aldehydes, which was originally designated as aldehyde decarbonylase [57]. The enzyme has been purified from a strain of Cyanobacterial species and catalyzes an oxygen dependent formation of an alkane and formate from an aldehyde as shown in Scheme 4.18 [58]. The enzyme has been shown to contain a diferrous cofactor, for which a hypothetical mechanism modeled on the CYP19A1- and CYP51A1-catalyzed acyl-carbon cleavage reaction is shown in Scheme 4.19. Here, the diferrous state of the enzyme, upon reaction with O_2, forms a peroxide anion that reacts with aldehyde to give (**54**). Homolytic cleavage and fragmentation produces a carbon radical. In the CYP19A1 and CYP51A1 conversions, such an intermediate (**57**) is neutralized by hydrogen transfer to Fe^{III}-$O^{\bullet}$, generating the resting state of the P450 and producing an olefinic linkage. In the case of the aldehyde deformylase, a further 2e + $2H^+$ are required to produce the saturated olefin (**59**).

4.7 Modulation of the Acyl-Carbon Bond Cleavage Activity of CYP17A1 by Cytochrome b_5

In the cases of aromatase and 14α-demethylase, there seems to be no regulation of the formation or consumption of the intermediates of the pathways. Thus, the metabolism of the hydroxy- and oxo- intermediates (e.g. **14** and **15**, Scheme 4.5) is regulated by the same factors that govern the initial hydroxylation process. This may be so because the intermediates have no other biological role than to be funneled into the production of the final demethylated products. With CYP17A1, the situation is quite different. Here, the initial hydroxylated steroid is required on the one hand for the production of an androgenic steroid and on the other for that of anti-inflammatory corticoids. Therefore, on intuitive grounds, there seems to be a need for some sort of regulation of the pathway, in particular for that of androgen production in the female. Indeed, during the course of studies on the elucidation of the chemical mechanism of CYP17A1, it was discovered that for the human enzyme, the androgen-producing lyase reaction was heavily dependent on the presence of the membrane-bound form of cytochrome b_5 [34, 59, 60]. Table 4.1 summarizes the essential data in terms of K_{cat}/K_m values. For the hydroxylation of

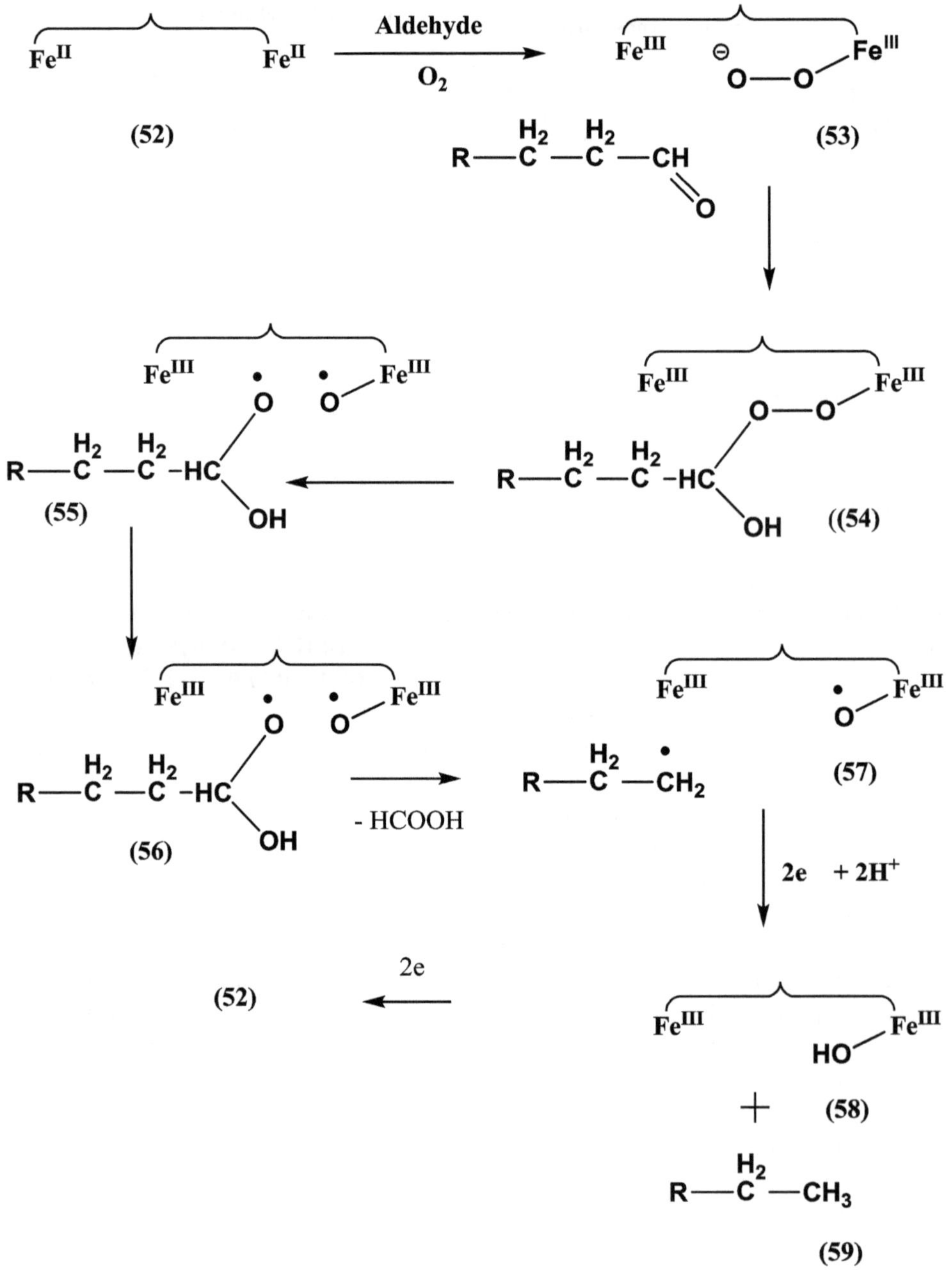

Scheme 4.19 Mechanism of deformylation by a diferrous enzyme modeled on the CYP19A1 and CYP51A1 reactions

pregnenolone (**30**), the parameter increased barely twofold in the presence cytochrome b_5, whereas the two side-chain cleavage reactions, from being non-existent in the absence of cytochrome b_5 reached a respectable level in its presence. Interestingly, with the Δ^4-steroid, progesterone (**31**), as a substrate, the direct cleavage represented by *Reaction 3* was undetectable, while the cleavage of the 17-α-hydroxylated progesterone (*Reaction 2*) was totally dependent on cytochrome b_5.

The proposed role of cytochrome b_5 in regulating the acyl-carbon bond cleavage activity of human CYP17A1 gains further support from

Table 4.1 Kinetic parameters of purified human CYP17A1[a]

	Kinetic parameters					
	Without cytochrome b_5			With cytochrome b_5		
	K_m	V_{max}	k_{cat}/K_m	K_m	V_{max}	k_{cat}/K_m
Δ^5-steroid						
Pregnenolone **30** → 17α-hydroxypregnenolone (of the type **32**)	0.65	2.20	5.6×10^4	0.65	3.90	1.0×10^5
Pregnenolone **30** → 5,16-diene (of the type **34**)	–	<0.02	–	0.65	0.39	1.0×10^4
17α-hydroxypregnenolone **32** → DHEA (of the type **33**)	–	<0.02	–	1.18	2.45	3.5×10^4
Δ^4-steroid						
Progesterone **31** → 17α-hydroxyprogesterone (partial structure **32**)	1.00	3.33	5.6×10^4	1.00	4.66	7.8×10^4
Progesterone **31** → 4,16-diene (partial structure **34**)	–	0.00	0.00	–	0.00	0.00
+						
17α-hydroxyprogesterone (partial **32**) → androstenedione (partial structure **33**)	–	0.00	0.00	5.00	1.80	6.0×10^3

[a]Activities are expressed as nmol of product formed/min/nmol P450. K_m data are in μM and k_{cat}/K_m units are $M^{-1}\,s^{-1}$ (Adapted from [33]. Structures referred to in Table 4.1 are illustrated in Scheme 4.10)

immunohistochemical studies [61]. The adrenal zona reticularis, the adrenocortical tissue layer that excretes androgens, was shown to express CYP17A1 and cytochrome b_5, whereas the zona fasciculata that excretes cortisol but not androgens expresses CYP17A1 only [62, 63]. In addition, in the human adrenal zona fasciculata and zona reticularis, the expression of CYP17A1 remains reasonably constant as a function of age, whereas the expression of cytochrome b_5 in the zona reticularis, at the onset of adrenarche, increases along with the acyl-carbon bond cleavage activity of CYP17A1 [64, 65]. Furthermore, high levels of cytochrome b_5 have been reported in human testes and in the adenomas of patients suffering from Cushing's Syndrome, which produce excessive androgens [62, 65–69].

To explore the nature of the interaction of CYP17A1 with cytochrome b_5, we performed mutagenesis of certain basic amino acid residues of CYP17A1 that we anticipated can form electrostatic interactions with acidic residues of cytochrome b_5. The basic residues targeted were selected from an amino acid linear sequence alignment of CYP17A1 with P-450$_{BM-3}$, being in or immediately adjacent to the P-450$_{BM-3}$ redox protein binding region [70]. Initially, residues targeted for mutation were changed to alanine. However, during the course of the work, single amino acid mutations, Arg^{347} → His and Arg^{358} → Gln, had been reported to result in the loss of the acyl-carbon cleavage activity of CYP17A1 and to cause sexual phenotype changes in 46XY human patients [71]. These disease-state mutants were then also constructed.

The mutant proteins were purified to homogeneity and detailed kinetic studies performed [72–75]. The two disease-state proteins and their alanine equivalents, as well as another mutant, Arg^{449} → Ala, exhibited the hydroxylation activity but were completely devoid of cytochrome b_5-dependent acyl-carbon cleavage activities. These mutants thus had been converted from multifunctional enzymes into conventional hydroxylases. Mutational experiments also showed that the interaction between CYP17A1 and cytochrome b_5 involves the cationic residues of the former, in particular Arg^{347}, Arg^{358} and possibly Arg^{349}. The anionic residues of cytochrome b_5 that pair with the three aforementioned cationic sites of CYP17A1 were not known, but a subsequent study on the mutation of cytochrome b_5 had shown that the substitution of Glu^{48} and Glu^{49} with Gly generated mutants that showed decreased activity to stimulate the side-chain cleaving activity of CYP17A1 [76].

The interaction of cytochrome b_5 with CYP17A1 has also been studied by NMR spectroscopy [77]. In these studies, the tail-less soluble form of cytochrome b_5 was used, which has been shown to be completely "inactive" in stimulating the side-chain cleavage activity of CYP17A1 [60]. Notwithstanding this, the NMR spectroscopy results by and large corroborate the kinetic studies in which the catalytic properties of CYP17A1 were studied. Thus, Glu^{48}-Ala and Glu^{49}-Ala mutants of the soluble form of cytochrome b_5 had impaired interaction with CYP17A1. Similarly, mutants of human CYP17A1 in which the cationic Arg^{347}, Arg^{358} and Arg^{449} residues, essential for cytochrome b_5-dependent stimulation of side-chain cleavage activity [74], had been replaced exhibited impaired NMR interaction with cytochrome b_5 [77]. The NMR spectroscopy data have also been interpreted to suggest that the binding of cytochrome b_5 and CPR to CYP17A1 is mutually exclusive. This conclusion is at variance with the kinetic data showing stimulation of cleavage activity of a mixture of CYP17A1 plus cytochrome b_5 by CPR, and of CPR plus CYP17A1 by cytochrome b_5, implying the involvement of a ternary complex comprising all three components as the catalytically competent entity. The interpretation of the NMR spectroscopy data will mean the involvement, presumably, in succession, of two binary complexes.

The role of cytochrome b_5 in regulating the various activities of P450s has been well documented [citations in 75] and it is argued that cytochrome b_5 may enhance the efficiency of electron transfer to a key iron-oxygen species involved in the catalytic P450 cycle [78]. Our results with an aldehyde analogue lacking the 21-methyl group (structure **30**, the 21-methyl group replaced by H), provided a forceful argument against the direct involvement of cytochrome b_5 in the electron transfer process during the side-chain cleavage catalyzed by CYP17A1 [74]. We had shown that deformylation of the aldehyde analogue occurred through the same chemical mechanism as the side-chain cleavage of the two physiological substrates [38]. However, this deformylation occurred at a rate that was fourfold faster than any physiological transformation catalyzed by CYP17A1, yet did not require, nor was enhanced by, cytochrome b_5. Furthermore, the mutant CYP17A1 proteins that had lost their physiological side-chain cleavage activities were able to cleave the formyl side chain. Thus, the oxidative cleavage of the acyl-carbon side chain by CYP17A1 does not depend on electron transfer properties of cytochrome b_5. Cumulatively, these findings bolstered our earlier hypothesis that the interaction of cytochrome b_5 with the CYP17A1-substrate complex causes protein conformational changes that culminate in directing the iron-peroxy species away from C-17 towards C-20 (Scheme 4.20). This facilitates a nucleophlic attack of the peroxide anion on the carbonyl carbon producing a tetrahedral adduct that follows the side-chain cleavage path. The resonance Raman spectroscopic studies cited above [41] would support the notion of a conformational change that stabilizes an iron-peroxide anion required for the cleavage reaction. This trend is more strongly noted when 17-α-hydroxypregnenolone is bound to CYP17A1 than when 17α-hydroxyprogesterone is bound. The data in Table 4.1 show that the former is a preferred substrate for cleavage.

Finally, we have shown that the kinetic parameters for the cytochrome b_5-dependent cleavage of the 17α-hydroxy-Δ^5-steroid (**30**) are more favorable than those for the cleavage of the 17α-hydroxy-Δ^4-steroid (**31**). When considered with the finding that the zona reticularis does not express 3β-hydroxysteroid dehydrogenase/$\Delta^{5\text{-}4}$ isomerase [79], it seems reasonable to infer that in vivo cleavage of 17α-hydroxypregnenolone produces dehydroepiandrosterone, which using the sulphotransferase activity of the adrenal cortex, is secreted as the inactive sulphoconjugate, which under normal conditions prevents the contribution toward the production of the androgenic steroids. These interrelated and coordinated transformations provide the mechanism by which the human adrenal gland avoids contributing to the unwanted production of active androgens, testosterone and androstenedione, enabling the female of the human species to escape from the physiological ramifications that could be promoted by the male hormone.

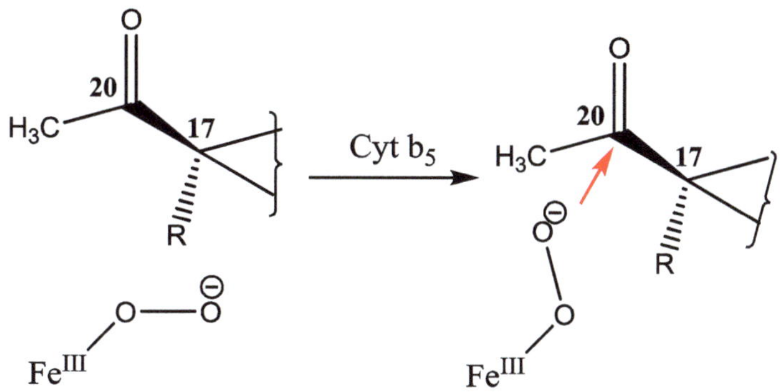

Scheme 4.20 Conformational role of cytochrome b_5 in modulating the cleavage activity of CYP17A1. The cartoon diagram shows that binding of cytochrome b_5 to CYP17A1 promotes protein conformation changes that direct the iron-oxygen species away from C-17 toward the carbonyl carbon at C-20 to allow the formation of a tetrahedral complex (shown by *red color*)

Reviews Apart from citations in the text [2, 5–8, 75], useful reviews dealing with specific aspects of the subject are available [80, 81].

Note added in proof Recently another alternative to the mechanism of Scheme 4.11 has been suggested (Yoshimoto FK, Guengerich FP, J Am Chem Soc 136: 15016–25).

Acknowledgments We acknowledge with pleasure the help of Dr. Peter Lee-Robichaud in writing the section on the interaction of CYP17A1 with cytochrome b_5, an area in which he has made seminal contributions.

References

1. Omura T, Sato RA (1962) New cytochrome in liver microsomes. J Biol Chem 237:1375–1376
2. Ortiz de Montellano PR (2005) Cytochrome P-450: structure, mechanism, and biochemistry, 3rd edn. Kluwer Academic/Plenum Press, New York
3. Akhtar M, Calder MR, Corina DL, Wright JN (1981) The status of oxygen atoms in the removal of C-19 in oestrogen biosynthesis. J Chem Soc Chem Commun:129–131
4. Akhtar M, Calder MR, Corina DL, Wright JN (1982) Mechanistic studies on C-19 demethylation in oestrogen biosynthesis. Biochem J 201:569–580
5. Stevenson DE, Wright JN, Akhtar M (1988) Mechanistic consideration of P-450 dependent enzymic reactions: studies on oestriol biosynthesis. J Chem Soc Perkin Trans 1:2043–2052
6. Akhtar M, Wright JN (1991) A unified mechanistic view of oxidative reactions catalysed by P-450 and related Fe-containing enzymes. Nat Prod Rep 8:527–551
7. Groves JT (2005) Models and mechanisms of cytochrome P450 action. In: Ortiz de Montellano PR (ed) Cytochrome P450: structure, mechanism, and biochemistry, 3rd edn. Kluwer Academic/Plenum Publishers, New York, pp 1–43
8. Makris TM, Denisov I, Schlichting I, Sligar SG (2005) Activation of molecular oxygen by cytochrome P450. In: Ortiz de Montellano PR (ed) Cytochrome P450: structure, mechanism, and biochemistry, 3rd edn. Kluwer Academic/Plenum Publishers, New York, pp 149–182
9. Sharrock M, Debrunner PG, Shultz C, Lipscomb JD, Marshall VP, Gunsalus IC (1976) Cytochrome P-450_{cam} and its complexes: mossbauer parameter of the haem-iron. Biochim Biophys Acta 420:8–26
10. Weiss JJ (1964) Nature of the iron-oxygen bond in oxyhaemoglobin. Nature 202:83–84
11. Pauling L (1964) Nature of the iron–oxygen bond in oxyhæmoglobin. Nature 203:182–183
12. Morato T, Hyano M, Dorfman RI, Axelrod LR (1961) The intermediate steps in the biosynthesis of estrogens from androgens. Biochem Biophys Res Commun 20:334–338
13. Ryan KJ (1959) Biological aromatisation of steroids. J Biol Chem 234:268–272
14. Akhtar M, Skinner SJM (1968) The intermediary role of a 19-oxoandrogen in the biosynthesis of oestrogen. Biochem J 109:318–321
15. Skinner SJM, Akhtar M (1969) The stereospecific removal of a C-19 hydrogen atom in oestrogen biosynthesis. Biochem J 114:75–81
16. Arigoni D, Battaglia R, Akhtar M, Smith T (1975) Stereospecificity of oxidation at C-19 in oestrogen biosynthesis. J Chem Soc Chem Commun 185–186
17. Osawa Y, Shibata K, Rohrer D, Week C, Duax WL (1975) Reassignment of the absolute configuration of 19-substituted 19-hydroxysteroids and stereomechanism of estrogen biosynthesis. J Am Chem Soc 97:4400–4402

18. Thompson EA, Siiteri PK (1974) Utilization of oxygen and reduced nicotinamide adenine dinucleotide phosphate by human placental microsomes during aromatization of androstenedione. J Biol Chem 249:5364–5374
19. Bloch K (1965) The biological synthesis of cholesterol. Science 150:19–28
20. Miller WL, Brady DR, Gaylor JL (1971) Investigation of the component reactions of oxidative demethylation of sterols: metabolism of 4α-hydroxymethyl steroids. J Biol Chem 246:5147–5153
21. Bloxham DP, Wilton DC, Akhtar M (1971) Studies on the mechanism and regulation of C-4 demethylation in cholesterol biosynthesis: the role of adenosine 3′,5′-cyclic monophosphate. Biochem J 125:625–634
22. Rahier A (2011) Dissecting the sterol C-4 demethylation process in higher plants. From structures and genes to catalytic mechanism. Steroids 76:350–352
23. Olson JA, Landberg M, Bloch K (1957) On the demethylation of lanosterol to cholesterol. J Biol Chem 228:941–956
24. Canonica L, Fiecchi A, Kienele MG, Scala A, Galli G, Paoletti EG, Paoletti RJ (1968) The fate of the 15-beta hydrogen of lanosterol in cholesterol biosynthesis. J Am Chem Soc 90:3597–3598
25. Akhtar M, Rahimtula AD, Watkinson IA, Wilton DC, Munday KA (1969) The status of C-6, C-7, C-15 and C-16 hydrogen atoms in cholesterol biosynthesis. Eur J Biochem 9:107–111
26. Gibson GF, Goad LJ, Goodwin TW (1968) J Chem Soc Chem Commun:1458–1460
27. Watkinson IA, Wilton DC, Munday KA, Akhtar M (1971) The formation and reduction of the 14,15-double bond in cholesterol biosynthesis. Biochem J 121:131–137
28. Akhtar M, Freeman CW, Wilton DC, Boar RB, Copsey DB (1977) The pathway for the removal of the 14α-methyl group of lanosterol: the role of lanost-8-ene-3β,32-diol in cholesterol biosynthesis. Bioorg Chem 6:473–481
29. Akhtar M, Alexander K, Boar RB, McGhie JF, Barton DHR (1978) Chemical and enzymic studies on the characterisation of intermediates during the removal of the 14α-methyl group in cholesterol biosynthesis. Biochem J 169:449–463
30. Shyadehi AZ, Lamb DC, Kelly SL, Kelly DE, Schunck WH, Wright JN, Corina D, Akhtar M (1996) The mechanism of the acyl-carbon bond cleavage reaction catalyzed by recombinant sterol 14-α-demethylase of *Candida albicans* (other names are: lanosterol 14α-demethylase, P-450_{14DM} and CYP51). J Biol Chem 271:12445–12450
31. Nakajin S, Hall SPF, Onoda M (1981) Testicular microsomal cytochrome P-450 for C_{21} steroid side chain cleavage. Spectral and binding studies. J Biol Chem 256:6134–6139
32. Nakajin S, Takahashi M, Shinoda M, Hall SPF (1985) Cytochrome b_5 promotes the synthesis of Δ^{16}–C19 steroids by homogeneous cytochrome P-450 C21 side-chain cleavage from pig testes. Biochem Biophys Res Commun 132:708–713
33. Lee-Robichaud P, Wright JN, Akhtar ME, Akhtar M (1995) Modulation of the activity of human 17-α-hydroxylase-17,20-lyase (CYP17) by cytochrome b_5: endocrinological and mechanistic implications. Biochem J 308:901–908
34. Gower BD, Holland KT, Mallet AI, Rennie PJ, Watkins WJ (1994) Comparison of 16-androstene steroid concentrations in sterile apocrine sweat and axillary secretions: interconversion of 16-androstenes by the axillary microflora–a mechanism for axillary odour production on man? J Steroid Biochem Mol Biol 48:409–418
35. Miller SL, Wright JN, Corina DL, Akhtar M (1991) Mechanistic studies on pregnene side-chain cleavage enzyme (17α-hydroxylase-17,20-lyase) using ^{18}O. J Chem Soc Chem Commun:157–159
36. Akhtar M, Corina DL, Miller SL, Shyadehi AZ, Wright JN (1994) Mechanism of the acyl-carbon cleavage and related reactions catalysed by multifunctional P-450s: studies on cytochrome P-$450_{17\alpha}$. Biochemistry 33:4410–4418
37. Akhtar M, Corina DL, Miller SL, Shyadehi AZ, Wright JN (1994) Incorporation of label from $^{18}O_2$ into acetate during side-chain cleavage catalysed by cytochrome P-$450_{17\alpha}$ (17α-hyroxylase-17,20-lyase). J Chem Soc Perkin Trans 1:263–267
38. Lee-Robichaud P, Shyadehi AZ, Wright JN, Akhtar ME, Akhtar M (1995) Mechanistic kinship between hydroxylation and desaturation reactions: acyl-carbon cleavage promoted by pig and human CYP17 (P-$450_{17\alpha}$; 17α-hydroxylase-17,20-lyase. Biochemistry 34:14104–14113
39. Imai T, Globerman H, Gertner JM, Kagawa N, Waterman MR (1993) Expression and purification of functional human 17α-hydroxylase/17,20-lyase (P450c17) in *Escherichia coli*. Use of this system for study of a novel form of combined 17-α-hydroxylase/17,20-lyase deficiency. J Biol Chem 268:19681–19689
40. Akhtar M, Corina DL, Pratt J, Smith T (1976) Studies on the removal of C-19 in oestrogen biosynthesis using 18O2. J Chem Soc Chem Commun:854–856
41. Gregory M, Mak PJ, Sligar SG, Kinkaid JR (2013) Differential hydrogen bonding in human CYP17 dictates hydroxylation versus lyase chemistry. Angew Chem Int Ed 52:5342–5345
42. Sen K, Hackett JC (2012) Coupled electron transfer and proton hopping in the final step of CYP19-catalyzed androgen aromatization. Biochemistry 51:3039–3049
43. Cheng Q, Sohl CD, Yoshimoto FK, Guengerich FP (2012) Oxidation of dihydrotestosterone by human cytochromes P450 19A1 and 3A4. J Biol Chem 287:29554–29567
44. Goto J, Fishman J (1977) Participation of a non-enzymic transformation in the biosynthesis of estrogens from androgens. Science 195:80–81

45. Caspi E, Wicha J, Arunachalam T, Nelson P, Spiteller G (1984) Estrogen biosynthesis. Concerning the obligatory intermediary of 2β-hydroxy-10-β-formylandrost-4-ene-3,17-dione. J Am Chem Soc 106:7282–7283
46. Cole PA, Bean JM, Robinson CH (1990) Conversion of a 3-deoxysteroid to 3-deoxyestrogen by human placental aromatase. Proc Natl Acad Sci U S A 87:2999–3003
47. Numazawa M, Nagoaka M, Sohtome N (2005) Aromatase reaction of 3-deoxyandrogens: steric mode of the C-19 oxygenation and cleavage of the C10-C19 bond by human placental aromatase. Biochemistry 44:10839–10485
48. Hackett JC, Brueggermeier RB, Hadad CM (2005) The final step of cytochrome P450 aromatase: a density functional theory study. J Am Chem Soc 127:5224–5287
49. Gosh D, Griswold J, Erman M, Pangborn W (2009) Structural basis for androgen specificity and oestrogen synthesis in human aromatase. Nature 457:219–223
50. Kadoharma N, Yarborough C, Zhou D, Chen S, Osawa Y (1992) Kinetic properties of aromatase mutants Pro308Phe, Asp309Asn and Asp309Ala and their interactions with aromatase inhibitors. J Steroid Biochem Mol Biol 43:693–701
51. Nagano S, Poulos TL (2005) Crystallographic study on the dioxygen complex of wild-type and mutant cytochrome P450cam. Implications for the dioxygen activation mechanism. J Biol Chem 280:31659–31663
52. Nagano S, Cupp-Vickery JR, Poulos TL (2005) Crystal structures of the ferrous dioxygen complex of wild-type cytochrome P450eryF and its mutants, A245S and A245T: investigation of the proton transfer system in P450eryF. J Biol Chem 280:22102–22107
53. Sen K, Hackett JC (2010) Peroxy-iron mediated deformylation in sterol 14α-demethylase catalysis. J Am Chem Soc 132:10293–10305
54. Robichaud P, Wright JN, Akhtar M (1994) Involvement of an O_2-derived nucleophilic species in acyl-carbon cleavage, catalysed by cytochrome P-45017α: implications for related P-450 catalysed fragmentation reactions. J Chem Soc Chem Commun:1501–1503
55. Roberts ES, Vaz ADN, Coon MJ (1991) Catalysis by cytochrome P-450 of an oxidative reaction in xenobiotic aldehyde metabolism: deformylation with olefin formation. Proc Natl Acad Sci U S A 88:8963–8966
56. Sivaramakrishnan S, Ouellet H, Matsumura H, Guan S, Moënne-Loccoz P, Burlingame AL, Ortiz de Montellano PR (2012) Proximal ligand electron donation and reactivity of the cytochrome P450 ferric–peroxo anion. J Am Chem Soc 134:6673–6684
57. Schirmer A, Rude MA, Li XZ, Popova E, del Cardayre SB (2010) Microbial biosynthesis of alkanes. Science 329:559–562
58. Pandelia ME, Li N, Nørgaard H, Warui DM, Rajakovich LJ, Chang W, Booker SJ, Krebs C, Bollinger JM Jr (2013) Substrate-triggered addition of dioxygen to the diferrous cofactor of aldehyde-deformylating oxygenase to form a diferric-peroxide intermediate. J Am Chem Soc 135:15801–15812
59. Katagiri M, Kagawa N, Waterman MR (1995) The role of cytochrome b_5 in the biosynthesis of androgens by human P-450c17. Arch Biochem Biophys 317:343–347
60. Lee-Robichaud P, Kaderbhai MA, Kaderbhai N, Wright JN, Akhtar M (1997) Interaction of human CYP17 (P-45017α, 17α-hydroxylase-17,20-lyase) with cytochrome b_5: importance of the orientation of the hydrophobic domain of cytochrome b_5. Biochem J 321:857–863
61. Suzuki T, Sasano H, Sawai T, Mason JI, Nagura H (1992) Immunohistochemistry and in situ hybridization of P-$450_{17\alpha}$ (17α-hydroxylase/17,20-lyase). J Histochem Cytochem 40:903–908
62. Dharia S, Slane A, Jian M, Connor M, Conley A, Parker CR Jr (2004) Colocalization of P450c17 and cytochrome b5 in androgen-synthesizing tissues of the human. Biol Reprod 71:83–88
63. Mapes S, Tarantal A, Parker CR Jr, Moran FM, Bahr JM, Pyter L, Conley A (2002) Adrenocortical cytochrome b5 expression during fetal development of the rhesus macaque. Endocrinology 143:1451–1458
64. Kaneko TC, Freije WA, Carr BR, Rainey WE (2000) Developmental changes in steroidogenic enzymes in human postnatal adrenal cortex: immunohistochemical studies. Clin Endocrinol 53:739–747
65. Mason JI, Estabrook RW, Purvis JL (1973) Testicular cytochrome P-450 and iron-sulphur protein as related to steroid metabolism. Ann NY Acad Sci 212:406–419
66. Sakai Y, Yanase T, Takayanagi R, Nakao R, Nishi Y, Haji M, Nawata H (1993) High expression of cytochrome b_5 in adrenocortical adenomas from patients with Cushing's syndrome associated with high secretion of adrenal androgens. J Clin Endocrinol Metab 76:1286–1290
67. Sakai Y, Yanase T, Hara T, Takayanagi R, Haji M, Nawata H (1994) In-vitro evidence for the regulation of 17,20-lyase activity by cytochrome b5 in adrenocortical adenomas from patients with Cushing's syndrome. Clin Endocrinol Oxf 40:205–209
68. Sakai Y, Yanase T, Hara T, Takayanagi R, Haji M, Nawata H (1994) Mechanism of abnormal production of adrenal androgens in patients with adrenocortical adenomas and carcinomas. J Clin Endocrinol Metab 78:36–40
69. Yanase T, Sasano H, Yubisui T, Sakai Y, Takayanagi R, Nawata H (1998) Immunohistochemical study of cytochrome b_5 in human adrenal gland and in adrenocortical adenomas from patients with Cushing's syndrome. Endocr J 45:89–95
70. Ravichandran KG, Boddupalli SS, Hasemann CA, Peterson JA, Deisenhofer J (1993) Crystal-structure of hemoprotein domain of P450BM-3, a prototype for microsomal P450s. Science 261:731–736
71. Geller DH, Auchus RJ, Mendonca BB, Miller WL (1997) The genetic and functional basis of isolated 17,20-lyase deficiency. Nat Genet 17:201–205

72. Lee-Robichaud P, Akhtar ME, Akhtar M (1998) Control of androgen biosynthesis in the human through the interaction of Arg^{347} and Arg^{358} of CYP17 with cytochrome b_5. Biochem J 332:293–296
73. Lee-Robichaud P, Akhtar ME, Akhtar M (1999) Lysine mutagenesis identifies cationic charges of human CYP17 that interact with cytochrome b_5 to promote male sex-hormone biosynthesis. Biochem J 342:309–312
74. Lee-Robichaud P, Akhtar ME, Wright JN, Sheikh QI, Akhtar M (2004) The cationic charges on Arg^{347}, Arg^{358} and Arg^{449} of human cytochrome P450c17 (CYP17) are essential for the enzyme's cytochrome b_5-dependent acyl-carbon cleavage activities. J Steroid Biochem Mol Biol 92:119–130
75. Akhtar M, Wright JN, Lee-Robichaud P (2011) A review of mechanistic studies on aromatase (CYP19) and 17α-hydroxylase-17,20-lyase (CYP17). J Steroid Biochem Mol Biol 125:2–12
76. Jacqueline L, Naffin-Olivos JL, Auchus RJ (2006) Human cytochrome b_5 requires residues E48 and E49 to stimulate the 17,20-lyase activity of cytochrome P450c17. Biochemistry 45:755–762
77. Estrada DF, Laurence JE, Scott EE (2013) Substrate-modulated cytochrome P450 17A1 and cytochrome b_5 interactions revealed by NMR. J Biol Chem 288:17008–17018
78. Pompon D, Coon MJ (1984) On the mechanism of action of cytochrome P-450: oxidation and reduction of the ferrous dioxygen complex of liver microsomal cytochrome P-450 by cytochrome b_5. J Biol Chem 259:15377–15385
79. Mapes S, Corbin CJ, Tarantal A, Conley A (1999) The primate adrenal zona reticularis is defined by expression of cytochrome b5, 17α-hydroxylase/17,20-lyase cytochrome P450 (P450c17) and NADPH-cytochrome P450 reductase (reductase) but not 3-β-hydroxysteroid dehydrogenase/Δ5-4-isomerase (3β-HSD). J Clin Endocrinol Metab 84:3382–3385
80. Hrycay EG, Bandiera SM (2012) The monooxygenase, peroxidase, and peroxygenase properties of cytochrome P450. Arch Biochem Biophys 522:71–89
81. Wertz DL, Valentine JS (2000) Nucleophilicity of iron-peroxo porphyrin complexes. Struct Bonding 97:37–60

5 Regioselective Versatility of Monooxygenase Reactions Catalyzed by CYP2B6 and CYP3A4: Examples with Single Substrates

Claudio A. Erratico, Anand K. Deo, and Stelvio M. Bandiera

Abstract

Hepatic microsomal cytochrome P450 (CYP) enzymes have broad and overlapping substrate specificity and catalyze a variety of monooxygenase reactions, including aliphatic and aromatic hydroxylations, *N*-hydroxylations, oxygenations of heteroatoms (N, S, P and I), alkene and arene epoxidations, dehalogenations, dehydrogenations and *N*-, *O*- and *S*-dealkylations. Individual CYP enzymes typically catalyze the oxidative metabolism of a common substrate in a regioselective and stereoselective manner. In addition, different CYP enzymes often utilize different monooxygenase reactions when oxidizing a common substrate. This review examines various oxidative reactions catalyzed by a CYP enzyme acting on a single substrate. In the first example, 2,2′,4,4′-tetrabromodiphenyl ether (BDE-47), a halogenated aromatic environmental contaminant, was oxidatively biotransformed by human CYP2B6. Nine different metabolites of BDE-47 were produced by CYP2B6 via monooxygenase reactions that included aromatic hydroxylation, with and without an NIH-shift, dealkylation and debromination. In the second example, lithocholic acid (3α-hydroxy-5β-cholan-24-oic acid), an endogenous bile acid, served as a substrate for human CYP3A4 and yielded five different metabolites via aliphatic hydroxylation and dehydrogenation reactions.

C.A. Erratico
Faculty of Pharmaceutical Sciences, The University of British Columbia, Vancouver, British Columbia V6T1Z3, Canada

Toxicology Center, University of Antwerp, Wilrijk 2610, Belgium

A.K. Deo
Faculty of Pharmaceutical Sciences, The University of British Columbia, Vancouver, British Columbia V6T1Z3, Canada

UCB Pharma, Briane l'Alleud 1420, Belgium

S.M. Bandiera (✉)
Faculty of Pharmaceutical Sciences, The University of British Columbia, Vancouver, British Columbia V6T1Z3, Canada
e-mail: bandiera@mail.ubc.ca

E.G. Hrycay and S.M. Bandiera (eds.), *Monooxygenase, Peroxidase and Peroxygenase Properties and Mechanisms of Cytochrome P450*, Advances in Experimental Medicine and Biology 851, DOI 10.1007/978-3-319-16009-2_5,

Keywords
Cytochrome P450 • Human CYP2B6 • Human CYP3A4 • Lithocholic acid • Oxidative biotransformation • Polybrominated diphenyl ethers • BDE-47 and BDE-99 • Hydroxylated metabolites

5.1 Introduction

One of the defining characteristics of cytochrome P450 (CYP[1]) enzymes is the large number and variety of substrates and the diversity of reactions catalyzed. This is especially true for the mammalian hepatic microsomal CYP enzymes involved in the oxidation of exogenous compounds. In contrast, the substrate selectivity of mitochondrial and most bacterial (microbial) CYP enzymes is much more restrictive.

The criteria governing substrate selectivity for human CYP enzymes are of considerable interest but are not well defined at present. It is well known that CYP enzymes have broad and overlapping substrate specificity and that many exogenous and endogenous compounds are substrates for more than one CYP enzyme, presumably because their structures fit the active sites of these enzymes. When several CYP enzymes participate in the metabolism of a substrate, they exhibit varying degrees of effectiveness. One CYP enzyme will often have higher catalytic efficiency, as reflected in differences in V_{max} and K_m values, compared with other CYP enzymes. These differences have important implications for in vivo and in vitro studies because the CYP enzyme primarily responsible for catalysis can change depending on the substrate concentration, the effect of induction and on other factors that alter CYP catalytic activity or expression. For example, the *O*-deethylation of ethoxyresorufin, a widely used catalytic probe for monitoring CYP1A induction (Fig. 5.1), is catalyzed by CYP1A, CYP2B and CYP2C enzymes, but the activity of purified rat CYP1A1 is approximately 10–20 times, 500 times and 30–40 times greater than that of purified rat CYP1A2, CYP2B1 and CYP2C6, respectively [1–3]. Owing to the low constitutive expression of CYP1A1, CYP1A2 and CYP2B enzymes, most of the ethoxyresorufin *O*-deethylase activity in livers of untreated rats is catalyzed by CYP2C enzymes, whereas the activity is catalyzed predominantly by CYP1A1 in livers of polycyclic aromatic hydrocarbon-pretreated rats [1–3].

In addition to differences in catalytic efficiency, CYP enzymes will often catalyze oxidative metabolism of a common substrate at different positions. For example, dextromethorphan, a non-opioid antitussive, undergoes extensive biotransformation in vivo and is metabolized by CYP2D6 and CYP3A4 in a position-dependent manner, with CYP2D6 catalyzing *O*-demethylation and CYP3A4 catalyzing *N*-demethylation [4] (Fig. 5.2). Structurally similar opioid compounds including codeine, morphine, oxycodone and tramadol also undergo *O*-demethylation by CYP2D6 and *N*-demethylation by CYP3A4 with different clinical outcomes [5]. While *O*-demethylation is the major pathway contributing to the in vivo clearance of dextromethorphan, it is a comparatively minor important pathway for the in vivo clearance of codeine [6].

Steroids are hydroxylated by CYP enzymes in a highly regioselective and stereospecific manner. Testosterone is a well-known example of a steroid substrate that undergoes hydroxylation at multiple sites by multiple CYP enzymes. Testosterone is hydroxylated at the 6β- and 15-β-positions by human CYP3A4 and rat

[1] *Abbreviations*: *BDE-47* 2,2′,4,4′-tetrabromodiphenyl ether, *BDE-99* 2,2′,4,4′,5-pentabromodiphenyl ether, *CAR* constitutive androstane receptor, *CYP* cytochrome P450, *lithocholic acid* 3α-hydroxy-5β-cholan-24-oic acid, *PBDEs* polybrominated diphenyl ethers, *PXR* pregnane X receptor, *QSAR* quantitative structure-activity relationship.

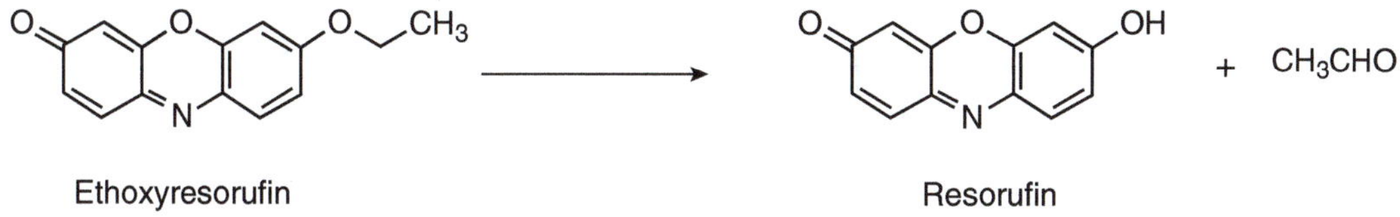

Fig. 5.1 CYP-catalyzed *O*-dealkylation of ethoxyresorufin

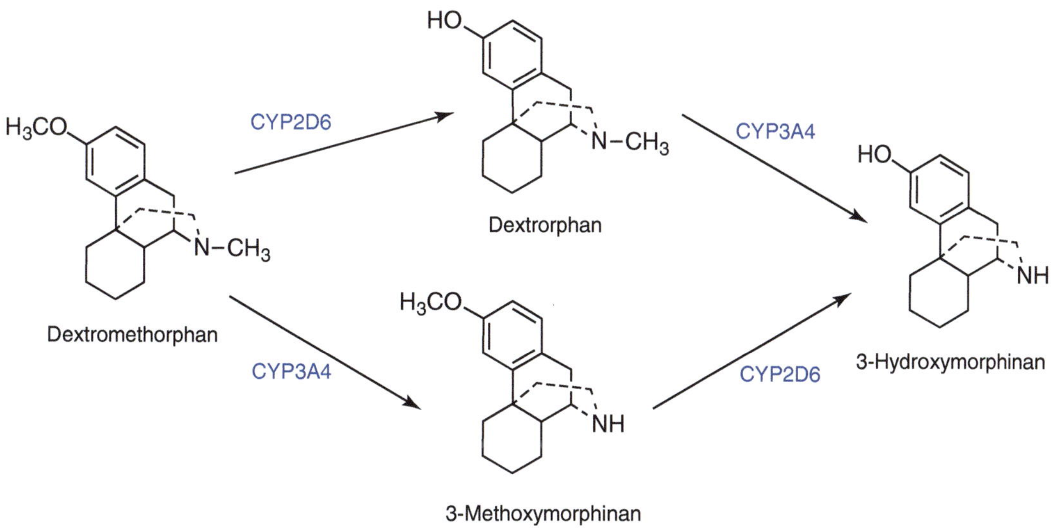

Fig. 5.2 Oxidative biotransformation of dextromethorphan by CYP2D6 and CYP3D4

CYP3A1, at the 7α-position by rat CYP2A1, at the 2α- and 16α-positions by rat CYP2C11, and at the 17β-position by human CYP2C9 and CYP2C19 and by rat CYP2B enzymes [7–9] (Fig. 5.3).

In contrast, some substrates are metabolized by a single CYP enzyme at multiple sites. For example, CYP2C9 metabolizes the anticoagulant, (*S*)-warfarin, but not (*R*)-warfarin or coumarin, by hydroxylating (*S*)-warfarin at the 6- and 7-positions [10, 11] (Fig. 5.4). These two metabolites account for 80–85 % of the in vivo clearance of (*S*)-warfarin [12]. Thus, regioselective and stereoselective oxidation of substrates by either single or multiple CYP enzymes is not uncommon.

One approach to understanding substrate selectivity is to assess the molecular and physical-chemical properties of substrates for each of the human CYP enzymes. Quantitative structure-activity relationship (QSAR) analysis can also be useful. QSAR evaluations have been conducted with CYP1A2, CYP2B6, CYP2C9, CYP2D6 and CYP3A4 [13, 14]. These studies show that a relatively small number of physical-chemical descriptors can be used to discriminate between substrates for the major human hepatic CYP enzymes. The availability of crystal structures for some CYP enzymes has allowed researchers to map the active site of the CYP enzymes, to establish the amino acids involved in substrate binding, to determine the geometry and flexibility of the active site, and to define the types of interactions between functional groups on the amino acids and substrates [15–18].

As outlined elsewhere in this book, hepatic microsomal CYP enzymes catalyze a variety of monooxygenase reactions including aliphatic and aromatic hydroxylations, *N*-hydroxylations, oxygenations of heteroatoms (N, S, P and I), alkene and arene epoxidations, dehalogenations, dehydrogenations and *N*-, *O*- and *S*-dealkylations. In this chapter, we illustrate some of the various oxidative reactions catalyzed by a CYP enzyme acting on a single substrate. Two

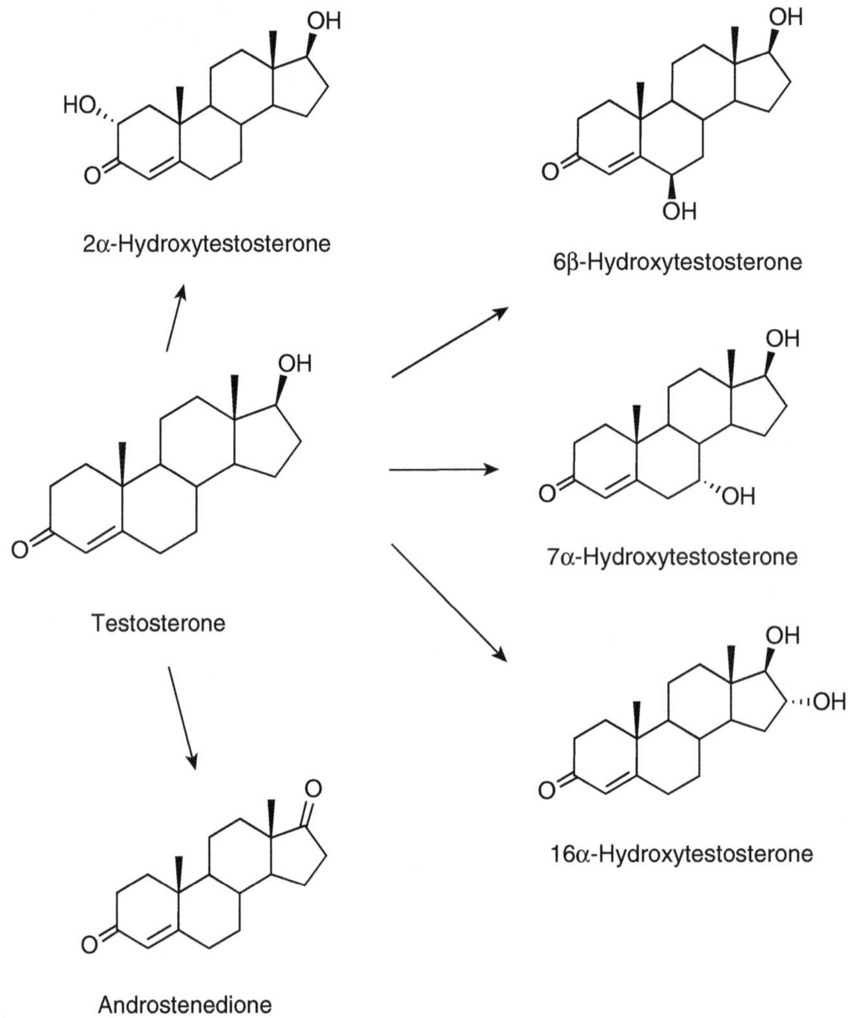

Fig. 5.3 Regioselective oxidation of testosterone

human CYP enzymes, CYP2B6 and CYP3A4, will be described. A halogenated aromatic compound, 2,2′,4,4′-tetrabromodiphenyl ether (BDE-47), is the substrate for CYP2B6 and an endogenous bile acid, 3α-hydroxy-5β-cholan-24-oic (lithocholic) acid, is the substrate for CYP3A4.

5.2 CYP2B6

CYP2B6 is the only enzyme belonging to the CYP2B subfamily in humans. A second CYP2B gene, CYP2B7, was identified as a splice variant of CYP2B6 but is not transcribed into protein [19]. CYP2B6 is expressed mostly in adult liver and has been quantified in liver samples of humans from all age groups, as well as in fetal liver samples from 10 weeks of gestation [20]. CYP2B6 mRNA or protein has also been detected in extrahepatic tissues including brain, kidney, intestine and skin [21–23]. The gene encoding CYP2B6 resides in the large multigene cluster, *CYP2ABFGST*, located on the long arm of chromosome 19 in humans [24]. CYP2B6 is a highly polymorphic enzyme with 37 alleles and more than 100 single nucleotide polymorphic variants [25]. Some CYP2B6 variants have decreased catalytic activity compared to the wild-type enzyme [25–28].

The expression of CYP2B6 in human liver is subject to interindividual variability and large

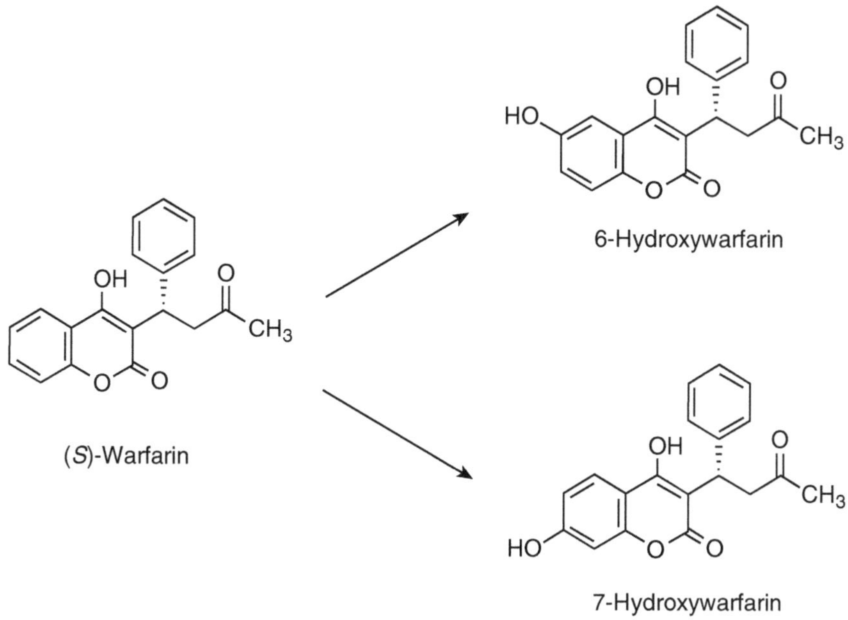

Fig. 5.4 Hydroxylation of (*S*)-warfarin

differences in human CYP2B6 protein levels and catalytic activities have been reported [13, 19, 29]. For example, CYP2B6 protein levels ranging from less than 1 to 70 pmol/mg microsomal protein, or from less than 1 to approximately 10 % of the total hepatic CYP content, have been determined for human liver samples [20, 30, 31]. The source of this variability is pretranslational as CYP2B6 mRNA levels vary widely among individual subjects [19, 32]. Among the genetic factors that can contribute to CYP2B6 variability are sex and ethnicity. It was shown that hepatic CYP2B6 mRNA and protein levels and catalytic activity are greater in women than men, and greater in Hispanic than Caucasian females [19]. Hepatic expression of CYP2B6 is also influenced by environmental factors as CYP2B6 is an inducible enzyme. For example, CYP2B6 mRNA and protein expression and CYP2B6 catalytic activity (measured as bupropion 4-hydroxylation) were induced in human primary hepatocytes pretreated with phenobarbital, dexamethasone or rifampicin, suggesting that the constitutive androstane receptor (CAR, gene designation NR1I3) and the pregnane X receptor (PXR, gene designation NR1I2) are involved in the regulation of CYP2B6 expression [33–36]. CYP2B6 mRNA levels were induced three to fivefold in human hepatocytes exposed to cigarette smoke extract [37]. In vivo, CYP2B6 protein levels were induced in regions of human brain by smoking (2.2- to 3.3-fold) and alcohol consumption (3.5- to 5.3-fold) [38].

A growing number of drugs, environmental pollutants and endogenous steroids with diverse chemical structures have been shown to be metabolized by CYP2B6 [14, 39, 40]. CYP2B6 is involved in the metabolism of several clinically useful drugs, including cyclophosphamide, ifosfamide, bupropion, antipyrine, diazepam, nevirapine, efavirenz, propofol and tamoxifen [29, 41–46]. Hydroxylation of bupropion, an antidepressant, to 4-hydroxybupropion is mediated almost exclusively by CYP2B6 and serves as a marker activity for CYP2B6 in human liver microsomes [29, 44]. CYP2B6 also metabolizes arachidonic acid, lauric acid and steroid hormones, including testosterone, estrone and 17β-estradiol [46–48]. In addition, CYP2B6 is involved in the biotransformation of insecticides such as methoxychlor, chlorpyrifos, diazinon and endosulfan [49–54], organohalogenated pollutants [55–59] and, to a lesser extent, polycyclic aromatic hydrocarbons such as benzo[*a*]pyrene, 6-aminochrysene, phenanthrene and dibenzo[*a*,*h*]anthracene [60–63].

Although the specific determinants of substrates for each CYP enzyme have not been precisely determined, general characteristics of CYP2B6 substrates include high lipid solubility, medium volume and neutral pKa [46, 64]. Substrates are non-planar (often V-shaped) molecules with hydrogen bond donors/acceptors [64–66].

5.3 CYP3A4

CYP3A enzymes are the most abundant CYP enzymes in human liver and intestine [67–69]. The human CYP3A subfamily consists of CYP3A4, CYP3A5, CYP3A7 and CYP3A43 enzymes. The four genes encoding these enzymes are clustered in tandem on human chromosome 7 [70]. CYP3A4 is the predominant CYP3A enzyme in human liver and small intestine and represents, on average, approximately 30 % and 60 % of the total CYP content, respectively, in these organs [69, 71, 72]. CYP3A4 protein expression has been quantified at levels ranging from 15 to more than 300 pmol/mg microsomal protein, or 3 % to more than 60 % of total CYP content, in human liver samples [73, 74]. CYP3A5 is expressed at much lower levels than CYP3A4 and could not be quantified in most human liver samples [69, 74, 75]. CYP3A5 shows overlapping catalytic selectivity with CYP3A4 [76] but has equal or lower activity than CYP3A4 for metabolism of many substrates [77]. CYP3A7 is expressed mainly in fetal liver, where it accounts for 30–50 % of the total hepatic CYP content, but has also been detected in some adult liver samples [70, 78]. CYP3A43 appears to be expressed in tissues at very low levels [74].

CYP3A4 is not polymorphic and the considerable interindividual variation in hepatic CYP3A4 expression appears to be largely genetically regulated. Sex has also been shown to be a determinant of CYP3A4 expression in human liver [73]. In addition, environmental factors are important contributors to the interindividual variability in CYP3A4 expression as CYP3A4 is highly inducible. Pretreatment with several drugs, including dexamethasone, phenobarbital, and rifampin and efavirenz, can upregulate CYP3A4 in vitro and in vivo [79–82]. Induction of CYP3A levels occurs via transcriptional activation, mainly through the activation of PXR [36, 79, 83]. CAR, the vitamin D receptor (VDR, gene designation NR1I1) and the glucocorticoid receptor are also thought to be involved in CYP3A induction [81, 84]. In contrast to induction, a relatively large number of xenobiotic compounds can inhibit CYP3A4 and thereby modify CYP3A4-mediated enzyme activity, without altering CYP3A4 mRNA or protein levels.

CYP3A enzymes are involved in the metabolism of a large number of clinically useful drugs, including midazolam, tamoxifen, quinidine and verapamil [85–89]. The important role of CYP3A4 in drug metabolism is accentuated by an estimate that CYP3A4 is responsible for the biotransformation of more than 50 % of prescription medications [83]. Steroid hormones, including cortisol, estradiol, progesterone and testosterone, are also metabolized by CYP3A4 [90].

CYP3A4 catalyzes the hydroxylation of alprazolam, cortisol, midazolam, testosterone and terfenadine, the *N*-dealkylation of benzphetamine, erythromycin, imipramine and tamoxifen, and the *N*-oxidation of nifedipine [86]. A diagnostic and widely used marker activity of CYP3A4 is testosterone 6β-hydroxylation. A common characteristic of CYP3A4 substrates is high molecular weight [64–66]. For instance, cyclosporine A and erythromycin are substrates of CYP3A4 and are among the largest CYP substrates known.

5.4 2,2′,4,4′-Tetrabromodiphenyl Ether (BDE-47)

A novel substrate of CYP2B6 was recently reported [91, 92]. BDE-47 is one of 209 possible isomers and congeners of polybrominated diphenyl ethers (PBDEs). PBDEs are man-made chemicals that were used as additive flame retardant chemicals on a variety of industrial and

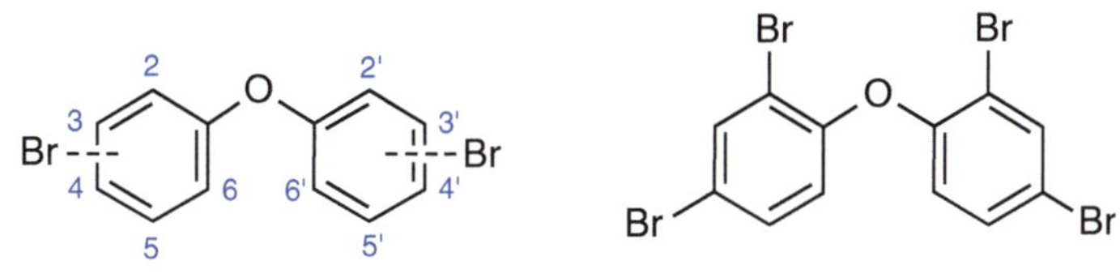

Fig. 5.5 General structure of PBDEs and structure of BDE-47

commercial products since 1965 [93, 94]. PBDEs were formulated as commercial mixtures known as penta-, octa-, and deca-BDE, according to their average bromine content. The penta-BDE mixture was used extensively in North America where it was applied to epoxy resins, textiles, paints and flexible polyurethane foam [95, 96]. The penta-BDE mixture consisted mostly of BDE-47 and 2,2′,4,4′,5-pentabromodiphenyl ether (BDE-99).

PBDEs are highly susceptible to release during manufacture, use, disposal and recycling of PBDE-containing products because they were not chemically bound to polymer components of the products [93, 94]. This factor, together with the high lipid solubility and chemical stability of PBDEs, has led to widespread contamination of the environment by PBDEs [93–96]. BDE-47, for example, has been detected in air, sediment, fish, marine mammals and in human blood, adipose tissue and breast milk, and is frequently the major congener detected in biotic samples [97–108]. Due to the persistence and environmental and human health concerns associated with BDE-47 and BDE-99, the manufacture of the penta-BDE mixture was voluntarily discontinued by the major U.S. manufacturer and in the European Union in 2004 [93–96] (Fig. 5.5).

5.4.1 CYP2B6-Mediated Biotransformation of BDE-47

Evidence from several in vivo and in vitro studies has shown that BDE-47 undergoes oxidative biotransformation to hydroxylated metabolites in rats and mice [109–114]. Formation of hydroxylated metabolites of BDE-47 by human liver preparations has also been reported [91, 115–117], but relatively few hydroxylated metabolites were structurally characterized and the CYP enzymes involved in BDE-47 oxidative metabolism were not determined.

We assessed the oxidative biotransformation of BDE-47 by human liver microsomes and by CYP enzymes using a liquid chromatography/mass spectrometry (LC/MS)-based method [118, 119]. Nine hydroxylated metabolites were formed when human liver microsomes were incubated with BDE-47 [92] (Fig. 5.6). Of the nine hydroxylated metabolites of BDE-47 produced by human liver microsomes, seven metabolites were identified using authentic standards. A monohydroxy-tetrabrominated and a dihydroxy-tetrabrominated metabolite were unidentified (called M1 and M2, respectively) [92]. Among a panel of human recombinant CYP enzymes, CYP2B6 was found to be the most active in the formation of all nine metabolites [92], indicating that CYP2B6 catalyzed the aromatic hydroxylation, dealkylation and dehalogenation of a single substrate.

The major metabolites produced by recombinant CYP2B6, as determined by kinetic analysis of the rates of metabolite formation (Table 5.1), were 4′-hydroxy-2,2′,4,5′-tetrabromodiphenyl ether (4′-OH-BDE-49), 5-hydroxy-2,2′,4,4′-tetrabromodiphenyl ether (5-OH-BDE-47), 6-hydroxy-2,2′,4,4′-tetrabromodiphenyl ether (6-OH-BDE-47), and possibly the unidentified monohydroxy-tetrabrominated metabolite [92]. These metabolites were produced by aromatic hydroxylation of the substrate, BDE-47. More specifically, hydroxylation of an unsubstituted carbon atom *meta* or *ortho* to the ether bond in BDE-47 led to formation of 5-OH-BDE-47 and 6-OH-BDE-47, respectively. The introduction of a hydroxyl group into an aromatic ring is thought to involve epoxidation of the aromatic ring followed by epoxide ring

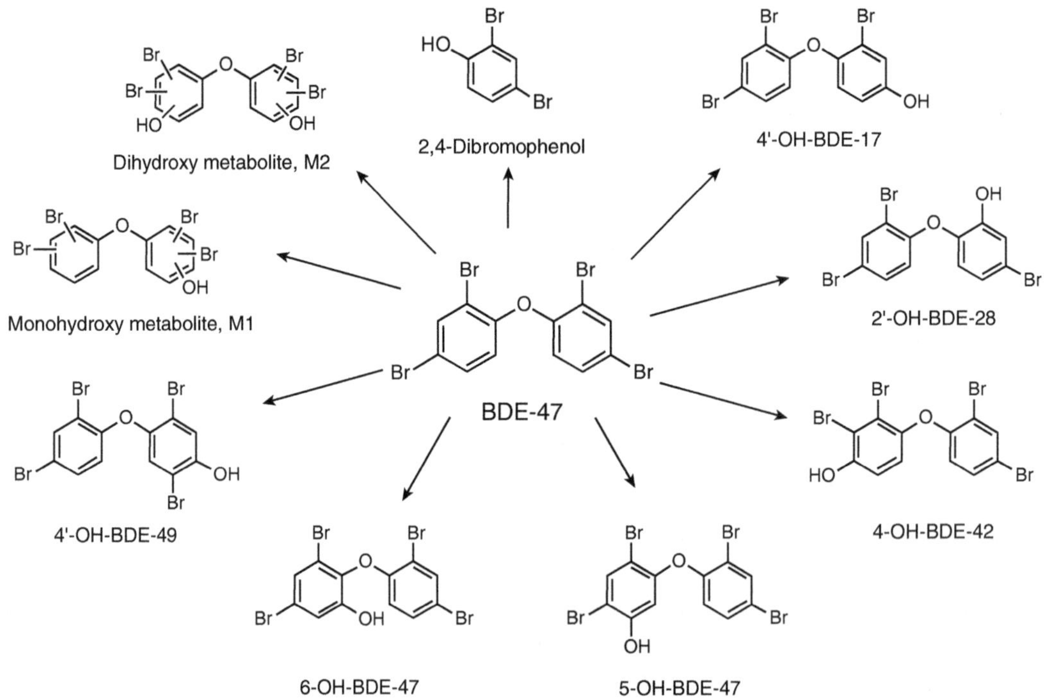

Fig. 5.6 Scheme showing the chemical structures of the hydroxylated metabolites of BDE-47 formed following incubation with human recombinant CYP2B6. General structures for the monohydroxylated tetrabromodiphenyl ether metabolite (*M1*) and the dihydroxylated tetrabromodiphenyl ether metabolite (*M2*) are also shown

Table 5.1 Kinetic parameters for the formation of hydroxylated metabolites of BDE-47 by human recombinant CYP2B6

Metabolite	V_{max} (pmol/min/nmol CYP)	V_{max} (response/min/nmol CYP)	K_m (μM)	K_i (μM)
4-OH-BDE-42	150 ± 49		2.7 ± 1.9	390 ± 150
5-OH-BDE-47	300 ± 27		5.8 ± 1.7	220 ± 75
6-OH-BDE-47	260 ± 37		2.8 ± 0.61	
4'-OH-BDE-49	270 ± 23		1.2 ± 0.10	540 ± 190
M1		570 ± 330	6.6 ± 6.6	

Rates of metabolite formation were determined over a substrate (BDE-47) concentration range of 0.5–200 μM, using 5 pmol of human recombinant CYP2B6/mL and an incubation time of 5 min, as described in Ref. [92]. Apparent V_{max}, K_m and K_i values were calculated using the Michaelis-Menten equation, the Hill equation or the substrate-inhibition equation, depending on which equation best fit the data, as described [92]. Values shown are the mean ± SD of three independent experiments. Rates of M1 formation are not expressed as pmol/min/nmol CYP because of the lack of authentic standards, and are therefore reported as response/min/nmol CYP

opening to yield a phenolic group. In some cases, the opening of the arene oxide is accompanied by intramolecular migration or shift of the hydrogen or substituent on one of the carbon atoms forming the epoxide to the adjacent carbon (i.e., the NIH shift). The exact chemical mechanism for aromatic hydroxylation remains unresolved [120, 121]. In the case of 5-OH-BDE-47 and 6-OH-BDE-47, both metabolites could have arisen from a common arene oxide intermediate involving carbon atoms at the 5 and 6 positions.

Formation of 4′-OH-BDE-49 or of 4-hydroxy-2,2′,3,4′-tetrabromodiphenyl ether (4-OH-BDE-42, another metabolite) demonstrates that BDE-47 was oxidized by CYP2B6 via hydroxylation at a substituted (i.e., brominated) *para* carbon atom accompanied by a NIH-shift of a bromine atom, possibly in concert with formation of an arene oxide intermediate. The V_{max} and K_m values for formation of 4′-OH-BDE-49, 5-OH-BDE-47 and 6-OH-BDE-47 by recombinant CYP2B6 were very similar, which implies that CYP2B6 did not exhibit a preference for *ortho*, *meta* or *para* hydroxylation, and catalyzed aromatic hydroxylation with and without the NIH-shift mechanism with equal efficiency.

Formation of 2,4-dibromophenol (2,4-DBP) resulted from *O*-dealkylation of BDE-47. Although 2,4-DBP was a minor CYP2B6-mediated metabolite of BDE-47, 2,4,5-tribromophenol was a major metabolite of BDE-99 [122]. Thus, *O*-dealkylation was a more important mechanism in the oxidative metabolism of BDE-99 by recombinant CYP2B6 [122], suggesting that small differences in the PBDE structure (i.e., presence or absence of a bromine on carbon 5) affects the rate of oxidative dealkyation in vitro. CYP-catalyzed *O*-dealkylation of ethers appears to involve direct hydroxylation of a carbon atom of the ether bond via a nonconcerted mechanism in which the carbon radical is transferred to the activated oxygen complex of CYP [123].

Formation of 4′-hydroxy-2,2′,4-tribromodiphenyl ether (4′-OH-BDE-17) and 2′-hydroxy-2,4,4′-tribromodiphenyl ether (2′-OH-BDE-28) could only have been produced by oxidative debromination of BDE-47 by CYP2B6. 4′-OH-BDE-17 and 2′-OH-BDE-28 are minor BDE-47 metabolites, suggesting that oxidative debromination was not a facile monooxygenase reaction for CYP2B6 under the experimental conditions used. A dihydroxy-tetrabrominated metabolite was also formed when recombinant CYP2B6 was incubated with BDE-47 but not with primary metabolites of BDE-47 [92]. Moreover, the inclusion of antibody against epoxide hydrolase in the reaction mixture had no effect on the formation of the di-OH-tetrabrominated-PBDE metabolite [92], suggesting that a stable epoxide intermediate was not involved in its formation and that it was produced directly from BDE-47 by P450-catalyzed dihydroxylation.

The variety of metabolites produced by CYP2B6 suggests that the active site of CYP2B6 can accommodate the binding of BDE-47 in various orientations, resulting in the oxidation of different carbon atoms of the BDE-47 molecule. The predominant role of CYP2B6 in the oxidative biotransformation of BDE-99 to multiple metabolites was recently reported [122], confirming that CYP2B6 is a versatile catalyst in the oxidative metabolism of BDE-47 and BDE-99. The flexibility of the CYP2B6 binding pocket is further reinforced by the diverse chemical structures of insecticide substrates. CYP2B6 has been shown to be an efficient catalyst for the conversion of malathion, an organophosphorothioate insecticide, to malaoxon via a sulfoxidation reaction [51], sulfoxidation of α-endosulfan, a chlorinated cyclodiene insecticide, but not of β-endosulfan [59], and formation of the oxon metabolite of chlorpyrifos, an organophosphorus insecticide [53]. In each case, the CYP2B6-mediated reaction leads to bioactivation of the insecticide.

5.5 Lithocholic Acid

Lithocholic acid is an endogenous bile acid and a physiological substrate of CYP3A4 [124]. Historically, bile acids (cholic acid, chenodeoxycholic acid, ursodeoxycholic acid, deoxycholic acid and lithocholic acid are the major bile acids in humans) have been thought of as liver-derived biological detergents that are important for the absorption of dietary fats and the excretion of lipid waste. The role of bile acids in mammalian physiology, however, is much broader and it is now recognized that bile acids are versatile signaling molecules that regulate their own synthesis and transport, and are involved in triglyceride, cholesterol, glucose and energy homeostasis [125–132].

Bile acids perform several functions in the body. First, bile acids provide the major driving force for canalicular bile flow, an osmotically driven process whereby water, organic solutes and electrolytes are secreted into bile [133–135]. Second, bile acids promote the elimination of cholesterol from the body through conversion of cholesterol to bile acids, a fraction of which are subsequently eliminated during enterohepatic circulation [133–136], and by active secretion of dietary and biliary cholesterol, together with phospholipid (predominantly phosphatidylcholine and phosphatidylethanolamine in humans) in the form of mixed micelles, from hepatocytes into bile [137–139]. Third, bile acid secretion aids the hepatobiliary elimination of lipophilic xenobiotic compounds, including many drugs, industrial organic compounds, environmental contaminants and their metabolites [140], and facilitates the excretion of excess metals such as iron, manganese, copper, zinc, magnesium and lead [134–136, 141]. Fourth, bile acids act as physiological emulsifiers that aid in the solubilization and absorption of dietary fats, including fatty acids, monoglycerides and lipid-soluble nutrients, including fat-soluble vitamins such as vitamins A, D, E, and K, by the intestinal epithelium [134–136, 141]. Fifth, bile acids serve as important signaling molecules with systemic effects. Bile acids are endogenous activators of a nuclear transcription factor, namely farnesoid xenobiotic receptor (FXR, gene designation NR1H4) [142, 143], which regulates the expression of several genes involved in cholesterol metabolism and bile acid synthesis and transport [125–132, 142–145]. Bile acids have also been identified as ligands for other nuclear transcription factors, including PXR [146] and VDR [147], and for a plasma membrane-bound G protein-coupled bile acid receptor, called TGR5 (also called GPBAR 1 and M-BAR) [130–132] that has been linked to triiodothyronine production, lipid homeostasis and thermogenesis [129, 130, 142].

Cholic acid and chenodeoxycholic acid, which comprise approximately 70 % of hepatic and biliary bile acids, are the most abundant bile acids in humans [148, 149]. Lithocholic acid is a more hydrophobic bile acid and represents approximately 4–5 % of hepatic and biliary bile acids in humans [148–150]. It has garnered a lot of attention because of its potential toxicity. Administration of lithocholic acid to experimental animals produces liver and biliary tract injury, including bile duct proliferation, multifocal necrosis and atrophy of hepatic lobules [148, 151, 152] (Fig. 5.7).

5.5.1 CYP3A4-Mediated Biotransformation of Lithocholic Acid

Bile acids can undergo hydroxylation at several positions on the steroid backbone as indicated in Fig. 5.8.

Oxidative biotransformation of lithocholic acid is a mechanism that can increase the clearance and elimination of lithocholic acid, thereby reducing its toxicity. In a study of the oxidative metabolism of lithocholic acid, we found that incubation of lithocholic acid with human liver microsomes yielded five metabolites identified as 3-ketocholanoic acid (3-oxo-5β-cholan-24-oic acid), hyodeoxycholic acid (3α,6α-dihydroxy-5-β-cholan-24-oic acid), ursodeoxycholic acid (3α,7β-dihydroxy-5β-cholan-24-oic acid), murideoxycholic acid (3α,6β-dihydroxy-5-β-cholan-24-oic acid) and 6-ketolithocholic acid (3α-hydroxy-6-oxo-5β-cholan-24-oic acid). Metabolites were analyzed using a LC/MS-based assay [124, 153] and the CYP enzymes involved in their formation were determined.

Recombinant CYP3A4 was the only enzyme, among a panel of human recombinant CYP enzymes, found to be active in lithocholic acid biotransformation. CYP3A4 catalyzed the oxidation of this steroid substrate at multiple positions, including aliphatic hydroxylation at the 6α-, 6β- and 7β-positions, and dehydrogenation at the 3 and 6 positions. The major metabolite was the 3-oxo metabolite, 3-ketocholanoic acid [124]. The 6α-hydroxy metabolite, hyodeoxycholic acid, was the second most abundant metabolite formed and ursodeoxycholic acid, murideoxycholic acid and 6-ketolithocholic acid were minor metabolites [124]. Based on the results obtained, a scheme for the biotransformation of lithocholic acid by CYP3A4 is proposed in Fig. 5.9.

Biotransformation of lithocholic acid by human liver microsomes and CYP3A4 occurred at three positions, C_3, C_6 and C_7. The preferred sites of lithocholic acid oxidation were $C_3 > C_6 > C_7$, in that order, indicating regioselective oxidation of this substrate by CYP3A4.

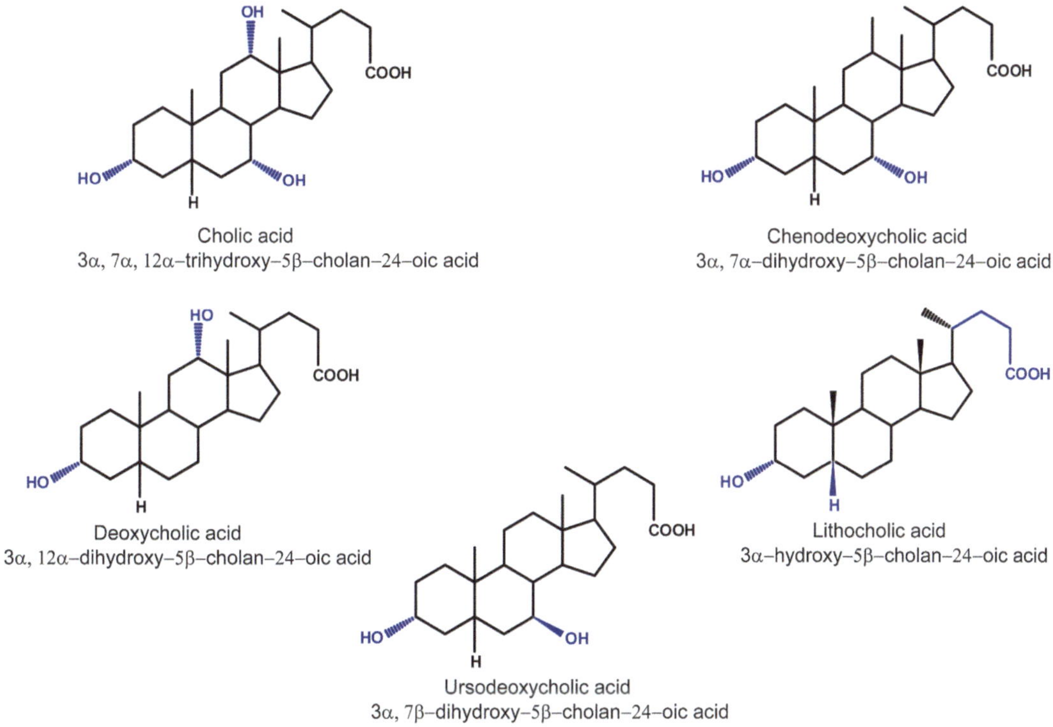

Fig. 5.7 Chemical structures of common bile acids in humans

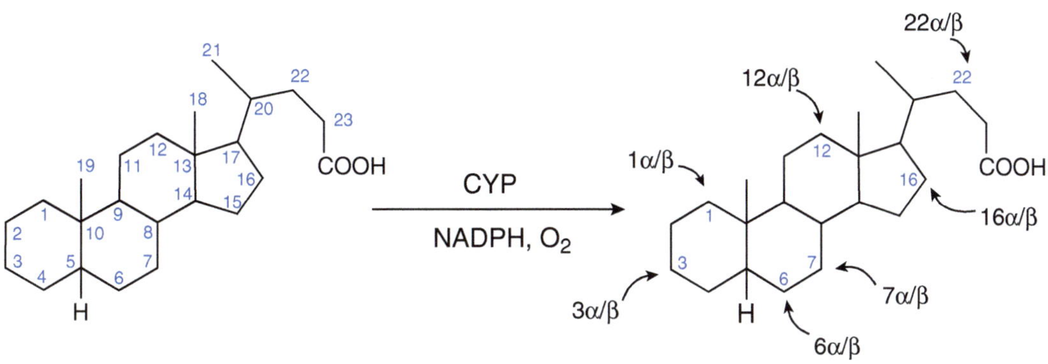

Fig. 5.8 Scheme showing possible hydroxylation sites of bile acids

Moreover, CYP3A4 also exhibited stereoselective metabolism in that hydroxylation at the 6α-position was preferred over hydroxylation at the 6β-position. Kinetic parameters for lithocholic acid metabolite formation by human recombinant CYP3A4 are shown in Table 5.2.

Formation of 3-ketocholanoic acid from lithocholic acid represents a dehydrogenation reaction and can occur via oxidation of the 3α-hydroxyl group, possibly through a geminal-diol intermediate, which can spontaneously rearrange to form a ketone group. A second possible mechanism involves 3β-oxidation followed by dehydration to form 3-ketocholanoic acid. Similar reactions are the oxidation of the 17-hydroxyl group of testosterone to the 17-keto group of androstenedione, which is catalyzed by rat CYP2B and human CYP2C enzymes [154], and

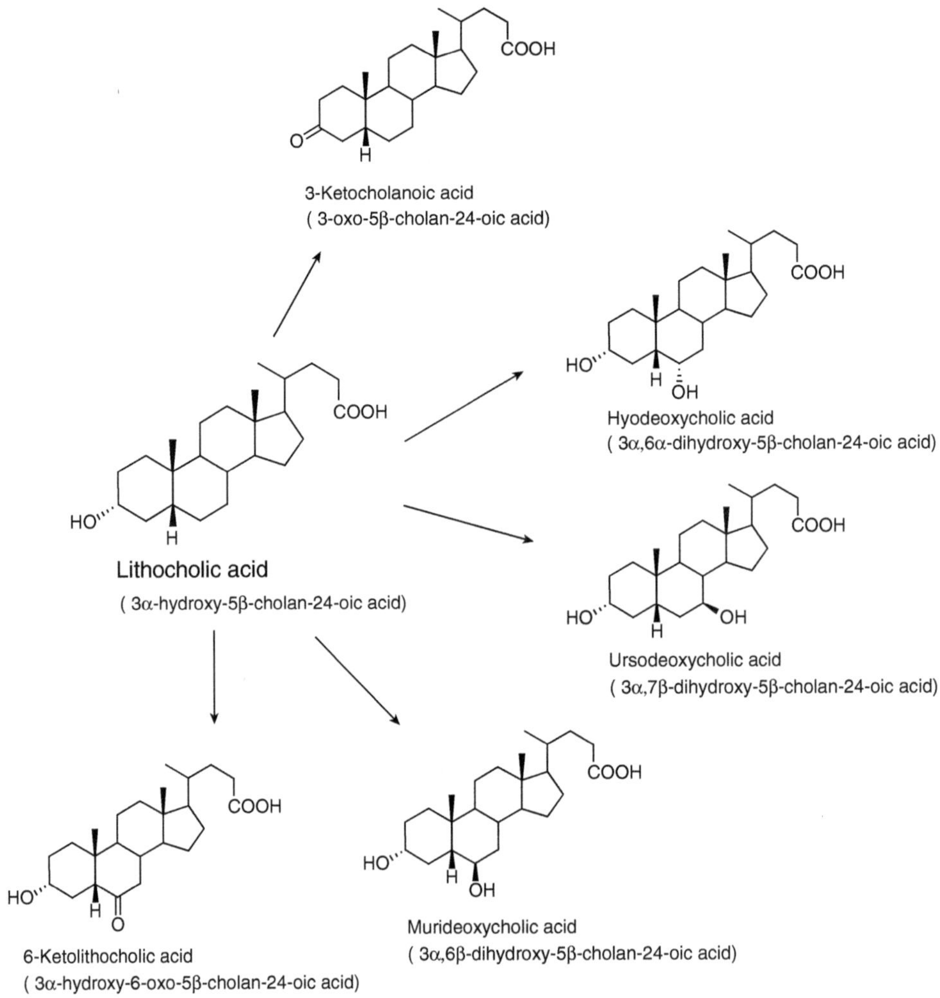

Fig. 5.9 Scheme showing lithocholic acid biotransformation by human recombinant CYP3A4

Table 5.2 Kinetic parameters for lithocholic acid metabolite formation by human recombinant CYP3A4

Metabolite	V_{max} (pmol/min/pmol CYP)	K_m (μM)	K' (μM)	n
3-Ketolithocholic acid	50 ± 2.5	31.9 ± 6.2		
Hyodeoxycholic acid	4.5 ± 0.4	45.4 ± 13.6		
Ursodeoxycholic acid	1.2 ± 0.1	40.1 ± 13.6		
Murideoxycholic acid	0.3 ± 0.0		35.4 ± 5.5	1.7 ± 0.3
6-Ketolithocholic acid	0.4 ± 0.0	29.3 ± 4.4		

Rates of metabolite formation were determined over a substrate (lithocholic acid) concentration range of 1–250 μM, using 30 pmol of human recombinant CYP3A4/mL and an incubation time of 30 min, as described in Ref. [124]. Kinetic parameters for formation of 3-ketolithocholic, hydodeoxycholic, ursodeoxycholic and 6-ketolithocholic acids were calculated using the Michaelis-Menten equation. Kinetic parameters for murideoxycholic acid formation were calculated using the Hill equation. Values shown are the mean ± SD of three independent experiments the oxidation of 6-hydroxyprogesterone to 6-ketoprogesterone by rat CYP2C13 [155].

A possible mechanism for the formation of 6-ketolithocholic acid, a minor metabolite, is the secondary oxidation of hyodeoxycholic acid. In this scenario, some of the hyodeoxycholic acid formed initially from lithocholic acid by CYP3A4 undergoes a subsequent dehydrogenation reaction via oxidation of the 6α-hydroxyl group to form a ketone group. As described above, the mechanism could involve a geminal-diol intermediate. In addition, loss of the α–facing hydroxyl group of the germinal-diol intermediate could result in formation of murideoxycholic acid, as a minor metabolite. Thus, formation of murideoxycholic acid and 3-ketolithocholic acid from hyodeoxycholic acid can occur by a stepwise process.

The variety of lithocholic acid metabolites produced by CYP3A4 suggests that the active site of CYP3A4 accommodates the binding of lithocholic acid in various orientations, resulting in the oxidation at different positions on the steroid molecule. Results of several studies support a model of CYP3A4 in which the active enzyme contains two or even three substrate or ligand binding sites [15–17], so that the enzyme can adopt multiple substrate-bound conformations and accommodate simultaneous binding of more than one substrate molecule and binding of relatively large compounds. These studies [15–17] suggest a remarkable flexibility in the CYP3A4 active site, which helps explain the promiscuity of this enzyme toward a range of substrates and ligands, as well as its ability to produce multiple metabolites from a single substrate such as lithocholic acid.

References

1. Burke MD, Mayer RT (1983) Differential effects of phenobarbitone and 3-methylcholanthrene induction on the hepatic microsomal metabolism and cytochrome P-450-binding of phenoxazone and a homologous series of its *n*-alkyl ethers (alkoxyresorufins). Chem Biol Interact 45:243–258
2. Burke MD, Thompson S, Elcombe CR, Halpert J, Haaparanta T, Mayer RT (1985) Ethoxy-, pentoxy- and benzyloxyphenoxazones and homologues: a series of substrates to distinguish between different induced cytochromes P-450. Biochem Pharmacol 34:3337–3345
3. Burke MD, Thompson S, Weaver RJ, Wolf CR, Mayer RT (1994) Cytochrome P450 specificities of alkoxyresorufin *O*-dealkylation in human and rat liver. Biochem Pharmacol 48:923–936
4. Jacqz-Aigrain E, Funck-Brentano C, Cresteil T (1993) CYP2D6- and CYP3A-dependent metabolism of dextromethorphan in humans. Pharmacogenetics 3:147–204
5. Smith HS (2011) The metabolism of opioid agents and the clinical impact of their active metabolites. Clin J Pain 27:824–838
6. Kharasch ED (2000) Opioid analgesics. In: Levy RH, Thummel KE, Trager WF, Hansten PD, Eichelbaum M (eds) Metabolic drug interactions. Lippincott Williams & Wilkins, Philadelphia, pp 297–319
7. Ryan D, Levin W (1990) Purification and characterization of hepatic microsomal cytochrome P-450. Pharmacol Ther 45:153–239
8. Rendic S, Nolteernsting E, Schänzer W (1999) Metabolism of anabolic steroids by recombinant human cytochrome P450 enzymes. J Chromatogr Biomed Appl 735:73–83
9. Choi MH, Skipper PL, Wishnok JS, Tannenbaum SR (2005) Characterization of testosterone 11-β-hydroxylation catalyzed by human liver microsomal cytochromes P450. Drug Metab Dispos 33:714–718
10. Rettie AE, Eddy AC, Heimark LD, Gibaldi M, Trager WF (1989) Characteristics of warfarin hydroxylation catalyzed by human liver microsomes. Drug Metab Dispos 17:265–270
11. Rettie AE, Korzekwa KR, Kunze KL, Lawrence RF, Eddy AC, Aoyama T, Gelboin HV, Gonzalez FJ, Trager WF (1992) Hydroxylation of warfarin by human cDNA-expressed cytochrome P-450: a role for P-4502C9 in the etiology of (*S*)-warfarin-drug interactions. Chem Res Toxicol 5:54–59
12. Trager WF (2000) Oral anticoagulants. In: Levy RH, Thummel KE, Trager WF, Hansten PD, Eichelbaum M (eds) Metabolic drug interactions. Lippincott Williams & Wilkins, Philadelphia, pp 403–413
13. Ekins S, Vandenbranden M, Ring BJ, Gillespie JS, Yang TJ, Gelboin HV, Wrighton SA (1998) Further characterization of the expression in liver and catalytic activity of CYP2B6. J Pharmacol Exp Ther 286:1253–1259
14. Ekins S, Wrighton SA (1999) The role of CYP2B6 in human xenobiotic metabolism. Drug Metab Rev 31:719–754
15. Yano JK, Wester MR, Schoch GA, Griffin KJ, Stout CD, Johnson EF (2004) The structure of human cytochrome P450 3A4 determined by x-ray crystallography to 2.05Å resolution. J Biol Chem 279:38091–38094
16. Ekroos M, Sjögren T (2006) Structural basis for ligand promiscuity in cytochrome P450 3A4. Proc Natl Acad Sci U S A 103:13682–13687

17. Sevrioukova IF, Poulos TL (2010) Structure and mechanism of the complex between cytochrome P450 3A4 and ritonavir. Proc Natl Acad Sci U S A 107:18422–18427
18. Dong D, Wu B (2012) Substrate selectivity of drug-metabolizing cytochrome P450s predicted from crystal structures and in silico modeling. Drug Metab Rev 44:1–17
19. Lamba V, Lamba J, Yasuda K, Strom S, Davila J, Hancock ML, Fackenthal JD, Rogan PK, Ring B, Wrighton SA, Schuetz EG (2003) Hepatic CYP2B6 expression: gender and ethnic differences and relationship to CYP2B6 genotype and CAR (constitutive androstane receptor) expression. J Pharmacol Exp Ther 307:906–922
20. Croom EL, Stevens JC, Hines RN, Wallace AD, Hodgson E (2009) Human hepatic CYP2B6 developmental expression: the impact of age and genotype. Biochem Pharmacol 78:184–190
21. Gervot L, Rochat B, Gautier JC, Bohnenstengel F, Kroemer H, de Berardinis V, Martin H, Beaune P, de Waziers I (1999) Human CYP2B6: expression, inducibility and catalytic activities. Pharmacogenetics 9:295–306
22. Janmohamed A, Dolphin CT, Phillips IR, Shephard EA (2001) Quantification and cellular localization of expression in human skin of genes encoding flavin-containing monooxygenases and cytochromes P450. Biochem Pharmacol 62:777–786
23. Ding X, Kaminsky LS (2003) Human extrahepatic cytochromes P450: function in xenobiotic metabolism and tissue-selective chemical toxicity in the respiratory and gastrointestinal tract. Annu Rev Pharmacol Toxicol 43:149–173
24. Hoffman SM, Nelson DR, Keeney DS (2001) Organization, structure and evolution of the CYP2 gene cluster on human chromosome 19. Pharmacogenetics 11:687–698
25. Lang T, Klein K, Fischer J, Nussler AK, Neuhaus P, Hofmann U, Eichelbaum M, Schwab M, Zanger UM (2001) Extensive genetic polymorphism in the human CYP2B6 gene with impact on expression and function in human liver. Pharmacogenetics 11:399–415
26. Jinno H, Tanaka-Kagawa T, Ohno A, Makino Y, Matsushima E, Hanioka N, Ando M (2003) Functional characterization of cytochrome P450 2B6 allelic variants. Drug Metab Dispos 31:398–403
27. Honda M, Muroi Y, Tamaki Y, Saigusa D, Suzuki N, Tomioka Y, Matsubara Y, Oda A, Hirasawa N, Hiratsuka M (2011) Functional characterization of CYP2B6 allelic variants in demethylation of antimalarian artemether. Drug Metab Dispos 39:1860–1865
28. Xu C, Ogburn ET, Guo Y, Dest Z (2012) Effects of the CYP2B6*6 allele on catalytic properties of inhibition of CYP2B6 in vitro: implication for the mechanism of reduced efavirenz metabolism and other CYP2B6 substrates in vivo. Drug Metab Dispos 40:717–725
29. Faucette SR, Hawke RL, Lecluyse EL, Shord SS, Yan B, Laethem RM, Lindley CM (2000) Validation of bupropion hydroxylation as a selective marker of human cytochrome P450 2B6 catalytic activity. Drug Metab Dispos 28:1222–1230
30. Mimura M, Baba T, Yamazaki H, Ohmori S, Inui Y, Gonzalez FJ, Guengerich FP, Shimada T (1993) Characterization of cytochrome P-450 2B6 in human liver microsomes. Drug Metab Dispos 21:1048–1056
31. Code EL, Crespi CL, Penman BW, Gonzalez FJ, Chang TKH, Waxman DJ (1997) Human cytochrome P450 2B6. Interindividual hepatic expression, substrate specificity, and role in procarcinogen activation. Drug Metab Dispos 25:985–992
32. Chang TKH, Bandiera SM, Chen J (2003) Constitutive androstane receptor and pregnane X receptor gene expression in human liver: interindividual variability and correlation with CYP2B6 mRNA levels. Drug Metab Dispos 31:7–10
33. Pascussi JM, Gerbal-Chaloin S, Fabre JM, Maurel P, Vilarem MJ (2000) Dexamethasone enhances constitutive androstane receptor expression in human hepatocytes: consequences on cytochrome P450 gene regulation. Mol Pharmacol 58:1441–1450
34. Goodwin B, Moore LB, Stoltz CM, McKee DD, Kliewer SA (2001) Regulation of the human CYP2B6 gene by the nuclear pregnane X receptor. Mol Pharmacol 60:427–431
35. Wang H, Faucette SR, Gilbert D, Jolley SL, Suseyoshi T, Negishi M, LeCluyse EL (2003) Glucucorticoid receptor enhancement of pregnane X receptor mediated CYP2B6 regulation in primary human hepatocytes. Drug Metab Dispos 31:620–630
36. Faucette SR, Zang T-C, Moore R, Sueyoshi T, Omiechinski CJ, LeCluyse EL, Negishi M, Wang H (2007) Relative activation of human pregnane X receptor versus constitutive androstane receptor defines distinct classes of CYP2B6 and CYP3A4 inducers. J Pharmacol Exp Ther 320:72–80
37. Washio I, Maeda M, Sugiura C, Shiga R, Yoshida M, Nonen S, Fujio Y, Azuma J (2011) Cigarette smoke extract induces CYP2B6 through constitutive androstane receptor in hepatocytes. Drug Metab Dispos 39:1–3
38. Mikyss S, Lerman C, Shields PG, Mash DC, Tyndale RF (2003) Smoking, alcoholism and genetic polymorphisms alter CYP2B6 levels in human brain. Neuropharmacology 45:122–132
39. Hodgson E, Rose RL (2007) The importance of cytochrome P450 2B6 in the human metabolism of environmental chemicals. Pharmacol Ther 113:420–428
40. Wang H, Tompkins LM (2008) CYP2B6: new insights into a historically overlooked cytochrome P450 enzyme. Curr Drug Metab 9:598–610
41. Chang TKH, Weber GF, Crespi CL, Waxman DJ (1993) Differential activation of cyclophosphamide and ifosphamide by cytochromes P-450 2B and

3A in human liver microsomes. Cancer Res 53:5629–5637
42. Styles JA, Davies A, Lim CK, De Matteis F, Stanley LA, White INH, Yuan ZX, Smith LL (1994) Genotoxicity of tamoxifen, tamoxifen epoxide and toremifene in human lymphoblastoid cells containing human cytochrome P450s. Carcinogenesis 15:5–9
43. Nakajima M, Yamamoto T, Nunoya KI, Yokoi T, Nagashima K, Inoue K, Funae Y, Shimada N, Kamataki T, Kuroiwa Y (1996) Role of human cytochrome P4502A6 in C-oxidation of nicotine. Drug Metab Dispos 24:1212–1217
44. Hesse LM, Venkatakrishnan K, Court MH, Von Moltke LL, Duan SX, Shader RI, Greenblatt DJ (2000) CYP2B6 mediates the in vitro hydroxylation of bupropion: potential drug interactions with other antidepressants. Drug Metab Dispos 28:1176–1183
45. Ward BA, Gorski JC, Jones DR, Hall SD, Flockhart DA, Desta Z (2003) The cytochrome P450 2B6 (CYP2B6) is the main catalyst of efavirenz primary and secondary metabolism: implication for HIV/AIDS therapy and utility of efavirenz as a substrate marker of CYP2B6 catalytic activity. J Pharmacol Exp Ther 306:287–300
46. Mo SL, Liu YH, Duan W, Wei MQ, Kanwar JR, Zhou SF (2009) Substrate specificity, regulation, and polymorphism of human cytochrome P450 2B6. Curr Drug Metab 10:730–753
47. Imaoka S, Yamada T, Hiroi T, Hayashi K, Sakaki T, Yabusaki Y, Funae Y (1996) Multiple forms of human P450 expressed in *Saccharomyces cerevisiae*. Systematic characterization and comparison with those of the rat. Biochem Pharmacol 51:1041–1050
48. Shou M, Korzekwa KR, Brooks EN, Krausz KW, Gonzalez FJ, Gelboin HV (1997) Role of human hepatic cytochrome P450 1A2 and 3A4 in the metabolic activation of estrone. Carcinogenesis 18:207–214
49. Tang J, Cao Y, Rose RL, Brimfield AA, Dai D, Goldstein JA, Hodgson E (2001) Metabolism of chlorpyrifos by human cytochrome P450 isoforms and human, mouse, and rat liver microsomes. Drug Metab Dispos 29:1201–1204
50. Buratti FM, Volpe MT, Meneguz A, Vittozzi L, Testai E (2003) CYP-specific bioactivation of four organophosphorthioate pesticides by human liver microsomes. Toxicol Appl Pharmacol 186:143–154
51. Buratti FM, D'Aniello A, Volpe MT, Meneguz A, Testai E (2005) Malathion bioactivation in the human liver: the contribution of different cytochrome P450 isoforms. Drug Metab Dispos 33:295–302
52. Sams C, Cocker J, Lennard MS (2004) Biotransformation of chlorpyrifos and diazinon by human liver microsomes and recombinant human cytochrome P450s (CYP). Xenobiotica 34:861–873
53. Foxemberg FJ, McGarrigle BP, Knaak JB, Kostinyak PJ, Olson JR (2007) Human hepatic cytochrome P450-specific metabolism of parathion and chlorpyrifos. Drug Metab Dispos 35:189–193
54. Ellison CA (2012) Human hepatic cytochrome P450-specific metabolism of the organophosphorus pesticides methyl parathion and diazinon. Drug Metab Dispos 40:1–5
55. Ariyoshi N, Oguri K, Koga N, Yoshimura H, Funae Y (1995) Metabolism of the highly persistent PCB congener, 2,4,5,2′,4′,5′-hexachlorobiphenyl by human CYP2B6. Biochem Biophys Res Commun 212:455–460
56. Stresser DM, Kupfer D (1999) Monospecific antipeptide antibody to cytochrome P450 2B6. Drug Metab Dispos 27:517–525
57. Hu Y, Kupfer D (2002) Metabolism of endocrine disruptor pesticide-methoxychlor by human P450s: pathways involving a novel catechol metabolite. Drug Metab Dispos 30:1035–1042
58. Hu YD, Kupfer D (2002) Enantioselective metabolism of the endocrine disruptor pesticide methoxychlor by cytochromes P450 (P450s): major differences in selective enantiomer formation by various P450 isoforms. Drug Metab Dispos 30:1329–1336
59. Lee HK, Moon JK, Chang CH, Choi H, Park HW, Park BS, Lee HS, Hwang EC, Lee YD, Liu KH, Kim JH (2006) Stereoselective metabolism of endosulfan by human liver microsomes and human cytochrome P450 isoforms. Drug Metab Dispos 34:1090–1095
60. Shou MG, Korzekwa KR, Crespi CL, Gonzalez FJ, Gelboin HV (1994) The role of 12 cDNA-expressed human, rodent, and rabbit cytochromes P450 in the metabolism of benzo[*a*]pyrene and benzo[*a*]pyrene *trans*-7,8-dihydrodiol. Mol Carcinog 10:159–168
61. Shou MG, Korzekwa KR, Krausz MS, Crespi CL, Gonzalez FJ, Gelboin HV (1994) Regioselective and stereoselective metabolism of phenanthrene by 12 cDNA-expressed human, rodent, and rabbit cytochromes P450. Cancer Lett 83:305–313
62. Yamazaki H, Mimura M, Oda Y, Gonzalez FJ, El Bayoumy K, Chae YH, Guengerich FP, Shimada T (1994) Activation of *trans*-1,2-dihydro-6-aminochrysene to genotoxic metabolites by rat and human cytochromes P450. Carcinogenesis 15:465–470
63. Cho TM, Rose RL, Hodgson E (2006) In vitro metabolism of naphthalene by human microsomal cytochrome P450 enzymes. Drug Metab Dispos 34:176–183
64. Lewis DFV (1996) Cytochromes P450: structure, function, and mechanism. Taylor & Francis, London
65. Lewis DFV (2000) On the recognition of mammalian microsomal cytochrome P450 substrates and their characteristics. Toward the prediction of human P450 substrate specificity and metabolism. Biochem Pharmacol 60:293–306
66. Lewis DFV (2001) Guide to cytochromes P450: structure and function. Taylor & Francis, London
67. Guengerich FP (2005) Human cytochrome P450 enzymes. In: Otiz de Montellano PR (ed) Cytochrome

P450: structure, mechanism, and biochemistry, 3rd edn. Kluwer Academic/Plenum Publishers, New York, pp 377–530
68. Martignoni M, Groothuis GMM, de Kanter R (2006) Species differences between mouse, rat, dog, monkey and human CYP-mediated drug metabolism, inhibition and induction. Expert Opin Drug Metab Toxicol 2:875–894
69. Paine MF, Hart HL, Ludington SS, Haining RL, Retie AE, Zeldin DC (2006) The human intestinal cytochrome P450 "pie". Drug Metab Dispos 34:880–886
70. Gellner K, Eiselt R, Hustert E, Arnold H, Koch I, Haberl M, Deglmann CJ, Burk O, Buntefuss D, Escher S, Bishop C, Koebe HG, Brinkmann U, Klenk HP, Kleine K, Meyer UA, Wojnowski L (2001) Genomic organization of the human CYP3A locus: identification of a new, inducible CYP3A gene. Pharmacogenetics 11:111–121
71. Wrighton SA, Brian WR, Sari MA, Iwasaki M, Guengerich FP, Raucy JL, Molowa DT, Van den Branden M (1990) Studies on the expression and metabolic capabilities of human liver cytochrome P450IIIA5 (HLp3). Mol Pharmacol 38:207–213
72. Shimada T, Yamazaki H, Mimura H, Inui Y, Guengerich FP (1994) Interindividual variations in human liver cytochrome P450 enzymes involved in the oxidation of drugs, carcinogens, and toxic chemicals: studies with liver microsomes of 30 Japanese and 30 Caucasians. J Pharmacol Exp Ther 270:414–423
73. Woldbold R, Klein K, Burk O, Nüssler AK, Neuhaus P, Eichelbaum M, Schwab M, Zamger UM (2003) Sex is a major determinant of CYP3A4 expression in human liver. Hepatology 38:978–988
74. Westlind-Johnsson A, Malmebo S, Johansson A, Otter C, Andersson TB, Johansson I, Edwards RJ, Boobis AR, Ingelman-Sundberg M (2003) Comparative analysis of CYP3A expression in human liver suggests only a minor role for CYP3A5 in drug metabolism. Drug Metab Dispos 31:755–761
75. Yamaori S, Yamazaki H, Iwano S, Kiyotani K, Matsumara K, Honda G, Nakagawa K (2004) CYP3A5 contributes significantly to CYP3A-mediated drug oxidations in liver microsomes from Japanese subjects. Drug Metab Pharmacokinet 19:120–129
76. Wrighton SA, Ring BJ, Watkins RB, Van den Branden M (1989) Identification of a polymorphically expressed member of the human cytochrome P-450III family. Mol Pharmacol 86:97–105
77. Williams JA, Ring BJ, Cantrell VE, Jones DR, Eckstein J, Ruterbories K, Hamman MA, Hall SD, Wrighton SA (2002) Comparative metabolic capabilities of CYP3A4, CYP3A5, and CYP3A7. Drug Metab Dispos 30:883–891
78. Daly AK (2006) Significance of the minor cytochrome P450 3A isoforms. Clin Pharmacokinet 45:13–31
79. Lehmann JM, McKee DD, Watson MA, Willson TM, Moore JT, Kliewer SA (1998) The human orphan nuclear receptor PXR is activated by compounds that regulate *CYP3A4* gene expression and cause drug interactions. J Clin Invest 102:1016–1023
80. Lu C, Li AP (2001) Species comparison in P450 induction: effects of dexamethasone, omeprazole, and rifampin on P450 isoforms 1A and 3A in primary cultured hepatocytes from man, Sprague–Dawley rat, minipig, and beagle dog. Chem Biol Interact 134:271–281
81. Wang K, Chen S, Xie W, Wan YJY (2008) Retinoids induce cytochrome P450 3A4 through the RXR/VDR-mediated pathway. Biochem Pharmacol 75:2204–2213
82. Woodland C, Huang TT, Gryz E, Bendayan R (2008) Expression, activity and regulation of CYP3A in human and rodent brain. Drug Metab Rev 40:149–168
83. Kliewer SA, Goodwin B, Willson TM (2002) The nuclear pregnane X receptor: a key regulator of xenobiotic metabolism. Endocr Rev 23:687–702
84. Burk O, Wojnowsk L (2004) Cytochrome 3A and their regulation, Naunyn-Schmiedeberg's Arch. Pharmacology 369:105–124
85. Nielsen TL, Rasmussen BB, Flinois JP, Beaune P, Brisen K (1999) In vitro metabolism of quinidine: the (3*S*)-3-hydroxylation of quinidine is a specific marker reaction for cytochrome P-4503A4 activity in human liver microsomes. J Pharmacol Exp Ther 289:31–37
86. Hrycay EG, Bandiera SM (2008) Cytochrome P450 enzymes. In: Gad SC (ed) Preclinical development handbook: ADME and biopharmaceutical properties. Wiley, Hoboken, pp 627–696
87. Christensen N, Mathiesen L, Postvoll LW, Winther B, Molden E (2009) Different enzyme kinetics of midazolam in recombinant CYP3A4 microsomes from human and insect sources. Drug Metab Pharmacokinet 24:261–268
88. Maekawa K, Yoshimura T, Saito Y, Fujimura Y, Aohara F, Emoto C, Iwasaki K, Hanioka N, Narimatsu S, Niwa T, Sawada J (2009) Functional characterization of CYP3A4: catalytic activities toward midazolam and carbamazepine. Xenobiotica 39:140–147
89. Hrycay EG, Bandiera SM (2009) Expression, function and regulation of mouse cytochrome P450 enzymes: comparison with human cytochrome P450 enzymes. Curr Drug Metab 10:1151–1183
90. Yamazaki H, Shimada T (1997) Progesterone and testosterone hydroxylation by cytochromes P450 2C19, 2C9, and 3A4 in human liver microsomes. Arch Biochem Biophys 346:161–169
91. Feo ML, Gross MS, McGarrigle BP, Eljarrat E, Barcelo D, Aga DS, Olson JR (2013) Biotransformation of BDE-47 to potentially toxic metabolites is predominantly mediated by human CYP2B6. Environ Health Perspect 121:440–446

92. Erratico CA, Szeitz A, Bandiera SM (2013) Biotransformation of 2,2′,4,4′-tetrabromodiphenyl ether (BDE-47) by human liver microsomes: identification of cytochrome P450 2B6 as the major enzyme involved. Chem Res Toxicol 26:721–731
93. Alaee M, Arias P, Sjödin A, Bergman A (2003) An overview of commercially used brominated flame retardants, their applications, their use patterns in different countries/regions and possible mode of release. Environ Int 29:683–689
94. LaGuardia AMJ, Hale RC, Harvey E (2006) Detailed polybrominated diphenyl ether (PBDE) congener composition of the widely used penta-, octa-, and deca-PBDE technical flame-retardant mixtures. Environ Sci Technol 40:6247–6254
95. Hale RC, LaGuardia MJ, Harvey E, Mainor TM (2002) Potential role of fire retardant-treated polyurethane foam as a source of brominated diphenyl ethers to the US environment. Chemosphere 46:729–735
96. Hites RA (2004) Polybrominated diphenyl ethers in the environment and in people: a meta-analysis of concentration. Environ Sci Technol 38:945–956
97. Darnerud PO, Eriksen GS, Jóhannesson T, Larsen PB, Viluksela M (2001) Polybrominated diphenyl ethers: occurrence, dietary exposure, and toxicology. Environ Health Perspect 109:49–68
98. Hale RC, La Guardia MJ, Harvey EP, Mainor TM, Duff MH, Gaylor MO (2001) Polybrominated diphenyl ether flame retardants in Virginia freshwater fishes (USA). Environ Sci Technol 35:4585–4591
99. Ikonomou M, Rayne S, Addison RF (2002) Exponential increase of brominated flame retardants, polybrominated diphenyl ethers, in the Canadian arctic from 1981 to 2000. Environ Sci Technol 36:1886–1892
100. Stapleton HM, Dodder NG, Offenberg JH, Schantz MM, Wise SA (2005) Polybrominated diphenyl ethers in house dust and clothes dryer lint. Environ Sci Technol 39:925–931
101. de Wit CA, Alaee M, Muir DCG (2006) Levels and trends of brominated flame retardants in the Arctic. Chemosphere 64:209–233
102. Gómara B, Herrero L, Ramos JJ, Mateo JR, Fernandez MA, Garcia JF, Gonzales MJ (2007) Distribution of polybrominated diphenyl ethers in human umbilical cord serum, paternal serum, maternal serum, placentas, and breast milk from Madrid population, Spain. Environ Sci Technol 41:6961–6968
103. Kelly BC, Ikonomou MG, Blair JD, Gobas FAPC (2008) Bioaccumulation behaviour of polybrominated diphenyl ethers (PBDEs) in a Canadian Arctic marine food web. Sci Total Environ 401:60–72
104. Vonderheide AP, Mueller KE, Meija J, Welsh L (2008) Polybrominated diphenyl ethers: causes of concern and knowledge gaps regarding environmental distribution, fate and toxicity. Sci Total Environ 400:425–436
105. Athanasiadou M, Cuadra SN, Marsh G, Bergman A, Jakobsson K (2008) Polybrominated diphenyl ethers (PBDEs) and bioaccumulative hydroxylated PBDE metabolites in young humans from Managua, Nicaragua. Environ Health Perspect 116:400–408
106. Sjödin A, Wong L-Y, Jones RS, Park A, Zhang Y, Hodge C, Dipietro E, McClure C, Turner W, Needham LL, Patterson DG Jr (2008) Serum concentrations of polybrominated diphenyl ethers (PBDEs) and polybrominated biphenyl (PBB) in the United States population: 2003–2004. Environ Sci Technol 42:1377–1384
107. Stapleton HM, Sjödin A, Jones RS, Niehuser S, Zhang Y, Patterson DG (2008) Serum levels of polybrominated diphenyl ethers (PBDEs) and foam recyclers and carpet installers working in the United States. Environ Sci Technol 42:3453–3458
108. Daniels JL, Pan I-J, Jones R, Anderson S, Patterson DG Jr, Needham LL, Sjödin A (2010) Individual characteristics associated with PBDE levels in U.S. milk samples. Environ Health Perspect 118:155–160
109. Örn U, Klasson-Wehler E (1998) Metabolism of 2,2′,4,4′-tetrabromodiphenyl ether in rat and mouse. Xenobiotica 28:199–211
110. Marsh G, Athanasiadou M, Athanassiadis XJ, Sandholm A (2006) Identification of hydroxylated metabolites in 2,2′,4,4′-tetrabromodiphenyl ether exposed rats. Chemosphere 63:690–697
111. Sanders JM, Chen LJ, Lebetkin EH, Burka LT (2006) Metabolism and disposition of 2,2′,4,4′-tetrabromodiphenyl ether following administration of single or multiple doses to rats and mice. Xenobiotica 36:103–117
112. Staskal DF, Hakk H, Bauer D, Diliberto JJ, Birnbaum LS (2006) Toxicokinetics of polybrominated diphenyl ether congeners 47, 99, 100, and 153 in mice. Toxicol Sci 94:28–37
113. Hamers T, Kamstra JH, Sonneveld E, Murk AJ, Visser TJ, Van Velzen MJ, Brouwer A, Bergman A (2008) Biotransformation of brominated flame retardants into potentially endocrine-disrupting metabolites, with special attention to 2,2′,4,4-′-tetrabromodiphenyl ether (BDE-47). Mol Nutr Food Res 52:284–298
114. Erratico CA, Moffatt SC, Bandiera SM (2011) Comparative oxidative metabolism of BDE-47 and BDE-99 by rat liver microsomes. Toxicol Sci 123:37–47
115. Lupton SJ, McGarrigle BP, Olson JR, Wood TD, Aga DS (2009) Human liver microsome-mediated metabolism of brominated diphenyl ethers 47, 99, and 153 and identification of their major metabolites. Chem Res Toxicol 22:1802–1809
116. Lupton SJ, McGarrigle BP, Olson JR, Wood TD, Aga DS (2010) Analysis of hydroxylated polybrominated diphenyl ether metabolites by liquid chromatography/atmospheric pressure chemical ionization tandem mass spectrometry. Rapid Commun Mass Spectrom 24:2227–2235

117. Marteau C, Chevolleau S, Jouanin I, Perdu E, De Sousa G, Rahmani R, Antignac JP, LeBizec B, Zalko D, Debrauwer L (2012) Development of a liquid chromatography/atmospheric pressure photoionization high resolution mass spectrometry analytical method for the simultaneous determination of polybrominated diphenyl ethers and their metabolites: application to BDE-47 metabolism in human hepatocytes. Rapid Commun Mass Spectrom 26:599–610
118. Moffatt S, Edwards PR, Szeitz A, Bandiera SM (2011) A validated liquid chromatography-mass spectrometry method for the detection and quantification of oxidative metabolites of 2,2′,4,4-′-tetrabromodiphenyl ether in rat hepatic microsomes. Am J Anal Chem 2:352–362
119. Erratico CA, Szeitz A, Bandiera SM (2010) Validation of a novel in vitro assay using ultra performance liquid chromatography-mass spectrometry (UPLC/MS) to detect and quantify hydroxylated metabolites of BDE-99 in rat liver microsomes. J Chromatogr B 878:1562–1568
120. Korzekwa KR, Swinney DC, Trager WT (1989) Isotopically labeled chlorobenzenes as probes for the mechanism of cytochrome P-450 catalyzed aromatic hydroxylation. Biochemistry 28:9019–9027
121. de Visser S, Shaik S (2003) A proton-shuttle mechanism mediated by the porphyrin in benzene hydroxylation by cytochrome P450 enzymes. J Am Chem Soc 125:7413–7424
122. Erratico CA, Szeitz A, Bandiera SM (2012) Oxidative metabolism of BDE-99 by human liver microsomes: predominant role of CYP2B6. Toxicol Sci 129:280–292
123. Ortiz de Montellano PR (1995) Oxygen activation and reactivity. In: Ortiz de Montellano PR (ed) Cytochrome P450: structure, mechanism, and biochemistry, 2nd edn. Plenum Press, New York, pp 245–304
124. Deo AK, Bandiera SM (2009) 3-Ketocholanoic acid is the major in vitro human hepatic microsomal metabolite of lithocholic acid. Drug Metab Dispos 37:1938–1947
125. Chiang JY (2004) Regulation of bile acid synthesis: pathways, nuclear receptors, and mechanisms. J Hepatol 40:539–551
126. Makishima M (2005) Nuclear receptors as targets for drug development: regulation of cholesterol and bile acid metabolism by nuclear receptors. J Pharmacol Sci 97:177–183
127. Kalaany NK, Mangelsdorf DJ (2006) LXRs and FXR: the yin and yang of cholesterol and fat metabolism. Annu Rev Physiol 68:159–191
128. Chiang JY (2009) Bile acids: regulation of synthesis. J Lipid Res 50:1955–1966
129. Nguyen A, Bouscarel B (2008) Bile acids and signal transduction: role in glucose homeostasis. Cell Signal 20:2180–2197
130. Fiorucci S, Mencarelli A, Palladino G, Cipriani S (2009) Bile acid-activated receptors: targeting TGR5 and farnesoid-X-receptor in lipid and glucose disorders. Trends Pharmacol Sci 30:570–580
131. Hylemon PB, Zhou H, Pandak WM, Ren S, Gil G, Dent P (2009) Bile acids as regulatory molecules. J Lipid Res 50:1509–1520
132. Lefebvre P, Cariou B, Lien F, Kuipers F, Staels B (2009) Role of bile acids and bile acid receptors in metabolic regulation. Physiol Rev 89:147–191
133. Hofmann AF (1999) Bile acids: the good, the bad, and the ugly. News Physiol Sci 14:24–29
134. Hofmann AF (1999) The continuing importance of bile acids in liver and intestinal disease. Arch Intern Med 159:2647–2658
135. Hofmann AF, Hagey LR (2008) Bile acids: chemistry, pathochemistry, biology, pathobiology, and therapeutics. Cell Mol Life Sci 65:2461–2483
136. Dawson PA, Schneider BL, Hofmann AF (2006) Bile formation and the enterohepatic circulation. In: Johnson LR (ed) Physiology of the gastrointestinal tract, vol 1, 4th edn. Academic, New York, pp 1437–1462
137. Small DM (2003) Role of ABC transporters in secretion of cholesterol from liver into bile. Proc Natl Acad Sci U S A 100:4–6
138. Borst P, Elferink RO (2002) Mammalian ABC transporters in health and disease. Annu Rev Biochem 71:537–592
139. Smit JJ, Schinkel AH, Oude Elferink RP, Groen AK, Wagenaar E, van Deemter DL, Mol CA, Ottenhoff R, van der Lugt NM, van Roon MA (1993) Homozygous disruption of the murine mdr2 P-glycoprotein gene leads to complete absence of phospholipid from bile and to liver disease. Cell 75:451–462
140. Lehman-McKeeman LD (2008) Absorption, distribution, and excretion of toxicants. In: Klaassen CD (ed) Casarett & Doull's toxicology: the basic science of poisons, 7th edn. McGraw Hill, Toronto, pp 131–159
141. Leuschner U (2009) Formation and secretion of bile and bilirubin metabolism. In: Dancygier H (ed) Clinical hepatology: principles and practice of hepatobiliary diseases, vol 1. Springer, Berlin, pp 103–126
142. Grober J, Zaghini I, Fujii H, Jones SA, Kliewer SA, Willson TM, Ono T, Besnard P (1999) Identification of a bile acid-responsive element in the human ileal bile acid-binding protein gene. Involvement of the farnesoid X receptor/9-cis-retinoic acid receptor heterodimer. J Biol Chem 274:29749–29754
143. Makishima M, Okamoto AY, Repa JJ, Tu H, Learned RM, Luk A, Hull MV, Lustig KD, Mangelsdorf DJ, Shan B (1999) Identification of a nuclear receptor for bile acids. Science 284:1362–1365
144. Sinal CJ, Tohkin M, Miyata M, Ward JM, Lambert G, Gonzalez FJ (2000) Targeted disruption

of the nuclear receptor FXR/BAR impairs bile acid and lipid homeostasis. Cell 102:731–744

145. Kim I, Ahn SH, Inagaki T, Choi M, Ito S, Guo GL, Kliewer SA, Gonzalez FJ (2007) Differential regulation of bile acid homeostasis by the farnesoid X receptor in liver and intestine. J Lipid Res 48:2664–2672
146. Staudinger JL, Goodwin B, Jones SA, Hawkins-Brown D, MacKenzie KI, LaTour A, Liu Y, Klaassen CD, Brown KK, Reinhard J, Willson TM, Koller BH, Kliewer SA (2001) The nuclear receptor PXR is a lithocholic acid sensor that protects against liver toxicity. Proc Natl Acad Sci U S A 98:3369–3374
147. Makishima M, Lu TT, Xie W, Whitfield GK, Domoto H, Evans RM, Haussler MR, Mangelsdorf DJ (2002) Vitamin D receptor as an intestinal bile acid sensor. Science 296:1313–1316
148. Hofmann AF (2002) Cholestatic liver disease: pathophysiology and therapeutic options. Liver 22(Suppl 2):14–19
149. Ridlon JM, Kang DJ, Hylemon PB (2006) Bile salt biotransformations by human intestinal bacteria. J Lipid Res 47:241–259
150. Setchell KD, Rodrigues CM, Clerici C, Solinas A, Morelli A, Gartung C, Boyer J (1997) Bile acid concentrations in human and rat liver tissue and in hepatocyte nuclei. Gastroenterology 112:226–235
151. Fickert P, Fuchsbichler A, Marschall HU, Wagner M, Zollner G, Krause R, Zatloukal K, Jaeschke H, Denk H, Trauner M (2006) Lithocholic acid feeding induces segmental bile duct obstruction and destructive cholangitis in mice. Am J Pathol 168:410–522
152. Beilke LD, Besselsen DG, Cheng Q, Kulkarni S, Slitt AL, Cherrington NJ (2008) Mininal role of hepatic transporters in the hepatoprotection against LCA-induced intrahepatic cholestasis. Toxicol Sci 102:196–204
153. Deo AK, Bandiera SM (2008) Biotransformation of lithocholic acid by rat hepatic microsomes: metabolite analysis by liquid chromatography/mass spectrometry. Drug Metab Dispos 36:442–451
154. Wood AW, Swinney DC, Thomas PE, Ryan DE, Hall PF, Levin W, Garland WA (1988) Mechanism of androstenedione formation from testosterone and epitestosterone catalyzed by purified cytochrome P-450b. J Biol Chem 263:17322–17332
155. Swinney DC, Ryan DE, Thomas PE, Levin W (1988) Evidence for concerted kinetic oxidation of progesterone by purified rat hepatic cytochrome P-450 g. Biochemistry 27:5461–5470

Cytochrome P450 Enzymes in the Bioactivation of Polyunsaturated Fatty Acids and Their Role in Cardiovascular Disease

6

Christina Westphal, Anne Konkel, and Wolf-Hagen Schunck

Abstract

Various members of the cytochrome P450 (CYP) superfamily have the capacity of metabolizing omega-6 and omega-3 polyunsaturated fatty acids (n-6 and n-3 PUFAs). In most mammalian tissues, CYP2C and CYP2J enzymes are the major PUFA epoxygenases, whereas CYP4A and CYP4F subfamily members function as PUFA hydroxylases. The individual CYP enzymes differ in their substrate specificities as well as regio- and stereoselectivities and thus produce distinct sets of epoxy and/or hydroxy metabolites, collectively termed CYP eicosanoids. Nutrition has a major impact on the endogenous CYP-eicosanoid profile. "Western diets" rich in n-6 PUFAs result in a predominance of arachidonic acid-derived metabolites, whereas marine foodstuffs rich in n-3 PUFAs shift the profile to eicosapentaenoic and docosahexaenoic acid-derived metabolites. In general, CYP eicosanoids are formed as second messengers of numerous hormones, growth factors and cytokines regulating cardiovascular and renal function, and a variety of other physiological processes. Imbalances in the formation of individual CYP eicosanoids are linked to the development of hypertension, myocardial infarction, maladaptive cardiac hypertrophy, acute kidney injury, stroke

Abbreviations: *AA* arachidonic acid, *ALA* alpha-linolenic acid, *COX* cyclooxygenase, *CYP* cytochrome P450, *DHA* docosahexaenoic acid, *EDHF* endothelium-derived hyperpolarizing factors, *EDP* epoxydocosapentaenoic acid, *EEQ* epoxyeicosatetraenoic acid, *EET* epoxyeicosatrienoic acid, *EPA* ecosapentaenoic acid, *HDoHE* hydroxydocosahexaenoic acid, *HEPE* hydroxyeicosapentaenoic acid, *HETE* hydroxyeicosatetraenoic acid, *HETrE* hydroxyeicosatrienoic acid, *I/R* ischemia-reperfusion, *KO* knockout, *LA* linoleic acid, *LOX* lipoxygenase, *PLA2* phospholipase A2, *PUFA* polyunsaturated fatty acid, *ROS* reactive oxygen species, *sEH* soluble epoxide hydrolase, *SHR* spontaneously hypertensive rat, *TAC* transverse aortic constriction, *TG* transgene, WT wild-type.

C. Westphal • A. Konkel • W.-H. Schunck (✉)
Max Delbrueck Center for Molecular Medicine, Robert-Rössle-Str. 10, Berlin 13125, Germany
e-mail: schunck@mdc-berlin.de

E.G. Hrycay and S.M. Bandiera (eds.), *Monooxygenase, Peroxidase and Peroxygenase Properties and Mechanisms of Cytochrome P450*, Advances in Experimental Medicine and Biology 851,
DOI 10.1007/978-3-319-16009-2_6, © Springer International Publishing Switzerland 2015

and inflammatory disorders. The underlying mechanisms are increasingly understood and may provide novel targets for the prevention and treatment of these disease states. Suitable pharmacological agents are under development and first proofs of concept have been obtained in animal models.

Keywords
Arachidonic acid • Eicosapentaenoic acid • Docosahexaenoic acid • Hydroxylases • Epoxygenases • Hypertension • Ischemia/reperfusion injury • Cardiac hypertrophy

6.1 Introduction: Discovery of the Third Branch of the Arachidonic Acid Cascade

The discovery chain leading to our current understanding of the pivotal role of cytochrome P450 (CYP) enzymes in the generation of biologically active metabolites of polyunsaturated fatty acids (PUFAs) was initiated with a series of seminal findings in the 1980s. In 1981, three laboratories demonstrated that liver and kidney microsomal as well as purified CYP enzymes catalyzed the oxygenation of arachidonic acid (AA; 20:4 n-6) [1–4]. Structural characterization of the metabolites indicated that CYP enzymes can metabolize AA via three reaction types [5, 6] (Fig. 6.1): (1) allylic oxidation to form cis,trans-conjugated "mid-chain" hydroxyeicosatetraenoic acids (5-, 8-, 9-, 11-, 12- and 15-HETE); (2) hydroxylation at or near the terminal methyl group (ω-/(ω−1)-hydroxylase reaction) yielding 20-, 19-, 18-, 17- and 16-HETE; and (3) olefin epoxidation (epoxygenase reaction) generating four regioisomeric epoxyeicosatrienoic acids (5,6-, 8,9-, 11,12- and 14,15-EET), each of which can be formed as either the *R,S* or the *S,R* enantiomer.

Subsequent studies revealed the presence of EETs as endogenous constituents in rat liver, rabbit kidney, and human urine providing the first proof for an active role of CYP enzymes in AA metabolism under in vivo conditions [7–10]. The biological tissues contained unique sets of regio- and stereoisomeric EETs substantiating the enzymatic origin of these metabolites and stimulating the search for the individual CYP enzymes involved in the regio- and enantioselective epoxidation of endogenous AA pools [11] (compare Sect. 6.2). Moreover, it became clear that CYP enzymes require free AA as a substrate suggesting that phospholipase-mediated AA release from membrane phospholipids provides the starting point for the formation and action of CYP-dependent AA metabolites under in vivo conditions (compare Sect. 6.3). Phospholipase activation is a common feature of the receptor-mediated actions of numerous vasoactive hormones, growth factors and cytokines. Accordingly, CYP-dependent AA metabolites were increasingly recognized to function as "second messengers". This concept became the key for our present understanding of how PUFA-metabolizing CYP enzymes are integrated into the regulation of a wide variety of physiological and pathophysiological processes.

Early studies on the potential physiological roles of CYP-dependent AA metabolites revealed effects on renal salt reabsorption and vascular tone and led to the hypothesis that alterations in the formation of 20-HETE and EETs contribute to the pathophysiology of hypertension [12]. Actually, it was then that the results obtained with two animal models of genetic hypertension until now have shaped our thinking and research about the renal and cardiovascular functions of 20-HETE and EETs

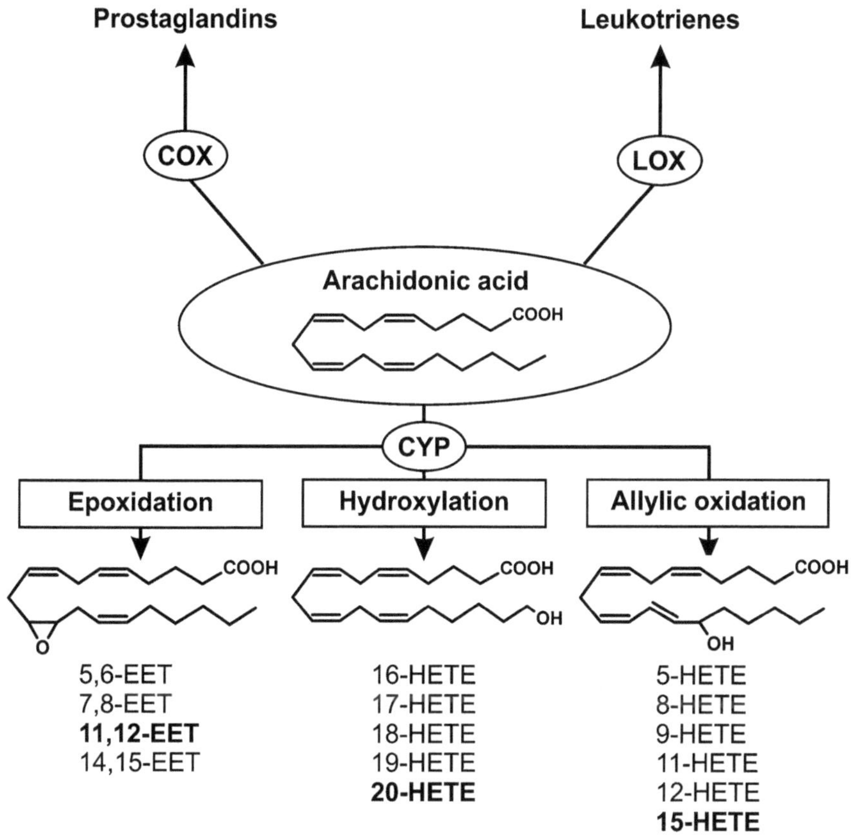

Fig. 6.1 Bioactivation of arachidonic acid (*AA*). Cyclooxygenases (*COX*), lipoxygenases (*LOX*), and CYP enzymes initiate the production of biologically active AA metabolites. CYP enzymes are able to metabolize AA by three different reaction types. Olefin epoxidation results in formation of regioisomeric epoxyeicosatrienoic acids (*EETs*). Hydroxylation at or near the terminal methyl group generates hydroxyeicosatetraenoic acids (*HETEs*) and allylic oxidation produces cis,trans-conjugated "mid-chain" HETEs (compare Sect. 6.1). The individual CYP enzymes involved in AA metabolism differ in their reaction specificities as well as regio- and stereoselectivities and thus produce enzyme specific sets of hydroxy- and epoxy metabolites (compare Sect. 6.2)

(compare Sect. 6.4). A study published in 1989 demonstrated that increased renal 20-HETE production contributes to the elevation of blood pressure in spontaneously hypertensive rats (SHR) [13]. Later on, the development of hypertension in salt-sensitive Dahl rats was attributed to the inability of this strain to upregulate renal EET biosynthesis in response to salt loading [14]. In conclusion, it appeared that hypertension can be caused by an imbalance of pro-(20-HETE) and antihypertensive (EETs) AA metabolites produced by CYP hydroxylases and CYP epoxygenases, respectively. Challenging this simplified view, a series of further studies showed that Dahl salt-sensitive rats exhibit not only a deficiency in EET formation but also in renal CYP hydroxylase expression and 20-HETE production [15]. The apparent paradox was solved after recognizing that the prohypertensive role of 20-HETE is related to its action as a potent vasoconstrictor in the vascular system of the kidney, whereas the antihypertensive role of 20-HETE is based on its capacity to inhibit sodium reabsorption in different segments of the nephron [16]. In contrast to the dual and site-specific role of 20-HETE, the vascular and tubular actions of EETs are apparently unidirectional and antihypertensive because they promote both vasodilation and salt excretion [16, 17].

Taken together, these early biochemical and pathophysiological studies established the CYP-dependent formation of biologically active hydroxy- and epoxymetabolites of AA as the so-called "third branch of the AA cascade" complementary to the previously discovered cyclooxygenase (COX) and lipoxygenase (LOX) initiated pathways of prostanoid and leukotriene formation [18] (Fig. 6.1). Collectively, the AA metabolites produced via all three pathways as well as nonenzymatic reactions are termed eicosanoids (from Greek eicosa = twenty, reflecting that these metabolites are derivatives of a 20 carbon fatty acid). In general and also in the present review, the term eicosanoid is used more broadly to also include related metabolites derived from other PUFAs. Currently, over a hundred different eicosanoids have been identified and the analysis of their specific biological functions has remained a highly active area of research [19].

Noteworthy in the historical context, the CYP branch of eicosanoid formation was discovered almost 50 years after recognizing the essentiality of PUFAs in the mammalian diet, 20 years after elucidating the enzymatic formation and structure of prostaglandins [20], 10 years after aspirin-like antiinflammatory and analgesic drugs were shown to act by inhibiting prostaglandin formation [21], and shortly after understanding the biosynthetic pathway of leukotrienes and their roles in inflammation and asthma [22]. Moreover, at that time, microsomal CYP enzymes were investigated primarily because of their recognized roles in drug and xenobiotic metabolism and their corresponding importance in pharmacology and toxicology. Thus, the discovery of CYP eicosanoids and their potential roles in the pathophysiology of hypertension indicated that microsomal CYP enzymes may be involved in important biological actions beyond drug metabolism and raised the hope of finding novel mechanisms regulating cardiovascular and renal function.

6.2 Reaction and Substrate Specificity of PUFA-Metabolizing CYP Enzymes in Human, Rat and Mouse

6.2.1 CYP Enzymes Involved in 20-HETE Generation

20-HETE is produced by ω-hydroxylation of AA. The capacity of catalyzing this reaction type is widespread among members of the CYP4A and CYP4F subfamilies [17, 23]. In addition to 20-HETE, typically minor amounts of 19-HETE are also generated. The resulting 20-HETE/19-HETE ratio may range from more than 20:1 to 8:1 and is an inherent feature of the individual CYP4A and CYP4F enzymes.

In the **human**, CYP4A11 and CYP4F2 contribute to renal and hepatic 20-HETE formation [24, 25]. A functional variant of CYP4A11 characterized by phenylalanine-to-serine substitution at amino acid position 434 is associated with essential hypertension (T8590C polymorphism of the CYP4A11 gene, compare Sect. 6.4.1) [26]. CYP4A22, the only other member of the human CYP4A subfamily, lacks hydroxylase activity, presumably due to an amino acid substitution at position 130 that is occupied by glycine in all other CYP4A enzymes but by serine in CYP4A22 [26]. However, recently discovered genetic polymorphisms of CYP4A22 include potential gain-of-function mutations (Gly130Ser) making this gene of particular interest for understanding interindividual differences in 20-HETE production [27, 28]. CYP4F3, originally identified as leukotriene B4 (LTB4) ω-hydroxylase in human blood cells, is a further interesting candidate for the production of 20-HETE in man. Alternative splicing of the CYP4F3 pre-mRNA occurs in the liver, kidney and other tissues resulting in a shift of substrate specificity of the mature enzyme from LTB4 to AA [29, 30]. In vitro, the corresponding CYP4F3B variant displayed significantly higher

AA ω-hydroxylase activity than CYP4A11 and CYP4F2 [31].

Recently, CYP2U1, a human CYP enzyme specifically expressed in the thymus and brain, was shown to function as an ω- and (ω−1)-hydroxylase of AA and other PUFAs [32], indicating that also CYP enzymes beyond the CYP4A and CYP4F subfamily members can contribute to 20-HETE production, in particular, in less investigated tissues and physiological conditions. However, exciting novel results can also be expected identifying the endogenous substrates and reaction specificities of CYP4V2 and CYP4F12, the still "orphan" members of the human CYP4 family. Polymorphisms in CYP4V2 and CYP4F12 genes are associated with ocular (Bietti's crystalline corneoretinal dystrophy) and skin disease (lamellar ichthyosis), respectively [33].

The **rat** genome encodes four members of the CYP4A subfamily. Among them, CYP4A1 is the most active AA ω-hydroxylase followed by CYP4A2, CYP4A3 and CYP4A8 [34, 35]. Rat CYP4F enzymes shown to generate 20-HETE include CYP4F1 and CYP4F2 [36]. Based on protein expression data and immunoinhibition experiments, it has been suggested that CYP4A1 is the major AA ω-hydroxylase in the rat heart and kidney, whereas CYP4A2 and/or CYP4F1/4 are the major 20-HETE producing enzymes in the rat lung and liver [37]. In the rat kidney, CYP4A1, CYP4A2 and CYP4A3 are expressed both in different segments of the nephron and in preglomerular arterioles [17, 38]. CYP4A8 was specifically localized to the renal and cerebral vasculature, where its enhanced expression is associated with androgen-induced hypertension in the normal rat and the severity of ischemic stroke in SHR, respectively [39, 40].

Compared to human and rat, the **mouse** genome contains the most extended cluster of CYP4A genes (http://drnelson.uthsc.edu/4ABX.2005.rat.pdf). The individual genes are located within the so-called Cyp4abx cluster on chromosome 4 [41]. Among the functional Cyp4a enzymes identified, Cyp4a12a is the predominant 20-HETE generating enzyme in the kidney of male mice [42]. In comparison, Cyp4a10 that is expressed in both male and female mice displays only a weak AA ω-hydroxylase activity. The female-specific Cyp4a14 lacks the ability of hydroxylating AA but shows significant ω-hydroxylase activity with lauric acid as substrate [42]. Surprisingly, Cyp4a14 gene disruption resulted in increased renal AA ω-hydroxylase activities and caused hypertension in male mice [43]. The mechanism obviously involves increased plasma androgen levels in the Cyp4a14 gene-disrupted mice followed by androgen-induced upregulation of the 20-HETE producing Cyp4a12. Subsequent studies proved that Cyp4a12a overexpression increases 20-HETE levels in preglomerular arterioles and is alone sufficient to elevate blood pressure in mice [44]. Providing a further example of the complex regulation of CYP-eicosanoid formation in mice, deletion of the Cyp4a10 gene caused salt-sensitive hypertension, associated with impaired regulation of the EET-generating Cyp2c44 and of the kidney epithelial sodium channel [45]. The mouse kidney also expresses a series of Cyp4f enzymes. However, their ability to metabolize AA has not yet been demonstrated [46].

6.2.2 CYP Enzymes Involved in EET Generation

Studies with purified or recombinant CYP enzymes demonstrated that, in particular, various members of the CYP2C and CYP2J subfamilies can function as AA epoxygenases [17, 47]. The CYP2C (compare: http://drnelson.uthsc.edu/rat2C.pdf) and CYP2J subfamilies (compare: http://drnelson.uthsc.edu/2Jrat.pdf) evolved differently in human, rat and mouse, making it difficult to identify orthologous genes and to transfer results from animal studies directly to human cardiovascular disease [41].

In the **human**, the CYP2C subfamily consists of four members (CYP2C8, CYP2C9, CYP2C18 and CYP2C19) and there is only a single CYP2J gene (CYP2J2). All corresponding CYP enzymes are able to produce EETs but differ in their catalytic activities, regio- and stereoselectivities as well as tissue specificities of expression. CYP2C8 and CYP2C9 have been considered as the major source of EETs in the human kidney and liver [48]. CYP2C8 generates 11,12- and 14,15-EET in a ratio of about 1.25:1 and preferentially produces the *R,S* enantiomers of both metabolites with a selectivity greater than 80 % [49, 50]. In porcine coronary arteries, antisense oligonucleotides downregulating a CYP2C8-related enzyme decreased bradykinin-induced EET formation and vascular relaxation [51]. This experiment provided direct evidence for the involvement of CYP2C enzymes in vascular EET formation and confirmed the concept that EETs function as endothelium-derived hyperpolarizing factors (EDHF) in various vascular beds [52–54]. Endothelial-specific overexpression of CYP2C8 lowers blood pressure and attenuates hypertension-induced renal injury in mice [55]. Surprisingly, the same CYP2C8 transgenic mice are more susceptible to myocardial infarction injury than wild-type (WT) mice [56]. This detrimental effect was explained by CYP2C8-mediated enhanced formation of reactive oxygen species (ROS) and cardiodepressive linoleic acid (LA) metabolites [56]. Compared to CYP2C8, CYP2C9 is less regio- and stereoselective and metabolizes AA to mixtures of 8(*S*),9(*R*)-, 11(*S*),12(*R*)- and 14(*R*),15(*S*)-EETs with optical purities of 66, 69 and 63 %, respectively [49, 50]. In addition to its ability of producing EETs, CYP2C9 was identified as a functionally significant source of ROS in coronary arteries [57]. In line with this finding, inhibition of CYP2C9 with sulfaphenazole improves endothelium-dependent, nitric oxide–mediated vasodilatation in patients with coronary artery disease [58]. The few studies with CYP2C18 and CYP2C19 show that these enzymes produce 8,9-, 11,12- and 14,15-EET [59, 60].

CYP2J2 has been identified as the major AA epoxygenase of the human heart [61] but is also expressed in other tissues including the vasculature, gastrointestinal tract and islets of Langerhans cells in the pancreas [62]. CYP2J2 generates all four regioisomeric EETs. The enzyme shows enantioselectivity in producing 14(*R*),15(*S*)-EET with an optical purity of 76 % but forms 8,9- and 11,12-EET as racemic mixtures [61]. Compared to CYP2C subfamily members, recombinant CYP2J2 displays rather weak enzymatic activities [60]. Nonetheless, transgenic mice with tissue-specific overexpression of CYP2J2 were developed as one of the most successful tools for studying the diverse beneficial effects of enhanced endogenous EET formation in cardiovascular disease [63, 64] (compare Sect. 6.4). Unlike CYP2C8 and CYP2C9, CYP2J2 is presumably not a relevant source of ROS [65].

The **rat** genome harbors 11 functional CYP2C genes. CYP2C23 has been identified as the predominant renal AA epoxygenase [66, 67]. This enzyme produces 8,9-, 11,12- and 14,15-EET in a ratio of 1:2:0.7. The enzyme shows a high degree of stereoselectivity and generates 8(*R*),9(*S*)-, 11(*R*),12(*S*)- and 14(*S*),15(*R*)-EET with optical purities of 95, 85, and 75 % [66]. CYP2C23 protein expression and activity is upregulated in the rat kidney upon excessive dietary salt intake [68]. A deficiency in CYP2C23-mediated renal EET formation is associated with the development of angiotensin II-induced hypertension and renal failure in the rat [69–71]. CYP2C11 was identified as the major AA epoxygenase in the liver of male rats [72]. However, CYP2C11 is also expressed in the heart, kidney and lung [37]. Moreover, CYP2C11 attracted particular interest as an EET-generating CYP enzyme in astrocytes and its potential role in the regulation of cerebral blood flow [73, 74]. Compared with CYP2C23, CYP2C11 is less regio- and stereoselective. CYP2C11 metabolizes AA to 8,9-, 11,12- and 14,15-EETs and also produces significant amounts of mid-chain HETEs [68].

The rat CYP2J gene cluster comprises five functional genes. Among them, CYP2J3 has been identified as a major AA epoxygenase in the heart [75]. Recombinant CYP2J3 metabolized AA to 14,15-, 11,12- and 8, 9-EETs and 19-HETE as the principal reaction products [75]. CYP2J4 is expressed in rat liver, intestine, olfactory mucosa, kidney, heart, and lung and can contribute to EET and HETE formation in these organs [76, 77].

Analysis of the **mouse** genome indicated the presence of 15 functional Cyp2c genes. Among them, Cyp2c44 is the enzyme most closely related to rat CYP2C23 [78]. Cyp2c44 metabolizes AA primarily to 8,9-, 11,12- and 14,15-EETs in a ratio of about 1:3:1 and shows a high stereoselectivity in producing the *R,S*-enantiomers of 8,9- and 11,12-EET with optical purities of 95 and 94 %, respectively [78]. Cyp2c44 is expressed in the liver, kidney and adrenals. Recent studies on Cyp2c44 knockout mice revealed an important role of this enzyme in the regulation of renal tubular salt reabsorption via EET-mediated inhibition of the epithelial sodium channel (ENaC) [79]. This function of Cyp2c44-derived EETs is essential for dopamine-induced natriuresis/diuresis and for preventing sodium reabsorption and hypertension in response to high dietary potassium intake [80, 81]. Other members of the mouse CYP2C subfamily shown to metabolize AA primarily to EETs include Cyp2c29, Cyp2c38, Cyp2c39, Cyp2c50 and Cyp2c54 [82, 83]. Cyp2c37 metabolizes AA to 12-HETE [82]. Cyp2c55 produces both EETs and HETEs [83]. Cyp2c40, a major Cyp2c enzyme expressed in the murine gastrointestinal tract, produces 16-HETE > 14,15-EET ≫ 8,9-EET > 11,12-EET in a moderate stereoselective manner with preference for 16(*R*)-HETE (66 %), 14(*R*),15(*S*)-EET (62 %), 11(*S*),12(*R*)-EET (70 %) and 8(*S*),9(*R*)-EET (86 %) [84]. The biological functions of most of these murine Cyp2c enzymes have not been characterized. Cyp2c29 apparently resembles human CYP2C9 regarding its capacity of producing both EETs and ROS in the vasculature [85]. Moreover, Cyp2c29 is involved in hypoxic pulmonary vasoconstriction [86]. Recently, Cyp2c knockout mice were developed by deleting the whole Cyp2c gene cluster [87]. This model can become important for studying the in vivo functions of Cyp2c genes and for establishing transgenic mice expressing selected human CYP2C enzymes.

Seven functional CYP2J genes (Cyp2j5, Cyp2j6, Cyp2j8, Cyp2j9, Cyp2j11, Cyp2j12 and Cyp2j13) are predicted by the sequence of the mouse genome [41]. Among them, Cyp2j5 has been most extensively characterized. Recombinant Cyp2j5 metabolizes AA to 14,15-, 11,12- and 8,9-EETs and 11- and 15-HETE [88]. Renal expression of Cyp2j5 is upregulated by androgens and downregulated by estrogens [89]. Female Cyp2j5 knockout mice show reduced plasma 17β-estradiol levels and increased blood pressure that can be normalized by estrogen replacement [90]. Recombinant Cyp2j6 was inactive with AA as substrate but metabolized benzphetamine [91]. Cyp2j9 was identified as an AA (ω−l)-hydroxylase predominantly expressed in the mouse brain [92]. Recently, the remaining four members of the murine Cyp2j subfamily (Cyp2j8, Cyp2j11, Cyp2j12 and Cyp2j13) were also cloned and heterologously coexpressed with the human NADPH-CYP oxidoreductase in insect cells [93]. The recombinant enzymes metabolized AA as well as LA (18:2 n-6) to enzyme-specific sets of epoxy and hydroxy metabolites [93].

Based on the studies summarized above, members of the CYP2C and CYP2J subfamilies are clearly the first candidates when searching for the identity of AA epoxygenases involved in the generation of biologically active EETs. However, it is important to note that other CYP enzymes share this catalytic ability [47]. Providing an example, Cyp2b19, a CYP enzyme specifically expressed in mouse skin keratinocytes, metabolizes AA and generates 14,15- and 11,12-EETs, and 11-, 12- and 15-HETEs. Cyp2b19-catalyzed AA metabolism is highly stereoselective for 11(*S*),12(*R*)- and 14(*S*),15(*R*)-EET, and 11(*S*)-, 12(*R*)- and 15(*R*)-HETE [94]. Cyp2b19 is the major source of endogenous EETs in mouse skin [95] and its enzymatic action can contribute to the regulation of epidermal cornification [96]. CYP2B12 is presumably the rat homolog of murine Cyp2b19 [97]. The human epidermis expresses various genes of the CYP1–4 families including CYP2B6, but the functional counterpart to Cyp2b19 remains to be identified [98]. In contrast to murine Cyp2b19, CYP2B6, the single representative of the CYP2B subfamily in humans, shows only a very weak AA epoxygenase activity [48]. Interestingly, however, CYP2B6 is remarkably active in epoxidizing the 14,15-and 11,12-double bonds of N-arachidonoylethanolamine (anandamide) [99]. Extending the uncertainties in predicting the reaction specificity of individual CYP enzymes solely based on their subfamily membership, human CYP1A2 also functions predominantly as an AA epoxygenase [48, 59]. Moreover, CYP2S1, one of the most recently discovered human CYP enzymes, is expressed in macrophages and metabolizes AA to EETs [100].

6.2.3 CYP Enzymes Involved in Subterminal AA Hydroxylation

The principle metabolites generated by subterminal hydroxylation are 16-, 17-, 18- and 19-HETE. Human CYP enzymes preferentially metabolizing AA to 19-HETE include CYP1A1 and CYP2E1. CYP1A1 generates 19-, 18-, 17- and 16-HETE in a ratio of 5:3:1:1.5, and also minor amounts of 14,15-EET [101]. CYP2E1 metabolizes AA predominantly to 19 and 18-HETE comprising 46 and 32 % of the total products formed [102]. CYP2E1 produces 19(*R*)- and 19(*S*)-HETE in a ratio of about 70:30 and 18 (*R*)-HETE with an optical purity of essentially 100 % [102]. 19-HETE counteracts the vasoconstrictory and proinflammatory effects of 20-HETE [103], suggesting that changes in vascular CYP1A1 and CYP2E1 expression may contribute to the regulation of blood pressure. In line with this hypothesis, the SHR model of genetic hypertension shows reduced CYP2E1 expression [104]. As mentioned above, murine Cyp2j9 provides a unique example of an enzyme that almost exclusively metabolizes AA to 19-HETE [92], whereas other CYP2J subfamily members function predominantly as epoxygenases and produce 19-HETE only as a minor product. CYP4F8 and CYP4F12, two human CYP enzymes primarily involved in prostaglandin metabolism, metabolize AA by ($\omega-2$)/($\omega-3$)-hydroxylation and produce 18-HETE as the main product [105]. Murine Cyp2c40 is currently the only CYP enzyme known to convert AA predominantly to 16-HETE [84].

6.2.4 CYP Enzymes Involved in the Generation of Mid-Chain HETEs

The enzymatic mechanism and biological significance of CYP-catalyzed AA conversion to mid-chain HETEs (5-, 8-, 9-, 11-, 12- and 15-HETE) is only partially understood. This class of CYP-dependent AA metabolites could be directly formed by hydroxylation with double bond migration or by bisallylic oxidation at C7, C10 or C13 followed by rearrangement to the corresponding dienols [106–108]. Mid-chain HETEs were identified as products of NADPH-dependent AA metabolism by liver microsomes as well as various recombinant CYP enzymes including CYP1A2, CYP2C8, CYP2C9 and CYP3A4 [109, 110]. In general, CYPs producing mid-chain HETEs function simultaneously as AA epoxygenases. A CYP enzyme exclusively catalyzing this reaction type has not yet been identified. Among the mid-chain HETEs generated by CYP enzymes, 12(*R*)-HETE attracted particular attention. 12-HETE can be further metabolized to a keto intermediate followed by a keto-reduction reaction yielding the dehydro-metabolite, 12-hydroxyeicosatrienoic acid (12-HETrE) [111]. 12-HETrE was detected in human tear film and follow-up studies in animal models implicated CYP4B1 and 12-HETrE as important components in corneal inflammation and neovascularization [112–114].

6.2.5 Long-Chain Omega-3 Fatty Acids as Alternative Substrates of AA Metabolizing CYP Enzymes

Traditionally, AA (20:4 n-6) has been considered as the main precursor of CYP eicosanoids. However, AA metabolizing CYP enzymes show rather broad substrate specificities and are able to function as hydroxylases or epoxygenases with virtually all PUFAs of both the n-6 and n-3 families (Fig. 6.2) [115]. Which of the various PUFAs becomes accessible and is actually metabolized largely depends on (1) the relative abundance of the individual PUFAs; (2) the substrate specificity of the phospholipases that release free PUFAs from membrane phospholipids and thus initiate their metabolism by CYP enzymes and other eicosanoid generating oxygenases; and (3) the substrate and reaction specificities of the CYP enzymes expressed in a given tissue.

To (1) Mammals unlike plants, marine phytoplankton and nematodes are unable to produce and interconvert n-6 and n-3 PUFAs [116–118]. Accordingly, these two PUFA families are essential components of the mammalian diet and the relative abundance of individual PUFAs in the body is determined by their dietary intake and subsequent tissue-specific mechanisms of metabolism, distribution and uptake [119]. "Western diets" typically contain n-6 and n-3 PUFAs in a ratio of about 15:1, whereas the genetic constitution of our ancestors presumably evolved in a nutritional environment with an n-6/n-3 PUFA ratio of nearly 1:1 [120]. Importantly, the relative deficiency of n-3 PUFAs in the modern human diet has been linked to an increased risk of cardiovascular disease and inflammatory disorders [116, 121–123].

AA is directly available from meat and dairy products or can be synthesized from linoleic acid (LA; 18:2 n-6) that is abundant in vegetable oils (Fig. 6.2). In line with the prevalence of n-6 PUFAs in the "Western diet", AA is indeed the predominant long-chain PUFA in most organs and tissues, except the brain and retina that are able to largely maintain high levels of docosahexaenoic acid (DHA; 22:n-6) even when the diet provides only small amounts of n-3 PUFAs [124–126]. Long-chain n-3 PUFAs, such as DHA and eicosapentaenoic acid (EPA; 20:5 n-3), can be synthesized from alpha-linolenic acid (ALA; 18:3 n-6) that is contained among others in leafy green vegetables (Fig. 6.2). However, the enzymatic steps converting ALA to EPA and further to DHA have limited efficiencies in human [119]. Fish oil and other seafood are a rich direct source of EPA and DHA due to the marine food chain starting with EPA/DHA producing phytoplankton [117]. Based on the accumulating evidence showing that EPA and DHA have beneficial effects in various cardiac disorders, the use of EPA/DHA supplements is recommended for the management of patients after myocardial infarction and for the treatment of hyperlipidemia [123, 127].

To (2) Under basal conditions, AA is predominantly esterified into the sn-2 position of membrane phospholipids and thus not accessible to CYP enzymes and other eicosanoid generating oxygenases. However, free AA becomes readily available in response to extracellular stimuli that activate phospholipases A2 (PLA2) that in turn release AA from the membrane stores [128]. In most tissues, extracellular signal-induced activation of the cytosolic calcium-dependent cPLA2 initiates AA release and eicosanoid formation. EPA and DHA are also incorporated into the sn-2 position of membrane phospholipids and thereby partially replace AA. The classical cPLA2 releases AA and EPA with almost equal efficiencies but is largely inactive in liberating DHA [129]. In the brain, AA and DHA are released by different mechanisms using cPLA2 for AA and a calcium-independent phospholipase A2 (most likely iPLA2β) for DHA [130, 131]. The identity of the PLA2 enzymes releasing DHA in other tissues remains to be elucidated. Recently, the endogenous levels of oxidized PUFA metabolites were compared in the livers of iPLA2γ knockout and WT mice. Interestingly, deletion of iPLA2γ was associated with a marked decrease of DHA- but not of LA- or AA-derived CYP epoxygenase metabolites [132].

To (3) The capacity of CYP enzymes to oxidize EPA and DHA was first shown with rat renal and hepatic microsomes [133, 134]. Recent studies with recombinant CYP enzymes clearly demonstrate that, in fact, all major AA metabolizing CYP enzymes accept these n-3 PUFAs as

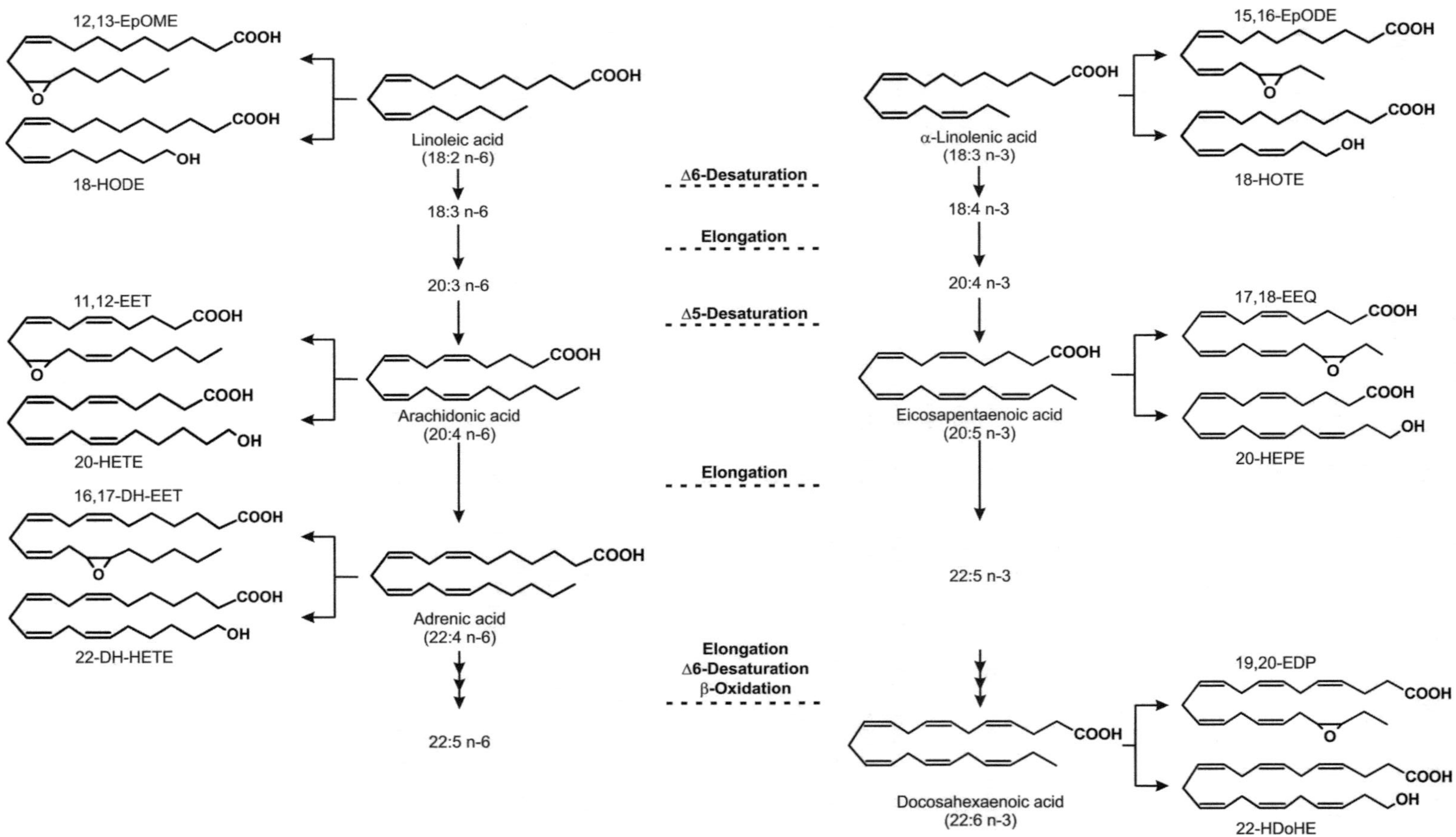

Fig. 6.2 Biosynthesis of long-chain n-6 and n-3 PUFAs. The biosynthetic pathway of long chain n-6 and n-3 polyunsaturated fatty acids provides a series of substrates that can be metabolized by CYP epoxýgenases and CYP hydroxylases (For further details, compare Sect. 6.2.5)

Table 6.1 AA-metabolizing CYP enzymes known to accept the fish-oil omega-3 fatty acids EPA and DHA as efficient alternative substrates

	Main metabolites			
Enzyme	AA	EPA	DHA	Refs.
Human				
CYP1A1	19-HETE	17,18-EEQ[a], 19-HEPE	19,20-EDP[a]	[59, 101, 137]
CYP1A2	11,12-EET	17,18-EEQ[a]	19,20-EDP[a]	[59, 137]
CYP2C8	11,12-EET, 14,15-EET	17,18-EEQ[b], 14,15-EEQ	19,20-EDP[b]	[59, 60, 136, 137]
CYP2C9	14,15-EET, 11,12-EET 8,9-EET	14,15-EEQ, 17,18-EEQ[a], 11,12-EEQ	10,11-EDP	[59, 60, 136, 137]
CYP2C18	8,9-EET, 11,12-EET, 14,15-EET	11,12-EEQ, 17,18-EEQ	19,20-EDP	[59]
CYP2C19	14,15-EET, 19-HETE	17,18-EEQ[a], 19-HEPE	7,8-EDP, 10,11-EDP, 19,20-EDP[a]	[59, 60, 137]
CYP2E1	19-HETE	19-HEPE, 17,18-EEQ[a]	21-HDoHE, 19,20-EDP[a]	[31, 59, 60, 137]
CYP2J2	14,15-EET	17,18-EEQ[a]	19,20-EDP[a]	[59, 60, 137]
CYP4A11	20-HETE	19-HEPE	22-HDoHE, 21-HDoHE	[31, 60]
CYP4F2	20-HETE	20-HEPE	22-HDoHE	[31, 60]
CYP4F3A	20-HETE	20-HEPE	22-HDoHE, 21-HDoHE	[31]
CYP4F3B	20-HETE	20-HEPE	22-HDoHE	[31, 276]
CYP4F8	18-HETE		19,20-EDP,	[105]
CYP4F12	18-HETE		19,20-EDP	[105]
CYP2S1	11,12-EET, 14,15-EET	14,15-EEQ, 17,18-EEQ		[100]
Rat				
CYP2C11	11,12-EET, 14,15-EET	17,18-EEQ[a]	10,11-EDP, 19,20-EDP	[60, 136]
CYP2C23	11,12-EET	17,18-EEQ[a]	10,11-EDP	[60, 136]
CYP4A1	20-HETE	19-HEPE, 20-HEPE, 17,18-EEQ[a]		[135]
Mouse				
Cyp4a12a	20-HETE	20-HEPE, 17,18-EEQ[a]	22-HDoHE	[42, 60]
Cyp4a12b	20-HETE	17,18-EEQ	20-HDoHE	[42, 60]
C. elegans				
CYP-33E2	11,12-EET, 19-HETE, 20-HETE	17,18-EEQ, 19-HEPE, 20-HEPE		[277]

[a]R,S-enantiomer
[b]S,R-enantiomer

efficient alternative substrates (Table 6.1) [31, 59, 60, 135, 136].

CYP2C and CYP2J enzymes that epoxidize AA to EETs also metabolize EPA to epoxyeicosatetraenoic acids (EEQs) and DHA to epoxydocosapentaenoic acids (EDPs). The (ω−3) double bond distinguishing EPA and DHA from AA is the preferred site of attack by most of the classical CYP epoxygenases. CYP2C enzymes are in general almost equally efficient when utilizing AA, EPA or DHA as substrates. Surprisingly, however, EPA is the clearly preferred substrate of CYP2J2, which is the predominant AA epoxygenase in the human heart. Moreover, CYP2J2 shows only a moderate regiospecificity when metabolizing AA but predominantly produces 17,18-EEQ from EPA.

CYP4A and CYP4F enzymes, hydroxylating AA to 20-HETE, metabolize EPA to 20-hydroxyeicosapentaenoic acid (20-HEPE) and DHA to 22-hydroxydocosahexaenoic acid (22-HDoHE). Human CYP4A11 is most active with EPA, whereas CYP4F2 prefers DHA over

AA and EPA [60]. CYP4A enzymes display remarkably increased (ω−1)-hydroxylase activities when metabolizing EPA or DHA instead of AA. Moreover, some of them even attack the (ω−3) double bond. For example, CYP4A11 metabolizes AA to 20-HETE and 19-HETE in a ratio of 82:18, EPA to 20-HEPE, 19-HEPE and 17,18-EEQ in a ratio of 28:62:10 and DHA to 22-HDoHE, 21-HDoHE and 19,20-EDP in a ratio of 48:44:8 [60]. Similarly, murine recombinant Cyp4a12a hydroxylates AA to 20-HETE and 19-HETE (80:20) but metabolizes EPA to 20-HEPE, 19-HEPE, and 17,18-EEQ in a ratio of 12:32:56 [42]. CYP1A1, CYP2E1 and other enzymes converting AA predominantly to 19-HETE or 18-HETE (CYP4F8 and CYP4F12) show pronounced (ω−3)-epoxygenase activities with EPA and DHA (for references, see Table 6.1).

Taken together, it can be concluded that the capacity of utilizing EPA and DHA as alternative substrates is shared by virtually all of the AA-metabolizing CYP enzymes belonging to the subfamilies 1A, 2C, 2E, 2J, 2U, 4A and 4F. The CYP enzymes generally respond to the altered double-bond structure and chain-length of their fatty acid substrates with remarkable changes in the regioselectivity and, in part, also in the type of the catalyzed oxygenation reaction. Moreover, 17,18-EEQ and 19,20-EDP, the unique epoxy metabolites of EPA and DHA, are formed with pronounced stereoselectivities [137]. CYP1A1, CYP1A2, CYP2E1, CYP2C9, CYP2C11, CYP2C19, CYP2C23 and CYP2J2 as well as murine Cyp4a12a and rat CYP4A1 preferentially generate the corresponding *R,S* enantiomers, whereas CYP2C8 and CYP2D6 show stereoselectivities in favor of producing 17(*S*),18(*R*)-EEQ and 19(*S*),20(*R*)-EDP [42, 135–137]. Noteworthy, CYP1A1 metabolizes AA to 19-HETE as the main product and epoxidizes EPA to 17(*R*),18(*S*)-EEQ with an optical purity greater than 98 % [101]. These substrate-dependent features of the PUFA metabolizing CYP enzymes may have important physiological implications, considering that the biological activities of CYP eicosanoids are dependent on the regio- and stereoisomeric position of their functional epoxy or hydroxy groups.

6.2.6 Effect of Dietary Omega-3 Fatty Acids on the Endogenous CYP-Eicosanoid Profile

First studies investigating the effects of marine omega-3 fatty acids (EPA and DHA) on eicosanoid formation were focused on potential changes in the production and activity of COX- and LOX-dependent metabolites. These studies were stimulated by the seminal observation in the 1970s of significantly lower myocardial infarction rates in Greenland Inuit's, who traditionally live on EPA/DHA-rich sea food, compared to Danish controls [138]. Subsequent world-wide epidemiological studies revealed the general existence of striking cardiovascular mortality differences between populations living on n-6 PUFA- versus n-3 PUFA-rich diets [121]. Giving first insight into the mechanisms that might explain the low myocardial infarction rate among Inuit's, EPA was shown to compete with AA yielding less proaggregatory (thromboxane A3 versus thromboxane A2) and less proinflammatory eicosanoids (leukotriene B5 versus leukotriene B4) via the COX- and LOX-dependent pathways [138, 139]. In contrast, prostacyclin I3, formed from EPA, acts with the same potency as vasodilator and inhibitor of platelet aggregation as its AA-derived counterpart prostacyclin I2. Indeed, a favorable shift of the thromboxane/prostacyclin ratio to a more antiaggregatory and vasodilatory state was shown in Intuits as well as in persons after long-term intake of high amounts of EPA (10–15 g/day) [140, 141]. Many of these studies were performed before the discovery of CYP eicosanoids and of other novel classes of lipid mediators, such as the resolvins, that could bring new twists in the search for EPA- and DHA-derived metabolites mediating the

Table 6.2 Effect of the dietary n-6/n-3 PUFA-ratio on the fatty acid and CYP eicosanoid profiles in different organs of the Rat

		Precursor PUFA			Epoxy metabolites			Hydroxy metabolites		
Organ	Diet	AA:	EPA:	DHA	EET:	EEQ:	EDP	HETE:	HEPE:	HDoHE
Left ventricle	n-6	87	0	13	86	0	14	42	1	57
	n-3	28	5	67	26	13	61	6	6	88
Kidney	n-6	95	0	5	93	0	7	61	0	39
	n-3	57	25	18	49	34	17	9	40	51
Cerebral cortex	n-6	45	0	55	43	0	57	41	0	59
	n-3	37	0	63	40	1	59	26	1	73
Lung	n-6	95	1	4	39	0	8	66	0	34
	n-3	34	24	42	23	36	41	8	8	84
Liver	n-6	88	0	12	82	1	17	32	0	68
	n-3	35	19	46	27	38	35	2	18	80
Pancreas	n-6	96	1	3	94	1	5	59	0	41
	n-3	28	41	31	18	55	27	2	11	87
Plasma	n-6	85	1	14	94	0	6	28	0	72
	n-3	29	23	48	15	47	38	1	13	86

Rats were fed either an n-6-rich diet or received a diet supplemented with the fish-oil n-3 PUFAs EPA and DHA for 3 weeks. Different tissues were analyzed as described previously [60]. Shown are the relative ratios of AA, EPA and DHA serving as potential CYP-eicosanoid precursors (AA:EPA:DHA) and the ratios of the corresponding epoxy- (EETs:EEQs:EDPs) and hydroxy metabolites (20-HETE:20-HEPE:22-HDoHE) formed in the different tissues

cardiovascular benefits of marine omega-3 fatty acids [142].

First evidence for the in vivo formation of EPA- and DHA-derived CYP epoxygenase metabolites was provided by the detection of EEQs and EDPs in human urine and plasma samples [143, 144]. Marked increases in the plasma levels of EPA- and DHA-derived epoxides and their vicinal diols were observed in healthy volunteers treated for 4 weeks with 4 g of an EPA/DHA-supplement [144] and in asthmatic patients who received for 3 weeks 4 g EPA + 2 g DHA per day [145]. Considering that AA remained the predominant long-chain PUFA despite EPA/DHA supplementation, these studies indicate that EPA and DHA were metabolized in vivo with significantly higher relative efficiencies compared to AA. Even without dietary intervention, the individual differences in the serum concentrations of EPA-derived CYP epoxygenase metabolites correlated well with the EPA content in red blood cells as shown in a recent study comparing the metabolite profiles in hyper- and normolipidemic humans [146].

We analyzed EPA/DHA-supplementation induced tissue-specific changes of the endogenous CYP-eicosanoid profile in the rat [60]. The animals received standard chow supplemented with 5 % sunflower oil (n-6 PUFA-rich diet), or additionally with 2.5 % OMACOR®-oil (a formulation of EPA/DHA-ethylesters containing 480 mg EPA and 360 mg DHA/g). The n-6 PUFA-rich diet resulted in a 10–20-fold excess of AA over EPA and DHA in most organs and tissues except the brain that maintained an almost 1:1 ratio of AA and DHA. In the heart, the AA content was about sevenfold higher than that of EPA + DHA. After EPA/DHA supplementation, the AA levels were generally reduced by 40–50 % and partially replaced by EPA and DHA in a tissue-specific manner. These changes in the relative PUFA levels correlated with marked changes in the endogenous CYP-eicosanoid profile (Table 6.2). For example, the ratio of EETs:EEQs:EDPs was shifted from 93:0:7 to 49:34:17 in the kidney and from 86:0:14 to 26:13:61 in the left ventricle. EPA/DHA-supplementation also modulated the endogenous formation of ω-hydroxylase products and resulted in a tissue-specific replacement of 20-HETE for 20-HEPE and 22-HDoHE (Table 6.2). The corresponding metabolite/precursor fatty acid ratios indicate that the CYP epoxygenases expressed in the different tissues metabolized EPA with a two to fourfold higher efficiency and DHA with almost the same efficiency, compared with AA [60]. Noteworthy, 17,18-EEQ and 19,20-EDP became the predominant CYP epoxygenase metabolites in most tissues. This finding is in line with the intrinsic feature of many individual CYP enzymes to catalyze preferentially the epoxidation of the (ω−3) double bond when having access to EPA and DHA as substrates (compare Sect. 6.2.5). Moreover, also a recent study in growing piglets identified the vicinal diols of 17,18-EEQ and 19,20-EDP as the epoxygenase-derived metabolites most markedly increased upon dietary n-3 PUFA supplementation [147].

Taken together, these studies demonstrate that the formation of endogenous CYP eicosanoids is highly susceptible to changes in the dietary n-6/n-3 PUFA ratio. Thus, the traditional view that AA is the main source of biologically active epoxy and hydroxy metabolites applies primarily to human populations and laboratory animals living on n-6 PUFA-rich ("Western") diets. However, EPA and DHA may readily become superior sources of CYP-dependent eicosanoids upon n-3 PUFA supplementation or a high dietary intake of fish oil and other marine foodstuffs.

The currently known biological activities of EPA- and DHA-derived CYP metabolites partially resemble those of their AA-derived counterparts, appear in part unique or can even produce opposite effects [148]. The epoxy metabolites of all three PUFAs share vasodilatory properties. However, the potencies of EEQs and EDPs may largely exceed those of EETs in some vascular beds [135, 149]. Interestingly, Cyp1a1 knockout mice display increased blood pressure presumably due to a reduced capacity of producing vasodilatory metabolites from n-3 PUFAs [150]. Potential candidates generated by Cyp1a1 are 17,18-EEQ and 19,20-EDP that efficiently relax murine aortic segments when added at picomolar concentrations [150]. Antiinflammatory effects were first revealed for 11,12- and 14,15-EET but are also exerted by EPA epoxides as exemplified by 17,18-EEQ [151, 152]. 17,18-EEQ and 19,20-EDP inhibit the Ca^{2+}- and isoproterenol-induced increased contractility of neonatal cardiomyocytes, indicating that these metabolites may act as endogenous antiarrhythmic agents [60]. Whereas certain EET regioisomers promote tumor angiogenesis and metastasis, 19,20-EDP and other regioisomeric DHA epoxides inhibit these crucial events in cancerogenesis [153, 154]. Moreover, CYP-dependent EPA- and DHA-derived epoxy metabolites were identified as potent antihyperalgesic agents in an animal model of pain [155]. These findings suggest that EPA- and DHA-derived CYP eicosanoids may serve as mediators in a variety of beneficial effects attributed to fish oil n-3 PUFAs, such as protection against cardiovascular disease, sudden cardiac death and tumor development [123, 156].

6.3 De Novo Biosynthesis and Metabolic Fate of CYP Eicosanoids

6.3.1 CYP Eicosanoids as Second Messengers

As already discussed in Sect. 6.2.5, AA, EPA or DHA become only accessible as substrates to the CYP enzymes after being released from membrane phospholipids. In this way, in vivo generation of CYP eicosanoids is normally strictly coupled to extracellular signals that trigger the activation of phospholipases, which in turn release the potential substrates from intracellular membrane stores. Accordingly, CYP eicosanoids are typically formed as second messengers of diverse hormones, cytokines and growth factors [17]. Examples include bradykinin-induced EET formation in endothelial cells as part of the vasodilatory response [53], VEGF-induced EET formation in angiogenesis [157], and angiotensin II-induced 20-HETE formation in vasoconstriction of renal arterioles [158]. After de novo

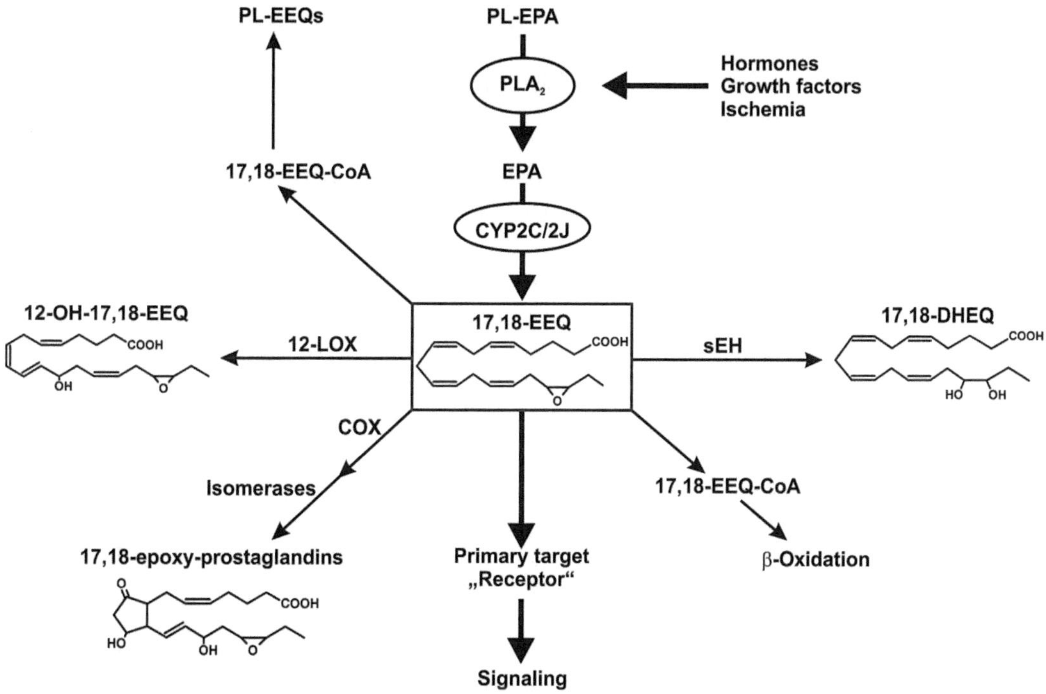

Fig. 6.3 Biosynthesis and metabolic fate of 17,18-EEQ. The biosynthesis of 17,18-EEQ is initiated by extracellular signals that activate phospholipases A2 (PLA2) in the given tissue. The PLA2 enzymes liberate eicosapentaenoic acid (*EPA*) from phospholipid (*PL*) stores and make free EPA accessible as substrate to the CYP enzymes. After its de novo synthesis, 17,18-EEQ triggers intracellular signaling pathways and can be further metabolized via various routes leading to its storage in membrane phospholipids, the formation of secondary metabolites with novel biological activities or to inactivation and degradation (For further details, see Sect. 6.3)

synthesis, CYP eicosanoids elicit cell type specific signaling pathways but are also subject to rapid further metabolism that may lead to (1) the generation of membrane pools of preformed CYP eicosanoids; (2) the formation of secondary metabolites with new biological activities; or (3) inactivation and degradation (Fig. 6.3).

6.3.2 Storage and Release

Unlike COX-dependent prostanoids, CYP-dependent hydroxy and epoxy metabolites are partially re-esterified into the sn-2 position of glycerophospholipids, generating a membrane pool of preformed CYP eicosanoids that is also accessible to PLA2 enzymes [159–161]. This unique feature of CYP eicosanoids is particularly important for their release and action in ischemia/reperfusion injury (compare Sect. 6.4).

6.3.3 Formation of Secondary Metabolites with New Biological Activities Through Actions of COX, LOX and CYP Enzymes

Several CYP eicosanoids such as 20-HETE, 5,6-EET or 17,18-EEQ still contain the double bond structure required for cyclooxygenation and can indeed serve as substrates of COX enzymes [16, 162, 163]. Depending on the COX enzymes and isomerases expressed in a given tissue, this route can result in the formation of 20-hydroxy, 5,6-epoxy or 17,18-epoxy analogs of diverse prostanoid subfamilies including prostaglandins, prostacyclins and thromboxanes. COX-dependent 20-HETE transformation has been proposed as an important mechanism in the regulation of renal microvascular tone [164]. A recent study showed that the proadipogenic effect of 20-HETE depends on its COX-2 mediated transformation to 20-OH-PGE2 [165]. The vasoactivity of 5,6-EET is dependent on the vascular bed and may consist of a vasodilator component of the primary metabolite and a vasoconstrictor component due to COX-dependent secondary metabolite formation [166, 167].

17,18-EEQ provides a thus far unique example of a CYP epoxygenase metabolite that can be further metabolized by LOX enzymes. Recently, 8-OH, 12-OH and 15-OH-17,18-EEQ were identified as endogenous metabolites in the peritoneal fluid of mice after feeding the animals an EPA-rich diet [168]. In vitro, the individual regioisomers can be enzymatically synthesized incubating 17,18-EEQ with purified 8-LOX, 12-LOX and 15-LOX, respectively. Among them, 12(*S*)-OH-17,18-EEQ (compare Fig. 6.3) displays highly potent antiinflammatory action by limiting neutrophil infiltration in experimental murine peritonitis. In vitro, 12(*S*)-OH-17,18-EEQ inhibits neutrophil chemotaxis with an EC_{50} of 0.6 nM [168]. These remarkable findings suggest that the combined actions of CYP epoxygenases and 12-LOX are an important component of the metabolic cascade mediating the antiinflammatory effects of dietary EPA intake. Another EPA-initiated antiinflammatory pathway uses 18-HEPE as a precursor and leads to formation of the E-series of resolvins that are highly potent mediators in the resolution of inflammation [169]. The enzymatic origin of 18-HEPE may include COX (after binding aspirin) or CYP enzymes, whereby the identity of the latter remains to be clarified in humans and mammals [170].

As shown in Fig. 6.4, CYP epoxygenases and CYP hydroxylases also can cooperate in producing secondary metabolites with unique biological activities. EETs are metabolized by CYP4A and CYP4F enzymes to hydroxy EETs (HEETs) and the same class of metabolites is also efficiently produced by CYP2C-catalyzed epoxidation of 20-HETE [71, 171, 172]. Among the CYP enzymes expressed in the rat kidney, CYP4A1 preferentially hydroxylates 11,12-EET [171], whereas CYP2C23 predominantly epoxidizes the 8,9 double bond of 20-HETE [71]. Importantly, the HEETs formed via both pathways act as high-affinity ligands of the peroxisome proliferator-activated receptor alpha (PPARα) that is involved in the regulation of lipid metabolism as well as the control of inflammation [71, 171].

6.3.4 Inactivation and Degradation

EETs and related epoxy metabolites derived from other PUFAs are rapidly degraded to the corresponding vicinal diols by the soluble epoxide hydrolase (sEH) [173, 174]. This mechanism leads to a loss of most of the biological activities attributed to EETs, although the vicinal diols can show, in part, overlapping net effects. EETs incorporated into membrane phospholipids or bound in the cytosol to fatty acid binding proteins are largely protected from enzymatic hydrolysis [175, 176]. The sEH enzyme is highly expressed in all major organs and throughout the cardiovascular system [174]. Its expression is further induced by angiotensin II and thus contributes to decreased EET levels in hypertension and cardiac disease [177, 178]. Over the last decade, pharmacological inhibition of sEH-mediated EET hydrolysis became a highly active field of research with great promise for the prevention and treatment of cardiovascular disease [179] (compare Sect. 6.4).

Resembling the metabolic fate of fatty acids, CYP eicosanoids can also become subject to peroxisomal and mitochondrial β-oxidation but also to chain elongation. Thereby, partial β-oxidation as well as chain elongation may produce metabolites with novel biological activities [180].

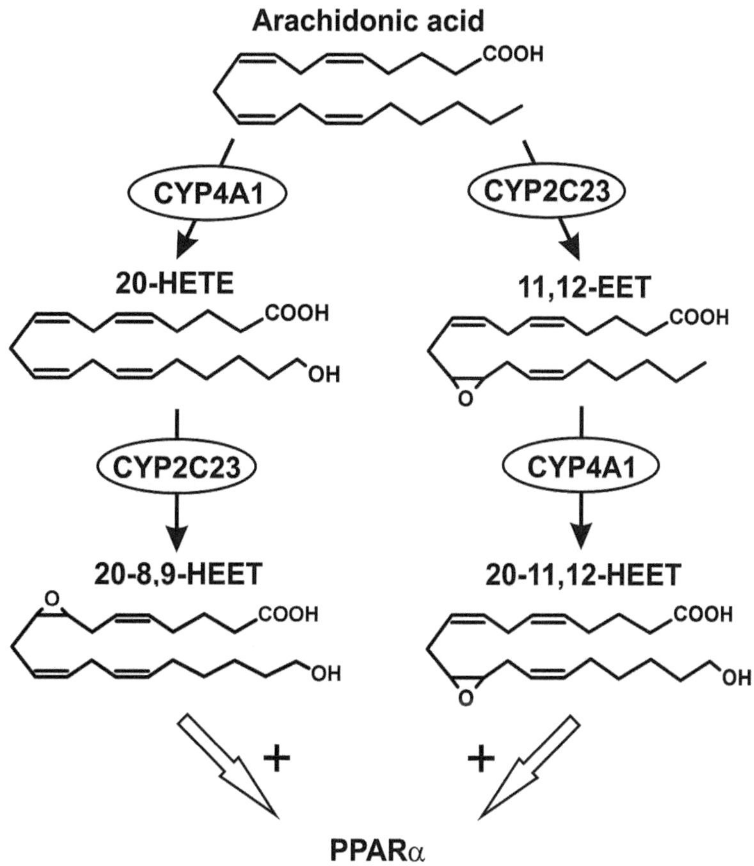

Fig. 6.4 Renal arachidonic acid (*AA*) metabolism by CYP4A1 and CYP2C23. CYP4A and CYP2C enzymes produce 20-HETE and EETs as primary products. However, they can also cooperate to generate secondary hydroxy-epoxy metabolites (HEETs) that function as high-affinity ligands of the transcription factor PPARα (For further details, compare Sect. 6.3)

6.4 CYP Eicosanoids in Cardiovascular Function and Disease

As described in the Introduction, studies in rat models of genetic hypertension led to the concept that imbalances in CYP eicosanoid formation contribute to the pathogenesis of hypertension and target organ damage. Subsequent studies proved this hypothesis in various other animal models of hypertension (see Table 6.3) and provided mechanistic insight into the partially opposing roles of EETs and 20-HETE in the regulation of vascular, renal and cardiac function [16, 17, 54, 180]. The basic concept was successfully extended and specified to a series of other disease conditions such as ischemia-induced injury of the heart, kidney and brain (Table 6.4), cardiac hypertrophy and arrhythmia (Table 6.5), inflammatory disorders, and atherosclerosis [179, 181–186].

6.4.1 Hypertension and Target Organ Damage

Trying to understand the mechanisms linking CYP-eicosanoid formation to blood pressure regulation, it is helpful to distinguish three major types of alterations in CYP-dependent AA metabolism that are associated with the development of hypertension (compare Table 6.3): (**1**) increased vascular CYP hydroxylase expression and 20-HETE formation resulting in vasoconstriction and vascular inflammation; (**2**) decreased renal tubular CYP4A expression and 20-HETE formation resulting in impaired renal function; and (**3**) decreased CYP epoxygenase and/or increased sEH expression

Table 6.3 Role of CYP eicosanoids in animal models of hypertension

Model	Genetic or pharmacological intervention	Effect	Refs.
Spontaneously hypertensive rats (SHR)	Treatment with $SnCl_2$	Inhibits 20-HETE formation and prevents the development of hypertension	[13]
	Adenovirus-mediated overexpression of CYP epoxygenases	Prevents development of hypertension	[278]
	sEH inhibition	Persistent reduction of blood pressure in female SHR when sEH inhibitor is administered in perinatal phase	[279]
Salt-sensitive hypertension	Salt-resistant Dahl rats -high dietary salt intake combined with CYP epoxygenase inhibitor or A2AR antagonist	Loss of the salt-resistant phenotype due to the inability of upregulating the adenosine-A2AR-EET axis	[14, 210, 211]
	Normal rats with high salt diet and CYP4A inhibitor	Normal rats are rendered salt-sensitive	[202]
Angiotensin II-induced hypertension	High salt diet combined with AngII-infusion in rat	Inability to upregulate renal CYP2C/EET expression is associated with hypertension and renal injury	[280]
	Rats overexpressing human renin and angiotensinogen	Reduction in renal microsomal AA epoxygenase and hydroxylase activities	[69]
	Rats overexpressing human renin and angiotensinogen treated with fenofibrate	Fenofibrate restores renal CYP2C23 expression and protects against hypertension and renal damage	[71]
	Renin transgenic rats treated with CYP4A- and/or sEH inhibitor	Attenuates the development of hypertension and target organ damage	[254]
	Pharmacologic sEH inhibition in mice	sEH-inhibition prevents and reverses angiotensin II-infusion hypertension	[212]
High fat diet and metabolic syndrome	High fat diet	Reduction of renal hydroxylase and epoxygenase activity	[281]
	Obese HO-2 KO mice treated with EET agonist or sEH inhibitor	Reduction in blood pressure and body weight gain, increased insulin sensitivity	[282]
Preeclampsia and Pregnancy	CYP-epoxygenase inhibition in rat models of preeclampsia	Amelioration of hypertension and endothelial dysfunction likely via reduced trophoblast-mediated 5,6-EET formation	[283]
	CYP epoxygenase inhibition in normal pregnancy of rat	Hypertension and impaired renal function	[284]
Androgen-induced hypertension	CYP4A inhibitor	Prevents androgen-induced 20-HETE overproduction and protects against hypertension and renal injury	[193, 194]
Cyclosporine-induced nephrotoxicity	CYP4A inhibitor	Amelioration of cyclosporine A-induced nephrotoxicity and hypertension	[285]
TG and KO mouse models of hypertension	CYP4F2 overexpression in mice	Enhanced 20-HETE production and increased blood pressure	[286, 287]
	Cyp4a14 KO and Cyp4a12 TG mice	Development of 20-HETE-dependent hypertension	[43, 288]
	Cyp2c44 KO mice	Impaired sodium reabsorption and hypertension in response to high dietary potassium intake	[81]

leading to decreased EET levels and resulting in impaired vasodilation and renal salt excretion.

Classical genetic models of hypertension or complex models of secondary hypertension such as angiotensin II-infusion hypertension show combinations of these three basic imbalances in CYP-eicosanoid formation. Moreover, these models frequently do not allow deriving the actual cause-and-effect relationships between the disease state and the associated changes in

Table 6.4 Role of CYP eicosanoids in animal models of ischemia/reperfusion injury

Model	Genetic or pharmacological intervention	Effect	Refs.
Heart			
Ex vivo global I/R in isolated rat hearts	11,12-EET, 14,15-EET, 19-HETE	Improved postischemic functional recovery after treatment with 11,12-EET, but not with 14,15-EET or 19-HETE	[75]
Left anterior descending (LAD) artery occlusion in dogs and rats	11,12-EET, 14,15-EET	Reduced infarct size	[225, 226]
LAD occlusion; ischemic preconditioning (IPC) in canine hearts	CYP ω-hydroxylase inhibitor; 20-HETE antagonist	Reduction in infarct size; synergistic beneficial effect with IPC	[222, 223]
In vivo I/R and ischemic pre- and postconditioning in rats	CYP epoxygenase inhibitor	Inhibition of CYP epoxygenase prevents the beneficial effect of postconditioning	[228]
In vivo I/R remote preconditioning of trauma (RPCT) in rats	CYP epoxygenase inhibitor; EET antagonist	Inhibition of EET formation or action abolishes the protective effects of RPCT	[229, 289]
Ex vivo global I/R in isolated mouse hearts	Cardiomyocyte- or endothelial cell-specific overexpression of CYP2J2	Improved functional recovery in mice with cardiomyocyte-, but not endothelial cell-specific overexpression of CYP2J2	[56, 182, 224]
Kidney			
Transient occlusion of renal artery and vein	CYP hydroxylase inhibitor; 20-HETE antagonist	Inhibition of 20-HETE formation or action ameliorates I/R-induced renal injury	[183]
Brain			
Middle cerebral artery occlusion (MCAO)	CYP ω-hydroxylase inhibitor	Reduction of ischemic infarct size	[234]
MCAO	sEH inhibitors in rats or sEH KO mice; Estradiol	Protective effect of estradiol is partially mediated by downregulation of cerebral sEH expression	[290]
Intracerebral hemorrhage in rats	Inhibitor of 20-HETE synthesis	Inhibition of 20-HETE synthesis reduced infarct size	[234]

Table 6.5 Role of CYP eicosanoids in animal models of cardiac hypertrophy

Model	Genetic or pharmacological intervention	Effect	Refs.
Transverse aortic constriction (TAC)	CYP2J2 TG mice	Prevention of ventricular connexin 43 delocalization and arrhythmia	[248]
	sEH inhibitor	Prevention and reversal of cardiac hypertrophy	[246]
	sEH knockout mice	Improved cardiac function	[247]
Chronic β-adrenergic stimulation by isoproterenol	CYP2J2 TG mice	Prevention of atrial fibrosis and atrial fibrillation	[248]
Angiotensin II-induced hypertrophy	sEH inhibitor	Reduction in left ventricular hypertrophy	[178]
	sEH knockout mice	Improved cardiac function	[247]
Doxorubicin-induced cardiotoxiticy	CYP2J2 TG mice	Reduced cardiotoxicity and improved cardiac function	[249]

CYP-eicosanoid formation. These problems have been partially overcome due to the recent progress in developing suitable pharmacological tools that specifically target the formation and action of 20-HETE and EETs. These tools include selective inhibitors of CYP hydroxylases [23],

CYP epoxygenases [187], and the sEH [179] as well as synthetic agonists and antagonists of 20-HETE [188, 189] and EETs [190–192]. Moreover, genetic engineering has been increasingly used to dissect the tissue-specific actions of CYP eicosanoids and to prove their significance in the development of cardiovascular disease (compare Table 6.3).

To (1) Androgen-induced hypertension provides a good example of how pharmacological and genetic interventions can be successfully combined for elucidating the prohypertensive and proinflammatory role of 20-HETE [193]. First, androgen treatment was shown to elevate blood pressure in rats. Indicating an important role of 20-HETE, androgen-induced hypertension was associated with increased vascular CYP4A expression and could be ameliorated by treating the animals with an inhibitor of CYP4A-mediated 20-HETE synthesis [194]. Moreover, adenovirus-mediated vascular overexpression of a 20-HETE generating CYP4A enzyme was alone sufficient to cause hypertension and renal injury in rats [195]. Partially explaining these in vivo observations, 20-HETE has been identified (1) as a potent vasoconstrictor by inhibiting calcium-activated potassium (BK) channels in vascular smooth muscle cells [196] and (2) to promote endothelial dysfunction by uncoupling endothelial nitric oxide synthase (eNOS) and activating the proinflammatory transcription factor NF-κB [197]. Beyond these mechanisms, 20-HETE is able to induce angiotensin-converting enzyme expression resulting in enhanced local and circulating angiotensin II-levels that contribute to the systemic prohypertensive effects of vascular 20-HETE overproduction [198, 199].

Other animal models, where hypertension may rely on similar 20-HETE-mediated mechanisms, include cyclosporine-induced hypertension in rats, androgen-induced hypertension in mice, and blood pressure elevation in Cyp4a14 knockout mice that is associated with an upregulation of androgen-inducible Cyp4a12 (for references, see Table 6.3). Increased urinary 20-HETE levels are associated with endothelial dysfunction in humans indicating that 20-HETE may also play an important role in human vascular pathophysiology [200]. Interestingly, the same study found significantly higher 20-HETE levels in men compared to women.

To (2) There are two major sites of 20-HETE generation in the kidney: (1) preglomerular microvessels where 20-HETE mediates vasoconstriction by inhibiting BK channels and (2) the renal tubule where 20-HETE promotes salt excretion by inhibiting Na^+-K^+-ATPase in proximal tubules and the Na^+-K^+-$2Cl^-$ cotransporter in the thick ascending loop of Henle [16, 17, 201]. Accordingly, renal tubular 20-HETE deficiency may contribute to the development of hypertension as first suggested based on studies with salt-sensitive Dahl rats [15]. Proving this hypothesis, treatment with a selective inhibitor of 20-HETE formation was sufficient to promote salt-sensitive hypertension in normal Sprague-Dawley rats [202]. Tubular 20-HETE deficiency is obviously also involved in the development of DOCA-salt induced hypertension in mice [203, 204] and some other animal models.

In humans, the T8590C polymorphism leads to the expression of a functional variant of CYP4A11 with reduced AA ω-hydroxylase activity [26]. Carriers of the C-allele show an increased risk of developing essential and salt-sensitive hypertension [26, 205]. Moreover, the CYP4A11 T8590C genotype was suggested to predict responses to medications that affect sodium homeostasis in hypertensive patients [206]. Functional polymorphisms exist also in the human CYP4F2 gene and were recently shown to associate with hypertension and other components of the metabolic syndrome [207].

To (3) Under physiological conditions, EETs are involved in the regulation of renal blood flow and salt excretion. EETs mediate vasodilator responses and represent the major EDHF in renal arterioles [53, 201, 208]. In distal tubules, EETs inhibit sodium reabsorption by reducing ENaC activity [79, 209]. Adenosine acting via the adenosine A2A receptor (A2AR) promotes renal EET formation in response to high dietary salt increase. The inability to upregulate this pathway is associated with the development of salt-sensitive hypertension in Dahl salt-sensitive rats [210]. Proving the importance of the adenosine-A2AR-EET axis, salt-resistant rats are rendered hypertensive inhibiting the key components of this pathway [211].

EET deficiency caused by downregulation of CYP epoxygenases and/or upregulation of sEH is also an important feature and mediator of angiotensin II-induced hypertension (Table 6.3). For example, in double transgenic rats overexpressing the human angiotensinogen and renin genes, fenofibrate restored CYP2C23-mediated renal EET formation and prevented the development of hypertension and renal injury [71]. Pharmacological inhibition of the sEH enzyme prevented and reversed angiotensin II-infusion hypertension in mice [212]. Direct evidence for the protective role of CYP epoxygenases in angiotensin II-induced hypertension comes from recent studies using transgenic mice with endothelial specific overexpression of the human enzymes CYP2C8 and CYP2J2 [55, 64].

In humans, circulating levels of 20-HETE are increased and those of EETs are decreased in renovascular disease, whereas the urinary excretion of 20-HETE is reduced [213]. Moreover, as reviewed by other authors, genetic association studies indicate that certain functional polymorphisms in the human CYP2J2 and sEH (EPHX2) genes may be linked to an increased risk of developing hypertension, coronary artery disease and stroke [214–216].

6.4.2 Ischemia-Reperfusion Injury

Ischemia-reperfusion (I/R) induced organ damage is a common feature of myocardial infarction, acute kidney injury and stroke. Early events initiating the pathophysiological cascade include ATP depletion and Ca^{2+}-overload followed by a rapid activation of PLA2 enzymes. Ischemia-induced PLA2 activation plays a critical role in I/R injury of the heart and brain [217–219] and has also been demonstrated in the kidney [220]. PLA2 activation results in the generation of potentially toxic lyso-phospholipids as well as accumulation of free AA that in turn may trigger disturbances in eicosanoid formation in the reperfusion phase. Moreover, the activated PLA2 is able to release preformed CYP eicosanoids from their membrane stores as shown for 20-HETE in the kidney [183]. I/R-induced excessive 20-HETE formation was also shown in the heart [221]. Accumulating evidence from various animal models suggests that 20-HETE plays a major detrimental role in I/R-injury, whereas measures increasing EET formation and action exert strong protective effects (Table 6.4).

Heart First studies leading to the recognition of the detrimental role of 20-HETE in myocardial infarction were performed in dogs. In canine hearts subjected to coronary artery ligation, inhibition of endogenous 20-HETE formation reduced infarct size, whereas exogenous 20-HETE administration exacerbated the injury [222]. Moreover, inhibition of 20-HETE formation enhances the beneficial effects of ischemic preconditioning (IPC) on the severity of myocardial infarction [223].

Initiating research on the protective role of EETs in the heart, exogenous EET administration to isolated perfused hearts was found to improve postischemic functional recovery and also to prevent electrocardiogram abnormalities in the reperfusion phase [75, 224]. EET pretreatments also efficiently reduced myocardial infarction size after transient coronary artery occlusion [182, 225, 226]. Further studies revealed an essential role of EETs in mediating the beneficial effects of pre- and postconditioning [227–229]. Mimicking the effects of exogenous EET administration, cardiomyocyte-specific overexpression of human CYP2J2 in transgenic mice improved recovery of pump function, and ventricular repolarization after ischemia [63]. In line with the cardioprotective effects of EETs, pharmacological inhibition of the sEH enzyme as well as sEH gene deletion ameliorated myocardial I/R-injury in mice [230]. Moreover, a synthetic EET analog was successfully used for protecting isolated murine hearts against global ischemia-induced loss of pump function and myocardial injury [231].

The potential mechanisms underlying the opposing roles of 20-HETE and EETs have been discussed in recent reviews [182, 232, 233]. Accordingly, the detrimental role of 20-HETE in myocardial I/R-injury is certainly multifactorial and involves vasoconstrictor and proinflammatory actions similar to those discussed above in the context of hypertension and vascular injury (compare Sect. 6.4.1). Moreover, 20-HETE induces inherent mechanisms of apoptosis in cardiomyocytes probably by inhibiting the mitochondrial ATP-sensitive potassium channel. EETs oppose the vasoconstrictor and proinflammatory action of 20-HETE and are able to induce prosurvival mechanisms in cardiomyocytes.

Brain Studies in rat models of brain I/R-injury demonstrated that blockade of 20-HETE synthesis ameliorates cerebral vasospasm following subarachnoid hemorrhage, and reduces infarct size in ischemic stroke [234, 235]. Pharmacological inhibition as well as genetic deletion of the sEH enzyme is protective in mouse models of ischemic stroke [236]. Interestingly, sex-specific expression of the sEH (male>female) has been linked to the pronounced sex difference in the extent of brain injury after cerebral artery occlusion in mice [237]. Based on these and further findings, sEH has been proposed as a novel therapeutic target in stroke [238]. A recent review gives further information on the potential mechanisms of cerebral I/R-injury that are beneficially modulated upon inhibiting 20-HETE or increasing EET levels by inhibiting the sEH [185].

Kidney I/R-induced acute kidney injury (AKI) leads to increased morbidity and mortality, particularly after cardiovascular surgery and kidney transplantation [239–241]. I/R-induced mechanisms in the kidney include persistent vasoconstriction, inflammation, endothelial dysfunction, and tubular injury [242, 243]. As analyzed in a rat model of AKI, 20-HETE released in the ischemic phase plays an important role in setting the stage for the subsequent events leading to renal failure. Inhibiting the formation or action of 20-HETE during ischemia improved the recovery of renal tissue perfusion and oxygenation in the early reperfusion phase and protected against subsequent inflammatory cell infiltration, tubular epithelial cell apoptosis and decline of renal function [183]. 20-HETE overproduction also exacerbates the cytotoxic and proapoptotic effects of chemical hypoxia on cultured primary renal tubular epithelial cells [244]. In contrast, protective effects of 20-HETE were observed in another rat model of AKI. In this model, systemic long-term inhibition of 20-HETE formation aggravated and antagonizing 20-HETE action in the reperfusion phase ameliorated renal I/R injury [184]. These apparently contradictory results probably reflect the unique dual role of 20-HETE in the kidney that unlike other organs requires 20-HETE for its normal function.

6.4.3 Cardiac Hypertrophy and Arrhythmia

Maladaptive cardiac hypertrophy is associated with structural and electrical remodeling eventually leading to heart failure and increased propensity to ventricular tachyarrhythmia and sudden cardiac death [245]. This disease may occur upon chronic pressure overload due to aortic stenosis but also develops frequently in more complex disease and stress conditions such as hypertension and myocardial infarction or adrenergic overdrive (Table 6.5).

Indicating an important protective role of CYP epoxy metabolites in pressure overload-induced cardiac hypertrophy, pharmacological sEH inhibition prevents and reverses left ventricular hypertrophy after transverse aortic constriction (TAC) in mice [246]. Pressure overload as well as angiotensin II-induced maladaptive cardiac hypertrophy is ameliorated in sEH knockout compared to wild-type mice [247]. Genetic analysis in rats identified the sEH gene (EPHX2) as a susceptibility factor for heart failure [247]. Interestingly, some of the rat strains used for experimental studies carry an EPHX2 promoter variant that decreases basal expression of the sEH enzyme and abolishes its angiotensin II-inducibility [247]. In mice, upregulation of cardiac sEH expression was shown to be essential for angiotensin II-induced cardiac hypertrophy [178].

Direct experimental evidence for a cardioprotective role of enhanced endogenous EET biosynthesis was provided comparing the development of TAC-induced cardiac hypertrophy in CYP2J2-transgenic mice and corresponding wild-type littermates [248]. Cardiomyocyte-specific overexpression of the human CYP epoxygenase markedly improved the survival of the animals and prevented the development of ventricular tachyarrhythmia vulnerability. Reduced arrhythmia susceptibility was related to CYP2J2-mediated protection against hypertrophy-induced delocalization of left ventricular connexin-43. CYP2J2 transgenic mice also displayed improved electrical remodeling in β-adrenergic stimulation-induced cardiac hypertrophy. In this model, CYP2J2 overexpression specifically prevented the development of fibrosis and atrial fibrillation susceptibility [248]. Other studies with CYP2J2 transgenic mice demonstrate that enhanced cardiac EET biosynthesis also protects against doxorubicin-induced cardiotoxicity [249] and the development of heart failure upon long-term infusion of angiotensin II or isoproterenol [250]. In vitro, exogenous administration of 14,15-EET inhibited the hypertrophic response of cultured cardiomyocytes to isoproterenol, whereas 20-HETE was alone sufficient to induce cellular hypertrophy [251].

Taken together, these studies revealed an important role of CYP eicosanoids in the pathogenesis of cardiac hypertrophy, heart failure and arrhythmia. The mechanisms are only partially understood but obviously include opposing roles of 20-HETE and EETs in mediating or suppressing prohypertrophic, proinflammatory and proapoptotic signaling pathways in cardiomyocytes [252]. Moreover, 20-HETE and EETs modulate ion channel activities in cardiomyocytes and thus influence cardiac electrophysiology and Ca^{2+}-handling [253]. Recently, combined inhibition of 20-HETE formation and of EET degradation was shown to attenuate hypertension and cardiac hypertrophy in Ren-2 transgenic rats [254]. This study provides an example for the role of 20-HETE and EETs in conditions of severe hypertension and end-organ damage and also of the promising therapeutic potential of approaches targeting the CYP-eicosanoid pathway in such complex disease states.

6.5 Conclusions

During the last three decades, the work of many laboratories improved our understanding of the physiological and pathophysiological relevance of the CYP-eicosanoid pathway. Extending the initial discoveries showing important roles in hypertension and renal failure, novel and previously unexpected implications have been revealed in myocardial infarction, maladaptive cardiac hypertrophy, acute kidney injury and stroke. Recent progress in CYP-eicosanoid profiling as well as genetic association studies suggest that much of what has been learned from animal experiments is also relevant to human cardiovascular disease. The list is steadily growing and we have to apologize for not explicitly covering in this review many other exciting findings regarding for example the role of CYP-eicosanoids in the gastrointestinal tract [255], lung [256, 257] and liver [258]. There is also significant progress in understanding the contribution of CYP-eicosanoids to the regulation of neurohormone release and pain sensation [259, 260]. Another important avenue of research led from recognizing EETs as stimulators of insulin secretion in isolated rat pancreatic islets [261] via the identification of CYP2J2 as a major epoxygenase in this cell type [262] to the hypothesis that EETs may be the link between endothelial dysfunction and insulin resistance [263]. The most recent paper in this chain demonstrates beneficial effects of sEH inhibition on glucose homeostasis and islet damage in a streptozotocin-induced diabetic mouse model [264].

Beyond their nowadays well-established roles in cardiovascular health and disease, CYP eicosanoids have been recently recognized as mediators of physiological and pathophysiological forms of angiogenesis [153, 265–269]. On the one hand, these novel findings improve our understanding of repair mechanisms and may open new opportunities for promoting wound healing. On the other hand, these findings indicate that alterations in the CYP-eicosanoid pathway may contribute to tumor proliferation and metastasis, age-related macular degeneration and other disease states associated with pathological angiogenesis. They also suggest that interventions into the CYP-eicosanoid pathway aimed at protecting vascular, cardiac and renal function may have detrimental side effects in promoting cancer progression. This concern was specifically raised against therapeutic strategies that increase the endogenous EET levels because, in particular, these AA-derived metabolites could function as "double-edged swords" [270].

The balance of n-6 and n-3 PUFAs in the diet has been recognized as one of the most important modifiable risk factors for the development of cardiovascular disease but also to influence cancerogenesis and pathologic neovascularization in ocular disease. Studies on the substrate and reaction specificity as well as on diet-induced changes in the endogenous CYP-eicosanoid profile clearly demonstrate that CYP enzymes do not metabolize only AA but also a wide range of other n-6 and n-3 PUFAs. In particular, the CYP-dependent metabolism of EPA and DHA generates sets of epoxy metabolites with superior vasodilatory, antiinflammatory and cardioprotective properties compared to the AA-derived counterparts. First studies also suggest that these n-3 PUFA-derived metabolites have unique biological activities in exerting antiarrhythmic effects [60] and suppressing tumor angiogenesis [154]. It is tempting to speculate but remains to be directly shown that the CYP-eicosanoid pathway mediates a variety of the beneficial effects attributed to diets rich in EPA and DHA.

However, there are also important gaps of knowledge. In particular, our understanding of the CYP-eicosanoid induced signaling pathways is incomplete and hampered by the fact that the primary cellular targets of the diverse epoxy and hydroxyl metabolites have not yet been identified. Eicosanoids generated via the COX- and LOX-dependent pathways exert their biological functions by activating G-protein coupled receptors [18]. Accumulating evidence suggests that there are also receptor-like membrane proteins specifically interacting with individual CYP eicosanoids [271–273] and it will be exciting to learn their molecular identities in the near future. Unexpected help in unraveling the components of CYP-eicosanoid mediated signaling pathways in mammals may also come from studies on PUFA-derived signaling in small animal models such as *Caenorhabditis elegans* and *Drosophila melanogaster* [118]. Indicating the existence of evolutionary conserved mechanisms, CYP-33E2, a CYP enzyme resembling the human cardiac epoxygenase CYP2J2, is expressed in *Caenorhabditis elegans* and contributes there to the regulation of pharynx activity, an organ continuously pumping in nematodes [274]. Moreover, CYP eicosanoids are obviously essential for mediating the behavioral response of *Caenorhabditis elegans* to hypoxia-reoxygenation perhaps via signaling pathways partially resembling those expressed in mammals and mediating the effects of CYP eicosanoids in ischemia-reperfusion injury of the heart, brain and kidney [275].

Acknowledgments This review is dedicated to the late John Charles (Jack) McGiff, MD and Professor of Pharmacology. Professor McGiff made seminal discoveries on the role of CYP eicosanoids in blood pressure regulation and he inspired many students and colleagues to follow his way of successfully combining biochemistry and pathophysiology for understanding the mechanisms of complex cardiovascular diseases. This work was supported by grants from the Deutsche Forschungsgemeinschaft (DFG): Schu822/5; FOR 1054 and Schu822/7-1; FOR 1368.

References

1. Capdevila J, Parkhill L, Chacos N, Okita R, Masters BS, Estabrook RW (1981) The oxidative metabolism of arachidonic acid by purified cytochromes P-450. Biochem Biophys Res Commun 101:1357–1363
2. Capdevila J, Chacos N, Werringloer J, Prough RA, Estabrook RW (1981) Liver microsomal cytochrome P-450 and the oxidative metabolism of arachidonic acid. Proc Natl Acad Sci U S A 78:5362–5366
3. Morrison AR, Pascoe N (1981) Metabolism of arachidonate through NADPH-dependent oxygenase of renal cortex. Proc Natl Acad Sci U S A 78:7375–7378
4. Oliw EH, Lawson JA, Brash AR, Oates JA (1981) Arachidonic acid metabolism in rabbit renal cortex. Formation of two novel dihydroxyeicosatrienoic acids. J Biol Chem 256:9924–9931
5. Capdevila JH, Falck JR, Dishman E, Karara A (1990) Cytochrome P-450 arachidonate oxygenase. Methods Enzymol 187:385–394
6. Capdevila JH, Falck JR, Estabrook RW (1992) Cytochrome P450 and the arachidonate cascade. FASEB J 6:731–736
7. Capdevila J, Pramanik B, Napoli JL, Manna S, Falck JR (1984) Arachidonic acid epoxidation: epoxyeicosatrienoic acids are endogenous constituents of rat liver. Arch Biochem Biophys 231:511–517
8. Falck JR, Schueler VJ, Jacobson HR, Siddhanta AK, Pramanik B, Capdevila J (1987) Arachidonate epoxygenase: identification of epoxyeicosatrienoic acids in rabbit kidney. J Lipid Res 28:840–846
9. Toto R, Siddhanta A, Manna S, Pramanik B, Falck JR, Capdevila J (1987) Arachidonic acid epoxygenase: detection of epoxyeicosatrienoic acids in human urine. Biochim Biophys Acta 919:132–139
10. Catella F, Lawson JA, Fitzgerald DJ, FitzGerald GA (1990) Endogenous biosynthesis of arachidonic acid epoxides in humans: increased formation in pregnancy-induced hypertension. Proc Natl Acad Sci U S A 87:5893–5897
11. Karara A, Dishman E, Blair I, Falck JR, Capdevila JH (1989) Endogenous epoxyeicosatrienoic acids. Cytochrome P-450 controlled stereoselectivity of the hepatic arachidonic acid epoxygenase. J Biol Chem 264:19822–19827
12. McGiff JC (1991) Cytochrome P-450 metabolism of arachidonic acid. Annu Rev Pharmacol Toxicol 31:339–369
13. Sacerdoti D, Escalante B, Abraham NG, McGiff JC, Levere RD, Schwartzman ML (1989) Treatment with tin prevents the development of hypertension in spontaneously hypertensive rats. Science 243:388–390
14. Makita K, Takahashi K, Karara A, Jacobson HR, Falck JR, Capdevila JH (1994) Experimental and/or genetically controlled alterations of the renal microsomal cytochrome P450 epoxygenase induce hypertension in rats fed a high salt diet. J Clin Invest 94:2414–2420
15. Roman RJ, Alonso-Galicia M, Wilson TW (1997) Renal P450 metabolites of arachidonic acid and the development of hypertension in Dahl salt-sensitive rats. Am J Hypertens 10:63S–67S
16. McGiff JC, Quilley J (1999) 20-HETE and the kidney: resolution of old problems and new beginnings. Am J Physiol 277:R607–R623
17. Roman RJ (2002) P-450 metabolites of arachidonic acid in the control of cardiovascular function. Physiol Rev 82:131–185
18. Funk CD (2001) Prostaglandins and leukotrienes: advances in eicosanoid biology. Science 294:1871–1875
19. Buczynski MW, Dumlao DS, Dennis EA (2009) Thematic review series: proteomics. An integrated omics analysis of eicosanoid biology. J Lipid Res 50:1015–1038
20. Bergstroem S, Ryhage R, Samuelsson B, Sjoevall J (1963) Prostaglandins and related factors. 15. The structures of prostaglandin E1, $F_{1\alpha}$, and $F_{1\beta}$. J Biol Chem 238:3555–3564
21. Vane JR (1971) Inhibition of prostaglandin synthesis as a mechanism of action for aspirin-like drugs. Nat New Biol 231:232–235
22. Samuelsson B (1983) Leukotrienes: mediators of immediate hypersensitivity reactions and inflammation. Science 220:568–575
23. Kroetz DL, Xu F (2005) Regulation and inhibition of arachidonic acid ω-hydroxylases and 20-HETE formation. Annu Rev Pharmacol Toxicol 45:413–438
24. Powell PK, Wolf I, Jin R, Lasker JM (1998) Metabolism of arachidonic acid to 20-hydroxy-5,8,11,14-eicosatetraenoic acid by P450 enzymes in human liver: involvement of CYP4F2 and CYP4A11. J Pharmacol Exp Ther 285:1327–1336
25. Lasker JM, Chen WB, Wolf I, Bloswick BP, Wilson PD, Powell PK (2000) Formation of 20-hydroxyeicosatetraenoic acid, a vasoactive and natriuretic eicosanoid, in human kidney. Role of Cyp4F2 and Cyp4A11. J Biol Chem 275:4118–4126
26. Gainer JV, Bellamine A, Dawson EP, Womble KE, Grant SW, Wang Y, Cupples LA, Guo CY, Demissie S, O'Donnell CJ, Brown NJ, Waterman MR, Capdevila JH (2005) Functional variant of CYP4A11 20-hydroxyeicosatetraenoic acid synthase is associated with essential hypertension. Circulation 111:63–69
27. Hiratsuka M, Nozawa H, Katsumoto Y, Moteki T, Sasaki T, Konno Y, Mizugaki M (2006) Genetic polymorphisms and haplotype structures of the CYP4A22 gene in a Japanese population. Mutat Res 599:98–104
28. Lino Cardenas CL, Renault N, Farce A, Cauffiez C, Allorge D, Lo-Guidice JM, Lhermitte M, Chavatte P, Broly F, Chevalier D (2011) Genetic polymorphism of CYP4A11 and CYP4A22 genes and *in silico*

insights from comparative 3D modelling in a French population. Gene 487:10–20
29. Christmas P, Jones JP, Patten CJ, Rock DA, Zheng Y, Cheng SM, Weber BM, Carlesso N, Scadden DT, Rettie AE, Soberman RJ (2001) Alternative splicing determines the function of CYP4F3 by switching substrate specificity. J Biol Chem 276:38166–38172
30. Corcos L, Lucas D, Le Jossic-Corcos C, Dreano Y, Simon B, Plee-Gautier E, Amet Y, Salaun JP (2012) Human cytochrome P450 4F3: structure, functions, and prospects. Drug Metabol Drug Interact 27:63–71
31. Fer M, Corcos L, Dreano Y, Plee-Gautier E, Salaun JP, Berthou F, Amet Y (2008) Cytochromes P450 from family 4 are the main omega hydroxylating enzymes in humans: CYP4F3B is the prominent player in PUFA metabolism. J Lipid Res 49:2379–2389
32. Chuang SS, Helvig C, Taimi M, Ramshaw HA, Collop AH, Amad M, White JA, Petkovich M, Jones G, Korczak B (2004) CYP2U1, a novel human thymus- and brain-specific cytochrome P450, catalyzes ω- and (ω-1)-hydroxylation of fatty acids. J Biol Chem 279:6305–6314
33. Kelly EJ, Nakano M, Rohatgi P, Yarov-Yarovoy V, Rettie AE (2011) Finding homes for orphan cytochrome P450s: CYP4V2 and CYP4F22 in disease states. Mol Interv 11:124–132
34. Nguyen X, Wang MH, Reddy KM, Falck JR, Schwartzman ML (1999) Kinetic profile of the rat CYP4A isoforms: arachidonic acid metabolism and isoform-specific inhibitors. Am J Physiol 276: R1691–R1700
35. Yamaguchi Y, Kirita S, Hasegawa H, Aoyama J, Imaoka S, Minamiyama S, Funae Y, Baba T, Matsubara T (2002) Contribution of CYP4A8 to the formation of 20-hydroxyeicosatetraenoic acid from arachidonic acid in rat kidney. Drug Metab Pharmacokinet 17:109–116
36. Xu F, Falck JR, Ortiz de Montellano PR, Kroetz DL (2004) Catalytic activity and isoform-specific inhibition of rat cytochrome P450 4F enzymes. J Pharmacol Exp Ther 308:887–895
37. El-Sherbeni AA, Aboutabl ME, Zordoky BN, Anwar-Mohamed A, El-Kadi AO (2013) Determination of the dominant arachidonic acid cytochrome P450 monooxygenases in rat heart, lung, kidney, and liver: protein expression and metabolite kinetics. AAPS J 15:112–122
38. Marji JS, Wang MH, Laniado-Schwartzman M (2002) Cytochrome P-450 4A isoform expression and 20-HETE synthesis in renal preglomerular arteries. Am J Physiol Renal Physiol 283:F60–F67
39. Singh H, Schwartzman ML (2008) Renal vascular cytochrome P450-derived eicosanoids in androgen-induced hypertension. Pharmacol Rep 60:29–37
40. Dunn KM, Renic M, Flasch AK, Harder DR, Falck J, Roman RJ (2008) Elevated production of 20-HETE in the cerebral vasculature contributes to severity of ischemic stroke and oxidative stress in spontaneously hypertensive rats. Am J Physiol Heart Circ Physiol 295:H2455–H2465
41. Nelson DR, Zeldin DC, Hoffman SM, Maltais LJ, Wain HM, Nebert DW (2004) Comparison of cytochrome P450 (CYP) genes from the mouse and human genomes, including nomenclature recommendations for genes, pseudogenes and alternative-splice variants. Pharmacogenetics 14:1–18
42. Muller DN, Schmidt C, Barbosa-Sicard E, Wellner M, Gross V, Hercule H, Markovic M, Honeck H, Luft FC, Schunck WH (2007) Mouse Cyp4a isoforms: enzymatic properties, gender- and strain-specific expression, and role in renal 20-hydroxyeicosatetraenoic acid formation. Biochem J 403:109–118
43. Holla VR, Adas F, Imig JD, Zhao X, Price E Jr, Olsen N, Kovacs WJ, Magnuson MA, Keeney DS, Breyer MD, Falck JR, Waterman MR, Capdevila JH (2001) Alterations in the regulation of androgen-sensitive Cyp 4a monooxygenases cause hypertension. Proc Natl Acad Sci U S A 98:5211–5216
44. Wu CC, Mei S, Cheng J, Ding Y, Weidenhammer A, Garcia V, Zhang F, Gotlinger K, Manthati VL, Falck JR, Capdevila JH, Schwartzman ML (2013) Androgen-sensitive hypertension associates with upregulated vascular CYP4A12-20-HETE synthase. J Am Soc Nephrol 24:1288–1296
45. Nakagawa K, Holla VR, Wei Y, Wang WH, Gatica A, Wei S, Mei S, Miller CM, Cha DR, Price E Jr, Zent R, Pozzi A, Breyer MD, Guan Y, Falck JR, Waterman MR, Capdevila JH (2006) Salt-sensitive hypertension is associated with dysfunctional Cyp4a10 gene and kidney epithelial sodium channel. J Clin Invest 116:1696–1702
46. Stec DE, Flasch A, Roman RJ, White JA (2003) Distribution of cytochrome P-450 4A and 4F isoforms along the nephron in mice. Am J Physiol Renal Physiol 284:F95–F102
47. Zeldin DC (2001) Epoxygenase pathways of arachidonic acid metabolism. J Biol Chem 276:36059–36062
48. Rifkind AB, Lee C, Chang TK, Waxman DJ (1995) Arachidonic acid metabolism by human cytochrome P450s 2C8, 2C9, 2E1, and 1A2: regioselective oxygenation and evidence for a role for CYP2C enzymes in arachidonic acid epoxygenation in human liver microsomes. Arch Biochem Biophys 320:380–389
49. Daikh BE, Lasker JM, Raucy JL, Koop DR (1994) Regio- and stereoselective epoxidation of arachidonic acid by human cytochromes P450 2C8 and 2C9. J Pharmacol Exp Ther 271:1427–1433
50. Zeldin DC, DuBois RN, Falck JR, Capdevila JH (1995) Molecular cloning, expression and characterization of an endogenous human cytochrome P450 arachidonic acid epoxygenase isoform. Arch Biochem Biophys 322:76–86

51. Fisslthaler B, Popp R, Kiss L, Potente M, Harder DR, Fleming I, Busse R (1999) Cytochrome P450 2C is an EDHF synthase in coronary arteries. Nature 401:493–497
52. Campbell WB, Gebremedhin D, Pratt PF, Harder DR (1996) Identification of epoxyeicosatrienoic acids as endothelium-derived hyperpolarizing factors. Circ Res 78:415–423
53. Campbell WB, Falck JR (2007) Arachidonic acid metabolites as endothelium-derived hyperpolarizing factors. Hypertension 49:590–596
54. Campbell WB, Fleming I (2010) Epoxyeicosatrienoic acids and endothelium-dependent responses. Pflugers Arch 459:881–895
55. Lee CR, Imig JD, Edin ML, Foley J, DeGraff LM, Bradbury JA, Graves JP, Lih FB, Clark J, Myers P, Perrow AL, Lepp AN, Kannon MA, Ronnekleiv OK, Alkayed NJ, Falck JR, Tomer KB, Zeldin DC (2010) Endothelial expression of human cytochrome P450 epoxygenases lowers blood pressure and attenuates hypertension-induced renal injury in mice. FASEB J 24:3770–3781
56. Edin ML, Wang Z, Bradbury JA, Graves JP, Lih FB, DeGraff LM, Foley JF, Torphy R, Ronnekleiv OK, Tomer KB, Lee CR, Zeldin DC (2011) Endothelial expression of human cytochrome P450 epoxygenase CYP2C8 increases susceptibility to ischemia-reperfusion injury in isolated mouse heart. FASEB J 25:3436–3447
57. Fleming I, Michaelis UR, Bredenkotter D, Fisslthaler B, Dehghani F, Brandes RP, Busse R (2001) Endothelium-derived hyperpolarizing factor synthase (Cytochrome P450 2C9) is a functionally significant source of reactive oxygen species in coronary arteries. Circ Res 88:44–51
58. Fichtlscherer S, Dimmeler S, Breuer S, Busse R, Zeiher AM, Fleming I (2004) Inhibition of cytochrome P450 2C9 improves endothelium-dependent, nitric oxide-mediated vasodilatation in patients with coronary artery disease. Circulation 109:178–183
59. Fer M, Dreano Y, Lucas D, Corcos L, Salaun JP, Berthou F, Amet Y (2008) Metabolism of eicosapentaenoic and docosahexaenoic acids by recombinant human cytochromes P450. Arch Biochem Biophys 471:116–125
60. Arnold C, Markovic M, Blossey K, Wallukat G, Fischer R, Dechend R, Konkel A, von Schacky C, Luft FC, Muller DN, Rothe M, Schunck WH (2010) Arachidonic acid-metabolizing cytochrome P450 enzymes are targets of ω-3 fatty acids. J Biol Chem 285:32720–31733
61. Wu S, Moomaw CR, Tomer KB, Falck JR, Zeldin DC (1996) Molecular cloning and expression of CYP2J2, a human cytochrome P450 arachidonic acid epoxygenase highly expressed in heart. J Biol Chem 271:3460–3468
62. Scarborough PE, Ma J, Qu W, Zeldin DC (1999) P450 subfamily CYP2J and their role in the bioactivation of arachidonic acid in extrahepatic tissues. Drug Metab Rev 31:205–234
63. Seubert J, Yang B, Bradbury JA, Graves J, Degraff LM, Gabel S, Gooch R, Foley J, Newman J, Mao L, Rockman HA, Hammock BD, Murphy E, Zeldin DC (2004) Enhanced postischemic functional recovery in CYP2J2 transgenic hearts involves mitochondrial ATP-sensitive K^+ channels and p42/p44 MAPK pathway. Circ Res 95:506–514
64. Askari A, Thomson SJ, Edin ML, Zeldin DC, Bishop-Bailey D (2013) Roles of the epoxygenase CYP2J2 in the endothelium. Prostaglandins Other Lipid Mediat 107:56–63
65. Spiecker M, Liao JK (2005) Vascular protective effects of cytochrome P450 epoxygenase-derived eicosanoids. Arch Biochem Biophys 433:413–420
66. Karara A, Makita K, Jacobson HR, Falck JR, Guengerich FP, DuBois RN, Capdevila JH (1993) Molecular cloning, expression, and enzymatic characterization of the rat kidney cytochrome P-450 arachidonic acid epoxygenase. J Biol Chem 268:13565–13570
67. Imaoka S, Wedlund PJ, Ogawa H, Kimura S, Gonzalez FJ, Kim HY (1993) Identification of CYP2C23 expressed in rat kidney as an arachidonic acid epoxygenase. J Pharmacol Exp Ther 267:1012–1016
68. Holla VR, Makita K, Zaphiropoulos PG, Capdevila JH (1999) The kidney cytochrome P-450 2C23 arachidonic acid epoxygenase is upregulated during dietary salt loading. J Clin Invest 104:751–760
69. Kaergel E, Muller DN, Honeck H, Theuer J, Shagdarsuren E, Mullally A, Luft FC, Schunck WH (2002) P450-dependent arachidonic acid metabolism and angiotensin II-induced renal damage. Hypertension 40:273–279
70. Zhao X, Pollock DM, Zeldin DC, Imig JD (2003) Salt-sensitive hypertension after exposure to angiotensin is associated with inability to upregulate renal epoxygenases. Hypertension 42:775–780
71. Muller DN, Theuer J, Shagdarsuren E, Kaergel E, Honeck H, Park JK, Markovic M, Barbosa-Sicard E, Dechend R, Wellner M, Kirsch T, Fiebeler A, Rothe M, Haller H, Luft FC, Schunck WH (2004) A peroxisome proliferator-activated receptor-α activator induces renal CYP2C23 activity and protects from angiotensin II-induced renal injury. Am J Pathol 164:521–532
72. Capdevila JH, Karara A, Waxman DJ, Martin MV, Falck JR, Guenguerich FP (1990) Cytochrome P-450 enzyme-specific control of the regio- and enantiofacial selectivity of the microsomal arachidonic acid epoxygenase. J Biol Chem 265:10865–10871
73. Alkayed NJ, Narayanan J, Gebremedhin D, Medhora M, Roman RJ, Harder DR (1996) Molecular characterization of an arachidonic acid epoxygenase in rat brain astrocytes. Stroke 27:971–979

74. Medhora M, Narayanan J, Harder D (2001) Dual regulation of the cerebral microvasculature by epoxyeicosatrienoic acids. Trends Cardiovasc Med 11:38–42
75. Wu S, Chen W, Murphy E, Gabel S, Tomer KB, Foley J, Steenbergen C, Falck JR, Moomaw CR, Zeldin DC (1997) Molecular cloning, expression, and functional significance of a cytochrome P450 highly expressed in rat heart myocytes. J Biol Chem 272:12551–12559
76. Zhang QY, Ding X, Kaminsky LS (1997) cDNA cloning, heterologous expression, and characterization of rat intestinal CYP2J4. Arch Biochem Biophys 340:270–278
77. Yaghi A, Bradbury JA, Zeldin DC, Mehta S, Bend JR, McCormack DG (2003) Pulmonary cytochrome P-450 2J4 is reduced in a rat model of acute *Pseudomonas* pneumonia. Am J Physiol Lung Cell Mol Physiol 285:L1099–L1105
78. DeLozier TC, Tsao CC, Coulter SJ, Foley J, Bradbury JA, Zeldin DC, Goldstein JA (2004) CYP2C44, a new murine CYP2C that metabolizes arachidonic acid to unique stereospecific products. J Pharmacol Exp Ther 310:845–854
79. Pidkovka N, Rao R, Mei S, Gong Y, Harris RC, Wang WH, Capdevila JH (2013) Epoxyeicosatrienoic acids (EETs) regulate epithelial sodium channel activity by extracellular signal-regulated kinase 1/2 (ERK1/2)-mediated phosphorylation. J Biol Chem 288:5223–5231
80. Zhang MZ, Wang Y, Yao B, Gewin L, Wei S, Capdevila JH, Harris RC (2013) Role of epoxyeicosatrienoic acids (EETs) in mediation of dopamine's effects in the kidney. Am J Physiol Renal Physiol 305:F1680–F1686
81. Sun P, Antoun J, Lin DH, Yue P, Gotlinger KH, Capdevila J, Wang WH (2012) Cyp2c44 epoxygenase is essential for preventing the renal sodium absorption during increasing dietary potassium intake. Hypertension 59:339–347
82. Luo G, Zeldin DC, Blaisdell JA, Hodgson E, Goldstein JA (1998) Cloning and expression of murine CYP2Cs and their ability to metabolize arachidonic acid. Arch Biochem Biophys 357:45–57
83. Wang H, Zhao Y, Bradbury JA, Graves JP, Foley J, Blaisdell JA, Goldstein JA, Zeldin DC (2004) Cloning, expression, and characterization of three new mouse cytochrome P450 enzymes and partial characterization of their fatty acid oxidation activities. Mol Pharmacol 65:1148–1158
84. Tsao CC, Foley J, Coulter SJ, Maronpot R, Zeldin DC, Goldstein JA (2000) CYP2C40, a unique arachidonic acid 16-hydroxylase, is the major CYP2C in murine intestinal tract. Mol Pharmacol 58:279–287
85. Sun D, Yang YM, Jiang H, Wu H, Ojaimi C, Kaley G, Huang A (2012) Roles of CYP2C29 and RXRγ in vascular EET synthesis of female mice. Am J Physiol Regul Integr Comp Physiol 298: R862–R869
86. Pokreisz P, Fleming I, Kiss L, Barbosa-Sicard E, Fisslthaler B, Falck JR, Hammock BD, Kim IH, Szelid Z, Vermeersch P, Gillijns H, Pellens M, Grimminger F, van Zonneveld AJ, Collen D, Busse R, Janssens S (2006) Cytochrome P450 epoxygenase gene function in hypoxic pulmonary vasoconstriction and pulmonary vascular remodeling. Hypertension 47:762–770
87. Scheer N, Kapelyukh Y, Chatham L, Rode A, Buechel S, Wolf CR (2012) Generation and characterization of novel cytochrome P450 Cyp2c gene cluster knockout and CYP2C9 humanized mouse lines. Mol Pharmacol 82:1022–1029
88. Ma J, Qu W, Scarborough PE, Tomer KB, Moomaw CR, Maronpot R, Davis LS, Breyer MD, Zeldin DC (1999) Molecular cloning, enzymatic characterization, developmental expression, and cellular localization of a mouse cytochrome P450 highly expressed in kidney. J Biol Chem 274:17777–17788
89. Ma J, Graves J, Bradbury JA, Zhao Y, Swope DL, King L, Qu W, Clark J, Myers P, Walker V, Lindzey J, Korach KS, Zeldin DC (2004) Regulation of mouse renal CYP2J5 expression by sex hormones. Mol Pharmacol 65:730–743
90. Athirakul K, Bradbury JA, Graves JP, DeGraff LM, Ma J, Zhao Y, Couse JF, Quigley R, Harder DR, Zhao X, Imig JD, Pedersen TL, Newman JW, Hammock BD, Conley AJ, Korach KS, Coffman TM, Zeldin DC (2008) Increased blood pressure in mice lacking cytochrome P450 2J5. FASEB J 22:4096–4108
91. Ma J, Bradbury JA, King L, Maronpot R, Davis LS, Breyer MD, Zeldin DC (2002) Molecular cloning and characterization of mouse CYP2J6, an unstable cytochrome P450 isoform. Biochem Pharmacol 64:1447–1460
92. Qu W, Bradbury JA, Tsao CC, Maronpot R, Harry GJ, Parker CE, Davis LS, Breyer MD, Waalkes MP, Falck JR, Chen J, Rosenberg RL, Zeldin DC (2001) Cytochrome P450 CYP2J9, a new mouse arachidonic acid ω-1 hydroxylase predominantly expressed in brain. J Biol Chem 276:25467–25479
93. Graves JP, Edin ML, Bradbury JA, Gruzdev A, Cheng J, Lih FB, Masinde TA, Qu W, Clayton NP, Morrison JP, Tomer KB, Zeldin DC (2013) Characterization of four new mouse cytochrome P450 enzymes of the CYP2J subfamily. Drug Metab Dispos 41:763–773
94. Keeney DS, Skinner C, Travers JB, Capdevila JH, Nanney LB, King LE Jr, Waterman MR (1998) Differentiating keratinocytes express a novel cytochrome P450 enzyme, CYP2B19, having arachidonate monooxygenase activity. J Biol Chem 273:32071–32079
95. Du L, Yermalitsky V, Ladd PA, Capdevila JH, Mernaugh R, Keeney DS (2005) Evidence that cytochrome P450 CYP2B19 is the major source of

epoxyeicosatrienoic acids in mouse skin. Arch Biochem Biophys 435:125–133
96. Ladd PA, Du L, Capdevila JH, Mernaugh R, Keeney DS (2003) Epoxyeicosatrienoic acids activate transglutaminases in situ and induce cornification of epidermal keratinocytes. J Biol Chem 278:35184–35192
97. Keeney DS, Skinner C, Wei S, Friedberg T, Waterman MR (1998) A keratinocyte-specific epoxygenase, CYP2B12, metabolizes arachidonic acid with unusual selectivity, producing a single major epoxyeicosatrienoic acid. J Biol Chem 273:9279–9284
98. Du L, Neis MM, Ladd PA, Lanza DL, Yost GS, Keeney DS (2006) Effects of the differentiated keratinocyte phenotype on expression levels of CYP1-4 family genes in human skin cells. Toxicol Appl Pharmacol 213:135–144
99. Sridar C, Snider NT, Hollenberg PF (2011) Anandamide oxidation by wild-type and polymorphically expressed CYP2B6 and CYP2D6. Drug Metab Dispos 39:782–788
100. Fromel T, Kohlstedt K, Popp R, Yin X, Awwad K, Barbosa-Sicard E, Thomas AC, Lieberz R, Mayr M, Fleming I (2013) Cytochrome P4502S1: a novel monocyte/macrophage fatty acid epoxygenase in human atherosclerotic plaques. Basic Res Cardiol 108:319–330
101. Schwarz D, Kisselev P, Ericksen SS, Szklarz GD, Chernogolov A, Honeck H, Schunck WH, Roots I (2004) Arachidonic and eicosapentaenoic acid metabolism by human CYP1A1: highly stereoselective formation of 17(R),18(S)-epoxyeicosatetraenoic acid. Biochem Pharmacol 67:1445–1457
102. Laethem RM, Balazy M, Falck JR, Laethem CL, Koop DR (1993) Formation of 19(*S*)-, 19(*R*)-, and 18(*R*)-hydroxyeicosatetraenoic acids by alcohol-inducible cytochrome P450 2E1. J Biol Chem 268:12912–12918
103. Cheng J, Ou JS, Singh H, Falck JR, Narsimhaswamy D, Pritchard KA Jr, Schwartzman ML (2008) 20-Hydroxyeicosatetraenoic acid causes endothelial dysfunction via eNOS uncoupling. Am J Physiol Heart Circ Physiol 294:H1018–H1026
104. Zhang F, Deng H, Kemp R, Singh H, Gopal VR, Falck JR, Laniado-Schwartzman M, Nasjletti A (2005) Decreased levels of cytochrome P450 2E1-derived eicosanoids sensitize renal arteries to constrictor agonists in spontaneously hypertensive rats. Hypertension 45:103–108
105. Stark K, Wongsud B, Burman R, Oliw EH (2005) Oxygenation of polyunsaturated long chain fatty acids by recombinant CYP4F8 and CYP4F12 and catalytic importance of Tyr-125 and Gly-328 of CYP4F8. Arch Biochem Biophys 441:174–181
106. Capdevila JH, Falck JR, Harris RC (2000) Cytochrome P450 and arachidonic acid bioactivation. Molecular and functional properties of the arachidonate monooxygenase. J Lipid Res 41:163–181
107. Oliw EH, Bylund J, Herman C (1996) Bisallylic hydroxylation and epoxidation of polyunsaturated fatty acids by cytochrome P450. Lipids 31:1003–1021
108. Brash AR, Boeglin WE, Capdevila JH, Yeola S, Blair IA (1995) 7-HETE, 10-HETE, and 13-HETE are major products of NADPH-dependent arachidonic acid metabolism in rat liver microsomes: analysis of their stereochemistry, and the stereochemistry of their acid-catalyzed rearrangement. Arch Biochem Biophys 321:485–492
109. Hornsten L, Bylund J, Oliw EH (1996) Dexamethasone induces bisallylic hydroxylation of polyunsaturated fatty acids by rat liver microsomes. Arch Biochem Biophys 332:261–268
110. Bylund J, Kunz T, Valmsen K, Oliw EH (1998) Cytochromes P450 with bisallylic hydroxylation activity on arachidonic and linoleic acids studied with human recombinant enzymes and with human and rat liver microsomes. J Pharmacol Exp Ther 284:51–60
111. Yamamoto S, Nishimura M, Conners MS, Stoltz RA, Falck JR, Chauhan K, Laniado-Schwartzman M (1994) Oxidation and keto reduction of 12-hydroxy-5,8,10,14-eicosatetraenoic acids in bovine corneal epithelial microsomes. Biochim Biophys Acta 1210:217–225
112. Mieyal PA, Dunn MW, Schwartzman ML (2001) Detection of endogenous 12-hydroxyeicosatrienoic acid in human tear film. Invest Ophthalmol Vis Sci 42:328–332
113. Mezentsev A, Mastyugin V, Seta F, Ashkar S, Kemp R, Reddy DS, Falck JR, Dunn MW, Laniado-Schwartzman M (2005) Transfection of cytochrome P4504B1 into the cornea increases angiogenic activity of the limbal vessels. J Pharmacol Exp Ther 315:42–50
114. Seta F, Patil K, Bellner L, Mezentsev A, Kemp R, Dunn MW, Schwartzman ML (2007) Inhibition of VEGF expression and corneal neovascularization by siRNA targeting cytochrome P450 4B1. Prostaglandins Other Lipid Mediat 84:116–127
115. Konkel A, Schunck WH (2011) Role of cytochrome P450 enzymes in the bioactivation of polyunsaturated fatty acids. Biochim Biophys Acta 1814:210–222
116. De Caterina R (2011) n-3 Fatty acids in cardiovascular disease. N Engl J Med 364:2439–2450
117. Martins DA, Custodio L, Barreira L, Pereira H, Ben-Hamadou R, Varela J, Abu-Salah KM (2013) Alternative sources of n-3 long-chain polyunsaturated fatty acids in marine microalgae. Mar Drugs 11:2259–2281
118. Vrablik TL, Watts JL (2013) Polyunsaturated fatty acid derived signaling in reproduction and development: insights from *Caenorhabditis elegans* and

Drosophila melanogaster. Mol Reprod Dev 80:244–259
119. Arterburn LM, Hall EB, Oken H (2006) Distribution, interconversion, and dose response of n-3 fatty acids in humans. Am J Clin Nutr 83:1467S–1476S
120. Simopoulos AP (2008) The importance of the ω-6/ω-3 fatty acid ratio in cardiovascular disease and other chronic diseases. Exp Biol Med (Maywood) 233:674–688
121. Lands WE (2005) Dietary fat and health: the evidence and the politics of prevention: careful use of dietary fats can improve life and prevent disease. Ann N Y Acad Sci 1055:179–192
122. Calder PC (2006) n-3 Polyunsaturated fatty acids, inflammation, and inflammatory diseases. Am J Clin Nutr 83:1505S–1519S
123. Kris-Etherton PM, Harris WS, Appel LJ (2002) Fish consumption, fish oil, ω-3 fatty acids, and cardiovascular disease. Circulation 106:2747–2757
124. SanGiovanni JP, Chew EY (2005) The role of ω-3 long-chain polyunsaturated fatty acids in health and disease of the retina. Prog Retin Eye Res 24:87–138
125. Uauy R, Hoffman DR, Peirano P, Birch DG, Birch EE (2001) Essential fatty acids in visual and brain development. Lipids 36:885–895
126. Rapoport SI, Igarashi M (2009) Can the rat liver maintain normal brain DHA metabolism in the absence of dietary DHA? Prostaglandins Leukot Essent Fatty Acids 81:119–123
127. Saravanan P, Davidson NC, Schmidt EB, Calder PC (2010) Cardiovascular effects of marine ω-3 fatty acids. Lancet 376:540–550
128. Leslie CC (2004) Regulation of arachidonic acid availability for eicosanoid production. Biochem Cell Biol 82:1–17
129. Wada M, DeLong CJ, Hong YH, Rieke CJ, Song I, Sidhu RS, Yuan C, Warnock M, Schmaier AH, Yokoyama C, Smyth EM, Wilson SJ, FitzGerald GA, Garavito RM, de Sui X, Regan JW, Smith WL (2007) Enzymes and receptors of prostaglandin pathways with arachidonic acid-derived versus eicosapentaenoic acid-derived substrates and products. J Biol Chem 282:22254–22266
130. Rosa AO, Rapoport SI (2009) Intracellular- and extracellular-derived Ca^{2+} influence phospholipase A_2-mediated fatty acid release from brain phospholipids. Biochim Biophys Acta 1791:697–705
131. Cheon Y, Kim HW, Igarashi M, Modi HR, Chang L, Ma K, Greenstein D, Wohltmann M, Turk J, Rapoport SI, Taha AY (2012) Disturbed brain phospholipid and docosahexaenoic acid metabolism in calcium-independent phospholipase A_2-VIA (iPLA$_2$ β-knockout mice. Biochim Biophys Acta 1821:1278–1286
132. Liu X, Moon SH, Mancuso DJ, Jenkins CM, Guan S, Sims HF, Gross RW (2013) Oxidized fatty acid analysis by charge-switch derivatization, selected reaction monitoring, and accurate mass quantitation. Anal Biochem 442:40–50
133. Van Rollins M, Frade PD, Carretero OA (1988) Oxidation of 5,8,11,14,17-eicosapentaenoic acid by hepatic and renal microsomes. Biochim Biophys Acta 966:133–149
134. VanRollins M (1995) Epoxygenase metabolites of docosahexaenoic and eicosapentaenoic acids inhibit platelet aggregation at concentrations below those affecting thromboxane synthesis. J Pharmacol Exp Ther 274:798–804
135. Lauterbach B, Barbosa-Sicard E, Wang MH, Honeck H, Kargel E, Theuer J, Schwartzman ML, Haller H, Luft FC, Gollasch M, Schunck WH (2002) Cytochrome P450-dependent eicosapentaenoic acid metabolites are novel BK channel activators. Hypertension 39:609–613
136. Barbosa-Sicard E, Markovic M, Honeck H, Christ B, Muller DN, Schunck WH (2005) Eicosapentaenoic acid metabolism by cytochrome P450 enzymes of the CYP2C subfamily. Biochem Biophys Res Commun 329:1275–1281
137. Lucas D, Goulitquer S, Marienhagen J, Fer M, Dreano Y, Schwaneberg U, Amet Y, Corcos L (2010) Stereoselective epoxidation of the last double bond of polyunsaturated fatty acids by human cytochromes P450. J Lipid Res 51:1125–1133
138. Dyerberg J, Bang HO, Stoffersen E, Moncada S, Vane JR (1978) Eicosapentaenoic acid and prevention of thrombosis and atherosclerosis? Lancet 2:117–119
139. Terano T, Salmon JA, Moncada S (1984) Biosynthesis and biological activity of leukotriene B5. Prostaglandins 27:217–232
140. Fischer S, Weber PC, Dyerberg J (1986) The prostacyclin/thromboxane balance is favourably shifted in Greenland Eskimos. Prostaglandins 32:235–241
141. von Schacky C, Fischer S, Weber PC (1985) Long-term effects of dietary marine ω-3 fatty acids upon plasma and cellular lipids, platelet function, and eicosanoid formation in humans. J Clin Invest 76:1626–1631
142. Calder PC (2009) Polyunsaturated fatty acids and inflammatory processes: new twists in an old tale. Biochimie 91:791–795
143. Knapp HR, Miller AJ, Lawson JA (1991) Urinary excretion of diols derived from eicosapentaenoic acid during n-3 fatty acid ingestion by man. Prostaglandins 42:47–54
144. Shearer GC, Harris WS, Pedersen TL, Newman JW (2010) Detection of ω-3 oxylipins in human plasma and response to treatment with ω-3 acid ethyl esters. J Lipid Res 51:2074–2081
145. Lundstrom SL, Yang J, Brannan JD, Haeggstrom JZ, Hammock BD, Nair P, O'Byrne P, Dahlen SE, Wheelock CE (2013) Lipid mediator serum profiles in asthmatics significantly shift following dietary supplementation with ω-3 fatty acids. Mol Nutr Food Res 57:1378–1389

146. Schuchardt JP, Schmidt S, Kressel G, Dong H, Willenberg I, Hammock BD, Hahn A, Schebb NH (2013) Comparison of free serum oxylipin concentrations in hyper- vs. normolipidemic men. Prostaglandins Leukot Essent Fatty Acids 89:19–29
147. Bruins MJ, Dane AD, Strassburg K, Vreeken RJ, Newman JW, Salem N Jr, Tyburczy C, Brenna JT (2013) Plasma oxylipin profiling identifies polyunsaturated vicinal diols as responsive to arachidonic acid and docosahexaenoic acid intake in growing piglets. J Lipid Res 54:1598–1607
148. Westphal C, Konkel A, Schunck WH (2011) CYP-eicosanoids—a new link between ω-3 fatty acids and cardiac disease? Prostaglandins Other Lipid Mediat 96:99–108
149. Ye D, Zhang D, Oltman C, Dellsperger K, Lee HC, Van Rollins M (2002) Cytochrome P-450 epoxygenase metabolites of docosahexaenoate potently dilate coronary arterioles by activating large-conductance calcium-activated potassium channels. J Pharmacol Exp Ther 303:768–776
150. Agbor LN, Walsh MT, Boberg JR, Walker MK (2012) Elevated blood pressure in cytochrome P4501A1 knockout mice is associated with reduced vasodilation to ω-3 polyunsaturated fatty acids. Toxicol Appl Pharmacol 264:351–360
151. Node K, Huo Y, Ruan X, Yang B, Spiecker M, Ley K, Zeldin DC, Liao JK (1999) Anti-inflammatory properties of cytochrome P450 epoxygenase-derived eicosanoids. Science 285:1276–1279
152. Morin C, Sirois M, Echave V, Albadine R, Rousseau E (2010) 17,18-Epoxyeicosatetraenoic acid targets PPARγ and p38 mitogen-activated protein kinase to mediate its anti-inflammatory effects in the lung: role of soluble epoxide hydrolase. Am J Respir Cell Mol Biol 43:564–575
153. Panigrahy D, Edin ML, Lee CR, Huang S, Bielenberg DR, Butterfield CE, Barnes CM, Mammoto A, Mammoto T, Luria A, Benny O, Chaponis DM, Dudley AC, Greene ER, Vergilio JA, Pietramaggiori G, Scherer-Pietramaggiori SS, Short SM, Seth M, Lih FB, Tomer KB, Yang J, Schwendener RA, Hammock BD, Falck JR, Manthati VL, Ingber DE, Kaipainen A, D'Amore PA, Kieran MW, Zeldin DC (2012) Epoxyeicosanoids stimulate multiorgan metastasis and tumor dormancy escape in mice. J Clin Invest 122:178–191
154. Zhang G, Panigrahy D, Mahakian LM, Yang J, Liu JY, Stephen Lee KS, Wettersten HI, Ulu A, Hu X, Tam S, Hwang SH, Ingham ES, Kieran MW, Weiss RH, Ferrara KW, Hammock BD (2013) Epoxy metabolites of docosahexaenoic acid (DHA) inhibit angiogenesis, tumor growth, and metastasis. Proc Natl Acad Sci U S A 110:6530–6535
155. Morisseau C, Inceoglu B, Schmelzer K, Tsai HJ, Jinks SL, Hegedus CM, Hammock BD (2010) Naturally occurring monoepoxides of eicosapentaenoic acid and docosahexaenoic acid are bioactive antihyperalgesic lipids. J Lipid Res 51:3481–3490
156. Lavie CJ, Milani RV, Mehra MR, Ventura HO (2009) ω-3 Polyunsaturated fatty acids and cardiovascular diseases. J Am Coll Cardiol 54:585–594
157. Webler AC, Michaelis UR, Popp R, Barbosa-Sicard-E, Murugan A, Falck JR, Fisslthaler B, Fleming I (2008) Epoxyeicosatrienoic acids are part of the VEGF-activated signaling cascade leading to angiogenesis. Am J Physiol Cell Physiol 295: C1292–C1301
158. Alonso-Galicia M, Maier KG, Greene AS, Cowley AW Jr, Roman RJ (2002) Role of 20-hydroxyeicosatetraenoic acid in the renal and vasoconstrictor actions of angiotensin II. Am J Physiol Regul Integr Comp Physiol 283:R60–R68
159. Karara A, Dishman E, Falck JR, Capdevila JH (1991) Endogenous epoxyeicosatrienoyl-phospholipids. A novel class of cellular glycerolipids containing epoxidized arachidonate moieties. J Biol Chem 266:7561–7569
160. Carroll MA, Balazy M, Huang DD, Rybalova S, Falck JR, McGiff JC (1997) Cytochrome P450-derived renal HETEs: storage and release. Kidney Int 51:1696–1702
161. Kaduce TL, Fang X, Harmon SD, Oltman CL, Dellsperger KC, Teesch LM, Gopal VR, Falck JR, Campbell WB, Weintraub NL, Spector AA (2004) 20-Hydroxyeicosatetraenoic acid (20-HETE) metabolism in coronary endothelial cells. J Biol Chem 279:2648–2656
162. Oliw EH (1991) 17R(18S)epoxyeicosatetraenoic acid, a cytochrome P-450 metabolite of 20:5n-3 in monkey seminal vesicles, is metabolized to novel prostaglandins. Biochem Biophys Res Commun 178:1444–1450
163. Oliw EH, Okamoto S, Hornsten L, Sato F (1992) Biosynthesis of prostaglandins from 17(18)epoxy-eicosatetraenoic acid, a cytochrome *P*-450 metabolite of eicosapentaenoic acid. Biochim Biophys Acta 1126:261–268
164. Cheng MK, McGiff JC, Carroll MA (2003) Renal arterial 20-hydroxyeicosatetraenoic acid levels: regulation by cyclooxygenase. Am J Physiol Renal Physiol 284:F474–F479
165. Kim DH, Puri N, Sodhi K, Falck JR, Abraham NG, Shapiro J, Schwartzman ML (2013) Cyclooxygenase-2 dependent metabolism of 20-HETE increases adiposity and adipocyte enlargement in mesenchymal stem cell-derived adipocytes. J Lipid Res 54:786–793
166. Yang W, Gauthier KM, Reddy LM, Sangras B, Sharma KK, Nithipatikom K, Falck JR, Campbell WB (2005) Stable 5,6-epoxyeicosatrienoic acid analog relaxes coronary arteries through potassium channel activation. Hypertension 45:681–686
167. Moreland KT, Procknow JD, Sprague RS, Iverson JL, Lonigro AJ, Stephenson AH (2007) Cyclooxygenase (COX)-1 and COX-2 participate in

5,6-epoxyeicosatrienoic acid-induced contraction of rabbit intralobar pulmonary arteries. J Pharmacol Exp Ther 321:446–454
168. Kubota T, Arita M, Isobe Y, Iwamoto R, Goto T, Yoshioka T, Urabe D, Inoue M, Arai H (2014) Eicosapentaenoic acid is converted via ω-3 epoxygenation to the anti-inflammatory metabolite 12-hydroxy-17,18-epoxyeicosatetraenoic acid. FASEB J 28:586–593
169. Serhan CN, Chiang N, Van Dyke TE (2008) Resolving inflammation: dual anti-inflammatory and pro-resolution lipid mediators. Nat Rev Immunol 8:349–361
170. Arita M, Clish CB, Serhan CN (2005) The contributions of aspirin and microbial oxygenase to the biosynthesis of anti-inflammatory resolvins: novel oxygenase products from ω-3 polyunsaturated fatty acids. Biochem Biophys Res Commun 338:149–157
171. Cowart LA, Wei S, Hsu MH, Johnson EF, Krishna MU, Falck JR, Capdevila JH (2002) The CYP4A isoforms hydroxylate epoxyeicosatrienoic acids to form high affinity peroxisome proliferator-activated receptor ligands. J Biol Chem 277:35105–35112
172. Le Quere V, Plee-Gautier E, Potin P, Madec S, Salaun JP (2004) Human CYP4F3s are the main catalysts in the oxidation of fatty acid epoxides. J Lipid Res 45:1446–1458
173. Morisseau C, Hammock BD (2013) Impact of soluble epoxide hydrolase and epoxyeicosanoids on human health. Annu Rev Pharmacol Toxicol 53:37–58
174. Harris TR, Hammock BD (2013) Soluble epoxide hydrolase: gene structure, expression and deletion. Gene 526:61–74
175. Widstrom RL, Norris AW, Spector AA (2001) Binding of cytochrome P450 monooxygenase and lipoxygenase pathway products by heart fatty acid-binding protein. Biochemistry 40:1070–1076
176. Widstrom RL, Norris AW, Van Der Veer J, Spector AA (2003) Fatty acid-binding proteins inhibit hydration of epoxyeicosatrienoic acids by soluble epoxide hydrolase. Biochemistry 42:11762–11767
177. Ai D, Fu Y, Guo D, Tanaka H, Wang N, Tang C, Hammock BD, Shyy JY, Zhu Y (2007) Angiotensin II up-regulates soluble epoxide hydrolase in vascular endothelium *in vitro* and *in vivo*. Proc Natl Acad Sci U S A 104:9018–9023
178. Ai D, Pang W, Li N, Xu M, Jones PD, Yang J, Zhang Y, Chiamvimonvat N, Shyy JY, Hammock BD, Zhu Y (2009) Soluble epoxide hydrolase plays an essential role in angiotensin II-induced cardiac hypertrophy. Proc Natl Acad Sci U S A 106:564–569
179. Imig JD, Hammock BD (2009) Soluble epoxide hydrolase as a therapeutic target for cardiovascular diseases. Nat Rev Drug Discov 8:794–805
180. Spector AA, Norris AW (2007) Action of epoxyeicosatrienoic acids on cellular function. Am J Physiol Cell Physiol 292:C996–C1012
181. Wu CC, Gupta T, Garcia V, Ding Y, Schwartzman ML (2014) 20-HETE and blood pressure regulation: clinical implications. Cardiol Rev 22:1–12
182. Seubert JM, Zeldin DC, Nithipatikom K, Gross GJ (2007) Role of epoxyeicosatrienoic acids in protecting the myocardium following ischemia/reperfusion injury. Prostaglandins Other Lipid Mediat 82:50–59
183. Hoff U, Lukitsch I, Chaykovska L, Ladwig M, Arnold C, Manthati VL, Fuller TF, Schneider W, Gollasch M, Muller DN, Flemming B, Seeliger E, Luft FC, Falck JR, Dragun D, Schunck WH (2011) Inhibition of 20-HETE synthesis and action protects the kidney from ischemia/reperfusion injury. Kidney Int 79:57–65
184. Regner KR, Zuk A, Van Why SK, Shames BD, Ryan RP, Falck JR, Manthati VL, McMullen ME, Ledbetter SR, Roman RJ (2009) Protective effect of 20-HETE analogues in experimental renal ischemia reperfusion injury. Kidney Int 75:511–517
185. Imig JD, Simpkins AN, Renic M, Harder DR (2011) Cytochrome P450 eicosanoids and cerebral vascular function. Expert Rev Mol Med 13:e7. doi:10.1017/S1462399411001773
186. Deng Y, Theken KN, Lee CR (2010) Cytochrome P450 epoxygenases, soluble epoxide hydrolase, and the regulation of cardiovascular inflammation. J Mol Cell Cardiol 48:331–341
187. Brand-Schieber E, Falck JF, Schwartzman M (2000) Selective inhibition of arachidonic acid epoxidation in vivo. J Physiol Pharmacol 51:655–672
188. Yu M, Cambj-Sapunar L, Kehl F, Maier KG, Takeuchi K, Miyata N, Ishimoto T, Reddy LM, Falck JR, Gebremedhin D, Harder DR, Roman RJ (2004) Effects of a 20-HETE antagonist and agonists on cerebral vascular tone. Eur J Pharmacol 486:297–306
189. Williams JM, Murphy S, Burke M, Roman RJ (2010) 20-hydroxyeicosatetraeonic acid: a new target for the treatment of hypertension. J Cardiovasc Pharmacol 56:336–344
190. Gauthier KM, Falck JR, Reddy LM, Campbell WB (2004) 14,15-EET analogs: characterization of structural requirements for agonist and antagonist activity in bovine coronary arteries. Pharmacol Res 49:515–524
191. Gauthier KM, Deeter C, Krishna UM, Reddy YK, Bondlela M, Falck JR, Campbell WB (2002) 14,15-Epoxyeicosa-5(Z)-enoic acid: a selective epoxyeicosatrienoic acid antagonist that inhibits endothelium-dependent hyperpolarization and relaxation in coronary arteries. Circ Res 90:1028–1036
192. Imig JD, Elmarakby A, Nithipatikom K, Wei S, Capdevila JH, Tuniki VR, Sangras B, Anjaiah S, Manthati VL, Reddy DS, Falck JR (2010) Development of epoxyeicosatrienoic acid analogs with in vivo anti-hypertensive actions. Front Physiol 1 (article 157):1–8

193. Wu CC, Schwartzman ML (2011) The role of 20-HETE in androgen-mediated hypertension. Prostaglandins Other Lipid Mediat 96:45–53
194. Singh H, Cheng J, Deng H, Kemp R, Ishizuka T, Nasjletti A, Schwartzman ML (2007) Vascular cytochrome P450 4A expression and 20-hydroxyeicosatetraenoic acid synthesis contribute to endothelial dysfunction in androgen-induced hypertension. Hypertension 50:123–129
195. Inoue K, Sodhi K, Puri N, Gotlinger KH, Cao J, Rezzani R, Falck JR, Abraham NG, Laniado-Schwartzman M (2009) Endothelial-specific CYP4A2 overexpression leads to renal injury and hypertension via increased production of 20-HETE. Am J Physiol Renal Physiol 297:F875–F884
196. Imig JD, Zou AP, Stec DE, Harder DR, Falck JR, Roman RJ (1996) Formation and actions of 20-hydroxyeicosatetraenoic acid in rat renal arterioles. Am J Physiol 270:R217–R227
197. Cheng J, Wu CC, Gotlinger KH, Zhang F, Falck JR, Narsimhaswamy D, Schwartzman ML (2010) 20-Hydroxy-5,8,11,14-eicosatetraenoic acid mediates endothelial dysfunction via IκB kinase-dependent endothelial nitric-oxide synthase uncoupling. J Pharmacol Exp Ther 332:57–65
198. Sodhi K, Wu CC, Cheng J, Gotlinger K, Inoue K, Goli M, Falck JR, Abraham NG, Schwartzman ML (2010) CYP4A2-induced hypertension is 20-hydroxyeicosatetraenoic acid- and angiotensin II-dependent. Hypertension 56:871–878
199. Cheng J, Garcia V, Ding Y, Wu CC, Thakar K, Falck JR, Ramu E, Schwartzman ML (2012) Induction of angiotensin-converting enzyme and activation of the renin-angiotensin system contribute to 20-hydroxyeicosatetraenoic acid-mediated endothelial dysfunction. Arterioscler Thromb Vasc Biol 32:1917–1924
200. Ward NC, Rivera J, Hodgson J, Puddey IB, Beilin LJ, Falck JR, Croft KD (2004) Urinary 20-hydroxyeicosatetraenoic acid is associated with endothelial dysfunction in humans. Circulation 110:438–443
201. Imig JD (2013) Epoxyeicosatrienoic acids, 20-hydroxyeicosatetraenoic acid, and renal microvascular function. Prostaglandins Other Lipid Mediat 104–105:2–7
202. Hoagland KM, Flasch AK, Roman RJ (2003) Inhibitors of 20-HETE formation promote salt-sensitive hypertension in rats. Hypertension 42:669–673
203. Honeck H, Gross V, Erdmann B, Kargel E, Neunaber R, Milia AF, Schneider W, Luft FC, Schunck WH (2000) Cytochrome P450-dependent renal arachidonic acid metabolism in desoxycorticosterone acetate-salt hypertensive mice. Hypertension 36:610–616
204. Zhou Y, Luo P, Chang HH, Huang H, Yang T, Dong Z, Wang CY, Wang MH (2008) Clofibrate attenuates blood pressure and sodium retention in DOCA-salt hypertension. Kidney Int 74:1040–1048
205. Laffer CL, Gainer JV, Waterman MR, Capdevila JH, Laniado-Schwartzman M, Nasjletti A, Brown NJ, Elijovich F (2008) The T8590C polymorphism of CYP4A11 and 20-hydroxyeicosatetraenoic acid in essential hypertension. Hypertension 51:767–772
206. Williams JS, Hopkins PN, Jeunemaitre X, Brown NJ (2011) CYP4A11 T8590C polymorphism, salt-sensitive hypertension, and renal blood flow. J Hypertens 29:1913–1918
207. Fava C, Montagnana M, Danese E, Sjogren M, Almgren P, Guidi GC, Hedblad B, Engstrom G, Minuz P, Melander O (2012) The functional variant V433M of the CYP4F2 and the metabolic syndrome in Swedes. Prostaglandins Other Lipid Mediat 98:31–36
208. Imig JD, Navar LG, Roman RJ, Reddy KK, Falck JR (1996) Actions of epoxygenase metabolites on the preglomerular vasculature. J Am Soc Nephrol 7:2364–2370
209. Wei Y, Sun P, Wang Z, Yang B, Carroll MA, Wang WH (2006) Adenosine inhibits ENaC via cytochrome P-450 epoxygenase-dependent metabolites of arachidonic acid. Am J Physiol Renal Physiol 290:F1163–F1168
210. Carroll MA (2012) Role of the adenosine$_{2A}$ receptor-epoxyeicosatrienoic acid pathway in the development of salt-sensitive hypertension. Prostaglandins Other Lipid Mediat 98:39–47
211. Liclican EL, McGiff JC, Falck JR, Carroll MA (2008) Failure to upregulate the adenosine$_{2A}$ receptor-epoxyeicosatrienoic acid pathway contributes to the development of hypertension in Dahl salt-sensitive rats. Am J Physiol Renal Physiol 295:F1696–F1704
212. Jung O, Brandes RP, Kim IH, Schweda F, Schmidt R, Hammock BD, Busse R, Fleming I (2005) Soluble epoxide hydrolase is a main effector of angiotensin II-induced hypertension. Hypertension 45:759–765
213. Minuz P, Jiang H, Fava C, Turolo L, Tacconelli S, Ricci M, Patrignani P, Morganti A, Lechi A, McGiff JC (2008) Altered release of cytochrome P450 metabolites of arachidonic acid in renovascular disease. Hypertension 51:1379–1385
214. Theken KN, Lee CR (2007) Genetic variation in the cytochrome P450 epoxygenase pathway and cardiovascular disease risk. Pharmacogenomics 8:1369–1383
215. Spiecker M, Liao J (2006) Cytochrome P450 epoxygenase CYP2J2 and the risk of coronary artery disease. Trends Cardiovasc Med 16:204–208
216. Zordoky BN, El-Kadi AO (2010) Effect of cytochrome P450 polymorphism on arachidonic acid metabolism and their impact on cardiovascular diseases. Pharmacol Ther 125:446–463
217. Bonventre JV, Huang Z, Taheri MR, O'Leary E, Li E, Moskowitz MA, Sapirstein A (1997) Reduced

fertility and postischaemic brain injury in mice deficient in cytosolic phospholipase A2. Nature 390:622–625
218. Tabuchi S, Uozumi N, Ishii S, Shimizu Y, Watanabe T, Shimizu T (2003) Mice deficient in cytosolic phospholipase A2 are less susceptible to cerebral ischemia/reperfusion injury. Acta Neurochir Suppl 86:169–172
219. Saito Y, Watanabe K, Fujioka D, Nakamura T, Obata JE, Kawabata K, Watanabe Y, Mishina H, Tamaru S, Kita Y, Shimizu T, Kugiyama K (2012) Disruption of group IVA cytosolic phospholipase A_2 attenuates myocardial ischemia-reperfusion injury partly through inhibition of TNF-α-mediated pathway. Am J Physiol Heart Circ Physiol 302: H2018–H2030
220. Nakamura H, Nemenoff RA, Gronich JH, Bonventre JV (1991) Subcellular characteristics of phospholipase A_2 activity in the rat kidney. Enhanced cytosolic, mitochondrial, and microsomal phospholipase A_2 enzymatic activity after renal ischemia and reperfusion. J Clin Invest 87:1810–1818
221. Nithipatikom K, DiCamelli RF, Kohler S, Gumina RJ, Falck JR, Campbell WB, Gross GJ (2001) Determination of cytochrome P450 metabolites of arachidonic acid in coronary venous plasma during ischemia and reperfusion in dogs. Anal Biochem 292:115–124
222. Nithipatikom K, Gross ER, Endsley MP, Moore JM, Isbell MA, Falck JR, Campbell WB, Gross GJ (2004) Inhibition of cytochrome P450 ω-hydroxylase: a novel endogenous cardioprotective pathway. Circ Res 95:e65–e71
223. Nithipatikom K, Endsley MP, Moore JM, Isbell MA, Falck JR, Campbell WB, Gross GJ (2006) Effects of selective inhibition of cytochrome P-450 ω-hydroxylases and ischemic preconditioning in myocardial protection. Am J Physiol Heart Circ Physiol 290:H500–H505
224. Batchu SN, Law E, Brocks DR, Falck JR, Seubert JM (2009) Epoxyeicosatrienoic acid prevents postischemic electrocardiogram abnormalities in an isolated heart model. J Mol Cell Cardiol 46:67–74
225. Gross GJ, Hsu A, Falck JR, Nithipatikom K (2007) Mechanisms by which epoxyeicosatrienoic acids (EETs) elicit cardioprotection in rat hearts. J Mol Cell Cardiol 42:687–691
226. Gross GJ, Gauthier KM, Moore J, Falck JR, Hammock BD, Campbell WB, Nithipatikom K (2008) Effects of the selective EET antagonist, 14,15-EEZE, on cardioprotection produced by exogenous or endogenous EETs in the canine heart. Am J Physiol Heart Circ Physiol 294:H2838–H2844
227. Gross GJ, Gauthier KM, Moore J, Campbell WB, Falck JR, Nithipatikom K (2009) Evidence for role of epoxyeicosatrienoic acids in mediating ischemic preconditioning and postconditioning in dog. Am J Physiol Heart Circ Physiol 297:H47–H52
228. Yu GG, Zeng XJ, Wang HX, Lu LQ, Zheng SP, Ma LQ, Chang J, Wang J, Zhang DM, Du FH, Zhang LK (2011) Cytochrome P450 2J3/epoxyeicosatrienoic acids mediate the cardioprotection induced by ischaemic post-conditioning, but not preconditioning, in the rat. Clin Exp Pharmacol Physiol 38:63–70
229. Gross GJ, Baker JE, Moore J, Falck JR, Nithipatikom K (2011) Abdominal surgical incision induces remote preconditioning of trauma (RPCT) via activation of bradykinin receptors (BK2R) and the cytochrome P450 epoxygenase pathway in canine hearts. Cardiovasc Drugs Ther 25:517–522
230. Motoki A, Merkel MJ, Packwood WH, Cao Z, Liu L, Iliff J, Alkayed NJ, Van Winkle DM (2008) Soluble epoxide hydrolase inhibition and gene deletion are protective against myocardial ischemia-reperfusion injury in vivo. Am J Physiol Heart Circ Physiol 295: H2128–H2134
231. Batchu SN, Lee SB, Qadhi RS, Chaudhary KR, El-Sikhry H, Kodela R, Falck JR, Seubert JM (2011) Cardioprotective effect of a dual acting epoxyeicosatrienoic acid analogue towards ischaemia reperfusion injury. Br J Pharmacol 162:897–907
232. Gross GJ, Falck JR, Gross ER, Isbell M, Moore J, Nithipatikom K (2005) Cytochrome P450 and arachidonic acid metabolites: role in myocardial ischemia/reperfusion injury revisited. Cardiovasc Res 68:18–25
233. Nithipatikom K, Gross GJ (2010) Review article: epoxyeicosatrienoic acids: novel mediators of cardioprotection. J Cardiovasc Pharmacol Ther 15:112–119
234. Miyata N, Seki T, Tanaka Y, Omura T, Taniguchi K, Doi M, Bandou K, Kametani S, Sato M, Okuyama S, Cambj-Sapunar L, Harder DR, Roman RJ (2005) Beneficial effects of a new 20-hydroxyeicosatetraenoic acid synthesis inhibitor, TS-011 [N-(3-chloro-4-morpholin-4-yl) phenyl-N-'-hydroxyimido formamide], on hemorrhagic and ischemic stroke. J Pharmacol Exp Ther 314:77–85
235. Renic M, Klaus JA, Omura T, Kawashima N, Onishi M, Miyata N, Koehler RC, Harder DR, Roman RJ (2009) Effect of 20-HETE inhibition on infarct volume and cerebral blood flow after transient middle cerebral artery occlusion. J Cereb Blood Flow Metab 29:629–639
236. Zhang W, Otsuka T, Sugo N, Ardeshiri A, Alhadid YK, Iliff JJ, DeBarber AE, Koop DR, Alkayed NJ (2008) Soluble epoxide hydrolase gene deletion is protective against experimental cerebral ischemia. Stroke 39:2073–2078
237. Zhang W, Iliff JJ, Campbell CJ, Wang RK, Hurn PD, Alkayed NJ (2009) Role of soluble epoxide hydrolase in the sex-specific vascular response to cerebral ischemia. J Cereb Blood Flow Metab 29:1475–1481
238. Zhang W, Koerner IP, Noppens R, Grafe M, Tsai HJ, Morisseau C, Luria A, Hammock BD, Falck JR, Alkayed NJ (2007) Soluble epoxide hydrolase: a

novel therapeutic target in stroke. J Cereb Blood Flow Metab 27:1931–1940
239. Lameire N, Van Biesen W, Vanholder R (2005) Acute renal failure. Lancet 365:417–430
240. Aydin Z, van Zonneveld AJ, de Fijter JW, Rabelink TJ (2007) New horizons in prevention and treatment of ischaemic injury to kidney transplants. Nephrol Dial Transplant 22:342–346
241. Karkouti K, Wijeysundera DN, Yau TM, Callum JL, Cheng DC, Crowther M, Dupuis JY, Fremes SE, Kent B, Laflamme C, Lamy A, Legare JF, Mazer CD, McCluskey SA, Rubens FD, Sawchuk C, Beattie WS (2009) Acute kidney injury after cardiac surgery: focus on modifiable risk factors. Circulation 119:495–502
242. Sutton TA, Fisher CJ, Molitoris BA (2002) Microvascular endothelial injury and dysfunction during ischemic acute renal failure. Kidney Int 62:1539–1549
243. Bonventre JV, Zuk A (2004) Ischemic acute renal failure: an inflammatory disease? Kidney Int 66:480–485
244. Nilakantan V, Maenpaa C, Jia G, Roman RJ, Park F (2008) 20-HETE-mediated cytotoxicity and apoptosis in ischemic kidney epithelial cells. Am J Physiol Renal Physiol 294:F562–F570
245. Wang Y, Hill JA (2010) Electrophysiological remodeling in heart failure. J Mol Cell Cardiol 48:619–632
246. Xu D, Li N, He Y, Timofeyev V, Lu L, Tsai HJ, Kim IH, Tuteja D, Mateo RK, Singapuri A, Davis BB, Low R, Hammock BD, Chiamvimonvat N (2006) Prevention and reversal of cardiac hypertrophy by soluble epoxide hydrolase inhibitors. Proc Natl Acad Sci U S A 103:18733–18738
247. Monti J, Fischer J, Paskas S, Heinig M, Schulz H, Gosele C, Heuser A, Fischer R, Schmidt C, Schirdewan A, Gross V, Hummel O, Maatz H, Patone G, Saar K, Vingron M, Weldon SM, Lindpaintner K, Hammock BD, Rohde K, Dietz R, Cook SA, Schunck WH, Luft FC, Hubner N (2008) Soluble epoxide hydrolase is a susceptibility factor for heart failure in a rat model of human disease. Nat Genet 40:529–537
248. Westphal C, Spallek B, Konkel A, Marko L, Qadri F, Degraff LM, Schubert C, Bradbury JA, Regitz-Zagrosek V, Falck JR, Zeldin DC, Muller DN, Schunck WH, Fischer R (2013) CYP2J2 overexpression protects against arrhythmia susceptibility in cardiac hypertrophy. PLoS One 8:e73490. doi:10.1371/journal.pone.0073490
249. Zhang Y, El-Sikhry H, Chaudhary KR, Batchu SN, Shayeganpour A, Jukar TO, Bradbury JA, Graves JP, DeGraff LM, Myers P, Rouse DC, Foley J, Nyska A, Zeldin DC, Seubert JM (2009) Overexpression of CYP2J2 provides protection against doxorubicin-induced cardiotoxicity. Am J Physiol Heart Circ Physiol 297:H37–H46
250. Wang X, Ni L, Yang L, Duan Q, Chen C, Edin ML, Zeldin DC, Wang DW (2014) CYP2J2-derived epoxyeicosatrienoic acids suppress endoplasmic reticulum stress in heart failure. Mol Pharmacol 85:105–115
251. Tse MM, Aboutabl ME, Althurwi HN, Elshenawy OH, Abdelhamid G, El-Kadi AO (2013) Cytochrome P450 epoxygenase metabolite, 14,15-EET, protects against isoproterenol-induced cellular hypertrophy in H9c2 rat cell line. Vascul Pharmacol 58:363–373
252. Alsaad AM, Zordoky BN, Tse MM, El-Kadi AO (2013) Role of cytochrome P450-mediated arachidonic acid metabolites in the pathogenesis of cardiac hypertrophy. Drug Metab Rev 45:173–195
253. Xiao YF (2007) Cyclic AMP-dependent modulation of cardiac L-type Ca^{2+} and transient outward K^+ channel activities by epoxyeicosatrienoic acids. Prostaglandins Other Lipid Mediat 82:11–18
254. Certikova Chabova V, Walkowska A, Kompanowska-Jezierska E, Sadowski J, Kujal P, Vernerova Z, Vanourkova Z, Kopkan L, Kramer HJ, Falck JR, Imig JD, Hammock BD, Vaneckova I, Cervenka L (2010) Combined inhibition of 20-hydroxyeicosatetraenoic acid formation and of epoxyeicosatrienoic acids degradation attenuates hypertension and hypertension-induced end-organ damage in Ren-2 transgenic rats. Clin Sci (Lond) 118:617–632
255. Zeldin DC, Foley J, Goldsworthy SM, Cook ME, Boyle JE, Ma J, Moomaw CR, Tomer KB, Steenbergen C, Wu S (1997) CYP2J subfamily cytochrome P450s in the gastrointestinal tract: expression, localization, and potential functional significance. Mol Pharmacol 51:931–943
256. Jacobs ER, Zeldin DC (2001) The lung HETEs (and EETs) up. Am J Physiol Heart Circ Physiol 280: H1–H10
257. Loot AE, Fleming I (2011) Cytochrome P450-derived epoxyeicosatrienoic acids and pulmonary hypertension: central role of transient receptor potential C6 channels. J Cardiovasc Pharmacol 57:140–147
258. Sacerdoti D, Gatta A, McGiff JC (2003) Role of cytochrome P450-dependent arachidonic acid metabolites in liver physiology and pathophysiology. Prostaglandins Other Lipid Mediat 72:51–71
259. Terashvili M, Tseng LF, Wu HE, Narayanan J, Hart LM, Falck JR, Pratt PF, Harder DR (2008) Antinociception produced by 14,15-epoxyeicosatrienoic acid is mediated by the activation of β-endorphin and met-enkephalin in the rat ventrolateral periaqueductal gray. J Pharmacol Exp Ther 326:614–622
260. Wagner K, Inceoglu B, Hammock BD (2011) Soluble epoxide hydrolase inhibition, epoxygenated fatty acids and nociception. Prostaglandins Other Lipid Mediat 96:76–83
261. Falck JR, Manna S, Moltz J, Chacos N, Capdevila J (1983) Epoxyeicosatrienoic acids stimulate

glucagon and insulin release from isolated rat pancreatic islets. Biochem Biophys Res Commun 114:743–749

262. Zeldin DC, Foley J, Boyle JE, Moomaw CR, Tomer KB, Parker C, Steenbergen C, Wu S (1997) Predominant expression of an arachidonate epoxygenase in islets of Langerhans cells in human and rat pancreas. Endocrinology 138:1338–1346
263. Mustafa S, Sharma V, McNeill JH (2009) Insulin resistance and endothelial dysfunction: are epoxyeicosatrienoic acids the link? Exp Clin Cardiol 14:e41–e50
264. Chen L, Fan C, Zhang Y, Bakri M, Dong H, Morisseau C, Maddipati KR, Luo P, Wang CY, Hammock BD, Wang MH (2013) Beneficial effects of inhibition of soluble epoxide hydrolase on glucose homeostasis and islet damage in a streptozotocin-induced diabetic mouse model. Prostaglandins Other Lipid Mediat 104–105:42–48
265. Cheranov SY, Karpurapu M, Wang D, Zhang B, Venema RC, Rao GN (2008) An essential role for SRC-activated STAT-3 in 14,15-EET-induced VEGF expression and angiogenesis. Blood 111:5581–5591
266. Fleming I (2011) The cytochrome P450 pathway in angiogenesis and endothelial cell biology. Cancer Metastasis Rev 30:541–555
267. Panigrahy D, Greene ER, Pozzi A, Wang DW, Zeldin DC (2011) EET signaling in cancer. Cancer Metastasis Rev 30:525–540
268. Pozzi A, Popescu V, Yang S, Mei S, Shi M, Puolitaival SM, Caprioli RM, Capdevila JH (2010) The anti-tumorigenic properties of peroxisomal proliferator-activated receptor α are arachidonic acid epoxygenase-mediated. J Biol Chem 285:12840–12850
269. Yang S, Wei S, Pozzi A, Capdevila JH (2009) The arachidonic acid epoxygenase is a component of the signaling mechanisms responsible for VEGF-stimulated angiogenesis. Arch Biochem Biophys 489:82–91
270. Wang D, Dubois RN (2012) Epoxyeicosatrienoic acids: a double-edged sword in cardiovascular diseases and cancer. J Clin Invest 122:19–22
271. Spector AA (2009) Arachidonic acid cytochrome P450 epoxygenase pathway. J Lipid Res 50(Suppl): S52–S56
272. Yang W, Tuniki VR, Anjaiah S, Falck JR, Hillard CJ, Campbell WB (2008) Characterization of epoxyeicosatrienoic acid binding site in U937 membranes using a novel radiolabeled agonist, 20-125i-14,15-epoxyeicosa-8(Z)-enoic acid. J Pharmacol Exp Ther 324:1019–1027
273. Chen Y, Falck JR, Manthati VL, Jat JL, Campbell WB (2011) 20-Iodo-14,15-epoxyeicosa-8(Z)-enoyl-3-azidophenylsulfonamide: photoaffinity labeling of a 14,15-epoxyeicosatrienoic acid receptor. Biochemistry 50:3840–3848
274. Kosel M, Wild W, Bell A, Rothe M, Lindschau C, Steinberg CE, Schunck WH, Menzel R (2011) Eicosanoid formation by a cytochrome P450 isoform expressed in the pharynx of *Caenorhabditis elegans*. Biochem J 435:689–700
275. Ma DK, Rothe M, Zheng S, Bhatla N, Pender CL, Menzel R, Horvitz HR (2013) Cytochrome P450 drives a HIF-regulated behavioral response to reoxygenation by *C. elegans*. Science 341:554–558
276. Harmon SD, Fang X, Kaduce TL, Hu S, Raj Gopal V, Falck JR, Spector AA (2006) Oxygenation of ω-3 fatty acids by human cytochrome P450 4F3B: effect on 20-hydroxyeicosatetraenoic acid production. Prostaglandins Leukot Essent Fatty Acids 75:169–177
277. Kulas J, Schmidt C, Rothe M, Schunck WH, Menzel R (2008) Cytochrome P450-dependent metabolism of eicosapentaenoic acid in the nematode *Caenorhabditis elegans*. Arch Biochem Biophys 472:65–75
278. Xiao B, Li X, Yan J, Yu X, Yang G, Xiao X, Voltz JW, Zeldin DC, Wang DW (2010) Overexpression of cytochrome P450 epoxygenases prevents development of hypertension in spontaneously hypertensive rats by enhancing atrial natriuretic peptide. J Pharmacol Exp Ther 334:784–794
279. Koeners MP, Wesseling S, Ulu A, Sepulveda RL, Morisseau C, Braam B, Hammock BD, Joles JA (2011) Soluble epoxide hydrolase in the generation and maintenance of high blood pressure in spontaneously hypertensive rats. Am J Physiol Endocrinol Metab 300:E691–E698
280. Zhao X, Pollock DM, Inscho EW, Zeldin DC, Imig JD (2003) Decreased renal cytochrome P450 2C enzymes and impaired vasodilation are associated with angiotensin salt-sensitive hypertension. Hypertension 41:709–714
281. Wang MH, Smith A, Zhou Y, Chang HH, Lin S, Zhao X, Imig JD, Dorrance AM (2003) Downregulation of renal CYP-derived eicosanoid synthesis in rats with diet-induced hypertension. Hypertension 42:594–599
282. Sodhi K, Inoue K, Gotlinger KH, Canestraro M, Vanella L, Kim DH, Manthati VL, Koduru SR, Falck JR, Schwartzman ML, Abraham NG (2009) Epoxyeicosatrienoic acid agonist rescues the metabolic syndrome phenotype of HO-2-null mice. J Pharmacol Exp Ther 331:906–916
283. Herse F, Lamarca B, Hubel CA, Kaartokallio T, Lokki AI, Ekholm E, Laivuori H, Gauster M, Huppertz B, Sugulle M, Ryan MJ, Novotny S, Brewer J, Park JK, Kacik M, Hoyer J, Verlohren S, Wallukat G, Rothe M, Luft FC, Muller DN, Schunck WH, Staff AC, Dechend R (2012) Cytochrome P450 subfamily 2J polypeptide 2 expression and circulating epoxyeicosatrienoic metabolites in preeclampsia. Circulation 126:2990–2999
284. Huang H, Chang HH, Xu Y, Reddy DS, Du J, Zhou Y, Dong Z, Falck JR, Wang MH (2006)

Epoxyeicosatrienoic acid inhibition alters renal hemodynamics during pregnancy. Exp Biol Med (Maywood) 231:1744–1752
285. Blanton A, Nsaif R, Hercule H, Oyekan A (2006) Nitric oxide/cytochrome P450 interactions in cyclosporin A-induced effects in the rat. J Hypertens 24:1865–1872
286. Fava C, Montagnana M, Melander O (2009) Overexpression of cytochrome P450 4F2 in mice increases 20-hydroxyeicosatetraenoic acid production and arterial blood pressure. Kidney Int 76:913, author reply 913–914
287. Liu X, Zhao Y, Wang L, Yang X, Zheng Z, Zhang Y, Chen F, Liu H (2009) Overexpression of cytochrome P450 4F2 in mice increases 20-hydroxyeicosatetraenoic acid production and arterial blood pressure. Kidney Int 75:1288–1296
288. Ding Y, Wu CC, Garcia V, Dimitrova I, Weidenhammer A, Joseph G, Zhang F, Manthati VL, Falck JR, Capdevila JH, Schwartzman ML (2013) 20-HETE induces remodeling of renal resistance arteries independent of blood pressure elevation in hypertension. Am J Physiol Renal Physiol 305:F753–F763
289. Gross GJ, Hsu A, Gross ER, Falck JR, Nithipatikom K (2013) Factors mediating remote preconditioning of trauma in the rat heart: central role of the cytochrome P450 epoxygenase pathway in mediating infarct size reduction. J Cardiovasc Pharmacol Ther 18:38–45
290. Koerner IP, Zhang W, Cheng J, Parker S, Hurn PD, Alkayed NJ (2008) Soluble epoxide hydrolase: regulation by estrogen and role in the inflammatory response to cerebral ischemia. Front Biosci 13:2833–2841

7 Monooxygenation of Small Hydrocarbons Catalyzed by Bacterial Cytochrome P450s

Osami Shoji and Yoshihito Watanabe

Abstract

Cytochrome P450s (P450s) catalyze the NAD(P)H/O_2-dependent monooxygenation of less reactive organic molecules under mild conditions. The catalytic activity of bacterial P450s is very high compared with P450s isolated from animals and plants, and the substrate specificity of bacterial P450s is also very high. Accordingly, their catalytic activities toward nonnative substrates are generally low especially toward small hydrocarbons. However, mutagenesis approaches have been very successful for engineering bacterial P450s for the hydroxylation of small hydrocarbons. On the other hand, "decoy" molecules, whose structures are very similar to natural substrates, can be used to trick the substrate recognition of bacterial P450s, allowing the P450s to catalyze oxidation reactions of nonnative substrates without any substitution of amino acid residues in the presence of decoy molecules. Thus, the hydroxylation of small hydrocarbons such as ethane, propane, butane and benzene can be catalyzed by P450BM3, a long-alkyl-chain hydroxylase, using substrate misrecognition of P450s induced by decoy molecules. Furthermore, a number of H_2O_2-dependent bacterial P450s can catalyze the peroxygenation of a variety of nonnative substrates through a simple substrate–misrecognition trick, in which catalytic activities and enantioselectivity are dependent on the structure of decoy molecules.

O. Shoji (✉)
Bioinorganic Chemistry Laboratory, Department of Chemistry, Nagoya University, Furo-cho, Chikusa-ku, Nagoya 464-8602, Japan
e-mail: shoji.osami@a.mbox.nagoya-u.ac.jp

Y. Watanabe
Bioinorganic Chemistry Laboratory, Research Center for Materials Science, Nagoya University, Furo-cho, Chikusa-ku, Nagoya 464-8602, Japan

E.G. Hrycay and S.M. Bandiera (eds.), *Monooxygenase, Peroxidase and Peroxygenase Properties and Mechanisms of Cytochrome P450*, Advances in Experimental Medicine and Biology 851,
DOI 10.1007/978-3-319-16009-2_7, © Springer International Publishing Switzerland 2015

Keywords
Bacterial P450s • Mutagenesis approaches • Gaseous alkanes • Decoy molecules • Hydrogen peroxide

7.1 Introduction

Cytochrome P450s (P450s) are a superfamily of hemeproteins that catalyze the monooxygenation of inert substrates in conjunction with the biosynthesis of steroids, drug metabolism and detoxification of xenobiotics [1, 2]. Because P450s efficiently catalyze the monooxygenation of less reactive substrate C–H bonds under mild conditions, P450s have attracted much attention as candidates of biocatalysts that are useful to synthetic chemistry [3–5]. Among the P450s reported thus far, bacterial P450s are regarded as promising candidates, because they have very high catalytic activities, and a practical amount of these enzymes can be obtained in soluble form using a typical *Escherichia coli* gene-expression system. However, bacterial P450s exhibit very high substrate specificity. To utilize bacterial P450s as biocatalysts, their substrate specificities must be changed to target substrates. This chapter describes the monooxygenation of nonnative substrates catalyzed by bacterial P450s.[1] The first topic is the reaction mechanisms of P450s including their switch-on mechanism. The second topic covers the hydroxylation of small alkanes by mutants of P450BM3 and wild-type P450BM3 with the assistance of decoy molecules. Finally, we describe the H_2O_2-dependent peroxygenation of a variety of nonnative substrates catalyzed by bacterial P450s.

7.2 Switch-on Mechanism of P450s

P450s activate molecular oxygen through a heme iron center with the thiolate of cysteine acting as the fifth ligand to generate the active oxidant species that consists of a porphyrin π radical cation ferryl species ($Por^{\bullet+}Fe^{IV}{=}O$) known as Compound I [6–8]. The catalytic monooxygenase cycle of P450 involves the following steps (Fig. 7.1): (1) substrate binding, which results in the removal of a water molecule ligated to the heme iron to cause a positive reduction–oxidation (redox) potential shift of the heme iron [9–11]; (2) reduction of Fe^{3+} (ferric) to Fe^{2+} (ferrous) by the first electron transfer from NAD(P)H through the reductase domain; (3) binding of molecular oxygen to the ferrous heme; (4) production of Compound I by reductive activation of molecular oxygen through the second electron transfer from NAD(P)H; and (5) monooxygenation of the bound substrate by Compound I [2]. In the formation of Compound I, two protons from outside of P450 are indispensable and are transferred via a P450 proton relay system. The aspartate-251 residue serves as a donor of the first proton to the distal oxygen atom of the ferriperoxo anion intermediate, while the conserved threonine-252 residue in the active site of P450 serves as a donor of the second proton to the distal oxygen atom of the ferrihydroperoxide intermediate [12]. In addition, mutations of the threonine to alanine or valine in P450cam can result in H_2O_2 production (uncoupling of P450) rather than monooxygenation of the substrate. According to the reaction mechanism of P450s, appropriate binding of the substrate to the active site of P450s is crucial for initiating the catalytic cycle. Therefore, substrate binding has the role of a switch to commence the reactions of P450s. This switch-on mechanism contributes to the high substrate specificities of bacterial P450s such as P450BM3 and P450cam. These P450s thus show very low catalytic activity toward nonnative substrates. Interestingly, a similar switch-on mechanism was observed in H_2O_2-dependent P450s such as $P450_{BS\beta}$ and $P450_{SP\alpha}$.

[1] A review of this scope cannot include all the references pertaining to the subject matter presented.

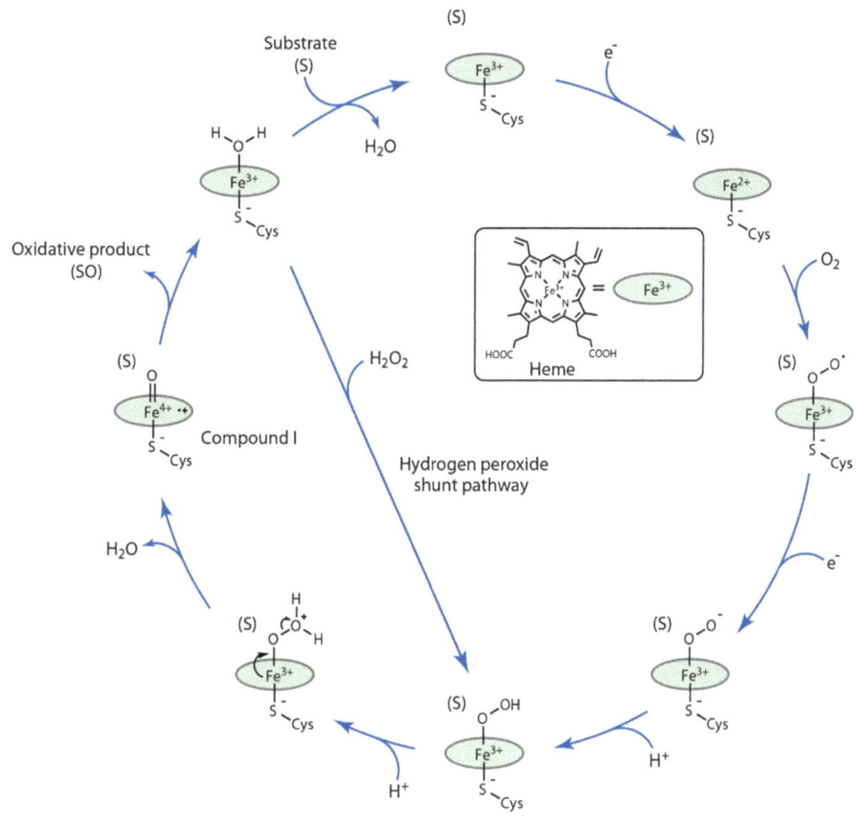

Fig. 7.1 The catalytic cycle of P450s including the hydrogen peroxide–shunt pathway

Although the general acid–base catalyst is essential for the generation of Compound I using H_2O_2, the H_2O_2-dependent P450s lack any general acid–base residue around the heme according to the crystal structures of $P450_{BS\beta}$ and $P450_{SP\alpha}$ in the palmitic acid–bound form (Fig. 7.2) [13, 14]. These findings indicate that the carboxylate group of palmitic acid placed at the distal side of the heme is crucial for the generation of Compound I. This substrate-assisted reaction mechanism confers the switch-on mechanism upon the H_2O_2-dependent P450s and thus contributes to the high substrate specificity. In fact, $P450_{SP\alpha}$ and $P450_{BS\beta}$ never oxidize substrates other than fatty acids. The switch-on mechanism of P450s is important to realize the efficient hydroxylation in the natural reaction system, but it is not advantageous in the viewpoint of the diversification of the substrate range for synthetic applications.

7.3 Hydroxylation of Small Hydrocarbons by P450BM3

7.3.1 P450BM3

P450BM3 (CYP102A1) isolated from *Bacillus megaterium* catalyzes the hydroxylation of long-alkyl-chain fatty acids at the ω-1, ω-2 and ω-3 positions (Fig. 7.3a) [15–17]. Because P450BM3 is a structurally self-sufficient P450 (i.e., the reductase domain is fused with the P450 domain on the same peptide chain), P450BM3 has an extremely high hydroxylation activity with a turnover number of more than 16,000 [18–20]. This is the highest oxidase activity reported among the P450s thus far. Because of its high catalytic activity for hydroxylation, P450BM3 is expected to be a biocatalyst for the hydroxylation of inert substrate C–H bonds. The crystal structure of

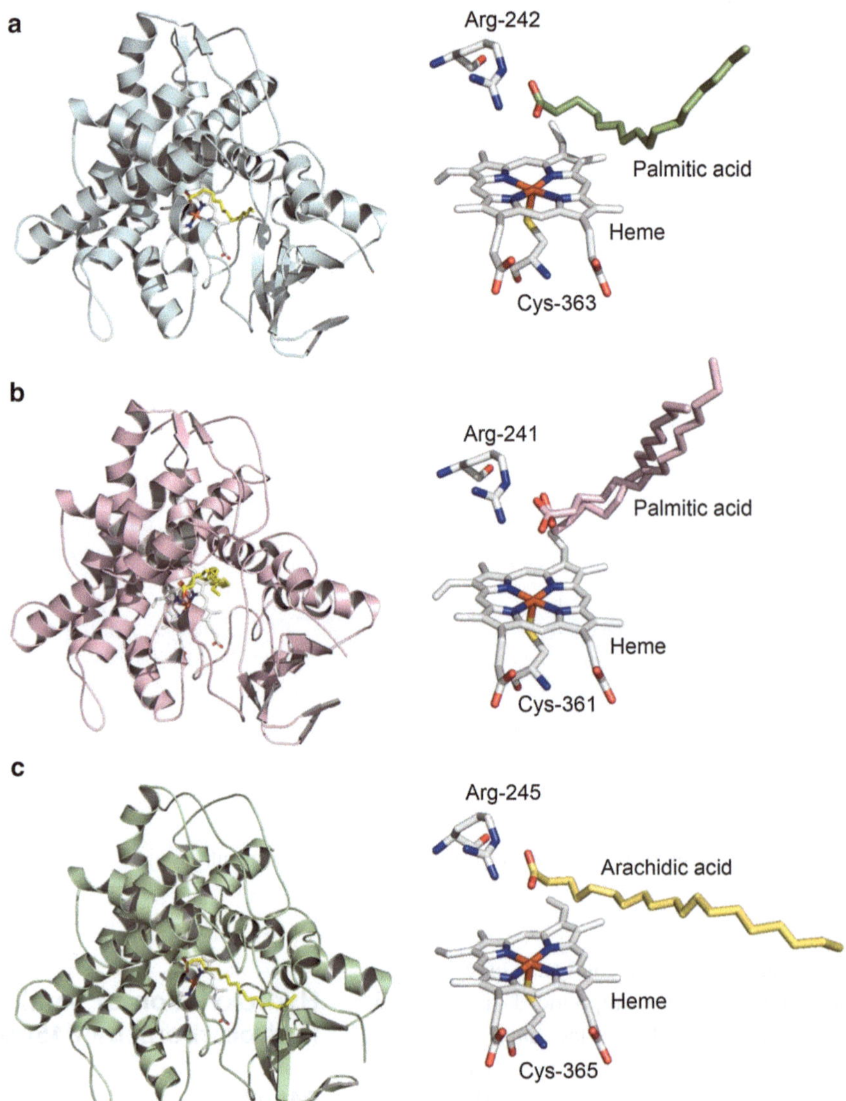

Fig. 7.2 Crystal structures (*left*) and active-site structures (*right*) of $P450_{BS\beta}$ (**a**), $P450_{SP\alpha}$ (**b**) and CYP152L1 (**c**). PDB codes for $P450_{BS\beta}$, $P450_{SP\alpha}$, and CYP152L1 are 1IZO, 3AWM and 4L40, respectively

P450BM3 with palmitoleic acid shows that palmitoleic acid is fixed by two major interactions (Fig. 7.3b) consisting of the ionic interaction of the substrate carboxylate group with Arg-47 and Tyr-51, and the hydrophobic interaction of the alkyl chain with amino acids at the substrate binding site [21]. As described in the previous section, the catalytic hydroxylation activity of bacterial P450BM3 was generally low for nonnative substrate oxidations, particularly for small alkanes. In Sects. 3.2 and 3.3, two approaches to alter the inherent substrate specificity to small alkanes are described.

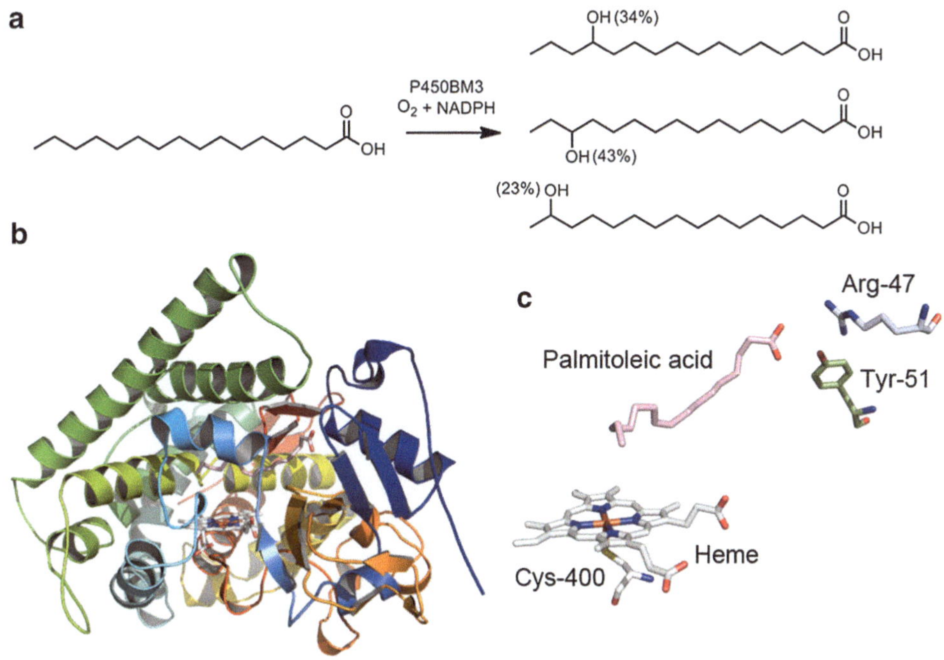

Fig. 7.3 Fatty acid hydroxylation catalyzed by P450BM3 (**a**). The overall structure (**b**) and the active site structure (**c**) of P450BM3 (PDB code: 1FAG)

7.3.2 Engineering of P450BM3 for the Hydroxylation of Small Hydrocarbons

According to the reaction mechanism of P450BM3, molecules having structures very different from those of native substrates, especially a small molecule, cannot start the first step of the catalytic cycle, resulting in very low or no catalytic activity. Because the substrate specificity in enzymatic reactions is governed by the local chemical environment of the enzyme active site, a variety of P450BM3 mutants [4] were prepared by site-directed mutagenesis as well as by random mutagenesis to alter their substrate specificities by redesigning the active site of P450BM3 for the hydroxylation of small alkanes such as gaseous alkanes [22–26] (Fig. 7.4). For ethane hydroxylation, a mutant with 17 amino acid substitutions (R47C, V78F, A82S, K94I, P142S, T175I, A184V, F205C, S226R, H236Q, E252G, R255S, A290V, A328F, L353V, E464G and I710T) named 35E11 was constructed. The rate of ethanol formation and total turnover by 35E11 were estimated to be 0.4 min^{-1} and 250, respectively [25]. The 35E11 mutant was also active in catalyzing propane hydroxylation and it was further developed by random mutagenesis. The resulting mutant named $P450_{PMO}R1$ had 24 amino acid substitutions (R47C, L52I, A74E, V78F, A82G, K94I, P142S, T175I, A184V, L188P, F205C, S226R, H236Q, E252G, R255S, A290V, A328F, L353V, I366V, G443A, E464G, P654K, I710T and E1037G). The rate of propanol formation and total turnover number of $P450_{PMO}R1$ were estimated to be 455 min^{-1} and 35,600, respectively [22]. Although mutations were introduced at various positions including the non-active site (Fig. 7.5), these mutations were assumed to provide space to accommodate gaseous alkanes in an appropriate manner because of the structural change in the active site. A similar strategy was used to prepare a P450cam mutant enzyme to

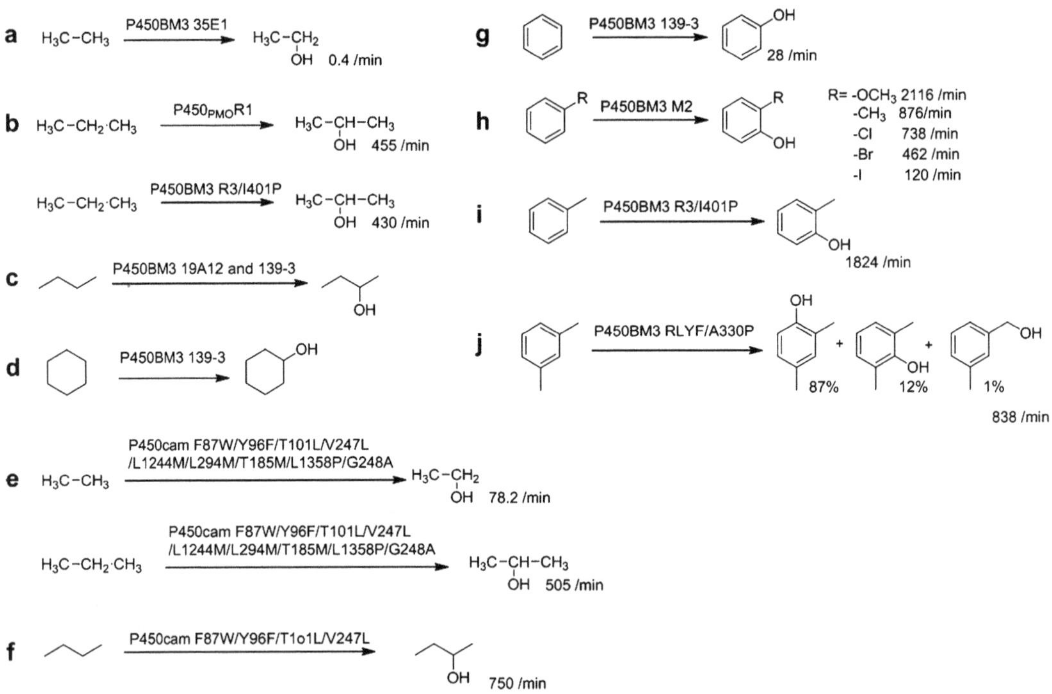

Fig. 7.4 Hydroxylation of small hydrocarbons catalyzed by P450s involving reductive activation of molecular oxygen: **a** ethane [25], **b** propane [22, 27], **c** butane [26, 28], **d** cyclohexane [28], **e** ethane and propane [29], **f** butane [30], **g** benzene [31], **h** monosubstituted benzenes [32], **i** toluene [27], **j** xylene [33]

catalyze the hydroxylation of gaseous alkanes [29, 30, 34]. For direct benzene hydroxylation, a mutant of P450BM3 named 139-3 [28] with 11 amino acid replacements (V78A, H138Y, T175I, V178I, A184V, H236Q, E252G, R255S, A290V, A295T and L353V) was shown to be active [31]. The initial rate of phenol formation was reported to be 28 min^{-1}. A variety of mutants were also constructed for the hydroxylation of substituted benzene substrates such as toluene (R3/I401P mutant with 4 amino acid replacements, R47S, Y51W, A330P, I401P) [27], *o*-xylene (RLYF/A330P mutant with 3 amino acid replacements, R47S, Y51W, A330P) [33], *p*-xylene and halogenated benzenes (M2 mutant with 3 amino acid replacements, R47L, Y51F, I401M) [32, 35]. These results clearly indicate that mutagenesis of P450s to construct a binding pocket suitable for nonnative substrates is a promising technique.

7.3.3 Substrate Misrecognition of P450BM3 for the Hydroxylation of Small Hydrocarbons

If the size and shape of the active site of P450s were reshaped by the addition of a small molecule that can bind to the active site, even wild-type P450BM3 would be able to provide a binding pocket for accommodating nonnative substrates and could oxidize them without being subjected to mutagenesis. A simple addition of perfluorocarboxylic acids (PFCs) as inert dummy substrates (decoy molecules) can turn wild-type P450BM3 into a small alkane hydroxylase without replacing any amino acid residues [36, 37]. Because the C–F bond dissociation energy is sufficiently high (116 kcal mol^{-1}), PFCs are never oxidized by P450BM3 [38]. We assumed that P450BM3 cannot

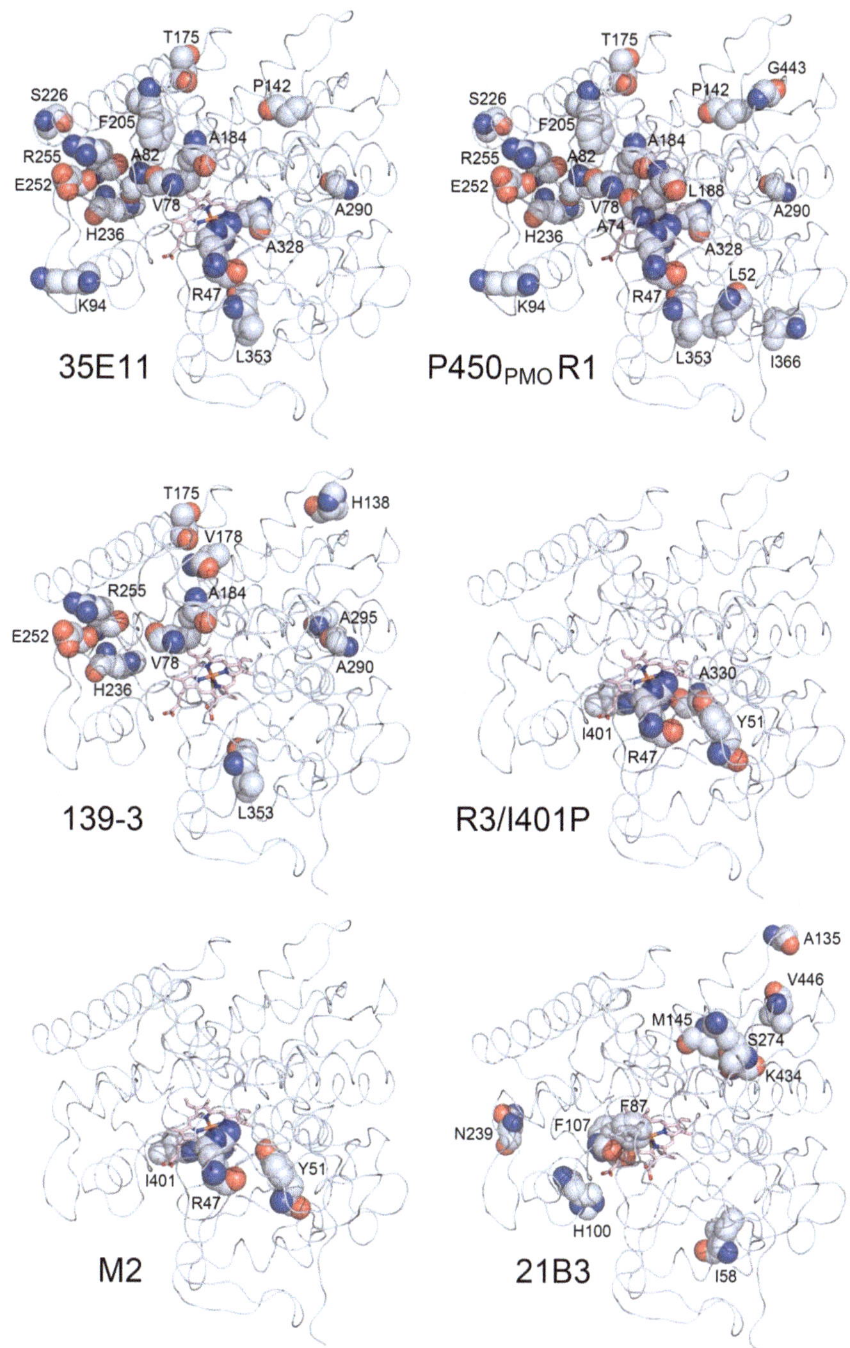

Fig. 7.5 Positions of the amino acids replaced in P450BM3 mutants. Mutated amino acids in the heme domain were represented as sphere models on the crystal structure of the substrate-free form of P450BM3 (PDB code: 1BU7)

distinguish between PFCs and fatty acids, because of the similar atomic radius of the fluorine and hydrogen atom. Thus, PFCs would bind to the active site of P450BM3 similarly to the natural substrate-binding manner and initiate activation of molecular oxygen in a similar

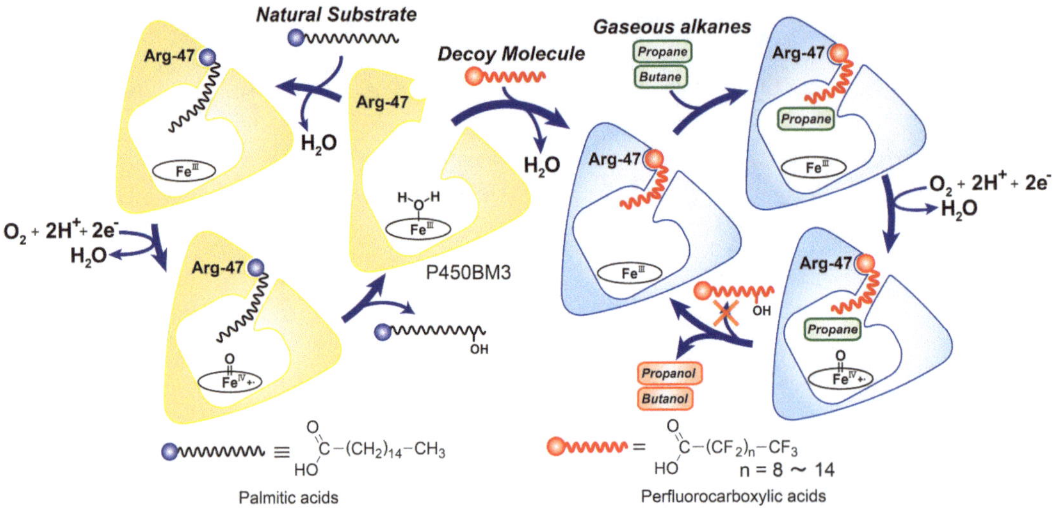

Fig. 7.6 Schematic representation of reaction mechanisms of the natural reaction system (*left*) and decoy molecule system (*right*) of P450BM3. Subterminal carbons of fatty acids are hydroxylated in the natural reaction (*left*). By simple addition of a decoy molecule such as PFC10, the hydroxylation reaction of a small alkane (for example, propane or butane) is catalyzed by P450BM3 because of substrate misrecognition of P450BM3 (*right*)

manner as do long-alkyl-chain fatty acids to afford Compound I (Fig. 7.6, right). PFCs bearing shorter alkyl chains of 8–14 carbon atoms (PFC8-PFC14) are expected to provide space for nonnative substrates because the structure of the substrate-binding site of P450BM3 is suitable for accommodating a fatty acid bearing 16 carbon atoms. In fact, propane, butane and cyclohexane were hydroxylated by P450BM3 in the presence of PFCs to yield 2-propanol, 2-butanol and cyclohexanol, respectively (Fig. 7.7). In sharp contrast, no products were detected in the reactions for any of the alkanes in the absence of PFCs, showing that addition of the decoy molecules to wild-type P450BM3 conferred the ability to catalyze the hydroxylation of gaseous alkanes. Interestingly, the rate of product formation was influenced significantly by the alkyl chain length of the PFCs. PFC10 showed the largest rate of product formation for propane hydroxylation (67 min^{-1}) and the highest coupling efficiency (18 %) among the PFCs examined. PFC9 and PFC10 were almost equally effective for butane hydroxylation and showed the fastest rate for the formation of 2-butanol (100–113 min^{-1}), whereas PFC9 demonstrated the highest coupling efficiency (57 %). In the hydroxylation of cyclohexane, PFC9 showed the fastest rate for cyclohexanol formation (110 min^{-1}) with a 35 % coupling efficiency. These results indicate that the larger alkanes tended to prefer the shorter alkyl-chain PFCs for efficient reactions and that the combination of the alkyl chain length of the PFCs and the size of the alkane substrate governs the efficiency of the hydroxylation reaction catalyzed by P450BM3. The P450BM3-decoy molecule system also catalyzes the hydroxylation of octane, hexane and branched alkanes [37]. The primary carbons of ethane were also hydroxylated under the pressure condition of 0.5 MPa ethane using PFC10 as a decoy molecule (40 h^{-1} $P450^{-1}$) [39], whereas the hydroxylation of methane gas under the pressure condition of 0.5 MPa methane gave no methanol. Furthermore, the hydroxylation of benzene and monosubstituted benzenes was also catalyzed by the P450BM3-

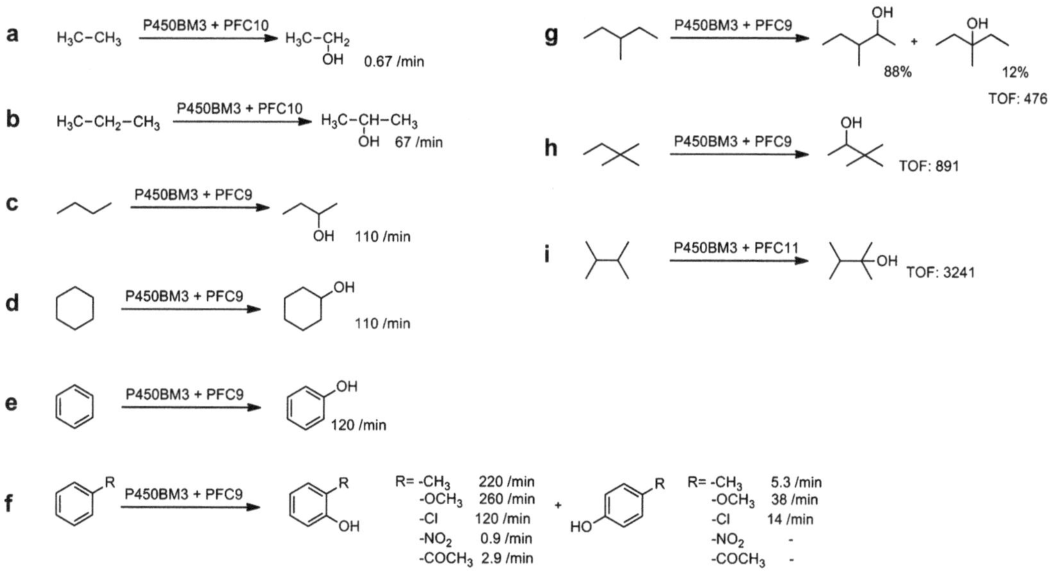

Fig. 7.7 Hydroxylation of small hydrocarbons catalyzed by P450BM3 with perfluorinated carboxylic acids as decoy molecules (TOF: turn over number) [36, 37, 39, 40]

decoy molecule system [40]. PFC9 afforded the turnover rate of 120 min^{-1} and a coupling efficiency of 24 %.

It is noteworthy to mention that the catalytic turnover rate and coupling efficiency were higher than those of the engineered P450BM3 mutants prepared by directed evolution [31]. Although the hydroxylation of benzene is generally accompanied by overoxidation products [41], the selective formation of phenol (more than 99 %) without any overoxidation products was observed in the P450BM3-decoy system. Under the reaction conditions, phenol is expected to escape rapidly from the active site of P450BM3 because of the heme cavity comprising hydrophobic amino acid residues and the hydrophobic nature of PFCs. Toluene was also hydroxylated and PFC9 gave the largest turnover rate (220 min^{-1}) for the selective *o*-hydroxylation and a coupling efficiency of 56 %. The selective *o*-hydroxylation was also observed in the hydroxylation of anisole, chlorobenzene, nitrobenzene and acetophenone, showing that the *o*-position of monosubstituted benzenes was hydroxylated selectively, irrespective of the substituents (Fig. 7.7).

7.4 Hydrogen Peroxide: Dependent Monooxygenation by P450s

7.4.1 Hydrogen Peroxide: Shunt Reaction of P450s

In place of molecular oxygen, two electrons and two protons, H_2O_2 can be used as an oxidant for the generation of the active species through a "shunt reaction" (see Fig. 7.1), whereby P450 ferric heme iron directly reacts with H_2O_2 to form a ferric-H_2O_2 complex that dissociates via a heterolytic cleavage of the peroxy O–O bond to give Compound I. The use of H_2O_2 consists of an attractive option for monooxygenation reactions catalyzed by P450s, because redox partner proteins such as NAD(P)H–P450 oxidoreductase are not required for generation of Compound I. Furthermore, a low cost for H_2O_2 allows researchers to perform monooxygenations through the shunt pathway on an industrial scale. The H_2O_2-dependent P450 reaction is considered to be a significant reaction pathway for practical applications. The H_2O_2-shunt reaction

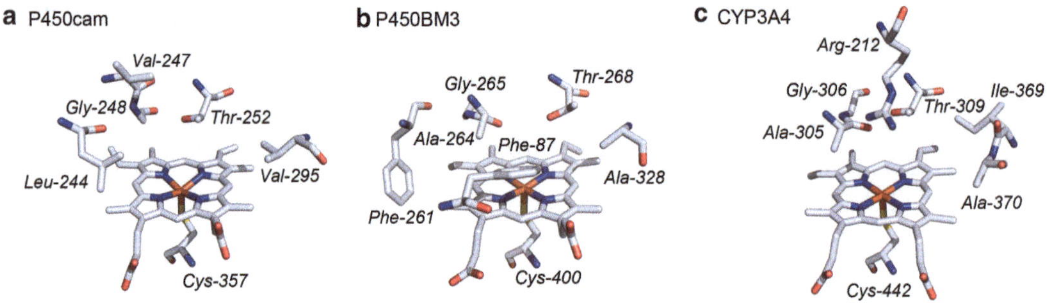

Fig. 7.8 Active site structures of **a** P450cam (PDB code: 2CPP) [57], **b** P450BM3 (PDB code: 1JPZ) [57] and **c** CYP3A4 (PDB code: 1TQN) [58]

has been applied to a variety of oxidations that are commonly catalyzed by the usual P450 reaction system [42–56]. However, the shunt reaction is generally inefficient and the catalytic activities of P450s are also lower than NAD(P)H/O_2-supported reactions including oxygen activation. The crystal structures of P450s reported so far have revealed that amino acid residues that serve as general acid–base catalysts in heme peroxidases and catalase (histidine, aspartic acid, glutamic acid) are not observed in the distal side of the heme of P450 enzymes (Fig. 7.8). General acid–base residues are crucial for the facile generation of Compound I (Fig. 7.9), which is responsible for H_2O_2-dependent peroxidation. Indeed, most H_2O_2-dependent heme enzymes such as horseradish peroxidase (HRP) [59], chloroperoxidase (CPO) [60, 61], aromatic peroxygenase (APO), now known as *Agrocybe aegerita* unspecific peroxygenase (*Aae*UPO) [62–64], cytochrome *c* peroxidase (C*c*P) [65], catalase-peroxidase (katG) [66] and catalase [67] have general acid–base residues close to the heme iron (Figs. 7.10 and 7.11). The critical role of the general acid–base residues has been studied using a myoglobin framework. When a general acid–base residue is introduced into an appropriate position of the distal side of myoglobin by point mutagenesis, the formation of myoglobin Compound I is accelerated and higher peroxidase and peroxygenase activities are observed [68–70]. The lack of any general acid–base residue in the active site of P450s and the high hydrophobicity of the heme pocket of P450s result in low efficiency for the formation of Compound I and the subsequent peroxidation.

7.4.2 Engineering of P450s for Construction of Artificial H_2O_2-Dependent P450s

A variety of P450 mutants have been prepared by site-directed mutagenesis and directed evolution (random mutagenesis) for the development of artificial H_2O_2-dependent P450s [71–75]. Among them, mutants of P450BM3 showed relatively high peroxygenase activity. The F87A mutant of P450BM3, the first artificial H_2O_2-dependent P450, was reported in 2001 [76] and succeeded in catalyzing the oxidation of myristic acid and *p*-nitrophenoxydodecanoic acid (12-pNCA) [77] with a turnover rate of 162 min^{-1}. However, its Michaelis–Menten constant (K_m) for H_2O_2 was estimated to be 24 mM. The F87V mutant of P450BM3 [78] was further improved by applying random mutation based on the F87A mutant. In the peroxygenation of 12-pNCA, the initial turnover rate of the resulting 21B3 mutant containing nine amino acid substitutions in addition to F87A (I58V, H100R, F107L, A135S, M145V, N239H, S274T, K434E and V446I) was more than 18 times higher than that of the F87A mutant (see Fig. 7.5) [79]. Its Michaelis–Menten constant (K_m) for H_2O_2 was also improved from 24 to 10 mM. The 21B3 mutant also catalyzed styrene epoxidation [80]. A stable double mutant of 21B3 (W96A/F405L) showed a larger

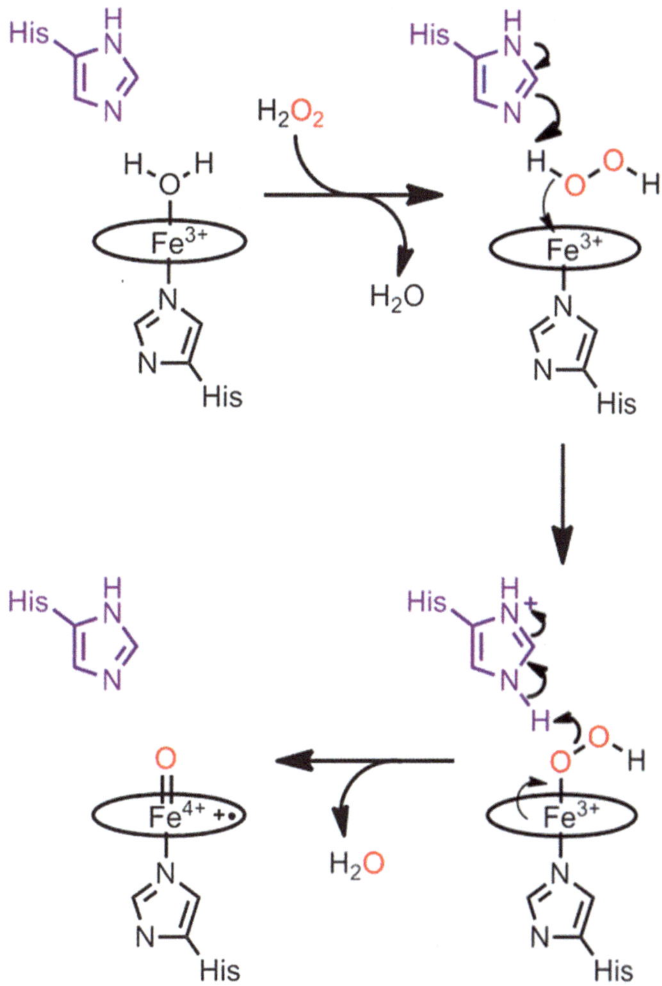

Fig. 7.9 General acid–base function of distal histidine in the formation of Compound I using H_2O_2 as an oxidant

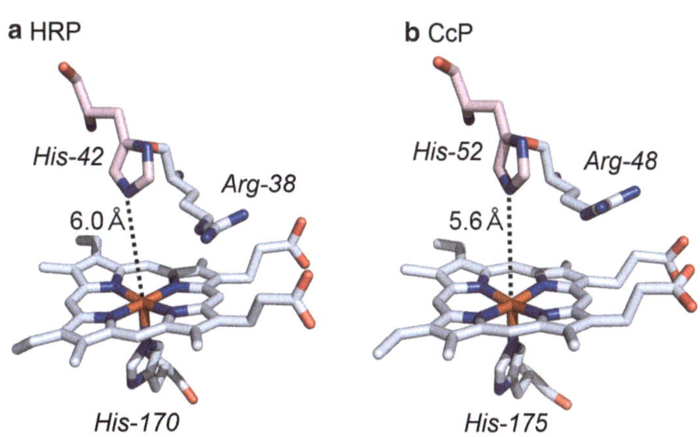

Fig. 7.10 Active site structures of **a** HRP (PDB code: 1ATJ) [59] and **b** CcP (PDB code: 1CCA) [65]

Fig. 7.11 Active site structures of **a** CPO (PDB code: 1CPO) [59, 60] and **b** APO (PDB code: 2YP1) [62]

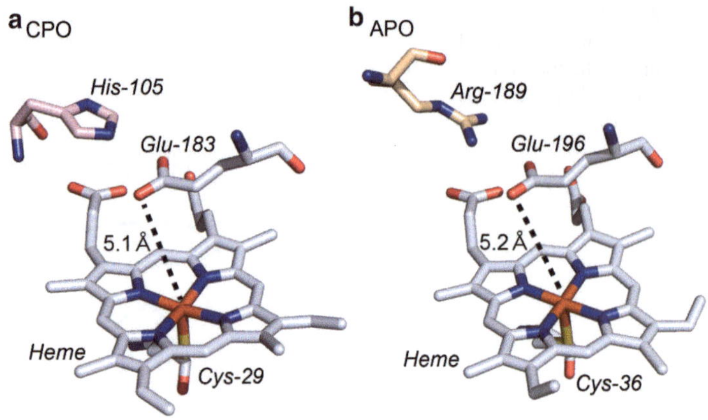

peroxidase activity than the 21B3 mutant. It is noteworthy that any mutation introducing general acid–base residue(s) is not involved in artificial H_2O_2-dependent P450s. Mutations were sometimes introduced in places other than the active site. Therefore, no general acid–base residue in the active site was required for the facile generation of Compound I in any artificial H_2O_2-dependent P450 enzyme. A possible explanation for this is that water molecules exist in the active site and may serve as a general acid–base catalyst.

7.4.3 CYP152 Family and Hydrogen Peroxide-Dependent Reactions

While most P450s use the reductive molecular oxygen activation process for monooxygenation, $P450_{SP\alpha}$ (CYP152B1) from *Sphingomonas paucimobilis* [81–86], $P450_{BS\beta}$ (CYP152A1) from *Bacillus subtilis* [14, 87–92] and $P450_{CLA}$ (CYP152A2) from *Clostridium acetobutylicum* [93] use H_2O_2 as the oxidant and catalyze the hydroxylation of long-alkyl-chain fatty acids with high catalytic activities (Fig. 7.12). According to the "Cytochrome P450 website" conducted by Dr. D. R. Nelson (http://drnelson.uthsc.edu/CytochromeP450.html), there were 30 genes that encoded P450 proteins classified as CYP152 family members (Jan. 2014). A genetic family tree of the CYP152 family is shown in Fig. 7.13. Among P450s classified in the CYP152 family, $P450_{SP\alpha}$ reported in 1994 [95] is the first P450 to be categorized as a H_2O_2-dependent P450. $P450_{SP\alpha}$ catalyzes the 100 %-α-selective hydroxylation of long-alkyl-chain fatty acids such as myristic acid. The crystal structure of $P450_{SP\alpha}$ (CYP152B1) at a resolution of 2.1 Å (PDB code 1IZO) reported in 2003 is the first crystal structure of a CYP152 family member (see Fig. 7.2a) [14]. $P450_{BS\beta}$ has 42 % amino acid identity with $P450_{SP\alpha}$ and oxidizes the α- and β-positions of fatty acids in a roughly 40:60 ratio (see Fig. 7.12a). The crystal structure of $P450_{BS\beta}$ revealed that the enzyme lacks general acid–base residues around the distal side of the heme, although general acid–base residues are highly conserved in heme peroxidases and peroxygenases, as described in Sect. 3.1. Instead of the general acid–base residues, the terminal carboxylate group of the bound fatty acid interacts with the guanidine group of Arg-242 located near the heme group (see Fig. 7.2a). The distance between an oxygen atom of the carboxylate group of palmitic acid and the heme iron is 5.3 Å, which is very close to that of APO (5.2 Å, see Fig. 7.11b) [62] and CPO (5.1 Å) (see Fig. 7.11) [60, 61]. Furthermore, the resulting salt bridge in the active site is very similar to that of APO [62]. This structure suggests the following unique catalytic mechanism (Fig. 7.14). The catalytic reaction begins with the fixation of a substrate through interaction of the terminal carboxyl group of the fatty acid with Arg-242, located near the heme. The general acid–base function of the fatty acid-Arg-242 salt bridge allows facile generation of the

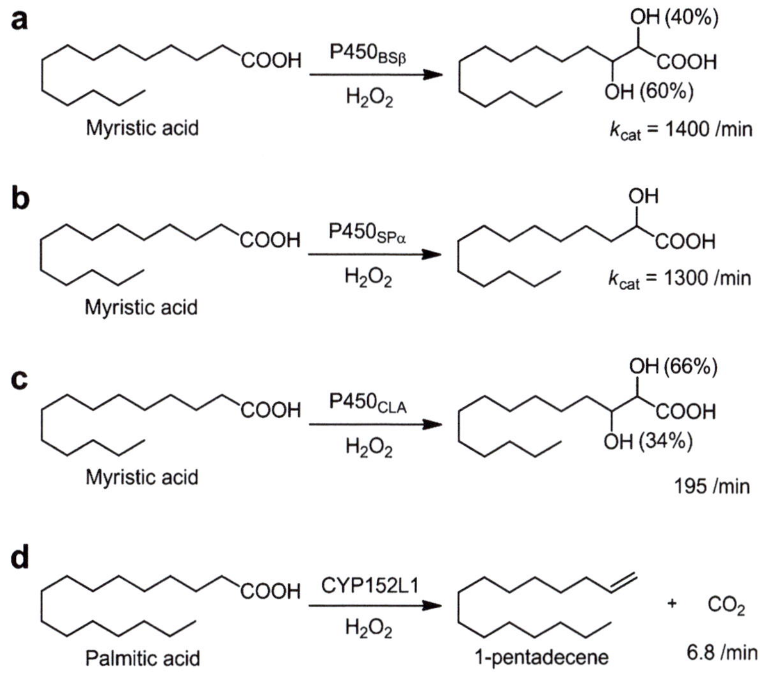

Fig. 7.12 Myristic acid hydroxylation catalyzed by H_2O_2-dependent P450s: **a** $P450_{BS\beta}$ [13], **b** $P450_{SP\alpha}$ [13], **c** $P450_{CLA}$ [93] and **d** CYP152L1 [94]

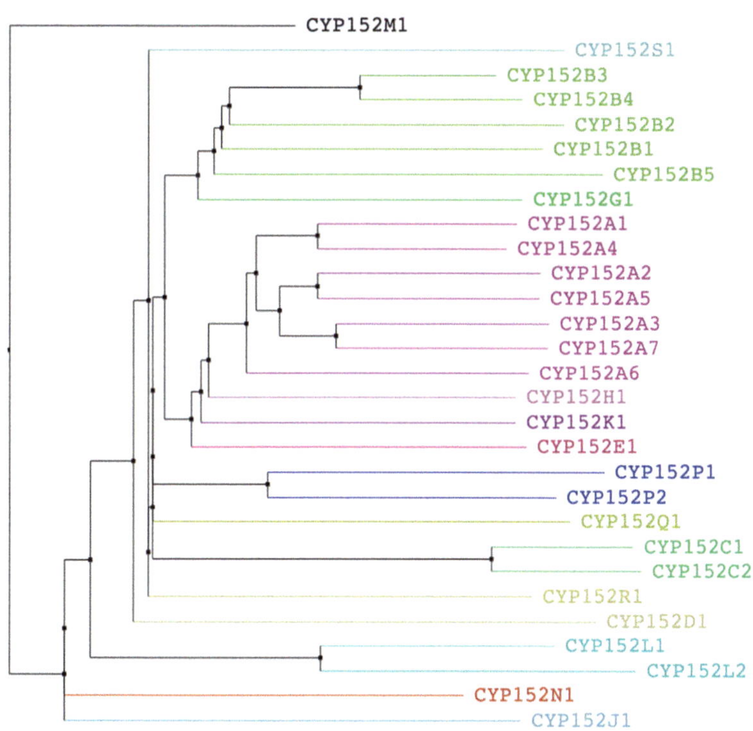

Fig. 7.13 Genetic family tree of CYP152 family

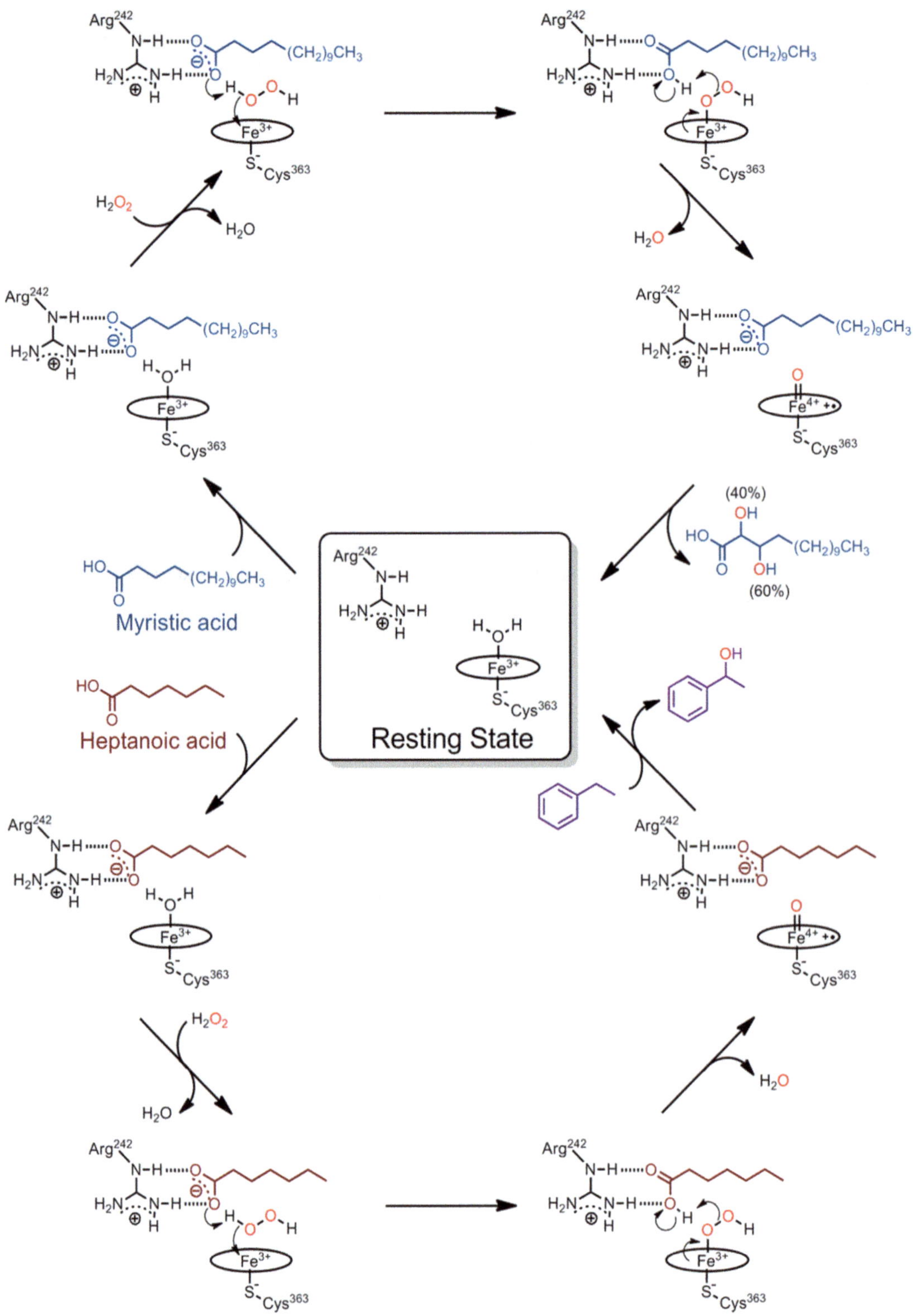

Fig. 7.14 Proposed catalytic hydroxylation mechanism for $P450_{BS\beta}$ and roles of the substrate carboxylate–Arg242 salt bridge (*upper*). The substrate misrecognition system in the oxidation of the nonnative ethylbenzene substrate in the presence of the heptanoic acid decoy molecule (*lower*)

active Compound I species to oxidize the substrate. Without the interaction of the carboxyl group, $P450_{BS\beta}$ does not start the reaction. Because this unique catalytic mechanism contributes to the high substrate specificity and regioselectivity of the hydroxylation, $P450_{BS\beta}$ never oxidizes substrates other than long-alkyl-chain fatty acids such as tetradecane, 1-tetradecanol, or tetradecanal. Similar to $P450_{BS\beta}$, $P450_{SP\alpha}$ also catalyzed the hydroxylation of long-alkyl-chain fatty acids using H_2O_2 as an oxidant. The crystal structure of the palmitic acid-bound form of $P450_{SP\alpha}$ revealed that the key interaction between the carboxylate of palmitic acid and the guanidine group of arginine near the heme are conserved in $P450_{SP\alpha}$ (see Fig. 7.2b, PDB code: 3AWM), indicating that the substrate-assisted reaction mechanism for the formation of Compound I is the same as that of $P450_{BS\beta}$ (see Fig. 7.14) [13]. Recently, the crystal structure of CYP152L1 (see Fig. 7.2c, PDB code: 4L40, 4L54) [96], which catalyzes decarboxylation of fatty acids using H_2O_2 (see Fig. 7.12) [94], was reported. Interestingly, the structure of the active site was not much different from $P450_{SP\alpha}$ and $P450_{BS\beta}$ (see Fig. 7.2).

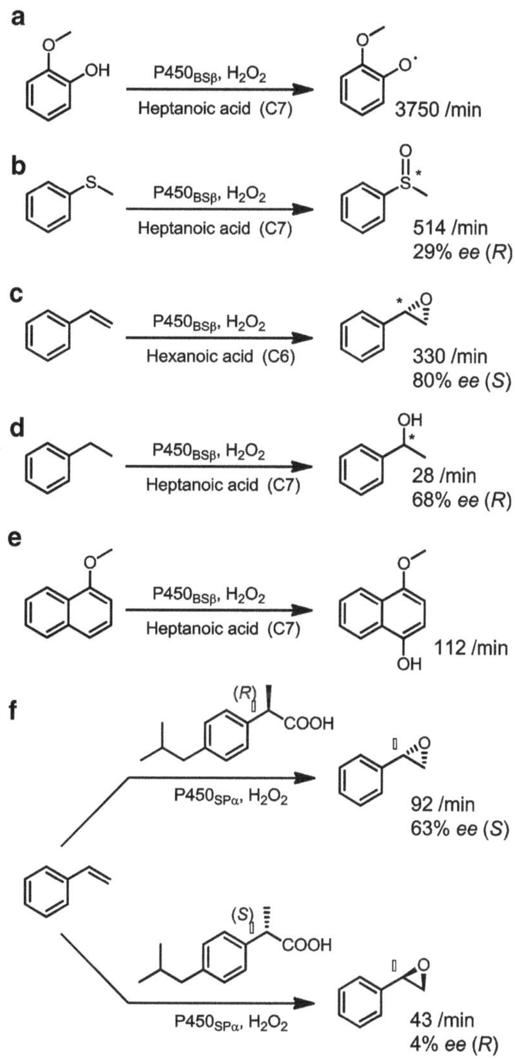

Fig. 7.15 Reactions catalyzed by H_2O_2-dependent P450s in the presence of decoy molecules: **a** guaiacol [97], **b** thioanisole [98], **c** styrene [97], **d** ethylbenzene [97], **e** 1–methoxynaphthalene [99], **f** styrene [100]

7.4.4 Substrate Misrecognition of H_2O_2-Dependent P450s

Using a series of short-alkyl-chain carboxylic acids (C4–C10) as decoy molecules, $P450_{BS\beta}$ oxidizes a wide variety of nonnative substrates (Fig. 7.15) [97]. Through this simple substrate--misrecognition trick using decoy molecules, oxidation (or peroxygenation) of nonnative substrates such as one-electron oxidation of guaiacol [97], sulfoxidation of thioanisole [98], epoxidation of styrene, C–H bond hydroxylation of ethylbenzene [97] and aromatic ring hydroxylation of 1-methoxynaphthalene [99] were catalyzed by $P450_{BS\beta}$ (see Fig. 7.15). Interestingly, the catalytic activities are highly dependent on the alkyl chain length of the decoy molecules. The enantioselective hydroxylation of ethylbenzene, for example, is also dependent on the structure of decoy molecules [97]. The peroxygenation of nonnative substrates never proceeded without decoy molecules, and thus these reactions were catalyzed by $P450_{BS\beta}$ with the aid of these short-alkyl-chain carboxylic acids. Apparently, $P450_{BS\beta}$ misrecognizes decoy molecules as its native substrates and the resulting salt bridge serves as the general acid–base catalyst to allow formation of Compound I, followed by the oxidation of nonnative substrates (see Fig. 7.14). The role of decoy molecules was confirmed by the crystal structure analysis of a heptanoic acid-bound form of $P450_{BS\beta}$ [101]. The inversion of enantioselectivity by noncovalent modification of the active site of $P450_{SP\alpha}$ was observed [100]. The enantioselectivity of styrene oxide formation was altered by the nature of the decoy molecules. (*R*)-Ibuprofen increased (*S*)-styrene oxide formation whereas (*S*)-ibuprofen preferentially yielded (*R*)-styrene oxide, showing that (*R*)-ibuprofen effectively increased (*S*)-selectivity and that the enantioselectivity could be controlled by simply selecting the (*R*)- or (*S*)-enantiomer of ibuprofen (see Fig. 7.15f). The crystal structure of $P450_{SP\alpha}$ containing (*R*)-ibuprofen revealed that (*R*)-ibuprofen was accommodated similarly to palmitic acid and that the carboxylate group of (*R*)-ibuprofen interacted with Arg-241. The docking simulation of styrene located in the active site of the (*R*)-ibuprofen-bound form of $P450_{SP\alpha}$ suggests the possible orientation of the vinyl group of styrene in the active site giving (*S*)-styrene oxide. These experimental observations demonstrated that a variety of oxidation reactions using small hydrocarbon substrates can be catalyzed by H_2O_2-dependent P450s that utilize decoy molecules.

7.5 Conclusions

Given their high hydroxylation activity toward inert substrate C–H bonds, bacterial P450s have been expected to be biocatalysts for synthetic applications. Their use as biocatalysts, however, has been limited because their substrate specificities are very high and thus they do not oxidize nonnative substrates efficiently, especially small hydrocarbons. The recent examples of engineered P450s highlighted here show that bacterial P450s with appropriate mutations can efficiently catalyze the hydroxylation of small hydrocarbons. For example, the catalytic reaction rate for propane hydroxylation with $P450_{PMO}R1$ reached more than 450 min^{-1}. Furthermore, the strategy for the activation of P450s using decoy molecules that induce substrate misrecognition of P450s have emerged as an alternative approach for the hydroxylation of small hydrocarbons without replacing any amino acid residues. Direct benzene hydroxylation yielding phenol catalyzed by P450BM3 and using perfluorononanoic acid proceeded even faster than that catalyzed by P450BM3 engineered by repeated mutagenesis. Combination of mutagenesis and decoy molecule approaches will overcome the limitation of the substrate specificity in bacterial P450s for the production of pharmaceuticals and fine chemicals, in which high enantioselectivity and regioselectivity as well as high oxidation activity are required.

Acknowledgments This work was supported, in part, by Grants-in-Aid for Scientific Research (S) to Y. W. (24225004) and for Young Scientists (A) to O. S. (21685018), and a Grant-in-Aid for Scientific Research on Innovative Areas "Molecular Activation Directed toward Straightforward Synthesis" to O. S. (25105724) from the Ministry of Education, Culture, Sports, Science and Technology (Japan).

References

1. Sono M, Roach MP, Coulter ED, Dawson JH (1996) Heme-containing oxygenases. Chem Rev 96:2841–2888
2. Ortiz de Montellano PR (2005) Cytochrome P450: structure, mechanism, and biochemistry, 3rd edn. Kluwer Academic/Plenum Publishers, New York
3. Denisov IG, Makris TM, Sligar SG, Schlichting I (2005) Structure and chemistry of cytochrome P450. Chem Rev 105:2253–2277
4. Whitehouse CJC, Bell SG, Wong LL (2012) $P450_{BM3}$ (CYP102A1): connecting the dots. Chem Soc Rev 41:1218–1260
5. Fasan R (2012) Tuning P450 enzymes as oxidation catalysts. ACS Catal 2:647–666
6. Dunford HB, Stillman JS (1976) Function and mechanism of action of peroxidases. Coord Chem Rev 19:187–251
7. Rittle J, Green MT (2010) Cytochrome P450 compound I: capture, characterization, and C-H bond activation kinetics. Science 330:933–937
8. Ost TWB, Clark J, Mowat CG, Miles CS, Walkinshaw MD, Reid GA, Chapman SK, Daff S (2003) Oxygen activation and electron transfer in flavocytochrome P450 BM3. J Am Chem Soc 125:15010–15020
9. Sligar SG (1976) Coupling of spin, substrate, and redox equilibria in cytochrome P450. Biochemistry 15:5399–5406

10. Murataliev MB, Feyereisen R (1996) Functional interactions in cytochrome P450BM3. Fatty acid substrate binding alters electron-transfer properties of the flavoprotein domain. Biochemistry 35:15029–15037
11. Daff SN, Chapman SK, Turner KL, Holt RA, Govindaraj S, Poulos TL, Munro AW (1997) Redox control of the catalytic cycle of flavocytochrome P-450 BM3. Biochemistry 36:13816–13823
12. Davydov R, Macdonald IDG, Makris TM, Sligar SG, Hoffman BM (1999) EPR and ENDOR of catalytic intermediates in cryoreduced native and mutant oxy-cytochromes P450cam: mutation-induced changes in the proton delivery system. J Am Chem Soc 121:10654–10655
13. Fujishiro T, Shoji O, Nagano S, Sugimoto H, Shiro Y, Watanabe Y (2011) Crystal structure of H_2O_2-dependent cytochrome $P450_{SP\alpha}$ with its bound fatty acid substrate. J Biol Chem 286:29941–29950
14. Lee DS, Yamada A, Sugimoto H, Matsunaga I, Ogura H, Ichihara K, Adachi S, Park SY, Shiro Y (2003) Substrate recognition and molecular mechanism of fatty acid hydroxylation by cytochrome P450 from *Bacillus subtilis*. Crystallographic, spectroscopic, and mutational studies. J Biol Chem 278:9761–9767
15. Narhi LO, Fulco AJ (1986) Characterization of a catalytically self-sufficient 119,000-Dalton cytochrome P-450 monooxygenase induced by barbiturates in *Bacillus megaterium*. J Biol Chem 261:7160–7169
16. Boddupalli SS, Pramanik BC, Slaughter CA, Estabrook RW, Peterson JA (1992) Fatty acid monooxygenation by $P450_{BM\text{-}3}$: product identification and proposed mechanisms for the sequential hydroxylation reactions. Arch Biochem Biophys 292:20–28
17. Ravichandran KG, Boddupalli SS, Hasemann CA, Peterson JA, Deisenhofer J (1993) Crystal-structure of hemoprotein domain of P450BM-3, a prototype for microsomal P450s. Science 261:731–736
18. Noble MA, Miles CS, Chapman SK, Lysek DA, Mackay AC, Reid GA, Hanzlik RP, Munro AW (1999) Roles of key active-site residues in flavocytochrome P450 BM3. Biochem J 339:371–379
19. Girvan HM, Marshall KR, Lawson RJ, Leys D, Joyce MG, Clarkson J, Smith WE, Cheesman MR, Munro AW (2004) Flavocytochrome P450 BM3 mutant A264E undergoes substrate-dependent formation of a novel heme iron ligand set. J Biol Chem 279:23274–23286
20. Girvan HM, Toogood HS, Littleford RE, Seward HE, Smith WE, Ekanem IS, Leys D, Cheesman MR, Munro AW (2009) Novel haem co-ordination variants of flavocytochrome P450 BM3. Biochem J 417:65–76
21. Li HY, Poulos TL (1997) The structure of the cytochrome P450BM-3 haem domain complexed with the fatty acid substrate, palmitoleic acid. Nat Struct Biol 4:140–146
22. Fasan R, Chen MM, Crook NC, Arnold FH (2007) Engineered alkane-hydroxylating cytochrome P450 (BM3) exhibiting nativelike catalytic properties. Angew Chem Int Ed 46:8414–8418
23. Fasan R, Meharenna YT, Snow CD, Poulos TL, Arnold FH (2008) Evolutionary history of a specialized P450 propane monooxygenase. J Mol Biol 383:1069–1080
24. Chen MMY, Snow CD, Vizcarra CL, Mayo SL, Arnold FH (2012) Comparison of random mutagenesis and semi-rational designed libraries for improved cytochrome P450 BM3-catalyzed hydroxylation of small alkanes. Protein Eng Des Sel 25:171–178
25. Meinhold P, Peters MW, Chen MMY, Takahashi K, Arnold FH (2005) Direct conversion of ethane to ethanol by engineered cytochrome P450BM3. ChemBioChem 6:1765–1768
26. Staudt S, Müller CA, Marienhagen J, Böing C, Buchholz S, Schwaneberg U, Gröger H (2012) Biocatalytic hydroxylation of *n*-butane with in situ cofactor regeneration at low temperature and under normal pressure. Beilstein J Org Chem 8:186–191
27. Whitehouse CJC, Yang W, Yorke JA, Rowlatt BC, Strong AJF, Blanford CF, Bell SG, Bartlam M, Wong LL, Rao ZH (2010) Structural basis for the properties of two single-site proline mutants of CYP102A1 (P450BM3). ChemBioChem 11:2549–2556
28. Glieder A, Farinas ET, Arnold FH (2002) Laboratory evolution of a soluble, self-sufficient, highly active alkane hydroxylase. Nat Biotechnol 20:1135–1139
29. Xu F, Bell SG, Lednik J, Insley A, Rao ZH, Wong LL (2005) The heme monooxygenase cytochrome P450cam can be engineered to oxidize ethane to ethanol. Angew Chem Int Ed 44:4029–4032
30. Bell SG, Stevenson JA, Boyd HD, Campbell S, Riddle AD, Orton EL, Wong LL (2002) Butane and propane oxidation by engineered cytochrome P450cam. Chem Commun 490–491
31. Farinas ET, Alcalde M, Arnold F (2004) Alkene epoxidation catalyzed by cytochrome P450 BM-3 139–3. Tetrahedron 60:525–528
32. Dennig A, Lülsdorf N, Liu H, Schwaneberg U (2013) Regioselective o-hydroxylation of monosubstituted benzenes by P450 BM3. Angew Chem Int Ed 52:8459–8462
33. Whitehouse CJC, Rees NH, Bell SG, Wong LL (2011) Dearomatisation of *o*-xylene by $P450_{BM3}$ (CYP102A1). Chem Eur J 17:6862–6868
34. Bell SG, Orton E, Boyd H, Stevenson JA, Riddle A, Campbell S, Wong LL (2003) Engineering cytochrome P450cam into an alkane hydroxylase. Dalton Trans 11:2133–2140

35. Dennig A, Marienhagen J, Ruff AJ, Guddat L, Schwaneberg U (2012) Directed evolution of P450 BM3 into a *p*-xylene hydroxylase. ChemCatChem 4:771–773
36. Kawakami N, Shoji O, Watanabe Y (2011) Use of perfluorocarboxylic acids to trick cytochrome P450BM3 into initiating the hydroxylation of gaseous alkanes. Angew Chem Int Ed 50:5315–5318
37. Zilly FE, Acevedo JP, Augustyniak W, Deege A, Hausig UW, Reetz MT (2011) Tuning a P450 enzyme for methane oxidation. Angew Chem Int Ed 50:2720–2724
38. Banks RE, Tatlow JC (1986) A guide to modern organofluorine chemistry. J Fluor Chem 33:227–346
39. Kawakami N, Shoji O, Watanabe Y (2013) Direct hydroxylation of primary carbons in small alkanes by wild-type cytochrome P450BM3 containing perfluorocarboxylic acids as decoy molecules. Chem Sci 4:2344–2348
40. Shoji O, Kunimatsu T, Kawakami N, Watanabe Y (2013) Highly selective hydroxylation of benzene to phenol by wild-type cytochrome P450BM3 assisted by decoy molecules. Angew Chem Int Ed 52:6606–6610
41. Zhai PM, Wang LQ, Liu CH, Zhang SC (2005) Deactivation of zeolite catalysts for benzene oxidation to phenol. Chem Eng J 111:1–4
42. Nordblom GD, White RE, Coon MJ (1976) Studies on hydroperoxide-dependent substrate hydroxylation by purified liver microsomal cytochrome P-450. Arch Biochem Biophys 175:524–533
43. Renneberg R, Scheller F, Ruckpaul K, Pirrwitz J, Mohr P (1978) NADPD and H_2O_2-dependent reactions of cytochrome P-450LM compared with peroxidase catalysis. FEBS Lett 96:349–353
44. Hrycay EG, Gustafsson JA, Ingelman–Sundberg M, Ernster L (1975) Sodium periodate, sodium chlorite, organic hydroperoxides, and H_2O_2 as hydroxylating agents in steroid hydroxylation reactions catalyzed by partially purified cytochrome P-450. Biochem Biophys Res Commun 66:209–216
45. Renneberg R, Capdevila J, Chacos N, Estabrook RW, Prough RA (1981) Hydrogen peroxide-supported oxidation of benzo[*a*]pyrene by rat liver microsomal fractions. Biochem Pharmacol 30:843–848
46. Holm KA, Engell RJ, Kupfer D (1985) Regioselectivity of hydroxylation of prostaglandins by liver microsomes supported by NADPH versus H_2O_2 in methylcholanthrene-treated and control rats: formation of novel prostaglandin metabolites. Arch Biochem Biophys 237:477–489
47. Koo LS, Tschirret-Guth RA, Straub WE, Moenne-Loccoz P, Loehr TM, Ortiz de Montellano PR (2000) The active site of the thermophilic CYP119 from *Sulfolobus solfataricus*. J Biol Chem 275:14112–14123
48. Ogura H, Nishida CR, Hoch UR, Perera R, Dawson JH, Ortiz de Montellano PR (2004) EpoK, a cytochrome P450 involved in biosynthesis of the anticancer agents epothilones A and B. Substrate-mediated rescue of a P450 enzyme. Biochemistry 43:14712–14721
49. Anari MR, Josephy PD, Henry T, O'Brien PJ (1997) Hydrogen peroxide supports human and rat cytochrome P450 1A2-catalyzed 2-amino-3-methylimidazo[4,5-*f*]-quinoline bioactivation to mutagenic metabolites: significance of cytochrome P450 peroxygenase. Chem Res Toxicol 10:582–588
50. Bui PH, Hankinson O (2009) Functional characterization of human cytochrome P450 2S1 using a synthetic gene-expressed protein in *Escherichia coli*. Mol Pharmacol 76:1031–1043
51. Niraula NP, Kanth BK, Sohng JK, Oh TJ (2011) Hydrogen peroxide-mediated dealkylation of 7-ethoxycoumarin by cytochrome P450 (CYP107AJ1) from *Streptomyces peucetius* ATCC27952. Enzym Microb Technol 48:181–186
52. Goyal S, Banerjee S, Mazumdar S (2012) Oxygenation of monoenoic fatty acids by CYP175A1, an orphan cytochrome P450 from *Thermus thermophilus* HB27. Biochemistry 51:7880–7890
53. Khan KK, He YA, He YQ, Halpert JR (2002) Site-directed mutagenesis of cytochrome P450eryF: implications for substrate oxidation, cooperativity, and topology of the active site. Chem Res Toxicol 15:843–853
54. Matsumura H, Wakatabi M, Omi S, Ohtaki A, Nakamura N, Yohda M, Ohno H (2008) Modulation of redox potential and alteration in reactivity via the peroxide shunt pathway by mutation of cytochrome P450 around the proximal heme ligand. Biochemistry 47:4834–4842
55. Zhang Z, Li Y, Stearns RA, Ortiz De Montellano PR, Baillie TA, Tang W (2002) Cytochrome P450 3A4-mediated oxidative conversion of a cyano to an amide group in the metabolism of pinacidil. Biochemistry 41:2712–2718
56. Gelb MH, Heimbrook DC, Malkonen P, Sligar SG (1982) Stereochemistry and deuterium isotope effects in camphor hydroxylation by the cytochrome P450cam monooxygenase system. Biochemistry 21:370–377
57. Poulos TL, Finzel BC, Howard AJ (1987) High-resolution crystal structure of cytochrome P450cam. J Mol Biol 195:687–700
58. Yano JK, Wester MR, Schoch GA, Griffin KJ, Stout CD, Johnson EF (2004) The structure of human microsomal cytochrome P450 3A4 determined by X-ray crystallography to 2.05 Å resolution. J Biol Chem 279:38091–38094
59. Gajhede M, Schuller DJ, Henriksen A, Smith AT, Poulos TL (1997) Crystal structure of horseradish peroxidase C at 2.15 Å resolution. Nat Struct Biol 4:1032–1038
60. Sundaramoorthy M, Terner J, Poulos TL (1995) The crystal structure of chloroperoxidase: a heme

peroxidase-cytochrome P450 functional hybrid. Structure 3:1367–1377
61. Sundaramoorthy M, Terner J, Poulos TL (1998) Stereochemistry of the chloroperoxidase active site: crystallographic and molecular-modeling studies. Chem Biol 5:461–473
62. Piontek K, Strittmatter E, Ullrich R, Grobe G, Pecyna MJ, Kluge M, Scheibner K, Hofrichter M, Plattner DA (2013) Structural basis of substrate conversion in a new aromatic peroxygenase: cytochrome P450 functionality with benefits. J Biol Chem 288:34767–34776
63. Wang XS, Peter S, Ullrich R, Hofrichter M, Groves JT (2013) Driving force for oxygen-atom transfer by heme-thiolate enzymes. Angew Chem Int Ed 52:9238–9241
64. Wang XS, Peter S, Kinne M, Hofrichter M, Groves JT (2012) Detection and kinetic characterization of a highly reactive heme-thiolate peroxygenase compound I. J Am Chem Soc 134:12897–12900
65. Goodin DB, Mcree DE (1993) The Asp-His-iron triad of cytochrome *c* peroxidase controls the reduction potential, electronic structure, and coupling of the tryptophan free-radical to the heme. Biochemistry 32:3313–3324
66. Yamada Y, Fujiwara T, Sato T, Igarashi N, Tanaka N (2002) The 2.0 Å crystal structure of catalase-peroxidase from *Haloarcula marismortui*. Nat Struct Biol 9:691–695
67. Ko TP, Day J, Malkin AJ, McPherson A (1999) Structure of orthorhombic crystals of beef liver catalase. Acta Crystallogr D 55:1383–1394
68. Watanabe Y, Ueno T (2003) Introduction of P450, peroxidase, and catalase activities into myoglobin by site-directed mutagenesis: diverse reactivities of compound I. Bull Chem Soc Jpn 76:1309–1322
69. Ozaki SI, Roach MP, Matsui T, Watanabe Y (2001) Investigations of the roles of the distal heme environment and the proximal heme iron ligand in peroxide activation by heme enzymes via molecular engineering of myoglobin. Acc Chem Res 34:818–825
70. Watanabe Y, Nakajima H, Ueno T (2007) Reactivities of oxo and peroxo intermediates studied by hemoprotein mutants. Acc Chem Res 40:554–562
71. Li QS, Ogawa J, Schmid RD, Shimizu S (2005) Indole hydroxylation by bacterial cytochrome P450 BM-3 and modulation of activity by cumene hydroperoxide. Biosci Biotechnol Biochem 69:293–300
72. Vidal-Limón A, Aguila S, Ayala M, Batista CV, Vazquez-Duhalt R (2013) Peroxidase activity stabilization of cytochrome P450BM3 by rational analysis of intramolecular electron transfer. J Inorg Biochem 122:18–26
73. Joo H, Lin ZL, Arnold FH (1999) Laboratory evolution of peroxide-mediated cytochrome P450 hydroxylation. Nature 399:670–673
74. Kumar S, Chen CS, Waxman DJ, Halpert JR (2005) Directed evolution of mammalian cytochrome P450 2B1: mutations outside of the active site enhance the metabolism of several substrates, including the anticancer prodrugs cyclophosphamide and ifosfamide. J Biol Chem 280:19569–19575
75. Kumar S, Liu H, Halpert JR (2006) Engineering of cytochrome P450 3A4 for enhanced peroxide-mediated substrate oxidation using directed evolution and site-directed mutagenesis. Drug Metab Dispos 34:1958–1965
76. Li QS, Ogawa J, Shimizu S (2001) Critical role of the residue size at position 87 in H_2O_2-dependent substrate hydroxylation activity and H_2O_2 inactivation of cytochrome P450BM-3. Biochem Biophys Res Commun 280:1258–1261
77. Schwaneberg U, Schmidt-Dannert C, Schmitt J, Schmid RD (1999) A continuous spectrophotometric assay for P450 BM-3, a fatty acid hydroxylating enzyme, and its mutant F87A. Anal Biochem 269:359–366
78. Cirino PC, Arnold FH (2002) Regioselectivity and activity of cytochrome P450 BM-3 and mutant F87A in reactions driven by hydrogen peroxide. Adv Synth Catal 344:932–937
79. Cirino PC, Arnold FH (2003) A self-sufficient peroxide-driven hydroxylation biocatalyst. Angew Chem Int Ed 42:3299–3301
80. Sanchez-Sanchez L, Roman R, Vazquez-Duhalt R (2012) Pesticide transformation by a variant of CYPBM3 with improved peroxygenase activity. Pestic Biochem Physiol 102:169–174
81. Matsunaga I, Sumimoto T, Ueda A, Kusunose E, Ichihara K (2000) Fatty acid-specific, regiospecific, and stereospecific hydroxylation by cytochrome P450 (CYP152B1) from *Sphingomonas paucimobilis*: substrate structure required for α-hydroxylation. Lipids 35:365–371
82. Imai Y, Matsunaga I, Kusunose E, Ichihara K (2000) Unique heme environment at the putative distal region of hydrogen peroxide-dependent fatty acid α-hydroxylase from *Sphingomonas paucimobilis* (peroxygenase $P450_{SP\alpha}$). J Biochem 128:189–194
83. Matsunaga I, Yamada M, Kusunose E, Miki T, Ichihara K (1998) Further characterization of hydrogen peroxide-dependent fatty acid α-hydroxylase from *Sphingomonas paucimobilis*. J Biochem 124:105–110
84. Matsunaga I, Sumimoto T, Kusunose E, Ichihara K (1998) Phytanic acid α-hydroxylation by bacterial cytochrome P450. Lipids 33:1213–1216
85. Matsunaga I, Yokotani N, Gotoh O, Kusunose E, Yamada M, Ichihara K (1997) Molecular cloning and expression of fatty acid α-hydroxylase from *Sphingomonas paucimobilis*. J Biol Chem 272:23592–23596
86. Matsunaga I, Yamada M, Kusunose E, Nishiuchi Y, Yano I, Ichihara K (1996) Direct involvement of hydrogen peroxide in bacterial α-hydroxylation of fatty acid. FEBS Lett 386:252–254

87. Lee DS, Yamada A, Matsunaga I, Ichihara K, Adachi SI, Park SY, Shiro Y (2002) Crystallization and preliminary X-ray diffraction analysis of fatty-acid hydroxylase cytochrome $P450_{BS\beta}$ from *Bacillus subtilis*. Acta Crystallogr D 58:687–689
88. Matsunaga I, Shiro Y (2004) Peroxide-utilizing biocatalysts: structural and functional diversity of heme-containing enzymes. Curr Opin Chem Biol 8:127–132
89. Matsunaga I, Sumimoto T, Ayata M, Ogura H (2002) Functional modulation of a peroxygenase cytochrome P450: novel insight into the mechanisms of peroxygenase and peroxidase enzymes. FEBS Lett 528:90–94
90. Matsunaga I, Ueda A, Fujiwara N, Sumimoto T, Ichihara K (1999) Characterization of the ybdT gene product of *Bacillus subtilis*: novel fatty acid β-hydroxylating cytochrome P450. Lipids 34:841–846
91. Matsunaga I, Ueda A, Sumimoto T, Ichihara K, Ayata M, Ogura H (2001) Site-directed mutagenesis of the putative distal helix of peroxygenase cytochrome P450. Arch Biochem Biophys 394:45–53
92. Matsunaga I, Yamada A, Lee DS, Obayashi E, Fujiwara N, Kobayashi K, Ogura H, Shiro Y (2002) Enzymatic reaction of hydrogen peroxide-dependent peroxygenase cytochrome P450s: kinetic deuterium isotope effects and analyses by resonance Raman spectroscopy. Biochemistry 41:1886–1892
93. Girhard M, Schuster S, Dietrich M, Durre P, Urlacher VB (2007) Cytochrome P450 monooxygenase from *Clostridium acetobutylicum*: a new α-fatty acid hydroxylase. Biochem Biophys Res Commun 362:114–119
94. Rude MA, Baron TS, Brubaker S, Alibhai M, Del Cardayre SB, Schirmer A (2011) Terminal olefin (1-alkene) biosynthesis by a novel P450 fatty acid decarboxylase from *Jeotgalicoccus* species. Appl Environ Microbiol 77:1718–1727
95. Matsunaga I, Kusunose E, Yano I, Ichihara K (1994) Separation and partial characterization of soluble fatty acid α-hydroxylase from *Sphingomonas paucimobilus*. Biochem Biophys Res Commun 201:1554–1560
96. Belcher J, McLean KJ, Matthews S, Woodward LS, Fisher K, Rigby SE, Nelson DR, Potts D, Baynham MT, Parker DA, Leys D, Munro AW (2014) Structure and biochemical properties of the alkene producing cytochrome P450 $OleT_{JE}$ (CYP152L1) from the *Jeotgalicoccus* sp. 8456 bacterium. J Biol Chem 289:6535–6550
97. Shoji O, Fujishiro T, Nakajima H, Kim M, Nagano S, Shiro Y, Watanabe Y (2007) Hydrogen peroxide dependent monooxygenations by tricking the substrate recognition of cytochrome $P450_{BS\beta}$. Angew Chem Int Ed 46:3656–3659
98. Fujishiro T, Shoji O, Watanabe Y (2010) Non-covalent modification of the active site of cytochrome P450 for inverting the stereoselectivity of monooxygenation. Tetrahedron Lett 52:395–397
99. Shoji O, Wiese C, Fujishiro T, Shirataki C, Wünsch B, Watanabe Y (2010) Aromatic C–H bond hydroxylation by P450 peroxygenases: a facile colorimetric assay for monooxygenation activities of enzymes based on Russig's blue formation. J Biol Inorg Chem 15:1109–1115
100. Fujishiro T, Shoji O, Kawakami N, Watanabe T, Sugimoto H, Shiro Y, Watanabe Y (2012) Chiral-substrate-assisted stereoselective epoxidation catalyzed by H_2O_2-dependent cytochrome $P450_{SP\alpha}$. Chem Asian J 7:2286–2293
101. Shoji O, Fujishiro T, Nagano S, Tanaka S, Hirose T, Shiro Y, Watanabe Y (2010) Understanding substrate misrecognition of hydrogen peroxide dependent cytochrome P450 from *Bacillus subtilis*. J Biol Inorg Chem 15:1331–1339

Use of Chemical Auxiliaries to Control P450 Enzymes for Predictable Oxidations at Unactivated C-H Bonds of Substrates

8

Karine Auclair and Vanja Polic

Abstract

Cytochrome P450 enzymes (P450s) have the ability to oxidize unactivated C-H bonds of substrates with remarkable regio- and stereoselectivity. Comparable selectivity for chemical oxidizing agents is typically difficult to achieve. Hence, there is an interest in exploiting P450s as potential biocatalysts. Despite their impressive attributes, the current use of P450s as biocatalysts is limited. While bacterial P450 enzymes typically show higher activity, they tend to be highly selective for one or a few substrates. On the other hand, mammalian P450s, especially the drug-metabolizing enzymes, display astonishing substrate promiscuity. However, product prediction continues to be challenging. This review discusses the use of small molecules for controlling P450 substrate specificity and product selectivity. The focus will be on two approaches in the area: (1) the use of decoy molecules, and (2) the application of substrate engineering to control oxidation by the enzyme.

Keywords

Biocatalysis • Decoy molecule • Fatty acid • Perfluorinated carboxylic acid • Carbolide • Chemical auxiliary • Theobromine • Molecularly imprinted polymer

8.1 Introduction

Cytochrome P450 enzymes, herein referred to as P450s, form a large family of heme-dependent monooxygenases. Ubiquitous in nature, they are found in bacteria, fungi, mammals and other organisms. In bacteria, P450s are involved typically in the biosynthesis of secondary metabolites, while in mammals they are responsible largely for the metabolism of xenobiotics (foreign compounds) such as drugs, carcinogens and environmental pollutants, as well as the biosynthesis and metabolism of endogenous or naturally-occurring bioactive compounds such as eicosanoids (eicosanoic acids, leukotrienes, prostaglandins), fatty acids, steroids and vitamins

K. Auclair (✉) • V. Polic
Department of Chemistry, McGill University,
801 Sherbrooke Street West, Montreal, Quebec H3A0B8, Canada
e-mail: karine.auclair@mcgill.ca

E.G. Hrycay and S.M. Bandiera (eds.), *Monooxygenase, Peroxidase and Peroxygenase Properties and Mechanisms of Cytochrome P450*, Advances in Experimental Medicine and Biology 851,
DOI 10.1007/978-3-319-16009-2_8,

[1]. One attribute that allows P450s to perform such diverse roles is their ability to oxidize unactivated C-H bonds of substrates with remarkable regio- and stereoselectivity [1–4]. It is this feature that has caught the attention of synthetic chemists. In this respect, there has been huge interest in potentially harnessing the biocatalytic prowess of P450s. Unfortunately, these enzymes are not without limitations. They often suffer from low stability in vitro, require expensive cofactors and demonstrate difficult product predictability.

8.2 Biocatalysis

Biocatalysis is defined as the use of biomolecules, whether naturally-occurring or engineered, for the purpose of facilitating a desired reaction. It can provide many of the benefits of chemical catalysis, adhering at the same time to many principles of green chemistry [5–8]. Typically, an ideal catalyst should be able to predictably catalyze chemo-, regio- and stereoselective reactions with high efficiency while maintaining a large substrate scope. Today, further economical and environmental constraints must also be considered [5, 8]. Although the field of biocatalysis has greatly expanded over the last few decades, there are still many obstacles to overcome, which vary on an enzyme-to-enzyme basis. Many enzymes typically suffer from low stability, narrow substrate scope, need for expensive cofactors and high production costs. Significant efforts by the research community to overcome these problems have led, over the past century, to major advances, yet more challenges lie ahead. The recent popularity of "green chemistry" is definitely positioning biocatalysis as a top research priority.

8.2.1 Chemical Catalysts for Oxidations at Unactivated C-H Bonds of Substrates

A variety of chemical catalysts are known for substrate oxidation at unactivated C-H bonds [9–11]. These catalysts can be separated into two general classes, either containing or lacking an active-site metal. Metal-containing catalysts typically have complex coordinating ligands responsible for tuning the reduction-oxidation (redox) potential of the metal, as well as imparting a degree of chemo-, regio- and stereoselectivity. For example, the metalloporphyrin catalysts were modeled after the prosthetic heme-iron group found in the active site of P450s [10]. In contrast to P450s, metalloporphyrin-based chemical catalysts can contain different metals such as cobalt, copper, iron, manganese, ruthenium and vanadium [10]. Although selectivity can be achieved with these catalysts, it is typically poor and unpredictable and must be determined empirically. In many cases, numerous oxidation products are obtained ranging from aliphatic and aromatic alcohols to carbonyls [12].

Reasonable predictability and good yields for oxidations at unactivated C-H bonds have been observed with non-porphyrin, metal-based catalysts [9–11]. The regio- and stereoselectivity of this second class of metal-catalysts is generally dependent on the steric and electronic properties of the substrate. Furthermore, the catalysts tend to favor oxidation at tertiary C-H bonds over secondary C-H bonds, or else over-oxidation and loss of chirality is the norm [13, 14]. Although excellent diastereoselectivity has been achieved with some chemical catalysts, to date there have been no reports of *enantioselective* metal catalysts performing hydroxylations of unactivated CH_2 groups. Chiral oxaziridine catalysts developed by Du Bois and coworkers [15] can favor enantioselective hydroxylations. However, these catalysts show a high preference for tertiary C-H bonds over less energetically favorable secondary C-H bonds. Moreover, they do not distinguish between multiple sterically and electronically similar tertiary C-H bonds. Finally, metal-free catalysts such as dioxiranes and oxaziridines have also been shown to oxidize unactivated C-H bonds and favoring reactions at tertiary carbons [15, 16].

Although there are a variety of chemical catalysts available for oxidation of unactivated C-H bonds, there are still important needs in this area. Overall, chemical catalysts suffer from poor selectivity and product predictability, and they often yield over-oxidation products. Furthermore, their selectivity is based on electronic and steric properties of the substrates, which often limits them to oxidation of more electronically favorable tertiary bonds and prevents discrimination between C-H bonds similar in these aspects.

8.2.2 P450 Biocatalysts

In parallel, interest has been invested into the design and use of P450s as biocatalysts in hopes of providing complementary reactivity to the available chemical toolbox. As mentioned previously, P450s are well known for their exceptional ability to selectively oxidize a large variety of substrates [2, 3, 17, 18]. This provides them with two attributes that are highly sought in a catalyst: scope and selectivity. As an example, human P450s can be credited for being involved in the metabolism of approximately 75 % of all clinical drugs [17]. Their impressive ability to selectively discern unactivated methylene bonds among many electronically and sterically similar groups is obvious when looking at the structure of drug metabolites [2, 3, 17, 18]. In addition, P450s have the ability to oxidize C-H bonds of varying strength, from methine to methyl groups, often without over-oxidation to the ketone. Despite these impressive attributes, a number of drawbacks still limit the applications of P450s as biocatalysts [19]. Poor stability and low turnover rates, as well as the need for expensive cofactors, are among their greater limitations. To overcome some of these issues, industrial processes, for example, typically use fermentation reactors [20].

In research settings, P450s of microbial origin are usually preferred because they are more easily expressed in high yields and in a soluble form, and have higher activities and coupling efficiencies. In contrast, mammalian P450s have a wide scope of reactivity and convoluted chemo-, regio- and stereoselectivity. Their ability to accept various substrates provides them with a clear advantage in a research context or with non-natural substrates.

8.2.3 The Promiscuity Paradox

While mammalian P450s may be attractive biocatalysts because of their high substrate promiscuity, predicting the structures of their products remains challenging. This promiscuity paradox is an important obstacle hindering the use of mammalian P450s as biocatalysts. Although a large diversity of substrates is recognized, particularly by xenobiotic-metabolizing P450s, many of these substrates are suboptimal and/or have multiple binding modes. As a result, increased uncoupling rates are often observed due to uncontrolled water access to the active site [21]. One strategy to overcome these issues involves protein engineering, which has been used to improve enzymatic activity, change substrate specificity and alter product distributions [22, 23]. As reviewed below, other research groups have attempted to control P450 selectivity and improve product predictability by using small molecules. There are generally two strategies by which this has been achieved. The first approach uses decoy molecules to control the site of oxidation by blocking a part of the active site. In contrast, the second strategy applies substrate engineering to control oxidation by the enzymes.

8.3 Decoy Molecules

Decoy molecules are inert dummy molecules that are structurally similar to the natural substrate of a specific enzyme [24]. The term was first coined by Watanabe and coworkers in 2007 [24] and was considerably explored with respect to $P450_{BS\beta}$. The general mode of action for decoy molecules involves partial filling up of the active-site space and consequent decrease of the degree of translational freedom allowed for the substrate while in the enzyme. In this respect, the decoys have the ability to increase the selectivity of the enzymatic reaction by prompting the substrate to bind in a consistent orientation. Careful consideration must be used when selecting a decoy molecule. It must either be inert and not be transformed by the enzyme, or not be in competition with the desired substrate. To this end, two types of decoy molecules have been explored for their P450-directing capabilities: (1) short-chain fatty acids, and (2) perfluorinated fatty acids.

8.3.1 Short-Chain Fatty Acids

In 2007, experiments reported by the Watanabe group enabled the recognition of unnatural substrates by P450s without the use of mutagenesis [24]. At the time, this strategy was unique in taking advantage of short-chain fatty acids to act as ligands of $P450_{BS\beta}$, mimicking part of the substrate and leaving a small portion of the substrate binding pocket available near the heme iron (Fig. 8.1). This allowed the transformation of the unnatural substrates guaiacol, styrene, ethylbenzene and thioanisole (Fig. 8.2) [24, 25]. In these studies, $P450_{BS\beta}$ was chosen as the biocatalytic system because it is soluble, does not require a redox partner and can efficiently use hydrogen peroxide instead of oxygen and the expensive NADPH cofactor. Isolated from *Bacillus subtilis*, $P450_{BS\beta}$ (also known as P450 152A1) normally catalyzes the hydroxylation of long-chain fatty acid substrates such as myristic acid to form α- and β-hydroxymyristic acid [26].

As part of their initial studies, the Watanabe group evaluated a series of decoy molecules for facilitating guaiacol oxidation by $P450_{BS\beta}$ [24]. Thus, carboxylic acids varying in length from two to ten carbons were added to the reaction mixtures and the products were quantified. Many of the acids allowed for guaiacol oxidation to occur, with the highest reaction rate being observed in the presence of heptanoic acid (3,750 turnovers min^{-1}). This rate is tenfold higher than that reported for the natural substrate, myristic acid, suggesting activation of the enzyme by the decoy molecule. This enhancement was attributed to the decoy molecules keeping the catalytic cycle "on", which long-chain fatty acid substrates normally do when they bind. The catalytic cycle is turned "off"

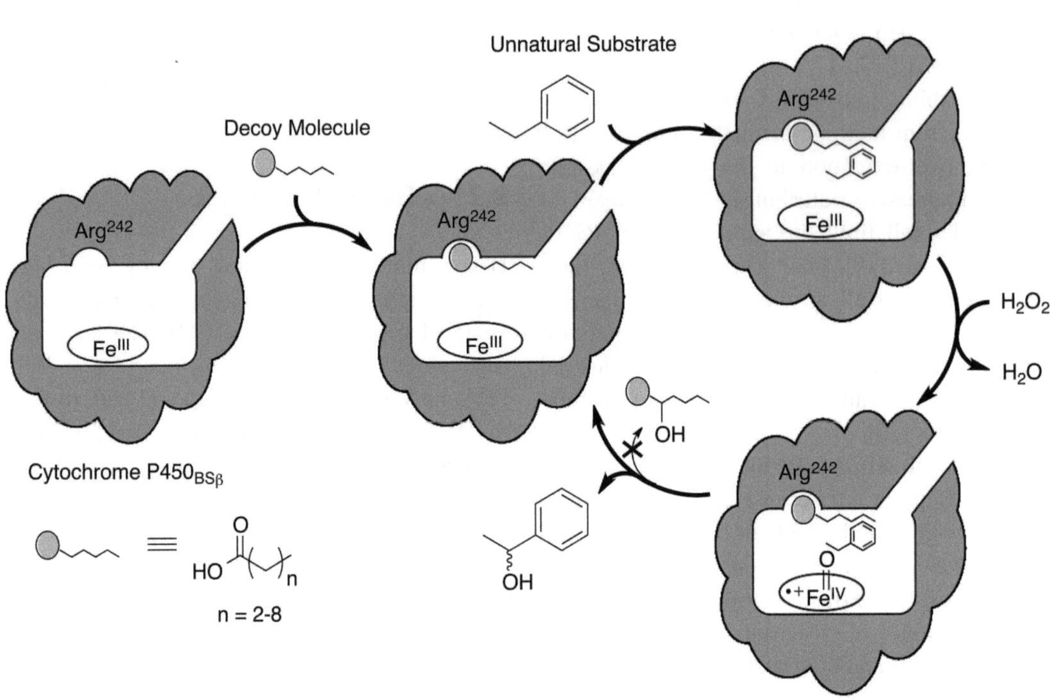

Fig. 8.1 Suggested catalytic cycle for the oxidation of unnatural substrates by $P450_{BS\beta}$ in the presence of decoy molecules [24]

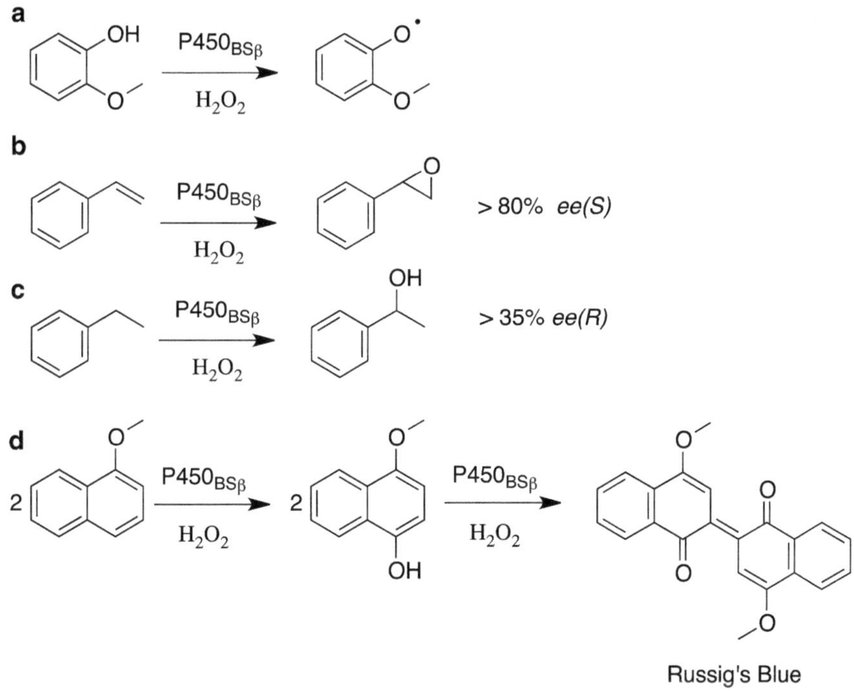

Fig. 8.2 Substrate scope for the oxidation of unnatural substrates by $P450_{BS\beta}$ in the presence of the decoy molecules, heptanoic or hexanoic acid. Substrates shown are **a** guaiacol, **b** styrene, **c** ethylbenzene and **d** 1-methoxynaphthalene, along with the major product of their oxidation by $P450_{BS\beta}$ [24, 28]

when the fatty acid dissociates [24]. As expected, in the presence of myristic acid, no oxidation of guaiacol occurred.

The concept was further investigated with styrene epoxidation by $P450_{BS\beta}$ in the presence of carboxylic acids with 4–8 carbon chain lengths [24]. Styrene epoxidation occurred in the presence of any decoy molecule examined, with both regioselectivity and enantioselectivity (*ee*) greater than 80 %. The maximum reaction rate of 334 turnovers min^{-1} was observed in the presence of hexanoic acid. Furthermore, under the same conditions, hydroxylation of ethylbenzene afforded the corresponding 2° alcohol [24]. Oxidation of the methylene carbon occurred with the highest enantioselectivity (68 % *ee* (*R*)) and highest reaction rate (28 turnovers min^{-1}) in the presence of heptanoic acid. Taken together, these results demonstrate that $P450_{BS\beta}$ can catalyze the oxidation of non-fatty acid substrates when tricked into misrecognizing short-chain fatty acids instead of the natural myristic acid substrate. To confirm the proposed mechanism, the authors obtained crystal structures of $P450_{BS\beta}$ complexed with heptanoic acid [27]. Using crystal structure analysis, they showed that fatty acids with carbon chain lengths less than ten carbon atoms are too short to extend into the hydrophobic substrate access channel [27]. This results in loose fixation of the hydrophobic tail, preventing hydroxylation of the decoy molecules and allowing for small molecules to enter and act as substrates. Comparison of the decoy molecule-bound crystal structure to the substrate-free crystal structure also demonstrated that binding of the decoy molecules did not induce large structural changes.

Following their initial success with decoy molecules, Watanabe and coworkers further investigated the scope of molecules that can be oxidized using $P450_{BS\beta}$. In 2010, they reported that $P450_{BS\beta}$ can catalyze the conversion of 1-methoxynaphthalene to 4,4′-dimethoxy-[2,2′]-binaphthalenylidene-1,1′-dione, also called Russig's blue dye (Fig. 8.2d) [28]. This study demonstrated that hydroxylation of an aromatic C-H bond (turnover rate of 112 min^{-1}) can be catalyzed by this P450, adding to the scope of the system. This was the first example of an efficient aromatic C-H bond oxidation by a hydrogen peroxide-dependent P450 enzyme. The expected *O*-demethylation product was not observed, yet the reaction was not as regioselective as with styrene or ethylbenzene. Additionally, by comparing the conversion rate of the intermediate 4-methoxy-1-naphthol to the conversion rate of 1-methoxynaphthalene, it was determined that the rate-limiting step is the first step, i.e., hydroxylation of the aromatic C-H bond of 1-methoxynaphthalene. With this in mind, the authors investigated the use of 1-methoxynaphthalene as the substrate in a colorimetric assay for the indirect observation of $P450_{BS\beta}$ catalytic activity. To test this idea, they compared the effects of carboxylic acids with lengths of 4–10 carbons on the turnover of 1-methoxynaphthalene by $P450_{BS\beta}$. Since the effect of the decoy molecules was similar for 1-methoxynaphthalene, ethylbenzene and styrene, they used the formation of Russig's blue to estimate activities of a series of $P450_{BS\beta}$ mutants generated in search of higher aromatic C-H bond oxidation activity [28]. Their colorimetric assay allowed them to set up high-throughput screening for the identification of many other types of decoy molecules. From this screen, they found that wild-type $P450_{BS\beta}$ in combination with pentanoic acid shows the best activity for aromatic C-H bond oxidation. Mutant V170F in combination with octanoic acid had comparable activity. Unexpectedly, they identified mono-*tert*-butyl succinate as an additional useful potential decoy molecule. However, its binding mode remains unclear.

Because the observed catalytic activity and selectivity depended on the length of the carboxylic acid carbon chain used as a decoy molecule, the authors hypothesized that structural changes to the decoy molecule might affect the stereochemical outcome of the reaction. Thus, sulfoxidation of thioanisole was examined in the presence of various decoy molecules (Table 8.1) [25]. The majority of the decoy molecules examined allowed for the sulfoxidation of thioanisole to occur without over-oxidation to the sulfone product, generally favoring *R*-stereoselectivity with modest *ee*. The highest turnover (514 min^{-1}) and stereoselectivity (29 % *ee*) was obtained once again with heptanoic acid. After testing a large scope of potential decoy molecules, it was concluded that branched carboxylic acids both with and without a chiral center could serve as decoy molecules.

Interestingly, *p*-substituted methylphenylacetic acid effectively inverted the stereoselectivity to favor the *S* enantiomer (11 % ee) while *m*- and *o*-substitutions maintained *R*-stereoselectivity. Molecular modeling simulations with phenylacetic acid (PAA), *p*-methylphenylacetic acid (*p*-MPAA), *o*-methylphenylacetic acid (*o*-MPAA) and *m*-methylphenylacetic acid (*m*-MPAA) showed that for *p*-MPAA, the methyl group is placed over the heme porphyrin ring resulting in a steric clash with thioanisole, while for *m*- and *o*-MPAA the methyl group is positioned away from the heme plane avoiding steric hindrance. After modeling thioanisole in the active site with minimized steric repulsion, it was clear that with PAA, *m*-MPAA and *o*-MPAA, the *pro-R* side of the sulfur lone pairs is closer to the heme iron, allowing for oxidation. For *p*-MPAA however, due to steric repulsion of the methyl group with the thioanisole phenyl ring, the *pro-S* lone pair is closer to the heme (Fig. 8.3). Although the *S*-enantioselectivity observed in the presence of the decoy molecules was not particularly high, this was the first example of a non-mutagenic method for inversion of stereochemistry in a P450.

8.3.2 Perfluorinated Fatty Acids

After it was established that $P450_{BS\beta}$ can be tricked into oxidizing unnatural substrates using decoy molecules, Watanabe and others directed their focus on another fatty acid-metabolizing P450. In 2011, groups led by Reetz and by Watanabe reported independently on the oxidation capabilities of $P450_{BM3}$ (P450 102A1) toward short alkanes with the aid of decoy molecules [29, 30]. Isolated from *Bacillus megaterium*, $P450_{BM3}$ has the highest catalytic

Table 8.1 Sulfoxidation of thioanisole catalyzed by $P450_{BS\beta}$ in the presence of carboxylic acid decoy molecules [25]

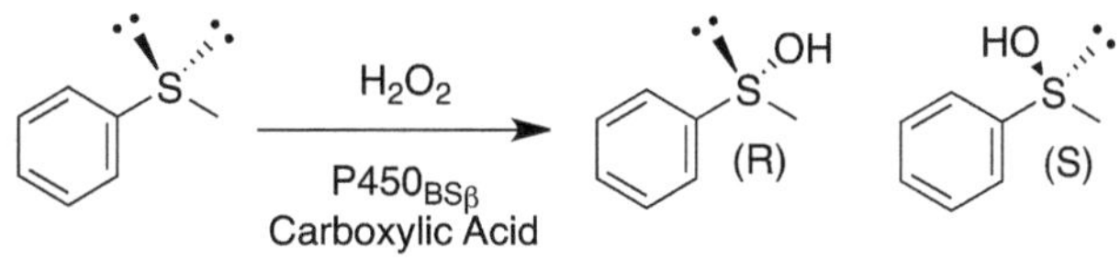

Decoy molecule	Turnover rate[a] (min^{-1})	% ee (*R* or *S*)
None	24 ± 2	4 ± 0 (*R*)
$CH_3(CH_2)_6COOH$	251 ± 20	21 ± 4 (*R*)
$CH_3(CH_2)_5COOH$	514 ± 72	29 ± 1 (*R*)
$CH_3(CH_2)_4COOH$	390 ± 53	21 ± 2 (*R*)
$CH_3(CH_2)_3COOH$	274 ± 48	16 ± 1 (*R*)
$CH_3(CH_2)_2COOH$	261 ± 75	18 ± 0 (*R*)
CH_3CH_2COOH	270 ± 21	20 ± 1 (*R*)
(S)-(+)-2-Methylbutyric acid	214 ± 27	19 ± 1 (*R*)
(±)-2-Methylbutyric acid	146 ± 22	13 ± 1 (*R*)
$PhCH_2CH_2CH_2COOH$	370 ± 39	19 ± 0 (*R*)
$PhCH_2CH_2COOH$	462 ± 37	21 ± 0 (*R*)
$PhCH_2COOH$ (PAA)	71 ± 10	14 ± 0 (*R*)
o-Me-$PhCH_2COOH$ (*o*-MPAA)	34 ± 9	9 ± 1 (*R*)
m-Me-$PhCH_2COOH$ (*m*-MPAA)	32 ± 8	12 ± 2 (*R*)
p-Me-$PhCH_2COOH$ (*p*-MPAA)	89 ± 14	11 ± 2 (*S*)
p-Isopropyl-$PhCH_2COOH$	55 ± 2	10 ± 1 (*S*)
PhCOOH	41 ± 5	12 ± 1 (*R*)

[a]The unit for catalytic activity is nmol product min^{-1} nmol $P450^{-1}$

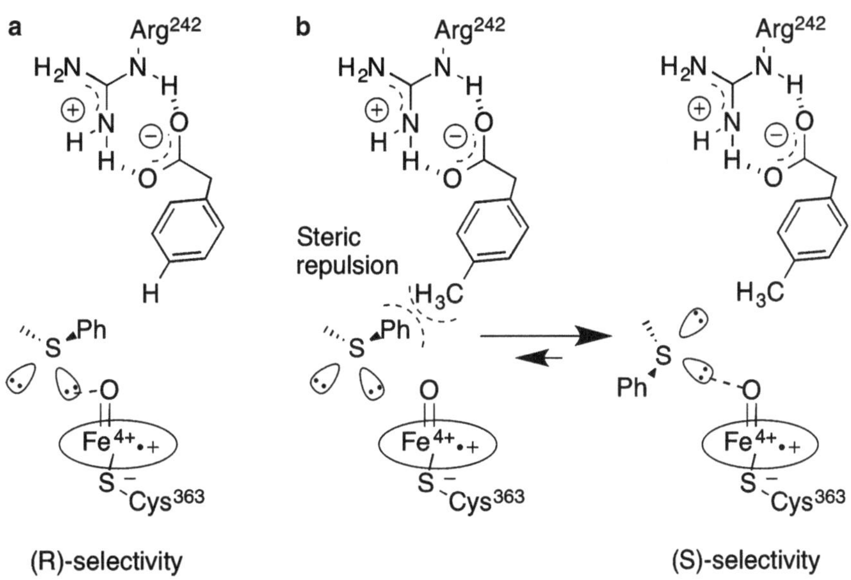

Fig. 8.3 Suggested transition states for stereochemical inversion during the sulfoxidation of thioanisole by $P450_{BS\beta}$ with **a** PAA and **b** *p*-MPAA [25]

rate reported for a P450, >15,000 turnovers min^{-1}, observed for the oxidation of arachidonic acid [30]. Its natural substrates include myristic and palmitic acid, which consist of 14 and 16 carbon chains, respectively [29, 30]. The high catalytic rate of $P450_{BM3}$ is attributed largely to the fact that its heme and reductase domains are located on the same polypeptide chain, resulting in efficient electron coupling [29, 30]. As with many other P450s, $P450_{BM3}$ is known for its large active site that is able to incorporate two molecules at a time [31–33]. Although this is ideal for achieving the cooperative effects desired with decoy molecules, problems can arise with small substrates due to a large number of binding orientations [29]. $P450_{BM3}$ catalyzes the hydroxylation of long-chain fatty acids at the $\omega-1$, $\omega-2$, or $\omega-3$ positions. Here again, molecules that are not fatty acids are typically not accepted as substrates because the carboxylate group is required for catalytic activity. Although many mutagenic studies have been performed with $P450_{BM3}$ with respect to broadening the substrate scope [34–44], this study demonstrated that perfluorocarboxylic acids (PFCs) can be used as decoy molecules to expand the substrate scope of $P450_{BM3}$ without resorting to mutagenesis. PFCs are a great alternative to fatty acids because the greater bond energy of C-F (116 kcal mol^{-1}) compared to that of C-H (95–99 kcal mol^{-1}) renders them inert toward oxidation [30]. Moreover, PFCs are considerably larger in size than fatty acids. For example, the CF_3 moiety is comparable in size to an ethyl group [45]. Overall, PFCs take up more space in the active site while maintaining the hydrophobic character of standard fatty acids [29].

The ability of PFCs to bind to $P450_{BM3}$ was first confirmed by monitoring changes in the UV/Vis spectrum [29, 30] and the consumption of NADPH in the presence of PFCs that were 8–14 carbons in length [30]. Reetz and coworkers conducted further spectral studies to characterize the effect of the PFCs in the active site. They concluded that PFCs not only reduce the space in the active site as do the original fatty acid molecules, but also activate the enzyme by displacing the distal water molecule and converting the heme iron from low-spin to high-spin [29]. Next, the directing effects of the PFCs on hydroxylation were examined with a scope of small alkane substrates ranging from methane to octane and including some constitutional isomers of hexane (Table 8.2) [29, 30]. Although the majority of the PFCs allowed for hydroxylation of the small molecules to occur, the length of the PFC chain had a significant effect on the rate of oxidation. Smaller molecules typically showed better turnover with longer PFCs. Moreover, in the absence of PFCs, no transformation occurred [29, 30]. In all cases, PFCs had a positive effect on the enzymatic turnover of the alkanes but little effect on regioselectivity and stereoselectivity [29]. This was attributed to the lack of functional groups on the alkanes, preventing consistent orientation. Interestingly, Watanabe demonstrated that $P450_{BM3}$ can successfully hydroxylate propane, butane and cyclohexane to 2-propanol, 2-butanol and cyclohexanol respectively in the presence of a PFC but did not oxidize methane or ethane [30]. As an example, propane was converted at a rate of 67 turnovers min^{-1} in the presence of PFC-C10, while cyclohexane was converted at a rate of 110 turnovers min^{-1} in the presence of PFC-C9 [30]. In contrast, Reetz and coworkers observed that *n*-butane had higher turnover with PFC-C7 (3,632 min^{-1}) while propane had the highest turnover with PFC-C11 (1,021 min^{-1}). The oxidation of methane to methanol is notoriously more challenging. Initial studies with $P450_{BM3}$ and PFCs by Reetz and coworkers suggested that methane might be converted to methanol [29]. However, further analysis of the data revealed that methane was not oxidized to any appreciable extent [46]. Although the turnover of propane reported by both groups is still lower than the turnover reported for $P450_{BM3}$ mutants, the PFC method is much simpler to apply than generation of mutant enzymes [29]. Finally, it is important to note that substitution of the hydrogen atoms for fluorines in the decoy molecule was crucial for the reactions to occur. Indeed, $P450_{BM3}$ did not convert propane to propanol in the presence of decanoic acid [30].

Table 8.2 Conversion rates of gaseous molecules by $P450_{BM3}$ in the presence of PFCs [29, 30]

Substrate	PFC[a]	Rate[b] (min^{-1})	Regioselectivity	Refs.
Methane	C10	2,472		[29]
Propane	C11	1,021		[29]
	C10	67		[30]
Butane	C7	3,632		[29]
	C10	113		[30]
Hexane	C11	525	2-/3-Hexanol = 77:23	[29]
Octane	C9	1,184	2-/3-/4-Octanol = 10:42:48	[29]
Cyclohexane	C9	110		[30]
3-Methylpentane	C9	476	2-/3-Hydroxy = 88:12	[29]
3,3-Dimethylbutane	C9	891	Only 2-Hydroxy	[29]
2,3-Dimethylbutane	C11	3,241	Only 2-Hydroxy	[29]

[a]Optimal perfluorinated carboxylic acid reported
[b]Observed turnover rates

Table 8.3 Hydroxylation of aromatic rings by $P450_{BM3}$ [47]

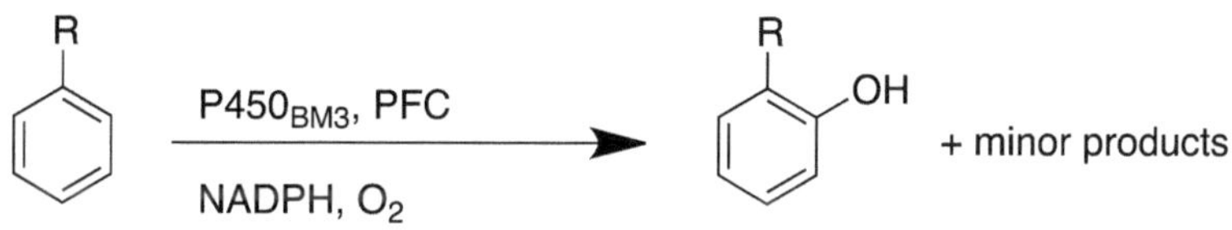

Substrate	PFC[a]	Rate (min^{-1})[b]	Major product
Benzene	C9	120 ± 9	Phenol
Toluene	C9	220 ± 7	*Ortho*
Anisole	C9	260 ± 4	*Ortho*
Chlorobenzene	C9	120 ± 4	*Ortho*
Nitrobenzene	C9	0.9 ± 0.05	*Ortho*
Acetophenone	C9	2.9 ± 0.1	*Ortho*

[a]The chain length of PFC that elicited the greatest effect
[b]Rates are given in units of min^{-1} per P450. Uncertainty is given as the standard deviation from at least three measurements

More recently, Watanabe, Shoji and colleagues published another study that utilizes $P450_{BM3}$ for the hydroxylation of aromatic rings such as those of benzene, toluene, anisole, chlorobenzene, nitrobenzene and acetophenone (Table 8.3) [47]. The reaction was successfully catalyzed using $P450_{BM3}$ and various PFCs with exclusive formation of phenols. The transformation of benzene proceeds most efficiently in the presence of PFC-C9, with a turnover rate of 120 min^{-1}. For toluene hydroxylation, the turnover rate was higher (220 min^{-1}) but diminished when electron withdrawing groups decorated the ring. Although PFCs had a positive effect on the rate, they did not affect regioselectivity. For all aromatic substrates examined, the major hydroxylation site was *ortho* to the existing substituent, even when the *meta* position is electronically favored. This suggests that the regioselectivity is controlled mainly by the residues in the active site and not by the PFC or the substrate [47].

In summary, the use of decoy molecules provides a simple and quick way for increasing the substrate scope of these biocatalysts without the need for protein mutagenesis. They can also be used to influence the regio- and stereoselectivity of P450-catalyzed reactions. This is attributed to the fact that $P450_{BS\beta}$ and $P450_{BM3}$

can accept multiple substrates into their relatively large active sites leading to positive cooperative effects [31–33]. Such methodology leaves room for further investigation of decoy molecule design and their effect on reaction selectivity and turnover. Finally, this elegant strategy remains to be expanded to P450s other than $P450_{BS\beta}$ and $P450_{BM3}$, and especially to drug-metabolizing P450s, many of which show cooperative behavior.

8.4 Substrate Engineering

While many research groups in the P450 area have focused on engineering the proteins for biocatalytic applications, other groups have instead explored the idea of engineering the substrates. There are many instances in the literature suggesting that the presence of specific functional groups can improve substrate acceptance by biocatalysts [48–53]. This concept is often referred to as substrate engineering and is typically applied to non-natural substrates that are unlikely to show high affinity for the enzyme and for which products may be difficult to predict [54, 55]. To apply the substrate-engineering concept, the desired substrate is first covalently linked to another molecule, termed the chemical auxiliary, which has specific recognition motifs (Fig. 8.4). The auxiliary serves to improve recognition by the enzyme and encourage productive binding, as well as control the binding orientation for selective transformation [54, 56]. Once the transformation is achieved, the auxiliary is cleaved off and the desired product isolated.

Weber and coworkers investigated the use of chemical auxiliaries that they termed docking/protecting (d/p) groups for improving the scope of hydroxylation by various monooxygenases [54, 55]. They evaluated various microorganisms as catalysts for the hydroxylation of non-natural substrates (Fig. 8.5). They first demonstrated that the selected substrates were not hydroxylated, or did not undergo side reactions, in the absence of the docking/protecting groups. However, covalent addition of a chemical auxiliary to the substrates

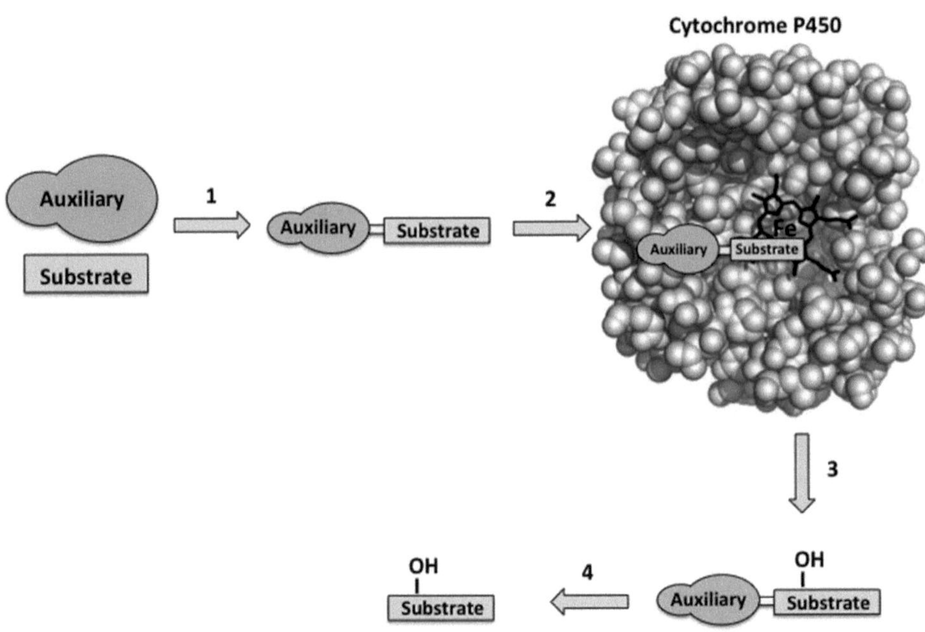

Fig. 8.4 A simplified depiction of the use of chemical auxiliaries to aid in transformation of unnatural substrates by P450s. Typical steps include *1* chemical coupling of the auxiliary to the substrate of interest, *2* incubation of the auxiliary-substrate with the enzyme, *3* oxidation of the non-natural substrate by the P450 enzyme and product release, and *4* chemical cleavage of the auxiliary

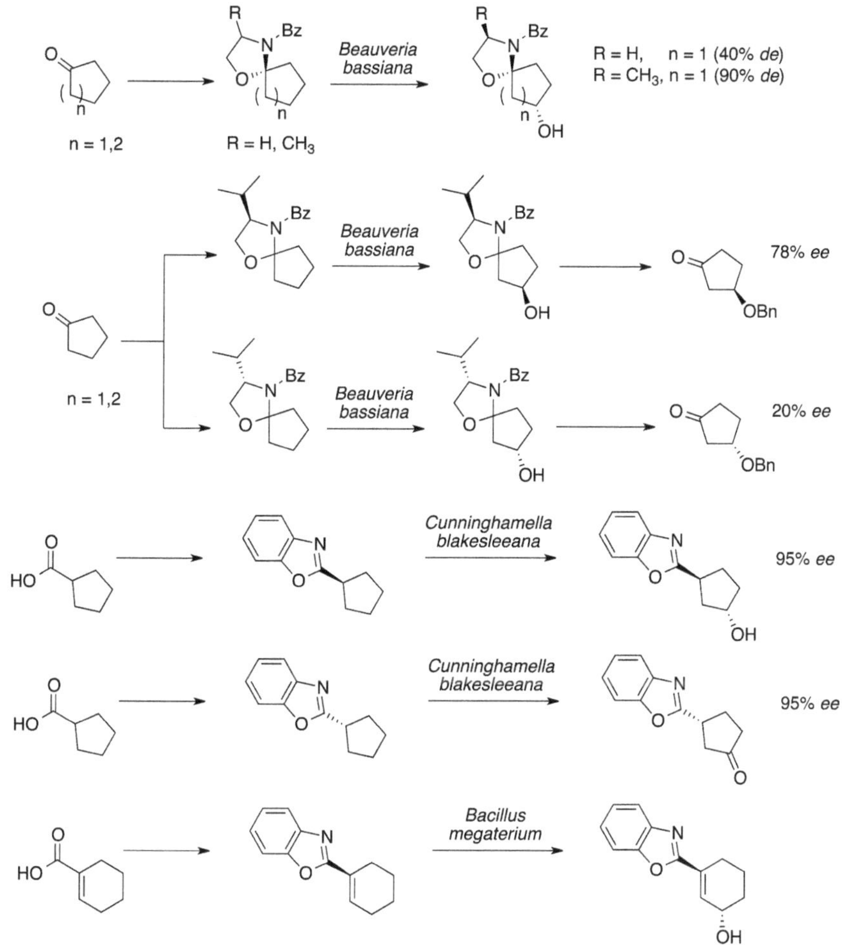

Fig. 8.5 Selected examples from the substrate scope explored by Griengl and coworkers using substrate engineering for biohydroxylation of cyclic alkanes [54, 55]

resulted in more selective hydroxylation by the microorganisms. Reduction, oxidation and other side reactions became minor [54]. A subsequent study demonstrated that regio- and enantioselectivity can be significantly altered by varying the nature of the auxiliary [55]. Although these studies were successful at narrowing down the number of products and favoring hydroxylation, whole-cell biocatalysts frequently yield multiple products. Purified enzymes offer a more defined system, often allowing structure-activity relationships to be drawn. Enzymes remain a popular choice for biocatalysts [57–59].

8.4.1 Bacterial P450s

Building on the work of Weber and coworkers, Munzer et al. [60] applied the 2-aminophenol chemical auxiliary for the expansion of the substrate scope of $P450_{BM3}$. Their work combined the use of substrate engineering and protein mutagenesis. Thus, 2-aminophenol was used as a chemical auxiliary for establishing cyclopentanoic acid as a substrate of various $P450_{BM3}$ mutants. Cyclopentanoic acid was not turned over in the absence of the auxiliary, and only two out of the four possible diastereoisomers were produced when the carboxylate was cyclized into a benzoxazole.

In a separate study by the same group, protecting-group manipulations at the anomeric carbon of globally-protected monosaccharides were used to control the regioselectivity of $P450_{BM3}$-catalyzed deprotection elsewhere on the ring (Fig. 8.6) [38]. For example, their work showed that benzylation at the anomeric position turns globally-methylated galactose and glucose into $P450_{BM3}$ substrates that are regioselectively deprotected at O-4 by $P450_{BM3}$ F87A or F87I mutants.

Sherman and coworkers investigated the potential of another P450 enzyme for hydroxylation of engineered substrates [56]. Their studies focused on the oxidation of carbocyclic rings covalently linked to a desosamine glycoside auxiliary. A mutant of the macrolide P450 monooxygenase PikC was used [56]. PikC is involved in the hydroxylation of the 12- and 14-membered ring macrolides, YC-17 and narbomycin, respectively, in the pikromycin biosynthetic pathway of *Streptomyces venezuelae* [61, 62]. Although PikC is not naturally self-sufficient, the authors created a self-sufficient fusion mutant, $PikC_{D50N}$-RhFRED, which is ~13 times more active than the wild-type enzyme, making it more attractive as a biocatalyst [63]. This self-sufficient P450 was used in the transformation of engineered substrates. Based on the X-ray crystal structure of PikC complexed with different macrolides, the authors reasoned that the desosamine moiety might position the macrolide substrates in the active site. This reasoning stemmed from the observation that the majority of hydrogen bonds and electrostatic interactions within the active site are formed by the desosamine group of the natural substrates, while the macrolactones are held in place merely through more general hydrophobic interactions [64, 65]. In support of their hypothesis, they first synthesized a series of carbocyclic rings ranging in size from 12 to 15 carbons and linked to a desosamine glycoside, altogether referred to as carbolides. Following reaction with PikC, the conversion yields ranged from 35 % to 65 %, decreasing as the ring size increased (Table 8.4). In each case, however, multiple hydroxylation products were detected. The authors next obtained X-ray crystal structures for two of their carbolides complexed with $PikC_{D50N}$. Based on structure analysis, synthetic standards of the most likely major products were synthesized and the LC-MS retention times were compared to those of the reaction products. This allowed the authors to assign the regioselectivity for the majority of the reaction products, showing that hydroxylation by PikC occurs primarily at sites most distant from the auxiliary. Interestingly, no over-oxidation was observed and no hydroxylation was detected on the auxiliary. Following this, other types of desosamine derivatives were synthesized, both cyclic and linear, and reacted with $PikC_{D50N}$-RhFRED. Transformation of smaller and more rigid cyclic derivatives was not observed, possibly due to their inability to reach the heme-iron oxidized species. The linear substrates produced multiple monohydroxylation products in modest

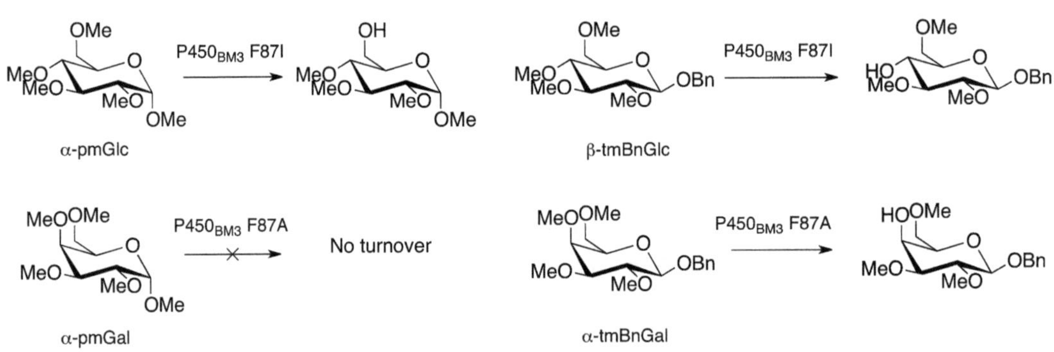

Fig. 8.6 Deprotection of globally-methylated glucose (glc) and galactose (gal) by $P450_{BM3}$ mutants [38]

Table 8.4 The scope and turnover of unnatural substrates by $PikC_{D50N}$-RhFRED [56]

Substrate	K_d (μM)	Yield (%)	No. of products
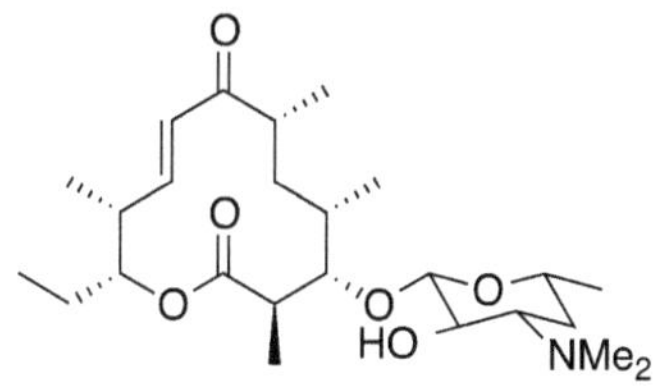	19	>99	2
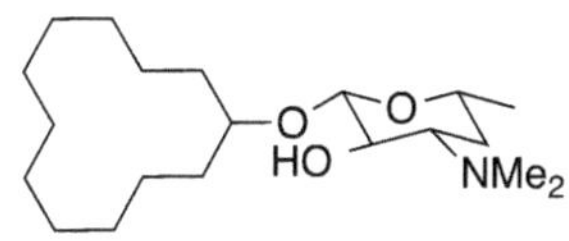	309	47	7
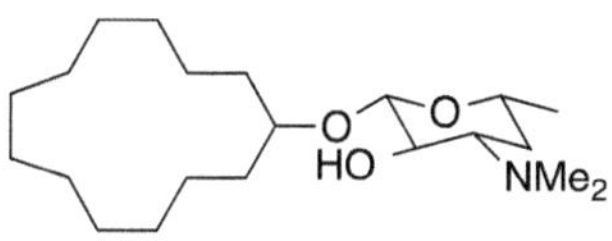	218	65	6
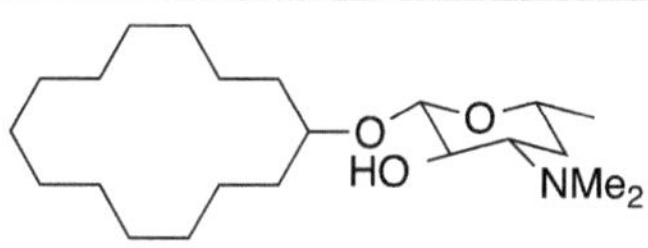	289	63	6
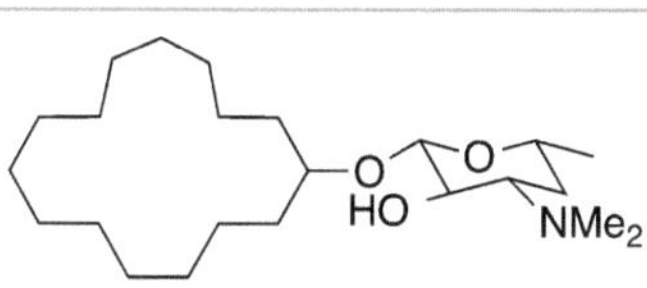	243	35	9
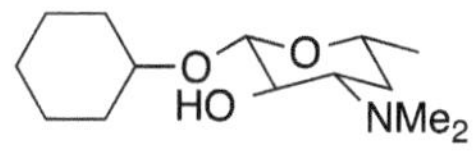	NB[a]	0	
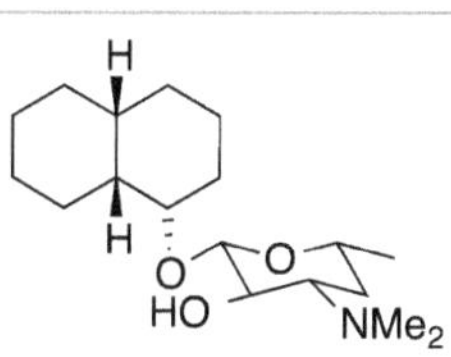	NB	0	0
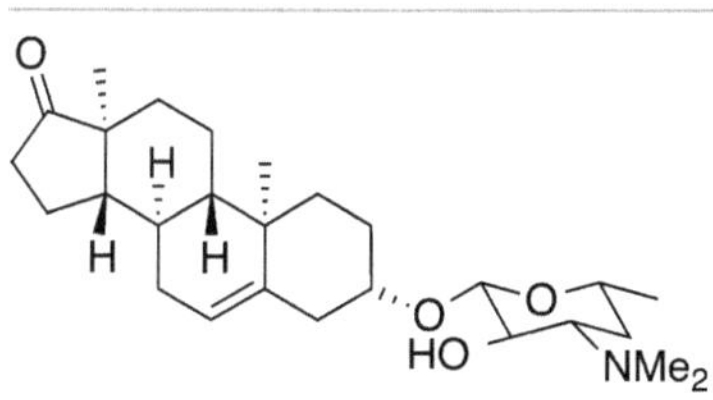	NB	<1	0

(continued)

Table 8.4 (continued)

Substrate	K_d (μM)	Yield (%)	No. of products
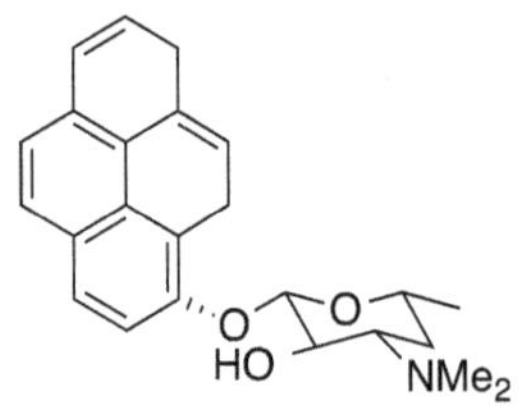	>5,000	4	1
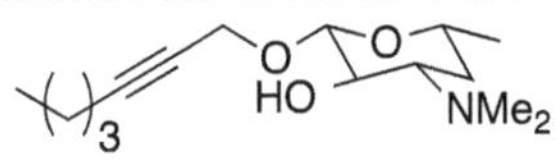	NB	0	0
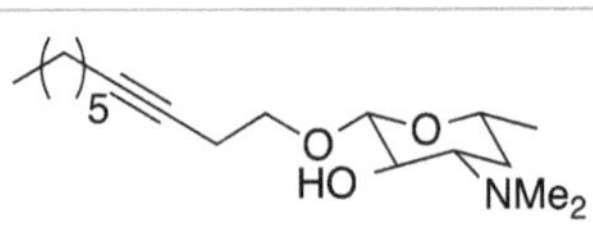	2,900	8	5
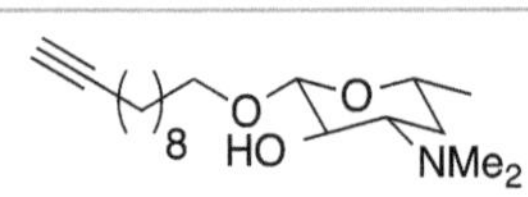	2,300	14	7

[a]*NB* no binding

yields, and an aromatic derivative was the only substrate to yield a single oxidation product albeit in fairly low yield. The crystal structures suggest that the inherent flexibility of the carbocycles can lead to multiple binding modes. Moreover, it was established that the desosamine moiety adopts two different binding modes that vary significantly, resulting in the flipping or rotating of the whole substrate, which can further explain the observed poor regio- and stereoselectivities.

8.4.2 Mammalian P450s

Most mammalian P450 enzymes show unequalled substrate promiscuity among P450s [2, 20, 66, 67]. This is an attractive quality for a useful and versatile biocatalyst, and can eliminate the need for protein mutagenesis to expand the substrate scope. While the studies reviewed above aimed at expanding the substrate scope of bacterial enzymes, more recent studies use substrate engineering to control the regio- and stereoselectivity of mammalian P450s in a predictable manner. This was successfully achieved with two human P450 enzymes, P450 3A4 and P450 2E1 [68, 69].

Human P450 3A4 is found mainly in the liver and accounts for the metabolism of about 50 % of all xenobiotics [70]. Highly promiscuous by nature due to its large and flexible active site, substrate prediction with this enzyme has been very challenging. The rules that govern substrate binding and oxidation selectivity have still not been fully defined. Auclair and coworkers have successfully applied substrate engineering using theobromine as the chemical auxiliary for controlling and predicting the regio- and stereoselectivity of P450 3A4-catalyzed oxidations [68].

Drawing inspiration from the known substrate lisofylline, it was envisaged that the theobromine moiety could serve as a chemical auxiliary for

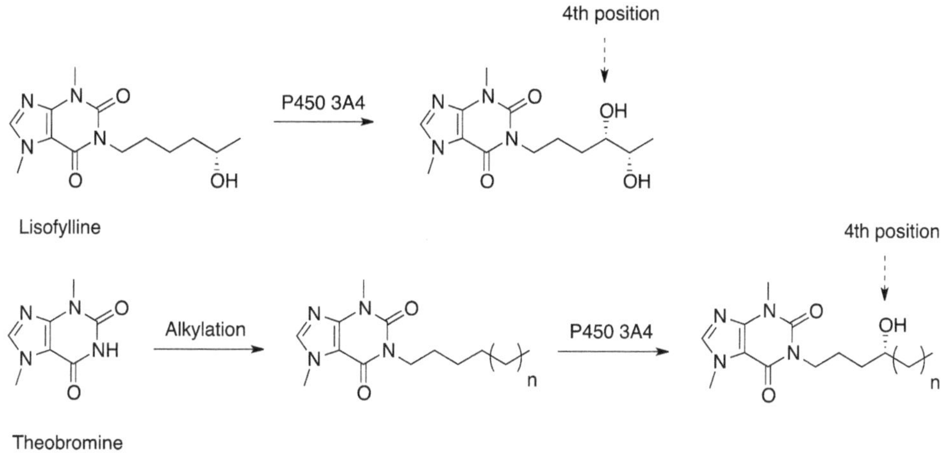

Fig. 8.7 Oxidation of lisofylline and theobromine derivatives by P450 3A4

the transformation of other substrates (Fig. 8.7). Lisofylline is metabolized in part by P450 3A4 via hydroxylation at the fourth carbon from the theobromine moiety [71]. It was therefore hypothesized that substrates covalently bound to theobromine would be hydroxylated at the fourth atom from theobromine.

To investigate the directing potential of theobromine, various small alkanes and alkenes were covalently linked to theobromine and the resulting substrate-auxiliaries were reacted with P450 3A4 in the presence of either CHP alone or NADPH, O_2 and the NADPH-P450 oxidoreductase (CPR) redox partner. CHP was used to bypass the consumption of the expensive NADPH cofactor and CPR, as previously reported [66]. Product distribution and stereoselectivity for the P450 3A4-catalyzed hydroxylation of theobromine-substrates were confirmed to be indistinguishable for both the NADPH/O_2/CPR and the CHP systems. CHP also has the advantage of improving the P450 3A4 initial catalytic rate by 30 % [66]. The conversion yields of theobromine derivatives and product structures were determined using LC-MS and MS fragmentation patterns, by comparison with authentic synthetic standards. As predicted, theobromine successfully directed hydroxylation at the fourth atom away from the auxiliary for all the compounds that had a fourth methylene carbon, or directed epoxidation where a double bond was present at the fourth carbon (Fig. 8.8). Additionally, there was a definite preference for *pro-R* facial selectivity during oxygen atom insertion.

Transformation proceeded with excellent regioselectivity for most substrates while enantiomeric ratios reached 75:25 for hydroxylations and >99:1 for epoxide formation. In all cases, no over-oxidation to the ketone and no oxidation of the auxiliary were observed. After optimization of the conditions, the conversion yields reached 70 %. This was the first example where products were reliably predicted for complex substrates reacted with a promiscuous P450. Further investigations examined functional group tolerance and showed that with the theobromine auxiliary, P450 3A4 was able to selectively oxidize methylene C-H bonds at the fourth position in the presence of nearby 3° C-H bonds, double bonds and heteroatoms (Fig. 8.8). Again, product distribution and stereoselectivity for theobromine-substrates were indistinguishable for both the NADPH/O_2/CPR and CHP systems. Although crystal structures of the theobromine-containing substrates complexed with P450 3A4 have yet to be obtained, in silico docking studies support the idea that theobromine may bind P450 3A4 in a mode conducive to reaction at the fourth carbon from the auxiliary. In contrast to previous studies, this work reports a chemical auxiliary that functions to position the substrates in the active

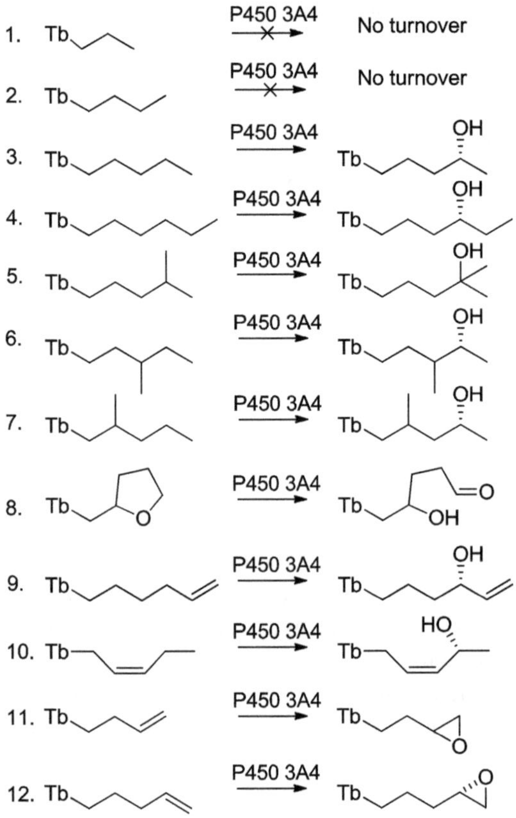

Fig. 8.8 The scope of functionalized theobromine substrates and their conversion products by P450 3A4. Oxidations were performed in the presence of either CHP or NADPH, O_2 and CPR; *Tb* theobromine [68]

site such that, like a ruler, the site of oxidation can be predicted based on the distance from the auxiliary.

While the use of theobromine as a chemical auxiliary shows promise for controlling the site of oxidation by P450 3A4, this system offers a number of other advantages. Theobromine is not only achiral, inexpensive and easily functionalized in high yields, but also contains a chromophore enabling easier purification and isolation of the starting material and products. Additionally, theobromine can penetrate cells allowing for the scale up of this method for fermentation purposes. With this in mind, Larsen et al. investigated yet another potential role for the theobromine auxiliary: facilitating product recovery from whole-cell reaction mixtures [72]. Due to the complex nature of the medium in whole-cell reactions, isolation of the desired product typically accounts for ~80 % of the cost. Extraction of the product often requires large volumes of organic solvents, followed by extensive purification, which is not only costly but also time consuming and generates large volumes of waste. To circumvent these issues, the authors turned to the aid of molecularly imprinted polymers (MIPs) designed to specifically recognize theobromine for isolation of starting materials and products of P450 3A4 reactions [72]. Imprinted polymers are generated by polymerization of monomers and cross-linkers in the presence of a template [73]. The MIPs are expected to have high affinity for molecules resembling the template and can serve to extract such molecules (referred to as target molecules) from a complex mixture [74–81]. The interactions between the polymer and the target molecules are non-covalent interactions, and the affinity is largely based on the complementarity of the size, shape, electrostatics and hydrogen bonding of the MIP with the target molecule. Thus, MIPs were synthesized using various theobromine derivatives as templates. In each case, the MIPs performed very well, with >85 % recovery of the theobromine derivatives and high reusability. The efficacy was maintained whether the target molecules were extracted from dilute, concentrated, aqueous or complex (LB broth) mixtures, and resulted in >90 % purity of the isolated product. In contrast, when the molecule used as the template to generate the MIP was structurally different from theobromine, as with isopropyl-β-D-1-thiogalactopyranoside (IPTG), the recovery of theobromine derivatives dropped below detection. In combination, these results show that chemical auxiliaries such as theobromine can find use not only in controlling the selectivity of P450 transformations but also in facilitating product recovery. This strategy should apply to other systems.

In a similar approach, the Auclair group used type II ligands as chemical auxiliaries to target unnatural substrates to the P450 2E1 active site [69]. Although type II ligands are typically viewed as P450 inhibitors due to stabilization of the low spin-state via coordination of the heme-iron to an aromatic nitrogen [69, 82, 83], studies have shown that in general, type II binding ligands have a higher affinity for the active site than similar but nitrogen-lacking substrates [84–86]. Moreover, a number of reports have demonstrated that P450s such as P450 3A4 can metabolize type II ligands, and in some cases to a greater extent than related type I ligand analogues [84, 87–91]. In hopes of uncovering complementary selectivity to that observed with the theobromine auxiliary, the authors investigated a variety of type II ligands as chemical auxiliaries for transformation of substrates by P450 2E1 [69]. After screening various pyridines, quinolone and imidazole derivatives, they focused further studies on nicotinate as the chemical auxiliary. Nicotinate looked promising because not only is methyl nicotinate a known type II ligand of P450 2E1 [85], but the naturally occurring P450 2E1 substrate, nicotine-derived nitrosamine ketone (NKK), contains a nicotinate group, positioned a few carbons away from the site of oxidation [92]. In addition, nicotinate is inexpensive, has a chromophore and is easily functionalized, reversibly, at the carboxylic acid group. A series of nicotinate esters were thus generated and reacted with P450 2E1 in the presence of the natural NADPH cofactor and CPR, or CHP [66]. From the results obtained, the authors confirmed the preference of P450 2E1 to oxidize aliphatic CH_2 or alkenyl CH groups furthest from the auxiliary ($\omega-1$). This selectivity has previously been observed for the hydroxylation of fatty acids such as lauric, myristic, palmitic and stearic acids by P450 2E1 [93]. In the absence of the auxiliary, however, none of the substrates tested by the authors were significantly transformed by the enzyme. These results show that type II binding can be used to favor and direct biocatalysis.

Although the use of chemical auxiliaries in biocatalysis requires two extra steps, this is often offset by a number of advantages, the key one being that the desired products might otherwise require several synthetic steps to be prepared. Moreover, the chemical auxiliaries can be used to facilitate detection and recovery of the products.

8.5 Conclusion

Overall, substrate engineering is largely unexplored in the area of P450 biocatalysis. However, the successful stories summarized here demonstrate the high potential of this strategy.

Acknowledgments Writing of this chapter and research in the area of P450 enzymes in the Auclair group have been funded by the National Science and Engineering Research Council of Canada (NSERC), the Center in Green Chemistry and Catalysis, Merck Frosst Canada Ltée, Boehringer Ingelheim Canada and AstraZeneca Canada. V.P. was supported by scholarships from the Dr. Richard H. Tomlinson Foundation, Walter C. Sumner Foundation and the Centre in Green Chemistry and Catalysis.

References

1. Guengerich FP (2005) Human cytochrome P450 enzymes. In: Ortiz de Montellano PR (ed) Cytochrome P450: structure, mechanism, and biochemistry, 3rd edn. Kluwer Academic/Plenum Publishers, New York, pp 377–530
2. Bernhardt R (2006) Cytochromes P450 as versatile biocatalysts. J Biotechnol 124:128–145
3. Gillam EMJ (2005) Exploring the potential of xenobiotic-metabolising enzymes as biocatalysts: evolving designer catalysts from polyfunctional cytochrome P450 enzymes. Clin Exp Pharmacol Physiol 32:147–152
4. Guengerich FP (2001) Common and uncommon cytochrome P450 reactions related to metabolism and chemical toxicity. Chem Res Toxicol 14:611–650
5. Manley JB, Anastas PT, Cue BW (2008) Frontiers in Green Chemistry: meeting the grand challenges for sustainability in R&D and manufacturing. J Clean Prod 16:743–750
6. Sheldon RA (2008) E factors, green chemistry and catalysis: an odyssey. Chem Commun 29:3352–3365
7. Sheldon RA (2007) The E factor: fifteen years on. Green Chem 9:1273–1283
8. Anastas P, Eghbali N (2010) Green chemistry: principles and practice. Chem Soc Rev 39:301–312
9. Godula K, Sames D (2006) C-H bond functionalization in complex organic synthesis. Science 312:67–72
10. Che C, Lo VK, Zhou C, Huang J (2011) Selective functionalisation of saturated C-H bonds with metalloporphyrin catalysts. Chem Soc Rev 40:1950–1975
11. Yamaguchi J, Yamaguchi AD, Itami K (2012) C-H bond functionalization: emerging synthetic tools for natural products and pharmaceuticals. Angew Chem Int Ed 51:8960–9009

12. Costas M (2011) Selective C-H oxidation catalyzed by metalloporphyrins. Coord Chem Rev 255:2912–2932
13. Chen MS, White MC (2007) A predictably selective aliphatic C-H oxidation reaction for complex molecule synthesis. Science 318:783–787
14. Chen MS, White MC (2010) Combined effects on selectivity in Fe-catalyzed methylene oxidation. Science 327:566–571
15. Brodsky BH, Du Bois J (2005) Oxaziridine-mediated catalytic hydroxylation of unactivated 3 ° C−H bonds using hydrogen peroxide. J Am Chem Soc 127:15391–15393
16. Rella MR, Williard PG (2007) Oxidation of peptides by methyl(trifluoromethyl)dioxirane: the protecting group matters. J Org Chem 72:525–531
17. Guengerich FP (2008) Cytochrome P450 and chemical toxicology. Chem Res Toxicol 21:70–83
18. Guengerich FP (1999) Cytochrome P-450 3A4: regulation and role in drug metabolism. Annu Rev Pharmacol Toxicol 39:1–17
19. Munro AW, Girvan HM, Mason AE, Dunford AJ, McLean KJ (2013) What makes a P450 tick? Trends Biochem Sci 38:140–150
20. Julsing MK, Cornelissen S, Bühler B, Schmid A (2008) Heme-iron oxygenases: powerful industrial biocatalysts? Curr Opin Chem Biol 12:177–186
21. Loida PJ, Sligar SG (1993) Engineering cytochrome P-450cam to increase the stereospecificity and coupling of aliphatic hydroxylation. Protein Eng 6:207–212
22. Gillam EMJ (2007) Extending the capabilities of nature's most versatile catalysts: directed evolution of mammalian xenobiotic-metabolizing P450s. Arch Biochem Biophys 464:176–186
23. Kumar S (2011) Engineering cytochrome P450 biocatalysts for biotechnology, medicine, and bioremediation. Expert Opin Drug Metab Toxicol 6:115–131
24. Shoji O, Fujishiro T, Nakajima H, Kim M, Nagano S, Shiro Y, Watanabe Y (2007) Hydrogen peroxide dependent monooxygenations by tricking the substrate recognition of cytochrome $P450_{BS\beta}$. Angew Chem Int Ed 46:3656–3659
25. Fujishiro T, Shoji O, Watanabe Y (2011) Non-covalent modification of the active site of cytochrome P450 for inverting the stereoselectivity of monooxygenation. Tetrahedron Lett 52:395–397
26. Lee DS, Yamada A, Sugimoto H, Matsunaga I, Ogura H, Ichihara K, Adachi SI, Park SY, Shiro Y (2003) Substrate recognition and molecular mechanism of fatty acid hydroxylation by cytochrome P450 from *Bacillus subtilis*: crystallographic, spectroscopic, and mutational studies. J Biol Chem 278:9761–9767
27. Shoji O, Fujishiro T, Nagano S, Tanaka S, Hirose T, Shiro Y, Watanabe Y (2010) Understanding substrate misrecognition of hydrogen peroxide dependent cytochrome P450 from *Bacillus subtilis*. J Biol Inorg Chem 15:1331–1339
28. Shoji O, Wiese C, Fujishiro T, Shirataki C, Wünsch B, Watanabe Y (2010) Aromatic C-H bond hydroxylation by P450 peroxygenases: a facile colorimetric assay for monooxygenation activities of enzymes based on Russig's blue formation. J Biol Inorg Chem 15:1109–1115
29. Zilly FE, Acevedo JP, Augustyniak W, Deege A, Häusig UW, Reetz MT (2011) Tuning a P450 enzyme for methane oxidation. Angew Chem 123:2772–2776
30. Kawakami N, Shoji O, Watanabe Y (2011) Use of perfluorocarboxylic acids to trick cytochrome P450BM3 into initiating the hydroxylation of gaseous alkanes. Angew Chem Int Ed 50:5315–5318
31. Ueng YF, Kuwabara T, Chun YJ, Guengerich FP (1997) Cooperativity in oxidations catalyzed by cytochrome P450 3A4. Biochemistry 36:370–381
32. Denisov IG, Baas BJ, Grinkova YV, Sligar SG (2007) Cooperativity in cytochrome P450 3A4: linkages in substrate binding, spin state, uncoupling, and product formation. J Biol Chem 282:7066–7076
33. Atkins WM (2005) Non-Michaelis-Menten kinetics in cytochrome P450-catalyzed reactions. Annu Rev Pharmacol Toxicol 45:291–310
34. Jung ST, Lauchli R, Arnold FH (2011) Cytochrome P450: taming a wild type enzyme. Curr Opin Biotechnol 22:809–817
35. Lewis JC, Coelho PS, Arnold FH (2011) Enzymatic functionalization of carbon-hydrogen bonds. Chem Soc Rev 40:2003–2021
36. Glieder A, Farinas ET, Arnold FH (2002) Laboratory evolution of a soluble, self-sufficient, highly active alkane hydroxylase. Nat Biotechnol 20:1135–1139
37. Lewis JC, Arnold FH (2009) Catalysts on demand: selective oxidations by laboratory-evolved cytochrome P450 BM3. CHIMIA Int J Chem 63:309–312
38. Lewis JC, Bastian S, Bennett CS, Fu Y, Mitsuda Y, Chen MM, Greenberg WA, Wong C, Arnold FH (2009) Chemoenzymatic elaboration of monosaccharides using engineered cytochrome P450 BM3 demethylases. Proc Natl Acad Sci U S A 106:16550–16555
39. Fasan R, Meharenna YT, Snow CD, Poulos TL, Arnold FH (2008) Evolutionary history of a specialized P450 propane monooxygenase. J Mol Biol 383:1069–1080
40. Meinhold P, Peters MW, Hartwick A, Hernandez AR, Arnold FH (2006) Engineering cytochrome P450 BM3 for terminal alkane hydroxylation. Adv Syn Catal 348:763–772
41. Lussenburg BMA, Babel LC, Vermeulen NPE, Commandeur JNM (2005) Evaluation of alkoxyresorufins as fluorescent substrates for cytochrome P450 BM3 and site-directed mutants. Anal Biochem 341:148–155
42. Van Vugt-Lussenburg BMA, Damsten MC, Maasdijk DM, Vermeulen NPE, Commandeur JNM (2006) Heterotropic and homotropic cooperativity by a drug-metabolising mutant of cytochrome P450 BM3. Biochem Biophys Res Commun 346:810–818

43. Li QS, Schwaneberg U, Fischer P, Schmid RD (2000) Directed evolution of the fatty-acid hydroxylase P450 BM-3 into an indole-hydroxylating catalyst. Chem Eur J 6:1531–1536
44. Whitehouse CJC, Bell SG, Yang W, Yorke JA, Blanford CF, Strong AJF, Morse EJ, Bartlam M, Rao Z, Wong LL (2009) A highly active single-mutation variant of P450BM3 (CYP102A1). ChemBioChem 10:1654–1656
45. Leroux F (2004) Atropisomerism, biphenyls, and fluorine: a comparison of rotational barriers and twist angles. ChemBioChem 5:644–649
46. Zilly FE, Acevedo JP, Augustyniak W, Deege A, Häusig UW, Reetz MT (2013) Corrigendum: tuning a P450 enzyme for methane oxidation. Angew Chem Int Ed 52:13503
47. Shoji O, Kunimatsu T, Kawakami N, Watanabe Y (2013) Highly selective hydroxylation of benzene to phenol by wild-type cytochrome P450BM3 assisted by decoy molecules. Angew Chem Int Ed 52:1–5
48. Pietz S, Wolker D, Haufe G (1997) Selectivity of the biooxygenation of *N*-phenylcarbamates by the fungus *Beauveria bassiana*. Tetrahedron 4020:17067–17078
49. Dawsonz MJ, Lawrence GC, Mayalll J, Noble D, Roberts SM, Turner MK, Wall WF (1986) Microbial hydroxylation of cyclohexylcyclohexane: synthesis of an analogue of leukotriene-b3. Tetrahedron Lett 27:1089–1092
50. Holland HL, Brown FM, Larsen BG, Zabic M (1995) Biotransformation of organic sulfides. Part 7 Formation of chiral isothiocyanato sulfoxides and related compounds by microbial biotransformation. Tetrahedron: Asym 6:1569–1574
51. Sundby E, Azerad R, Anthonsen T (1998) 2,2-Dimethyl-1,3-propanediol as protective group promotes microbial hydroxylation of cis-bicyclo [3.3.0]octane-3,7-dione. Biotechnol Lett 20:337–340
52. Vigne B, Archelas A, Fustoss R (1991) Microbial transformations 18. Regiosepecific *para*-hydroxylation of aromatic carbamates mediated by the fungus *Beauveria sulfurescens*. Tetrahedron 47:1447–1458
53. Holland HL, Morris TA, Nava PJ, Zabic M (1999) A new paradigm for biohydroxylation by *Beauveria bassiana* ATCC7159. Tetrahedron 55:7441–7460
54. Braunegg G, de Raadt A, Feichtenhofer S, Griengl H, Kopper I, Lehmann A, Weber H (1999) The concept of docking/protecting groups in biohydroxylation. Angew Chem Int Ed 38:2763–2766
55. De Raadt A, Griengl H, Weber H (2001) The concept of docking and protecting groups in biohydroxylation. Chemistry 7:27–31
56. Li S, Chaulagain MR, Knauff AR, Podust LM, Montgomery J, Sherman DH (2009) Selective oxidation of carbolide C-H bonds by an engineered macrolide P450 mono-oxygenase. Proc Natl Acad Sci U S A 106:18463–18468
57. De Raadt A, Griengl H, Petsch M, Plachota P, Schoo N, Weber H, Braunegg G, Kopper I, Kreiner M, Zeiser A, Kieslich K (1996) Microbial hydroxylation of 2-cycloalkylbenzoxazoles. Part I. Product spectrum obtained from *Cunninghamella blakesleeana* DSM 1906 and *Bacillus megaterium* DSM 32. Tetrahedron: Asym 7:467–472
58. De Raadt A, Griengl H, Petsch M, Plachota P, Schoo N, Weber H (1996) Microbial hydroxylation of 2-cycloalkylbenzoxazoles. Part II Determination of product structures and enhancement of enantiomeric excess. Tetrahedron: Asym 7:473–490
59. De Raadt A, Griengl H, Petsch M, Plachota P, Schoo N, Weber H, Braunegg G, Kopper I, Kreiner M, Zeiser A (1996) Microbial hydroxylation of 2-cycloalkylbenzoxazoles. Part III Determination of product enantiomeric excess and cleavage of benzoxazoles. Tetrahedron: Asym 7:491–496
60. Münzer DF, Meinhold P, Peters MW, Feichtenhofer S, Griengl H, Arnold FH, Glieder A, De Raadt A (2005) Stereoselective hydroxylation of an achiral cyclopentanecarboxylic acid derivative using engineered P450s BM-3. Chem Commun 28(20):2597–2599
61. Xue Y, Zhao L, Liu HW, Sherman DH (1998) A gene cluster for macrolide antibiotic biosynthesis in *Streptomyces venezuelae*: architecture of metabolic diversity. Proc Natl Acad Sci U S A 95:12111–12116
62. Xue Y, Wilson D, Zhao L, Sherman DH (1998) Hydroxylation of macrolactones YC-17 and narbomycin is mediated by the pikC-encoded cytochrome P450 in *Streptomyces venezuelae*. Chem Biol 5:661–667
63. Li S, Podust LM, Sherman DH (2007) Engineering and analysis of a self-sufficient biosynthetic cytochrome P450 PikC fused to the RhFRED reductase domain. J Am Chem Soc 129:12940–12941
64. Sherman DH, Li S, Yermalitskaya LV, Kim Y, Smith JA, Waterman MR, Podust LM (2006) The structural basis for substrate anchoring, active site selectivity, and product formation by P450 PikC from *Streptomyces venezuelae*. J Biol Chem 281:26289–26297
65. Li S, Ouellet H, Sherman DH, Podust LM (2009) Analysis of transient and catalytic desosamine-binding pockets in cytochrome P-450 PikC from *Streptomyces venezuelae*. J Biol Chem 284:5723–5730
66. Chefson A, Zhao J, Auclair K (2006) Replacement of natural cofactors by selected hydrogen peroxide donors or organic peroxides results in improved activity for CYP3A4 and CYP2D6. ChemBioChem 7:916–919
67. Podust LM, Sherman DH (2012) Diversity of P450 enzymes in the biosynthesis of natural products. Nat Prod Rep 29:1251–1266
68. Larsen AT, May EM, Auclair K (2011) Predictable stereoselective and chemoselective hydroxylations and epoxidations with P450 3A4. J Am Chem Soc 133:7853–7858
69. Menard A, Fabra C, Huang Y, Auclair K (2012) Type II ligands as chemical auxiliaries to favor enzymatic transformations by P450 2E1. ChemBioChem 13:2527–2536
70. Redlich G, Zanger UM, Riedmaier S, Bache N, Giessing ABM, Eisenacher M, Stephan C, Meyer HE, Jensen ON, Marcus K (2008) Distinction between

human cytochrome P450 (CYP) isoforms and identification of new phosphorylation sites by mass spectrometry. J Proteome Res 7:4678–4688
71. Shin HS, Slattery JT (1998) CYP3A4-mediated oxidation of lisofylline to lisofylline 4,5-diol in human liver microsomes. J Pharm Sci 87:390–393
72. Larsen AT, Lai T, Polic V, Auclair K (2012) Dual use of a chemical auxiliary: molecularly imprinted polymers for the selective recovery of products from biocatalytic reaction mixtures. Green Chem 14:2206–2211
73. Cormack PAG, Elorza AZ (2004) Molecularly imprinted polymers: synthesis and characterisation. J Chromatog B 804:173–182
74. He C, Long Y, Pan J, Li K, Liu F (2007) Application of molecularly imprinted polymers to solid-phase extraction of analytes from real samples. J Biochem Biophys Methods 70:133–150
75. Kist TBL, Mandaji M (2004) Separation of biomolecules using electrophoresis and nanostructures. Electrophoresis 25:3492–3497
76. Qiao F, Sun H, Yan H, Row KH (2006) Molecularly imprinted polymers for solid phase extraction. Chromatographia 64:625–634
77. Turiel E, Martín-Esteban A (2010) Molecularly imprinted polymers for sample preparation: a review. Anal Chim Acta 668:87–99
78. Jin Y, Row KH (2007) Solid-phase extraction of caffeine and catechin compounds from green tea by caffeine molecular imprinted polymer. Bull Korean Chem Soc 28:276–280
79. Theodoridis G, Manesiotis P (2002) Selective solid-phase extraction sorbent for caffeine made by molecular imprinting. J Chromatog A 948:163–169
80. Tse Sum Bui B, Haupt K (2010) Molecularly imprinted polymers: synthetic receptors in bioanalysis. Anal Bioanal Chem 398:2481–2492
81. Cirillo G, Curcio M, Parisi OI, Puoci F, Iemma F, Spizzirri UG, Restuccia D, Picci N (2011) Molecularly imprinted polymers for the selective extraction of glycyrrhizic acid from liquorice roots. Food Chem 125:1058–1063
82. Vasaitis TS, Bruno RD, Njar VCO (2011) CYP17 inhibitors for prostate cancer therapy. J Steroid Biochem Mol Biol 125:23–31
83. Mercer EI (1991) Sterol biosynthesis inhibitors: their current status and modes of action. Lipids 26:584–597
84. Peng CC, Pearson JT, Rock DA, Joswig-Jones CA, Jones JP (2010) The effects of type II binding on metabolic stability and binding affinity in cytochrome P450 CYP3A4. Arch Biochem Biophys 497:68–81
85. Jones JP, Joswig-Jones CA, Hebner M, Chu Y, Koop DR (2011) The effects of nitrogen-heme-iron coordination on substrate affinities for cytochrome P450 2E1. Chem Biol Interact 193:50–56
86. Peng CC, Cape JL, Rushmore T, Crouch GJ, Jones JP (2008) Cytochrome P450 2C9 type II binding studies on quinoline-4-carboxamide analogues. J Med Chem 51:8000–8011
87. Pearson JT, Hill JJ, Swank J, Isoherranen N, Kunze KL, Atkins WM (2006) Surface plasmon resonance analysis of antifungal azoles binding to CYP3A4 with kinetic resolution of multiple binding orientations. Biochemistry 45:6341–6353
88. Dahal UP, Joswig-Jones C, Jones JP (2012) Comparative study of the affinity and metabolism of type I and type II binding quinoline carboxamide analogues by cytochrome P450 3A4. J Med Chem 55:280–290
89. Pearson J, Dahal UP, Rock D, Peng CC, Schenk JO, Joswig-Jones C, Jones JP (2011) The kinetic mechanism for cytochrome P450 metabolism of type II binding compounds: evidence supporting direct reduction. Arch Biochem Biophys 511:69–79
90. Chiba M (2001) P450 interaction with HIV protease inhibitors: relationship between metabolic stability, inhibitory potency, and P450 binding spectra, Drug Metab. Drug Metab Dispos 29:1–3
91. Hutzler JM, Melton RJ, Rumsey JM, Schnute ME, Locuson CW, Wienkers LC (2006) Inhibition of cytochrome P450 3A4 by a pyrimidineimidazole: evidence for complex heme interactions. Chem Res Toxicol 19:1650–1659
92. Krishnan S, Wasalathanthri D, Zhao L, Schenkman JB, Rusling JF (2011) Efficient bioelectronic actuation of the natural catalytic pathway of human metabolic cytochrome P450s. J Am Chem Soc 133:1459–1465
93. Adas F, Salaün JP, Berthou F, Picart D, Simon B, Amet Y (1999) Requirement for ω and (ω-1)-hydroxylations of fatty acids by human cytochromes P450 2E1 and 4A11. J Lipid Res 40:1990–1997

Cytochrome P450 Enzymes and Electrochemistry: Crosstalk with Electrodes as Redox Partners and Electron Sources

9

Victoria V. Shumyantseva, Tatiana Bulko, Evgeniya Shich, Anna Makhova, Alexey Kuzikov, and Alexander Archakov

Abstract

The functional significance of cytochrome P450 (P450) enzymes includes their ability to catalyze the biotransformation of xenobiotics (foreign compounds) and endogenous compounds. P450 enzymes play an important role in the detoxification of exogenous bioactive compounds and hydrophobic xenobiotics (e.g. carcinogens, drugs, environment pollutants, food supplements, medicines, plant products) and in the biotransformation of endogenous bioactive compounds (e.g. amino acids, cholesterol, eicosanoids, saturated/unsaturated fatty acids, melatonin, steroid hormones). Electrode/P450 systems are analyzed in terms of the mechanisms underlying P450-catalyzed reactions. Bioelectrocatalysis-based screening of potential substrates or inhibitors of P450 enzymes, the stoichiometry of the electrocatalytic cycle, oxidation-reduction (redox) thermodynamics, and the peroxide shunt pathway are described. Electrochemical techniques are utilized for investigating the influence of (1) the vitamin B group, (2) vitamins (e.g. vitamins A and B) and antioxidants (e.g. taurine), and (3) drugs and antioxidants (e.g. mexidol, ethoxidol) on biocatalysis using P450 enzymes, and on the metabolism of drugs catalyzed by P450 3A4. The characteristics, performance and potential applications of P450 electrochemical systems are also discussed.

Keywords

Cytochrome P450 • Electrochemistry • Enzyme electrode • Thermodynamics • Antioxidants

V.V. Shumyantseva (✉) • T. Bulko • A. Kuzikov
Laboratory of Bioelectrochemistry, Institute of Biomedical Chemistry, Pogodinskaya St. 10, Moscow 119121, Russia
e-mail: viktoria.shumyantseva@ibmc.msk.ru

E. Shich • A. Makhova
The First Moscow State Medical University, Trubetskaya St. 8, Moscow 119991, Russia

A. Archakov
Institute of Biomedical Chemistry, Pogodinskaya St. 10, Moscow 119121, Russia

E.G. Hrycay and S.M. Bandiera (eds.), *Monooxygenase, Peroxidase and Peroxygenase Properties and Mechanisms of Cytochrome P450*, Advances in Experimental Medicine and Biology 851,
DOI 10.1007/978-3-319-16009-2_9,

9.1 Introduction

Cytochrome P450 (P450) enzymes are widely distributed throughout the biological kingdoms and have numerous important catalytic functions. For example, these ubiquitous and versatile biocatalysts (1) act as electron carriers by reducing diverse substrates; (2) function as monooxygenases by oxygenating a variety of organic substrates using O_2 as a cosubstrate and NAD(P)H; (3) generate reactive oxygen species (ROS) (e.g. superoxide anion, hydrogen peroxide) during monooxygenase reactions that are not coupled efficiently to NAD(P)H utilization; (4) act as four-electron oxidases via the NAD(P)H oxidase pathway; and (5) function as peroxidases and peroxygenases by utilizing organic hydroperoxides and H_2O_2 as cosubstrates [1–5]. In spite of such multifunctionality, the main catalytic function of P450 enzymes is monooxygenation (e.g. hydroxylation of organic substrates) performed toward drugs and foreign chemicals entering the body. From the medical viewpoint, in vitro biocatalysis with P450s is a promising technique for conducting clinical medicine studies, analyzing drug-drug interactions, and performing substrate/inhibitor studies. From the chemical viewpoint, this diversity of functions is based on redox properties of the P450 heme iron. From the electrochemical viewpoint, the most important characteristic feature of hemeproteins is their ability to execute direct electron transfer from the heme iron to the electrode surface [6, 7].

Investigation of the catalytic activities of isolated cytochrome P450 enzymes requires the obligatory presence of redox partners and electron donors such as NAD(P)H [8]. However, redox partners are not obligatory upon electrochemical reduction of P450 family hemeproteins, and the catalytic system is essentially simplified [6, 7, 9, 10]. The electrochemical approach is especially important in cases of unknown physiological partners (e.g. P450 51 MT, systematic name P450 51B1) [11–13], P450 106A2 [14], or P450 109D1 [15]. Electrochemical systems execute the dual function of substituting partner proteins and serving as sources of electrons for redox enzymes. P450 enzymes perform direct electron transfer from the heme iron to the electrode surface, a reaction represented by the appearance on the voltammogram of a pair of peaks corresponding to the oxidation and reduction processes. The electrode reaction of the heme can be described by the equation (eqn)

$$\mathrm{Fe(III)} + e^- \leftrightarrow \mathrm{Fe(II)} \qquad (9.1)$$

The relevance of such an approach is apparent. Indeed, P450-based enzyme electrodes can be used as biosensors in personalized medicine, high-throughput screening and drug interference studies. From this viewpoint, electrochemical systems based on recombinant P450 enzymes are most promising because they enable researchers to standardize the analysis format [6, 7, 9, 10, 16–19]. A very interesting modification of a P450-based electrode is using microsomes instead of P450 enzymes [20]. The authors discuss the future applicability of such systems as biosensors in therapeutic drug monitoring.

As of Aug. 13th, 2013, 1,261 P450 families containing genes that encode P450 proteins have been identified. In addition, a total of 21,039 P450 sequences have been named in all groups of organisms [21]. Studies aimed at the search for novel drugs, estimation of their toxicity and drug-drug interactions have demonstrated that P450 enzymes are the most significant preparations for use in practical clinical medicine. The clinical significance of P450s is connected with their functions in the metabolism of exogenous and endogenous compounds. P450 enzymes play an important role in living organisms by detoxifying hydrophobic xenobiotics or biologically-active compounds [22]. Endogenous or biologically-active substrates of P450 enzymes include arachidonic acid, saturated/unsaturated fatty acids, bile acids, eicosanoids such as prostaglandins, steroid hormones, uroporphyrogens, retinoids and vitamins D and E [3]. Exogenous substrates of human P450 enzymes include medicinal drugs and externally-penetrating compounds (e.g., carcinogens, herbicides, pesticides, plant components). Xenobiotic metabolism proceeds with participation of human P450 families P450 1, P450 2, P450 3 and, to a lesser degree, P450

4. For instance, P450 1A2 catalyzes the metabolism of over 106 substrates, of which 72 are medicinal preparations (e.g., caffeine, paracetamol, phenacetin) (http://cpd.ibmh.msk.su; knowledge data on P450s). Five human P450 enzymes (P450s 1A2, 2C9, 2C19, 2D6 and 3A4) were responsible for the metabolism of over 87 % of medicinal drugs that were in use in 2008 [23]. Here, we review the recent progress made in the electrochemical approach for analyzing the catalytic activities of P450 enzymes and discuss the opportunity of regulating this class of drug-metabolizing biocatalysts.

9.2 Analysis of Current-Voltage Characteristics of Electrodes with Immobilized P450s for Screening of Substrates and Inhibitors

One of the major challenges of engineering (technical) nanotechnology lies in coupling nano-sized bio-objects with measuring analytical devices. The preparation of hybrid bioorganic nanocomposite materials provides a "bridge" for such coupling. Because P450-based test systems are in great demand, various methods for high-throughput screening of substrates and inhibitors are being actively developed [24, 25]. Electrochemical approaches are particularly useful for studying enzyme-substrate interactions due to their high sensitivity [6, 7, 9, 10]. A special feature of P450-based electrochemical sensors is the usage of nanostructured electrodes for improvement of the sensitivity of analyses. Electrochemical therapeutic drug monitoring was described in several publications devoted to characteristics of P450-modified electrodes. An amperometric P450 2B4-based electrochemical biosensor was proposed for determination of phenobarbital as a substrate for P450 (detection limit, 0.289 μM). Phenobarbital is a medicinal drug used in the treatment of epilepsy [26]. P450 2B6 was incorporated into chitosan-modified colloidal gold nanoparticles films. P450 2B6 electrodes were utilized for the investigation of electrocatalytic behaviors of P450 2B6 towards lidocaine, bupropion and cyclophosphamide [27]. Continuous monitoring of naproxen by an electrode modified with carbon nanotubes/microsomal P450 1A2 was studied using cyclic voltammetry [20]. The limit of detection was estimated to be 16 ± 1 μM. Direct electrochemistry of P450 6A1 was studied with an edge-plane graphite electrode, and catalytic activity with aldrin was demonstrated using cyclic voltammetry (CV) [28]. Attempts to develop a potentiometric analysis of enzyme-substrate interaction proved to be futile because no direct correlation between the shifts of reduction potential with and without substrate was established. Besides, such an approach was ineffective in the search for P450 inhibitors. Inhibitors of P450 enzymes may be potential medical preparations by influencing one or an alternate metabolic pathway and by lowering the activity of one or an alternate hemeprotein responsible for drug metabolism [29]. P450s also serve as target objects during the development of novel anticancer drugs in hormone-dependent cancer diseases [30]. Potential inhibitors of P450 19A1 (aromatase) are used in the treatment of breast cancer in estrogen-dependent tumors. Inhibitors of P450 17A1 (17α-hydroxylase) are finding applications in the treatment of androgen-dependent prostate cancer [31].

To develop the algorithm in the search for potential substrates and inhibitors of P450 enzymes, we tested various electrochemical methods. Gold nanoparticles, stabilized by the synthetic membrane-like surfactant didodecyldimethylammonium bromide (DDAB), provide effective electron transport between the graphite electrode and P450 heme. A synthetic lipid membrane with colloidal gold, DDAB/Au, contains a sufficient amount of water to maintain the hemeprotein structure and to secure fixation of P450 enzymes on graphite screen-printed electrodes. P450s 1A2, 2B4, 3A4, 11A1 (P450scc), 17A1 and 51B1 were studied in the presence of substrates and/or inhibitors of the enzymes [6, 7, 18, 32–35].

To investigate electroanalytic characteristics, current-voltage electrode responses were used. Registration was accomplished by using CV and voltammetric analysis (square wave and differential pulse voltammetry). Substrates for appropriate P450 enzymes cause substantial enhancement of catalytic current at controlled potential, while inhibitors do not alter or even lower the current responses. Based on results of amperometry, CV, differential pulse voltammetry (DPV) and square-wave voltammetry (SWV), it is possible to conduct a search and investigate kinetic parameters for potential substrates and inhibitors of P450 enzymes. The proposed electrochemical approach is a sort of a bio bar code for determining P450 substrate/inhibitor competence. Electrochemical methods can be coupled with tandem liquid chromatography/mass spectrometry (LC/MS/MS). Analysis of electrochemical reaction products using LC/MS demonstrated the presence of the same substances as revealed in enzymatic reactions in the presence of cofactors and partner proteins [27, 33].

P450 3A4 plays the major functional role in metabolizing substrates compared with other P450 enzymes. Data retrieved from a website database (http://cpd.ibmh.msk.su) indicated that P450 3A4 metabolized 225 substrates of which 191 were therapeutic drugs [36]. Among the 97 inhibitors of this enzyme, 87 compounds were therapeutic drugs. This database is dedicated to the P450 superfamily. Information on P450-catalyzed reactions, substrate preferences and peculiarities of induction and inhibition is available through this management system. Remarkably, P450 3A4 was involved in the metabolism of ~55 % of therapeutic drugs that were in use in 2005 [37]. Analysis of the interaction of P450 3A4 with testosterone using CV and DPV revealed the presence of catalytic current in enzyme-substrate interactions (see Fig. 9.1). The ratio of maximal amplitudes of reductive currents in the presence of testosterone and without substrate may be expressed as $I(O_2 + Ts)/I(O_2) = 1.7$. The inhibitor ketoconazole (36 mM) lowers this ratio to 1.3.

An apparent Michaelis-Menten constant K_M may be calculated from electrochemical data based on the Michaelis-Menten eqn (9.2) and its electrochemical form (9.3).

$$v = \frac{v_{max} \cdot [S]}{K_M + [S]} \tag{9.2}$$

$$I = \frac{I_{max} \cdot [S]}{K_M{}^{app} + [S]} \tag{9.3}$$

The calculated K_M values for different electrochemical systems are presented in Table 9.1, and are comparable with earlier published data obtained with biochemical methods (Fig. 9.2).

The electrochemical reduction of human recombinant P450 17A1 was investigated with the aim of searching for potential inhibitors of this hemeprotein [34]. Inhibitors of P450 17A1 are used in clinical oncology for treatment of

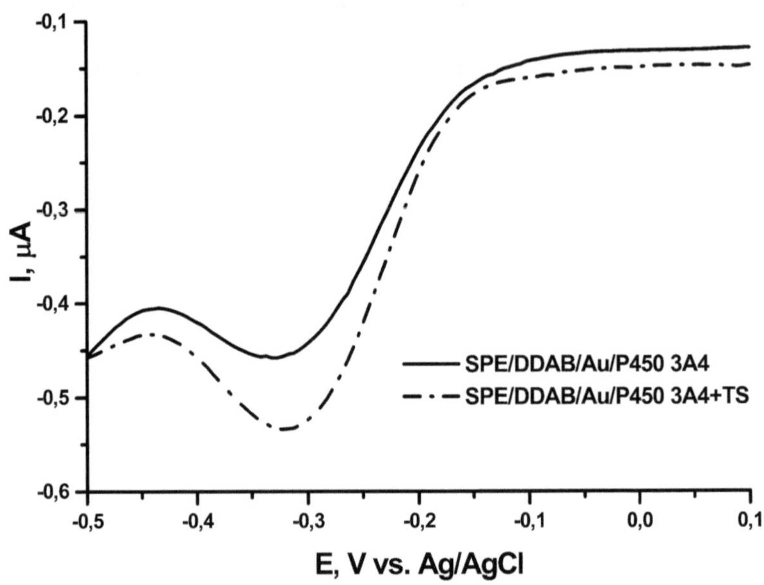

Fig. 9.1 Reductive DPV of screen-printed DDAB/Au/P450 3A4 electrode before (-) and after the addition of 200 µM testosterone (TS) (-·-). Electrolyte volume is 1 ml 100 mM potassium phosphate buffer plus 50 mM NaCl, pH 7.4

Table 9.1 Electrochemical Michaelis-Menten constants (K_M) calculated from the results of amperometric titration of DDAB/Au/P450 or DDAB/P450 electrodes [6]

Electrode	K_M
DDAB/Au/P450 2B4 + benzphetamine	13 μM
DDAB/Au/P450 11A1 (P450scc) + cholesterol	830 μM (upon titration of 14 mM standard cholesterol solution)
	17 μM (upon titration of 10 mM cholesterol in ethanol)
DDAB/Au/P450 51B1 + lanosterol	30 μM
DDAB/P450 3A4 + testosterone	67 μM
DDAB/P450 3A4 + diclofenac	40 μM

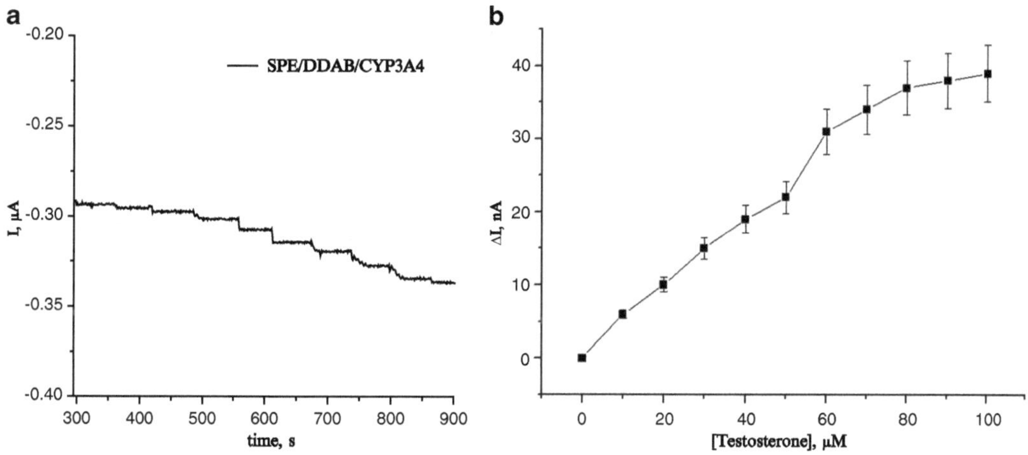

Fig. 9.2 (**a**) Chronoamperometry experiments of the SPE/DDAB/P450 3A4 upon testosterone titration (10 mM stock solution in ethanol). The potential of the working electrode was kept at −0.5 V (vs. Ag/AgCl). (**b**) Michaelis-Menten plot based on the current measurement

prostate cancer [29]. P450 17A1 was immobilized on an electrode modified with DDAB/Au. Analysis of the electrochemical behavior of P450 17A1 was conducted in the presence of the substrate pregnenolone, (1) the inhibitor ketoconazole, (2) and synthetic derivatives of pregnenolone: acetylpregnenolone, (3) cyclopregnenolone, (4) and tetrabrompregnenolone, (5). Ketoconazole, an azole inhibitor of P450s, blocked catalytic current in the presence of the substrate pregnenolone. Compounds 3–5 did not demonstrate substrate properties according to absence of catalytic current. Compound (3) did not influence catalytic activity with pregnenolone as the substrate, whereas compounds (4) and (5) demonstrated inhibitor properties at the micromole level. Electrochemical reduction of P450 17A1 may serve as an adequate substitution for the reconstituted system that requires additional redox partners for demonstration of catalytic activity of P450 hemeproteins (Fig. 9.3).

9.3 Stoichiometry of the Electrocatalytic Cycle of P450 2B4

The stoichiometry of the electrocatalytic cycle of P450 2B4 was studied in kinetic mode according to the bielectrode scheme [38]. Graphite screen-printed electrodes with immobilized P450 2B4 were used as working electrodes (at the potential E = −0.450 V, vs. Ag/AgCl) and electrodes, modified with cytochrome *c* (E = −0.05 V) or Prussian blue (E = 0 V), as measuring electrodes (for H_2O_2) and Clark-type electrode (for O_2). The rate of benzphetamine *N*-demethylation was 17 ± 3 nmol/nmol enzyme/min; peroxide

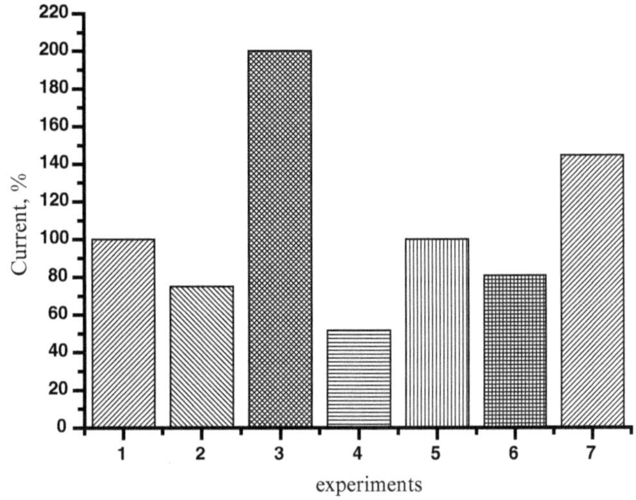

Fig. 9.3 Peak intensity of reductive DPV of screen-printed electrodes in aerobic buffer (with baseline correction): (*1*) DDAB/Au/CYP17A1, (*2*) DDAB/Au/CYP17A1 + **3** (10 μM), (*3*) DDAB/Au/CYP17A1 + **3** (10 μM) + **1** (10 μM), (*4*) DDAB/Au/CYP17A1 + **4** (10 μM), (*5*) DDAB/Au/CYP17A1 + **4** (10 μM) + **1** (10 μM), (*6*) DDAB/Au/CYP17A1 + **5** (10 μM), (*7*) DDAB/Au/CYP17A1 + **5** (10 μM) + **1** (10 μM)

production was 4.8 ± 0.7 nmol/nmol enzyme/min (substrate-free system) and 3.3 ± 0.6 nmol/nmol enzyme/min (0.5 mM benzphetamine); oxygen consumption rate by P450 2B4 was 19.4 ± 0.6 nmol/nmol enzyme/min (in the presence of benzphetamine) and 4.8 ± 0.4 nmol/nmol enzyme/min (without substrate). The bielectrode scheme enables registering oxygen consumption and hydrogen peroxide formation in kinetic mode during the reduction of hemeprotein. Based on the stoichiometry of P450 electrocatalysis, the adequacy of electrochemical reduction of P450 and the P450 monooxygenase system was revealed. Stoichiometric ratios were determined for electrocatalytic reduction of P450 2B4 without substrate ($\Delta O_2 : \Delta H_2O_2 = 1 : 0.92$) and in the presence of substrate ($\Delta O_2 : \Delta H_2O_2 :$ formaldehyde $= 1 : 0.15 : 0.8$). Based on results of reaction rate measurement data and the value of the stoichiometrical coefficient, it was concluded that electrode-immobilized P450 2B4 behaves in essentially the same way as does the monooxygenase microsomal system in solution.

9.4 Redox Thermodynamics of P450 2B4

The temperature dependence of the redox potential $E^{0'}$ of P450 2B4 allows the determination of thermodynamic parameters of the electron transfer process, in particular, the standard entropy ($\Delta S^{0'}{}_{rc}$) and enthalpy $\Delta H^{0'}{}_{rc}$ changes associated with reduction of the oxidized hemeprotein. Thermodynamics parameters of the electrochemical cycle of P450s are an important factor that determines the driving force for effective electron transfer between electrode and heme iron. Therefore, studies on the thermodynamics behavior of P450s allow gaining a better insight into the electron transfer kinetics and properties of hemeprotein molecules. Dynamic electrochemistry such as CV is a particularly useful tool for determining electrochemical as well as thermodynamic characteristics of metalloenzymes [6].

The dependence of the redox potential of P450 2B4 on temperature was determined in the temperature range of 5–30 °C, thus enabling thermodynamic parameters of P450 2B4 to be calculated. Studies on the redox thermodynamics of P450s can provide a better understanding of the electron transfer mechanism and, hence, the improvement of sensitivity of electrochemical sensing systems. Electrochemically-obtained partition of the enthalpic and entropic contributions to the redox potential of P450 will be helpful to the elucidation of molecular mechanisms and factors that influence the protein's redox kinetics and electrocatalysis.

Reduction enthalpy is influenced by metal-ligand bonding and electrostatic interactions between the heme iron and the surrounding polypeptide chain on the one hand and by electrostatic heme-solvent interaction on the other. Reduction entropy depends on conformational changes occurring during electron transfer, and influencing the dynamic properties of the protein in the two redox states [39]. Electrochemical methods permit determining the factorization of the formal potential $E^{0'}$ into enthalpic and entropic contributions, each of which can be obtained through direct electrochemical experimentation at variable temperatures using screen-printed nanostructured electrodes and P450 2B4.

The rate constants (k_s) for electron transfer between the absorbed protein and the electrode were determined from the scan-rate dependence of the anodic and cathodic peak potentials following the Laviron's model of surface-controlled diffusionless electrochemical systems [40, 41] (Table 9.2). The decrease of the heterogeneous electron-transfer rate constant with temperature can reflect complicated processes on the electrode surface associated with partial conformational mobility of protein.

There are several factors regulating the redox potential $E^{0'}$. These are heme exposure to the solvent, the nature of the solvent medium, heme-protein interactions and the ligand binding effect. In the case of P450, the influence of organic substrate and oxygen binding is also a factor [1, 3]. Measurements of $E^{0'}$ as a function of temperature and the application of Eq. (9.4) allow the estimation of $\Delta S^{0'}_{rc}$ and $\Delta H^{0'}_{rc}$

$$E^{0'} = \frac{-\Delta H^{0'}}{nF} + \frac{T\Delta S^{0'}}{nF} \tag{9.4}$$

where n is the number of electrons and F is Faraday's constant, 96,485 °C/mol. The reaction entropy for the reduction of oxidized P450 2B4 is calculated from $\Delta S^{0'}_{rc} = nF\,dE^{0'}/dT$; thus, $\Delta S^{0'}_{rc}$ was determined from the slope of $E^{0'}$ vs. temperature (Fig. 9.4). A linear decrease in $E^{0'}$ with increasing temperature (temperature range of 5–30 °C) was observed. Such dependence is typical for metalloproteins as was

Table 9.2 Electrochemical and kinetic parameters of the reduction reaction for P450 2B4 on DDAB/Au electrode

T, K	ΔE, mV	$E^{0'}$, V(vs. SHE)	I_{ox}, A	I_{red}, A	I_{ox}/I_{red}	k_s, s^{-1}
278	0.164	−0.038	$0.101 \cdot 10^{-6}$	$0.097 \cdot 10^{-6}$	1.04	0.70
284	0.166	−0.037	$0.105 \cdot 10^{-6}$	$0.14 \cdot 10^{-6}$	0.75	0.68
288	0.160	−0.042	$0.082 \cdot 10^{-6}$	$0.091 \cdot 10^{-6}$	0.9	0.67
293	0.173	−0.052	$0.079 \cdot 10^{-6}$	$0.062 \cdot 10^{-6}$	1.27	0.66
299	0.162	−0.061	$0.110 \cdot 10^{-6}$	$0.093 \cdot 10^{-6}$	1.18	0.65

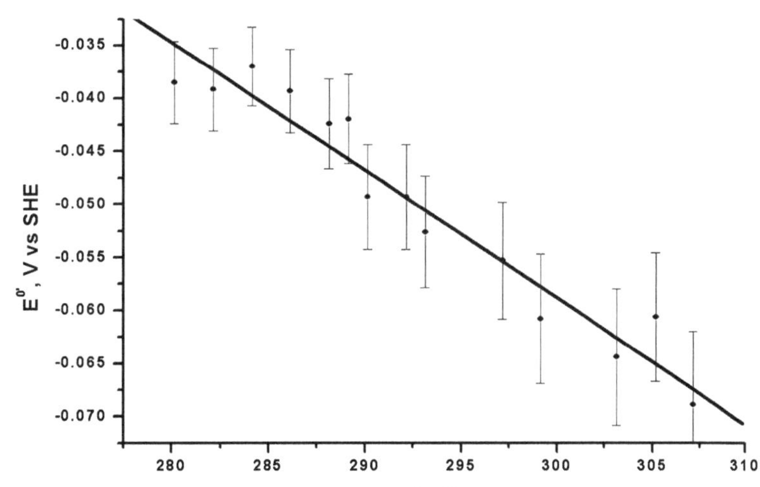

Fig. 9.4 $E^{0'}$ vs. T plot of DDAB/Au/P450 2B4/Nafion electrodes in anaerobic conditions. Scan rate 50 mV/s. Electrolyte volume is 1 ml, 100 mM potassium phosphate buffer, containing 50 mM NaCl, pH 7.4

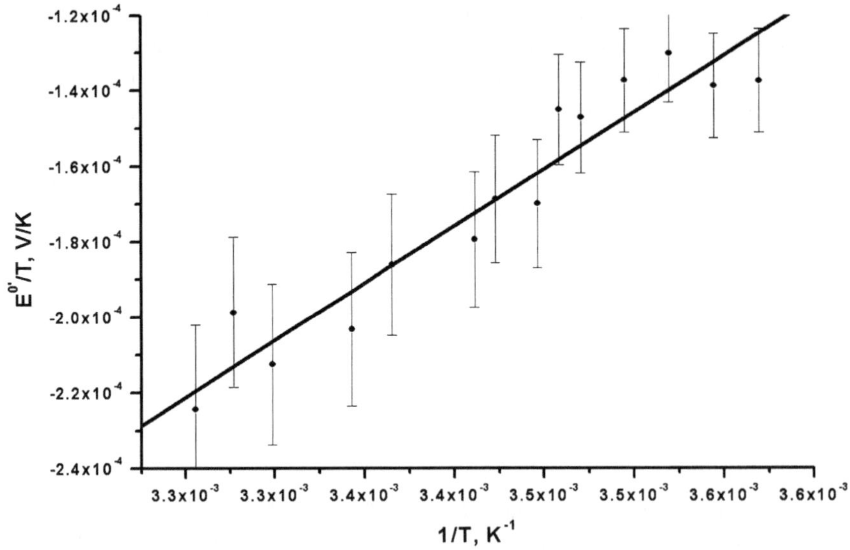

Fig. 9.5 $E^{0'}/T$ vs. 1/T of DDAB/Au/P450 2B4/Nafion electrodes in anaerobic conditions. Scan rate 50 mV/s. Electrolyte volume is 1 ml, 100 mM potassium phosphate buffer, containing 50 mM NaCl, pH 7.4

confirmed for myoglobin [39], hemoglobin [42] and cytochrome *c* [43, 44].

The enthalpy change $\Delta H^{0'}_{rc}$ was obtained from the Gibbs-Helmholtz equation, namely, as a negative slope of the $E^{0'}/T$ vs. 1/T plot (Fig. 9.5).

The thermodynamics parameters for P450 2B4 reduction, – $\Delta H^{0'}_{rc}$ and $\Delta S^{0'}_{rc}$ –, are both negative, which is typical for the hemeprotein's redox thermodynamics [39, 42–44]. From the commonly accepted viewpoint, the enthalpy loss is due to predominance of the reduced heme of P450 2B4 over its oxidized ferric form and, very likely, is due to the limited accessibility of the solvent in the reduced state [42, 44]. The entropic contribution and entropy loss upon reduction are accounted for by conformational variations during the electrochemical cycle.

In the presence of oxygen as a cosubstrate of P450, the CV reduction peaks increased while CV oxidation peaks vanished completely [45–47]. This phenomenon provides evidence for the electrocatalytic activity of P450, with additional conformation being based on the rapid binding of oxygen to reduced ferrous P450 ($k > 10^6\ M^{-1}\ s^{-1}$) [3].

In summary, a biocompatible organic-inorganic (DDAB-gold-nanoparticles) nanocomposite was applied for P450 2B4 immobilization and direct electron-transfer investigation. Redox thermodynamics of P450 2B4 in such a microenvironment and the factorization of the redox potential $E^{0'}$ into enthalpic and entropic contributions can give insight into comprehension of molecular mechanisms of electron transfer in such heterogeneous systems as electrode/protein complexes. The thermodynamics of mammalian P450 2B4 was studied in DDAB surfactant films on a basal plane pyrolytic graphite electrode [47, 48]. Electrochemical experiments and thermodynamic parameters described as enthalpy and entropy are very similar to those found in earlier publications [6] (Table 9.3).

9.5 Light-Driven Biocatalysis with P450 Peroxygenases via In Situ Generation of Hydrogen Peroxide

The most common reaction catalyzed by P450s is a monooxygenase reaction, i.e., insertion of one atom of oxygen into an organic substrate, while

Table 9.3 Thermodynamic parameters of electrochemical reduction of hemeproteins

	Parameter		
Protein	$\Delta S^{0'}$ (J mol^{-1} K^{-1})	$\Delta H^{0'}$ (kJ mol^{-1})	Refs.
Hemoglobin	−349.28	−97.18	[40]
Cytochrome c	−44÷−75	−34÷−43.1	[41]
Myoglobin	−51.1	−1.3	[36]
P450 2B4	−115.7	−29.1	[7]
P450 2B4	−151	−46	[46]
P450 BM3	−163	−47	[48]

the other oxygen atom is reduced to water, according to the following eqn.

$$AH + O_2 + 2H^+ + 2e^- \rightarrow AOH + H_2O \quad (9.5)$$

where AH signifies the organic substrate and AOH represents the monooxygenated product. Manifestation of catalytic activity of isolated cytochromes from the P450 superfamily requires obligatory presence of redox partners (e.g. NADPH-P450 oxidoreductase) and electron donors such as NAD(P)H [1–5, 8], which restricts application of the P450 monooxygenase reaction for routine analysis or screening of potential P450 substrates or inhibitors. Redox partners are not required for electron transfer between the heme of P450 family hemeproteins and electrode, essentially simplifying the catalytic system and thus overcoming the multicomponent nature of P450 monooxygenase systems [6]. In addition, P450s have an alternative route for mono-oxygenation via the "peroxide shunt" in the presence of H_2O_2 [2, 13] or organic hydroperoxides such as cumene hydroperoxide or *tert*-butyl hydroperoxide. The mechanism of this reaction presupposes substrate-heme-peroxide complex formation followed by insertion of an oxygen atom into a substrate molecule [2], as shown in the eqn below

$$\begin{aligned} &\mathrm{Fe(III)} + \mathrm{R\text{-}O\text{-}O\text{-}H} + \mathrm{AH} \rightarrow (\mathrm{AH})- \\ &\mathrm{Fe(III)} - \mathrm{R\text{-}O\text{-}OH} \rightarrow (\mathrm{AH}) - \mathrm{Por}\bullet^{+}\mathrm{Fe(IV)} \\ &\quad = \mathrm{O} + \mathrm{ROH} \rightarrow \mathrm{Fe(III)} + \mathrm{ROH} + \mathrm{AOH} \end{aligned} \quad (9.6)$$

where Fe(III) represents ferric P450, R-O-O-H is the hydroperoxide, R denotes an organic substituent or H atom, AH depicts the substrate, $Por\bullet^{+}Fe(IV){=}O$ designates Compound I, $Por\bullet^{+}$ signifies the porphyrin radical cation group, ROH denotes the reduced hydroperoxide or H_2O, and AOH depicts the monooxygenated product [2]. In this process, P450s do not require electrons for heme iron reduction [2, 13]. The catalytic activity of peroxide-dependent reactions of P450 enzymes is usually (but not always) exceeded by that of NAD(P)H/O_2-dependent reactions, and peroxide-dependent reactions usually require high concentrations of peroxides [48, 49]. In practice, the "peroxide shunt" utility is limited by rapid enzyme inactivation.

Bacterial P450 peroxygenases (e.g. P450 152 family) [2, 50] consume hydrogen peroxide via the peroxide shunt pathway and demonstrate catalytic activity without NAD(P)H as electron donor. P450s 152A1 and 152A2 have high synthetic potential towards polycyclic aromatic hydrocarbons, styrene, ethylbenzene, guaicol and unsaturated medium-chain fatty acids (C_{10}–C_{16}). However, these P450 peroxygenases have poor operational stability in the presence of hydrogen peroxide. To avoid this drawback, a light-driven approach was proposed [49] that utilized riboflavin, flavin mononucleotide (FMN) or flavin adenine dinucleotide (FAD) to generate hydrogen peroxide in situ. During visible light–excited flavins reduction by electrons from ethylenediaminetetraacetic acid and with subsequent interaction of reactive intermediates with O_2, hydrogen peroxide was produced. Reactions for the conversion of myristic acid using P450s 152A1 and 152A2 were performed and in situ generation of H_2O_2 represented an efficient way of reconstituting catalytic activity displayed by these bacterial P450 peroxygenases. Myristic acid was hydroxylated at the α- and β-carbon atoms by P450s 152A1 or 152A2. The

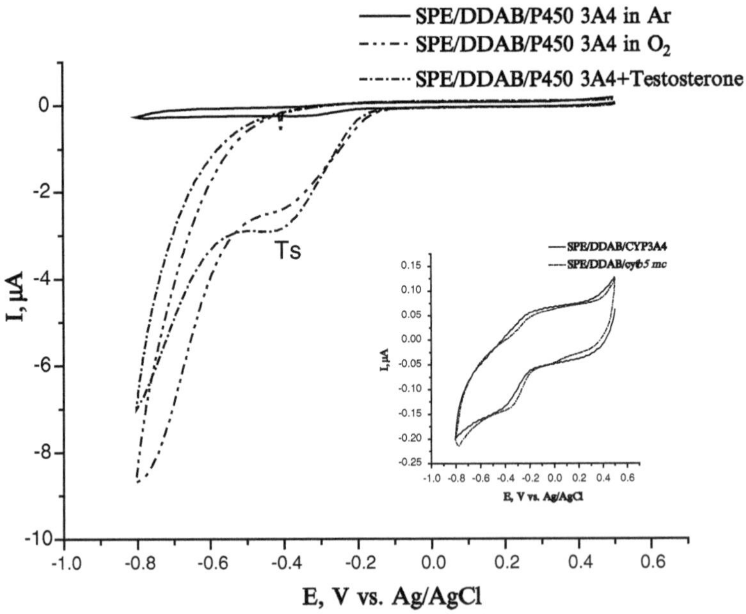

Fig. 9.6 Cyclic voltammograms (CV) of DDAB/CYP3A4 in argon saturated 0.1 M potassium phosphate buffer, plus 0.05 M NaCl; pH 7.4 (-); in O_2 saturated buffer (-¨-); in O_2 saturated buffer +2 mM testosterone, Ts (-·-). The scan rate 0.05 V/s. Inset: DDAB/CYP3A4 and DDAB/b_5(*mc*) in argon saturated 0.1 M potassium phosphate buffer plus 0.05 M NaCl, pH 7.4

ratio of α:β hydroxylated myristic acid was 1:2 for P450 152A1 and 24:1 for P450 152A2 using a gas chromatography assay [50]. This approach is an effective way for biotechnological application of P450 peroxygenase enzymes.

9.6 Electrochemical Registration of P450 Interaction with Redox Partner(s)

Because electron transfer is one of the basic characteristics of the P450 superfamily, studies focused on the peculiarities of P450 3A4 interactions with microsomal cytochrome b_5(*mc*) within the P450 monooxygenase system. The membrane-like synthetic surfactant, didodecyldimethylammonium bromide (DDAB) [47], was used for immobilization of P450 3A4 and cytochrome b_5(*mc*) on a screen-printed graphite electrode surface. We compared the electro- analytical characteristics of P450 3A4 and cytochrome b_5(*mc*) studied by means of CV responses. A single redox couple was observed at $E^0 = -0.302$ V (vs. Ag/AgCl), ($E^0 = $ (Ered + Eox)/2) for P450 3A4 and $E^0 = -0.278$ V for cytochrome b_5(*mc*) (vs. Ag/AgCl) (Fig. 9.6, Inset). In the presence of molecular oxygen, the one-electron reduction is followed by rapid oxygen binding, which manifested in only one reduction peak in CV [14, 45]. In oxygenated electrolyte buffer, the electrocatalytic properties of hemeproteins toward oxygen and testosterone as a substrate are clearly pronounced (Fig. 9.6).

We also compared the formal reduction potential of P450 3A4 and cytochrome b_5(*mc*) in aerobic conditions. The reduction potential of P450 3A4: Ered P450 3A4 $= -0.403$ V (vs. Ag/AgCl), is shifted to the cathodic area in comparison to cytochrome b_5(*mc*): Ered $= -0.350$ V (vs. Ag/AgCl). The observed data testify to the hardly probable direct electron transfer from cytochromes b_5(*mc*) to P450 3A4, owing to the close proximity of their reduction potential values. Substrate binding may act as the thermodynamic trigger facilitating electron transfer from protein redox partners to P450 [46]. In our experiments, however, we did not register a pronounced shift of redox potential during substrate binding (Fig. 9.6). The many roles of cytochrome b_5 in P450 monooxygenase reactions are discussed from various viewpoints [51–53].

The influence of cytochromes b_5(*mc*) on electrochemically-driven P450 3A4 catalysis towards testosterone was investigated. The ratio of the catalytic current, Ic (with testosterone, P450 3A4 + Ts) to the current in oxygenated buffer, I (without substrate, P450 3A4), Ic/I, corresponds to 2.0 ± 0.12. The Ic/I ratio for P450 3A4 + cytochrome $b_5(mc)$ (1:1) system corresponds to 3.99 ± 0.15, testifying to the influence of cytochrome $b_5(mc)$ on electrocatalysis. The data obtained in these experiments provide evidence for the stimulatory effect of cytochrome $b_5(mc)$ on P450–induced electrocatalysis.

Cytochrome $b_5(mc)$ appeared to be capable of stimulating the electrocatalytic activity of P450 3A4 towards testosterone. The stimulation afforded by cytochrome $b_5(mc)$ can be explained in terms of this effector's role as an allosteric regulator. Or possibly, its role involves a direct input of the second of the two electrons required for monooxygenase reactions catalyzed by P450. The electrochemical approach can clarify not only the mechanism of enzyme/substrate/inhibitor interactions, but also protein/protein interactions from the electron transfer processes point of view.

9.7 Influence of the Vitamin B Group on Electrocatalytic Activity of P450 3A4

The problem of drug-drug interactions for personalized therapy is intensively studied from different viewpoints [20, 54–59]. At the same time, the influence of other biologically-active compounds on drug-metabolizing enzymes and, also, on pharmacological functions, pharmacokinetics and pharmacodynamics of therapeutic agents has high clinical significance and offers a fertile field for pharmacological research.

P450 3A4 recognizes and metabolizes a broad range of structurally diverse therapeutic agents [5, 60]. As a consequence, many clinically relevant drug-drug interactions are associated with inhibition and/or induction of this enzyme [55]. Thus, influence of the vitamin B group on the pharmacokinetics of diclofenac (voltaren), a non-steroidal anti-inflammatory drug used for the treatment of arthritis, ankylosing spondylitis and acute muscle pain, was studied by means of electrochemical methods [61]. To investigate the electroanalytic characteristics of electrochemical systems with P450 immobilized on the electrode surface, we have resorted to cyclic voltammetry and voltammetric analysis (square wave voltammetry and differential pulse voltammetry). Electrochemical studies were performed to examine the influence of the vitamin B group on catalytic activity of P450 3A4 towards diclofenac. Thiamine (water soluble vitamin B1) alone did not demonstrate substrate properties (at 1.5 mM concentration), but blocked catalytic activity of P450 3A4 towards diclofenac. Diclofenac produced a 137 % increase of catalytic current without thiamine but in the presence of thiamine, only 67 % of catalytic current can be measured (Fig. 9.7). Since thiamine has nitrogen with an unshared pair of electrons in a pyrimidine

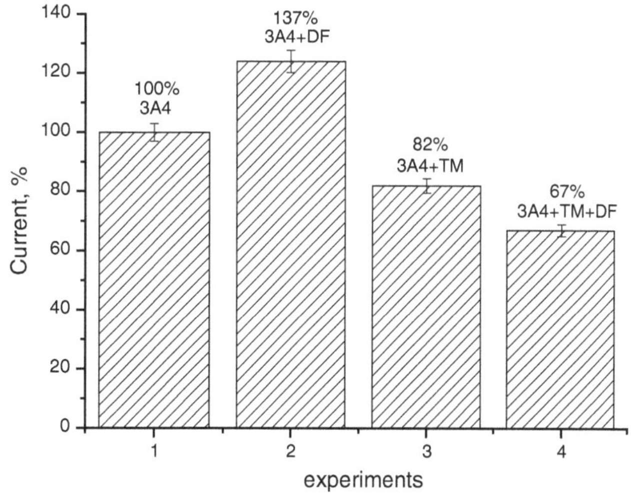

Fig. 9.7 Peak intensity (%) of reductive SWV of screen-printed electrodes in aerobic buffer (with baseline correction): DDAB/Au/P450 3A4 (*1*); DDAB/Au/P450 3A4 + 100 μM diclofenac (DF) (*2*); DDAB/Au/P450 3A4 + 1.5 mM thiamine (TM) (*3*); DDAB/Au/P450 3A4 + 1.5 mM thiamine (TM), then 100 μM DF (*4*). The relative standard deviation (R.S.D.) of 3 % was calculated for n = 5

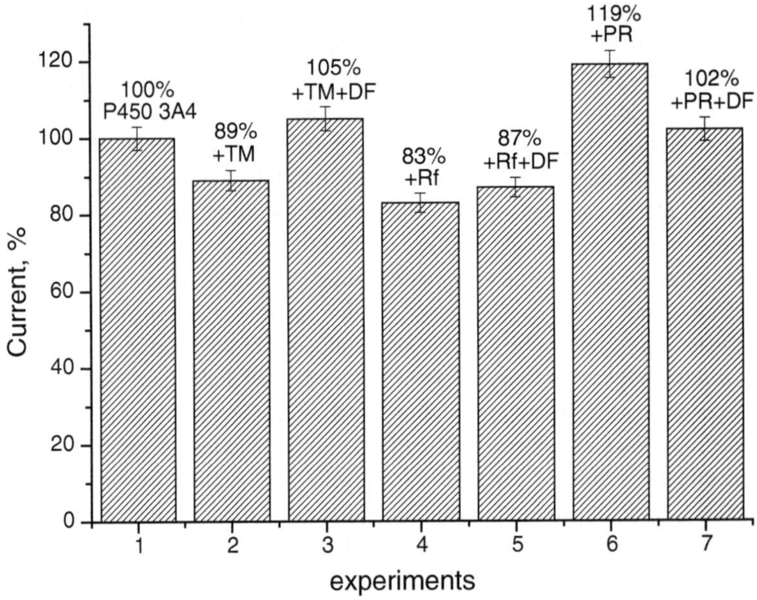

Fig. 9.8 Peak intensity (%) of reductive SWV of screen-printed electrodes in aerobic buffer (with baseline correction): DDAB/Au/P450 3A4 (*1*); DDAB/Au/P450 3A4 + TM (0.3 mM) (*2*); DDAB/Au/P450 3A4 + TM (0.3 mM, then DF (*3*); DDAB/Au/P450 3A4 + riboflavin (Rf) (0.3 mM) (*4*); DDAB/Au/P450 3A4 + Rf (0.3 mM) then DF (*5*); DDAB/Au/P450 3A4+ pyridoxine (PR) (0.3 mM) (*6*); DDAB/Au/P450 3A4 + PR (0.3 mM) then DF (*7*). The relative standard deviation (R.S.D.) of 3 % was calculated for n = 5

heterocycle and sulphur in the thiasole ring, it is reasonable to suggest that thiamine interacts with the heme iron of P450, similar to the azole inhibitors ketoconazole and itraconazole [3], while competing with molecular oxygen. Oxygen is a cosubstrate of P450 enzymes and binds to the heme iron during the catalytic cycle. Thiamine acts as a noncompetitive inhibitor against the organic substrate diclofenac with an electrochemical inhibition constant of Ki = 0.45 ± 0.15 mM. Therefore, thiamine reduces the activity of P450 3A4 towards diclofenac. These results suggest that thiamine can influence, i.e. slow down, the metabolism of diclofenac in a concentration-dependent manner.

Comparative investigation of the influence of thiamine (vitamin B1), riboflavin (vitamin B2) and pyridoxine (vitamin B6) at a 300 μM concentration was conducted on the electrochemical DDAB/Au/P450 3A4/diclofenac system (Fig. 9.8). Thiamine (300 μM) inhibited the electrocatalytic $Fe^{+3} \leftrightarrow Fe^{+2}$ process of P450 3A4 and lowered catalytic current with diclofenac as a substrate (Fig. 9.8, experiments 1–3). Riboflavin (300 μM), in accordance with SWV data, also inhibited the catalytic activity of P450 3A4 toward diclofenac (Fig. 9.8, experiments 4, 5). Importantly, pyridoxine (300 μM) did not diminish electrochemical reduction of P450 3A4, but rather promoted this process (Fig. 9.8, experiment 6). However, catalytic current corresponding to diclofenac in the presence of pyridoxine (102 %) did not exceed the level of diclofenac itself (137 %), without pyridoxine (Fig. 9.7, experiment 2; Fig. 9.8, experiment 7). These results suggest that pyridoxine regulates the catalytic activity of P450 3A4 with mediation of electron transfer to the heme iron, but interferes with diclofenac monooxygenation. We propose that pyridoxine action is possibly associated with the antioxidant properties of vitamin B6 (pyridoxine).

Our results demonstrate that, based on the electrochemical behavior of the electrode/P450 3A4 system, it is possible to assess the influence of the vitamin B group on catalytic activity of P450 alone and to observe the substrate/inhibitor competence of this enzyme.

9.8 The Dose-Dependent Influence of Vitamins and Drugs with Antioxidant Properties on Electrochemically-Driven P450 Catalysis

According to the catalytic cycle of P450 enzymes, one of the oxygen atoms is involved in the oxidation of organic substrates and the other is reduced to water (eqn. 9.5). However, in some cases, a reaction does not correspond to the stoichiometry of the above equation (see also Sect. 9.3).

A portion of the electrons is spent on the reduction of oxygen without substrate conversion (eqns. 9.8.1, 9.8.2 and 9.8.3). This phenomenon is known as uncoupling. A portion of the redox equivalents participates in oxidative side reactions. In the course of the reaction, ROS such as hydrogen peroxide and superoxide anion radicals (as well as water) are formed (Scheme 9.1) [2, 3, 19–21].

ROS interact with P450 to cause an inactivation of protein [1, 62–64]. Antioxidants can diminish the level of ROS by a radical-scavenging effect that modulates the activities of P450 enzymes, which are known to generate reactive intermediates (Scheme 9.1). Scavenging substances are essential in the antioxidant defense against ROS, and can influence the catalytic functions of the hemeprotein. We investigated the role of antioxidants in P450-mediated catalysis and have shown that vitamins A, C and E exhibiting antioxidant properties influence catalytic activities of P450 3A4 [65]. Electrochemically-driven P450 catalysis is also accompanied by ROS generation [38]. Therefore, the influence of free radical-scavenging substances (ROS "traps") on electrocatalysis may be reasonably expected.

Antioxidant compounds are routinely included in pharmaceutical formulations to minimize the oxidative degradation of the active pharmaceutical ingredient(s). To minimize drug-drug interactions, it is necessary to choose safe drug-drug or drug-vitamin combination regimens and adjust drug dosage appropriately. Based on analysis of electrochemical parameters of P450, the algorithm that allows elucidating the properties of antioxidants was developed.

Vitamin C (in the range 0.03–1 mM) and vitamins A and E (in the range 10–100 μM) stimulated the dose-dependent growth of the cathodic peak current of P450 3A4, corresponding to heme reduction according to the equation Fe (III) + 1 e → Fe(II) (Figs. 9.9 and 9.10). To better understand the mechanism of the stimulating effect of vitamins with antioxidant properties in P450 electrocatalysis, we tested the influence of *tert*-butyl alcohol, a well-known ROS scavenger [66], on the cathodic reductive current corresponding to heme reduction. *tert*-butyl alcohol (0.05 M) stimulated the reductive current in the SPE/DDAB/Au/P450 3A4 electrode system (138 ± 10 %) and neutralized the antioxidant effect of vitamin C. These experiments confirm the participation of ROS in the electrocatalysis of P450 3A4.

Our findings are in line with previously shown experimental data that vitamin C and cytochrome *c* could enhance electron transfer in reaction-

Scheme 9.1 RH is the substrate for P450, ROH is the product of the monooxygenase reaction and AO is the antioxidant

$$RH + O_2 \rightarrow ROH + H_2O \quad (9.8)$$

$$O_2 + 2H^+ + 2e \rightarrow H_2O_2 \quad (9.8.1)$$

$$O_2 + H^+ + e \rightarrow HO_2 \rightarrow O_2^- + H^+ \quad (9.8.2)$$

$$HO_2 + H^+ + e \rightarrow H_2O_2 \rightarrow HO_2^- + H^+ \quad (9.8.3)$$

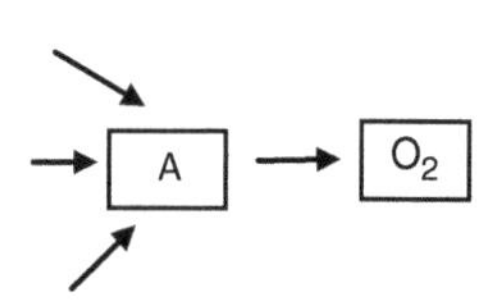

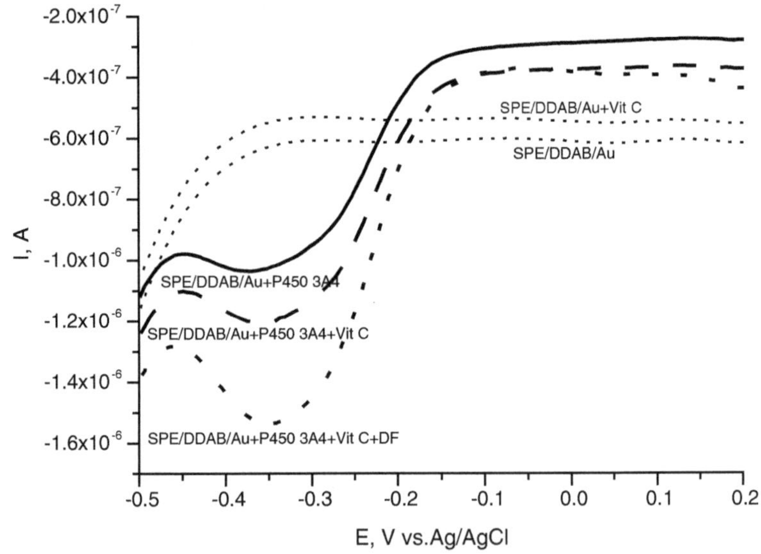

Fig. 9.9 Reductive square wave voltammograms of screen-printed DDAB/Au/P450 3A4 electrode (—); DDAB/Au/P450 3A4 + vitamin C (0.3 mM) (---); DDAB/Au/P450 3A4 + vitamin C (0.3 mM), then 100 μM diclofenac (DF) (. . .), DDAB/Au (. . .) and DDAB/Au + vitamin C (0.3 mM) (. . .). Electrolyte volume is 1 ml of 100 mM potassium phosphate buffer plus 50 mM NaCl, pH 7.4

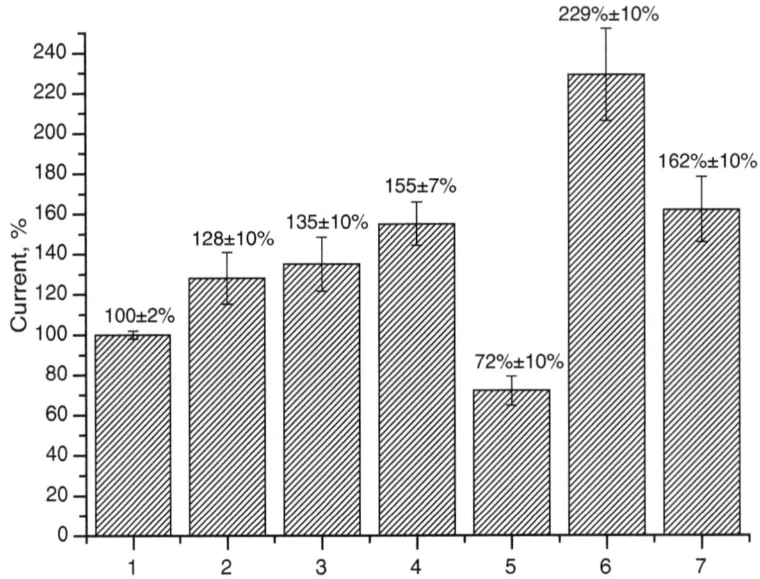

Fig. 9.10 Peak intensity (%) of reductive square wave voltammograms of screen-printed electrodes in aerobic buffer (with baseline correction): DDAB/Au/P450 3A4 (*1*); DDAB/Au/P450 3A4 + 100 μM diclofenac (DF) (*2*); DDAB/Au/P450 3A4 + vitamin C (0.6 mM) (*3*); DDAB/Au/P450 3A4 + vitamin C (0.6 mM), then 100 μM DF (*4*). DDAB/Au/P450 3A4 + vitamin C (1.7 mM) (*5*); DDAB/Au/P450 3A4 + 100 μM vitamin A (*6*); DDAB/Au/P450 3A4 + 100 μM vitamin E (*7*)

mediated redox processes by serving as nonspecific redox-activity facilitators for heme peroxidases such as chloroperoxidase and horseradish peroxidase [67]. It was also shown that vitamin C, being a strong antioxidant, is capable of scavenging ROS in low concentration ranges, and provides prooxidant capacity in high concentration ranges [68].

In electrochemical experiments, vitamins-antioxidants serve as modulating and/or stimulating additives with respect to P450

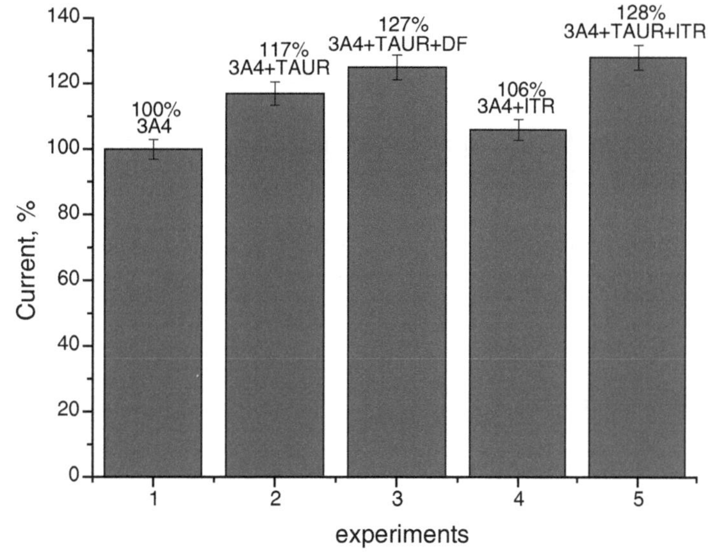

Fig. 9.11 Peak intensity (%) of reductive SWV of screen-printed electrodes in aerobic buffer (with baseline correction): DDAB/Au/P450 3A4 (*1*); DDAB/Au/P450 3A4 + taurine (TAUR) (50 μM) (*2*); DDAB/Au/P450 3A4 + TAUR (50 μM, then DF (*3*); DDAB/Au/P450 3A4 + itraconazole (ITR) (10 μM) (*4*); DDAB/Au/P450 3A4+ TAUR (50 μM), then ITR (*5*). The relative standard deviation (R.S.D.) of 3 % was calculated for n = 5

electrochemical activity, due to their free-radical scavenging, antihypoxants properties or electron mediator features.

We also studied the influence of taurine as an antioxidant on electrochemical activity of P450 3A4. Taurine or 2-aminoethane sulfonic acid is a major free β-amino acid. The functions of taurine include osmoregulation, cell membrane stabilization, antioxidation, detoxification and neuromodulation [69]. In our experiments, taurine enhanced the cathodic reduction current of hemeproteins up to 117 %, as was shown with the SWV technique (Fig. 9.11). Antioxidant properties of taurine are well established and we propose that taurine influences the P450 reduction step by interfering with ROS produced during heme reduction [2, 70]. In the presence of taurine, the catalytic current of diclofenac does not exceed its maximum value (137 %) and amounts to only 127 % (Fig. 9.11, experiment 3).

Itraconazole is an antifungal drug used in therapy of different mycoses [71] and acts as an inhibitor of P450 3A4 [56, 57]. In the presence of taurine, the inhibitory properties of itraconazole are not so pronounced, and the P450 3A4 reduction current corresponds to 128 % in comparison with 106 % for itraconazole itself (Fig. 9.11). This means that in the presence of taurine, the inhibitory properties of itraconazole are weakened.

9.9 Conclusion

The electrochemistry of P450 in combination with nanotechnologies enables researchers to miniaturize electrodes and measuring devices for potential application of the "laboratory-on-the-chip", as well as microarrays and point-of-care, biosensors. Further development of methods for high-throughput screening of potential substrates and inhibitors of P450 enzymes is connected with miniaturization of devices, automation of processes, as well as cutting down expenses for reagents, assay time and analysis steps, which in turn leads to reduction of research costs. The electrochemically-driven P450 catalysis is an alternative model system for pharmacological research and drug-drug interaction studies. The electrochemical experiments have elucidated the possible mechanism of dose-dependent interaction of vitamins exhibiting antioxidant properties with clinical drugs. These findings provide primary data for future clinical risk prediction studies, especially for those devoted to the interaction of drugs with antioxidants. Regulation and modulation of P450 3A4 activity through the action of vitamins-antioxidants, upon their appointment in a combination with clinical drugs metabolized by P450s, will probably become an essential requirement in clinical routine practice. Antioxidants intake can lead to alteration in pharmacodynamic efficiency, which demands special attention from doctors because the prescribed medical product can bring about changes in an efficiency/safety profile.

Acknowledgements The work is done in the framework of the State Academies of Sciencies fundamenal research program for 2013–2020.

References

1. Archakov AA, Bachmanova GI (1990) Cytochrome P450 and active oxygen. Taylor & Francis, London
2. Hrycay EG, Bandiera SM (2012) The monooxygenase, peroxidase, and peroxygenase properties of cytochrome P450. Arch Biochem Biophys 522:71–89
3. Lewis DFV (2001) Guide to cytochromes P450: structure and function. Taylor & Francis, London
4. Ortiz de Montellano PR (2005) Cytochrome P450: structure, mechanism, and biochemistry, 3rd edn. Kluwer Academic/Plenum Publishers, New York
5. Zanger U, Schwab M (2013) Cytochrome P450 enzymes in drug metabolism: regulation of gene expression, enzyme activities, and impact of genetic variation. Pharmacol Ther 138:103–141
6. Shumyantseva V, Bulko T, Suprun E, Chalenko Y, Vagin M, Rudakov Y, Shatskaya M, Archakov A (2011) Electrochemical investigations of cytochromes P450. Biochim Biophys Acta 1814:94–101
7. Shumyantseva V, Suprun E, Bulko T, Chalenko Y, Archakov A (2012) Electrochemical sensor systems for medicine. In: Rozlosnik N (ed) Nanomedicine in diagnostics, 1st edn. CRC Press/Taylor & Francis, St. Helier, pp 68–95
8. Pandey A, Flück C (2013) NADPH P450 oxidoreductase: structure, function, and pathology of diseases. Pharmacol Ther 138:229–254
9. Schneider E, Clark D (2013) Cytochrome P450 (CYP) enzymes and the development of CYP biosensors. Biosens Bioelectron 39:1–13
10. Yarman A, Wollenberger U, Scheller FW (2013) Sensors based on cytochrome P450 and CYP mimicking systems. Electrochim Acta 110:63–72
11. Bellamine A, Mangla A, Nes WD, Waterman MR (1999) Characterization and catalytic properties of the sterol 14α-demethylase from *Mycobacterium tuberculosis*. Proc Natl Acad Sci U S A 96:8937–8942
12. Pikuleva IA (2006) Cytochrome P450s and cholesterol homeostasis. Pharmacol Ther 112:761–773
13. Hlavica P (2009) Assembly of non-natural electron transfer conduits in the cytochrome P450 system: a critical assessment and update of artificial redox constructs amenable to exploitation in biotechnological areas. Biotechnol Adv 27:103–121
14. Colas H, Ewen K, Hannemann F, Bistolas N, Wollenberger U, Bernhardt R, de Oliveira P (2012) Direct and mediated electrochemical response of the cytochrome P450 106A2 from *Bacillus megaterium* ATCC 13368. Bioelectrochemistry 87:71–77
15. Khatri Y, Girhard M, Romankiewicz A, Urlacher VB, Bernhardt R (2010) Regioselective hydroxylation of norisoprenoids by CYP109D1 from *Sorangium cellulosum* So ce56. Appl Microbiol Biotechnol 88:485–495
16. Estabrook RW, Faulkner KM, Shet MS, Fisher CW (1996) Application of electrochemistry for P450-catalyzed reactions. Methods Enzymol B 272:44–51
17. Bistolas N, Wollenberger U, Jung C, Scheller FW (2005) Cytochrome P450 biosensors – a review. Biosens Bioelectron 20:2408–2423
18. Shumyantseva V, Bulko T, Archakov A (2005) Electrochemical reduction of cytochrome P450 as an approach to the construction of biosensors and bioreactors. J Inorg Biochem 99:1051–1063
19. Udit AK, Gray HB (2005) Electrochemistry of heme-thiolate proteins. Biochem Biophys Res Commun 338:470–476
20. Baj-Rossi C, Rezzonico Jost T, Cavallini A, Grassi F, De Michelli G, Carrara S (2014) Continuous monitoring of Naproxen by a cytochrome P450-based electrochemical sensor. Biosens Bioelectron 53:283–287
21. http://drnelson.uthsc.edu/P450.statsfile.html
22. Nebert DW, Russel DW (2002) Clinical importance of the cytochromes P450. Lancet 360:1155–1162
23. Guengerich FP (2008) Cytochrome P450 and chemical toxicology. Chem Res Toxicol 21:70–83
24. Persson KP, Ekehed S, Otter C, Lutz M, VcPheat J, Masomirembwa CM, Andersson TB (2006) Evaluation of human liver slices and reporter gene assays as systems for predicting the cytochrome P450 induction potential of drugs *in vivo* in humans. Pharmacol Res 23:56–66
25. Tupeinen M, Jouko U, Jorma J, Olavi P (2005) Multiple P450 substrates in a single run: rapid and comprehensive in vitro interaction assay. Eur J Pharm Res 24:123–132
26. Alonso-Lomillo MA, Gonzalo-Ruiz J, Domínguez-Renedo O, Muñoz FJ, Arcos-Martínez MJ (2008) CYP450 biosensors based on gold chips for antiepileptic drugs determination. Biosens Bioelectron 23:1733–1736
27. Liu S, Peng L, Yang X, Wu Y, He L (2008) Electrochemistry of cytochrome P450 enzyme on nanoparticle-containing membrane-coated electrode and its application for drug sensing. Anal Biochem 375:209–216
28. Zhang L, Liu X, Wang C, Liu X, Cheng G, Wu Y (2010) Expression, purification and direct electrochemistry of cytochrome P450 6A1 from the house fly, *Musca domestica*. Protein Expr Purif 71:74–78

29. Handratta VD, Vasaitis TS, Njar VC, Gediya LK, Kataria R, Chopra P, Newman D, Farquhar R, Guo Z, Qiu Y, Brodie AM (2005) Novel C-17-heteroaryl steroidal CYP17 inhibitors/antiandrogens: synthesis, in vitro biological activity, pharmacokinetics, and antitumor activity in the LAPC4 human prostate cancer xenograft model. J Med Chem 48:2972–2984
30. Bruno RD, Njar CO (2007) Targeting cytochrome P450 enzymes: a new approach in anti-cancer drug development. Bioorg Med Chem 15:5047–5060
31. Sadeghi S, Ferrero S, Di Nardo G, Gilardi G (2012) Drug-drug interactions and cooperative effects detected in electrochemically driven human cytochrome P450 3A4. Bioelectrochemistry 86:87–91
32. Carrara S, Shumyantseva V, Archakov A, Samorì B (2008) Screen-printed electrodes based on carbon nanotubes and cytochrome P450scc for highly-sensitive cholesterol. Biosens Bioelectron 24:148–150
33. Shumyantseva VV, Bulko TV, Kuznetsova GP, Lisitsa AV, Ponomarenko EA, Karuzina II, Archakov AI (2007) Electrochemical reduction of sterol-14-α-demethylase from *Mycobacterium tuberculosis* (CYP51b1). Biochem Mosc 72:658–663
34. Shumyantseva VV, Bulko TV, Misharin A, Archakov AI (2011) Screening of potential substrates or inhibitors of cytochrome P450 17A1 (CYP17A1) by electrochemical methods. Biochem (Mosc) Suppl B Biomed Chem 5:55–59
35. Shumyantseva VV, Bulko TV, Rudakov YO, Kuznetsova GP, Samenkova NF, Lisitsa AV, Karuzina II, Archakov AI (2007) Electrochemical properties of cytochromes P450 using nanostructured electrodes: direct electron transfer and electrocatalysis. J Inorg Biochem 101:859–865
36. Lisitsa AV, Gusev SA, Karusina II, Archakov AI, Koymans L (2001) Cytochrome P450 database. SAR QSAR. Environ Res 12:359–366
37. Foti RS, Rock DA, Wienkers LC, Wahlstrom JL (2010) Selection of alternative CYP3A4 probe substrates for clinical drug interaction studies using in vitro data and in vivo simulation. Drug Metab Dispos 38:981–987
38. Rudakov YO, Shumyantseva VV, Bulko TV, Suprun EV, Kuznetsova GP, Samenkova NF, Archakov AI (2008) Stoichiometry of electrocatalytic cycle of cytochrome P450 2B4. J Inorg Biochem 102:2020–2025
39. Liu X, Huang Y, Zhang W, Fan G, Fan C, Li G (2005) Electrochemical investigation of redox thermodynamics of immobilized myoglobin: ionic and ligation effects. Langmuir 21:375–378
40. Laviron E (1979) General expression of the linear potential sweep voltammogram in the case of diffusionless electrochemical systems. J Electroanal Chem 101:19–28
41. Liu J, Guo C, Li CM, Li Y, Chi Q, Huang X, Liao L, Yu T (2009) Carbon-decorated ZnO nanowire array: a novel platform for direct electrochemistry of enzymes and biosensing applications. Electrochem Commun 11:202–205
42. He X, Zhu L (2006) Direct electrochemistry of hemoglobin in cetylpyridinium bromide film: redox thermodynamics and electrocatalysis to nitric oxide. Electrochem Commun 8:615–620
43. Battistuzzi G, Borsari M, Rossi G, Sola M (1998) Effects of solvent on the redox properties of cytochrome c: cyclic voltammetry and 1H NMR experiments in mixed water-dimethylsulfoxide solutions. Inorg Chim Acta 272:168–175
44. Borsari M, Bellei M, Tavagnacco C, Peressini S, Millo D, Costa G (2003) Redox thermodynamics of cytochrome c in mixed water–organic solvent solutions. Inorg Chim Acta 349:182–188
45. Shumyantseva VV, Ivanov YD, Bistolas N, Scheller FW, Archakov AI, Wollenberger U (2004) Direct electron transfer of cytochrome P450 2B4 at electrodes modified with non-ionic detergent and colloidal clay nanoparticles. Anal Chem 76:6046–6052
46. Johnson DL, Lewis BC, Elliot DJ, Miners JO, Martin LL (2005) Electrochemical characterization of the human cytochrome P450 CYP2C9. Biochem Pharmacol 69:1533–1541
47. Joseph S, Rusling JF, Lvov YM, Friedberg T, Fuhr U (2003) An amperometric biosensor with human CYP3A4 as a novel drug screening tool. Biochem Pharmacol 65:1817–1826
48. Hagen KD, Gillan J, Im SC, Landefeld S, Meal G, Hiley M, Waskell L, Hill M, Udit A (2013) Electrochemistry of mammalian cytochrome P450 2B4 indicates tunable thermodynamic parameters in surfactant films. J Inorg Biochem 129:30–34
49. Kanaeva IP, Dedinskii IR, Scotselyas ED, Krainev IG, Guleva IV, Sevryukova IF, Koen YM, Kuznetsova GP, Bachmanova GI, Archakov AI (1992) Comparative study of monomeric reconstituted and membrane microsomal monooxygenase systems of the rabbit liver: I. Properties of NADPH-cytochrome P450 reductase and cytochrome P450 LM2 (2B4) monomers. Arch Biochem Biophys 298:395–402
50. Girhard M, Kunigk E, Tihovsky S, Shumyantseva VV, Urlacher VB (2013) Light-driven biocatalysis with cytochrome P450 peroxygenases. Biotechnol Appl Biochem 60:112–118
51. Schenkman JB, Jansson I (2003) The many roles of cytochrome b_5. Pharmacol Ther 97:139–152
52. Im SC, Waskell L (2011) The interaction of microsomal cytochrome P450 2B4 with its redox partners, cytochrome P450 reductase and cytochrome b_5. Arch Biochem Biophys 507:144–153
53. Storbeck KH, Swart A, Goosen P, Swart P (2013) Cytochrome b_5: novel roles in steroidogenesis. Mol Cell Endocrinol 371:87–99
54. Carrara S, Cavallini A, Erokhin V, Albini GD, De Micheli G (2011) Multi-panel drugs detection in human serum for personalized therapy. Biosens Bioelectron 26:3914–3919

55. Zhou S, Xue C, Yu X, Li C, Wang G (2007) Clinically important drug interactions potentially involving mechanism-based inhibition of cytochrome P450 3A4 and the role of therapeutic drug monitoring. Ther Drug Monit 29:687–708
56. Hisata A, Ohno Y, Yamamoto T, Suzuki H (2010) Prediction of pharmacokinetic drug–drug interaction caused by changes in cytochrome P450 activity using in vivo information. Pharmacol Ther 125:230–248
57. Zhang L, Reynolds KS, Zhao P, Huang SM (2010) Drug interactions evaluation: an integrated part of risk assessment of therapeutics. Toxicol Appl Pharmacol 243:134–145
58. Fantuzzi A, Capria E, Mak L, Dodhia HS, Sadeghi J, Collins S, Somers G, Huq E, Gilardi G (2010) An electrochemical microfluidic platform for human P450 drug metabolism profiling. Anal Chem 82:10222–10227
59. Fantuzzi A, Mak L, Capria E, Dodhia V, Panicco P, Collins S, Gilardi G (2011) A new standardized electrochemical array for drug metabolic profiling with human cytochromes P450. Anal Chem 83:3831–3839
60. Shen S, Marchick MR, Davis MR, Doss GA, Pohl LR (1999) Metabolic activation of diclofenac by human cytochrome P450 3A4: role of 5-hydroxydiclofenac. Chem Res Toxicol 12:214–222
61. Makhova A, Shumyantseva V, Shich E, Bulko T, Kukes V, Sizova O, Ramenskaya G, Usanov S, Archakov A (2011) Electroanalysis of cytochrome P450 3A4 catalytic properties with nanostructured electrodes: the influence of vitamin B group on diclofenac metabolism. BioNanoSci 1:46–52
62. Yasui H, Hayashi S, Sakurai H (2005) Possible involvement of singlet oxygen species as multiple oxidants in P450 catalytic reactions. Drug Metab Pharmacokinet 20:1–13
63. Guengerich FP (1978) Destruction of heme and hemoproteins mediated by liver microsomal reduced nicotinamide adenine dinucleotide phosphate-cytochrome P-450 reductase. Biochemistry 17:3633–3639
64. Bondy S, Naderi S (1994) Contribution of hepatic cytochrome P450 systems to the generation of reactive oxygen species. Biochem Pharmacol 48:155–159
65. Shumyantseva V, Makhova A, Bulko T, Kuzikov A, Shich E, Suprun E, Kukes V, Usanov S, Archakov A (2013) The dose-dependent influence of vitamins with antioxidant properties on electrochemically-driven cytochromes P450 catalysis. Oxid Antioxid Med Sci 2:113–117
66. Jeong J, Kim C, Yoon J (2009) The effect of electrode material on the generation of oxidants and microbial inactivation in the electrochemical disinfection processes. Water Res 43:895–901
67. Gade S, Bhattacharya S, Manoj K (2012) Redox active molecules cytochrome c and vitamin C enhance heme-enzyme peroxidations by serving as non-specific agents for redox relay. Biochem Biophys Res Commun 419:211–214
68. Bian C, Xiong H, Zhang X, Ye Y, Gu H, Wang S (2012) Electrochemical detection of BSA damage induced by Fenton reagents in room temperature ionic liquid. Sensors Actuators B Chem 169:368–373
69. Matsuda H, Kinoshita K, Shimida A, Takahashi K, Fukuen S, Fukuda T, Takahashi K, Yamamoto I, Azuma J (2002) Taurine modulates induction of cytochrome P450 3A4 mRNA by rifampicin in the HepG2 cell line. Biochim Biophys Acta 1593:93–98
70. Nakamura T, Ogasawara M, Koyama I, Nemoto M, Yoshida T (1993) The protective effect of taurine on the biomembrane against damage produced by oxygen radicals. Biol Pharm Bull 16:970–972
71. Welsh O, Vera-Cabrera L, Welsh E (2010) Onychomycosis. Clin Dermatol 28:151–159

Mechanistic Basis of Electron Transfer to Cytochromes P450 by Natural Redox Partners and Artificial Donor Constructs

10

Peter Hlavica

Abstract

Cytochromes P450 (P450s) are hemoproteins catalyzing oxidative biotransformation of a vast array of natural and xenobiotic compounds. Reducing equivalents required for dioxygen cleavage and substrate hydroxylation originate from different redox partners including diflavin reductases, flavodoxins, ferredoxins and phthalate dioxygenase reductase (PDR)-type proteins. Accordingly, circumstantial analysis of structural and physicochemical features governing donor-acceptor recognition and electron transfer poses an intriguing challenge. Thus, conformational flexibility reflected by togging between closed and open states of solvent exposed patches on the redox components was shown to be instrumental to steered electron transmission. Here, the membrane-interactive tails of the P450 enzymes and donor proteins were recognized to be crucial to proper orientation toward each other of surface sites on the redox modules steering functional coupling. Also, mobile electron shuttling may come into play. While charge-pairing mechanisms are of primary importance in attraction and complexation of the redox partners, hydrophobic and van der Waals cohesion forces play a minor role in docking events. Due to catalytic plasticity of P450 enzymes, there is considerable promise in biotechnological applications. Here, deeper insight into the mechanistic basis of the redox machinery will permit optimization of redox processes via directed evolution and DNA shuffling. Thus, creation of hybrid systems by fusion of the modified heme domain of P450s with proteinaceous electron carriers helps obviate the tedious reconstitution procedure and induces novel activities. Also, P450-based amperometric biosensors may open new vistas in pharmaceutical and clinical implementation and environmental monitoring.

P. Hlavica (✉)
Walther-Straub-Institut für Pharmakologie und Toxikologie der LMU, Goethestrasse 33, 80336 München, Germany
e-mail: hlavica@lrz.uni-muenchen.de

E.G. Hrycay and S.M. Bandiera (eds.), *Monooxygenase, Peroxidase and Peroxygenase Properties and Mechanisms of Cytochrome P450*, Advances in Experimental Medicine and Biology 851,
DOI 10.1007/978-3-319-16009-2_10,

Keywords
P450 • Redox machinery • Key determinants • Genetic engineering • Biotechnological exploitation

10.1 Introduction

Cytochrome P450 (CYP or P450) enzymes, occurring in organisms from all domains of life [1–5], represent a superfamily of ever-growing *b*-type heme-thiolate proteins [6]. The metalloenzymes are of major importance in both the biosynthesis of endogenous compounds [7, 8] and oxidative clearance of a vast array of drugs, toxins and environmental pollutants characterized by high structural diversity [9, 10]. These processes require the consecutive delivery of two electrons to the ferric P450 catalysts to convert the unreactive atmospheric dioxygen via a generally accepted O-O bond activation cycle to a high-valent iron-oxo species capable of attacking C-H entities and heteroatoms in substrate molecules [11, 12]. Apart from this consensus mechanism, recent data spark particular interest in a "multi-oxidant" concept, providing a rationale for the striking catalytic diversification of P450s [13–15].

Although it would appear that the plethora of *CYP* genes evolved from a common ancestor [16], there exist variations in the nature of the intermediate carrier systems bridging NAD(P)H-derived reducing equivalents to specific terminal P450 acceptors [17]. Thus, in class I P450s comprising bacterial and eukaryotic mitochondrial hemoproteins, a flavin-containing ferredoxin reductase (FdR) usually operates in conjunction with an [Fe_2-S_2] cluster-bearing ferredoxin (Fdx) to shuttle electrons from the reduced cofactor to the heme iron [18]. Noteworthy, in the CYP107H1- and CYP176A1-dependent microbial electron transport chains, unusual FMN-carrying flavodoxins act as functional substitutes for ferredoxins [19–24]. In the class II monooxygenase apparatus comprising microsomal P450 proteins, FAD/FMN prosthetic components in the structure of NADPH-cytochrome P450 oxidoreductase (POR) foster swift electron delivery to the various candidates; here, the NADH-driven cytochrome b_5 (b_5)/b_5 oxidoreductase pair can serve as an alternate redox partner [18]. On the other hand, the unique CYP55A1 enzyme, promoting reductive conversion of nitric oxide to the gaseous nitrous oxide, utilizes NADH as a direct electron supplier without the need for any auxiliary mediator [25]. With other P450s such as human CYP2S1 or bacterial CYP152A/B, the typical $O_2/2e^-/2H^+$ proteinaceous systems fail to stimulate catalytic activity, while utilization of H_2O_2 or fatty acid hydroperoxides permits efficient substrate turnover via the peroxygenase main route based on homolytic peroxy O-O bond scission [26–28]. Similarly, biocatalysts such as CYP5A or CYP74, bringing about rearrangement of endoperoxides and hydroperoxides, respectively, require neither oxygen nor an NAD(P)H-type electron source [29, 30].

As can be readily seen, the pronounced P450-dependent specification of the redox machinery creates the challenging task of more detailed analysis of the structural and functional characteristics of the diverse electron transfer entities to improve our understanding of the observed electrochemical phenomena. In this respect, molecular modeling of composite 3D P450 constructs on the basis of the crystal structure, chemical modification and genetic engineering of a broad spectrum of hemoproteins provided an appreciable picture of both the overall topology of key determinants dictating donor docking/orientation and the nature of the driving forces supporting these events [31, 32]; this also helped assess the redox dynamics of the systems [33]. Circumstantial insight into these processes will be beneficial to the development of novel strategies serving to simplify transmission of reducing power, such as efficient installation of the peroxide shunt pathway to overcome the prohibitive costs for NAD(P)H as the constant electron donor or curtailing of the complex electron transfer conduits [34]; this will give an impetus to exploitation of more flexible P450s in biotechnological areas encompassing the production of fine chemicals, drug processing or degradation of environmental pollutants [35, 36]. The present chapter thus highlights significant breakthroughs in our knowledge about the mechanistic basis of donor/acceptor interactions in the functionally diversified domain of P450s, paving the way for innovative tailoring of versatile redox modules.

10.2 Mechanistic Principles of Electron Transport by Natural Redox Partners of P450s

10.2.1 NADPH-Cytochrome P450 Oxidoreductase (POR)

10.2.1.1 Evolutionary History

Microsomal POR represents a prototypic member of the fairly small family of diflavin redox proteins. The enzyme bears one molecule each of FAD and FMN as cofactors and favors electron transfer from NADPH to eukaryotic P450s or cytochrome *c* as the ultimate acceptors [37, 38]. Precursors of the 78-kDa POR proteins have been hypothesized to arise from ancestral fusion of genes encoding an FMN-binding bacterial flavodoxin and a plant-type FAD-complexed ferredoxin-NADP$^+$ reductase. Subsequent evolutionary steps helped create an α-helical interdomain linker, allowing efficient functional coupling of the two flavins and an N-terminal membrane anchor region (Fig. 10.1) [39–41]. Flavodoxins as such operate in photosynthetic processes or participate in nitrate reduction as well as in methionine and biotin producing pathways [42]. Similarly, ferredoxin-NADP$^+$ reductases display high functional plasticity in supporting auto- and heterotrophic reactions [43].

Analysis of the genetic code for representative PORs from taxonomically diverse eukaryotic species mostly points at the involvement of a single gene in protein expression. Thus, the human gene, located on chromosome 7, contains 16 exons and has been found to be highly polymorphic [44–46]. Likewise, the rat gene carries 16 exons, 15 of which are coding exons. Organization of the latter strictly correlates with functional or structural domains [47]. Moreover, cytogenetic mapping of insect and fungal oxidoreductases suggests them to be single-copy products [48, 49]. Exception to this rule is given by the widespread polyploidy in plants, giving rise to gene duplication and divergence. In this respect, about 54 gene sequences encoding PORs derived from a total of 35 different plant species have as of now been identified, most of the paralogous enzymes at least partially complementing each other [50, 51]. Multiple-alignment studies revealed the majority of full-length POR proteins isolated from mammalian, insect, fungal and plant phyla to share 33–38 % amino acid sequence homology [49].

10.2.1.2 Electrochemical Features of Electron Transfer

The family of POR proteins mediates electron transfer in the NADPH→FAD→FMN→P450 redox system. Here, the flavin cofactors have a vital function in the step-down process from the obligatory two-electron donor NADPH to the one-electron acceptor P450. Using rabbit POR as a probe, flavins were shown to exist as one-electron reduced air-stable blue (neutral) semiquinones (FMN/FMNH$^{\bullet}$, $E_0^{'} = -110\,\text{mV}$ and FAD/FADH$^{\bullet}$, $E_0^{'} = -290\,\text{mV}$) or two-electron fully reduced red (anionic) forms (FMNH$^{\bullet}$/FMNH$_2$, $E_0^{'} = -270\,\text{mV}$ and FADH$^{\bullet}$/FADH$_2$, $E_0^{'} = -365\,\text{mV}$) equilibrating between these states [52–55]. Noteworthy, no shift from the blue di-semiquinone (FMNH$^{\bullet}$, FADH$^{\bullet}$) toward the red species is observed upon increasing the pH of the reaction media [56]. However, the lipid bilayer of membrane-tethered POR was found to impact the redox potential of the FMN/FAD prosthetic groups: application of anionic phospholipids was shown to drive the $E_0^{'}$ for the red forms of both cofactors to more negative values, favoring

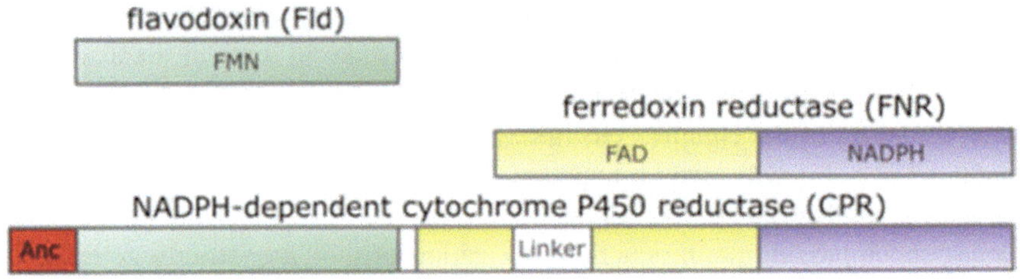

Fig. 10.1 Molecular evolution of NADPH-cytochrome P450 oxidoreductase (POR). The mammalian diflavin protein POR originates from ancestral fusion of the genes of flavodoxin and ferredoxin oxidoreductase with the subsequent creation of a flexible interdomain linker and a membrane anchor serving in proper orientation of the electron donor toward P450s (Data taken from Ref. [51])

electron transfer to P450s [57]. Similarly, structural aberrations in the flavin-binding domains of reductases from different taxa may dramatically affect the redox parameters and abilities to support the catalytic activity of P450s. For example, the redox potential of the $FMNH^{\bullet}/FMNH_2$ pair in yeast POR was shown to be more positive than that for the blue couple. This behavior contrasts the situation in the rat and plant POR species and is similar to the inversion observed with the reductase moiety of the bacterial flavocytochrome CYP102A1 [58, 59].

Elucidation of the precise pathway of flavin-driven redox cycling during electron donation to heme catalytic centers has been fuelled by techniques such as deflavination and reconstitution [60] or dissection of PORs into their component domains [61], permitting more detailed studies on the kinetic and thermodynamic properties of the enzymes. Thus, triggering of the cycle is thought to be brought about by the stable, 635 nm-absorbing $FAD\text{-}FMNH^{\bullet}$ semiquinone potentially generated during a priming reaction [62]. Upon hydride transfer from NADPH ($E_0' = -320$ mV) to the latter species, interflavin electron flow proceeds from $FADH^{-}\text{-}FMNH^{\bullet}$ to yield $FADH^{\bullet}\text{-}FMNH^{-}$. At high molar excess of NADPH, this process is reversible [63]. However, under in vivo conditions, the $FMNH^{-}$ entity acts as the major one-electron supplier to the ferric heme iron of P450s, thereby returning to the resting semiquinone form in the $FADH^{\bullet}\text{-}FMNH^{\bullet}$ duo [52, 64]. Electron cycling between the essentially equipotential members of this redox couple to regenerate $FMNH^{-}$ seems fairly unfavorable and, indeed, occurs as a single-exponential process at a modest rate of 55 s^{-1}, even dropping to a value of 11 s^{-1} when dithionite substitutes for NADPH as the reductant. This suggests cofactor binding to play a pivotal role in regulating internal electron flux [65]. Fully reduced FMN released in the gated electron transfer event serves in P450 reduction via a one-electron step. In summation, microsomal PORs usually cycle in a 1-3-2-1 sequence, denoting the total number of electrons carried by the flavins (Fig. 10.2) [64, 66]. Opposite to this, the microbial CYP102A1 fusion protein undergoes a reduction cycle of 0-2-1-0 lacking any priming reaction [67].

10.2.1.3 Structural Elements Governing Intramolecular Electron Transfer

A drastic step forward in the study of functional domains in POR proteins was made by comparison of the full-length amino acid sequences derived from a broad spectrum of species to unveil highly conserved signature motifs amenable to circumstantial analysis by genetic engineering [49, 50]. Moreover, availability of crystallographic data for human, rat and yeast PORs [68–70] enabled three-dimensional modeling of critical enzyme structures [71, 72]. Thus, investigation of the N-terminal α-helical signal anchor segments of mammalian oxidoreductases disclosed the carboxy termini to be located on the cytoplasmic side of the endoplasmic reticulum, with the first 55–56 amino acid residues being sufficient for stable membrane insertion/retention, proper orientation and maintenance of catalytic efficiency [73–75].

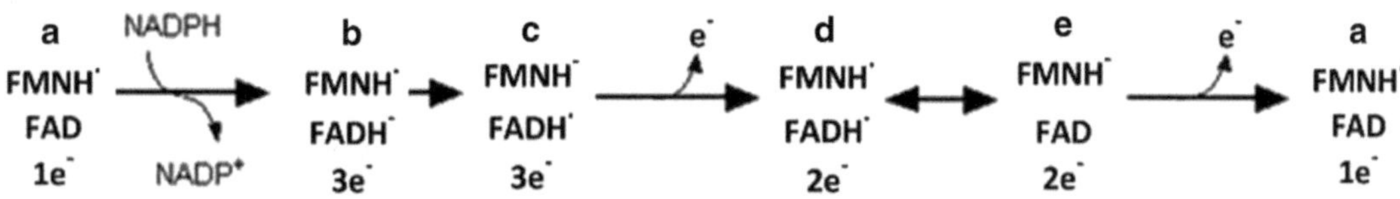

Fig. 10.2 POR-supported intra- and intermolecular electron conduction to P450s. Hydride transfer from NADPH to the stable semiquinone (**a**) elicits sequential formation of the fully reduced (**b**, **c**) intermediates to enable one-electron supply to ferric P450 associated with release of a resting semiquinone duo (**d**). Electron swapping between the latter redox couple regenerates a fully reduced cofactor (**e**) again permitting electron donation to P450s. The mammalian oxidoreductase thus cycles between the 1- and 3-electron reduced states

Despite the fairly low overall sequence identity (33–43 %) of POR enzymes from various organisms [49, 50], the FMN-binding domains exhibit a high degree of conservation [76]. Chain tracing in the scaffolds of human and rat proteins revealed an α-β-α architecture composed of a five-stranded parallel β-sheet in the core fold flanked by a variable number of α-helices, with the FMN cofactor positioned at the tip of the C-terminal side of the β-sheet [69, 71, 77]. Using the human enzyme (hPOR; NCBI reference sequence NP_000932.3) as a template, molecular docking and site-directed mutagenesis experiments suggested a set of residues such as Q90, T91, T142, H183 and N185 to be involved in FMN fixation, though Y143 and Y181 obviously act as key players [76, 78, 79]. The aromatic side chains of the two tyrosines, sitting on the *re*- and *si*-face, respectively, of the isoalloxazine ring, clasp the FMN unit at nearly the same distance of 3.5 Å [71, 77]. Both positions are conserved in the rat homolog [69] sharing 94 % sequence identity with the human counterpart [79], and Y→D exchange in the rodent protein was shown to indeed block efficient electron transfer [80]. Moreover, replacement in the human catalyst of F184, lying close to the pyrimidine tail of the cofactor, with leucine or glutamine caused a 40- to 50-fold increase in the K_d value for FMN association. This was interpreted to reflect a vital role of F184 in stabilization of the electron carrier [78]. Strikingly, L86 and L219, deeply buried in two hydrophobic cores of the FMN domain of POR from *Anopheles minimus*, aligns with F86 and F219 in the human analog. Experimental introduction into the insect enzyme of phenylalanine residues in place of the leucines proved to be beneficial to FMN docking and protein folding [75]. Of note, X-ray crystallography of oxidoreductase from the yeast *Saccharomyces cerevisiae* helped discover a second FMN-binding region at the interface of the linker and standard cofactor-bearing domain. The novel site displays low conservation throughout the gene family, with only two residues, namely, T71 and D187 corresponding to T93 and D211 in hPOR, being invariant [70]. It has been hypothesized that a single FMN molecule shuttles between the structural doublet associated with semiquinone transition from the neutral to the anionic state [70].

A fragment spanning about 40 amino acid residues of predominantly polar character bridges the gap between the FMN/FAD-harboring loci. This linker is speculated to serve in proper orientation of the flavins, the isoalloxazine rings of which make an angle of about 150° to each other and reside at a minimum distance of 3.5 Å [41, 69, 79]. In fact, mutations in the short random-coil hinge, preceding the FAD-connecting unit composed of residues G235 to R246 in the hPOR structure, induce drastic rearrangement of the FMN/FAD topology impacting intramolecular electron transfer [81]. Crystallographic analysis of the FAD-docking region in the rat protein showed the isoalloxazine entity to be hosted at the boundary between the cofactor- and NADPH-binding site, with the remainder of the molecule extending to the interface between the FAD-binding pocket and the connecting domain [69]. Site-directed mutagenesis was used to verify the functional importance of the various determinants. This demonstrated Y456 to make van der Waals contact with the *si*-side of the FAD isoalloxazine structure and to hydrogen-bond to the ribityl 4′-hydroxyl, while amino groups in the side chains of R454 (equivalent to R457 in hPOR), G488 and T491 stabilize the negatively charged pyrophosphate. Moreover, the aromatic nucleus of Y478 stacks on one side of the adenine moiety [68, 69, 82]. Interestingly, the interplay of the S457/D675/C630 triad may have a dual role in the control of the flavin redox potential and stabilization of the transition state to facilitate hydride transfer [83, 84].

A set of homologous residues lining the NADPH-binding cavity, constituted of alternating α-helices and β-strands, operate in fixation and orientation of the cofactor in a bipartite mode. Thus, highly conserved amino acids encompassing C566, S596, R597, K602 and Y604 (rat POR numbering) make up a specific motif attracting the 2′-phosphate of NADPH via H-bonding or salt-bridging, such as to cause discrimination against NADH [79, 85–87]. In accord with this, introduction of hydrophobic elements in place of the positively charged arginine and lysine residues at positions 597 and 602 to create the triple mutant R597M/K602W/W677A resulted in a 170-fold increase in the apparent binding affinity for NADH compared to the wild-type enzyme, paralleled by an IC_{50} value for inhibition by $NADP^+$ that was 50-times higher than that of the parent enzyme [88]. Similarly, simple W676A exchange in hPOR allowed the NADH-dependent reductive potency to become equivalent to that of the NADPH-driven event [87]. On the other hand, this manipulation was found to compromise NADPH-promoted reduction beyond the two-electron level owing to slow release of $NADP^+$ from the active site upon first hydride transfer, suggestive of a vital function of the C-terminal tryptophan in electrochemical processes [89]. The mechanistic basis of such an action relies on the assumption that the π-stacking indole of W677, building a lid above the *re*-face of the FAD isoalloxazine, moves away to permit direct contact of the flavin moiety with the nicotinamide ring of NADPH required for efficient hydride transfer [69, 72]. Here, local movement of the short G631-N635 loop may be beneficial to NADPH/$NADP^+$ binding/release [90]. An overall diagram of the POR polypeptide fold disclosing the topology of the diverse cofactor-binding domains is presented in Fig. 10.3.

It should be pointed out that POR enzymes are highly polymorphic proteins. Currently, about 48 missense mutations have been identified in the human reductase (www.cypalleles.ki.se/por.

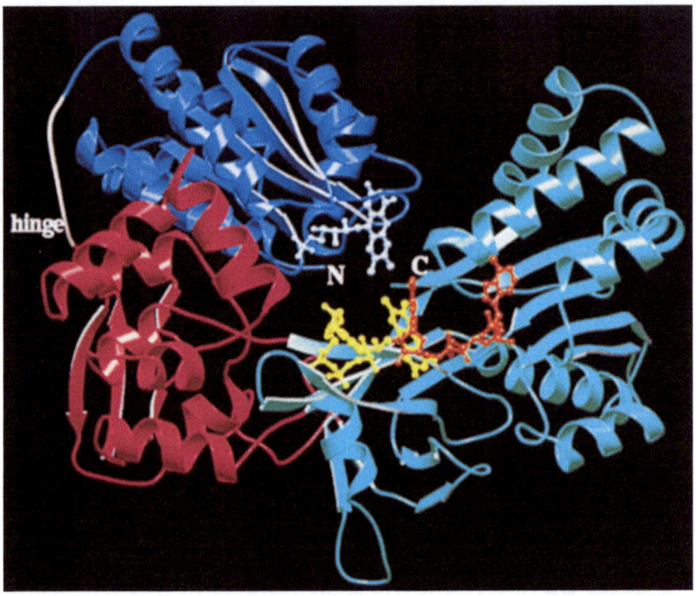

Fig. 10.3 Ribbon diagram illustrating the overall polypeptide fold and topology of POR. The FMN-binding domain is given in *blue*, while the NADP(H)- and FAD-docking sites are presented in *green*. The connecting interdomain fragment is depicted in *red*. Cofactors are presented in the ball and stick mode (PDB ID: 1AMO) (Reproduced from Ref. [69])

htm), part of which overlap with conserved residues critical for FMN/FAD/NADPH fixation [45, 79, 91]. Examples of this are found in the most important variants T142A, Y181D, R457H, Y459H, V492E, C569Y, R600W and Y607C [45, 79]. These amino acid substitutions have been recognized to be deleterious to electron transfer and, consequently, cause disordered steroidogenesis along with skeletal malformations when the allelic hPORs are to serve as obligatory donors to CYP17A1 and CYP19A1 [91, 92]. Moreover, the altered phenotypes may impact the P450-catalyzed metabolic biotransformation of both curative drugs/prodrugs and toxins [45, 46, 93]. Noteworthy, the Y181D-induced perturbation of electron flow/P450 activity has been reported to undergo restoration upon the addition of excess FMN to the assay media [76, 94]. In this case, the potential existence of a second FMN-binding site (see above) might permit the exogenous cofactor to act as a bypass.

10.2.1.4 Structural Features Steering Functional POR Docking to P450s

To bring about efficient electron shuttling from POR enzymes to P450s, a large-scale conformational rearrangement of the FMN domain is required. Available 3D structures of mammalian reductases display a closed conformation of the core region with the isoalloxazine ring of FMN being shielded by the FAD cofactor at a distance ranging from 4 to 5 Å [68, 69]. Hence, electron donation to P450s necessitates concerted movement of the domains leading to an open state associated with exposure of the FMN moiety to the solvent to enable contact with the hemoproteins. In this regard, the "closed-open" transition, as studied with free or membrane-anchored POR, was recognized to be conducted by the flexible hinge motif adjusting the distance between the FAD/FMN entities to 29–60 Å [81, 95, 96]. More detailed analysis by sophisticated spectroscopic techniques revealed the POR molecule to, indeed, toggle between a multiplicity of closed and open conformations in solution [97, 98]. Generally, opening is driven by flavin reduction, whereas closure predominates in the oxidized enzyme and is supported by NADPH binding to facilitate loading of reducing equivalents [99, 100]. These findings are in line with the "swinging" model of POR-mediated electron transfer from the nicotinamide coenzyme to the heme iron of P450s [101].

Based on the construction of model complexes between the redox partners, a docking area of ~870 $Å^2$ was calculated to guide productive encounter of the solvent-exposed FMN domain with P450s [81]. This patch, located on the surface of the extended reductase molecule, bears an electronegative profile arising from accommodation of three clusters of putative salt-bridging residues encompassing E92, E93, D113, E115, E116, E142, D144, D147 and D208 (rat POR numbering), speculated to provide a rationale for snuggly fit of the electropositive proximal face of the different P450s obviously binding in a very similar fashion [69, 71, 81, 102, 103]. The negatively charged elements surrounding the FMN moiety were predicted to form a cleft allowing a minimal distance of ~12 Å between the cofactor and the heme group [81]. The hypothetical acidic contact sites were substantiated by genetic engineering: Mutation of D113, E115 and E116 to alanine disclosed the residues to stabilize the CYP2B1/POR adduct on the one hand and open new avenues to more efficient electron transfer to the hemoprotein partner on the other [103]. Moreover, replacement of hPOR amino acids corresponding to E142, D144 and D147 in the rat homolog with the less bulky polar serine or glycine substitutes was found to moderately impinge on the catalytic efficiency (k_{cat}/K_m) of CYP2D6, while D208N exchange caused a drastic fall in P450-dependent activities [71, 104]. Two thirds of the determinants examined display 60–90 % conservation across the multitude of taxonomically diverse reductase species, the rest being invariant [76].

10.2.2 Cytochrome b_5

Cytochrome b_5 (b_5), occurring in a wide range of phyla, is a membrane-anchored amphipathic hemoprotein operating in concert with POR or NADH-cytochrome b_5 oxidoreductase as electron donor to desaturating systems involved in fatty acid synthesis and plasmalogen-producing enzymes [105]. Of note, soluble forms of human b_5 and NADH-cytochrome b_5 oxidoreductase found in erythrocytes were shown to be capable of reducing methemoglobin [106, 107]; here, deletion of codon 298 in the gene of the flavoprotein component was detected to cause functional deficiency associated with hereditary methemoglobinemia [108]. Moreover, the ferrihemoglobin-coupled redox triad brought about O_2-dependent substrate turnover in a monooxygenase-type reaction [109]. In parallel, a considerable number of P450s were recognized to have substrate-specific obligatory requirement for electron supply by b_5 [110, 111].

10.2.2.1 Topology of the Membrane-Spanning and Heme-Binding Domains of Cytochrome b_5

Two mammalian b_5 isoforms were identified, one inserted into the endoplasmic reticulum and the other bound to the outer membrane of mitochondria. These proteins arise from different genes [112]. The hydrophobic membrane anchor of the microsomal homolog, functioning as a static retention signal, was shown to span the bilayer of the endoplasmic reticulum such that the carboxy-terminus extends to the lumen of the organelle [113, 114]. However, mutation of the C-terminal L124/M125/Y126 triad in rat b_5 to alanine was found to induce location of the engineered hemoprotein in both the cytosol and microsomal membrane, suggestive of the existence of loosely- and firmly-integrated forms differing by the overall content of α-helical structure [115, 116]. In fact, the membrane-interactive tail of b_5 has been detected to function as a stop-transfer sequence giving rise to inversion of protein orientation in the endoplasmic reticulum to permit versatile processing of nascent precytochrome b_5 during topogenesis, resulting in final positioning of the integral electron carrier in the N_{out}-C_{in} mode [117]. Another triad of potential interest embedded in the 43-amino-acid membrane-binding domain of b_5 refers to tryptophan residues at locations 108, 109 and 112. However, studies with the W108L/W112L double mutant failed to disclose any impact on electron transfer to CYP2B4 as a probe acceptor [118]. Finally, attention was drawn to P115, forming a 26° kink in a helix when occurring in the *trans* conformation. Surprisingly, P→A exchange resulted in normal insertion of the mutant into the membrane and a wild-type enzyme level of activity in a P450 test system [119].

Microsomal b_5, being 60 % α-helical, is a fairly small polypeptide composed of 134 amino acid residues, with the cytosolic heme-containing region showing ~92 % sequence identity throughout the different mammalian isoforms [105, 120]. Availability of the crystal and solution structure of the protein permitted insight into the architecture of the heme-binding pocket. Thus, the prosthetic group was shown to reside in a hydrophobic crevice, the iron atom being coordinated to histidines at positions 39 and 63; the latter reactant has some exposure to solvent via a water channel [121–123]. Dependence of the heme-holding stability on the histidine axial ligation was confirmed by H39S/C mutations, also affecting the spin state of the heme iron [124]. Apart from steric factors, changes in hydrophobicity of the heme microenvironment may modulate the electrochemical properties of the hemoprotein [125]. In accord with this, V45H/E substitutions were found to shift the redox potential of the wild-type protein ($E_0^{'} = -10$ mV) to values of +8 mV and −26 mV, respectively [126]. Similarly, manipulation of hydrophobicity by replacement of V61 with histidine revealed to influence interaction of the heme with its pocket, resulting in broadening of the latter; this moved $E_0^{'}$ of the mutant by +21 mV [127]. Special interest focuses on the interplay of the F35/F58 duo, stabilizing heme docking through π-stacking overlap with the porphyrine macrocycle [128, 129]. Moreover, phenylalanine-35 is part of a hydrophobic patch of 350 Å^2 on the surface of b_5 [130] and member of a network that includes Y74 and the axial H39 being in direct van der Waals and electrostatic contact with the heme [131, 132]. Apart from this, the conserved F35 is pivotal to fine tuning of the redox potential: F35Y exchange was discerned to make E_0' 66 mV more negative compared to the parent protein [133]. Finally, P40, another component of the surface patch producing a sharp γ-bend in the polypeptide chain, is believed to significantly contribute to a fixed folding pattern of the heme pocket due to its rotational restriction [134]. It should be kept in mind that, in contrast to the crystalline state, b_5 is heterogenous in solution due to the presence of two isomers, differing with respect to 180° rotation of the heme plane around the axis defined by α,γ-meso protons [135]. A diagram of key structural motifs in the b_5 backbone chain is given in Fig. 10.4.

10.2.2.2 Interaction of Cytochrome b_5 with Electron Donors

The microsomal FAD-containing NADH-cytochrome b_5 oxidoreductase acts as a physiological electron donor to the ferric b_5. Anaerobic photoreduction of the FAD moiety was observed to form the red anionic semiquinone being in equilibrium with the blue neutral species. The latter turned out to be the primary intermediate in the NADH-driven hydride transfer process [136, 137]. Here, the conserved T66 entity in the reductase structure was shown to participate in modulation of the rate-limiting interconversion of the semiquinone forms [137]. Furthermore, mutation experiments verified the importance of the specific arrangement of R63, Y65 and S99 in the β-sheet barrel core of the flavoprotein in maintenance of FAD docking by electrostatic and H-bonding attraction of the *si*-face of the isoalloxazine ring [138, 139]. Similarly, the backbone amide nitrogen of M126 forms a hydrogen bond to the phosphate oxygen of the cofactor [140]. In addition, a series of residues including K110, S127, G179 and P275 were predicted to participate in the anchoring and proper positioning of the NADH electron donor [140–145]. In this regard, the active-site C273 was considered to be critical for accurate orientation of the nicotinamide nucleus prior to hydride transfer [146]. Most interestingly, G179

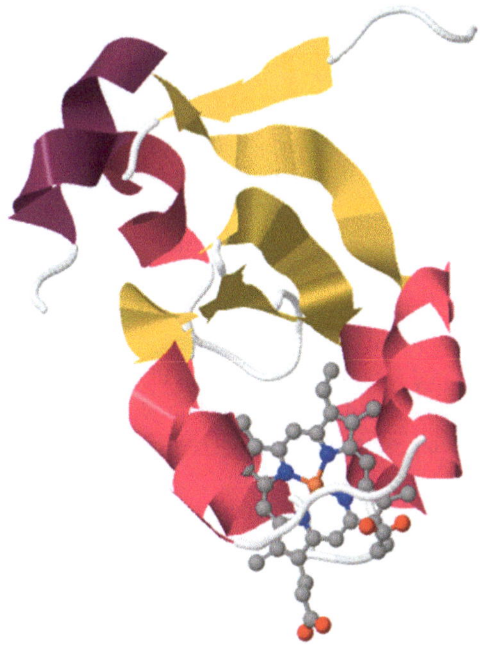

Fig. 10.4 Schematic structure of bovine cytochrome b_5. The approximately cylindrical molecule (PDB ID: 1CYO) houses α-helices 2–5 (*in red*) clustering around the prosthetic heme group given in *grey* balls and sticks. A five-stranded β-sheet (in *yellow*) in the center of the amphiphilic polypeptide separates the heme-binding pocket from a more peripheral helical segments 1 and 6 (in *violet*) (Data taken from Ref. [122])

and D239 were recognized to be required for efficient NADH/NADPH selectivity [144, 147].

Rapid electron transfer from NADH-cytochrome b_5 oxidoreductase to b_5 was shown to require N-terminal myristoylation of the flavoprotein to stabilize its orientation in the endoplasmic reticular membrane as a prerequisite for optimal productive encounter of the redox partners [148]. Here, circumstantial analysis implicated three reductase lysine residues hosted at positions 41, 125 and 163 in complementary charge pairing with the single exposed porphyrine propionate and a cluster of glutamate carboxyl groups at locations 43, 48, 49 and 53 (rat hemoprotein numbering) in the b_5 polypeptide, surrounding the heme edge at a distance of ~12 Å [149–152]. Qualitatively, the same b_5 carboxyls were recognized to be essential for electrostatic interaction with prospective cationic groups in the alternate POR electron donor

[153]. Employing the covalently cross-linked diflavoprotein/b_5 heterodimer as a model for a functional electron-transfer complex, FMN depletion unveiled the cofactor-binding POR domain to be the active center for responsiveness to the hemoprotein. Here, lysines at positions 72, 74 and 75 are likely candidates for charge pairing with the b_5 carboxyls [154]. As evidenced by the detrimental effect of removal of the N-terminus of POR on the functional coupling with b_5, the intact hydrophobic tails of both redox proteins are required for efficient cross-talk between the partners facilitated by free lateral movement in the plane of the membrane [155]. Here, the nature of the system utilized for reconstitution of the matrix may steer the kinetics of intra- and intermolecular electron transfer [156]. Noteworthy, flash-induced b_5 reduction by POR was found to proceed at a first-order rate about 10 % that measured with NADH-cytochrome b_5 oxidoreductase [157, 158].

10.2.2.3 Characteristics of the Catalytic Cytochrome b_5/Cytochrome P450 Redox Adduct

Cytochrome b_5 plays a supportive role as a modifier of NADPH/POR-driven monooxygenations depending on the type of substrate and P450 species involved. For example, the presence of b_5 invariably improves efficiency of product formation from methoxyflurane by CYP2B4 [159], fosters mephenytoin and chlorzoxazone turnover by CYP2C19 and CYP2E1 [160], and stimulates testosterone biotransformation by CYP3A4 [161].

The mechanism by which b_5 impacts P450 activity has been extensively studied. When bound to ferric P450, the intermediate carrier elicits a low-to-high spin transition in the iron coordination sphere of the heme chromophore of the terminal pigment [162, 163]. Owing to the unfavorable discrepancy in the midpoint potential between the Fe^{3+}/Fe^{2+} couples of b_5 (−2.6 to +5.1 mV) and substrate-free ferric P450 (~ −400 to −300 mV), acceptance by the latter species of the first electron from the presumed donor protein can be excluded [33, 125]. In contrast, E_0' for the labile oxyferrous P450 form is raised to 50 mV [164], permitting introduction of the second electron by ferrous b_5 [111, 165] at a rate faster than autoxidation of the Fe^{3+}-O_2^- intermediate associated with H_2O_2 release. This is expected to enhance economy of product formation at the expense of superoxide [166, 167]. It has to be noted that b_5 may also exert non-redox, conformational effects on P450s. Thus, the modifier was shown to increase the steady-state level of the substrate-bound iron-oxo complex through lowering the energy of activation [162, 168] and to influence the rate of regeneration of ferric P450 from the oxygenated precursor as an index of the velocity of oxidative substrate turnover [169]. Precedent to this kinetic behavior is given by the action of apocytochrome b_5 on the rate of productive decay of substrate-bound oxyferrous CYP101A1 [170]. Moreover, interaction of holo-/apo-b_5 with CYP17 triggers rearrangement of the iron-dioxygen ligand necessary to awaken lyase activity [171, 172] or promotes repositioning of substrate to favor 16-α-hydroxylation [173]. Generally, incorporation of heme-depleted b_5 into reconstituted systems containing members of the CYP2 and CYP3 families was found to enhance typical catalytic activities to differing extents [174, 175]. Specific studies with the CYP4A7 species suggested apo-b_5 to possibly alter the conformation of the substrate-binding pocket and/or accelerate product release [176]. In summation, these findings point at a dual role of b_5 as an electron donor on the one hand and an allosteric effector on the other [177, 178].

The hydrophobic α-helical, membrane-spanning domain of b_5 was demonstrated to play a dominant role in productive association with CYP2B4 [179]. However, introduction of alanines into the membrane anchor, expected to cause all amino acids distal to the insertion to undergo a 100° rotation, failed to disrupt any specific helix-helix interactions. This was interpreted to mean that b_5/P450 binding proceeds through a nonspecific mechanism [180]. In contrast, the S90-D104 fragment, linking the heme domain of b_5 with the C-terminal hydrophobic sequence, was postulated to restrict orientation of the donor/

acceptor heme regions, facilitating formation of a functional complex [181]. Stability of the latter was shown to be granted by electrostatic and H-bonding attraction of complementary P450 residues by the invariant b_5 amino acids Y30, E44, E48, D60 and T65, cooperating with the exposed heme propionate group [182–184].

10.2.3 Ferredoxins

Ferredoxins are small, soluble iron-sulfur proteins mediating electron transfer to P450s and other proteins such as nitrate and sulfite reductase. Classification of the intermediate carriers comprises different prototypes depending on the total number as well as Fe/S-proportion of the prosthetic clusters defining the active-site structure of the various electron shuttles [185]. In this regard, [Fe_2-S_2]-bearing ferredoxins, occurring in plants, bacteria and vertebrates, are of special interest [186]; the latter category includes both pro- and eukaryotic representatives [187]. Here, most extensive studies focus on the mammalian mitochondrial adrenodoxin (Adx) and the microbial putidaredoxin (Pdx) [188, 189], donating electrons to class I P450s [18]. Electron bridging requires prior transfer of reducing equivalents to ferredoxins by FAD-carrying NAD(P)H-ferredoxin reductases generally belonging to distinct types of unrelated protein families [40]. With respect to this, NADPH-adrenodoxin reductase (AdR) and NADH-putidaredoxin reductase (PdR) were uniformly assigned glutathione reductase-type redox proteins [190, 191].

10.2.3.1 Recognition of Adrenodoxin by Redox Partners

Site-directed mutagenesis experiments revealed the core domain of Adx, housing a single [Fe_2-S_2] cluster, to be mandatory for Adx/AdR association, while a second, acidic interaction site encompassing residues at positions 56–90 serves in docking of both AdR and P450s [192]. In accord with this, D72, D76 and D79 of Adx build up a tight H-bonding network with R211, R240 and R244 of AdR [193], but equally well operate in fixation of the steroidogenic CYP11A1 [194]. Noteworthy, the salt bridge between the invariant E74/R89 residues turned out to exert a principal stabilizing force impacting the orientation and redox properties of the iron-sulfur motif in parallel to AdR and CYP11A1 binding [195]. Genetic engineering of Y82 suggested the amino acid to be of importance in complex formation of Adx with CYP11A1 and CYP11B1, but to leave electron transfer unaffected [196]. In contrast, histidine at position 56 was recognized to control the integrity and ligand field of the protein region surrounding the [Fe_2-S_2] cluster [197]. Thus, H56T exchange was found to shift the redox potential of the wild-type Adx (−274 mV) to a value of −340 mV, causing a ~2.3-fold increase in the rate of CYP11A1 reduction [198]. Similarly, the vicinal T54 was recognized to modulate the protein's redox state: conservative T→S replacement lowered $E_0^{'}$ by ~60 mV as compared to the native ferredoxin without affecting AdR coupling and CYP11A1 reduction, though there was a marginal decrease in K_d for spectral binding of the hemoprotein [199, 200]. Of note, C-terminal truncation (Δ113–128) of Adx followed by S112W substitution was found to cause an 11-fold increase in the rate of CYP11A1 reduction associated with a 60-fold rise in the enzyme's catalytic efficiency [200]. Finally, sequential deletion of residues E47, G48, T49, L50 and A51, located in a surface loop covering the iron-sulfur center, disclosed the domain to be crucial to regulation of the redox potential and functional coupling of AdR and CYP11A1 [201, 202].

As can be readily seen, the extensive spacial overlap of the interaction sites of Adx for AdR and P450 makes formation of a ternary complex improbable [203]. This view is substantiated by results from carbodiimide-mediated covalent crosslinking of Adx carboxylates to lysines on either AdR or CYP11A1. Structure-based assessment of the individual crosslink positions excluded a cluster model, but unequivocally suggested the ferredoxin to act as a mobile electron shuttle [204, 205]. Here, transport of reducing equivalents was hypothesized to proceed via both monomeric or dimeric Adx species [206]. The architecture of Adx-Adx assembly resulting in an asymmetric dimer was disclosed by crystal-based molecular modeling [207].

10.2.3.2 Molecular Recognition of Putidaredoxin by Redox Partners

Major driving forces in Pdx/PdR recognition were proven to encompass steric complementarity along with hydrophobicity and polarity. Thus, modeling studies combined with crystal-based mutagenesis experiments to modify both bulkiness of prospective key amino acids and their efficiency in charge pairing or van der Waals contacts identified the voluminous Y33 and R66 of Pdx, flanking the 365 Å^2 protein-protein interface, to bind to R65/T66 and E335, respectively, in PdR. Substitution of the two ferredoxin residues with amino acids of lower molecular mass significantly increased the binding affinity of mutated Pdx to PdR, but drastically diminished k_{cat} for electron transfer to the iron-sulfur cluster in view of moderate effects on E_0' [208, 209]. This was interpreted to mean that the bulky side chains of tyrosine and arginine prevent tight docking of Pdx, so that transfer of reducing equivalents may occur via alternate pathways. In fact, optimal orientation for swift electron flow from FAD to the [Fe_2-S_2] center was predicted to be provided by interaction of W310 of PdR with D38 of the intermediate carrier [208, 209]. Moreover, ion pairing of the two residues is expected to lower the activation free energy for reduction of the metal cluster [189]. Evaluation of mutation and crosslinking data suggested the α-helical E72 of Pdx to form a salt bridge with K409 of PdR serving to establish and stabilize the electron transfer complex [209, 210], while the adjacent C73 seems to not only modulate the ferredoxin's redox potential but to also define spacial approach of the subunits of the redox partners [208, 211]. Owing to flexibility of its aromatic ring, the C-terminal W106 of Pdx, oriented toward the center of the groove close to W330 of PdR [208], is thought to play a mediating and/or regulating role in the electron transfer process [212].

Importantly, the tryptophan at position 106 is of dominant importance in functional coupling of Pdx with the camphor-hydroxylating bacterial CYP101A1. Here, W106 is of higher relevance to transfer of the second electron to the oxyferrous hemoprotein than to donation of the first reducing equivalent to the ferric enzyme. This was argued to arise from the fact that the bulky, rigid indole ring of the tryptophan residue is apt to penetrate deep enough to approach the heme-binding loop of CYP101A1 [213] and induce structural changes required for acceleration of dioxygen activation, thus assisting the role of Pdx as an allosteric effector [189, 214]. It thus appears that the essential tryptophan exists in a conformational microheterogeneity [215]. In addition, D38 of the ferredoxin component was recognized to represent another hot spot in the two-step reductive event [214]. Starting from 3D modeling and molecular dynamics simulations, a series of amino acids such as D34 of the intermediate carrier were hypothesized to be likely candidates for intermolecular salt bridge formation, affording fixation of the Pdx/CYP101A1 complex [216]. In fact, D34N mutation was shown to depress catalytic efficiency (V_{max}/K_m) of the P450 system to a level 44 % of that found with the wild-type Pdx species [217]. Moreover, S42C exchange in the polypeptide clearly impacted donor/acceptor interaction [211].

Comparative evaluation of the general docking mode of the redox partners disclosed partial overlap of the proposed binding areas for PdR and CYP101A1 on the surface of the Pdx molecule, suggesting that the reductase and the hemoprotein cannot simultaneously interact with the ferredoxin [211]. This view seems to be in contrast to the competent function of a ternary PdR-Pdx-CYP101A1 fusion protein reported by others. Mobility of the fixed Pdx subunit of the latter construct appeared to be, nonetheless, high enough to pass electrons to exogenous native CYP101A1 introduced into the assay mixture [218]. In accord with this, analysis by optical biosensor techniques demonstrated the covalently immobilized three-component complex to exhibit only loose arrangement between Pdx and the terminal acceptor [219]. Also, studies on the kinetic behavior in dependence on the molar proportion of the individual redox partners supported the notion that Pdx acts as an electron transfer shuttle between PdR and CYP101A1 in analogy to the Adx-promoted route [220]. This raises the question as to what extent the bacterial CYP101A1-dependent system might be comparable to the mitochondrial CYP11A1-steered redox chain. Thus, inspection of the superimposed Adx/Pdx 3D structures, no doubt, permits one to discern significant homology (Fig. 10.5). Despite this, the two ferredoxins cannot substitute for each other in the two catalytic pathways owing to pronounced discrepancies in a series of functional determinants: (1) none of the acidic residues of Pdx corresponding to those vital to Adx fixation to CYP11A1 participate in intermolecular interactions with CYP101A1; (2) while T49 of Adx controls the redox dynamics of the iron-sulfur cluster, the equivalent S44 in Pdx fails to play such a role; (3) whereas the C-terminal aromatic tryptophan of Pdx is pivotal to tight CYP101A1 docking, the extended analogous region of Adx is deficient in such a P450-binding element. The interplay of these shortcomings causes Pdx and Adx to be unable to donate the second electron to the oxyferrous forms of the heterologous hemoproteins [221].

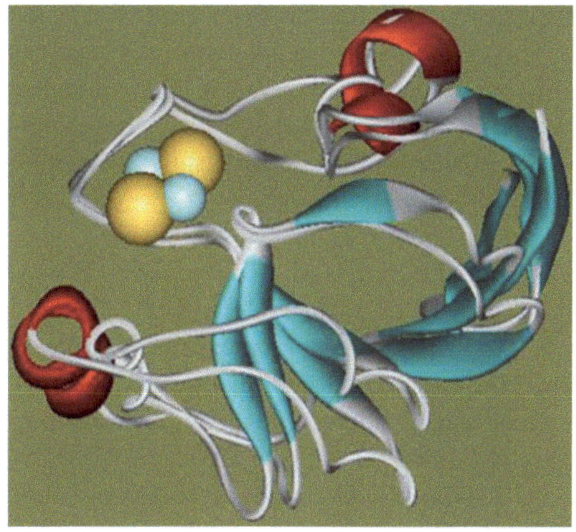

Fig. 10.5 Superposition of the three-dimensional structures of adrenodoxin (*top*) and putidaredoxin (*bottom*). The Adx (PDB ID: 1AYF) and Pdx (PDB ID: 1PUT) proteins, representing typical examples for vertebrate-type ferredoxins characterized by a sequence homology of ~35 %, display a similar planar geometry of the iron-sulfur cluster-containing region (spheres in *yellow* and *blue*). Generally, the overall folding topology of the α-helical and β-sheet elements shows a high degree of identity, with a 1.64 Å r.m.s. deviation between the two electron carriers (Reproduced from Ref. [221])

10.2.4 Unorthodox Electron Transfer Chains

Though P450s usually receive reducing equivalents from their dedicated redox partners, nonconventional electron transfer chains are frequently constructed to facilitate in vitro reconstitution of the donor/acceptor modules. For this purpose, the vertebrate-type ferredoxin/ferredoxin reductase components belong to the most frequently used surrogates of native intermediate carriers. Thus, the mitochondrial AdR/Adx couple turned out to interact with intact microsomal CYP1A1 such as to support erythromycin *N*-demethylation at higher efficiency compared to the inherent electron donor [222]. Similarly, Adx was demonstrated to cross-react with CYP2B enzymes, N-terminal hemoprotein truncation eliciting balanced reductive potency between the ferredoxin-promoted and P450 reductase-driven systems [223, 224]. Also, the truncated microsomal CYP17A1 and CYP21A2 proteins show higher steroid 17α-hydroxylase and 21-hydroxylase activity, respectively, with AdR/Adx compared to POR as the electron supplier [225]. Interestingly, CYP46A1, predominantly functional in cholesterol 24-hydroxylation in the brain, was found to interact with Adx as a redox component [226]. The mitochondrial carrier also sustains electron donation to the bacterial steroid 15β-hydroxylase CYP106A2 from *Bacillus megaterium* [227]. The system even operates at elevated catalytic capacity when AdR is replaced with NADPH-flavodoxin reductase from *Escherichia coli* to establish a novel robust redox chain [228]. Noteworthy, a mitochondrial ferredoxin reductase/ferredoxin unit from the fission yeast *Schizosaccharomyces pombe* was found to have >50 % sequence similarity with the mammalian AdR/Adx counterpart. The redox pair supports CYP11A1-catalyzed biotransformation of 7-dehydrocholesterol [229].

In addition, [Fe_2-S_2] proteins from non-mitochondrial sources have been demonstrated to transfer electrons to heterologous P450s. For example, plant-type ferredoxin and NADPH-ferredoxin reductase from spinach chloroplasts promote oxidative substrate turnover by microsomal CYP1A2 and CYP3A4 [230, 231] as well as 25-hydroxylation of vitamin D_2 by CYP105A1 from *Streptomyces griseolus* [232]. Furthermore, bacterial electron transport systems such as the PdR/Pdx pair proved to foster β-carotene hydroxylation by the thermostable CYP175A1 species [233]. When working in concert with PdR, the microbial palustrisredoxin A factor readily feeds reducing equivalents to CYP199A2, preferentially metabolizing four-substituted benzoates [234]. Also, reconstitution of linredoxin and linredoxin reductase from a soil pseudomonad with CYP2B4 yields a collective efficiently metabolizing benzphetamine [235].

In some instances, *Escherichia coli* flavodoxin/flavodoxin reductase was detected to provide a basis for facile electron donation to microsomal P450s such as CYP1A2 [230], CYP3A4 [231] and CYP17A1 [236], but to equally-well pass electrons via a ping-pong mechanism to the microbial fatty acid oxidases CYP102A1 and CYP152A2 [237, 238]. Of note, a catalytically active system could also be established by employing flavodoxin reductase together with the unusual flavodoxin cindoxin from *Citrobacter braakii* as redox partners for CYP107H1 from *Bacillus subtilis*, having a role in biotin biosynthesis [19]. Finally, electron supply by POR from the yeast *Candida apicola* to the myristate-metabolizing CYP109B1 seems to be a unique case, where a eukaryotic diflavin reductase acts as a versatile electron donor to a bacterial hemoprotein [239]. In summation, cross-reactivity of electron carriers with a diversity of heterologous P450s can be reconciled with evolutionary conservation of a common functional domain architecture steering donor/acceptor interactions [32].

10.3 Topology of Critical Regions in P450s Dictating Interaction with Natural Redox Partners

10.3.1 Docking of NADPH-Cytochrome P450 Oxidoreductase

Data from chemical/immunochemical modification, molecular modeling and targeted mutagenesis were collated to generate an overall picture of key determinants in P450s responsible for POR fixation. Here, the N-terminal membrane-spanning signal anchor sequence of microsomal P450s seems to have a general role in protein-protein association: deletion of the membrane-immersed portion of CYP1A2 drastically decreases affinity for POR [230]. Genetic tailoring of a (Δ2–27)-variant of CYP2B4 proved to be detrimental to POR binding, resulting in a pronounced drop in the efficiency of electron transfer to the recipient [179, 240]. Chemical modification of the enzyme's N-terminal α-amino group through covalent attachment of fluorescein isothiocyanate was recognized to compromise reductase docking via motional perturbation of the fluorophore-labeled region, eliciting a long-range effect on some distant patch involved in productive POR complexation [241]. Similarly, truncation of CYP2D6 was found to increase the K_d value for reductase binding by a factor of 11 [242]. Surprisingly, analogous manipulation of CYP2C3 and CYP2E1 failed to impede fixation of the flavoprotein [243, 244]. In contrast, the N-terminus of CYP6B33 from the insect *Papilio multicaudatus* is likely to maintain a protein fold obviously instrumental to communication with POR [245]. Also, construction of the (Δ1–66)-derivative of CYP52A3 from the yeast *Candida maltosa* was found to diminish reactivity toward POR [246].

To assess critical residues in P450s involved in the functional coupling of POR, CYP1A1 was covalently modified through treatment with acetic anhydride or an azido analog of benzphetamine. Selective blockage of four lysines at positions 97, 271, 279 and 407 was found to eliminate POR-dependent enzyme activity [247, 248]. This finding corresponds to results from studies with antibody targeted against a K271/K279-containing fragment of the hemoprotein, disclosing inhibition of metabolic turnover as the potential consequence of a rise in K_m for POR [249]. In addition, attachment of 4,4′-dithiodipyridine to C293 in the CYP1A1 polypeptide was shown to be reversible upon incorporation of POR into the assay media, suggesting the residue to be located close to the reductase-binding motif [250]. This view receives support from antibody-directed suppression of substrate turnover following blockage of a region in the CYP1A2 homolog aligning with positions C293 to N301 in CYP1A1 [251]. Moreover, nitration of Y243 and Y271 in the CYP1A2 molecule was found to slow down electron transfer from POR to the acceptor [252]. Finally, site-directed mutagenesis helped verify prospective key players: Replacement in CYP1A1 of lysine at positions 271 and 279 with isoleucine caused a severe loss of responsiveness to POR for the hemoprotein [253]. Similarly, there was a 2- to 4-fold increase in the K_d value for POR anchoring when the basic lysines occurring at positions 94, 99, 440 and 453 in CYP1A2 were exchanged for an acidic residue [254, 255].

Within the plethora of drug-metabolizing P450s, inhibition of CYP2B1-mediated substrate oxidation by immunoprecipitation of the enzyme's K122 to T231 sequence was shown to be less pronounced when antipeptide was added after reconstitution of the system with POR, proposing the epitope to be most likely engaged in POR association [256]. This concept is in line with R125 obviously having a critical role in this event [257]. Further lysine residues in CYP2B1, putatively serving as candidates for reductase recognition, reside at positions 251, 384, 422 and 433 [258]. Kinetic analysis of the chemically modified CYP2B4 analog in the absence and presence of protective amounts of POR disclosed lysines 139, 144, 251 and 384 to be in presumptive contact with the electron donor at a distance of about 3–4 Å [259]. Moreover, substitution of predominantly basic amino acids, hosted in the polypeptide fragment spanning residues R122 to K139, with the hydrophobic alanine entity drastically increased the K_d value for reductase binding to CYP2B4 [260]. Additional positively charged elements in the surface structure of the hemoprotein, identified by genetic engineering to promote electron flow from POR, include K225, H226, R232, R253 and H285 along with R422, K433 and R443 located in the vicinity of the heme edge [260–262]. On the other hand, a series of aromatic and hydrophobic amino acids such as F223, F227, F244, V267 and L270 were uncovered to participate in π-π-stacking and H-bonding interactions with POR [261, 263]. Of note, charge-reversal mutation K139E in the polymorphic CYP2B6.8 variant was found to impair functional complexation with POR [264]. This finding agrees with data from cross-linking experiments with the wild-type enzyme, disclosing competition of the latter with the synthetic D134-R140 peptide in reductase capture [265]. It should be mentioned that arginines at positions 139, 144 and 442, hypothesized to be beneficial to contacts with the electron donor in allelic CYP2C8 and CYP2C9 proteins as well as in CYP2C19, coincide with corresponding patches on CYP2B members [266–268]. Also, homology modeling of CYP2E1 in parallel with chemical inactivation by nitration of a series of tyrosines elucidated a close relationship between the FMN domain of POR and Y422 [269]. Noteworthy, C98W mutation in CYP3A4 was found to significantly hamper affinity to POR, associated with a 41 % diminution in the maximum rate of electron flow between the P450 and flavoprotein [270]. In addition, molecular modeling revealed the neighboring Y99 to be in close proximity to the cofactor-binding region of POR, while Y430 forms a hydrogen bond with an acidic reductase residue at a distance of 2.3 Å [271].

Inspection of microsomal P450s involved in the biosynthesis of natural products helped rescue further information about the architecture of donor/acceptor complexes. Thus, chemical and genetic modification of CYP17A1, lying at the crossroad of androgen and corticoid formation, unveiled the positively charged amino acids K326, K327, R347 and K358 to constitute part of the POR-contacting area [272, 273]. Similarly, a set of missense mutations at K121, R339, R341 and R356 provided clues to better understanding of the mode of interaction of POR with the steroid 21-hydroxylase CYP21A2 [274–276]. Finally, construction of a molecular model of the lanosterol 14α-demethylase CYP51F1 from yeast allowed identification of unique residues such as H101, K358, R426 and K433, serving as prospective sites for reductase association [277]. A compilation of the topological data for POR docking to the diverse P450s is given in Table 10.1.

10.3.2 Docking of Cytochrome b_5

In cases where P450s exhibit an obligatory requirement for electron donation by b_5 to maintain optimal rates of substrate turnover, structural integrity of the hydrophobic tail portion of the oxidases seems to be pivotal to productive donor/acceptor coupling. For example, optical biosensor studies with CYP2B4 lacking amino acids 2–27 disclosed removal of the signal anchor to result in defective binding of the intermediate carrier accompanied by a pronounced drop in the reductive force [179, 240]. Apart from this, circumstantial exploration of a set of CYP2 members helped ascertain an array of critical b_5-docking entities sitting remote from the enzymes' N-terminus. Thus, strongly perturbed donor anchoring upon generation of the R129S derivative of CYP2A5 suggested the RRFS fragment in the polypeptide chain to be a key recognition motif [278]. This conclusion nicely coincides with the fact that the homologous CYP2A4, bearing a R129S point mutation, fails to stimulate substrate oxidation [279]. Interestingly, site-specific attack on K122, R125 and S128 in the CYP2B1 polypeptide by immunochemical manipulation or protein kinase-mediated phosphorylation was found to be competitively antagonized by the presence of b_5, suggesting these residues to be in contact with the electron donor [280, 281]. Moreover, elements R122, R126, R133, F135, M137, K139, H226 and K433, selected by computer docking of a CYP2B4 model, were substituted with alanine to evaluate the function of the amino acid side chain distal to the β-carbon. All the mutants tested displayed diminished ability to bind b_5 [260]. Genetic engineering was also employed to confirm the biological importance of K428 and K434 in CYP2E1/b_5 complexation [282].

Studies extended to other P450 families verified sites K127 and K421 on CYP3A4 to be essential for efficient b_5 coupling [283]. Moreover, impairment of the fundamental chemistry by introduction of mutations at positions 83, 88, 347, 358 and 449 in CYP17A1 was demonstrated to hamper propensity for 17,20-lyase activity by disrupting responsiveness to the b_5 component [171, 273, 284]. Table 10.1 provides a synopsis of key amino acids in P450s governing interaction with b_5.

10.3.3 Docking of Ferredoxins

Use of a specific fluorescence probe localized the heme group of the mitochondrial CYP11A1 protein ~26 Å remote from the binding surface for adrenodoxin (Adx) [285]. Here, basic residues K73, K109, K110, K126, K145, K267, K270, K338 and K342 on the mature hemoprotein form were substantiated to govern reactivity toward Adx by the ferredoxin's ability to act as an almost complete protector against the lysine-modifying agents, succinic anhydride or fluorescein isothiocyanate, employed for enzyme engineering [286, 287]. Two additional lysines corresponding to K377 and K381 in the precursor form of steroidogenic CYP11A1 were identified by site-directed mutagenesis as also being vital to Adx association. Estimated K_d values for donor docking increased about 150- to 600-fold compared to the wild-type enzyme depending on the particular lysine substitute [288]. This finding fits data from specific chemical labeling of lysines in the peptide comprising amino acids M369 to K381 in the CYP11A1 molecule, eliciting a

Table 10.1 Prospective key amino acids of P450 enzymes governing interactions with redox partners: representative results from molecular modeling, genomic analyses and site-directed mutagenesis

				Interacting redox partner			
CYP enzyme	Residue modified	Alignment position[a]	Location in 2° structure[b]	POR	b_5	Fdx[c]	Refs.
51F1	H101	57	αB	+			[277]
1A1	K97	59	αB	+			[247]
1A2	K94	59	αB	+			[254]
17A1	K83	59	αB		+		[171]
101A1	R72	59	αB		+	+	[298]
11A1	K73	63	αB			+	[286]
101D1	R77	63	αB			+	[304]
1A2	K99	64	αB	+			[254, 255]
17A1	K88	64	αB		+		[171]
2B1	K122	97	αC		+		[280]
2B4	R122	97	αC	+	+		[260]
11A1	K109	97	αC			+	[286]
101A1	R109	97	αC			+	[299]
119	R109	97	αC			+	[303]
3A4	K127	98	αC		+		[283]
11A1	K110	98	αC			+	[286]
2B1	R125	100	αC	+	+		[257]
101A1	R112	100	αC			+	[299–301]
101D1	R113	100	αC			+	[304]
2A5	R129	101	αC		+		[278, 279]
2B4	R126	101	αC	+	+		[260]
21A2	K121	101	αC	+			[274]
2B1	S128	103	αC		+		[281]
2B2	S128	103	αC		+		[281]
2B4	S128	103	αC		+		[281]
102A1	L104	104	αC	+			[354]
2B4	R133	108	αC1	+	+		[260]
2B4	F135	110	αC1	+	+		[260]
2B4	M137	112	αC1	+	+		[260]
2B4	K139	113	αC1	+	+		[259, 260]
2B6	K139	113	αC1	+			[264]
11A1	K126	114	αC1			+	[286]
2C9	R144	118	αD	+			[267]
11A1	K145	133	αD			+	[286]
2B4	H226	197	αG	+	+		[260, 261]
2B4	F227	198	αG	+			[261]
2B4	R232	203	αG	+			[261]
101A1	K197	209	αG			+	[302]
2B4	F244	215	αG	+			[261]
1A1	K271	221	αG	+			[247]
2B1	K251	222	αG	+			[258]
2B4	K251	222	αG	+			[259]
1A2	Y271	224	αG	+			[252]
2B4	R253	224	αG	+			[261]
1A1	K279	229	αG-αH	+			[253]
2B4	V267	236	αH	+			[263]
2B4	L270	239	αH	+			[263]
1A1	C292	241	αH-αI	+			[250]

(continued)

Table 10.1 (continued)

CYP enzyme	Residue modified	Alignment position[a]	Location in 2° structure[b]	Interacting redox partner			Refs.
				POR	b_5	Fdx[c]	
11A1	K267	245	αH-αI			+	[286]
11A1	K270	248	αH-αI			+	[286]
2B4	H285	250	αI	+			[262]
17A1	K326	288	αJ	+			[272]
17A1	R347	308	αJ‘	+	+		[273, 284]
21A2	R339	308	αJ‘	+			[276]
21A2	R341	310	αJ‘	+			[276]
51F1	K358	310	αJ‘	+			[277]
11A1	K338	315	αK			+	[286, 287]
27A1	K354	315	αK			+	[296]
199A2	R285	315	αK			+	[305]
11A1	K342	319	αK			+	[286]
17A1	R358	319	αK	+	+		[273, 284]
27A1	K358	319	αK			+	[296]
21A2	R356	325	αK	+			[276]
1A1	K407	349	β2(2)-β1(3)	+			[247]
2B1	K384	349	β2(2)-β1(3)	+			[258]
2B4	K384	349	β2(2)-β1(3)	+			[259]
51F1	R426	377	MR	+			[277]
3A4	K421	380	MR		+		[283]
27A1	R418	380	MR			+	[296]
11A1	K405	383	MR			+	[292]
101A1	K344	383	MR		+	+	[298]
51F1	K433	384	MR	+			[277]
2B1	K422	386	HBR	+			[258]
2B4	R422	386	HBR	+			[260]
2E1	Y422	386	HBR	+			[269]
102A1	Q387	387	HBR	+			[354]
1A2	K440	388	HBR	+			[254]
3A4	Y430	388	HBR	+			[271]
2E1	K428	391	HBR		+		[282]
1A2	K453	397	HBR	+			[255]
2B1	K433	397	HBR	+			[258]
2B4	K433	397	HBR	+	+		[260]
2E1	K434	397	HBR		+		[282]
1A2	R455	399	HBR	+			[254]
11A1	R426	404	αL			+	[292]
24A1	R466	404	αL			+	[294]
2B4	R443	407	αL	+			[260]
2C19	R442	407	αL	+			[268]
17A1	R449	407	αL		+		[284]
101D1	R371	407	αL			+	[304]
199A2	L369	408	αL			+	[305]

[a]Positions were determined by screening the sequences of the target P450 enzymes against the crystal structure of substrate-bound CYP102A1 (PDB ID: 1ZO9) as described previously [277, 307]

[b]Allocation of the alignment positions to definite domains of α-helical or β-sheet structure is based on the CYP102A1 architecture [314]. *MR* meander region, *HBR* heme-binding region

[c]The category of ferredoxins includes Adx, Arx, Pdx and Pux

drastic fall in responsiveness to the electron-supplying factor [289]. Of note, point mutation R→C at position 366 in CYP11B1, corresponding to K377 in the CYP11A1 congener, was detected to give rise to breakdown of the catalytic efficiency of 11β-hydroxylation to a level ~25 % that of the native protein. This has been interpreted to mean that a change to cysteine eliminates a positive charge and leaves a cove on the enzyme's surface, most likely impacting Adx fixation [290]. Interest also focused on residue C264 lying proximate to K267 in the so-called "hinge" region. Indeed, chemical blockage of the surface cysteine was found to hamper CYP11A1-promoted turnover through curtailing the catalyst's capacity to interact with Adx [291]. Moreover, biochemical and molecular modeling studies based on the crystal structure of the redox partner jointly supported the concept that K405 and R426 (numbering of the mature hemoprotein form) participate in electrostatic contacts with Adx [292]. There seems to exist an interplay between the latter amino acid and the conserved vicinal E429 residue responsible for fine tuning of the stability of the assembled complex [293]. Crystallographic analysis of the 24-hydroxylase CYP24A1 from rat again revealed structural elements K378 and K382, aligning with lysines at positions 377 and 381 in bovine CYP11A1, to operate as key players in Adx recognition. In addition, the invariant R466, located 8–10 Å remote from the conserved lysines, was assigned a dominant function in ferredoxin-driven electron transfer [294]. The critical arginine aligns with R426 in CYP11A1 and R458 in the murine CYP27B1. In fact, R458Q substitution was detected to induce a 36-fold rise in the apparent K_m value for Adx associated with a drastic decrease in electron pressure [295]. Finally, introduction of the K354A/K358A/R418S triad into CYP27A1 involved in bile acid biosynthesis was shown to be destructive to ferredoxin binding [296].

Epitope mapping, carried out with bacterial CYP101A1 from *Pseudomonas putida* in the presence of a set of antigenic peptides directed against areas presumed to be of functional relevance, suggested regions spanning residues 63–72 and 108–117, respectively, to potentially participate in putidaredoxin docking [297]. This view was underpinned by the severe loss of reactivity toward Pdx upon creation of hemoprotein variants bearing non-ionic amino acids in place of the positively charged arginine at positions 72, 109 and 112 [217, 298, 299]. It should be emphasized that ferredoxin binding to R112 has been recognized to also be beneficial to intracomplex electron transfer to the ferric heme iron-oxo species [300, 301]. Moreover, the ability of the intermediate carrier to shield K197 in the P450 molecule from attack by chemical modifiers qualifies the lysine residue as part of the Pdx recognition site [302]. Similarly, reversal of the cationic charge by K344E mutation was demonstrated to cause perturbation of donor docking [298]. Interestingly, CYP119A1 from thermophilic *Sulfolobus acidocaldarius* utilizes Pdx as the electron supplier. Here, D77R mutation of the hemoprotein was detected to markedly enhance fixation of the redox partner and stimulate electron flow by a factor of about 5 compared to the parent enzyme, obviously eliminating a potentially repulsive protein-protein interaction [303]. The repellent effect of D77 thus might serve in proper Pdx orientation.

Other bacterial P450s receive electrons via [Fe_2-S_2]-type ferredoxins genomically associated with the individual oxidases. For example, evaluation of the electrostatic surface profile of CYP101D1 from the oligotrophic *Novosphingobium aromaticivorans* suggested amino acids such as R77, R113 and R371 to contribute to specificity in [2Fe-2S]-type ferredoxin (Arx) recognition [304]. Furthermore, the benzoic acid-oxidizing CYP199A2 from *Rhodopseudomonas palustris* was recognized to carry two surface hot spots presumed to be significant factors in steering cross-reactivity of ferredoxins. Thus, the presence of R285 as well as charge reversal at L369 were hypothesized to be responsible for preferential functional coupling of the physiological redox partner palustrisredoxin compared to the heterologous Pdx [305, 306]. A summary of data for ferredoxin docking to vertebrate-type P450s is presented in Table 10.1.

10.3.4 Overall Architecture of Redox Domains and Mechanism of Electron Donor Docking

Increasing interest in elucidation of the molecular mechanism of electron transfer in P450 systems creates fundamental demand for visualization of the architecture of donor-binding sites ruling catalytic potency of the enzymes. To accomplish this goal, homology modeling has to be carried out using sophisticated strategies for construct building. Judging from data for root mean square (r.m.s.) deviation of critical C_α atoms and φ/ψ-angle distribution, comparative alignment by knowledge-based techniques of P450s from different phyla with the bacterial CYP102A1, having its 3D structure determined, suggested the microbial enzyme to be a robust template for elucidation of structure-function relationships [277, 307]. Thus, mapping of key amino acid residues from microsomal, mitochondrial and bacterial hemoprotein species recognized to contribute to redox partner interactions (Table 10.1) onto the CYP102A1 scaffold yielded a scenario (Fig. 10.6) describing the general spatial distribution of contact sites [32].

As can be seen, the majority of points presumed to dictate contact with redox proteins cluster close to the center of the proximal face of the hemoprotein model. Highest density of binding sites, amounting to 47 % of the total number of key players, is found in the triad formed by α-helical structures C/C1, bordering the core fold on the top, the G-helical fragment, located more in the periphery of the P450 molecule, and the heme-binding region. The remaining interaction sites appear to be of minor importance, each housing but 5–8 % of the overall volume of anchoring elements (Table 10.1). Surprisingly, the population of functional determinants in the various target P450s, residing in the three preeminent donor-docking epitopes, displays a very low to moderate extent of conservation ranging from 9 % to 27 %. This might arise from the need for conformational flexibility to enable encounter with heterologous redox proteins. Indeed, ~38 % of the contact sites harbored in helices C/C1/G and the heme-binding domain have overlap of POR fixation with b_5 recognition. This behavior agrees with the ability of increasing amounts of b_5, integrated into assay media containing a constant level of POR, to gradually transform the biphasic kinetic tracings, prototypic of NADPH-driven P450 reduction, to a sluggish monophasic reaction as is characteristic of electron donation by b_5 [308, 309]. This lends support to the notion of a functional antagonism between the two redox proteins. On the other hand, b_5 fails to interfere with nonproductive physical anchoring of reductase to P450, as evidenced by visible difference spectrometry [308]. This seems to hint at functional diversification of the POR-docking loci [310] potentially acting in substrate-induced cooperativity [241]. Though overlap of regions involved in POR and ferredoxin association is lacking, the redox domain architecture (Fig. 10.6) displays epitopes fostering binding of the two carrier species to cluster in close proximity to each other in helices B/C and the H-I interhelical loop, possibly caused by certain analogy in the structural organization of the electron transfer proteins [185]. Similarly, evolutionary commonality induces joint contact points for b_5 and ferredoxins on the proximal face of P450s constituted by portions of helices B, C, K and the meander stretch (Table 10.1). It should be mentioned that the crystal structure of an archetypal b_5 homolog isolated from a bacterial strain has been identified [311]. This prompts one to speculate that b_5-type proteins may act as natural electron donors also to certain microbial P450s.

Evaluation of the array of data summarized in Table 10.1 disclosed 81 % of the overall population of amino acids predicted to operate in redox partner binding in the various target P450s to belong to the category of positively charged entities, about two thirds of the reactants being represented by lysine residues and one third by arginines. Indeed, calculation of the electrostatic surface potential for a series of P450s showed the dipole moment of the hemoproteins to be oriented such as to help direct the intermediate

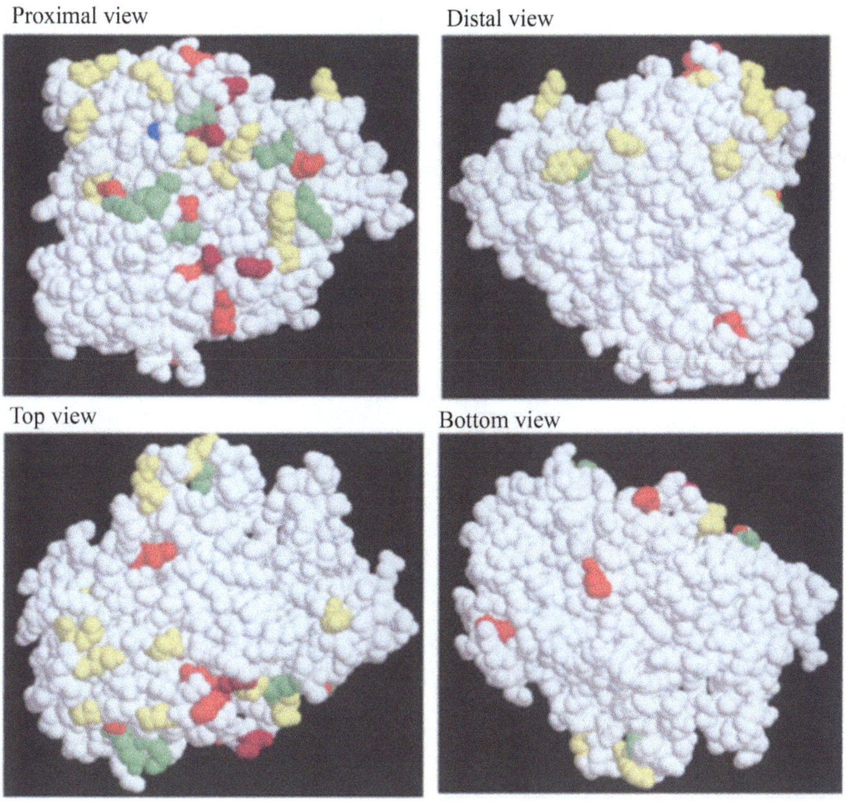

Fig. 10.6 Generalized molecular model featuring critical surface sites in P450s operating in recognition and binding of redox proteins. The composite profile was built by mapping the topological data of key determinants steering electron donor fixation onto the substrate-bound CYP102A1 template. The color code denotes: *yellow spheres*, POR-binding sites; *blue spheres*, b_5-binding sites; *green spheres*, sites common to POR and b_5; *red spheres*, Fdx-binding sites; *purple spheres*, sites common to Fdx and b_5. For *top* and *bottom* views, the coordinates of the images were rotated by 90° in the x-axis (Data taken from Ref. [32])

carriers toward a patch of positively charged elements on the proximal face [312–314]. This behavior underpins salt-bridge formation with carboxylates in the donor proteins (see Sect. 10.2) to be the most salient driving force in complexation, as exemplified by the allocation of interfacial residues involved in the CYP3A4-b_5 encounter [283] depicted in Fig. 10.7. In agreement with this principle, charge shielding by high concentrations of mobile ions was shown to elicit disintegration of donor/acceptor anchoring associated with a drop in electron flow [255, 293, 298]. Moreover, ~10 % of the key players bear a polar side group serving in generation of a flexible H-bonding link to some basic group(s) in the intermediate carriers, with tyrosines presumably being favored mediators of weakly polar inter-residue contacts [252, 269, 271].

Since electrostatic phenomena, no doubt, prevail in functional coupling of redox partners, it does not seem surprising that only a minority (~9 %) of the total of sites attracting electron donors can be assigned to the class of lipophilic amino acids largely accommodated in helices C1 and G. Here, aromatic and aliphatic representatives cooperate in π-π-stacking and van der Waals interactions with reactants in the diverse redox proteins [260, 261, 263, 305].

10.3.5 Factors Impacting Organization of Protein-Protein Association

10.3.5.1 The Role of Phospholipids

The phospholipid matrix serving in insertion and assembly of the components of the P450-

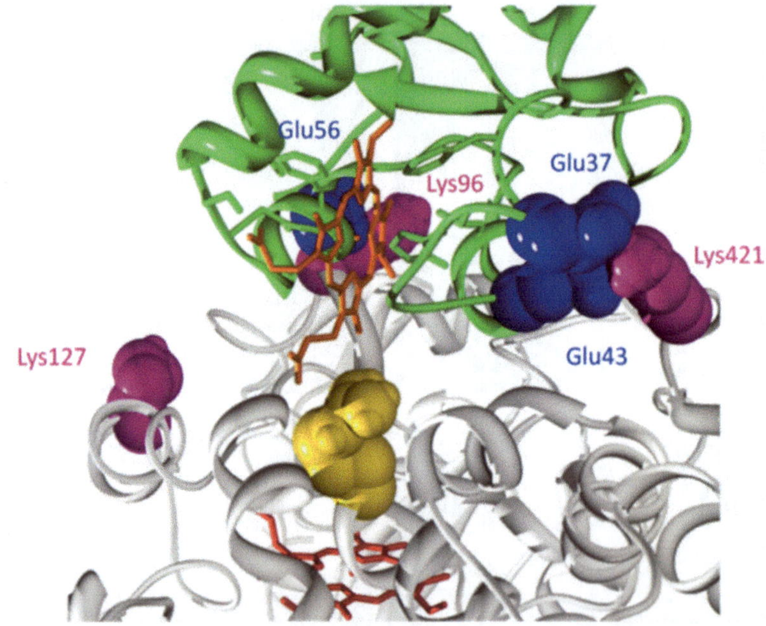

Fig. 10.7 Functional importance of electrostatic interactions between the protein surfaces of cytochrome b_5 and CYP3A4. The monooxygenase and the electron carrier are presented in *white* and *green*, with their heme groups being shown in *red* and *orange*, respectively. The interacting residues on the redox partners are depicted in *magenta* and *blue*, while the critical R446 is colored *golden*. As is evident, b_5 approaches the B-B′ loop region and helix C of CYP3A4 via helices α4 and α5. Protein domains on the oxygenase far from the docking surface are truncated (Reproduced from Ref. [283])

dependent redox machinery was detected to strongly impact efficiency of electron transport in microsomal and mitochondrial systems by providing structural features permitting improved recognition, ordering and alignment of redox partners [315]. In this respect, the synthetic dilauroyl phosphatidylcholine as well as natural phospholipids were shown to decrease the apparent dissociation constant for P450/POR complexes to an extent depending on both the chain length of the lipids and the mode of reconstitution, yielding either micellar or vesicular systems [316–318]. Owing to predominance of positive charges on the proximal profile of P450s (see above), mixtures containing anionic lipids such as phosphatidylserine were found to favor P450/POR association by forcing the proteins into correct orientation toward each other [319–321]. Facilitated donor/acceptor binding appears to generally require prior phospholipid-induced relaxation of the tight multimeric P450 aggregates [322]. Indeed, displacement of the P450 oligomerization equilibrium toward monomers by use of a nanoscale construct bearing a palmitoyl-oleoyl phosphatidylcholine bilayer drastically improved flavoprotein-promoted P450 reducibility [323]. Collectively, phospholipids were recognized to act as allosteric effectors eliciting conformational alterations in P450s associated with an increase in α-helical content of the hemoproteins. This fosters functional coupling of different types of electron carriers [324, 325]. Thus, lipid was demonstrated to also modulate affinity of b_5 for CYP2B4 [326]. Conversely, b_5 binding to the enzyme caused a ~2-fold rise in reactivity of phosphatidylcholine to CYP2B4 [327]. Similarly, cholesterol lowers K_d for Adx docking to cardiolipin-saturated CYP11A1 by a factor of up to 20, while the ferredoxin, in its turn, improves cholesterol binding to steroidogenic CYP11A1 [328].

10.3.5.2 The Role of P450-P450 Aggregation

Formation of hetero-oligomers of P450 is well documented and may play a decisive role in regulatory mechanisms of electron transfer [329]. In fact, the combined presence of CYP1A2 and CYP2B4 reconstituted with POR in the same phosphatidylcholine vesicles provides conclusive evidence from changes in the enzyme-specific monooxygenase activities that the CYP1A2 moiety of the heteromeric P450 complex generates a high-affinity reductase adduct more effectively competing for the redox protein than CYP2B4 [330]. Comparable results were obtained when the CYP2E1/CYP2B4 pair was embedded into a phospholipid matrix in the presence of POR to probe competition for the reductant. Here, low levels of CYP2E1 turned out to cause a 23-fold increase in the apparent K_m value of CYP2B4 for the donor protein, while the analogous K_m of CYP2E1 for POR decreased significantly, allowing CYP2E1 to outact CYP2B4 [331]. Of note, CYP2A6/CYP2E1/POR co-expression in microsomal membranes disclosed the presence of a prototypic CYP2A6 substrate to impair electron flow to CYP2E1, suggestive of a regulatory function of substrate in P450 aggregation [332]. This view is substantiated by drug-drug interactions occurring as the output of competition for the ancillary POR enzyme of co-reconstituted P450 couples such as CYP2C9/CYP2C19 or CYP2D6/CYP3A4 [333, 334]. Furthermore, the formation in liposomal membranes of an equimolar complex between the mitochondrial CYP11A1 and CYP11B1 enzymes was found to have a stimulatory effect on the CYP11B1-dependent 11β-hydroxylase activity as the consequence of a conformational alteration, corresponding to changes in the K_m value for Adx [335]. A mathematical model taking account of the possible existence of multiple types of P450-based dimer formations was developed to explore the most probable mechanism(s) of such interactions in more detail [336].

One would be remiss without mentioning that a fraction of P450s integrated into membranous systems may also exist as homo-oligomers. Here, formation of large aggregates causes P450 immobilization to an extent depending on the lipid-to-protein ratio [337, 338]. Interestingly, incorporation of POR or b_5 was shown to readily disrupt the aggregation state of P450s, when in a membrane, via transient complexation with the monooxygenases. This might influence the amount of productive donor/acceptor adducts determining catalytic activity [339, 340]. Again, substrate may interfere with the docking events to modulate reactivity of the redox partners [33, 177].

10.4 P450/Redox Partner Fusion Enzymes

10.4.1 Natural Fusion Proteins

Among fusion enzymes, the cytosolic CYP102A1 from *Bacillus megaterium* represents a unique self-sufficient flavohemoprotein catalyzing (ω–n)-hydroxylation of medium- to long-chain saturated fatty acids [341]. Owing to its soluble nature and applicability as an excellent paradigm for the understanding of structure/function relationships in class II-type P450s, CYP102A1 represents the most extensively studied member of the CYP102A subfamily consisting of a large number of relatives, though only four additional homologs, namely CYP102A2/A3/A5 and A7 from diverse *Bacillus* strains, have so far been characterized. Here, comparison of the polypeptide structures revealed some deviations in active-site architecture [342–344].

The CYP102A1 enzyme is composed of an N-terminal heme domain connected via a short protein linker with a eukaryotic-like diflavin reductase module bearing one equivalent each of FAD and FMN [345]. Availability of the crystal structure of the FAD/NADPH-binding domain helped identify sites involved in NADPH fixation such as S965, R966, K972 and Y974 [346]. Noteworthy, the side chain of W1046 shields the FAD isoalloxazine ring from

NADPH, and motion of this residue drives pyridine nucleotide specificity to the formation of an $FADH_2$-$NAD(P)^+$ charge-transfer intermediate [347]. Aromatic stacking with W854 and Y860 was shown to stabilize the FAD cofactor, while amino acids at positions 729–743 have the potential to make contacts with the cognate FMN domain [346]. Due to the dimeric nature of CYP102A1, the obligatory electron tunneling route traverses both constituents of the dimer during a single turnover by switching from the FAD-binding site of one monomer to the FMN domain of the other one prior to passing on to the terminal acceptor [348]. Interestingly, modulation of the electrostatic microenvironment of the FMN-docking pocket, housing critical residues Y536 and G570 [349], by unusual integration of positively charged lysines at positions 572 and 580 as well as decreased flexibility of the short cofactor-binding loop were presumed to be jointly responsible for the observed repression of the neutral, blue FMN semiquinone radical paralleled by stabilization of the red, anionic hydroquinone species. This was shown to be coupled with a change in the E_0' values of the redox pairs securing electron flow to the heme unit [350–352]. In this regard, W574, located in the FMN domain, was demonstrated to provide a direct through-bond electron transfer pathway including P382 and C400 in the heme-binding peptide [349, 350], while the highly conserved W96 turned out to have a function in heme association and control of the spin state of the iron [353]. Moreover, the area around L104 and Q387 in the intact heme/FMN-binding fragment (Fig. 10.8) revealed to be vital to efficient functional association of the partners [354]. It has to be mentioned that a soluble form of microsomal b_5 was found to also undergo tight binding to CYP102A1, eliciting a low-to-high spin transition in the enzyme's heme iron. This suggested the electron donor to occupy a contact site on the proximal face of the P450 heme that overlaps with that for the FMN domain of the diflavin reductase [355].

Genetic exploration of the fungus *Fusarium oxysporum* revealed the existence of a loosely membrane-associated, self-sufficient flavocytochrome termed CYP505A1, sharing ~41 % sequence identity with the P450 moiety of the bacterial CYP102A1 counterpart [356]. The enzyme catalyzes pyridine nucleotide-driven (ω-1)- to (ω-3)-hydroxylation of saturated C9 to C16 fatty acids [357]. The reductase unit in the primary CYP505A1 structure, having 35 % sequence identity with that of CYP102A1, was shown to be fixed to the heme region via a linker consisting of 20 amino acids of mainly hydrophilic character [356]. Noteworthy, 28 % of the residues hosted in fractions forming the NADPH/FAD- and FMN-binding domains disclosed to be invariant, with hydrophilicity clearly prevailing in the bond-making events [356]. Here, electron transfer was shown to be strongly stimulated by the presence of substrate [358]. Another self-sufficient member of the CYP505 family, classified CYP505B1,

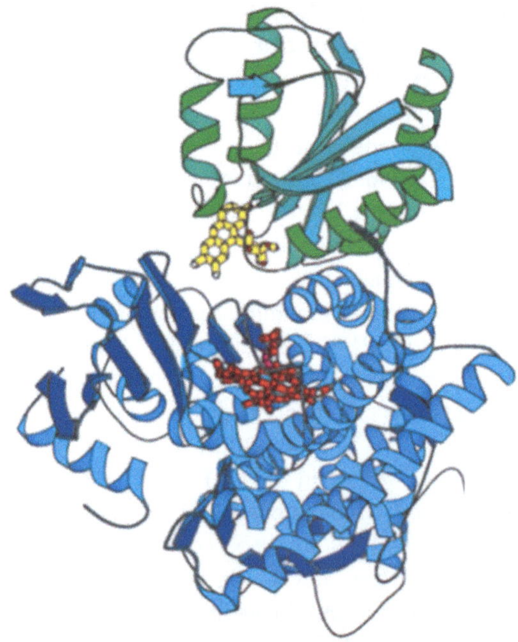

Fig. 10.8 View of the 3D structure of the complex between the heme- and FMN-binding domains of bacterial CYP102A1. The flavin-binding domain (*green*) hosting the FMN cofactor (*yellow*) is physically linked on the same polypeptide to the region (*blue*) surrounding the iron-porphyrin macrocycle (*red*). In this complex, the dimethylbenzene ring of FMN is oriented perpendicular to the heme plane at a distance of ~18 Å (Data taken from Ref. [350])

was isolated from the ascomycete *Fusarium verticillioides* and found to participate in the biosynthesis of the polyketide mycotoxin fumonisin. The flavohemoprotein displays 41 % sequence identity to CYP505A1 and 33 % identity to CYP102A1, with the putative cofactor-docking regions being arranged in the same order as in the homologs cited [359].

In addition, new types of P450-redox partner fusions have been unveiled. An example is bacterial P450 XplA (CYP177A1) from the *Rhodococcus rhodochrous* strain 11Y, catalyzing reductive denitration of the military explosive hexahydro-1,3,5-trinitro-1,3,5-triazine (RDX) [360]. The enzyme has an unusual structural organization comprising the N-terminal P450 heme domain fused to a flavodoxin unit [361, 362]. The latter was shown to contain most of the elements of a signature typical for FMN binding [360]. Unexpected features include the low affinity of the non-covalently bound FMN to its docking site (K_d = 1.09 μM) compared to reactivity of the cofactor toward flavodoxins from other bacterial sources and the strong positive shift of the redox potential of the FMN semiquinone/hydroquinone couple $\left(E_0' = -172\ \text{mV}\right)$, being the likely electron donor to the XplA heme [363].

Efficient RDX degradation was shown to require expression of the partnering reductase XplB. The FAD-containing carrier transfers reducing equivalents to the XplA-FMN in a 1:1 complex with high specificity for NADPH as the electron source [362, 363]. Here, collision of the two subunits represents a rate-limiting step. Of note, the deduced amino acid sequence of XplB has 42 % similarity to bovine mitochondrial AdR [360]. In accord with this, the XplA flavodoxin domain is capable of receiving electrons also from ferredoxin reductase before transferring them to the P450 heme [364].

Circumstantial exploration of the bacterial genome sequence library uncovered a completely novel class of self-sufficient P450 systems, representing a distinctive community of enzymes C-terminally fused to a phthalate dioxygenase reductase (PDR) module as the redox partner. The latter is folded into three domains involved in NADH/FMN binding and docking of the [Fe_2-S_2] cluster [365]. Thus, CYP116B1 from *Cupriavidus metallidurans*, a thiocarbamate herbicide-oxygenating fusion protein, displays stoichiometric binding of both FMN and the iron-sulfur center, electron transfer being supported by NAD(P)H with clear dominance of the triphosphopyridine nucleotide [366]. Similarly, CYP116B2 from a *Rhodococcus* species was demonstrated to be composed of an N-terminal P450 moiety separated by a short segment of about 16 amino acids from the reductase-like fragment, sharing 34 % sequence identity with the PDR family [367]. Closer investigation of the electron-supplying subunit predicted P578 to contact pyridine nucleotides, with NADPH having a ~500-fold preference over NADH in terms of the estimated K_d values [367, 368]. Moreover, the stretch spanning residues S532 to S536 was found to conform to the consensus motif for binding of the phosphate group of FMN, while a cluster of four highly conserved cysteines at positions 722, 727, 730 and 760 constitutes a [Fe_2-S_2] ferredoxin-type center [367]. The reduction potentials of the FMN semiquinone/hydroquinone and FeS entities were calculated to be approximately −270 mV and −214 mV, respectively [369]. Substrate screening for CYP116B2 revealed the enzyme to mediate dealkylation of substituted aromatic alkyl ethers, catalytic efficiency being higher with compounds bearing a shorter alkyl chain [370]. In addition, a new self-sufficient member of the CYP116 family was identified in *Rhodococcus ruber*. The fusion protein was recognized to have >90 % amino acid sequence identity to CYP116B2 and to consist of a heme domain, an FMN-hosting region and an iron-sulfur unit. In the presence of NADPH, the enzyme shows hydroxylase activity toward polycyclic aromatic hydrocarbons such as naphthalene or fluorene [371].

10.4.2 Artificial Self-Sufficient Fusion Proteins

The catalytic diversity of P450s has high potential for biotechnological exploitation. However, industrial application is frustrated by the need of costly NAD(P)H cofactors, cumbersome reconstitution of auxiliary electron donor systems and fairly low metabolic turnover. To simplify the procedure and improve the catalytic outcome, the natural P450/redox partner fusion proteins addressed above may represent excellent paradigms for the sculpturing of man-made chimeric analogs of desired autonomic electron transport [372]. This requires cDNA shuffling to design assembly of redox chain building blocks on a “molecular Lego” principle [373]. Also, a versatile “drop-in” vector for rapid creation of self-sufficient P450s has been developed [374].

10.4.2.1 P450/Diflavin Reductase Fusion Enzymes

A useful tool for the generation of simplistic P450 redox systems is featured by covalent fixation of the hemoprotein portion to a POR-like unit with the aim to construct the most suitable fusion [375]. In this way, the (Δ1–41)-truncated reductase moiety from yeast was genetically attached to rat CYP1A1. The construct displayed rotational mobility of the P450 fragment higher than that of CYP1A1 alone [376]. In the presence of NADPH, the rate of reduction of the substrate-bound fusion enzyme was found to be $>50\ s^{-1}$, suggesting that electrons were rapidly transferred from the cofactor through FAD and FMN to the heme iron [377]. Similarly, human CYP1A1 was connected to N-terminally truncated rat POR via a Ser-Thr dipeptide linker. Activity toward resorufins was shown to be 11- to 22-fold higher compared to the control [378]. Of note, human CYP1A2 genetically engineered with yeast reductase proved to be about twice as efficient in oxidative ethoxyresorufin biotransformation relative to the CYP1A1 fusion system [379]. Moreover, a fused construct derived from the cDNA for canine CYP2B11 in tandem with the code for the modified rat oxidoreductase exhibited an androstenedione metabolite profile very similar to that found with the reconstituted components [380]. The same approach was adapted to arrange chimeras produced by joining the C-terminus of mammalian CYP2C11 or CYP2D6 to the cytoplasmic domain of the cognate flavoproteins via a dipeptide linker. In either case, molecular organization seemed to be suboptimal, as judged from comparison of the k_{cat} values for substrate turnover with those of the non-fused systems [381, 382]. Similar observations were made upon linkage of human CYP3A4 to rat POR. Addition of excess exogenous POR and b_5 to the reaction mixtures were found to drastically enhance the rate of testosterone 6β-hydroxylation [161]. Deficiency in catalytic capacity of the fused construct may arise from the fairly short linker region restricting flexibility in orientation toward each other of the functional interfaces of the CYP3A4 and POR modules required to permit swift electron transfer. Optimization was achieved by engineering a number of triple adducts, among which the CYP3A4/reductase/b_5 product turned out to reflect the most appropriate ordering for high activity compared to the reconstitution premixes [383]. Though fusion of rat CYP4A1 with the native reductase unit gave a biocatalyst mediating lauric acid ω-hydroxylation at a rate threefold higher than that determined in reconstitution assays, metabolic capacity was not fully exhausted: supplementation with purified flavoprotein/b_5 strongly stimulated fatty acid consumption by potentially increasing collision frequency of the redox partners [384, 385].

The fusion strategy was also extended to microsomal steroidogenic P450s. Thus, the multifunctional CYP17A1 domain of different mammalian species was connected to a truncated form of yeast or rat reductase to yield a self-contained unit characterized by 17α-hydroxylase and 17,20-lyase activity, promoting biotransformation of progesterone and pregnenolone to the corresponding C19-derivatives [386, 387]. Here, the length and amino acid sequence of the hinge region between the redox components was demonstrated to play a decisive role in efficient intramolecular electron transfer [388]. Of interest, donation to the fused adduct of reducing equivalents by exogenous b_5 was shown to boost the lyase pathway to an extent depending on the genetic ancestry of the CYP17A1 moiety examined [389]. Similarly, linkage of bovine

CYP21A2 to N-terminally modified yeast POR gave a flavohemoprotein driving conversion of 17α-hydroxyprogesterone to 11-deoxycortisol at a catalytic efficiency two- to four-times greater than that found with the reconstituted CYP21A2 redox chain [390]. Surprisingly, a mitochondrial form of rat CYP27A1, when fixed to a heterologous POR motif, displayed notable potency for 27-hydroxylation of 5β-cholestanetriol in the absence of its native electron suppliers [391].

A suit of vectors for the expression of fungal and plant P450s as non-natural genetic fusions with various reductase isoforms have been developed. Thus, engineering of CYP51 from *Saccharomyces cerevisiae* to allow connection with its cognate redox partner resulted in swift 3-hydroxylanostenol demethylation [392]. Also, chimeric plant P450s CYP71B1, CYP73A and CYP76B1 were shown to work with higher overall capacity when plant reductases were permitted to act as fusion partners. Here, cinnamate 4-hydroxylation, a key reaction in phenylpropanoid biosynthesis, induces swift production of relevant secondary metabolites steering plant development, while catabolism of recalcitrant herbicides such as chlortoluron is of major importance in defense reactions [393–396].

Efforts were also undertaken to evaluate exploitation of the *Bacillus megaterium* (CYP102A1) reductase component (BMR) as a surrogate of POR, having 35 % sequence identity with the microbial analog [397]. In this way, a series of soluble self-sufficient CYP2C chimeras, generated by gene-fused assembly of the N-terminally modified P450s with BMR via a Pro-Ser-Arg linker, displayed activities toward prototypic marker substrates comparing favorably with those reported for the wild-type enzymes [381, 398, 399]. Similar observations were made with the CYP2E1/BMR and CYP3A4/BMR constructs, though coupling levels between product formation and NADPH consumption did not exceed 8–15 % [398, 400, 401]. Furthermore, swapping of the oxidoreductase module of the P450-like self-sufficient neuronal nitric oxide synthase for BMR was found to give rise to a manipulated multi-domain construct of low stability, nevertheless displaying appreciable oxygenase activity prone to the regulatory action of the calmodulin messenger protein [402]. Finally, fusion-mediated development of a reaction host for efficient 3′-hydroxylation of 4′,7-dihydroxyisoflavone (daidzein) was carried out by cross-linking CYP105D7 via a 20 amino acid peptide to the BMR-like reductase fragment of the self-contained CYP102D1 from *Streptomyces avermitilis*. The engineered enzyme metabolized daidzein at a k_{cat}/K_m value 24-fold higher than that measured with CYP105D7 reconstituted with Pdx/PdR [403].

10.4.2.2 P450/Ferredoxin/Ferredoxin Reductase Fusion Enzymes

A novel type of architecture was tested for utility in simplifying the P450-dependent redox machinery. Thus, microsomal rat CYP1A1 was manipulated by gene fusion to obtain a triple adduct encompassing Fdx and FdR from plant chloroplasts. Here, the CYP1A1/Fdx/FdR order revealed to permit the most efficient oxidative turnover of 7-ethoxycoumarin and the herbicide chlortoluron [404]. Similarly, mammalian mitochondrial CYP11A1, CYP11B1 and CYP27A1 enzymes were tethered to their native accessory redox partners via the production of a series of expression cassettes. Again, arrangement of the ligated modules was recognized to have a pivotal impact on the catalytic potency of the individual constructs, with the P450/AdR/Adx congener being superior to other species. This suggested Adx to be a key factor in determining the reaction rate [405–407].

Moreover, bacterial CYP101A1 from *Pseudomonas putida* was fixed to its dedicated electron donors to yield a tandem linear fusion enzyme. Of note, highest NADH-promoted camphor turnover was attained with an assembly, where the PdR/Pdx duo, linked by peptides of variable length, preceded the P450 domain, though activity as such was but 30 % that of the reconstituted wild-type system [218]. In contrast to this, a novel site-specific, branched CYP101A1 fusion protein with spatially equal geometry of the three-redox-component adduct was created to minimize structural constraints. To this end, the P450 module cross-linked with PdR via a peptide, including a reactive glutamine residue and Pdx attached to a lysine-bearing tag at the C-terminus, were associated with each other by the help of transglutaminase. This product displayed tenfold higher activity compared to the simple chimera described above [408]. Using PCNA, a DNA sliding clamp, as a scaffold in the engineering of a ring-shaped heterotrimeric complex of CYP101A1 tightly juxtaposed to its attendant electron donors, catalytic activity of the resulting construct could be raised to a level two orders of magnitude higher than that of the P450 alone [409]. Notably, a thermostable system was modeled by linkage of CYP175A1 from *Thermus thermophilus* to a new type of FdR and Fdx, with five small amino acids being inserted as a hinge between each component to increase flexibility. The fused protein displayed full NADPH-driven reactivity toward *β*-carotene even at 70 °C [410].

10.4.2.3 P450/Dioxygenase Reductase-Like Fusion Enzymes

Stimulus was given by the CYP116B2 precedent to mimicking the fusion organization of the enzyme's redox center. Thus, a plant-bacterial chimera was created by ligating the P450 domain of CYP93C1 from the soybean *Glycine max* to the PDR-like reductase module of the rhodococcal monooxygenase, catalyzing naringenin-to-genistein transformation at improved efficiency compared to hemoprotein mated with a usual plant reductase [395]. The same procedure was employed to engender genetically engineered merging of the FMN/Fe_2S_2-containing carrier moiety with the C-terminal heme unit of CYP101A1, CYP153A or CYP203A, yielding biocatalysts avidly attacking a diversity of compounds such as *d*-camphor, alkanes and 4-hydroxybenzoate [374, 411–413]. Also, interest focused on harnessing improved catalytic potency and broadening of substrate spectra upon fusion of the native or mutated, macrolide biosynthetic CYP107L1 protein with the PDR-type building block [414, 415]. Moreover, strategies were developed to attach the isolated heme domain of the explosive-degrading CYP177A1(XplA) via a 16-amino-acid-linker to the modified C-terminal reductase partner of CYP116B2. The artificial adduct revealed substrate specificities comparable to those of the wild-type enzyme with a K_d value for RDX docking of ~5 μM [362, 374]. It should be noted that a rare bacterial reductase has been purified from *Nocardia farcinica* bearing some resemblance to the molecular organization of PDR, though carrying an NADPH/FAD-binding module and an Fe_4S_4 cluster. Fusion of the electron donor with CYP51 gave a chimera that demethylated lanosterol at a 35-fold higher efficiency relative to the P450 unit alone [416].

10.5 Procedures to Evade Requirements for Supporting Redox Proteins and Cofactor Utilization

10.5.1 The Peroxide Shunt Pathway

The peroxide shunt serves in driving P450-catalyzed monooxygenations in the absence of an NADPH-dependent redox partner by reacting ferric hemoprotein with H_2O_2 or organic peroxides to generate the Fe^{3+}-OOH^- intermediate, protonation of which leads to release of water and formation of the high-energy iron-oxene species [417]. In this regard, the single-component bacterial peroxygenases CYP152B1 from *Sphingomonas paucimobilis*, CYP152A1 from *Bacillus subtilis* and CYP152A2 from *Clostridium acetobutylicum*, primarily catalyzing α- and β-hydroxylation of long-chain fatty acids, may be ideal model systems [28, 238, 418]; here, salt bridge formation between the fatty acid's carboxylate and an arginine located near the heme was shown to be essential to H_2O_2 ligation and proton delivery to initiate facile O-O bond cleavage [419]. More recently, a new member of the CYP152 family was purified from a *Jeotgalicoccus* species, operating in the peroxide-dependent biosynthesis of 1-alkenes via fatty acid decarboxylation [420]. Moreover, the microbial CYP107AJ1 from *Streptomyces peuceticus* has been ascribed to a putative peroxygenase class of P450s owing to lack of reactivity toward NADPH-driven redox partners, contrasting with the high catalytic efficiency in

H_2O_2-supported 7-ethoxycoumarin dealkylation [421]. Similarly, human CYP2S1 was found to be resistant to reduction by POR, while mediating swift oxidative metabolism of a series of environmental carcinogens in the presence of hydrogen peroxide, cumene hydroperoxide or fatty acid hydroperoxides [26, 27]. In analogy, mammalian brain CYP2D18 was detected to support conversion of dopamine to aminochrome exclusively in a peroxygenase mode [422].

Apart from this specific behavior, P450s usually accepting reducing equivalents from a natural redox partner may, nevertheless, exploit peroxides as alternative oxygen donors in substrate biotransformations. For instance, CYP2B4 was demonstrated to utilize cumene hydroperoxide or fatty acid hydroperoxides to bring about N-oxidation of 4-chloroaniline. Albeit, turnover was found to occur at a rate not exceeding 25 % of that observed with the pyridine nucleotide-driven process [423, 424]. Notably, exchange of the enzyme's highly conserved T302 for alanine was recognized to accelerate inactivation of the mutant through peroxide-induced denaturation of the apoprotein matrix and degradation of the heme macrocycle, pointing at a function of the threonine residue in P450 stabilization or diminution of the level of free reactive oxidant [425]. Employing CYP2D6 and CYP3A4 as probe catalysts because of their high substrate promiscuity, efficiency of the peroxygenase-like metabolic route relative to that determined by the action of natural cofactors was shown to largely rely on the type of "oxygen surrogate" employed [231, 426]. To maximize productive interactions of P450s with peroxides and minimize oxidative hemoprotein damage, substantial work was done by genetic enzyme engineering. Thus, CYP3A4 was subjected to random and site-directed mutagenesis to engender formation of a F228I/T309A variant characterized by a V_{max}/K_m for cumene hydroperoxide-supported 7-benzyloxyquinoline debenzylation 11-fold higher than the value observed with the wild-type enzyme. However, k_{cat} as such only amounted to ~18 % the level measured with CYP3A4 fortified with NADPH [427]. Among bacterial P450s, three random mutants that showed improved capacity for H_2O_2-dependent naphthalene oxidation were generated from CYP101A1. Here, DNA sequencing revealed that amino acid substitutions C242F, R280L and E331K potentially interfered with peroxide binding [428]. Kinetic analysis of the F87A mutant of full-length CYP102A1 unveiled the modified enzyme to be a somewhat more efficient utilizer of H_2O_2 in medium-chain fatty acid hydroxylation compared to the parental species, where peroxide-supported activity is hardly detectable, but to shift C-H bond functionalization away from the terminal position [429]. Applying sequential rounds of random mutagenesis, peroxidative catalyst performance of the F87A-modified heme domain of CYP102A1 was drastically improved by evolution of an allelic variant carrying nine additional amino acid substitutions dispersed throughout the protein scaffold, with exception of the active-site cavity and substrate access channel [430]. Enhanced H_2O_2-driven peroxygenase activity was shown to extend to fatty acid substrates and styrene, although major limitations of this system are rapid suicide inactivation as the result of peroxide-mediated heme destruction and significant decrease in thermostability [430]. Here, quantum mechanical and molecular mechanical calculations allowed rational identification of key oxidizable targets, permitting replacement of the latter with less sensitive entities. In fact, the double mutant W96A/F405L gave a more stable construct [431]. Moreover, thermostabilization of the laboratory-evolved heme-domain peroxygenase variant was achieved by further directed evolution, leading to the introduction of eight new amino acid substitutions [432].

10.5.2 Photo- and Electrochemical Manipulation of the P450 System

Innovative approaches to supersede the obligatory proteinaceous redox chains in the P450 toolbox include light-induced electron transfer to the heme iron via photoactivatable mediators or direct delivery of reducing equivalents from electrodes to promote substrate metabolism [34, 433]. In this way, light-induced reductive dehalogenation of environmental pollutants such as pentachloroethane was brought about in reaction mixtures containing EDTA/proflavin and CYP101A1. Activity remained unaffected upon the addition of exogenous Pdx [434]. Great promise was also shown by the construction of hybrid CYP102A1 heme domains consisting of Ru(II)-diimine photosensitizers attached to the single cysteine residues of the K97C, Q109C, Q397C and L407C mutants, strategically positioned in close proximity to the heme. Continuous irradiation of the systems with visible light permitted hydroxylation of lauric acid with variable total turnover numbers, with the L407C variant being the most efficient catalyst despite some degradation due to oxidative damage [435]. Similarly, cadmium sulfide semiconductor nanoparticles, frequently referred to as quantum dots (QDs), have attracted interest due to their unique size-tunable properties and high photostability during the light-dependent generation of free superoxide and hydroxyl radical species [436]. Adsorption of the positively charged hexahistidine-tagged CYP152A1 on the mercaptoacetic acid-capped QD surface was recognized to yield nanohybrids of differential spatial conformation [437]. UV light-induced triggering of the hemoprotein's peroxygenase activity was shown to cause α- and β-hydroxylation of myristic acid as well as conversion of *N*-acetyl-3,7-dihydroxyphenoxazine to resorufin at a rate 50 % of that found with H_2O_2 as the oxidant [438, 439].

Interfacing of P450s to viable amperometric devices to obtain highly efficient catalysis through direct mediator-free transfer of reducing equivalents requires modification of electrodes with agents that facilitate electron flow, prevent protein denaturation and cause appropriate orientation of the enzymes. To attain this goal, different types of bioelectrocatalysts have been developed [440]. Thus, riboflavin-bearing CYP1A2, CYP2B4 and CYP11A1 enzymes entrapped in a phospholipid vesicular system were cross-linked via glutaraldehyde to screen-printed (SP) thick film rhodium-graphite working electrodes, poised at −500 mV vs. Ag/AgCl reference electrodes. Rates of biosensor-driven *p*-hydroxylation of aniline, *N*-demethylation of aminopyrine and cholesterol side-chain cleavage were close to those obtained with NAD(P)H as the electron source [441]. Alternatively, CYP2B4 was adsorbed onto SP electrodes coated with colloidal gold nanoparticles stabilized with didodecyldimethylammonium bromide (DDAB) in the presence of the Nafion ionomer to improve film permeability. The construct adequately mediated benzphetamine *N*-dealkylation [442]. Modifying the immobilization scheme, studies were carried out with monomerized CYP2B4 incorporated into thin layers of non-ionic detergent and montmorillonite, a member of the mineral group of clays, on glassy carbon (GC) electrodes. Here, k_{cat} for aminopyrine turnover was shown to be comparable to the value of the microsomal system [443]. Moreover, a biocompatible film containing colloidal gold nanoparticles and chitosan was used to encapsulate CYP2B6 on GC sensors. Product analysis confirmed C-hydroxylation and heteroatom release from bupropion, lidocaine and cyclophosphamide to be the main pathways of drug oxidation [444]. Studies were also conducted with carbon cloth (CC) electrodes coated by immersion into DDAB dispersions embedding bacterial CYP101A1. Electrolyses performed under aerobic conditions in the presence of styrene and *cis*-*β*-methylstyrene as the probe substrates revealed styrene oxide and *trans*-*β*-methylstyrene oxide to be the major products resulting from oxidative attack by the P450, while some byproducts were speculated to rather arise from H_2O_2-driven reactions [445].

Substantial progress was achieved by construction of enzyme films of predesigned architecture via layer-by-layer self-assembly of hemoproteins and oppositely charged polyions on the surface of electrodes. Applying this regimen, CC sensors elaborated by casting CYP1A2/poly(styrenesulfonate) (PSS) microemulsions onto the solid supporters displayed good electrocatalytic performance of O_2 reduction to hydrogen peroxide, mediating epoxidation of styrene faster than CYP101A1 [446].

Electrochemical exploration was extended to assembly of enzyme films on the surface of derivatized gold electrodes by alternate adsorption of a P450 layer on top of a poly (diallyldimethyl-ammonium) (PDDA) layer. In this way, immobilization of CYP2E1 and CYP3A4 resulted in sensor devices mediating oxidative turnover of *p*-nitrophenol and midazolam, respectively, at fairly low catalytic rates [447, 448]. Here, covalent enzyme linkage to gold electrodes via flexible spacer molecules, bearing both thiol and disulfide groups as well as anchors to the proteins, was shown to increase metabolic efficiency. Making use of this strategy, the exposed C261 and C268 residues of CYP2E1 were intimately connected with cystamine-maleimide on gold biosensors. This procedure stimulated conversion of *p*-nitrophenol to *p*-nitrocatechol by a factor of 22 relative to the Au/PDDA array [447]. Similarly, human CYP2C9 was bonded to a gold electrode by the aid of its N-terminal lysine fixed to an 11-mercaptoundecanoic acid and octanethiol self-assembled monolayer. Electron transfer was calculated to proceed at a rate ranging from 6 to 31 s^{-1}, with warfarin being metabolized to the 7-hydroxy derivative at an apparent K_m of 3 μM [449].

10.6 Conclusions and Future Prospects

The present review focuses on the description of electron transfer events with emphasis on topological and functional features in the P450-dependent redox chain to gain a more detailed, structure-based insight into fundamental molecular principles steering donor-acceptor interactions. Improved comprehension may permit engineering to introduce more efficient electrochemical properties into the system such as facilitated redox partner association and intermolecular electron flow [200, 303], but equally-well may pave the way for the development of technologically viable hemoprotein species for extensive exploitation as versatile biocatalysts [35]. Here, directed evolution and DNA shuffling may be useful in the sculpturing of self-sufficient fusion proteins for preselected metabolic implementation [376, 395, 404] or in the development of peroxygenase-like P450s with upgraded resistance toward oxidative destruction [430, 431] to obviate the tedious reconstitution procedure. Also, artificial photo- and electrocatalytic devices might help avoid costly NAD(P)H utilization [434, 442].

Despite conspicuous biotechnological advances [450], industrial large-scale production of fine chemicals is presently limited to a fairly low number of processes making preferential use of microbial whole-cell catalysts harboring recombinant P450s co-expressed with an appropriate electron donor [451]. Relevant examples include hydrocortisone production via $P450_{lun}$-mediated 11β-hydroxylation of 11-deoxycortisol in a fungal bioreactor [452] or manufacture of the cholesterol-lowering drug pravastatin by CYP105A3-driven attack on compactin, employing the *Streptomyces* sp. Y-110 as the host [453]. Moreover, CYP71AV1-promoted three-step oxidation of amorphadiene to artemisinic acid, the immediate precursor of the antimalarial drug artemisinin, permitted industrial scale-up due to high productivity of the engineered *Saccharomyces cerevisiae* factory [454]. Similarly, long-chain α,ω-dicarboxylic acids, widely used as raw materials for the synthesis of products such as perfumes, hot-melting adhesives, engineering plastics or high quality lubricants, have been generated on a commercial scale from *n*-alkanes by fungal fermentation catalyzed by *Candida tropicalis*, housing CYP52A1 in conjunction with POR as the redox machinery [455].

Specialized exploitation of manipulated hemoproteins was recognized to be of high interest in gene-directed enzyme prodrug therapy (GDEPT) of cancer. Here, introduction of tumor-selective retroviral vectors, encoding P450s characterized by high metabolic potency in the reductase-supported intratumoral conversion of anticarcinogenic compounds such as cyclophosphamide or ifosfamide to their active intermediates, displayed a substantial therapeutic progress [456]. This has given an impetus to improvement of reactivity of CYP2B enzymes toward the oxazaphosphorines [457, 458]. In this regard, creation of the 114V/477W double mutant of human CYP2B6 increased the catalytic efficiency of cyclophosphamide oxidation by a factor of 4 [459]. Also, there is clear opportunity to expedite therapeutic efficacy by utilization of the fusion gene of the self-sufficient CYP2B6/POR chimera for infection of tumor cells [460]. The current advances lend confidence that this novel strategy may be promoted by the development of more sophisticated vector and promotor systems.

The engineered P450 redox machinery may also be exploited in phyto- and bioremediation processes. Thus, expression in tobacco plants of human CYP1A proteins or CYP76B1 from *Helianthus tuberosus* as hybrid enzymes fused with POR increased herbicide resistance toward a series of phenylureas such as chlortoluron due to swift detoxification [379, 396]. Similarly, transduction of rice plants with human CYP2B6 or CYP2C19 enhanced the ability to remove atrazine and metolachlor herbicides from soil [461]. Interest has also arisen in transgenic *Arabidopsis* plants producing the bizarre fusion protein CYP177A1(XplA) for targeted degradation of the widespread military explosive RDX, a priority pollutant contaminating liquid culture and soil leachate [462]. One factor complicating introduction of such technologies for environmental clean-up may be concerns about field application of genetically modified organisms, possibly entailing certain risks. Nevertheless, remediation of the biotope needs a robust catalytic apparatus capable of killing off hazardous anthropogenic toxicants via more flexible, pollutant-specific oxyfunctionalization. Here, polycyclic aromatic hydrocarbons such as phenanthrene, naphthalene, fluorene or benzo[*a*]pyrene proved to be targets for optimized biotransformation by chimeric CYP1A1/POR, mutated CYP102A1, fused CYP116B3 or modified CYP5136A3, expressed in microbial recombinant cells [371, 463–465]. However, inoculation and efficient maintenance of the population density of the engineered microbial biomass in terrestrial habitats still need improvement [466].

Finally, exploitation of P450-based electroanalytical techniques may become of increasing interest to enable more practical applications. Given appreciable sensitivity and recognition selectivity, miniaturized amperometric biosensors might be utilized for the determination of compounds important in pharmaceutical industry, clinical practice and environmental monitoring [467, 468]. Also, microfluidic devices were developed to improve analytical performance by decreasing analysis time and increasing reliability through automation [440]. This may foster high-throughput screening during the search for structural features of dynamic molecules having potential for therapeutic implementation [469].

Collectively, catalytic versatility, no doubt, adds the P450 redox system to the enzymatic armory for large-scale exploitation in a vast array of biotechnological areas. Despite huge progress in recent years, members of the hemoprotein family are, nevertheless, thought of as relatively fragile biocatalysts prone to spontaneous structural disruption or rapid inactivation at temperatures >40 °C [430]. Hence, future engineering strategies will have to focus on erasure of these shortcomings and evolution of novel activities to allow widespread application in metabolic processes.

Reviews *Abbreviations*: *AdR* NADPH-adrenodoxin reductase, *Adx* adrenodoxin, *Arx* [2Fe-2S]-type ferredoxin, b_5 cytochrome b_5, *BMR Bacillus megaterium* (CYP102A1) reductase component, *CC* carbon cloth electrode, *CYP or P450* cytochrome P450, *FdR* NAD(P)H-ferredoxin reductase, *Fdx* ferredoxin, *GC* glassy carbon electrode, *PCNA* proliferating cell nuclear antigen, *PdR* NADH-putidaredoxin reductase, *PDR* phthalate dioxygenase reductase, *Pdx* putidaredoxin, *POR* NADPH-P450 oxidoreductase, *Pux* palustrisredoxin, *r.m.s.* root mean square deviation, *SP* screen-printed electrode.

References

1. Nelson DR (2006) Cytochrome P450 nomenclature, 2004. Methods Mol Biol 320:1–10
2. Hlavica P (2011) Insect cytochromes P450: topology of structural elements predicted to govern catalytic versatility. J Inorg Biochem 105:1354–1364
3. Bak S, Beisson F, Bishop G, Hamberger B, Höfer R, Paquette S, Werck-Reichhart D (2011) Cytochromes p450. Arabidopsis Book 9:e0144

4. Hlavica P (2013) Evaluation of structural features in fungal cytochromes P450 predicted to rule catalytic diversification. Biochim Biophys Acta 1834:205–220
5. Lewis DFV, Wiseman A (2005) A selective review of bacterial forms of cytochrome P450 enzymes. Enzyme Microb Technol 36:377–384
6. Omura T, Sato R (1964) The carbon monoxide-binding pigment of liver microsomes. I. Evidence for its hemoprotein nature. J Biol Chem 239:2370–2378
7. Bernhardt R, Waterman MR (2007) Cytochrome P450 and steroid hormone biosynthesis. Met Ions Life Sci 3:361–396
8. Hlavica P, Lehnerer M (2010) Oxidative biotransformation of fatty acids by cytochromes P450: predicted key structural elements orchestrating substrate specificity, regioselectivity and catalytic efficiency. Curr Drug Metab 11:85–104
9. Hlavica P (2006) Functional interaction of nitrogenous organic bases with cytochrome P450: a critical assessment and update of substrate features and predicted key active-site elements steering the access, binding, and orientation of amines. Biochim Biophys Acta 1764:645–670
10. Brown CM, Reisfeld B, Mayeno AN (2008) Cytochrome P450: a structure-based summary of biotransformations using representative substrates. Drug Metab Rev 40:1–100
11. Meunier B, de Visser SP, Shaik S (2004) Mechanism of oxidation reactions catalyzed by cytochrome P450 enzymes. Chem Rev 104:3947–3980
12. Denisov IG, Makris TM, Sligar SG, Schlichting I (2005) Structure and chemistry of cytochrome P450. Chem Rev 105:2253–2277
13. Newcomb M, Hollenberg PF, Coon MJ (2003) Multiple mechanisms and multiple oxidants in P450-catalyzed hydroxylations. Arch Biochem Biophys 409:72–79
14. Hlavica P (2004) Models and mechanisms of O-O bond activation by cytochrome P450. A critical assessment of the potential role of multiple active intermediates in oxidative catalysis. Eur J Biochem 271:4335–4360
15. Hrycay EG, Bandiera SM (2012) The monooxygenase, peroxidase, and peroxygenase properties of cytochrome P450. Arch Biochem Biophys 522:71–89
16. Lewis DFV, Watson E, Lake BG (1998) Evolution of the cytochrome P450 superfamily: sequence alignments and pharmacogenetics. Mutat Res 410:245–270
17. Hannemann F, Bichet A, Ewen KM, Bernhardt R (2007) Cytochrome P450 systems – biological variations of electron transport chains. Biochim Biophys Acta 1770:330–344
18. Paine MJI, Scrutton NS, Munro AW, Gutierrez A, Roberts GCK, Wolf CR (2005) Electron transfer partners of cytochrome P450. In: Ortiz de Montellano PR (ed) Cytochrome P450: structure, mechanism, and biochemistry, 3rd edn. Kluwer Academic/Plenum Publishers, New York, pp 115–148
19. Stok JE, De Voss JJ (2000) Expression, purification, and characterization of BioI: a carbon-carbon bond cleaving cytochrome P450 involved in biotin biosynthesis in *Bacillus subtilis*. Arch Biochem Biophys 384:351–360
20. Lawson RJ, von Wachenfeldt C, Haq I, Perkins J, Munro AW (2004) Expression and characterization of the two flavodoxin proteins of *Bacillus subtilis*, YkuN and YkuP: biophysical properties and interactions with cytochrome P450 BioI. Biochemistry 43:12390–12409
21. Hawkes DB, Adams GW, Burlingame AL, Ortiz de Montellano PR, De Voss JJ (2002) Cytochrome $P450_{cin}$ (CYP176A), isolation, expression, and characterization. J Biol Chem 277:27725–27732
22. Hawkes DB, Slessor KE, Bernhardt PV, De Voss JJ (2010) Cloning, expression and purification of cindoxin, an unusual FMN-containing cytochrome P450 redox partner. Chembiochem 11:1107–1114
23. Kimmich N, Das A, Sevrioukova I, Meharenna Y, Sligar SG, Poulos TL (2007) Electron transfer between cytochrome $P450_{cin}$ and its FMN-containing redox partner, cinredoxin. J Biol Chem 282:27006–27011
24. Yeom J, Park W (2012) Biochemical characterization of ferredoxin-$NADP^+$ reductase interaction with flavodoxin in *Pseudomonas putida*. BMB Rep 45:476–481
25. Oshima R, Fushinobu S, Su F, Zhang L, Takaya N, Shoun H (2004) Structural evidence for direct hydride transfer from NADH to cytochrome $P450_{nor}$. J Mol Biol 342:207–217
26. Bui PH, Hankinson O (2009) Functional characterization of human cytochrome P450 2S1 using a synthetic gene-expressed protein in *Escherichia coli*. Mol Pharmacol 76:1031–1043
27. Bui PH, Hsu EL, Hankinson O (2009) Fatty acid hydroperoxides support cytochrome P450 2S1-mediated bioactivation of benzo[*a*]pyrene-7,8-dihydrodiol. Mol Pharmacol 76:1044–1052
28. Lee DS, Yamada A, Sugimoto H, Matsunaga I, Ogura H, Ichihara K, Adachi S, Park SY, Shiro Y (2003) Substrate recognition and molecular mechanism of fatty acid hydroxylation by cytochrome P450 from *Bacillus subtilis*. J Biol Chem 278:9761–9767
29. Chen CYK, Poole EM, Ulrich CM, Kulmacz RJ, Wang LH (2012) Functional analysis of human thromboxane synthase polymorphic variants. Pharmacogenet Genomics 22:653–658
30. Song WV, Funk CD, Brash AR (1993) Molecular cloning of an allene oxide synthase: a cytochrome P450 specialized for the metabolism of fatty acid hydroperoxides. Proc Natl Acad Sci U S A 90:8519–8523
31. Lewis DFV, Hlavica P (2000) Interactions between redox partners in various cytochrome P450 systems:

functional and structural aspects. Biochim Biophys Acta 1460:353–374
32. Hlavica P, Schulze J, Lewis DFV (2003) Functional interaction of cytochrome P450 with its redox partners: a critical assessment and update of the topology and predicted contact regions. J Inorg Biochem 96:279–297
33. Hlavica P (2007) Control by substrate of the cytochrome P450-dependent redox machinery: mechanistic insights. Curr Drug Metab 8:594–611
34. Hlavica P (2009) Assembly of non-natural electron transfer conduits in the cytochrome P450 system: a critical assessment and update of artificial redox constructs amenable to exploitation in biotechnological areas. Biotechnol Adv 27:103–121
35. Bernhardt R (2006) Cytochromes P450 as versatile biocatalysts. J Biotechnol 124:128–145
36. Kumar S (2010) Engineering P450 biocatalysts for biotechnology, medicine, and bioremediation. Expert Opin Drug Metab Toxicol 6:115–131
37. Murataliev MB, Feyereisen R, Walker FA (2004) Electron transfer by diflavin reductases. Biochim Biophys Acta 1698:1–26
38. Iyanagi T, Xia C, Kim JJP (2012) NADPH-cytochrome P450 oxidoreductase: prototypic member of the diflavin reductase family. Arch Biochem Biophys 528:72–89
39. Jenkins CM, Waterman MR (1999) Flavodoxin as a model for the P450-interacting domain of NADPH-cytochrome P450 reductase. Drug Metab Rev 31:195–203
40. Aliverti A, Pandini V, Pennati A, de Rosa M, Zanetti G (2008) Structural and functional diversity of ferredoxin-NADP$^+$ reductase. Arch Biochem Biophys 474:283–291
41. Porter TD, Kasper CB (1986) NADPH-cytochrome P-450 oxidoreductase: flavin mononucleotide and flavin adenine dinucleotide domains evolved from different flavoproteins. Biochemistry 25:1682–1687
42. Sancho J (2006) Flavodoxins: sequence, folding, binding, function and beyond. Cell Mol Life Sci 63:855–864
43. Ceccarelli EA, Arakaki AK, Cortez N, Carrillo N (2004) Functional plasticity and catalytic efficiency in plant and bacterial ferredoxin-NADP(H) reductases. Biochim Biophys Acta 1698:155–165
44. Yamano S, Aoyama T, McBride OW, Hardwick JP, Gelboin HV, Gonzalez FJ (1989) Human NADPH-P450 oxidoreductase: complementary DNA cloning, sequence and Vaccinia virus-mediated expression and localization of the *CYPOR* gene to chromosome 7. Mol Pharmacol 36:83–88
45. Hart SN, Zhong X (2008) P450-oxidoreductase: genetic polymorphism and implications for drug metabolism and toxicity. Expert Opin Metab Toxicol 4:439–452
46. Hu L, Zhuo W, He YJ, Zhou HH, Fan L (2012) Pharmacogenetics of P450 oxidoreductase: implications in drug metabolism and therapy. Pharmacogenet Genomics 22:812–819
47. Porter TD, Beck TW, Kasper CB (1990) NADPH-cytochrome P-450 oxidoreductase gene organization correlates with structural domains of the proteins. Biochemistry 29:9814–9818
48. Hovemann BT, Sehlmeyer F, Malz J (1997) *Drosophila melanogaster* NADPH-cytochrome P450 oxidoreductase: pronounced expression in antennae may be related to odorant clearance. Gene 189:213–219
49. Yadav JS, Loper JC (2000) Cytochrome P450 oxidoreductase gene and its differentially terminated cDNAs from the white rot fungus *Phanerochaete chrysosporium*. Curr Genet 37:65–73
50. Durst F, Nelson DR (1995) Diversity and evolution of plant P450 and P450-reductases. Drug Metabol Drug Interact 12:189–206
51. Jensen KL, Møller B (2010) Plant NADPH-cytochrome P450 oxidoreductases. Phytochemistry 71:132–141
52. Iyanagi T, Makino N, Mason HS (1974) Redox properties of the reduced nicotinamide adenine dinucleotide phosphate-cytochrome P450 and reduced nicotinamide adenine dinucleotide-cytochrome b_5 reductases. Biochemistry 13:1701–1710
53. Oprian DD, Coon MJ (1982) Oxidation-reduction states of FMN and FAD in NADPH-cytochrome P-450 reductase during reduction by NADPH. J Biol Chem 257:8935–8944
54. Massey V, Palmer G (1966) On the existence of spectrally distinct classes of flavoprotein semiquinones. A new method for quantitative production of flavoprotein semiquinones. Biochemistry 5:3181–3189
55. Müller F, Brüstlein M, Hemmerich P, Massey V, Walker WH (1972) Light-absorption studies on neutral flavin radicals. Eur J Biochem 25:573–580
56. Brenner S, Hay S, Munro AW, Scrutton NS (2008) Inter-flavin electron transfer in cytochrome P450 reductase – effects of solvent and pH identify hidden complexity in mechanism. FEBS J 275:4540–4557
57. Das A, Sligar SG (2009) Modulation of the cytochrome P450 reductase redox potential by the phospholipid bilayer. Biochemistry 48:12104–12112
58. Louerat-Oriou B, Perret A, Pompon D (1998) Differential redox and electron-transfer properties of purified yeast, plant and human NADPH-cytochrome P-450 reductases highly modulate cytochrome P-450 activities. Eur J Biochem 258:1040–1049
59. Daff SN, Chapman SK, Turner KL, Holt RA, Govindaraj S, Poulos TL, Munro AW (1997) Redox control of the catalytic cycle of flavocytochrome P-450BM3. Biochemistry 36:13816–13823
60. Hefti MH, Vervoort J, van Berkel WJH (2003) Deflavination and reconstitution of flavoproteins. Tackling fold and function. Eur J Biochem 270:4227–4242
61. Wolthers KR, Basran J, Munro AW, Scrutton NS (2003) Molecular dissection of human synthase reductase: determination of the flavin redox

potentials in full-length enzyme and isolated flavin-binding domains. Biochemistry 42:3911–3920
62. Iyanagi T (2007) Molecular mechanism of phase I and phase II drug-metabolizing enzymes: implications for detoxification. Int Rev Cytol 260:35–112
63. Gutierrez A, Paine M, Wolf CR, Scrutton NS, Roberts GCK (2002) Relaxation kinetics of cytochrome P450 reductase: internal electron transfer is limited by conformational changes and regulated by coenzyme binding. Biochemistry 41:4626–4637
64. Munro AW, Noble MA, Robledo L, Daff SN, Chapman SK (2001) Determination of the redox properties of human NADPH-cytochrome P450 reductase. Biochemistry 40:1956–1963
65. Gutierrez A, Munro AW, Grunau A, Wolf CR, Scrutton NS, Roberts GCK (2003) Interflavin electron transfer in human cytochrome P450 reductase is enhanced by coenzyme binding. Eur J Biochem 270:2612–2621
66. Vermilion JL, Ballou DP, Massey V, Coon MJ (1981) Separate roles of FMN and FAD in catalysis by liver microsomal NADPH-cytochrome P-450 reductase. J Biol Chem 256:266–277
67. Murataliev MB, Klein M, Fulco A, Feyereisen R (1997) Functional interactions in cytochrome P450BM3: flavin semiquinone intermediates, role of NADP(H), and mechanism of electron transfer by the flavoprotein domain. Biochemistry 36:8401–8412
68. Xia C, Panda SP, Marohnic CC, Martasek P, Masters BS, Kim JJP (2011) Structural basis for human NADPH-cytochrome P450 oxidoreductase deficiency. Proc Natl Acad Sci U S A 108:13486–13491
69. Wang M, Roberts DL, Paschke R, Shea TM, Masters BSS, Kim JJP (1997) Three-dimensional structure of NADPH-cytochrome P450 reductase: prototype for FMN- and FAD-containing enzymes. Proc Natl Acad Sci U S A 94:8411–8416
70. Lamb DC, Kim Y, Yermalitskaya LV, Yermalitsky VN, Lepesheva GI, Kelly SL, Waterman MR, Podust LM (2006) A second FMN binding site in yeast NADPH-cytochrome P450 reductase suggests a mechanism of electron transfer by diflavin reductases. Structure 14:51–61
71. Zhao Q, Modi S, Smith G, Paine M, McDonagh PD, Wolf CR, Tew D, Lian LY, Roberts GCK, Driessen HPC (1999) Crystal structure of the FMN-binding domain of human cytochrome P450 reductase at 1.93 Å resolution. Protein Sci 8:298–306
72. Hubbard PA, Shen AL, Paschke R, Kasper CB, Kim JJP (2001) NADPH-cytochrome P450 oxidoreductase. Structural basis for hydride and electron transfer. J Biol Chem 276:29163–29170
73. Kida Y, Ohgiya S, Mihara K, Sakaguchi M (1998) Membrane topology of NADPH-cytochrome P450 reductase on the endoplasmic reticulum. Arch Biochem Biophys 351:175–179
74. Bonina TA, Gilep AA, Estabrook RW, Usanov SA (2005) Engineering of proteolytically stable NADPH-cytochrome P450 reductase. Biochemistry (Mosc) 70:357–365
75. Sarapusit S, Xia C, Misra I, Rongnoparut P, Kim JJP (2008) NADPH-cytochrome P450 oxidoreductase from the mosquito *Anopheles minimus*: kinetic studies and the influence of Leu86 and Leu219 on cofactor binding and protein stability. Arch Biochem Biophys 477:53–59
76. Nicolo C, Flück CE, Mullis PE, Pandey AV (2010) Restoration of mutant cytochrome P450 reductase activity by external flavin. Mol Cell Endocrinol 321:245–252
77. Barsukov I, Modi S, Lian LY, Sze KH, Paine MJI, Wolf CR, Roberts GCK (1997) ^{1}H, ^{15}N and ^{13}C NMR resonance assignment, secondary structure and global fold of the FMN-binding domain of human cytochrome P450 reductase. J Biomol NMR 10:63–75
78. Paine MJI, Ayivor S, Munro A, Tsan P, Lian LY, Roberts GCK, Wolf CR (2001) Role of the conserved phenylalanine 181 of NADPH-cytochrome P450 oxidoreductase in FMN binding and catalytic activity. Biochemistry 40:13439–13447
79. Flück CE, Mullis PE, Pandey AV (2009) Modeling of human P450 oxidoreductase structure by in silico mutagenesis and MD simulation. Mol Cell Endocrinol 313:17–22
80. Shen AL, Porter TD, Wilson TE, Kasper CB (1989) Structural analysis of the FMN- binding domain of NADPH-cytochrome P-450 oxidoreductase by site-directed mutagenesis. J Biol Chem 264:7584–7589
81. Hamdane D, Xia C, Im SC, Zhang H, Kim JJP, Waskell L (2009) Structure and function of an NADPH-cytochrome P450 oxidoreductase in an open conformation capable of reducing cytochrome P450. J Biol Chem 284:11374–11384
82. Shen AL, Kasper CB (2000) Differential contributions of NADPH-cytochrome P450 oxidoreductase FAD binding site residues to flavin binding and catalysis. J Biol Chem 275:41087–41091
83. Shen AL, Kasper CB (1996) Role of Ser457 of NADPH-cytochrome P450 oxidoreductase in catalysis and control of FAD oxidation-reduction potential. Biochemistry 35:9451–9459
84. Shen AL, Sem DS, Kasper CB (1999) Mechanistic studies on the reductive half-reaction of NADPH-cytochrome P450 oxidoreductase. J Biol Chem 274:5391–5398
85. Shen AL, Christensen MJ, Kasper CB (1991) NADPH-cytochrome P-450 oxidoreductase. The role of cysteine 566 in catalysis and cofactor binding. J Biol Chem 266:19976–19980
86. Sem DS, Kasper CB (1993) Interaction with arginine 597 of NADPH-cytochrome P-450 oxidoreductase is

a primary source of the uniform binding energy used to discriminate between NADPH and NADH. Biochemistry 32:11548–11558
87. Döhr O, Paine MJI, Friedberg T, Roberts GCK, Wolf CR (2001) Engineering of a functional human NADH-dependent cytochrome P450 system. Proc Natl Acad Sci U S A 98:81–86
88. Elmore CL, Porter TD (2002) Modification of the nucleotide cofactor-binding site of cytochrome P-450 reductase to enhance turnover with NADH in vivo. J Biol Chem 277:48960–48964
89. Gutierrez A, Doehr O, Paine M, Wolf R, Scrutton NS, Roberts GCK (2000) Trp-676 facilitates nicotinamide coenzyme exchange in the reductive half-reaction of human cytochrome P450 reductase: properties of the soluble W676H and W676A mutant reductases. Biochemistry 39:15990–15999
90. Xia C, Hamdane D, Shen AL, Choi V, Kasper CB, Pearl NM, Zhang H, Im SC, Waskell L, Kim JJP (2011) Conformational changes of NADPH-cytochrome P450 oxidoreductase are essential for catalysis and cofactor binding. J Biol Chem 286:16246–16260
91. Huang N, Pandey AV, Agrawal V, Reardon W, Lapunzina PD, Mowat D, Jabs EW, Van Vliet G, Sack J, Flück CE, Miller WL (2005) Diversity and function of mutations in P450 oxidoreductase in patients with Antley-Bixler syndrome and disordered steroidogenesis. Am J Hum Genet 76:729–749
92. Pandey AV, Kempna P, Hofer G, Mullis PE, Flück CE (2007) Modulation of human CYP19A1 activity by mutant NADPH-P450 oxidoreductase. Mol Endocrinol 21:2579–2595
93. Chen X, Pan LQ, Naranmandura H, Zeng S, Chen SQ (2012) Influence of various polymorphic variants of cytochrome P450 oxidoreductase (POR) on drug metabolic activity of CYP3A4 and CYP2B6. PLoS One 7:e38495
94. Marohnic CC, Panda SP, McCammon K, Rueff J, Masters BSS, Kranendonk M (2010) Human cytochrome P450 oxidoreductase deficiency caused by the Y181D mutation: molecular consequences and rescue of defect. Drug Metab Dispos 38:332–340
95. Aigrain L, Pompon D, Morera S, Truan G (2009) Structure of the open conformation of a functional chimeric NADPH-cytochrome P450 reductase. EMBO Rep 10:742–747
96. Wadsäter M, Laursen T, Singha A, Hatzakis NS, Stamou D, Barker R, Mortensen K, Feidenhans R, Lindberg Møller B, Cardenas M (2012) Monitoring shifts in the conformation equilibrium of the membrane protein cytochrome P450 reductase (POR) in nanodiscs. J Biol Chem 287:34596–34603
97. Hay S, Brenner S, Khara B, Quinn AM, Rigby SEJ, Scrutton NS (2010) Nature of the energy landscape for gated electron transfer in a dynamic redox protein. J Am Chem Soc 132:9738–9745
98. Pudney CR, Heyes DJ, Khara B, Hay S, Rigby SEJ, Scrutton NS (2012) Kinetic and spectroscopic probes of motions and catalysis in the cytochrome P450 reductase family of enzymes. FEBS J 279:1534–1544
99. Pudney CR, Khara B, Johannissen LO, Scrutton NS (2011) Coupled motions direct electrons along human microsomal P450 chains. PLoS Biol 9: e1001222
100. Vincent B, Morellet N, Fatemi F, Aigrain L, Truan G, Guittet E, Lescop E (2012) The closed and compact domain organization of the 70-kDa human cytochrome P450 reductase in its oxidized state as revealed by NMR. J Mol Biol 420:296–309
101. Laursen T, Jensen K, Lindberg Møller B (2011) Conformational changes of the NADPH-dependent cytochrome P450 reductase in the course of electron transfer to cytochromes P450. Biochim Biophys Acta 1814:132–138
102. Hong Y, Li H, Yuan YC, Chen S (2010) Sequence-function correlation of aromatase and its interaction with reductase. J Steroid Biochem Mol Biol 118:203–206
103. Jang HH, Jamakhandi AP, Sullivan SZ, Yun CH, Hollenberg PF, Miller GP (2010) Beta sheet 2-alpha helix C loop of cytochrome P450 reductase serves as a docking site for redox partners. Biochim Biophys Acta 1804:1285–1293
104. Shen AL, Kasper CB (1995) Role of acidic residues in the interaction of NADPH- cytochrome P450 oxidoreductase with cytochrome P450 and cytochrome *c*. J Biol Chem 270:27475–27480
105. Schenkman JB, Jansson I (2003) The many roles of cytochrome b_5. Pharmacol Ther 97:139–152
106. Hultquist DE, Dean RT, Douglas RH (1974) Homogenous cytochrome b_5 from human erythrocytes. Biochem Biophys Res Commun 60:28–34
107. Bando S, Takano T, Yubisui T, Shirabe K, Takeshita M, Nakagawa A (2004) Structure of human erythrocyte NADH-cytochrome b_5 reductase. Biol Crystallogr 60:1929–1934
108. Shirabe K, Fujimoto Y, Yubisui T, Takeshita M (1994) An in-frame deletion of codon 298 in the NADH-cytochrome b_5 reductase gene results in hereditary methemoglobinemia type II (generalized type). A functional implication for the role of the COOH-terminal region of the enzyme. J Biol Chem 269:5952–5957
109. Golly I, Hlavica P (1983) The role of hemoglobin in the N-oxidation of 4-chloroaniline. Biochim Biophys Acta 760:69–76
110. Imai Y (1981) The roles of cytochrome b_5 in reconstituted monooxygenase systems containing various forms of hepatic microsomal cytochrome P-450. J Biochem 89:351–362
111. Pompon D, Coon MJ (1984) On the mechanism of action of cytochrome P-450. Oxidation and reduction of the ferrous dioxygen complex of liver microsomal cytochrome P-450 by cytochrome b_5. J Biol Chem 259:15377–15385

112. Lederer F, Ghrir R, Guiard B, Cortial S, Ito A (1983) Two homologous cytochromes b_5 in a single cell. Eur J Biochem 132:95–102
113. Vergeres G, Ramsden J, Waskell L (1995) The carboxyl-terminus of the membrane binding domain of cytochrome b_5 spans the bilayer of the endoplasmic reticulum. J Biol Chem 270:3414–3422
114. Honsho M, Mitoma J, Ito A (1998) Retention of cytochrome b_5 in the endoplasmic reticulum is transmembrane and luminal domain-dependent. J Biol Chem 273:20860–20866
115. Tanaka S, Kinoshita J, Kuroda R, Ito A (2003) Integration of cytochrome b_5 into endo- plasmic reticulum membrane: participation of carboxy-terminal portion of the transmembrane domain. J Biochem 133:247–251
116. Hanlon MR, Begum RR, Newbold RJ, Whitford D, Wallace BA (2000) *In vitro* membrane-inserted conformation of the cytochrome b_5 tail. Biochem J 352:117–124
117. Kaderbhai MA, Morgan R, Kaderbhai NN (2003) The membrane-interactive tail of cytochrome b_5 can function as a stop-transfer sequence in concert with a signal sequence to give inversion of protein topology in the endoplasmic reticulum. Arch Biochem Biophys 412:259–266
118. Vergeres G, Waskell L (1995) Cytochrome b_5, its functions, structure and membrane topology. Biochimie 77:604–620
119. Vergeres G, Waskell L (1992) Expression of cytochrome b_5 in yeast and characterization of mutants of the membrane-anchoring domain. J Biol Chem 267:12583–12591
120. Cowley AB, Altuve A, Kuchment O, Terzyan S, Zhang X, Rivera M, Benson DR (2002) Toward engineering the stability and hemin-binding properties of microsomal cytochrome b_5 into rat outer mitochondrial membrane cytochrome b_5: examining the influence of residues 25 and 71. Biochemistry 41:11566–11581
121. Mathews FS, Levine M, Argos P (1972) Three-dimensional fourier synthesis of calf liver cytochrome b_5 at 2.8 Å resolution. J Mol Biol 64:449–464
122. Durley RCE, Mathews FS (1996) Refinement and structural analysis of bovine cytochrome b_5 at 1.5 Å resolution. Acta Cryst D52:65–76
123. Konopka K, Waskell L (1988) Modification of trypsin-solubilized cytochrome b_5, apo-cytochrome b_5, and liposome-bound cytochrome b_5 by diethylpyrocarbonate. Arch Biochem Biophys 261:55–63
124. Wang WH, Lu J, Yao P, Xie Y, Huang ZX (2003) The distinct heme coordination environments and heme-binding stabilities of His39Ser and His39Cys mutants of cytochrome b_5. Protein Eng 16:1047–1054
125. Aono T, Sakamoto Y, Miura M, Takeuchi F, Hori H, Tsubaki M (2010) Direct electrochemical analysis of human cytochromes b_5 with a mutated heme pocket showed a good correlation between their midpoint and half wave potentials. J Biomed Sci 17:90–104
126. Cao C, Zhang Q, Wang ZQ, Wang YF, Wang YH, Wu H, Huang ZX (2003) ^{1}H NMR studies of the effect of mutation at valine 45 on heme microenvironment of cytochrome b_5. Biochimie 85:1007–1016
127. Cao C, Zhang Q, Xue LL, Ma J, Wang YH, Wu H, Huang ZX (2003) The solution structure of the oxidized bovine microsomal cytochrome b_5 mutant V61H. Biochem Biophys Res Commun 307:600–609
128. Dangi B, Sarma S, Yan C, Banville DL, Guiles RD (1998) The origin of differences in the physical properties of the equilibrium forms of cytochrome b_5 revealed through high-resolution NMR structures and backbone dynamic analyses. Biochemistry 37:8289–8302
129. Shan L, Lu JX, Gan JH, Wang YH, Huang ZX, Xia ZX (2005) Structure of the F58W mutant of cytochrome b_5: the mutation leads to multiple conformations and weakens stacking interactions. Acta Cryst D61:180–189
130. Mathews FS, Czerwinski EW (1985) Cytochrome b_5 and cytochrome b_5 reductase from the chemical and X-ray diffraction viewpoint. In: Martonosi AN (ed) The enzymes of biological membranes, 4th edn. Springer, New York, pp 235–300
131. Vergeres G, Chen DY, Wu FF, Waskell L (1993) The function of tyrosine 74 of cytochrome b_5. Arch Biochem Biophys 305:231–241
132. Yao P, Wu J, Wang YH, Sun BY, Xia ZX, Huang ZX (2002) X-ray crystallography, CD and kinetic studies revealed the essence of the abnormal behaviors of the cytochrome b_5 Phe35→Tyr mutant. Eur J Biochem 269:4287–4296
133. Yao P, Xie Y, Wang YH, Sun YL, Huang ZX, Xiao GT, Wang SD (1997) Importance of the conserved phenylalanine-35 of cytochrome b_5 to the protein's stability and redox potential. Protein Eng 10:575–581
134. Wang ZQ, Wang YH, Qian W, Wang HH, Chunyu LJ, Xie Y, Huang ZX (1999) Methanol-induced unfolding and refolding of cytochrome b_5 and its P40V mutant monitored by UV-visible, CD, and fluorescence spectra. J Protein Chem 18:547–555
135. Banci L, Bertini I, Rosato A, Scacchieri S (2000) Solution structure of oxidized microsomal rabbit cytochrome b_5. Factors determining the heterologous binding of the heme. Eur J Biochem 267:755–766
136. Kobayashi K, Iyanagi T, Ohara H, Hayashi K (1988) One-electron reduction of hepatic NADH-cytochrome b_5 reductase as studied by pulse radiolysis. J Biol Chem 263:7493–7499
137. Kimura S, Kawamura M, Iyanagi T (2003) Role of Thr66 in porcine NADH-cytochrome b_5 reductase in catalysis and control of the rate-limiting step in electron transfer. J Biol Chem 278:3580–3589

138. Nishida H, Inaka K, Miki K (1995) Specific arrangement of three amino acid residues for flavin-binding barrel structures in NADH-cytochrome b_5 reductase and the other flavin-dependent reductases. FEBS Lett 361:97–100
139. Kimura S, Nishida H, Iyanagi T (2001) Effects of flavin-binding motif amino acid mutations in the NADH-cytochrome b_5 reductase catalytic domain on protein stability and catalysis. J Biochem 130:481–490
140. Bewley MC, Marohnic CC, Barber MJ (2001) The structure and biochemistry of NADH-dependent cytochrome b_5 reductase are now consistent. Biochemistry 40:13574–13582
141. Strittmatter P, Kittler JM, Coghill JE (1992) Characterization of the role of lysine 110 of NADH-cytochrome b_5 reductase in the binding and oxidation of NADH by site-directed mutagenesis. J Biol Chem 267:20164–20167
142. Fujimoto Y, Shirabe K, Nagai T, Yubisui T, Takehita M (1993) Role of Lys-110 of human NADH-cytochrome b_5 reductase in NADH binding as probed by site-directed mutagenesis. FEBS Lett 322:30–32
143. Yubisui T, Shirabe K, Takeshita M, Kobayashi Y, Fukumaki Y, Sakaki Y, Takano T (1991) Structural role of serine 127 in the NADH-binding site of human NADH-cytochrome b_5 reductase. J Biol Chem 266:66–70
144. Roma GW, Crowley LJ, Davis CA, Barber MJ (2005) Mutagenesis of glycine 179 modulates both catalytic efficiency and reduced pyridine nucleotide specificity in cytochrome b_5 reductase. Biochemistry 44:13467–13476
145. Percy MJ, Crowley LJ, Boudreaux J, Barber MJ (2006) Expression of a novel P275L variant of NADH-cytochrome b_5 reductase gives functional insight into the conserved motif important for pyridine nucleotide binding. Arch Biochem Biophys 447:59–67
146. Shirabe K, Yubisui T, Nishino T, Takeshita M (1991) Role of cysteine residues in human NADH-cytochrome b_5 reductase studied by site-directed mutagenesis. CYS-273 and CYS-283 are located close to the NADH-binding site but are not catalytically essential. J Biol Chem 266:7531–7536
147. Marohnic CC, Bewley MC, Barber MJ (2003) Engineering and characterization of a NADPH-utilizing cytochrome b_5 reductase. Biochemistry 42:11170–11182
148. Ozols J, Carr SA, Strittmatter P (1984) Identification of the NH_2-terminal blocking group of NADH-cytochrome b_5 reductase as myristic acid and the complete amino acid sequence of the membrane-binding domain. J Biol Chem 259:13349–13354
149. Strittmatter P, Hackett CS, Korza G, Ozols J (1990) Characterization of the covalent cross-links of the active sites of amidinated cytochrome b_5 and NADH-cytochrome b_5 reductase. J Biol Chem 265:21709–21713
150. Strittmatter P, Kittler J, Goghill JE, Ozols J (1992) Characterization of lysyl residues of NADH-cytochrome b_5 reductase implicated in charge-pairing with active-site carboxyl residues of cytochrome b_5 by site-directed mutagenesis of an expression vector for the flavoprotein. J Biol Chem 267:2519–2523
151. Dailey HA, Strittmatter P (1979) Modification and identification of cytochrome b_5 carboxyl groups involved in protein-protein interaction with cytochrome b_5 reductase. J Biol Chem 254:5388–5396
152. Kawano M, Shirabe K, Nagai T, Takeshita M (1998) Role of carboxyl residues surrounding heme of human cytochrome b_5 in electrostatic interaction with NADH-cytochrome b_5 reductase. Biochem Biophys Res Commun 245:666–669
153. Dailey HA, Strittmatter P (1980) Characterization of the interaction of amphipathic cytochrome b_5 with stearyl coenzyme A desaturase and NADPH-cytochrome P450 reductase. J Biol Chem 255:5184–5189
154. Nisimoto Y, Otsuka-Murakami H (1988) Cytochrome b_5, cytochrome c, and cytochrome P-450 interactions with NADPH-cytochrome P-450 reductase in phospholipid vesicles. Biochemistry 27:5869–5876
155. Enoch HG, Strittmatter P (1979) Cytochrome b_5 reduction by NADPH-cytochrome P-450 reductase. J Biol Chem 254:8976–8981
156. Guengerich FP (2005) Reduction of cytochrome b_5 by NADPH-cytochrome P450 reductase. Arch Biochem Biophys 440:204–211
157. Bhattacharyya AK, Hurley JK, Tollin G, Waskell L (1994) Investigation of the rate limiting step for electron transfer from NADPH-cytochrome P450 reductase to cytochrome b_5: laser flash-photolysis study. Arch Biochem Biophys 310:318–324
158. Meyer TE, Shirabe K, Yubisui T, Takeshita M, Bes MT, Cusanovich MA, Tollin G (1995) Transient kinetics of intracomplex electron transfer in the human cytochrome b_5 reductase-cytochrome b_5 system: NAD^+ modulates protein-protein binding and electron transfer. Arch Biochem Biophys 318:457–464
159. Gruenke L, Konopka K, Cadieu M, Waskell L (1995) The stoichiometry of the cyto- chrome P-450-catalyzed metabolism of methoxyfurane and benzphetamine in the presence and absence of cytochrome b_5. J Biol Chem 270:24707–24718
160. Peng HM, Auchus RJ (2013) The action of cytochrome b_5 on CYP2E1 and CYP2C19 activities requires anionic residues D58 and D65. Biochemistry 52:210–220
161. Shet MS, Faulkner KM, Holmans PL, Fisher CW, Estabrook RW (1995) The effects of cytochrome b_5, NADPH-P450 reductase, and lipid on the rate of 6β-hydroxylation of testosterone as catalyzed by a

human P450 3A4 fusion protein. Arch Biochem Biophys 318:314–321
162. Hlavica P (1984) On the function of cytochrome b_5 in the cytochrome P-450-dependent oxygenase system. Arch Biochem Biophys 228:600–608
163. Tamburini PP, Gibson GG (1983) Thermodynamic studies on the protein-protein interactions between cytochrome P-450 and cytochrome b_5. Evidence for a central role of the cytochrome P-450 spin state in the coupling of substrate and cytochrome b_5 binding to the terminal hemoprotein. J Biol Chem 258:13444–13452
164. Guengerich FP (1983) Oxidation-reduction properties of rat liver cytochromes P-450 and NADPH-cytochrome P-450 reductase related to catalysis in reconstituted systems. Biochemistry 22:2811–2820
165. Noshiro M, Ullrich V, Omura T (1981) Cytochrome b_5 as electron donor for oxycytochrome P-450. Eur J Biochem 116:521–526
166. Ingelman-Sundberg M, Johansson I (1980) Cytochrome b_5 as electron donor to rabbit liver cytochrome P-450LM2 in reconstituted phospholipid vesicles. Biochem Biophys Res Commun 97:582–589
167. Gorsky LD, Coon MJ (1986) Effects of conditions for reconstitution with cytochrome b_5 on the formation of products in cytochrome P-450-catalyzed reactions. Drug Metab Dispos 14:89–96
168. Golly I, Hlavica P (1987) Regulative mechanisms in NADH- and NADPH-supported N-oxidation of 4-chloroaniline catalyzed by cytochrome b_5-enriched rabbit liver microsomal fractions. Biochim Biophys Acta 913:219–227
169. Guengerich FP, Ballou DP, Coon MJ (1976) Spectral intermediates in the reaction of oxygen with purified liver microsomal cytochrome P-450. Biochem Biophys Res Commun 70:951–956
170. Lipscomb JD, Sligar SG, Namtvedt MJ, Gunsalus IC (1976) Autoxidation and hydroxylation reactions of oxygenated cytochrome P-450_{cam}. J Biol Chem 251:1116–1124
171. Lee-Robichaud P, Akhtar ME, Akhtar M (1998) Control of androgen biosynthesis in the human through the interaction of Arg347 and Arg358 of CYP17 with cytochrome b_5. Biochem J 332:293–296
172. Auchus RJ, Lee TC, Miller WL (1998) Cytochrome b_5 augments the 17,20-lyase activity of human P450c17 without direct electron transfer. J Biol Chem 273:3158–3165
173. Storbeck KH, Swart AC, Goosen P, Swart P (2013) Cytochrome b_5: novel roles in steroidogenesis. Mol Cell Endocrinol 371:87–99
174. Yamazaki H, Shimada T, Martin MV, Guengerich FP (2001) Stimulation of cytochrome P450 reactions by apo-cytochrome b_5. Evidence against transfer of heme from cytochrome P450 3A4 to apo-cytochrome b_5 or heme oxygenase. J Biol Chem 276:30885–30891
175. Yamazaki H, Nakamura M, Komatsu T, Ohyama K, Hatanaka N, Asahi S, Shimada N, Guengerich FP, Shimada T, Nakajima M, Yokoi T (2002) Roles of NADPH-cytochrome P450 reductase and apo- and holo-cytochrome b_5 on xenobiotic oxidations catalyzed by 12 recombinant human cytochrome P450s expressed in membranes of *Escherichia coli*. Protein Expr Purif 24:329–337
176. Loughran PA, Roman LJ, Miller T, Masters BSS (2001) The kinetic and spectral characterization of the *E. coli*-expressed mammalian CYP4A7: cytochrome b_5 effects vary with substrate. Arch Biochem Biophys 385:311–321
177. Hlavica P, Lewis DFV (2001) Allosteric phenomena in cytochrome P450-catalyzed monooxygenations. Eur J Biochem 268:4817–4832
178. Porter TD (2002) The roles of cytochrome b_5 in cytochrome P450 reactions. J Biochem Mol Toxicol 16:311–316
179. Ivanov YD, Kanaeva IP, Kuznetsov VY, Lehnerer M, Schulze J, Hlavica P, Archakov AI (1999) The optical biosensor studies on the role of hydrophobic tails of NADPH-cytochrome P450 reductase and cytochromes P450 2B4 and b_5 upon productive complex formation with a monomeric reconstituted system. Arch Biochem Biophys 362:87–93
180. Mulrooney SB, Meinhardt DR, Waskell L (2004) The α-helical membrane spanning domain of cytochrome b_5 interacts with cytochrome P450 via nonspecific interactions. Biochim Biophys Acta 1674:319–326
181. Clarke TA, Im SC, Bidwai A, Waskell L (2004) The role of the length and sequence of the linker domain of cytochrome b_5 in stimulating cytochrome P450 2B4 catalysis. J Biol Chem 279:36809–36818
182. Hlavica P, Kellermann J, Golly I, Lehnerer M (1994) Chemical modification of Tyr34 and Tyr129 in rabbit liver microsomal cytochrome b_5 affects interaction with cytochrome P-450 2B4. Eur J Biochem 224:1039–1046
183. Stayton PS, Poulos TL, Sligar SG (1989) Putidaredoxin competitively inhibits cytochrome b_5-cytochrome $P450_{cam}$ association: a proposed molecular model for a cytochrome $P450_{cam}$ electron-transfer complex. Biochemistry 28:8201–8205
184. Stayton PS, Fisher MT, Sligar SG (1988) Determination of cytochrome b_5 association reactions. Characterization of metmyoglobin and cytochrome P-450_{cam} binding to genetically engineered cytochrome b_5. J Biol Chem 263:13544–13548
185. Degtyarenko KN, Kulikova TA (2001) Evolution of bioinorganic motifs in P450- containing systems. Biochem Soc Trans 29:139–147
186. Grindberg AV, Hannemann F, Schiffler B, Müller J, Heinemann U, Bernhardt R (2000) Adrenodoxin: structure, stability, and electron transfer properties. Proteins 40:590–612
187. Kostic M, Pochapsky SS, Obenauer J, Mo H, Pagani GM, Pejchal R, Pochapsky TC (2002) Comparison

of functional domains in vertebrate-type ferredoxins. Biochemistry 41:5978–5989
188. Ewen KM, Kleser M, Bernhardt R (2011) Adrenodoxin: the archetype of vertebrate-type [2Fe-2S] cluster ferredoxins. Biochim Biophys Acta 1814:111–125
189. Sevrioukova IF, Poulos TL (2011) Structural biology of redox partner interactions in $P450_{cam}$ monooxygenase: a fresh look at an old system. Arch Biochem Biophys 507:66–74
190. Ziegler GA, Vonrhein C, Hanukoglu I, Schulz GE (1999) The structure of adrenodoxin reductase of mitochondrial P450 systems: electron transfer for steroid biosynthesis. J Mol Biol 289:981–990
191. Sevrioukova IF, Poulos TL (2002) Putidaredoxin reductase, a new function for an old protein. J Biol Chem 277:25831–25839
192. Heinz A, Hannemann F, Müller JJ, Heinemann U, Bernhardt R (2005) The interaction domain of the redox protein adrenodoxin is mandatory for binding of the electron acceptor CYP11A1, but is not required for binding of the electron donor adrenodoxin reductase. Biochem Biophys Res Commun 228:491–498
193. Müller JJ, Lapko A, Bourenkov G, Ruckpaul K, Heinemann U (2001) Adrenodoxin reductase-adrenodoxin complex structure suggests electron transfer path in steroid biosynthesis. J Biol Chem 276:2786–2789
194. Coghlan VM, Vickery LE (1991) Site-specific mutations in human ferredoxin that affect binding to ferredoxin reductase and cytochrome $P450_{scc}$. J Biol Chem 266:18606–18612
195. Grindberg AV, Bernhardt R (2001) Contribution of a salt bridge to the thermostability of adrenodoxin determined by site-directed mutagenesis. Arch Biochem Biophys 396:25–34
196. Beckert V, Dettmer R, Bernhardt R (1994) Mutations of tyrosine 82 in bovine adrenodoxin that affect binding to cytochromes P450 11A1 and P450 11B1 but not electron transfer. J Biol Chem 269:2568–2573
197. Beckert V, Schrauber H, Bernhardt R, van Dijk AA, Kakoschke C, Wray V (1995) Mutational effects on the spectroscopic properties and biological activities of oxidized bovine adrenodoxin, and their structural implications. Eur J Biochem 231:226–235
198. Beckert V, Bernhardt R (1997) Specific aspects of electron transfer from adrenodoxin to cytochromes $P450_{scc}$ and $P450_{11\beta}$. J Biol Chem 272:4883–4888
199. Uhlmann H, Bernhardt R (1995) The role of threonine 54 in adrenodoxin for the properties of its iron-sulfur cluster and its electron function. J Biol Chem 270:29959–29966
200. Schiffler B, Kiefer M, Wilken A, Hannemann F, Adolph HW, Bernhardt R (2001) The interaction of bovine adrenodoxin with CYP11A1 (cytochrome $P450_{scc}$) and CYP11B1 (cytochrome $P450_{11\beta}$). Acceleration of reduction and substrate conversion by site-directed mutagenesis of adrenodoxin. J Biol Chem 276:36225–36232
201. Hannemann F, Rottmann M, Schiffler B, Zapp J, Bernhardt R (2001) The loop region covering the iron-sulfur cluster in bovine adrenodoxin comprises a new interaction site for redox partners. J Biol Chem 276:1369–1375
202. Zöllner A, Hannemann F, Lisurek M, Bernhardt R (2002) Deletions in the loop surrounding the iron-sulfur cluster of adrenodoxin severely affect the interactions with its native redox partners adrenodoxin reductase and cytochrome $P450_{scc}$ (CYP11A1). J Inorg Biochem 91:644–654
203. Vickery LE (1997) Molecular recognition and electron transfer in mitochondrial steroid hydroxylase systems. Steroids 62:124–127
204. Lambeth JD, Geren LM, Millett F (1984) Adrenodoxin interaction with adrenodoxin reductase and cytochrome $P\text{-}450_{scc}$. Cross-linking of protein complexes and effects of adrenodoxin modification by 1-ethyl-3-(3-dimethylaminopropyl)carbodiimide. J Biol Chem 259:10025–10029
205. Müller EC, Lapko A, Otto A, Müller JJ, Ruckpaul K, Heinemann U (2001) Covalently crosslinked complexes of bovine adrenodoxin with adrenodoxin reductase and cytochrome $P450_{scc}$. Mass spectrometry and Edman degradation of complexes of the steroidogenic hydroxylase system. Eur J Biochem 268:1837–1843
206. Beilke D, Weiss R, Löhr F, Pristovsek P, Hannemann F, Bernhardt R, Rüterjans H (2002) A new electron transport mechanism in mitochondrial steroid hydroxylase systems based on structural changes upon the reduction of adrenodoxin. Biochemistry 41:7969–7978
207. Müller A, Müller JJ, Muller YA, Uhlmann H, Bernhardt R, Heinemann U (1998) New aspects of electron transfer revealed by the crystal structure of a truncated bovine adrenodoxin, Adx(4-108). Structure 6:269–280
208. Kuznetsov VY, Blair E, Farmer PJ, Poulos TL, Pifferitti A, Sevrioukova IF (2005) The putidaredoxin reductase-putidaredoxin electron transfer complex. Theoretical and experimental studies. J Biol Chem 280:16135–16142
209. Sevrioukova IF, Poulos TL, Churbanova IY (2010) Crystal structure of the putidaredoxin reductase-putidaredoxin electron transfer complex. J Biol Chem 285:13616–13620
210. Aoki M, Ishimori K, Morishima I (1998) Roles of negatively charged surface residues of putidaredoxin in interactions with redox partners in $P450_{cam}$ monooxygenase system. Biochim Biophys Acta 1386:157–167
211. Holden M, Mayhew M, Bunk D, Roitberg A, Vilker V (1997) Probing the interactions of putidaredoxin

with redox partners in camphor P450 5-monooxygenase by mutagenesis of surface residues. J Biol Chem 272:21720–21725
212. Sevrioukova IF, Garcia C, Li H, Bhaskar B, Poulos TL (2003) Crystal structure of putidaredoxin, the [2Fe-2S] component of the $P450_{cam}$ monooxygenase system from *Pseudomonas putida*. J Mol Biol 333:377–392
213. Zhang W, Pochapsky SS, Pochapsky TC, Jain NU (2008) Solution NMR structure of putidaredoxin-cytochrome $P450_{cam}$ complex via a combined residual dipolar coupling-spin labeling approach suggests a role for Trp106 of putidaredoxin in complex formation. J Mol Biol 384:349–363
214. Kuznetsov VY, Poulos TL, Sevrioukova IF (2006) Putidaredoxin-to-cytochrome $P450_{cam}$ electron transfer: differences between the two reductive steps required for catalysis. Biochemistry 45:11934–11944
215. Stayton PS, Sligar SG (1991) Structural microheterogeneity of a tryptophan residue required for efficient biological electron transfer between putidaredoxin and cytochrome $P\text{-}450_{cam}$. Biochemistry 30:1845–1851
216. Pochapsky TC, Lyons TA, Kazanis S, Arakaki T, Ratnaswamy G (1996) A structure-based model for cytochrome $P450_{cam}$-putidaredoxin interactions. Biochimie 78:723–733
217. Roitberg AE, Holden MJ, Mayhew MP, Kurnikov IV, Beratan DN, Vilker VL (1998) Binding and electron transfer between putidaredoxin and cytochrome $P450_{cam}$. Theory and experiments. J Am Chem Soc 120:8927–8932
218. Sibbesen O, De Voss JJ, Ortiz de Montellano PR (1996) Putidaredoxin reductase-putidaredoxin-cytochrome $P450_{cam}$ triple fusion protein. Construction of a self-sufficient *Escherichia coli* catalytic system. J Biol Chem 271:22462–22469
219. Ivanov YD, Kanaeva IP, Karuzina II, Archakov AI, Hui Bon Hoa G, Sligar SG (2001) Molecular recognition in the $P450_{cam}$ monooxygenase system: direct monitoring of protein-protein interactions by using optical biosensor. Arch Biochem Biophys 391:255–264
220. Purdy MM, Koo LS, Ortiz de Montellano PR, Klinman JP (2004) Steady-state kinetic investigation of cytochrome $P450_{cam}$: interaction with redox partners and reaction with molecular oxygen. Biochemistry 43:271–281
221. Schiffler B, Bernhardt R (2003) Bacterial (CYP101) and mitochondrial P450 systems – how comparable are they? Biochem Biophys Res Commun 312:223–228
222. Anandatheerthavarada HK, Addya S, Mullick J, Avadhani NG (1998) Interaction of adrenodoxin with P450 1A1 and its truncated form P450MT2 through different domains: differential modulation of enzyme activities. Biochemistry 37:1150–1160
223. Lehnerer M, Schulze J, Bernhardt R, Hlavica P (1999) Some properties of mitochondrial adrenodoxin associated with its nonconventional electron donor function toward rabbit liver microsomal P450 2B4. Biochem Biophys Res Commun 254:83–87
224. Lehnerer M, Schulze J, Petzold A, Bernhardt R, Hlavica P (1995) Rabbit liver cytochrome P-450 2B5: high-level expression of the full-length protein in *Escherichia coli*, purification, and catalytic activity. Biochim Biophys Acta 1245:107–115
225. Pechurskaya T, Harnastai IN, Grabovec IP, Gilep AA, Usanov SA (2007) Adrenodoxin supports reactions catalyzed by microsomal steroidogenic cytochrome P450s. Biochem Biophys Res Commun 353:598–604
226. Liao WL, Dodder NG, Mast N, Pikuleva IA, Turko IV (2009) Steroid and protein ligand binding to cytochrome P450 46A1 as assessed by hydrogen-deuterium exchange and mass spectrometry. Biochemistry 48:4150–4158
227. Hannemann F, Virus C, Bernhardt R (2006) Design of an *Escherichia coli* system for whole cell mediated steroid synthesis and molecular evolution of steroid hydroxylases. J Biotechnol 124:172–181
228. Ringle M, Khatri Y, Zapp J, Hannemann F, Bernhardt R (2012) Application of a new versatile electron transfer system for cytochrome P450-based *Escherichia coli* whole-cell bioconversions. Appl Microbiol Biotechnol 97:7741–7754
229. Ewen KM, Schiffler B, Uhlmann-Schiffler H, Bernhardt R, Hannemann F (2008) The endogenous adrenodoxin reductase-like flavoprotein arh1 supports heterologous cytochrome P450-dependent substrate conversions in *Schizosaccharomyces pombe*. FEMS Yeast Res 8:432–441
230. Dong MS, Yamazaki H, Guo Z, Guengerich FP (1996) Recombinant human cytochrome P450 1A2 and an N-terminal-truncated form: construction, purification, aggregation properties, and interactions with flavodoxin, ferredoxin, and NADPH-cytochrome P450 reductase. Arch Biochem Biophys 327:11–19
231. Yamazaki H, Ueng YF, Shimada T, Guengerich FP (1995) Roles of divalent metal ions in oxidations catalyzed by recombinant cytochrome P450 3A4 and replacement of NADPH-cytochrome P450 reductase with other flavoproteins, ferredoxin, and oxygen surrogates. Biochemistry 34:8380–8389
232. Sawada N, Sakaki T, Yoneda S, Kusudo T, Shinkyo R, Ohta M, Inouye K (2004) Conversion of vitamin D_3 to 1α,25-dihydroxyvitamin D_3 by *Streptomyces griseolus* cytochrome P450SU-1. Biochem Biophys Res Commun 320:156–164
233. Momoi K, Hofmann U, Schmid RD, Urlacher VB (2006) Reconstitution of β-carotene hydroxylase activity of thermostable CYP175A1 monooxygenase. Biochem Biophys Res Commun 339:331–336
234. Bell SG, Hoskins N, Xu F, Caprotti D, Rao Z, Wong LL (2006) Cytochrome P450 enzymes from the

metabolically diverse bacterium *Rhodopseudomonas palustris*. Biochem Biophys Res Commun 342:191–196
235. Bernhardt R, Gunsalus IC (1992) Reconstitution of cytochrome P450 2B4 (LM2) activity with camphor and linalool monooxygenase electron donors. Biochem Biophys Res Commun 187:310–317
236. Jenkins CM, Waterman MR (1994) Flavodoxin and NADPH-flavodoxin reductase from *Escherichia coli* support bovine cytochrome P450c17 hydroxylase activities. J Biol Chem 269:27401–27408
237. McIver L, Leadbeater C, Campopiano DJ, Baxter RL, Daff SN, Chapman SK, Munro AW (1998) Characterization of flavodoxin $NADP^+$ oxidoreductase and flavodoxin; key components of electron transfer in *Escherichia coli*. Eur J Biochem 257:577–585
238. Girhard M, Schuster S, Dietrich M, Dürre P, Urlacher VB (2007) Cytochrome P450 monooxygenase from *Clostridium acetobutylicum*: a new α-fatty acid hydroxylase. Biochem Biophys Res Commun 362:114–119
239. Girhard M, Tieves F, Weber E, Smit MS, Urlacher VB (2013) Cytochrome P450 reductase from *Candida apicola*: versatile redox partner of bacterial P450s. Appl Microbiol Biotechnol 97:1625–1635
240. Lehnerer M, Schulze J, Pernecky SJ, Lewis DFV, Eulitz M, Hlavica P (1998) Influence of mutation of the amino-terminal signal anchor sequence of cytochrome P450 2B4 on the enzyme structure and electron transfer processes. J Biochem 124:396–403
241. Hlavica P, Golly I, Wolf J (1987) Influence of N, N-dimethylaniline on the association of phenobarbital-induced cytochrome P450 and NADPH-cytochrome c (P450) reductase in a reconstituted rabbit liver microsomal enzyme system. Biochim Biophys Acta 915:28–36
242. Hanna IH, Kim MS, Guengerich FP (2001) Heterologous expression of cytochrome P450 2D6 mutants, electron transfer, and catalysis of bufuralol hydroxylation: the role of aspartate 30 in structural integrity. Arch Biochem Biophys 393:255–261
243. von Wachenfeldt C, Richardson TH, Cosme J, Johnson EF (1997) Microsomal P450 2C3 is expressed as a soluble dimer in *Escherichia coli* following modifications of its N-terminus. Arch Biochem Biophys 339:107–114
244. Voznesensky AI, Schenkman JB, Pernecky SJ, Coon MJ (1994) The NH_2-terminal region of rabbit CYP2E1 is not essential for interaction with NADPH-cytochrome P450 reductase. Biochem Biophys Res Commun 203:156–161
245. Mao W, Berenbaum MR, Schuler MA (2008) Modifications in the N-terminus of an insect cytochrome P450 enhance production of catalytically active protein in baculovirus-Sf9 cell expression systems. Insect Biochem Mol Biol 38:66–75
246. Scheller U, Kraft R, Schröder KL, Schunck WH (1994) Generation of a soluble and functional cytosolic domain of microsomal cytochrome P450 52A3. J Biol Chem 269:12779–12783
247. Shen S, Strobel HW (1992) The role of cytochrome P450 lysine residues in the interaction between cytochrome P450 1A1 and NADPH-cytochrome P450 reductase. Arch Biochem Biophys 294:83–90
248. Cvrk T, Hodek P, Strobel HW (1996) Identification and characterization of cytochrome P450 1A1 amino acid residues interacting with a radiolabeled photoaffinity diazido-benzphetamine analogue. Arch Biochem Biophys 330:142–152
249. Shen S, Strobel HW (1995) Functional assessment of specific amino acid residues of cytochrome P450 1A1 using anti-peptide antibodies. Arch Biochem Biophys 320:162–169
250. Parkinson A, Ryan DE, Thomas PE, Jerina DM, Sayer JM, van Bladeren PJ, Haniu M, Shively JE, Levin W (1986) Chemical modification and inactivation of rabbit liver microsomal cytochrome P-450c by 2-bromo-4′-nitroacetophenone. J Biol Chem 261:11478–11486
251. Edwards RJ, Singleton AM, Murray BP, Sesardic D, Rich KJ, Davies DS, Boobis AR (1990) An anti-peptide antibody targeted to a specific region of rat cytochrome P-450 1A2 inhibits enzyme activity. Biochem J 266:497–504
252. Jänig GR, Kraft R, Blanck J, Ristau O, Rabe H, Ruckpaul K (1987) Chemical modification of cytochrome P-450LM4. Identification of functionally linked tyrosine residues. Biochim Biophys Acta 916:512–523
253. Cvrk T, Strobel HW (2001) Role of Lys271 and Lys279 residues in the interaction of cytochrome P450 1A1 with NADPH-cytochrome P450 reductase. Arch Biochem Biophys 385:290–300
254. Shimizu T, Tateishi T, Hatano M, Fuji-Kuriyama Y (1991) Probing the role of lysines and arginines in the catalytic function of cytochrome P450d by site-directed mutagenesis. J Biol Chem 266:3372–3375
255. Mayuzumi H, Sambongi C, Hiroya K, Shimizu T, Tateishi T, Hatano M (1993) Effect of mutations of ionic amino acids of cytochrome P450 1A2 on catalytic activities toward 7-ethoxycoumarin and methanol. Biochemistry 32:5622–5628
256. Frey AB, Waxman DJ, Kreibich G (1985) The structure of phenobarbital-inducible rat liver cytochrome P-450 isozyme PB-4. J Biol Chem 260:15253–15265
257. Omata Y, Dai R, Smith SV, Robinson RC, Friedman FK (2000) Synthetic peptide mimics of a predicted topographical interaction surface: the cytochrome P450 2B1 recognition domain for NADPH-cytochrome P450 reductase. J Protein Chem 19:23–32
258. Shen S, Strobel HW (1993) Role of lysine and arginine residues of cytochrome P450 in the interaction between cytochrome P450 2B1 and NADPH-

cytochrome P450 reductase. Arch Biochem Biophys 304:257–265

259. Bernhardt R, Kraft R, Otto A, Ruckpaul K (1988) Electrostatic interaction between cytochrome P-450LM2 and NADPH-cytochrome P-450 reductase. Biomed Biochim Acta 47:581–592
260. Bridges A, Gruenke L, Chang YT, Vakser IA, Loew G, Waskell L (1998) Identification of the binding site on cytochrome P450 2B4 for cytochrome b_5 and cytochrome P450 reductase. J Biol Chem 273:17036–17049
261. Lehnerer M, Schulze J, Achterhold K, Lewis DFV, Hlavica P (2000) Identification of key residues in rabbit liver microsomal cytochrome P450 2B4: importance in interactions with NADPH-cytochrome P450 reductase. J Biochem 127:163–169
262. Schulze J, Tschöp K, Lehnerer M, Hlavica P (2000) Residue 285 in cytochrome P450 2B4 lacking the NH_2-terminal hydrophobic sequence has a role in the functional association of NADPH-cytochrome P450 reductase. Biochem Biophys Res Commun 270:777–781
263. Kenaan C, Zhang H, Shea EV, Hollenberg PF (2011) Uncovering the role of hydrophobic residues in cytochrome P450-cytochrome P450 reductase interactions. Biochemistry 50:3957–3967
264. Zhang H, Sridar C, Kenaan C, Amunugama H, Ballou DP, Hollenberg PF (2011) Polymorphic variants of cytochrome P450 2B6 (CYP2B6.4-CYP2B6.9) exhibit altered rates of metabolism for bupropion and efavirenz: a charge-reversal mutation in the K139E variant (CYP2B6.8) impairs formation of a functional cytochrome P450-reductase complex. J Pharmacol Exp Ther 338:803–809
265. Bumpus NN, Hollenberg PF (2010) Cross-linking of human cytochrome P450 2B6 to NADPH-cytochrome P450 reductase: identification of a potential site of interaction. J Inorg Biochem 104:485–488
266. Kaspera R, Naraharisetti SB, Evangelista EA, Marciante KD, Psaty BM, Totah RA (2011) Drug metabolism by CYP2C8.3 is determined by substrate dependent interactions with cytochrome P450 reductase and cytochrome b_5. Biochem Pharmacol 82:681–691
267. Crespi CI, Miller VP (1997) The R144C change in the CYP2C9.2 allele alters interaction of the cytochrome P450 with NADPH:cytochrome P450 oxidoreductase. Pharmacogenetics 7:203–210
268. Wada Y, Mitsuda M, Ishihara Y, Watanabe M, Iwasaki M, Asahi S (2008) Important amino acid residues that confer CYP2C19 selective activity to CYP2C9. J Biochem 144:323–333
269. Lin H, Myshkin E, Waskell L, Hollenberg PF (2007) Peroxynitrite inactivation of human cytochrome P450s 2B6 and 2E1: heme modification and site-specific nitrotyrosine formation. Chem Res Toxicol 20:1612–1622
270. Wen B, Lampe JN, Roberts AG, Atkins WM, Rodrigues AD, Nelson SD (2006) Cysteine 98 in CYP3A4 contributes to conformational integrity required for P450 interaction with CYP reductase. Arch Biochem Biophys 454:42–54
271. Lin H, Kenaan C, Zhang H, Hollenberg PF (2012) Reaction of human cytochrome P450 3A4 with peroxinitrite: nitrotyrosine formation on the proximal side impairs its interaction with NADPH-cytochrome P450 reductase. Chem Res Toxicol 25:2642–2653
272. Nikfarjam L, Izumi S, Yamazaki T, Kominami S (2006) The interaction of cytochrome P450 17α with NADPH-cytochrome P450 reductase, investigated using chemical modification and MALDI-TOF mass spectrometry. Biochim Biophys Acta 1764:1126–1131
273. Geller DH, Auchus RJ, Miller WL (1999) P450c17 mutations R347H and R358Q selectively disrupt 17,20-lyase ctivity by disrupting interactions with P450 oxidoreductase and cytochrome b_5. Mol Endocrinol 13:167–175
274. Riepe FG, Hiort O, Grötzinger J, Sippell WG, Krone N, Holterhus PM (2008) Functional and structural consequences of a novel point mutation in the *CYP21A2* gene causing congenital adrenal hyperplasia: potential relevance of helix C for P450 oxidoreductase-21-hydroxylase interaction. J Clin Endocrinol Metab 93:2891–2895
275. Lajic S, Levo A, Nikoshkov A, Lundberg Y, Partanen J, Wedell A (1997) A cluster of missense mutations at Arg356 of human steroid 21-hydroxylase may impair redox partner interaction. Hum Genet 99:704–709
276. Robins T, Carlsson J, Sunnerhagen M, Wedell A, Person B (2006) Molecular model of human CYP21 based on mammalian CYP2C5: structural features correlate with clinical severity of mutations causing congenital adrenal hyperplasia. Mol Endocrinol 20:2946–2964
277. Ji H, Zhang W, Zhou Y, Zhang M, Zhu J, Song Y, Lü J, Zhu J (2000) A three-dimesional model of lanosterol 14α-demethylase of *Candida albicans* and its interaction with azole antifungals. J Med Chem 43:2493–2505
278. Juvonen RO, Iwasaki M, Negishi M (1992) Roles of residues 129 and 209 in the alteration by cytochrome b_5 of hydroxylase activities in mouse 2A P450S. Biochemistry 31:11519–11523
279. Honkakoski P, Linnala-Kankkunen A, Usanov SA, Lang MA (1992) Highly homologous cytochromes P-450 and b_5: a model to study protein-protein interactions in a reconstituted monooxygenase system. Biochim Biophys Acta 1122:6–14
280. Omata Y, Sakamoto H, Robinson RC, Pincus MR, Friedman FK (1994) Interaction between cytochrome P450 2B1 and cytochrome b_5: inhibition by synthetic peptides indicates a role for P450 residues

Lys-122 and Arg-125. Biochem Biophys Res Commun 201:1090–1095
281. Epstein PM, Curti M, Jansson I, Huang CK, Schenkman JB (1989) Phosphorylation of cytochrome P450: regulation by cytochrome b_5. Arch Biochem Biophys 271:424–432
282. Gao Q, Doneanu CE, Shaffer SA, Adman ET, Goodlett DR, Nelson SD (2006) Identification of the interactions between cytochrome P450 2E1 and cytochrome b_5 by mass spectrometry and site-directed mutagenesis. J Biol Chem 281:20404–20417
283. Zhao C, Gao Q, Roberts AG, Shaffer SA, Doneanu CE, Xue S, Goodlett DR, Nelson SD, Atkins WM (2012) Cross-linking mass spectrometry and mutagenesis confirm the functional importance of surface interactions between CYP3A4 and holo/apo cytochrome b_5. Biochemistry 51:9488–9500
284. Lee-Robichaud P, Akhtar ME, Akhtar M (1999) Lysine mutagenesis identifies cationic charges of human CYP17 that interact with cytochrome b_5 to promote male sex-hormone biosynthesis. Biochem J 342:309–312
285. Tuls J, Geren L, Lambeth JD, Millett F (1987) The use of a specific fluorescence probe to study the interaction of adrenodoxin with adrenodoxin reductase and cytochrome P-450$_{scc}$. J Biol Chem 262:10020–10025
286. Adamovich TB, Pikuleva IA, Chashchin VL, Usanov SA (1989) Selective chemical modification of cytochrome P-450$_{scc}$ lysine residues. Identification of lysines involved in the interaction with adrenodoxin. Biochim Biophys Acta 996:247–253
287. Tuls J, Geren L, Millett F (1989) Fluorescein isothiocyanate specifically modifies lysine 338 of cytochrome P-450$_{scc}$ and inhibits adrenodoxin binding. J Biol Chem 264:16421–16425
288. Wada A, Waterman MR (1992) Identification by site-directed mutagenesis of two lysine residues in cholesterol side chain cleavage cytochrome P450 that are essential for adrenodoxin binding. J Biol Chem 267:22877–22882
289. Tsubaki M, Iwamoto Y, Hiwatashi A, Ichikawa Y (1989) Inhibition of electron transfer from adrenodoxin to cytochrome P-450$_{scc}$ by chemical modification with pyridoxal 5′-phosphate: identification of adrenodoxin-binding site of cytochrome P-450$_{scc}$. Biochemistry 28:6899–6907
290. Parajes S, Loidi L, Reisch N, Dhir V, Rose IT, Hampel R, Quinkler M, Conway GS, Castro-Feijoo-L, Araujo-Vilar D, Pornbo M, Dominguez F, Williams EL, Cole TR, Kirk JM, Kaminsky E, Rumsby G, Arlt W, Krone N (2010) Functional consequences of seven novel mutations in the *CYP11B1* gene: four mutations associated with nonclassic and three mutations causing classic 11β-hydroxylase deficiency. J Clin Endocrinol Metab 95:779–788
291. Chernogolov A, Usanov S, Kraft R, Schwarz D (1994) Selective chemical modification of Cys 264 with diiodofluorescein iodacetamide as a tool to study the membrane topology of cytochrome P450$_{scc}$ (CYP11A1). FEBS Lett 340:83–88
292. Usanov SA, Graham SE, Lepesheva GI, Azeva TN, Strushkevich NV, Gilep AA, Estabrook RW, Peterson JA (2002) Probing the interaction of bovine cytochrome P450$_{scc}$ (CYP11A1) with adrenodoxin: evaluating site-directed mutations by molecular modeling. Biochemistry 41:8310–8320
293. Strushkevich NV, Harnastai IN, Usanov SA (2010) Mechanism of steroidogenic electron transport: role of conserved Glu 429 in destabilization of CYP11A1-adrenodoxin complex. Biochemistry (Mosc) 75:570–578
294. Annalora AJ, Goodin DB, Hong WX, Zhang Q, Johnson EF, Stout CD (2010) Crystal structure of CYP24A1, a mitochondrial cytochrome P450 involved in vitamin D metabolism. J Mol Biol 396:441–451
295. Urushino N, Yamamoto K, Kagawa N, Ikushiro S, Kamakura M, Yamada S, Kato S, Inouye K, Sakaki T (2006) Interaction between mitochondrial CYP27B1 and adrenodoxin: role of arginine 458 of mouse CYP27B1. Biochemistry 45:4405–4412
296. Pikuleva IA, Cao C, Waterman MR (1999) An additional electrostatic interaction between adrenodoxin and P450c27 (CYP27A1) results in tighter binding than between adrenodoxin and P450$_{scc}$ (CYP11A1). J Biol Chem 274:2045–2052
297. Kolesanova EF, Kozin SA, Rumyantsev AB, Jung C, Hui Bon Hoa G, Archakov AI (1997) Epitope mapping of cytochrome P450$_{cam}$ (CYP101). Arch Biochem Biophys 341:229–237
298. Stayton PS, Sligar SG (1990) The cytochrome P-450$_{cam}$ binding surface as defined by site-directed mutagenesis and electrostatic modeling. Biochemistry 29:7381–7386
299. Shimada H, Nagano S, Hori H, Ishimura Y (2001) Putidaredoxin-cytochrome P450$_{cam}$ interaction. J Inorg Biochem 83:255–260
300. Unno M, Shimada H, Toba Y, Makino R, Ishimura Y (1996) Role of Arg 112 of cytochrome P450$_{cam}$ in the electron transfer from reduced putidaredoxin. J Biol Chem 271:17869–17874
301. Nagano S, Shimada H, Tarumi A, Hishiki T, Kimata-Ariga Y, Egawa T, Suematsu M, Park SY, Adachi S, Shiro Y, Ishimura Y (2003) Infrared spectroscopic and mutational studies on putidaredoxin-induced conformational changes in ferrous CO-P450$_{cam}$. Biochemistry 42:14507–14514
302. Bernhardt R, Kraft R, Alterman M, Otto A, Schrauber H, Gunsalus IC, Ruckpaul K (1992) Common mechanism of interaction between cytochrome P-450 and electron donors in different monooxygenase systems. In: Archakov AI, Bachmanova GI (eds) Cytochrome P-450:

biochemistry and biophysics. INCO-TNC, Moscow, pp 204–209
303. Koo LS, Immoos CE, Cohen MS, Farmer PJ, Ortiz de Montellano PR (2002) Enhanced electron transfer and lauric acid hydroxylation by site-directed mutagenesis of CYP119. J Am Chem Soc 124:5684–5691
304. Yang W, Bell SG, Wang H, Zhou W, Hoskins N, Dale A, Bartlam M, Wong LL, Rao Z (2010) Molecular characterization of a class I P450 electron transfer system from *Novosphingobium aromaticivorans* DSM 12444. J Biol Chem 285:27372–27384
305. Bell SG, Xu F, Forward I, Bartlam M, Rao Z, Wong LL (2008) Crystal structure of CYP199A2, a *para*-substituted benzoic acid oxidizing cytochrome P450 from *Rhodopseudomonas palustris*. J Mol Biol 383:561–574
306. Bell SG, Xu F, Johnson EOD, Forward IM, Bartlam M, Rao Z, Wong LL (2010) Protein recognition in ferredoxin-P450 electron transfer in the class I CYP199A2 system from *Rhodopseudomonas palustris*. J Biol Inorg Chem 15:315–328
307. Williams PA, Cosme J, Sridhar V, Johnson EF, McRee DE (2000) Microsomal cytochrome P450 2C5: comparison to microbial P450s and unique features. J Inorg Biochem 81:183–190
308. Golly I, Hlavica P, Schartau W (1988) The functional role of cytochrome b_5 reincorporated into hepatic microsomal fractions. Arch Biochem Biophys 260:232–240
309. Fisher GJ, Gaylor JL (1982) Kinetic investigation of rat liver microsomal electron transport from NADH to cytochrome P450. J Biol Chem 257:7449–7455
310. Golly I, Hlavica P (1993) Inactivation of phenobarbital-inducible rabbit-liver microsomal cytochrome P-450 by allylisopropylacetamide: impact on electron transfer. Biochim Biophys Acta 1142:74–82
311. Kostanjevecki V, Leys D, Van Driessche G, Meyer TE, Cusanovich MA, Fischer U, Guisez Y, Van Beeumen J (1999) Structure and characterization of *Ectothiorhodospira vacuolata* cytochrome b_{558}, a prokaryotic homologue of cytochrome b_5. J Biol Chem 274:35614–35620
312. Williams PA, Cosme J, Sridhar V, Johnson EF, McRee DE (2000) Mammalian microsomal cytochrome P450 monooxygenase: structural adaptations for membrane binding and functional diversity. Mol Cell 5:121–131
313. Geller DH, Auchus RJ, Mendonca BB, Miller WL (1997) The genetic and functional basis of isolated 17,20-lyase deficiency. Nat Genet 17:201–205
314. Hasemann CA, Kurumbail RG, Boddupalli SS, Peterson JA, Deisenhofer J (1995) Structure and function of cytochromes P450: a comparative analysis of three crystal structures. Structure 2:41–62
315. Ingelman-Sundberg M (1986) Cytochrome P450 organization and membrane interactions. In: Ortiz de Montellano PR (ed) Cytochrome P450: structure, mechanism, and biochemistry. Plenum Press, New York, pp 119–160
316. Müller-Enoch D, Churchill P, Fleischer S, Guengerich FP (1984) Interaction of liver microsomal cytochrome P-450 and NADPH-cytochrome P-450 reductase in the presence and absence of lipid. J Biol Chem 259:8174–8182
317. Balvers WG, Boersma MG, Veeger C, Rietjens MCM (1993) Kinetics of cytochromes P-450 IA1 and IIB1 in reconstituted systems with dilauroyl- and distearoyl-glycerophosphocholine. Eur J Biochem 215:373–381
318. Blanck J, Jänig GR, Schwarz D, Ruckpaul K (1989) Role of lipid in the electron transfer between NADPH-cytochrome P-450 reductase and cytochrome P-450 from mammalian liver cells. Xenobiotica 19:1231–1246
319. Balvers WG, Boersma MG, Vervoort J, Ouwehand A, Rietjens MCM (1993) A specific interaction between NADPH-cytochrome reductase and phosphatidyl serine and phosphatidyl inositol. Eur J Biochem 218:1021–1029
320. Ingelman-Sundberg M, Blanck J, Smettan G, Ruckpaul K (1983) Reduction of cytochrome P-450LM$_2$ by NADPH in reconstituted phospholipid vesicles is dependent on membrane charge. Eur J Biochem 134:157–162
321. Imaoka S, Imai Y, Shimada T, Funae Y (1992) Role of phospholipids in reconstituted cytochrome P450 3A form and mechanism of their activation of catalytic activity. Biochemistry 31:6063–6069
322. Causey KM, Eyer CS, Backes WL (1990) Dual role of phospholipid in the reconstitution of cytochrome P-450LM$_2$-dependent activities. Mol Pharmacol 38:134–142
323. Davydov DR, Sineva EV, Sistla S, Davydova NY, Frank DJ, Sligar SG, Halpert JR (2010) Electron transfer in the complex of membrane-bound human cytochrome P450 3A4 with the flavin domain of P450BM-3: the effect of oligomerization of the heme protein and intermittent modulation of the spin equilibrium. Biochim Biophys Acta 1797:378–390
324. Ingelman-Sundberg M (1977) Phospholipids and detergents as effectors in the liver microsomal hydroxylase system. Biochim Biophys Acta 488:225–234
325. Yun CH, Ahn T, Guengerich FP (1998) Conformational change and activation of cytochrome P450 2B1 induced by salt and phospholipid. Arch Biochem Biophys 356:229–238
326. Bendzko P, Usanov SA, Pfeil W, Ruckpaul K (1982) Role of the hydrophobic tail of cytochrome b_5 in the interaction with cytochrome P450LM$_2$. Acta Biol Med Ger 41:K1–K8
327. Golly I, Hlavica P (1987) Influence of cytochrome b_5 on electron flow from NADPH-cytochrome *c* (P-450) reductase to cytochrome P450. In: Benford DJ, Bridges JW, Gibson GG (eds) Drug metabolism

– from molecules to man. Taylor & Francis, London, pp 468–472
328. Pember SO, Powell GL, Lambeth JD (1983) Cytochrome P-450$_{scc}$-phospholipid interactions. Evidence for a cardiolipin binding site and thermodynamics of enzyme interactions with cardiolipin, cholesterol, and adrenodoxin. J Biol Chem 258:3198–3206
329. Reed JR, Backes WL (2012) Formation of P450-P450 complexes and their effect on P450 function. Pharmacol Ther 133:299–310
330. Reed JR, Eyer M, Backes WL (2010) Functional interactions between cytochromes P450 1A2 and 2B4 require both enzymes to reside in the same phospholipid vesicle. Evidence for physical complex formation. J Biol Chem 285:8942–8952
331. Kenaan C, Shea EV, Lin H, Zhang H, Pratt-Hyatt MJ, Hollenberg PF (2013) Interactions between CYP2E1 and CYP2B4: effects on affinity for NADPH-cytochrome P450 reductase and substrate metabolism. Drug Metab Dispos 41:101–110
332. Tan Y, Patten CJ, Smith T, Yang CS (1997) Competitive interactions between cytochromes P450 2A6 and 2E1 for NADPH-cytochrome P450 oxidoreductase in the microsomal membranes produced by a baculovirus expression system. Arch Biochem Biophys 342:82–91
333. Hazai E, Kupfer D (2005) Interactions between CYP2C9 and CYP2C19 in reconstituted binary systems influence their catalytic activity: possible rationale for the inability of CYP2C19 to catalyze methoxychlor demethylation in human liver microsomes. Drug Metab Dispos 33:157–164
334. Li DN, Pritchard MP, Hanlon SP, Burchell B, Wolf CR, Friedberg T (1999) Competition between cytochrome P-450 isozymes for NADPH-cytochrome P-450 oxidoreductase affects drug metabolism. J Pharmacol Exp Ther 289:661–667
335. Ikushiro S, Kominami S, Tekemori S (1992) Adrenal P-450$_{scc}$ modulates activity of P-450$_{11\beta}$ in liposomal and mitochondrial membranes. J Biol Chem 267:1464–1469
336. Hazai E, Bikadi Z, Simonyi M, Kupfer D (2005) Association of cytochrome P450 enzymes is a determining factor in their catalytic activity. J Comput Aided Mol Des 19:271–285
337. Kawato S, Gut J, Cherry RJ, Winterhalter KH, Richter C (1982) Rotation of cytochrome P-450. I. Investigations of protein-protein interactions of cytochrome P-450 in phospholipid vesicles and liver microsomes. J Biol Chem 257:7023–7029
338. Gut J, Richter C, Cherry RJ, Winterhalter KH, Kawato S (1982) Rotation of cytochrome P-450. II. Specific interactions of cytochrome P-450 with NADPH-cytochrome P-450 reductase in phospholipid vesicles. J Biol Chem 257:7030–7036
339. Gut J, Richter C, Cherry RJ, Winterhalter KH, Kawato S (1983) Rotation of cytochrome P-450. Complex formation of cytochrome P-450 with NADPH-cytochrome P-450 reductase in liposomes demonstrated by combining protein rotation with antibody-induced cross-linking. J Biol Chem 258:8588–8594
340. Yamada M, Ohta Y, Bachmanova GI, Nishimoto Y, Archakov AI, Kawato S (1995) Dynamic interactions of rabbit liver cytochromes P450 IA2 and P450 IIB4 with cytochrome b_5 and NADPH-cytochrome P450 reductase in proteoliposomes. Biochemistry 34:10113–10119
341. Whitehouse CJC, Bell SG, Wong LL (2012) P450BM3 (CYP102A1): connecting the dots. Chem Soc Rev 41:1218–1260
342. Gustafsson MCU, Roitel O, Marshall KR, Noble MA, Chapman SK, Pessegueiro A, Fulco AJ, Cheesman MR, von Wachenfeldt C, Munro AW (2004) Expression, purification, and characterization of *Bacillus subtilis* cytochromes P450 CYP102A2 and CYP102A3: flavocytochrome homologues of P450BM3 from *Bacillus megaterium*. Biochemistry 43:5474–5487
343. Chowdhary PK, Alemseghed M, Haines DC (2007) Cloning, expression and characterization of a fast self-sufficient P450: CYP102A5 from *Bacillus cereus*. Arch Biochem Biophys 468:32–43
344. Dietrich M, Eiben S, Asta C, Do TA, Pleiss J, Urlacher VB (2008) Cloning, expression and characterization of CYP102A7, a self-sufficient P450 monooxygenase from *Bacillus licheniformis*. Appl Microbiol Biotechnol 79:931–940
345. Narhi LO, Fulco AJ (1987) Identification and characterization of two functional domains in cytochrome P-450BM-3, a catalytically self-sufficient monooxygenase induced by barbiturates in *Bacillus megaterium*. J Biol Chem 262:6683–6690
346. Joyce MG, Ekanem IS, Roitel O, Dunford AJ, Neeli R, Girvan HM, Bakler GJ, Curtis RA, Munro AW, Leys D (2012) The crystal structure of the FAD/NADPH-binding domain of flavocytochrome P450BM3. FEBS J 279:1694–1706
347. Neeli R, Roitel O, Scrutton NS, Munro AW (2005) Switching pyridine nucleotide specificity in P450BM3. Mechanistic analysis of the W1046H and W1046A enzymes. J Biol Chem 280:17634–17644
348. Kitazume T, Haines DC, Estabrook RW, Chen B, Peterson JA (2007) Obligatory intermolecular electron-transfer from FAD to FMN in dimeric P450BM-3. Biochemistry 46:11892–11901
349. Klein ML, Fulco AJ (1993) Critical residues involved in FMN binding and catalytic activity in cytochrome P450BM-3. J Biol Chem 268:7553–7561
350. Sevrioukova IF, Li H, Zhang H, Peterson JA, Poulos TL (1999) Structure of a cytochrome P450-redox partner electron-transfer complex. Proc Natl Acad Sci U S A 96:1863–1868
351. Hanley SC, Ost TWB, Daff S (2004) The unusual redox properties of flavocytochrome P450BM3

flavodoxin domain. Biochem Biophys Res Commun 325:1418–1423
352. Chen HC, Swenson RP (2008) Effect of the insertion of a glycine residue into the loop spanning residues 536–541 on the semiquinone state and redox properties of the flavin- mononucleotide-binding domain of flavocytochrome P450BM-3 from *Bacillus megaterium*. Biochemistry 47:13788–13799
353. Munro AW, Malarkey K, McKnight J, Thomson AJ, Kelly SM, Prince NC, Lindsay JG, Coggins JR, Miles JS (1994) The role of tryptophan 97 of cytochrome P450BM3 from *Bacillus megaterium* in catalytic function. Biochem J 303:423–428
354. Sevrioukova IF, Hazzard JT, Tollin G, Poulos TL (1999) The FMN to heme electron transfer in cytochrome P450BM-3. Effect of chemical modification of cysteines engineered at the FMN-heme domain interaction site. J Biol Chem 274:36097–36106
355. Noble MA, Girvan HM, Smith SJ, Smith WE, Murataliev M, Guzov VM, Feyereisen R, Munro AW (2007) Analysis of the interactions of cytochrome b_5 with flavocytochrome P450BM3 and its domains. Drug Metab Rev 39:599–617
356. Kitazume T, Takaya N, Nakayama N, Shoun H (2000) *Fusarium oxysporum* fatty-acid subterminal hydroxylase (CYP505) is a membrane-bound eukaryotic counterpart of *Bacillus megaterium* cytochrome P450BM3. J Biol Chem 275:39734–39740
357. Kitazume T, Tanaka A, Takaya N, Nakamura A, Matsuyama S, Suzuki T, Shoun H (2002) Kinetic analysis of hydroxylation of saturated fatty acids by recombinant $P450_{foxy}$ produced by an *Escherichia coli* expression system. Eur J Biochem 269:2075–2082
358. Nakayama N, Takemae A, Shoun H (1996) Cytochrome $P450_{foxy}$, a catalytically self-sufficient fatty acid hydroxylase of the fungus *Fusarium oxysporum*. J Biochem 119:435–440
359. Seo JA, Proctor RH, Plattner RD (2001) Characterization of four clustered and coregulated genes associated with fumonisin biosynthesis in *Fusarium verticillioides*. Fungal Genet Biol 34:155–165
360. Seth-Smith HMB, Rosser SJ, Basran A, Travis ER, Dabbs ER, Nicklin S, Bruce NC (2002) Cloning, sequencing, and characterization of the hexahydro-1,3,5-trinitro-1,3,5-triazine degradation gene cluster from *Rhodococcus rhodochrous*. Appl Environ Microbiol 68:4764–4771
361. Sabbadin F, Jackson R, Haider K, Tampi G, Turkenburg JP, Hart S, Bruce NC, Grogan G (2009) The 1.5-Å structure of XplA-heme, an unusual cytochrome P450 heme domain that catalyzes reductive biotransformation of royal demolition explosive. J Biol Chem 284:28467–28475
362. Rylott EL, Jackson RG, Sabbadin F, Seth-Smith HMB, Edwards J, Chong CS, Strand SE, Grogan G, Bruce NC (2011) The explosive-degrading cytochrome P450 XplA: biochemistry, structural features and prospects for bioremediation. Biochim Biophys Acta 1814:230–236
363. Bui SH, McLean KJ, Cheesman MR, Bradley JM, Rigby SEJ, Levy CW, Leys D, Munro AW (2012) Unusual spectroscopic and ligand binding properties of the cytochrome P450-flavodoxin fusion enzyme XplA. J Biol Chem 287:19699–19714
364. Rylott EL, Jackson RG, Edwards J, Womack GL, Seth-Smith HMB, Rathbone DA, Strand SE, Bruce NC (2006) An explosive-degrading cytochrome P450 activity and its targeted application for the phytoremediation of RDX. Nat Biotechnol 24:216–219
365. Gassner GT, Ludwig ML, Gatti DL, Correll CC, Ballou DP (1995) Structure and mechanism of the iron-sulfur flavoprotein phthalate dioxygenase reductase. FASEB J 9:1411–1418
366. Warman AJ, Robinson JW, Luciakova D, Lawrence AD, Marshall KR, Warren MJ, Cheesman MR, Rigby EJ, Munro AW, McLean KJ (2012) Characterization of *Cupriavidus metallidurans* CYP116B1 – a thiocarbamate herbicide oxygenating P450-phthalate dioxygenase reductase fusion protein. FEBS J 279:1675–1693
367. Roberts GA, Grogan G, Greter A, Flitsch SL, Turner NJ (2002) Identification of a new class of cytochrome P450 from a *Rhodococcus* sp. J Bacteriol 184:3898–3908
368. Roberts GA, Celik A, Hunter DJB, Ost TWB, White JH, Chapman SK, Turner NJ, Flitsch SL (2003) A self-sufficient cytochrome P450 with a primary structural organization that includes a flavin domain and a [2Fe-2S] redox center. J Biol Chem 278:48914–48920
369. Hunter DJB, Roberts GA, Ost TWB, White JH, Müller S, Turner NJ, Flitsch SL, Chapman SK (2005) Analysis of the domain properties of the novel cytochrome P450 RhF. FEBS Lett 579:2215–2220
370. Celik A, Roberts GA, White JH, Chapman SK, Turner NJ, Flitsch SL (2006) Probing the substrate specificity of the catalytically self-sufficient cytochrome P450 RhF from a *Rhodococcus* sp. Chem Commun 43:4492–4494
371. Liu L, Schmid RD, Urlacher VB (2006) Cloning, expression, and characterization of a self-sufficient cytochrome P450 monooxygenase from *Rhodococcus ruber* DSM 44319. Appl Microbiol Biotechnol 72:876–882
372. McLean KJ, Girvan HM, Munro AW (2007) Cytochrome P450/redox partner fusion enzymes: biotechnological and toxicological prospects. Expert Opin Drug Metab Toxicol 3:847–863
373. Gilardi G, Meharenna YT, Tsotsou GE, Sadeghi SJ, Fairhead M, Giannini S (2002) Molecular Lego: design of molecular assemblies of P450 enzymes for nanobiotechnology. Biosens Bioelectron 17:133–145
374. Sabbadin F, Hyde R, Robin A, Hilgarth EM, Delenne M, Flitsch S, Turner N, Grogan G, Bruce NC (2010) LICRED: a versatile drop-in vector for rapid generation of redox-self-sufficient cytochrome P450s. Chembiochem 11:987–994

375. Yabusaki Y (1995) Artificial P450/reductase fusion enzymes: what can we learn from their structures? Biochimie 77:594–603
376. Yamada M, Ohta Y, Sakaki T, Yabusaki Y, Ohkawa H, Kawato S (1999) Dynamic mobility of genetically expressed fusion protein between cytochrome P450 1A1 and NADPH- cytochrome P450 reductase in yeast microsomes. Biochemistry 38:9465–9470
377. Sakaki T, Kominami S, Takemori S, Ohkawa H, Akiyoshi-Shibata M, Yabusaki Y (1994) Kinetic studies on a genetically engineered fusion enzyme between rat cytochrome P450 1A1 and yeast NADPH-P450 reductase. Biochemistry 33:4933–4939
378. Chun YJ, Jeong TC, Roh JK, Guengerich FP (1997) Characterization of a fusion protein between human cytochrome P450 1A1 and rat NADPH-P450 oxidoreductase in *Escherichia coli*. Biochem Biophys Res Commun 230:211–214
379. Shiota N, Kodama S, Inui H, Ohkawa H (2000) Expression of human cytochromes P450 1A1 and P450 1A2 as fused enzymes with yeast NADPH-cytochrome P450 oxidoreductase in transgenic tobacco plants. Biosci Biotechnol Biochem 64:2025–2033
380. Harlow GR, Halpert JR (1996) Mutagenesis study of Asp-290 in cytochrome P450 2B11 using a fusion protein with rat NADPH-cytochrome P450 reductase. Arch Biochem Biophys 326:85–92
381. Helvig C, Capdevila JH (2000) Biochemical characterization of rat P450 2C11 fused to rat or bacterial NADPH-P450 reductase domains. Biochemistry 39:5196–5205
382. Deeni YY, Paine MJ, Ayrton AD, Clarke SE, Chenery R, Wolf CR (2001) Expression, purification, and biochemical characterization of a human cytochrome P450 CYP2D6-NADPH cytochrome P450 reductase fusion protein. Arch Biochem Biophys 396:16–24
383. Inui H, Maeda A, Ohkawa H (2007) Molecular characterization of specifically active recombinant fused enzymes consisting of CYP3A4, NADPH-cytochrome P450 oxidoreductase, and cytochrome b_5. Biochemistry 46:10213–10221
384. Chaurasia CS, Alterman MA, Lu P, Hanzlik RP (1995) Biochemical characterization of lauric acid ω-hydroxylation by a CYP4A1/NADPH-cytochrome P450 reductase fusion protein. Arch Biochem Biophys 317:161–169
385. Shet MS, Fisher CW, Holmans PL, Estabrook RW (1996) The ω-hydroxylation of lauric acid: oxidation of 12-hydroxylauric acid to dodecanedioic acid by a purified recombinant fusion protein containing P450 4A1 and NADPH-P450 reductase. Arch Biochem Biophys 330:199–208
386. Shet MS, Fisher CW, Arlotto MP, Shackleton CHL, Holmans PL, Martin-Wixtrom CA, Salki Y, Estabrook RW (1994) Purification and enzymatic properties of a recombinant fusion protein expressed in *Escherichia coli* containing the domains of bovine P450 17A and rat NADPH-P450 reductase. Arch Biochem Biophys 311:402–417
387. Shet MS, Fisher CW, Estabrook RW (1997) The function of recombinant cytochrome P450s in intact *Escherichia coli* cells: the 17α-hydroxylation of progesterone and pregnenolone by P450c17. Arch Biochem Biophys 339:218–225
388. Shibata M, Sakaki T, Yabusaki Y, Murakami H, Ohkawa H (1990) Genetically engineered P450 monooxygenases: construction of bovine P450c17/yeast reductase fused enzymes. DNA Cell Biol 9:27–36
389. Shet MS, Fisher CW, Tremblay Y, Belanger A, Conley AJ, Mason JI, Estabrook RW (2007) Comparison of the 17α-hydroxylase/C17,20-lyase activities of porcine, guinea pig and bovine P450c17 using purified recombinant fusion proteins containing P450c17 linked to NADPH-P450 reductase. Drug Metab Rev 39:289–307
390. Sakaki T, Shibata M, Yabusaki Y, Murakami H, Ohkawa H (1990) Expression of bovine cytochrome P450c21 and its fused enzymes with yeast NADPH-cytochrome P450 reductase in *Saccharomyces cerevisiae*. DNA Cell Biol 9:603–614
391. Sakaki T, Kominami S, Hayashi K, Akiyoshi-Shibata M, Yabusaki Y (1996) Molecular engineering study on electron transfer from NADPH-P450 reductase to rat mitochondrial P450c27 in yeast microsomes. J Biol Chem 271:26209–26213
392. Venkateswarlu K, Kelly DE, Kelly SL (1997) Characterization of *Saccharomyces cerevisiae* CYP51 and a CYP51 fusion protein with NADPH-cytochrome P-450 oxidoreductase expressed in *Escherichia coli*. Antimicrob Agents Chemother 41:776–780
393. Lamb SB, Lamb DC, Kelly SL, Stuckey DC (1998) Cytochrome P450 immobilization as a route to bioremediation/biocatalysis. FEBS Lett 431:343–346
394. Hotze M, Schröder G, Schröder J (1995) Cinnamate 4-hydroxylase from *Catharanthus roseus*, and a strategy for the functional expression of plant cytochrome P450 proteins as translational fusions with P450 reductase in *Escherichia coli*. FEBS Lett 374:345–350
395. Schückel J, Rylott EL, Grogan G, Bruce NC (2012) A gene-fusion approach to enabling plant cytochromes P450 for biocatalysis. Chembiochem 13:2758–2763
396. Didierjean L, Gondet L, Perkins R, Lau SMC, Schaller H, O'Keefe DP, Werck-Reichhart D (2002) Engineering herbicide metabolism in tobacco and Arabidopsis with CYP76B1, a cytochrome P450 enzyme from Jerusalem artichoke. Plant Physiol 130:179–189
397. Porter TD (1991) An unusual yet strongly conserved flavoprotein reductase in bacteria and mammals. Trends Biochem Sci 16:154–158
398. Dodhia VR, Fantuzzi A, Gilardi G (2006) Engineering human cytochrome P450 enzymes into

catalytically self-sufficient chimeras using molecular Lego. J Biol Inorg Chem 11:903–916
399. Rua F, Sadeghi SJ, Castrignano S, Di Nardo G, Gilardi G (2012) Engineering *Macaca fascicularis* cytochrome P450 2C20 to reduce animal testing for new drugs. J Inorg Biochem 117:277–284
400. Fairhead M, Giannini S, Gillam EMJ, Gilardi G (2005) Functional characterization of an engineered multidomain human P450 2E1 by molecular Lego. J Biol Inorg Chem 10:842–853
401. Degregorio D, Sadeghi SJ, Di Nardo G, Gilardi G, Solinas SP (2011) Understanding uncoupling in the multiredox centre P450 3A4-BMR model system. J Biol Inorg Chem 16:109–116
402. Fuziwara S, Sagami I, Rozhkova E, Craig D, Noble MA, Munro AW, Chapman SK, Shimizu T (2002) Catalytically functional flavocytochrome chimeras of P450BM3 and nitric oxide synthase. J Inorg Biochem 91:515–526
403. Choi KY, Jung E, Jung DH, An BR, Pandey BP, Yun H, Sung C, Park HY, Kim BG (2012) Engineering of daidzein 3′-hydroxylase P450 enzyme into catalytically self-sufficient cytochrome P450. Microb Cell Factories 11:81–90
404. Lacour T, Ohkawa H (1999) Engineering and biochemical characterization of the rat micro- somal cytochrome P450 1A1 fused to ferredoxin and ferredoxin-NADP^{+} reductase from plant chloroplasts. Biochim Biophys Acta 1433:87–102
405. Harikrishna JA, Black SM, Szklarz GD, Miller WL (1993) Construction and function of fusion enzymes of the human cytochrome P450$_{scc}$ system. DNA Cell Biol 12:371–379
406. Cao P, Bülow H, Dumas B, Bernhardt R (2000) Construction and characterization of a catalytic fusion protein system: P-450$_{11\beta}$-adrenodoxin reductase-adrenodoxin. Biochim Biophys Acta 1476:253–264
407. Dilworth FJ, Black SM, Guo YD, Miller WL, Jones G (1996) Construction of a P450c27 fusion enzyme: a useful tool for analysis of vitamin D$_3$ 25-hydroxylase activity. Biochem J 320:267–271
408. Hirakawa H, Kamiya N, Tanaka T, Nagamune T (2007) Intramolecular electron transfer in a cytochrome P450$_{cam}$ system with a site-specific branched structure. Protein Eng Des Sel 20:453–459
409. Hirakawa H, Nagamune T (2010) Molecular assembly of P450 with ferredoxin and ferredoxin reductase by fusion to PCNA. Chembiochem 11:1517–1520
410. Mandai T, Fujiwara S, Imaoka S (2009) Construction and engineering of a thermostable self-sufficient cytochrome P450. Biochem Biophys Res Commun 384:61–65
411. Kubota M, Nodate M, Yasumoto-Hirose M, Uchiyama T, Kagami O, Shizuri Y, Misawa N (2005) Isolation and functional analysis of cytochrome P450 CYP153A genes from various environments. Biosci Biotechnol Biochem 69:2421–2430
412. Bordeaux M, Galarneau A, Fajula F, Drone J (2011) A regioselective biocatalyst for alkane activation under mild conditions. Angew Chem Int Ed 50:2075–2079
413. Nodate M, Kubota M, Misawa N (2006) Functional expression system for cytochrome P450 genes using the reductase domain of self-sufficient P450RhF from *Rhodococcus* sp. NCIMB 9784. Appl Microbiol Biotechnol 71:455–462
414. Li S, Podust LM, Sherman DH (2007) Engineering and analysis of a self-sufficient biosynthetic cytochrome P450 PikC fused to the RhFRED reductase domain. J Am Chem Soc 129:12940–12941
415. Li S, Chaulagain M, Knauff AR, Podust LM, Montgomery J, Sherman DH (2009) Selective oxidation of carbolide C-H bonds by an engineered macrolide P450 mono-oxygenase. Proc Natl Acad Sci U S A 106:18463–18468
416. Choi KY, Jung EO, Jung DH, Pandey BP, Lee N, Yun H, Park H, Kim BG (2012) Novel iron-sulfur containing NADPH-reductase from *Nocardia farcinica* IFM 10152 and fusion construction with CYP51 lanosterol demethylase. Biotechnol Bioeng 109:630–636
417. Ortiz de Montellano PR (1995) Oxygen activation and reactivity. In: Ortiz de Montellano PR (ed) Cytochrome P450: structure, mechanism, and biochemistry, 2nd edn. Plenum Press, New York, pp 245–303
418. Fujishiro T, Shoji O, Nagano S, Sugimoto H, Shiro Y, Watanabe Y (2011) Crystal structure of H_2O_2-dependent cytochrome P450$_{SP\alpha}$ with its bound fatty acid substrate. Insight into the regioselective hydroxylation of fatty acids at the α-position. J Biol Chem 286:29941–29950
419. Matsunaga I, Yamada A, Lee DS, Obayashi E, Fujiwara N, Kobayashi K, Ogura H, Shiro Y (2002) Enzymatic reaction of hydrogen peroxide-dependent peroxygenase cytochrome P450s: kinetic deuterium isotope effects and analyses by resonance Raman spectroscopy. Biochemistry 41:1886–1892
420. Rude MA, Baron TS, Brubaker S, Alibhai M, Del Cardayre SB, Schirmer A (2011) Terminal olefin (1-alkene) biosynthesis by a novel P450 fatty acid decarboxylase from *Jeotgalicoccus* species. Appl Environ Microbiol 77:1718–1727
421. Niraula NP, Kanth BK, Sohng JK, Oh TJ (2011) Hydrogen peroxide-mediated dealkylation of 7-ethoxycoumarin by cytochrome P450 (CYP107AJ1) from *Streptomyces peucetius* ATCC 27952. Enzyme Microb Technol 48:181–186
422. Thompson CM, Capdevila JH, Strobel HW (2000) Recombinant cytochrome P450 2D18 metabolism of dopamine and arachidonic acid. J Pharmacol Exp Ther 294:1120–1130
423. Hlavica P, Golly I, Mietaschk J (1983) Comparative studies on the cumene hydroperoxide- and NADPH-supported N-oxidation of 4-chloroaniline. Biochem J 212:539–547

424. Golly I, Hlavica P, Wolf J (1984) The role of lipid peroxidation in the N-oxidation of 4-chloroaniline. Biochem J 224:415–421
425. Yoshigae Y, Kent UM, Hollenberg PF (2013) Role of the highly conserved threonine in cytochrome P450 2E1: prevention of H_2O_2-induced inactivation during electron transfer. Biochemistry 52:4636–4647
426. Chefson A, Zhao J, Auclair K (2006) Replacement of natural cofactors by selected hydrogen peroxide donors or organic peroxides results in improved activity for CYP3A4 and CYP2D6. Chembiochem 7:916–919
427. Kumar S, Liu H, Halpert JR (2006) Engineering of cytochrome P450 3A4 for enhanced peroxide-mediated substrate oxidation using directed evolution and site-directed mutagenesis. Drug Metab Dispos 34:1958–1965
428. Joo H, Lin Z, Arnold FH (1999) Laboratory evolution of peroxide-mediated cytochrome P450 hydroxylation. Nature 399:670–673
429. Cirino PC, Arnold FH (2002) Regioselectivity and activity of cytochrome P450BM-3 and mutant F87A in reactions driven by hydrogen peroxide. Adv Synth Catal 344:932–937
430. Cirino PC, Arnold FH (2003) A self-sufficient peroxide-driven hydroxylation biocatalyst. Angew Chem Int Ed 42:3299–3301
431. Vidal-Limon A, Aguila S, Ayala M, Batista CV, Vasquez-Duhalt R (2013) Peroxidase activity stabilization of cytochrome P450BM3 by rational analysis of intramolecular electron transfer. J Inorg Biochem 122:18–26
432. Salazar O, Cirino PC, Arnold FH (2003) Thermostabilization of a cytochrome P450 peroxygenase. Chembiochem 4:891–893
433. Holtmann D, Schrader J (2007) Approaches to recycling and substituting NAD(P)H as a CYP cofactor. In: Schmid RD, Urlacher VB (eds) Modern biooxidation: enzymes, reactions and applications. Wiley-VCH Verlag, Weinheim, pp 265–290
434. Li S, Wackett LP (1993) Reductive dehalogenation by cytochrome $P450_{cam}$: substrate binding and catalysis. Biochemistry 32:9355–9361
435. Tran NH, Huynh N, Chavez G, Nguyen A, Dwaraknath S, Nguyen TA, Nguyen M, Cheruzel L (2012) A series of hybrid P450BM3 enzymes with different catalytic activity in the light-initiated hydroxylation of lauric acid. J Inorg Biochem 115:50–56
436. Ipe BI, Lehnig M, Niemeyer CM (2005) On the generation of free radical species from quantum dots. Small 7:706–709
437. Ipe BI, Shukla A, Lu H, Zou B, Rehage H, Niemeyer CM (2006) Dynamic light-scattering analysis of the electrostatic interaction of hexahistidine-tagged cytochrome P450 enzyme with semiconductor quantum dots. Chemphyschem 7:1112–1118
438. Fruk L, Rajendran V, Spengler M, Niemeyer CM (2007) Light-induced triggering of peroxidase activity using quantum dots. Chembiochem 8:2195–2198
439. Gandubert VJ, Torres E, Niemeyer CM (2008) Investigation of cytochrome P450- modified cadmium sulfide quantum dots as photocatalysts. J Mater Chem 18:3824–3830
440. Sadeghi SJ, Fantuzzi A, Gilardi G (2011) Breakthrough in P450 bioelectrochemistry and future perspectives. Biochim Biophys Acta 1814:237–248
441. Shumyantseva VV, Bulko TV, Usanov SS, Schmid RD, Nicolini C, Archakov AI (2001) Construction and characterization of bioelectrocatalytic sensors based on cytochrome P450. J Inorg Biochem 87:185–190
442. Rudakov YO, Shumyantseva VV, Bulko TV, Suprun EV, Kuznetsova GP, Samenkova NF, Archakov AI (2008) Stoichiometry of electrocatalytic cycle of cytochrome P450 2B4. J Inorg Biochem 102:2020–2025
443. Shumyantseva VV, Ivanov YD, Bistolas N, Scheller FW, Archakov AI, Wollenberger U (2004) Direct electron transfer of cytochrome P450 2B4 at electrodes modified with nonionic detergent and colloidal clay nanoparticles. Anal Chem 76:6046–6052
444. Liu S, Peng L, Yang X, Wu Y, He L (2008) Electrochemistry of cytochrome P450 enzyme on nanoparticle-containing membrane-coated electrode and its applications for drug sensing. Anal Biochem 375:209–216
445. Zu X, Lu Z, Zhang Z, Schenkman JB, Rusling JF (1999) Electroenzyme-catalyzed oxidation of styrene and *cis-β*-methylstyrene using thin films of cytochrome $P450_{cam}$ and myoglobin. Langmuir 15:7372–7377
446. Estavillo C, Lu Z, Jansson I, Schenkman JB, Rusling JF (2003) Epoxidation of styrene by human cytochrome P450 1A2 by thin film electrolysis and peroxide activation compared to solution reactions. Biophys Chem 104:291–296
447. Fantuzzi A, Fairhead M, Gilardi G (2004) Direct electrochemistry of immobilized human cytochrome P450 2E1. J Am Chem Soc 126:5040–5041
448. Joseph S, Rusling JF, Lvov YM, Friedberg T, Fuhr U (2003) An amperometric biosensor with human CYP3A4 as a novel drug screening tool. Biochem Pharmacol 65:1817–1826
449. Yang M, Kabulski JL, Wollenberg L, Chen X, Subramanian M, Tracy TS, Lederman D, Gannett PM, Wu N (2009) Electrocatalytic drug metabolism by CYP2C9 bonded to a self-assembled monolayer-modified electrode. Drug Metab Dispos 37:892–899
450. Urlacher VB, Girhard M (2012) Cytochrome P450 monooxygenases: an update on perspectives for synthetic application. Trends Biotechnol 30:26–36
451. Julsing MK, Cornelissen S, Bühler B, Schmid A (2008) Heme-iron oxygenases: powerful industrial biocatalysts? Curr Opin Chem Biol 12:177–186
452. Bureik M, Bernhardt R (2007) Steroid hydroxylation: microbial steroid biotransformations using cytochrome P450 enzymes. In: Schmid RD,

Urlacher VB (eds) Modern biooxidation: enzymes, reactions and applications. Wiley-VCH Verlag, Weinheim, pp 155–176
453. Park JW, Lee JK, Kwon TJ, Yi DH, Kim YJ, Moon SH, Suh HH, Kang SM, Park YI (2003) Bioconversion of compactin into pravastatin by *Streptomyces* sp. Biotechnol Lett 25:1827–1831
454. Ro DK, Paradise EM, Ouellet M, Fisher KJ, Newman KL, Ndungu JM, Ho KA, Eachus RA, Ham TS, Kirby J, Chang CY, Withers ST, Shiba Y, Sarpong R, Keasling JD (2006) Production of the antimalarial drug precursor artemisinic acid in engineered yeast. Nature 440:940–943
455. Liu S, Li C, Fang X, Cao Z (2004) Optimal pH control strategy for high-level production of long-chain α, ω-dicarboxylic acid by Candida tropicalis. Enzym Microb Technol 34:73–77
456. Roy P, Waxman DJ (2006) Activation of oxazaphosphorines by cytochrome P450: application to gene-directed enzyme prodrug therapy of cancer. Toxicol In Vitro 20:176–186
457. Kumar S, Chen CS, Waxman DJ, Halpert JR (2005) Directed evolution of mammalian cytochrome P450 2B1. Mutations outside of the active site enhance the metabolism of several substrates, including the anti-cancer prodrugs cyclophosphamide and ifosfamide. J Biol Chem 280:19569–19575
458. Sun L, Chen CS, Waxman DJ, Liu H, Halpert JR, Kumar S (2007) Re-engineering cytochrome P450 2B11dH for enhanced metabolism of several substrates including anti-cancer prodrugs cyclophosphamide and ifosfamide. Arch Biochem Biophys 458:167–174
459. Nguyen TA, Tychpoulos M, Bichat F, Zimmermann C, Flinois JP, Diry M, Ahlberg E, Delaforge M, Corcos L, Beaune P, Dansette P, Andre F, de Waziers I (2008) Improvement of cyclophosphamide activation by CYP2B6 mutants: from in silico to ex vivo. Mol Pharmacol 73:1122–1133
460. Tychopoulos M, Corcos L, Genne P, Beaune P, de Waziers I (2005) A virus-directed enzyme prodrug therapy (VDEPT) strategy for lung cancer using a CYP2B6/NADPH-cytochrome P450 reductase fusion protein. Cancer Gene Ther 12:497–508
461. Kawahigashi H, Hirose S, Ohkawa H, Ohkawa Y (2006) Phytoremediation of the herbicides atrazine and metolachlor by transgenic rice plants expressing human CYP1A1, CYP2B6, and CYP2C19. J Agric Food Chem 54:2985–2991
462. Jackson RG, Rylott EL, Fournier D, Hawari J, Bruce NC (2007) Exploring the biochemical properties and remediation applications of the unusual explosive-degrading P450 system XplA/B. Proc Natl Acad Sci U S A 104:16822–16827
463. Chun YJ, Shimada T, Guengerich FP (1996) Construction of a human cytochrome P450 1A1–rat NADPH-cytochrome P450 reductase fusion protein cDNA and expression in *Escherichia coli*, purification, and catalytic properties of the enzyme in bacterial cells and after purification. Arch Biochem Biophys 330:48–58
464. Carmichael AB, Wong LL (2001) Protein engineering of *Bacillus megaterium* CYP102. The oxidation of polycyclic aromatic hydrocarbons. Eur J Biochem 268:3117–3125
465. Syed K, Porollo A, Miller D, Yadav JS (2013) Rational engineering of the fungal P450 monooxygenase CYP5136A3 to improve its oxidizing activity toward polycyclic aromatic hydrocarbons. Protein Eng Des Sel 26:553–557
466. Kanaly RA, Harayama S (2000) Biodegradation of high-molecular-weight polycyclic aromatic hydrocarbons by bacteria. J Bacteriol 182:2059–2067
467. Bistolas N, Wollenberger U, Jung C, Scheller FW (2005) Cytochrome P450 biosensors – a review. Biosens Bioelectron 20:2408–2423
468. Wang Y, Xu H, Zhang J, Li G (2008) Electrochemical sensors for clinic analysis. Sensors 8:2043–2081
469. Ansede JH, Thakker DR (2004) High-throughput screening for stability and inhibitory activity of compounds toward cytochrome P450-mediated metabolism. J Pharm Sci 93:239–255

Biological Diversity of Cytochrome P450 Redox Partner Systems

11

Kirsty J. McLean, Dominika Luciakova, James Belcher, Kang Lan Tee, and Andrew W. Munro

Abstract

Cytochrome P450 enzymes (P450s or CYPs) catalyze an enormous variety of oxidative reactions in organisms from all major domains of life. Their monooxygenase activity relies on the reductive scission of molecular oxygen (O_2) bound to P450 heme iron, and thus on the delivery of two electrons to the heme iron at discrete points in the catalytic cycle. Early studies suggested that P450 redox partner machinery fell into only two major classes: either the eukaryotic diflavin enzyme NADPH-cytochrome P450 oxidoreductase, or bacterial/mitochondrial NAD(P)H-ferredoxin reductase and ferredoxin partners. However, more recent studies, aided by genome sequence data, reveal a much more complex scenario. Several new types of P450 redox partner systems have now been characterized, including P450s naturally linked to their redox partners, or to a component protein of their P450 electron delivery system. Other P450s have evolved to bypass requirements for redox partners, and instead react directly with hydrogen peroxide or NAD(P)H to facilitate oxidative or reductive catalysis. Further P450s are fused to non-redox partner enzymes and can catalyse consecutive reactions in a common pathway. This chapter describes the biochemistry and the enormous natural diversity of P450 redox systems, including descriptions of novel P450s fused to non-redox partner proteins.

Keywords

Cytochrome P450 • Compound I • P450 redox systems • P450 oxidoreductase • P450 BM3 • P450-flavodoxin fusion protein • P450-ferredoxin fusion protein

K.J. McLean • D. Luciakova • J. Belcher
K.L. Tee • A.W. Munro (✉)
Manchester Institute of Biotechnology, The University of Manchester, 131 Princess Street, Manchester M1 7DN, UK
e-mail: andrew.munro@manchester.ac.uk

E.G. Hrycay and S.M. Bandiera (eds.), *Monooxygenase, Peroxidase and Peroxygenase Properties and Mechanisms of Cytochrome P450*, Advances in Experimental Medicine and Biology 851,
DOI 10.1007/978-3-319-16009-2_11,

11.1 Introduction

Cytochrome P450 enzymes (P450s or CYPs) are prodigious catalysts of oxidation reactions in Nature. P450s are heme *b*-containing enzymes that bind dioxygen (O_2) to a ferrous heme iron, and then further reduce and protonate the iron-oxo species to form reactive intermediates capable of oxygen addition chemistry on a substrate bound close to the P450 heme iron [1]. The key species responsible for substrate oxidation reactions is the ferryl-oxo porphyrin radical cation species known as compound I. This transient species was definitively characterized for the first time in 2010 by Rittle and Green, who presented both the spectroscopic characteristics of compound I and a clear demonstration of its catalytic potency in substrate oxidation using a thermophilic P450 enzyme as a model system [2]. To achieve oxygen activation and substrate oxidation, the canonical P450 catalytic cycle (Fig. 11.1) involves two sequential single electron reduction reactions that first convert a ferric heme iron to the ferrous state (enabling the binding of O_2), and then result in the further reduction

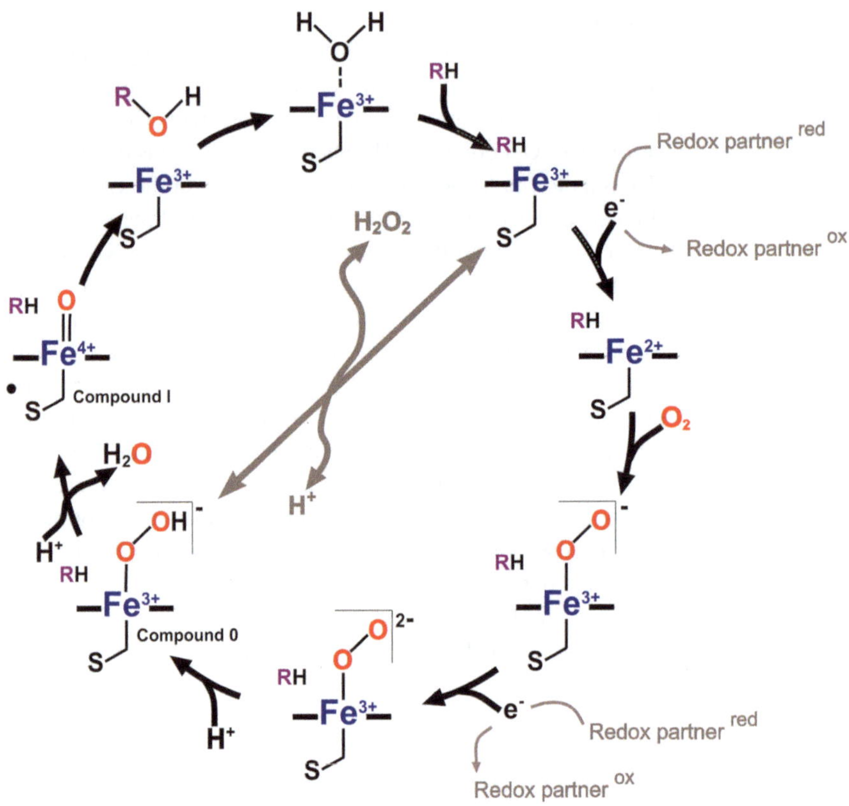

Fig. 11.1 The cytochrome P450 catalytic cycle. Intermediate species in the canonical P450 catalytic cycle are shown. At the *top* of the cycle is the resting state species where the ferric P450 heme iron is proximally coordinated by cysteine sulfur (S) in its thiolate form. Binding of substrate (RH) displaces the distal water ligand and often shifts the heme iron equilibrium toward high spin. Electron delivery from a redox partner reduces the heme iron to the ferrous form, which can then bind to molecular oxygen (O_2), and convert to the ferric-superoxo form. The delivery of a second electron from the redox partner forms the ferric-peroxo species, which becomes protonated to the reactive ferric-hydroperoxo (compound 0) form. A further protonation results in loss of a water molecule and production of the catalytically relevant compound I (ferryl-oxo porphyrin radical cation). Compound I abstracts a hydrogen atom from the substrate, and then "rebounds" a hydroxyl group to the substrate radical, forming a hydroxylated product (ROH), the dissociation of which enables rebinding of water in the distal position and restoration of the starting form

of the ferrous oxo (formally ferric superoxo) form to the ferric peroxo state. To facilitate the formation of compound I, successive protonation reactions then occur to produce first the transient ferric hydroperoxo state (compound 0), followed by formation of compound I following the loss of a water molecule after the second protonation step. To achieve substrate hydroxylation (as shown in the model catalytic cycle in Fig. 11.1), compound I abstracts a hydrogen atom from the substrate bound close to the heme (forming a transient compound II or ferryl-hydroxo species), and then "rebounds" the hydroxyl to the substrate radical, according to the radical rebound model of P450 catalysis developed by Groves [3].

The requirement for electrons in P450 catalysis is almost invariably met by the cofactors NADH and/or NADPH, which deliver electrons by hydride transfer to a flavin cofactor in a P450 redox partner enzyme. Until relatively recently, the types of P450 redox systems used were thought to be quite limited (to two major types). However, studies in recent years have revealed a more complex arrangement in nature, with diverse systems used in different organisms and even cases in which protein-based redox systems are no longer used. These different P450 redox systems are described in more detail in the sections below.

11.2 Class I and II P450 Redox Systems

Several early studies on P450 systems focused on mammalian P450 enzymes, and led to the identification of the diflavin (FAD- and FMN-containing) enzyme cytochrome P450 reductase (also known as CPR, and more recently referred to as POR for P450 oxidoreductase) as a redox partner for the membrane-anchored P450 enzymes found in the endoplasmic reticulum in the liver and other tissues [4]. CPR, like its eukaryotic P450 partners, is attached to the membrane by an N-terminal "anchor" region consistent with a single membrane spanning alpha helical segment. Recent studies from Monk and coworkers provided the first structure of an intact P450 enzyme (the *Saccharomyces cerevisiae* lanosterol 14α-sterol demethylase, or CYP51), showing the conformation of the membrane-spanning segment and its influence on the orientation of the P450 itself, resulting in the positioning of P450 substrate access regions toward the membrane [5]. Structural studies on the rat CPR (using a soluble version from which the N-terminal transmembrane segment is removed) confirmed its predicted structural relationships with FAD/NAD(P)H-binding ferredoxin reductases (FDRs) and FMN-binding flavodoxins (FLDs) [6]. The CPR crystal structure revealed a modular arrangement of the CPR with a N-terminal FLD domain orientated toward the C-terminal FDR domain such that the two flavin cofactors face toward each other and are separated by only ~4 Å between the dimethyl groups at the ends of their respective FMN and FAD cofactors [7]. In this "closed" conformation, there is clearly scope for direct electron transfer between FAD and FMN cofactors. However, while this orientation is clearly effective for inter-flavin electron transfer, it is probably not conducive to electron transfer to the P450 partner. As shown in Fig. 11.2, a reorientation of the FMN domain to a more "open" conformation is necessary to position the reduced FMN cofactor close to the heme-binding site in the P450, with the docking of the FMN domain proposed to occur on the P450 protein surface at the proximal side of heme – i.e. on the opposite side of the heme from the substrate-binding cavity, but close to the cysteine thiolate bond to the P450 heme iron.

CPR binds NADPH tightly and positions the redox active nicotinamide ring of the cofactor close to the FAD. Mutagenesis of several residues in CPR orthologues from different organisms has enabled identification of amino acids important for the binding of NADP(H) and for the progression of electron transfer to the FAD. The aromatic residues Trp677 (the indole ring thereof) and Tyr456 stack with the FAD isoalloxazine ring on opposite sides of the molecule in the rat CPR (a common arrangement in other CPRs) [7]. The displacement of the

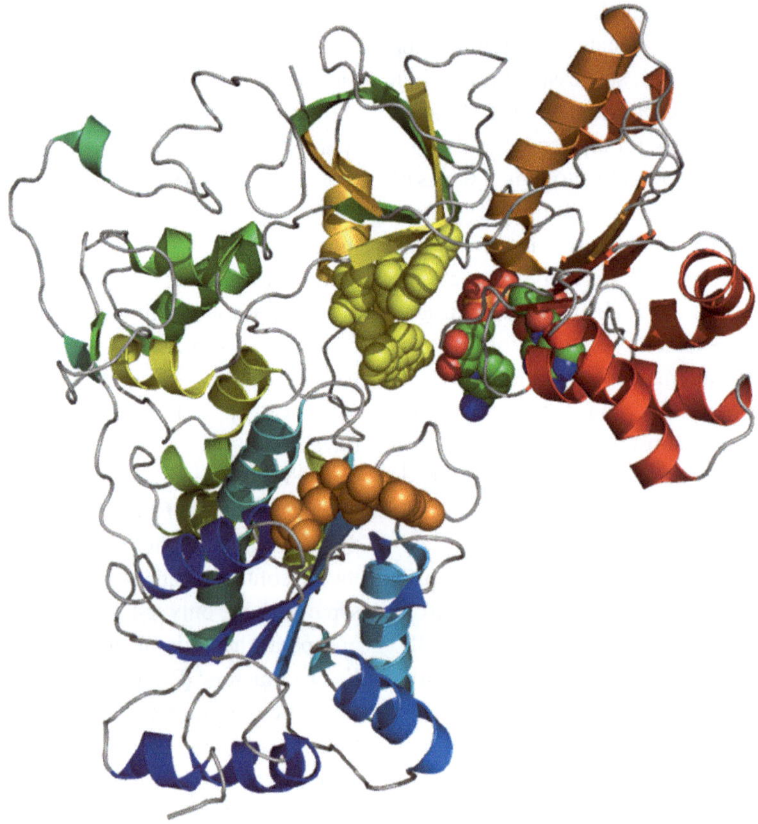

Fig. 11.2 The structure of cytochrome P450 reductase. The structure of rat CPR (PDB 3ES9) is shown in an open conformation in which the FAD/NADP(H)-binding and FMN-binding domains are separated, such that the FMN domain may be able to interact with and pass an electron to a P450 partner protein. In a more "closed" conformation, the domains would be more compactly arranged with the isoalloxazine rings of the flavins placed adjacent to one another to enable direct inter-flavin electron transfer [7]. The FAD/NADP(H) domain is shown in *red-to-green* colour with the FAD-binding region in *yellow/green* and the NADP(H) binding portion in *red/orange*, and with the FAD cofactor in *yellow* spheres. The FMN domain is in *blue* with the FMN in *orange* spheres. The $NADP^+$ cofactor is bound in the FAD/NADP(H) domain, close to the FAD cofactor. The $NADP^+$ is also shown in spheres with a *green* carbon backbone [107]

tryptophan side chain is an important catalytic step required for the reduced nicotinamide ring of NADPH to access the FAD and to transfer electrons (in the form of a hydride ion) to the flavin. The return of the conserved Trp side chain to its original position is likely also important to trigger the displacement of the oxidized $NADP^+$ and to allow the enzyme to perform inter-flavin electron transfer as a prerequisite for reduction of the P450 partner [8, 9]. It is considered that the resting form of CPR in vivo is one in which the FAD cofactor is fully oxidized, while the more positive potential FMN cofactor is in a single electron reduced and relatively air-stable blue semiquinone form [10]. In this model, NADPH-dependent electron transfer results in a 3-electron reduced form in which the FAD is fully reduced (the hydroquinone or $FADH_2$ form) and the FMN is a neutral semiquinone (FMNH). An internal electron transfer from $FADH_2$ to the FMN produces the FMN hydroquinone, which is the electron donor to the P450 partner. The first electron is transferred from the FMN hydroquinone to the ferric heme iron, restoring the FMN semiquinone and forming ferrous heme iron, which is then competent to bind O_2. A second electron is passed from the FAD to the FMN (restoring the resting, oxidized state of the FAD and forming the FMN hydroquinone again). The FMN hydroquinone then passes the second electron to the ferric-superoxo P450 species (forming the ferric peroxo state and enabling progression through the later protonation steps in the catalytic cycle) (Fig. 11.1). In this way, there is temporal control over electron transfer to the P450 (enabling dioxygen binding to heme iron to occur between the two individual electron transfer steps from CPR) and the CPR redox partner undergoes a 1-3-2-1 cycle, where the digits indicate the total number of electrons held on the CPR flavins at different stages in the process of P450 reduction.

The CPR-type redox partner system is widespread in eukaryotes and is often referred to as a class II (or E-class for eukaryotic-type) P450 redox partner. Its counterpart in early studies of P450 enzymes was the redox partner system supporting the activity of the cytochrome P450cam (CYP101A1) camphor 5-*exo* hydroxylase, an important member of the P450 enzyme superfamily and the first P450 for which a 3-dimensional structure was determined using X-ray crystallography [11]. P450cam is encoded by one of a suite of genes encoded on a transmissible plasmid in a strain of *Pseudomonas putida*, and the P450 plays a crucial role in initiating the breakdown of camphor as a source of energy for cell growth [12]. Also encoded on the same (CAM) plasmid are the redox partner proteins that supply electrons for P450cam catalysis. In this class I (or B-class for bacterial-type) system, electrons from NADH reduce the FAD-binding putidaredoxin reductase (PDR), which then transfers electrons one at a time to a ferredoxin (putidaredoxin or PD) [13]. PD binds a 2Fe-2S cluster which acts as a 1-electron carrier in this system. NADH reduces the PDR to its $FADH_2$ form, which then passes single electrons in two successive reactions to the PD. The PD reduces the P450cam in two steps, essentially as described above for the class II system. Thus, PD is an electron "shuttling" protein that interacts in turn with reduced PDR and then P450cam to facilitate catalysis in this class I system. Figure 11.3 shows the crystal structure for a complex between the PDR/PD proteins [14, 15]. The flavodoxin domain in CPR enzymes plays a similar role to that of PD, but in this case is fused to its NADPH dehydrogenase (FDR-like) domain.

Several bacterial P450 systems operate a class I redox system, including the *Pseudomonas* sp. terpineol oxidase P450terp (CYP108A1) and the *Mycobacterium tuberculosis* sterol demethylase CYP51B1 [16–18]. The ferredoxin component of the CYP51B1 redox system binds a 3Fe-4S iron-sulfur cluster [16, 17]. However, many such systems use 2Fe-2S cluster ferredoxins. These include the P450terp partner terpredoxin, with the most notable example being P450cam with its partner putidaredoxin [14, 18]. Figure 11.4 shows the crystal structure of a complex between P450cam and PD, illustrating how the ferredoxin docks on the proximal face of the P450 protein and places the iron-sulfur cluster close to the cysteine thiolate-heme iron bond and to the heme cofactor itself [19]. However, P450 redox systems using 4Fe-4S and even 7Fe cluster ferredoxins have been reported, with the *Mycobacterium smegmatis* FdxA protein shown to be a 7Fe ferredoxin (binding both a 3Fe-4S and a 4Fe-4S cluster) and suggested to support activity of one or more of the numerous P450s in this bacterium, and in the pathogen *M. tuberculosis*, which encodes 20 P450s [20–23]. The eukaryotic mitochondrial P450 systems also use a class I redox system, likely reflecting the prokaryotic origins of the mitochondrion. In these cases, the redox partners are the FAD-binding, NADPH-dependent adrenodoxin reductase and the 2Fe-2S ferredoxin adrenodoxin [24, 25]. However, it should also be noted that the FMN-binding flavodoxin proteins are also potential redox partners for prokaryotic P450s – as best demonstrated in the case of the *Citrobacter braakii* P450cin system (CYP176A1), where the flavodoxin cindoxin is the natural redox partner for the P450 and is itself reduced by a FAD-dependent reductase. P450cin can then oxidize the terpene cineole to enable the bacterium to use this molecule as a sole source of carbon for growth [26].

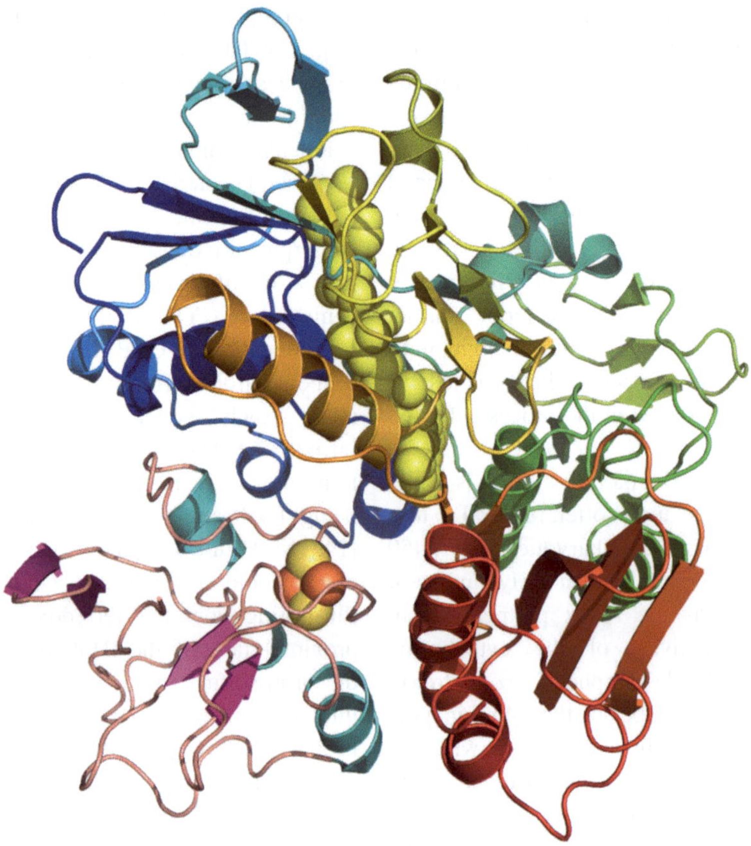

Fig. 11.3 The crystal structure of the putidaredoxin reductase: putidaredoxin complex. The image shows the crystal structure for a covalently linked complex between the P450cam putidaredoxin reductase (PDR) and putidaredoxin (PD) (PDB 3LB8). The complex was formed by cross-linking the transiently formed PDR/PD complex using 1-ethyl 3-[3-(dimethylamino)propyl] carbodiimide (EDC). The PD is shown in cyan and magenta with the 2Fe-2S cluster in atom coloured spheres. The PD iron-sulfur cluster is docked close to the FAD cofactor in the PDR protein. The PDR is shown in multicolour with the FAD cofactor in *yellow* spheres [14]

11.3 Flavocytochrome P450 BM3

The discovery of the *Bacillus megaterium* P450 BM3 enzyme by Armand Fulco and coworkers led to the characterization of the first "outlier" redox partner system from the class I/class II paradigm [27]. The 119 kDa P450 BM3 (CYP102A1) is a natural fusion of a soluble fatty acid hydroxylase P450 (at the N-terminus) to a soluble CPR module (at the C-terminus). The BM3 CPR domain lacks the membrane anchor region found in the eukaryotic CPRs, and is instead linked to the P450 domain by a short peptide linker region [28] (Fig. 11.5). BM3 is a high activity fatty acid hydroxylase, which hydroxylates a wide range of long chain fatty acids at the ω−1, ω−2 and ω−3 positions [28]. BM3 has the highest reported oxidase activity for a P450 (e.g. ~285 s^{-1} with arachidonic acid), facilitated by rapid electron transfer from the CPR module [29, 30]. Early studies on the BM3 enzyme demonstrated that the enzyme could oligomerize [31]. More recent work in this area has indicated that the dimeric form of the flavocytochrome is likely the catalytically relevant form, with electron transfer occurring between the CPR domain of one monomer and the P450 domain of the other [32–34]. A similar interdomain electron transfer process is thought

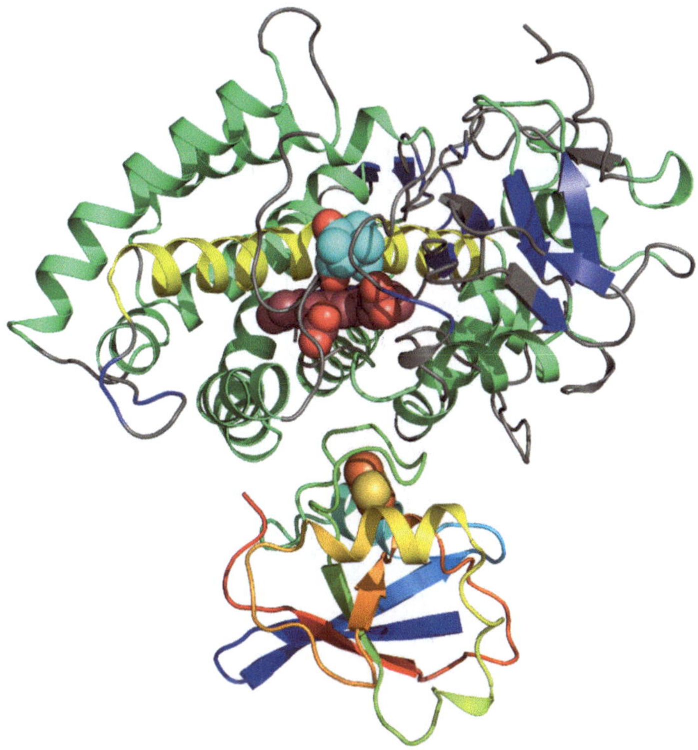

Fig. 11.4 The crystal structure of the putidaredoxin: P450cam complex. The *Pseudomonas putida* P450cam/putidaredoxin complex (PDB 4JX1) is shown in cartoon representation with the P450cam alpha helices in *green* and beta sheets in *blue*. The I-helix is highlighted in *yellow* and the heme shown in *purple* spheres with the oxygen atoms in *red*. The hydroxylated product 5-*exo*-hydroxycamphor (also in spheres) is shown above the heme plane in the active site cavity with carbon atoms coloured in cyan. Putidaredoxin (in multicolour) docks on the proximal face of the P450cam and positions its 2Fe-2S cluster (atom coloured spheres) in close vicinity to the P450 heme and to the cysteine thiolate-heme iron bond. The complex was formed using the homobifunctional maleimide cross-linker 1,6-bismaleimidohexane, and using a D19C mutant of PD with a K344C mutant of P450cam to facilitate the cross-linking [19]

to occur in the nitric oxide synthase (NOS) enzymes, which also use a CPR-like reductase domain to reduce a fused heme domain. In this case, the heme domain is another cysteine thiolate ligated hemoprotein that catalyzes successive oxidations of *L*-arginine to form first *N*-hydroxyl-*L*-arginine, and then *L*-citrulline and nitric oxide (NO) [35].

The binding of fatty acid substrates to P450 BM3 results in a substantial change in ferric heme iron spin-state from low-spin toward high-spin, and a concomitant large shift in P450 heme iron potential from −368 to −239 mV when bound to arachidonic acid [36]. This change favours electron transfer to the ferric heme iron and helps to ensure that electron transfer occurs only when substrate is available for oxidation. The catalytic cycle for the BM3 CPR is different to that for the eukaryotic CPRs, due mainly to alterations in the properties of the FMN-binding (flavodoxin) domain in BM3. The resting form of the BM3 CPR has both flavins (FAD and FMN) fully oxidized. Reduction with NADPH results in complete reduction of the FAD (to $FADH_2$), and successive single electron transfers occur through the FMN, with the FMN semiquinone species passing electrons to the heme iron at the same reaction stages as described above (and as shown in Fig. 11.1). The BM3 reductase thus undergoes a 0-2-1-0 redox cycle during P450 catalysis [37]. Transient kinetic studies of the reduction of the

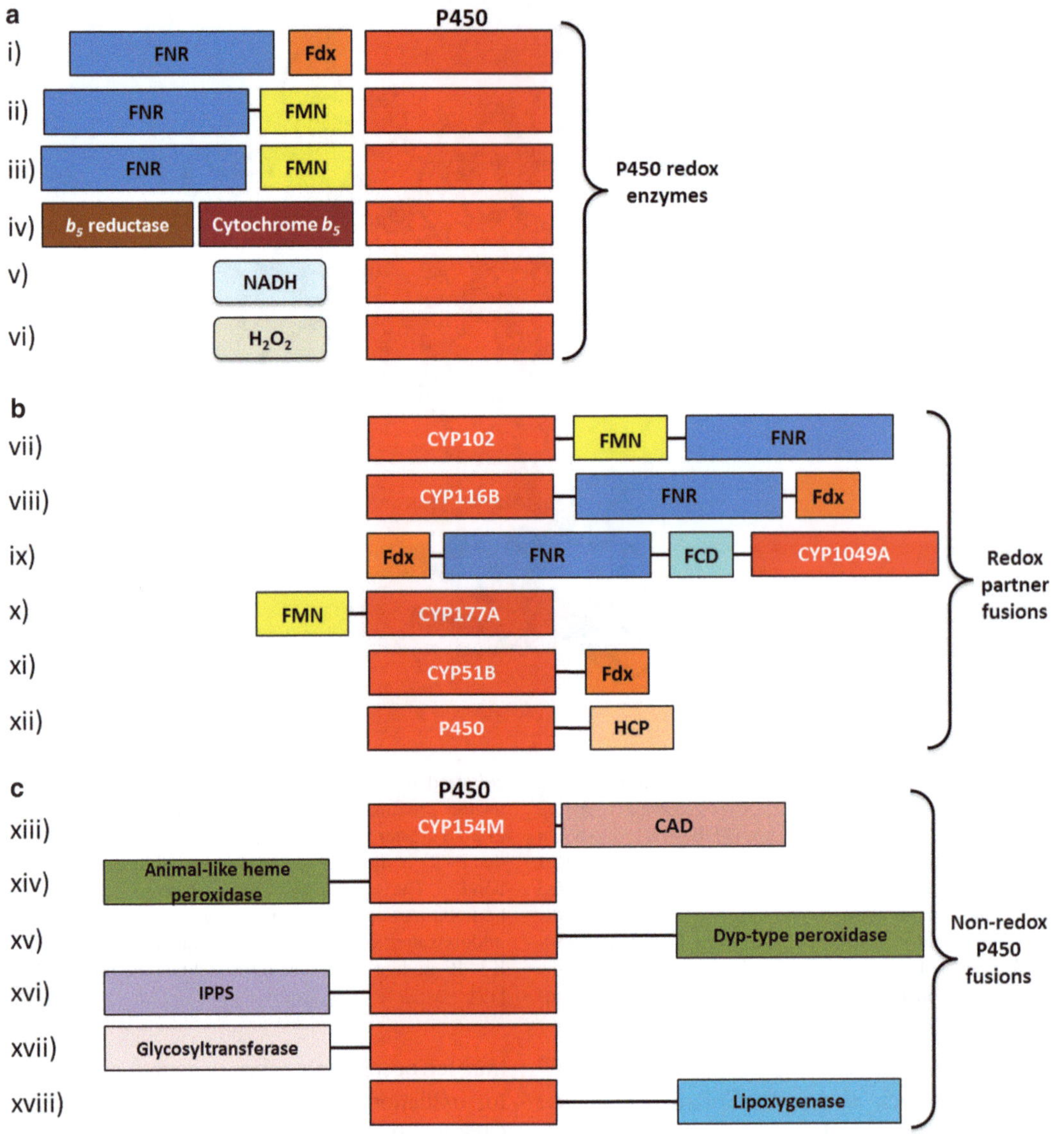

Fig. 11.5 Biological diversity of P450 redox partners, redox partner fusions and non-redox partner P450 fusion enzymes. A schematic overview of a number of selected P450 redox systems where P450s interact with separate redox partners or other chemical entities to drive catalysis (**a**); or where P450s are parts of fusions in which the other protein is either a redox partner system or displays activity unrelated to P450 reduction (**b** and **c**, respectively). **Section A** shows "standard" class I and II redox systems (*i* and *ii*), where both CPR and P450 components of the class II system contain N-terminal membrane anchor domains. (*iii*) is the P450cin type system where a flavodoxin replaces a ferredoxin [26]. (*iv*) is the yeast/fungal system that requires only cytochrome b_5 reductase and b_5 partners [100]. (*v*) and (*vi*) show the NADH-dependent P450nor and H_2O_2-dependent peroxygenase-type P450s [76, 79, 80]. **Section B** shows a variety of P450-redox partner fusion enzymes. (*vii*) is the BM3-type CPR fusion and (*viii*) the CYP116B-type PDOR-P450 fusion enzyme [28, 29, 45, 46]. (*ix*) is a benzoate dioxygenase reductase-P450 fusion (CYP1049 family), (*x*) a XplA-type flavodoxin-P450 fusion [56, 60], (*xi*) the *M. capsulatus* MCCYP51FX fusion [63] and (*xii*) an example of a P450-HCP fusion, where HCP is a hybrid cluster protein family member and potential P450 redox partner [108]. **Section C** shows various examples of P450s fused to non-redox partner proteins. (*xiii*) is a CYP154 family P450 fused to a cinammyl alcohol dehydrogenase-type module, (*xiv*) and (*xv*) are P450 fusions to different types of peroxidases, with the animal-like peroxidase fusions typified by the *Aspergillus nidulans* Ppo proteins [67]. (*xvi–xviii*) show uncharacterized P450 fusions to IPPS (*trans*-isoprenyl diphosphate synthase family member), glycosyltranferase and lipoxygenase like modules

isolated BM3 FMN domain with sodium dithionite showed formation of an anionic semiquinone, consistent with preceding EPR data from BM3 enzyme samples freeze-quenched during turnover, and pointing to the involvement of the BM3 FMN anionic semiquinone as the electron donor to the heme iron, as opposed to the FMN hydroquinone in the case of the eukaryotic CPRs [37, 38].

Several orthologues of P450 BM3 have been identified in other microbes, including two fatty acid hydroxylases (CYP102A2 and A3) in *B. subtilis* [39, 40]. Membrane-associated fatty acid hydroxylase activity was also found to be associated with CYP505A1 (P450foxy) from the fungus *Fusarium oxysporum*, a member of a distinct family of eukaryotic P450-CPR fusion enzymes [41] (Fig. 11.5). As with P450 BM3, P450foxy also catalyzes the $\omega-1$ to $\omega-3$ hydroxylation of fatty acids, and is also catalytically active with shorter chain fatty acids (C9 and C10) that are not efficient substrates for P450 BM3 [42].

11.4 The CYP116B Enzymes

Further complexity in the redox systems supporting P450 activities became clear with the identification of the CYP116B enzyme family, in which soluble bacterial P450s are fused to redox partner modules with structural similarity to the FMN and 2Fe-2S cluster binding phthalate dioxygenase reductase (PDOR) [43]. Analysis of microbial genomes provided the first evidence for these fusion enzymes, in which a redox partner distinct from the class I/II systems was recognized for the first time [44]. In the soluble bacterial CYP116B enzymes, the N-terminal P450 domain is fused to the C-terminal PDOR domain to form an ~86 kDa fusion enzyme [45] (Fig. 11.5). The CYP116B1 and CYP116B2 enzymes have been expressed and characterized, including studies to dissect the enzymes into their component P450 and PDOR domains, including the individual 2Fe-2S- and FMN-binding domains in the case of CYP116B2 from *Rhodococcus* sp. NCIMB 978 [45, 46]. Electron transfer in these enzymes proceeds from NAD(P)H through the FMN and 2Fe-2S clusters in the PDOR module, and then onto the P450 heme iron.

The related class I bacterial P450 CYP116A1 (from *Rhodococcus erythropolis* NI86/21) was previously shown to catalyze *N*-dealkylation reactions with the thiocarbamate herbicides vernolate (*S*-propyldipropylthiocarbamate) and EPTC (*S*-ethyldipropylthiocarbamate), as well as acting on the triazine herbicide atrazine (1-chloro-3-ethylamino-5-isopropylamino-2,4,6-triazine) [47, 48]. In view of the structural similarities between CYP116A1 and the heme domain of the *Cupriavidus metallidurans* CYP116B1, the thiocarbamate herbicides were tested for activity with CYP116B1, and both herbicides were shown to be hydroxylated on propyl chains, with the reaction progressing to *N*-dealkylation in the case of vernolate [45]. For the *Rhodococcus* sp. CYP116B2, activity in dealkylation of 7-ethoxycoumarin was demonstrated, along with hydroxylation and *O*-dealkylation of various alkyl aryl ethers [49], while the *Rhodococcus ruber* DSM 44319 CYP116B3 was also purified and shown to catalyze NADPH-dependent oxidation of a range of chemicals – including hydroxylation of naphthalene, fluorene, toluene and ethylbenzene [50]. A combination of rational and directed evolution approaches was also used to improve CYP116B3-dependent oxidative demethylation and deethylation reactions with 7-methoxycoumarin and 7-ethoxycoumarin, respectively [51]. Numerous protein engineering studies of P450 BM3 have demonstrated the catalytic versatility of mutants of this enzyme (e.g. [52, 53]), and studies on the CYP116B enzymes are now also beginning to point to their potential as biotechnologically important enzyme catalysts.

In other studies, the CYP116B PDOR modules have been used as surrogate electron donors for heterologous P450 enzymes, with potential applications in generating robust, catalytically self-sufficient P450-PDOR fusion enzymes [54]. Such fusions benefit from the relatively fast electron transfer kinetics of the PDOR modules, with NADPH-dependent reduction of the FMN/2Fe-2S centres in CYP116B1 occurring with a limiting rate constant of ~72 s^{-1} at 25 °C [45].

11.5 A Biotechnologically Important P450-Flavodoxin Fusion Protein

The discovery of an unusual cytochrome P450-flavodoxin fusion protein was made in *Rhodococcus rhodochrous* strain 11Y, isolated from soil contaminated with explosives. In this case, the flavodoxin forms the N-terminal domain of the enzyme. This P450 system (XplA or CYP177A1) was shown to catalyze the degradation of RDX (Royal Demolition eXplosive; hexahydro-1,3,5-trinitro-1,3,5-triazine) and to enable the bacterium to use this explosive as a sole source of nitrogen for growth. Electrons for the reaction are supplied by the adrenodoxin reductase-like XplB, encoded by the *xplB* gene in the same gene cluster as *xplA*. Thus, the electron transfer pathway in this system is from NADPH to the XplB FAD cofactor, and then on to the XplA FMN and finally to the P450 heme iron [55]. XplA degrades RDX under aerobic or aerobic conditions, with enzyme-catalyzed reductive denitration likely followed by hydration reactions to form products nitrite and formaldehyde, and either 4-nitro-2,4-diazabutanal (NDAB, under aerobic conditions) or methylenedinitramine (MEDINA, under anaerobic conditions) [56]. For applications in bioremediation, *Arabidopsis thaliana* was engineered to express the *xplA* gene, and plants grown in RDX-contaminated soil were shown to exhibit resistance to the explosive. Further studies with *A. thaliana* transformed with both *xplA* and *xplB*, along with the *Enterobacter cloacae* nitroreductase *nsfl* gene, showed that these plants could remove both RDX and the explosive 2,4,6-trinitrotoluene (TNT) from soil and allow plant growth at levels of these explosives that proved inhibitory to the growth of *A. thaliana* that expressed only *xplA* [57].

The structure of the XplA P450 domain was determined at 1.5 Å resolution, revealing a typical P450 fold and a compact active site with imidazole present as a ligand in the 6th (distal) position on the heme iron [58]. The structure highlighted the absence of the acid-alcohol pair (e.g. Asp251/Thr252 in the well-studied camphor hydroxylase P450cam) in XplA – which instead has Met394/Ala395 in the relevant positions in the P450 I helix [59]. The absence of amino acids with functions in stabilizing and protonating the P450 heme iron-oxo complexes is consistent with XplA having a predominantly reductive role (i.e. nitro group reduction). Other novel features of XplA relate to the unusually positive reduction potential of its flavodoxin FMN semiquinone/hydroquinone couple (−172 mV vs the normal hydrogen electrode, NHE) and the relatively weak binding of FMN to the protein (K_d = 1.09 μM, compared to, for example, the *Desulfovibrio vulgaris* flavodoxin with a K_d = 0.24 nM). These adaptations are likely made through evolution to facilitate a mainly reductive (rather than oxidative) role for this unusual P450 system [60, 61]. Approximately 30 different *xplA* genes have now been identified in different microbial genomes – with the majority of these being in *Rhodococci* [62].

11.6 A P450-Ferredoxin Fusion Protein

A further unusual P450 fusion protein was discovered in the methanotrophic bacterium *Methylococcus capsulatus*. In this case, the soluble P450 domain is at the N-terminal, with a ferredoxin linked at the C-terminal end via an alanine-rich linker [63]. The P450 (MCCYP51FX) has ~49 % identity between its heme domain and the CYP51B1 sterol demethylase enzyme from *M. tuberculosis*, while the ferredoxin domain has ~42 % identity to the 3Fe-4S cluster ferredoxin that is encoded by the *Rv0763c* gene located directly adjacent to the gene encoding CYP51B1 (*Rv0764c*) [17, 18]. Activity as a lanosterol demethylase was shown when MCCYP51FX was reconstituted with a heterologous (spinach) ferredoxin reductase, despite the observation that addition of lanosterol produced a type II P450 spectral change (Soret absorption shift to longer wavelength) that is more often associated with inhibitory interactions of a ligand with the P450 heme iron. Other studies indicated a propensity of the MCCYP51FX to self-aggregate, possibly indicating the formation of an oligomeric state in solution [63]. To date, this is the only characterized example of a P450-ferredoxin fusion enzyme, although other types of P450 fusion enzymes (to both redox and non-redox partners) have been identified, as described in the sections below.

11.7 Other Potential Catalytically Self-Sufficient P450s

Bioinformatics tools can be used to probe available genome sequences in order to identify P450s fused to other proteins. In a number of cases, these fusion partners are potential redox partners in view of their similarity to known redox enzymes. At the end of 2014, the CDART (Conserved Domain Architecture Retrieval Tool) program was used to identify potential P450 fusion enzymes and to assess whether these fused partners were likely to be electron donor systems for the P450, or else might play distinct roles in relation to the P450 [64]. Figure 11.5 shows selected results for the domain architecture of a range of P450 systems in which redox partners (either potential or proven) are found either as separate entities or are fused to the P450. Figure 11.5 also shows a number of P450 fusion enzymes in which the partner protein is not an electron transferase to the P450. In addition to the characterized proteins listed in the preceding sections, a number of uncharacterized potential P450-redox partner fusions are identified – including P450s fused to modules identified as (1) short chain dehydrogenases/reductases; (2) HCPs (hybrid cluster proteins) that contain a 4Fe-4S cluster; (3) Rieske-type 2Fe-2S iron cluster binding proteins; and (4) benzoate dioxygenase reductases – comprising 2Fe-2S ferredoxin, FAD-binding ferredoxin reductase and FCD (putative transcriptional regulator) genes. In the latter case, we have expressed and purified one such enzyme (CYP1049A1 from a *Burkholderia* sp.), confirming the predicted cofactor content of the reductase module and NAD(P)H-dependent reduction of the reductase portion of the enzyme (Luciakova et al. unpublished data). However, there are few other data available for the systems mentioned here, or for virtually all the rest of the other potentially new catalytically self-sufficient P450 systems identified from genome searches.

11.8 Non-redox Partner P450 Fusions

Of equal interest to the various likely P450-redox partner fusions identified in numerous prokaryotic and eukaryotic genomes are the several other P450 fusion enzymes in which the apparent fused partners are evidently not electron transfer modules. These protein modules may instead have catalytic functions that are associated with the activity of the P450 itself, e.g. providing a substrate for the linked P450, or using the P450 product as a substrate. Figure 11.5 shows examples for P450s fused (at either their N- or C-terminus) to cinammyl alcohol dehydrogenase (CAD), and to different types of peroxidase, glycosyltransferase, lipoxygenase and IPPS (isoprenyl diphosphate synthase) modules. Other types of P450 fusions from genome databases include those predicted to be fused to putative solute transporters (in ascomycete fungi), esterases/lipases, ubiquitin-like proteins, 2-oxoglutarate/Fe(II)-dependent oxygenase-type proteins and mannosyltransferases.

As is the case for the P450-redox partner fusion enzymes, very few of these non-redox partner fusion enzymes have been expressed and characterized. However, the fungal Ppo enzymes have been studied and are functional fusions of a N-terminal peroxidase/dioxygenase domain to a C-terminal P450. In *Aspergillus nidulans*, the Ppo's are involved in synthesis of psi factors – molecules derived from oleic acid and linoleic acid, and which play key roles in regulation of the fungal life cycle, as well as being implicated in the synthesis of mycotoxins [65, 66]. The *A. nidulans* PpoA enzyme was demonstrated to catalyze the oxidation of linoleic acid to 8*R*-HPODE (8*R*-hydroperoxyoctadecadienoic acid) in the peroxidase/dioxygenase domain, with the product then isomerized to 5,8-dihydroxyoctadecadienoic acid by the P450 domain [67]. The *A. nidulans* PpoC enzyme was similarly shown to catalyze dioxygenation of linoleic acid, forming 10-HPODE. However, in this case the product is not further isomerized by the P450 domain, in which the heme coordinating cysteine thiolate is replaced by a glycine, leading to a heme-depleted P450 domain [68]. A further characterized example of a P450/non-redox partner fusion enzyme is the *Penicillium brevicompactum* P450-hydrolase fusion protein encoded by the *mpaDE* gene. The MpaDE enzyme catalyzes key steps in the production of mycophenolic acid (MPA), a molecule with important immunosuppressant, antimicrobial and antitumour activities [69]. Specifically, the P450 component (MpaD) is predicted to catalyze methyl hydroxylation of the precursor 5-methylorsellinic acid (5-MOA) to form 5,7-dihydroxy-4-methylphthalide, followed by a lactonization reaction on this product catalyzed by the Zn-dependent hydrolase domain (MpaE) to form 5,7-dihydroxy-4-methylphthalide (DHMP) as the next intermediate in the MPA synthesis pathway [69].

11.9 P450s that Bypass Redox Partners

Not all P450s are dependent on electron transfer from redox partner, and two particular classes of P450s are stand-alone enzymes that perform reductive or oxidative transformations of their substrates without the participation of any partner proteins. The peroxygenase P450s use hydrogen peroxide directly to form the reactive compound 0 state, which is then transformed via a further protonation and dehydration to generate the catalytically relevant compound I (Fig. 11.1). The first characterized examples of this enzyme class were P450 BSβ from *Bacillus subtilis* (CYP152A1) and P450 SPα from *Sphingomonas paucimobilis* (CYP152B1) [70, 71]. CYP152B1 was found to produce predominantly alpha-hydroxy fatty acid products from a range of fatty acid substrates, with C14 and C15 saturated fatty acids (myristic and pentadecanoic acids), and the polyunsaturated arachidonic acid being particularly good substrates, and with the *S*-enantiomeric hydroxy fatty acids being formed at levels of >98 % of the overall products formed [72]. CYP152A1 produces both α- and β-hydroxylated fatty acids, with the greater proportion being the β-hydroxy products [70, 73]. CYP152A1 and CYP152B1 have similar steady-state rate constants for fatty acid hydroxylation of around 1,000 min^{-1} with their preferred substrates [70, 73, 74]. A more recently studied member of the CYP152 family is the CYP152L1 enzyme from a *Jeotgalicoccus* species (OleT). Although strongly related to the CYP152A1/B1 enzymes, OleT catalyzes predominantly the oxidative decarboxylation of long chain fatty acids, producing terminal alkenes as the major products (with much smaller amounts of α- and β-hydroxylated fatty acids). CYP152A1 was also shown to produce 1-pentadecene from the saturated C16 fatty acid palmitic acid, but this product was not detected in parallel studies with CYP152B1. However, CYP152 enzymes from three other microbes (*Corynebacterium efficiens*, *Kocuria rhizophila* and *Methylobacterium populi*) were also shown to produce 1-pentadecene from palmitic acid [75]. The crystal structure of OleT was determined, confirming the strong structural similarity of this P450 to CYP152A1/B1. However, the factors determining preference for H_2O_2-dependent decarboxylation over hydroxylation in these enzymes remain uncertain. Nonetheless, the apparent rate constant for OleT-dependent oxidation of fatty acids is very fast at ~167 s^{-1} with 200 μM H_2O_2 [76]. Other studies used redox partner systems (the *E. coli* flavodoxin reductase [FLDR] and flavodoxin [FLD], and the P450 BM3 CPR domain) to drive NADPH-dependent catalysis in CYP152A1 and the *Clostridium acetobutylicum* CYP152A2 ($P450_{CLA}$) enzymes. These studies showed that fatty acid hydroxylation could also be catalyzed by electron transfer using heterologous redox partner proteins [77]. NADPH-dependent fatty acid decarboxylation was also shown for OleT using either the fused PDOR domain of CYP116B1, or the *E. coli* FLDR/FLD redox partner combination [78].

The other major group of P450s that has evolved to function without the use of redox partners are the CYP55A P450 subfamily enzymes. The prototype enzyme of this class is CYP55A1 (P450nor) from the fungus *Fusarium oxysporum* (which also produces P450foxy), which catalyzes a purely reductive reaction in which two molecules of nitric oxide (NO) are converted to dinitrogen oxide (N_2O) according to the reaction: $2NO + NADH + H^+ \rightarrow N_2O + NAD^+ + H_2O$. Electrons are delivered directly by NADH without the participation of any other proteins [79, 80]. This reaction is the final step in the respiratory conversion of nitrite/nitrate to N_2O in the fungus [80]. Two distinct isoforms of CYP55A1 are produced that result from translation using (1) the first start codon (P450norA, which is then directed to the mitochondrion via an N-terminal targeting sequence), or (2) the subsequent start codon (P450norB that is cytoplasmically located) [81]. Orthologues of P450nor were also reported in other fungi – notably *Trichosporum cutaneum* (CYP55A4) and *Cylindrocarpon tonkinense*. As for the *F. oxysporum* P450nor, two different forms of the enzyme are found in *C. tonkinense*. However, in this case they originate from distinct genes – *CYP55A2* (which encodes P450nor1) and *CYP55A3* (which encodes P450nor2). Interestingly, P450nor1 is quite NADH-specific, whereas P450nor2 can use both NADH and NADPH, albeit with a higher affinity for NADPH [82].

Mechanistically, the P450nor reaction involves the binding of a molecule of NO to the heme iron of the P450 to form a ferric-NO complex that is then reduced to an intermediate with a red-shifted heme absorption maximum (at ~444 nm) that is considered to be a ferric-hydroxylamine radical species. The short lived intermediate then reacts with the second molecule of NO to generate the N_2O product, with release of a molecule of water and NAD^+ [83–85].

11.10 Cytochrome b_5 as a Redox Partner for P450 Enzymes

The cytochromes b_5 are small heme binding proteins found predominantly in eukaryotes, although b_5-like proteins and domains have now also been identified in bacteria, including in the purple bacterium *Ectothiorhodospira vacuolata* [86]. The b_5 proteins are small, membrane proteins (usually ~135 amino acids) with *bis*-His coordinated heme iron. Electron transfer to and from b_5 proteins occurs via an exposed edge of their heme cofactor, and the b_5 proteins shuttle between ferric and ferrous forms for single electron transfer reactions. Two distinct types of b_5 are found in mammalian tissues – one locating to the outer mitochondrial membrane, and the other to the endoplasmic reticulum membrane [87]. In the context of P450-dependent drug metabolism reactions, the b_5 isoform in the ER is of particular relevance, and is located on the cytoplasmic side of the ER membrane, attached via a C-terminal membrane anchor domain. The b_5 proteins are reduced by their partner cytochrome b_5 reductase, an NADH-dependent and FAD-binding protein that shares the same cellular location [88].

The relationships between b_5 and P450s are complex and remain incompletely understood. The b_5 heme iron has a relatively positive reduction potential (typically around 0 mV versus the normal hydrogen electrode, NHE) [89, 90]. This suggests that they should not be able to deliver the first of the two electrons required for P450 monooxygenation reactions, where the ferric P450 heme iron (in either low-spin or high-spin form) typically has a much more negative potential. However, delivery of the second electron from b_5 to a P450 ferric-superoxo heme species should be much more thermodynamically favourable. Early research on the roles of b_5 in P450 metabolism highlighted that addition of NADH could stimulate NADPH-dependent metabolism of drugs, and thus that a secondary system of electron transport involving cytochrome b_5 reductase and b_5 could provide the 2nd electron for P450 catalysis in reactions such as the activation of the anti-cancer drug ellipticine and the improved oxidation of nifedipine and testosterone by the major human drug metabolizing P450 CYP3A4 [91, 92]. In the case of human CYP2B4, the delivery of a 2nd electron to the P450 via b_5 was reported to occur much faster than from CPR, leading to more efficient catalysis in which NADPH oxidation was more tightly coupled to P450 product formation [93]. The b_5 protein also has a crucial role in the metabolism of steroid hormones. CYP17A1 in the adrenal cortex catalyzes the 17α-hydroxylation of both pregnenolone and progesterone, but 17-hydroxypregnenolone can be further oxidized by the same P450 to generate dehydroepiandrosterone in an acyl (17,20-lyase) bond cleavage reaction. 17-Hydroxypregnenolone can undergo a similar (but less efficient) lyase reaction to form androstenedione. The lyase reaction that forms androgens is enhanced by b_5, although there is still controversy relating to whether b_5's role involves electron transfer or is achieved through imparting conformational change on the P450. In the latter case, studies suggest that b_5 may influence the binding mode of the substrate such that the iron-oxo species is orientated toward the C20 position (and away from the C17 position), favouring the lyase reaction [94–96]. Other recent in vivo studies using a conditional deletion of b_5 in mouse revealed diminished activities of P450s in the 3A and 2C families, as well as decreased CYP17A 17,20-lyase activity as a consequence of diminished production of b_5 [97, 98].

Studies in yeast and fungal systems have indicated that NADH-cytochrome b_5 reductase and b_5 could act as a functional redox partner system supporting (1) CYP51 (sterol demethylase)-dependent oxidative demethylation of 24-methylene-24,25-dihydrolanosterol in *Candida albicans*, and (2) hydroxylation of 4-propylbenzoic acid in the lignin degrading *Phanaerochaete chrysosporium* [99, 100]. However, given the thermodynamic considerations, the most likely mechanism would appear to be delivery of the first electron via the b_5 reductase, and the second (to the ferric-superoxo intermediate) from b_5 [101, 102].

11.11 Conclusions

The cytochromes P450 are an enormously divergent enzyme superfamily, members of which are found in all of the major domains of life. For most eukaryotes (and for several prokaryotes) there are large numbers of different *CYP* genes in the genome [103]. Early studies on P450 enzymes suggested a relatively simple set of redox partner systems from which these enzymes obtain the electrons required for activation of molecular oxygen and substrate monooxygenation. However, as genome sequence data continue to accumulate and novel types of P450s are characterized, it has become increasingly clear that there is considerable diversity in the types of enzyme machinery that can reduce different P450 enzymes, including various P450s that are fused to some or all of the accessory enzyme components required to drive P450 substrate oxidation [104]. These findings have inspired the creation of artificial fusions between P450s and heterologous redox partners to produce catalytically self-sufficient P450s for biotechnological applications [105, 106]. Other P450s have apparently evolved as fusions to non-redox partner proteins, and this may instead enable more catalytically efficient biochemical transformations if the two fused partners are component enzymes in the same pathway [65–67]. However, other P450s have also evolved to forego any requirement for redox partners – by using naturally the "peroxide shunt" mechanism for catalysis, or by deriving electrons directly from NAD(P)H [76, 84]. Undoubtedly, these recent discoveries on the complexity of redox processes in P450 enzymes are only the tip of the iceberg, and coming years will inevitably see identification of further diversity in the redox processes used in the P450 superfamily.

References

1. Guengerich FP, Munro AW (2013) Unusual cytochrome P450 enzymes and reactions. J Biol Chem 288:17065–17073
2. Rittle J, Green MT (2010) Cytochrome P450 compound I: capture, characterization, and C-H bond activation kinetics. Science 330:933–937
3. Groves JT (2006) High-valent iron in chemical and biological oxidations. J Inorg Biochem 100:434–437
4. Pandey AV, Flück CE (2013) NADPH P450 oxidoreductase: structure, function, and pathology of diseases. Pharmacol Ther 138:229–254
5. Monk BC, Tomasiak TM, Keniya MV, Huschmann FU, Tyndall JD, O'Connell JD III, Cannon RD, McDonald JG, Rodriguez A, Finer-Moore JS, Stroud RM (2014) Architecture of a single membrane spanning cytochrome P450 suggests constraints that orient the catalytic domain relative to the bilayer. Proc Natl Acad Sci U S A 111:3865–3870
6. Porter TD (1991) An unusual yet strongly conserved flavoprotein reductase in bacteria and mammals. Trends Biochem Sci 16:154–158
7. Wang M, Roberts DL, Paschke R, Shea TM, Masters BS, Kim JJ (1997) Three-dimensional structure of NADPH-cytochrome P450 reductase: prototype for FMN- and FAD-containing enzymes. Proc Natl Acad Sci U S A 94:8411–8416
8. Gutierrez A, Doehr O, Paine M, Wolf CR, Scrutton NS, Roberts GC (2000) Trp-676 facilitates nicotinamide coenzyme exchange in the reductive half-reaction of human cytochrome P450 reductase: properties of the soluble W676H and W676A mutant reductases. Biochemistry 39:15990–15999
9. Gutierrez A, Munro AW, Grunau A, Wolf CR, Scrutton NS, Roberts GC (2003) Interflavin electron transfer in human cytochrome P450 reductase is enhanced by coenzyme binding. Relaxation kinetic studies with coenzyme analogues. Eur J Biochem 270:2612–2621
10. Iyanagi T, Xia C, Kim JJ (2012) NADPH-cytochrome P450 oxidoreductase: prototypic member of the diflavin reductase family. Arch Biochem Biophys 528:72–89
11. Poulos TL, Finzel BC, Howard AJ (1987) High-resolution crystal structure of cytochrome P450cam. J Mol Biol 195:687–700
12. Rheinwald JG, Chakrabarty AM, Gunsalus IC (1973) A transmissible plasmid controlling camphor oxidation in *Pseudomonas putida*. Proc Natl Acad Sci U S A 70:885–889
13. Sevrioukova IF, Garcia C, Li H, Bhaskar B, Poulos TL (2003) Crystal structure of putidaredoxin, the [2Fe-2S] component of the P450cam monooxygenase system from *Pseudomonas putida*. J Mol Biol 333:377–392
14. Sevrioukova IF, Poulos TL, Churbanova IY (2010) Crystal structure of the putidaredoxin reductase•putidaredoxin electron transfer complex. J Biol Chem 285:13616–13620
15. Churbanova IY, Poulos TL, Sevrioukova IF (2010) Production and characterization of a functional putidaredoxin reductase-putidaredoxin covalent complex. Biochemistry 49:58–67
16. Peterson JA, Lu JY, Geisselsoder J, Graham-Lorence S, Carmona C, Witney F, Lorence MC (1992) Cytochrome P-450terp. Isolation and purification of the protein and cloning and sequencing of its operon. J Biol Chem 267:14193–14203
17. Bellamine A, Mangla AT, Nes WD, Waterman MR (1999) Characterization and catalytic properties of the sterol 14α-demethylase from *Mycobacterium tuberculosis*. Proc Natl Acad Sci U S A 96:8937–8942
18. McLean KJ, Warman AJ, Seward HE, Marshall KR, Girvan HM, Cheesman MR, Waterman MR, Munro AW (2006) Biophysical characterization of the sterol demethylase P450 from *Mycobacterium tuberculosis*, its cognate ferredoxin, and their interactions. Biochemistry 45:8427–8443
19. Tripathi S, Li H, Poulos TL (2013) Structural basis for effector control and redox partner recognition in cytochrome P450. Science 340:1227–1230
20. Green AJ, Munro AW, Cheesman MR, Reid GA, von Wachenfeldt C, Chapman SK (2003) Expression, purification and characterisation of a *Bacillus subtilis* ferredoxin: a potential electron transfer donor to cytochrome P450 BioI. J Inorg Biochem 93:92–99
21. Puchkaev AV, Ortiz de Montellano PR (2005) The *Sulfolobus solfataricus* electron donor partners of thermophilic CYP119: an unusual non-NAD(P)H-dependent cytochrome P450 system. Arch Biochem Biophys 434:169–177
22. Ricagno S, de Rosa M, Aliverti A, Zanetti G, Bolognesi M (2007) The crystal structure of FdxA, a 7Fe ferredoxin from *Mycobacterium smegmatis*. Biochem Biophys Res Commun 360:97–102
23. McLean KJ, Munro AW (2008) Structural biology and biochemistry of cytochrome P450 systems in *Mycobacterium tuberculosis*. Drug Metab Rev 40:427–446
24. Ewen KM, Kleser M, Bernhardt R (2011) Adrenodoxin: the archetype of vertebrate-type [2Fe-2S] cluster ferredoxins. Biochim Biophys Acta 1814:111–125
25. Müller JJ, Lapko A, Bourenkov G, Ruckpaul K, Heinemann U (2001) Adrenodoxin reductase-complex structure suggests electron transfer path in steroid biosynthesis. J Biol Chem 276:2786–2789
26. Hawkes DB, Slessor KE, Bernhardt PV, De Voss JJ (2010) Cloning, expression and purification of cindoxin, an unusual FMN-containing cytochrome P450 redox partner. ChemBioChem 11:1107–1114
27. Narhi LO, Fulco AJ (1986) Characterization of a catalytically self-sufficient 119,000-dalton cytochrome P450 monooxygenase induced by

barbiturates in *Bacillus megaterium*. J Biol Chem 261:7160–7169
28. Narhi LO, Fulco AJ (1987) Identification and characterization of two functional domains in cytochrome P-450 BM-3, a catalytically self-sufficient monooxygenase induced by barbiturates in *Bacillus megaterium*. J Biol Chem 262:6683–6690
29. Noble MA, Miles CS, Chapman SK, Lysek DA, Mackay AC, Reid GA, Hanzlik RP, Munro AW (1999) Roles of key active-site residues in flavocytochrome P450 BM3. Biochem J 339:371–379
30. Munro AW, Daff S, Coggins JR, Lindsay JG, Chapman SK (1996) Probing electron transfer in flavocytochrome P-450 BM3 and its component domains. Eur J Biochem 239:403–409
31. Black SD, Martin ST (1994) Evidence for conformational dynamics and molecular aggregation in cytochrome P450 102 (BM-3). Biochemistry 33:12056–12062
32. Neeli R, Girvan HM, Lawrence A, Warren MJ, Leys D, Scrutton NS, Munro AW (2005) The dimeric form of flavocytochrome P450 BM3 is catalytically functional as a fatty acid hydroxylase. FEBS Lett 579:5582–5588
33. Kitazume T, Haines DC, Estabrook RW, Chen B, Peterson JA (2007) Obligatory intermolecular electron-transfer from FAD to FMN in dimeric P450BM-3. Biochemistry 46:11892–11901
34. Girvan HM, Dunford AJ, Neeli R, Ekanem IS, Waltham TN, Joyce MG, Leys D, Curtis RA, Williams P, Fisher K, Voice MW, Munro AW (2011) Flavocytochrome P450 BM3 mutant W1046A is a NADH-dependent fatty acid hydroxylase: implications for the mechanism of electron transfer in the P450 BM3 dimer. Arch Biochem Biophys 507:75–85
35. Siddhanta U, Presta A, Fan B, Wolan D, Rouseau DL, Stuehr DJ (1998) Domain swapping in inducible nitric-oxide synthase. Electron transfer occurs between flavin and heme groups located on adjacent subunits in the dimer. J Biol Chem 273:18950–18958
36. Daff SN, Chapman SK, Turner KL, Holt RA, Govindaraj S, Poulos TL, Munro AW (1997) Redox control of the catalytic cycle of flavocytochrome P-450 BM3. Biochemistry 36:13816–13823
37. Murataliev MB, Klein M, Fulco AJ, Feyereisen R (1997) Functional interactions in cytochrome P450 BM3: flavin semiquinone intermediates, role of NADP(H), and mechanism of electron transfer by the flavoprotein domain. Biochemistry 36:8401–8412
38. Hanley SC, Ost TW, Daff S (2004) The unusual redox properties of flavocytochrome P450 BM3 flavodoxin domain. Biochem Biophys Res Commun 325:1418–1423
39. Gustafsson MC, Roitel O, Marshall KR, Noble MA, Chapman SK, Pessegueiro A, Fulco AJ, Cheesman MR, von Wachenfeldt C, Munro AW (2004) Expression, purification, and characterization of *Bacillus subtilis* cytochromes P450 CYP102A2 and CYP102A3: flavocytochrome homologues of P450 BM3 from *Bacillus megaterium*. Biochemistry 43:5474–5487
40. Chowdhary PK, Alemseghed M, Haines DC (2007) Cloning, expression and characterization of a fast self-sufficient P450: CYP102A5 from *Bacillus cereus*. Arch Biochem Biophys 468:32–43
41. Nakayama N, Takemae A, Shoun H (1996) Cytochrome P450foxy, a catalytically self-sufficient fatty acid hydroxylase of the fungus *Fusarium oxysporum*. J Biochem 119:435–440
42. Kitazume T, Tanaka A, Takaya N, Nakamura A, Matsuyama S, Suzuki T, Shoun H (2002) Kinetic analysis of hydroxylation of saturated fatty acids by recombinant P450foxy produced by an *Escherichia coli* expression system. Eur J Biochem 269:2075–2082
43. Correll CC, Batie CJ, Ballou DP, Ludwig ML (1992) Phthalate dioxygenase reductase: a modular structure for electron transfer from pyridine nucleotides to [2Fe-2S]. Science 258:1604–1610
44. De Mot R, Parrey AHA (2002) A novel class of self-sufficient cytochrome P450 monooxygenases in prokaryotes. Trends Microbiol 10:502–508
45. Warman AJ, Robinson JW, Luciakova D, Lawrence AD, Marshall KR, Warren MJ, Cheesman MR, Rigby SE, Munro AW, McLean KJ (2012) Characterization of *Cupriavidus metallidurans* CYP116B1 – a thiocarbamate herbicide oxygenating P450-phthalate dioxygenase reductase fusion protein. FEBS J 279:1675–1693
46. Hunter DJ, Roberts GA, Ost TW, White JH, Muller S, Turner NJ, Flitsch SL, Chapman SK (2005) Analysis of the domain properties of the novel cytochrome P450 RhF. FEBS Lett 579:2215–2220
47. Nagy I, Compernolle F, Ghys K, Vanderleyden J, De Mot R (1995) A single cytochrome-P-450 system is involved in degradation of the herbicides EPTC (S-ethyl dipropylthiocarbamate) and atrazine by *Rhodococcus* sp. strain ni86/21. Appl Environ Microbiol 61:2056–2060
48. Nagy I, Schoofs G, Compernolle F, Proost P, Vanderleyden J, De Mot R (1995) Degradation of the thiocarbamate herbicide EPTC (S-ethyl dipropylcarbamothioate) and biosafening by *Rhodococcus* sp. strain ni86/21 involves an inducible cytochrome P-450 system and aldehyde dehydrogenase. J Bacteriol 177:676–687
49. Celik A, Roberts GA, White JH, Chapman SK, Turner NJ, Flitsch SL (2006) Probing the substrate specificity of the catalytically self-sufficient cytochrome P450RhF from a *Rhodococcus* sp. Chem Commun 2006:4492–4494

50. Liu L, Schmid RD, Urlacher VB (2006) Cloning, expression, and characterization of a self-sufficient cytochrome P450 monooxygenase from *Rhodococcus ruber* DSM 44319. Appl Microbiol Biotechnol 72:876–882
51. Liu L, Schmid RD, Urlacher VB (2010) Engineering cytochrome P450 monooxygenase CYP116B3 for high dealkylation activity. Biotechnol Lett 32:841–845
52. Coelho PS, Wang ZJ, Ener ME, Baril SA, Kannan A, Arnold FH, Brustad EM (2013) A serine-substituted P450 catalyzes highly efficient carbene transfer to olefins in vivo. Nat Chem Biol 9:485–487
53. Butler CF, Peet C, Mason AE, Voice MW, Leys D, Munro AW (2013) Key mutations alter the cytochrome P450 BM3 conformational landscape and remove inherent substrate bias. J Biol Chem 288:25387–25399
54. Sabbadin F, Grogan G, Bruce NC (2013) LICRED: a versatile drop-in vector for rapid generation of redox-self-sufficient cytochromes P450. Methods Mol Biol 987:239–249
55. Seth-Smith HM, Rosser SJ, Basran A, Travis ER, Dabbs ER, Nicklin S, Bruce NC (2002) Cloning, sequencing, and characterization of the hexahydro-1,3,5-trinitro-1,3,5-triazine degradation gene cluster from *Rhodococcus rhodochrous*. Appl Environ Microbiol 68:4764–4771
56. Jackson RG, Rylott EL, Fournier D, Hawari J, Bruce NC (2007) Exploring the biochemical properties and remediation applications of the unusual explosive-degrading P450 system XplA/B. Proc Natl Acad Sci U S A 104:16822–16827
57. Rylott EL, Jackson RG, Sabbadin F, Seth-Smith HM, Edwards J, Chong CS, Strand SE, Grogan G, Bruce NC (2011) The explosive-degrading cytochrome P450 XplA: biochemistry, structural features and prospects for bioremediation. Biochim Biophys Acta 1814:230–236
58. Sabbadin F, Jackson R, Haider K, Tampi G, Turkenburg JP, Hart S, Bruce NC, Grogan G (2009) The 1.5-Å structure of XplA-heme, an unusual cytochrome P450 heme domain that catalyzes reductive biotransformation of royal demolition explosive. J Biol Chem 284:28467–28475
59. Raag R, Martinis SA, Sligar SG, Poulos TL (1991) Crystal structure of the cytochrome P-450CAM active site mutant Thr252Ala. Biochemistry 30:11420–11429
60. Bui SH, McLean KJ, Cheesman MR, Bradley JM, Rigby SE, Levy CW, Leys D, Munro AW (2012) Unusual spectroscopic and ligand binding properties of the cytochrome P450-flavodoxin fusion enzyme XplA. J Biol Chem 287:19699–19714
61. McCarthy AA, Walsh MA, Verma CS, O'Connell DP, Reinhold M, Yalloway GN, D'Arcy D, Higgins TM, Voordouw G, Mayhew SG (2002) Crystallographic investigation of the role of aspartate 95 in the modulation of the redox potentials of *Desulfovibrio vulgaris*. Biochemistry 41:10950–10962
62. Seth-Smith HM, Edwards J, Rosser SJ, Rathbone DA, Bruce NC (2008) The explosive-degrading cytochrome P450 system is highly conserved among strains of *Rhodococcus* spp. Appl Environ Microbiol 74:4550–4552
63. Jackson CJ, Lamb DC, Marczylo TH, Warrilow AG, Manning NJ, Lowe DJ, Kelly DE, Kelly SL (2002) A novel sterol 14α-demethylase/ferredoxin fusion protein from *Methylococcus capsulatus* represents a new class of the cytochrome P450 superfamily. J Biol Chem 277:46959–46965
64. Geer LY, Domrachev M, Lipman DJ, Bryant SH (2002) CDART: protein homology by domain architecture. Genome Res 12:1619–1623
65. Tsitsigiannis DI, Kowieski TM, Zarnowski R, Keller NP (2005) Three putative oxylipin biosynthetic genes integrate sexual and asexual development in *Aspergillus nidulans*. Microbiology 151:1809–1821
66. Tsitsigiannis DI, Keller NP (2006) Oxylipins act as determinants of natural product biosynthesis and seed colonization in *Aspergillus nidulans*. Mol Microbiol 59:882–892
67. Brodhun F, Gobel C, Hornung E, Feussner I (2009) Identification of PpoA from *Aspergillus nidulans* as a fusion protein of a fatty acid heme dioxygenase/peroxidase and a cytochrome P450. J Biol Chem 284:11792–11805
68. Brodhun F, Schneider S, Gobel C, Hornung E, Feussner I (2010) PpoC from *Aspergillus nidulans* is a fusion protein with only one active haem. Biochem J 425:553–565
69. Hansen BG, Mnich E, Nielsen KF, Nielsen JB, Nielsen MT, Mortensen UH, Larsen TO, Patil KR (2012) Involvement of a natural fusion of a cytochrome P450 and a hydrolase in mycophenolic acid biosynthesis. Appl Environ Microbiol 78:4908–4913
70. Lee DS, Yamada A, Sugimoto H, Matsunaga I, Ogura H, Ichihara K, Adachi S, Park SY, Shiro Y (2003) Substrate recognition and molecular mechanism of fatty acid hydroxylation by cytochrome P450 from *Bacillus subtilis* – crystallographic, spectroscopic, and mutational studies. J Biol Chem 278:9761–9767
71. Fujishiro T, Shoji O, Nagano S, Sugimoto H, Shiro Y, Watanabe Y (2011) Crystal structure of H_2O_2-dependent cytochrome $P450_{SP\alpha}$ with its bound fatty acid substrate: insight into the regioselective hydroxylation of fatty acids at the α position. J Biol Chem 286:29941–29950
72. Matsunaga I, Sumimoto T, Ueda A, Kusunose E, Ichihara K (2000) Fatty acid-specific, regiospecific, and stereospecific hydroxylation by cytochrome P450 (CYP152B1) from *Sphingomonas paucimobilis*: substrate structure required for α-hydroxylation. Lipids 35:365–371
73. Matsunaga I, Yokotani N, Gotoh O, Kusunose E, Yamada M, Ichihara K (1997) Molecular cloning

and expression of fatty acid α-hydroxylase from *Sphingomonas paucimobilis*. J Biol Chem 272:23592–23596

74. Matsunaga I, Ueda A, Fujiwara N, Sumimoto T, Ichihara K (1999) Characterization of the ybdT gene product of *Bacillus subtilis*: novel fatty acid β-hydroxylating cytochrome P450. Lipids 34:841–846
75. Rude MA, Baron TS, Brubaker S, Alibhai M, Del Cardayre SB, Schirmer A (2011) Terminal olefin (1-alkene) biosynthesis by a novel P450 fatty acid decarboxylase from *Jeotgalicoccus* species. Appl Environ Microbiol 77:1718–1727
76. Belcher J, McLean KJ, Matthews S, Woodward LS, Fisher K, Rigby SE, Nelson DR, Potts D, Baynham MT, Parker DA, Leys D, Munro AW (2014) Structure and biochemical properties of the alkene producing cytochrome P450 OleT $_{JE}$ (CYP152L1) from the *Jeotgalicoccus* sp. 8456 bacterium. J Biol Chem 289:6535–6550
77. Girhard M, Schuster S, Dietrich M, Dürre P, Urlacher VB (2007) Cytochrome P450 monooxygenase from *Clostridium acetobutylicum*: a new α-fatty acid hydroxylase. Biochem Biophys Res Commun 362:114–119
78. Liu Y, Wang C, Yan J, Zhang W, Guan W, Lu X, Li S (2014) Hydrogen peroxide-independent production of α-alkenes by OleT $_{JE}$ P450 fatty acid decarboxylase. Biotechnol Biofuels 7:28
79. Daiber A, Shoun H, Ullrich V (2005) Nitric oxide reductase (P450nor) from *Fusarium oxysporum*. J Inorg Biochem 99:185–193
80. Shoun H, Tanimoto T (1991) Denitrification by the fungus *Fusarium oxysporum* and involvement of cytochrome P-450 in the respiratory nitrite reduction. J Biol Chem 266:11078–11082
81. Usuda K, Toritsuka N, Matsuo Y, Kim DH, Shoun H (1995) Denitrification by the fungus *Cylindrocarpon tonkinense*: anaerobic cell growth and two isozyme forms of cytochrome P-450nor. Appl Environ Microbiol 61:883–889
82. Shoun H, Fushinobu S, Jiang L, Kim SW, Wakagi T (2012) Fungal denitrification and nitric oxide reductase cytochrome P450nor. Phil Trans R Soc B Biol Sci 367:1186–1194
83. Shiro Y, Fujii M, Iizuka T, Adachi S, Tsukamoto K, Nakahara K, Shoun H (1995) Spectroscopic and kinetic-studies on reaction of cytochrome P450nor with nitric oxide – implication for its nitric oxide reduction mechanism. J Biol Chem 270:1617–1623
84. Lehnert N, Praneeth VKK, Paulat F (2006) Electronic structure of iron(II)-porphyrin nitroxyl complexes: molecular mechanism of fungal nitric oxide reductase (P450nor). J Comput Chem 27:1338–1351
85. Kostanjevecki V, Leys D, Van Driessche G, Meyer TE, Cusanovich MA, Fischer U, Guisez Y, Van Beeumen J (1999) Structure and characterization of *Ectothiorhodospira vacuolata* cytochrome b_{558}, a prokaryotic homologue of cytochrome b_5. J Biol Chem 274:35614–35620
86. Altuve A, Silchenko S, Lee KH, Kuczera K, Terzyan S, Zhang X, Benson DR, Rivera M (2001) Probing the differences between rat liver outer mitochondrial membrane cytochrome b_5 and microsomal cytochromes b_5. Biochemistry 40:9469–9483
87. Spatz L, Strittmatter P (1973) A form of reduced nicotinamide adenine dinucleotide cytochrome b_5 reductase containing both the catalytic site and an additional hydrophobic membrane-binding segment. J Biol Chem 248:793–799
88. Guzov VM, Houston HL, Murataliev MB, Walker FA, Feyereisen R (1996) Molecular cloning, overexpression in *Escherichia coli*, structural and functional characterization of house fly cytochrome b_5. J Biol Chem 271:26637–26645
89. Funk WD, Lo TP, Mauk MR, Brayer GD, MacGillivray RT, Mauk AG (1990) Mutagenic, electrochemical, and crystallographic investigation of the cytochrome b_5 oxidation-reduction equilibrium: involvement of asparagine-57, serine-64, and heme propionate-7. Biochemistry 29:5500–5508
90. Hildebrandt A, Estabrook RW (1971) Evidence for the participation of cytochrome b_5 in hepatic microsomal mixed-function oxidation reactions. Arch Biochem Biophys 143:66–79
91. Correia MA, Mannering GJ (1973) Reduced diphosphopyridine nucleotide synergism of the reduced triphosphopyridine nucleotide-dependent mixed-function oxidase system of hepatic microsomes. II. Role of the type I drug-binding site of cytochrome P-450. Mol Pharmacol 9:470–485
92. Sang-Choul I, Waskell L (2011) The interaction of microsomal cytochrome P450 2B4 with its redox partners, cytochrome P450 reductase and cytochrome b_5. Arch Biochem Biophys 507:144–153
93. Katagiri M, Kagawa N, Waterman MR (1995) The role of cytochrome b_5 in the biosynthesis of androgens by human P450C17. Arch Biochem Biophys 317:343–347
94. Akhtar M, Wright JN, Lee-Robichaud P (2011) A review of mechanistic studies on aromatase (CYP19) and 17α-hydroxylase-17,20-lyase (CYP17). J Steroid Biochem Mol Biol 125:2–12
95. Storbeck KH, Swart AC, Goosen P, Swart P (2013) Cytochrome b_5: novel roles in steroidogenesis. Mol Cell Endocrinol 371:87–99
96. Finn RD, McLaughlin LA, Ronseaux S, Rosewell I, Houston JB, Henderson CJ, Wolf CR (2008) Defining the in vivo role for cytochrome b_5 in cytochrome P450 function through the conditional hepatic deletion of microsomal cytochrome b_5. J Biol Chem 283:31385–31393
97. McLaughlin LA, Ronseaux S, Finn RD, Henderson CJ, Wolf CR (2010) Deletion of microsomal cytochrome b_5 profoundly affects hepatic and extrahepatic drug metabolism. Mol Pharmacol 78:269–278

98. Ichinose H, Wariishi H (2012) Heterologous expression and mechanistic investigation of a fungal cytochrome P450 (CYP5150A2): involvement of alternative redox partners. Arch Biochem Biophys 518:8–15
99. Syed K, Kattamuri C, Thompson TB, Yadav JS (2011) Cytochrome b_5 reductase-cytochrome b_5 as an active P450 redox enzyme system in *Phanerochaete chrysosporium*: atypical properties and in vivo evidence of electron transfer capability to CYP63A2. Arch Biochem Biophys 509:26–32
100. Henderson CJ, McLaughlin LA, Wolf CR (2013) Evidence that cytochrome b_5 and cytochrome b_5 reductase can act as sole electron donor to the hepatic cytochrome P450 system. Mol Pharmacol 83:1209–1217
101. Noble MA, Girvan HM, Smith SJ, Smith WE, Murataliev M, Guzov VM, Feyereisen R, Munro AW (2007) Analysis of the interactions of cytochrome b_5 with flavocytochrome P450 BM3 and its domains. Drug Metab Rev 39:599–617
102. Nelson DR, Goldstone JV, Stegeman JJ (2013) The cytochrome P450 genesis locus: the origin and evolution of animal cytochrome P450s. Philos Trans R Soc Lond B Biol Sci 368(1612):20120474
103. Munro AW, Girvan HM, McLean KJ (2007) Variations on a (t)heme – novel mechanisms, redox partners and catalytic functions in the cytochrome P450 superfamily. Nat Prod Rep 24:585–609
104. Sadeghi SJ, Gilardi G (2013) Chimeric P450 enzymes: activity of artificial redox fusions driven by different reductases for biotechnological applications. Biotechnol Appl Biochem 60:102–110
105. Munro AW, Girvan HM, McLean KJ (2007) Cytochrome P450 – redox partner fusion enzymes. Biochim Biophys Acta 1770:345–359
106. Hamdane D, Xia C, Im SC, Zhang H, Kim JJ, Waskell L (2009) Structure and function of an NADPH-cytochrome P450 oxidoreductase in an open conformation capable of reducing cytochrome P450. J Biol Chem 284:11374–11378
107. Filenko N, Spiro S, Browning DF, Squire D, Overton TW, Cole J, Constantinidou C (2007) The NsrR regulon of *Escherichia coli* K-12 includes genes encoding the hybrid cluster protein and the periplasmic, respiratory nitrite reductase. J Bacteriol 189:4410–4417
108. Macedo S, Mitchell EP, Romão CV, Cooper SJ, Coelho R, Liu MY, Xavier AV, LeGall J, Bailey S, Garner DC, Hagen WR, Teixeira M, Carrondo MA, Lindley P (2002) Hybrid cluster proteins (HCPs) from *Desulfovibrio desulfuricans* ATCC 27774 and *Desulfovibrio vulgaris* (Hildenborough): X-ray structures at 1.25 Å resolution using synchrotron radiation. J Biol Inorg Chem 7:514–525

Cytochrome P450$_{cin}$ (CYP176A1) 12

Jeanette E. Stok, Kate E. Slessor, Anthony J. Farlow,
David B. Hawkes, and James J. De Voss

Abstract

Cytochrome P450$_{cin}$ (P450$_{cin}$) (CYP176A1) is a bacterial P450 enzyme that catalyses the enantiospecific hydroxylation of 1,8-cineole to (1*R*)-6β-hydroxycineole when reconstituted with its natural reduction-oxidation (redox) partner cindoxin, *E. coli* flavodoxin reductase, and NADPH as a source of electrons. This catalytic system has become a useful tool in the study of P450s as not only can large quantities of P450$_{cin}$ be prepared and rates of oxidation up to 1,500 min^{-1} achieved, but it also displays a number of unusual characteristics. These include an asparagine residue in P450$_{cin}$ that has been found in place of the usual conserved threonine residue observed in most P450s. In general, this conserved threonine controls oxygen activation to create the potent ferryl (Fe(IV=O) porphyrin cation radical required for substrate oxidation. Another atypical characteristic of P450$_{cin}$ is that it utilises an FMN-containing redoxin (cindoxin) rather than a ferridoxin as is usually observed with other bacterial P450s (e.g. P450$_{cam}$). This chapter will review what is currently known about P450$_{cin}$ and how this enzyme has provided a greater understanding of P450s in general.

Keywords

Cytochrome P450$_{cin}$ • CYP176A1 • Cindoxin • Cindoxin reductase • Ferric hydroperoxy species • Ferryl porphyrin radical species

J.E. Stok • K.E. Slessor • A.J. Farlow • D.B. Hawkes
J.J. De Voss (✉)
School of Chemistry and Molecular Biosciences,
University of Queensland, St. Lucia, Brisbane 4072,
Australia
e-mail: j.devoss@uq.edu.au

E.G. Hrycay and S.M. Bandiera (eds.), *Monooxygenase, Peroxidase and Peroxygenase Properties and Mechanisms of Cytochrome P450*, Advances in Experimental Medicine and Biology 851,
DOI 10.1007/978-3-319-16009-2_12,

12.1 Introduction

Cytochrome P450s (P450s[1]) are a vast group of enzymes that typically catalyse the hydroxylation of an unactivated carbon atom within a molecule. Over 12,000 different P450s have been identified via sequencing and are widely distributed across both prokaryotes and eukaryotes. The 57 human P450s associated with xenobiotic metabolism and endogenous biosynthesis have, of course, drawn much attention because of their importance in drug metabolism and activation and in various human diseases [1]. However, our understanding of the mechanism of these enzymes has been greatly facilitated by investigations of a number of bacterial P450s that were experimentally more tractable.

Generally, P450s withdraw an oxygen atom that has been derived from molecular oxygen and insert it into a substrate. This requires the substrate to bind to the P450 active site, concomitantly displacing the water molecule typically coordinated to the P450 heme iron with a concomitant shifting of the Fe(III) from a low-spin to high-spin state (Fig. 12.1**A** to **B**). The P450 enzyme then accepts the first electron from its reduction-oxidation (redox) partner, which reduces the heme Fe(III) to Fe(II) and allows dioxygen coordination (Fig. 12.1**B** to **D**). The delivery of a second electron to yield the ferric peroxo species is followed by addition of a proton to form a ferric hydroperoxy (Fe(III)-OOH) complex (Fig. 12.1**D** to **F**) [2]. It has been proposed that in some cases, the ferric-peroxo or the ferric hydroperoxy species may be responsible for substrate oxidation but in general, the oxidation of the substrate is performed by the ferryl (Fe(IV=O) porphyrin cation radical species (Fig. 12.1**G**). To generate this ferryl species, the distal (furthest from ferric iron) oxygen atom of the ferric hydroperoxy (Fe(III)-OOH) species is protonated, leading to the release of a water molecule and scission of the oxygen-oxygen bond. The extremely powerful oxidising species thus produced then inserts an oxygen atom into the substrate and the newly formed product is subsequently released (Fig. 12.1**G** to **A**) [2]. It is thought that a highly conserved threonine in the P450 active site is instrumental in directing and controlling the protonation of the ferric hydroperoxy species (Fig. 12.1**F** to **G**) by facilitating proton delivery to the distal rather than the proximal (adjacent to ferric iron) oxygen atom [2].

One of the first P450s to be extensively characterised, and the first for which a crystal structure was available, was $P450_{cam}$ (CYP101A1) isolated from *Pseudomonas putida* [3, 4]. Together with $P450_{BM3}$ (CYP102A1), the study of $P450_{cam}$ has been crucial to the general mechanistic understanding of P450s. $P450_{cam}$ was an ideal model system as it was readily available in large quantities, was catalytically very active, utilised a defined substrate ((1*R*)-camphor), and yielded a three-dimensional crystal structure that was obtained relatively early in its study. $P450_{cam}$ was therefore instrumental in facilitating the elucidation of the reaction mechanisms of P450s [5] with a spectacular series of low temperature X-ray structures providing snapshots of the catalytic cycle [6]. In addition, the study of $P450_{cam}$ served to establish the "universal" features of P450s [7].

A second bacterial P450 enzyme that has also been important in both our structural and mechanistic understanding of P450s is $P450_{BM3}$ [8]. This enzyme provides a unique perspective on P450s as the heme-containing domain is fused to the redox partner domain. This enzymatic structural fusion characteristic delivers the electrons required by $P450_{BM3}$ to activate molecular oxygen and oxidise its substrate. This is an especially interesting bacterial system both because of its high catalytic activity and because the redox domain contains the two flavin cofactors commonly found in eukaryotic systems.

Overall, both $P450_{cam}$ and $P450_{BM3}$ have provided significant insights into common features of P450 structure and reaction mechanisms. However, a number of features specific to each P450 enzyme system were also found, such as a unique potassium binding site in $P450_{cam}$ [9]. With more exemplar P450s, our overall understanding of their reaction mechanisms is enhanced and truly universal

[1] *Abbreviations*: *P450 or CYP* cytochrome P450, *Fld E. coli* flavodoxin, *FdR E. coli* flavodoxin reductase, *GC/MS* gas chromatography/mass spectrometry, *Pdx* putidaredoxin, *PdR* putidaredoxin reductase, *CPR* NADPH-cytochrome P450 oxidoreductase, *CdR* cindoxin reductase, *Cdx* cindoxin, *HPLC* high-performance liquid chromatography, E_1 quinone/semiquinone, E_2 semiquinone/hydroquinone, *EPR* electron paramagnetic resonance, *ENDOR* electron nuclear double resonance.

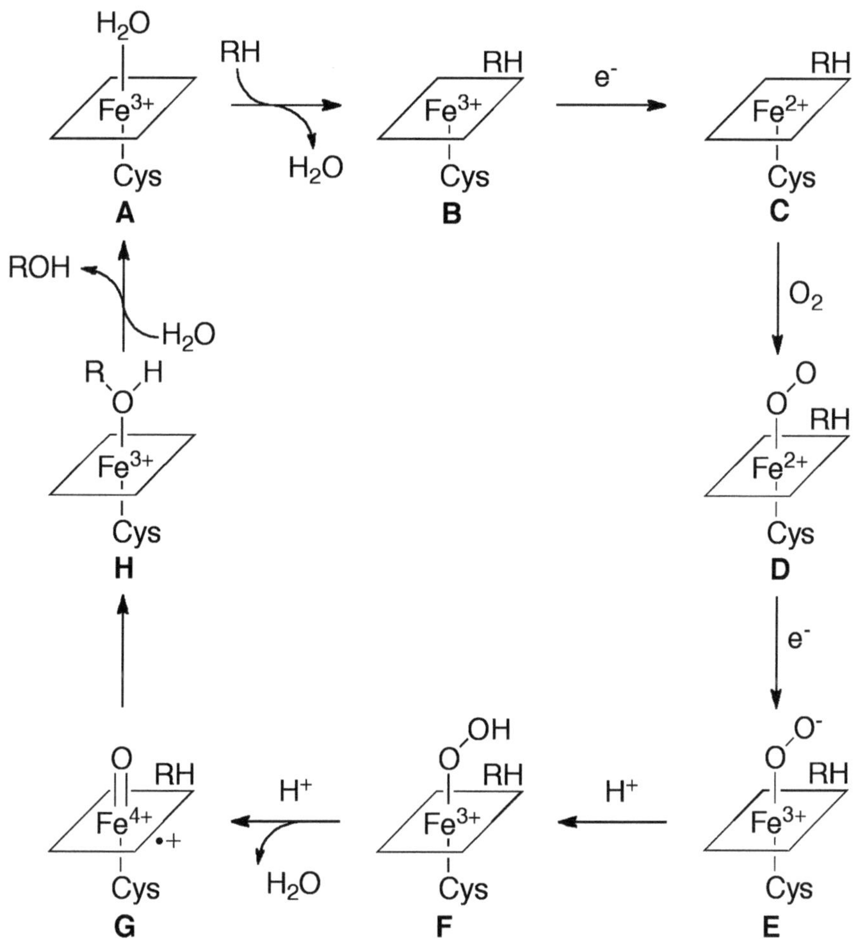

Fig. 12.1 Catalytic cycle for the oxidation reaction catalysed by P450 enzymes. *Abbreviations*: *Cys* conserved cysteine with thiolate coordinated to the heme iron, *RH* substrate, e^- electron

characteristics of this fascinating family of enzymes can be determined.

$P450_{cin}$ (CYP176A1) is a bacterial P450 enzyme that catalyses the enantiospecific hydroxylation of 1,8-cineole to (1*R*)-6-β-hydroxycineole[2] **2a** (Fig. 12.2). $P450_{cin}$ was originally isolated to explore its potential as an additional model P450 system for comparison with $P450_{cam}$ and $P450_{BM3}$, and subsequently $P450_{cin}$ has become a valuable tool in the investigation of P450 oxidation chemistry. *Citrobacter braakii*, the organism from which $P450_{cin}$ was isolated, can survive on cineole **1** as its sole carbon and energy source. As a consequence, $P450_{cin}$ was also specifically

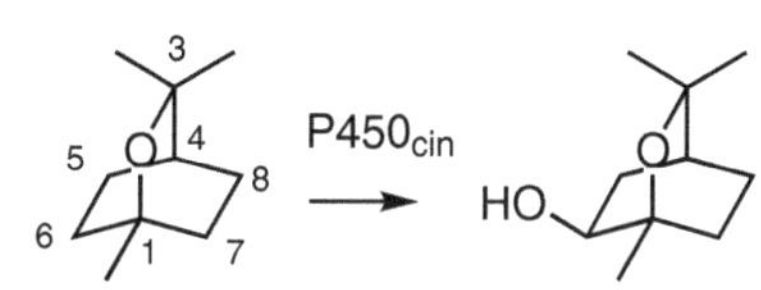

1 2a

Fig. 12.2 The enantiospecific hydroxylation of 1,8-cineole **1** to (1*R*)-6β-hydroxycineole **2a** as catalysed by $P450_{cin}$

[2] There is no consistency for naming hydroxycineoles in the literature. In order to discuss stereochemistry we use descriptors *α* and *β*. Employing a plane that passes through C5, C6, C7 and C8 of the molecule (See Fig. 12.2), we term any substituents that lie below the plane *α* and any substituents that are above this plane *β*. Three new stereogenic centres at C1, C4 and C6 are created following the hydroxylation of the *meso* cineole at either one of the carbons that lie adjacent to the C1 bridgehead. To differentiate the entantiomers formed following oxidation we have described the *pro-R* carbon as the carbon atom that leads to the *R*-C1 isomer and the pro-S carbon as that leading to the *S*-C1 isomer.

Table 12.1 General comparison of $P450_{cam}$ and $P450_{cin}$

		$P450_{cam}$ [67]	$P450_{cin}$ [10]
Organism		*Pseudomonas putida*	*Citrobacter braakii*
Abs. max. (nm)	No substrate	418	415
	Plus substrate	392	392
Substrate		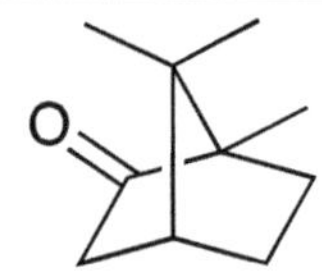	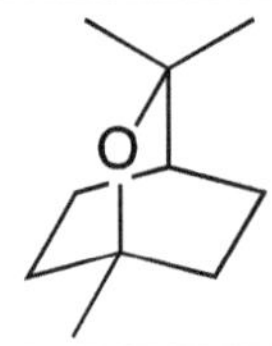
Binding constant (K_d) (μM)		1.6 ± 0.3 [41]	0.7
NAD(P)H consumption[a] (μM/min/μM P450)		847 [46]	218–891 [15]
Coupling[b]		95 [41]	80 ± 2 [12]
H-bond (regio/stereo control)		Tyr96 [68]	Asn242 [13]
H-bond (oxygen activation)		Thr252 [11]	Unknown [13]
Redox partners (cofactors)		$Pd(Fe_2S_2)$/PdR(FAD)	Cd(FMN)/CdR(FAD)[c]
Redox potentials (mV)	No substrate	$E_{m7.0}$ −307 [32]	$E_{m7.4}$ −182 [33]
			$E_{m7.4}$ −330 [14]
	Plus substrate	$E_{m7.0}$ −173 [32]	$E_{m7.4}$ −167 [33]
			$E_{m7.4}$ −202 [14]

[a]Ratio- $P450_{cam}$:PdR:Pdx (1:2:10); $P450_{cin}$:FdR:Cdx (1:2:8)–(1:10:20)

[b]Ratio of the amount of NADPH consumed compared to the amount of product formed

[c]Putative; from sequence homology

examined to determine its role in the biodegradation of cineole. In addition, as $P450_{cin}$ is a soluble, bacterial P450 that catalyses the hydroxylation of a small molecule, it was an ideal enzyme system in which to evaluate P450 biocatalytic capabilities. Overall, $P450_{cin}$ provides a distinct view on the chemistry of P450s and interestingly demonstrates a number of exceptions to what are considered universal P450 characteristics.

12.2 Initial Discoveries

$P450_{cin}$ was isolated from *C. braakii*, a gram-negative bacterial species that was found in a soil sample taken from beneath eucalyptus trees [10]. Initially, $P450_{cin}$ was purified from *C. braakii* that had been grown in minimal media containing cineole as its sole carbon source and it was demonstrated to be a P450 via carbon monoxide difference spectroscopy. Most P450s with a specific substrate, such as biosynthetic or biodegradative enzymes, are observed to undergo a shift in their resting UV-Visible absorbance from 417 (low-spin Fe(III)) to 392 nm (high-spin Fe(III)) upon substrate binding as a result of displacement of the water molecule bound to the heme iron. This characteristic shift (417–392 nm) was observed upon the addition of cineole to purified $P450_{cin}$ and provided the first indication that cineole was the natural substrate for this enzyme. In addition, the calculated affinity of $P450_{cin}$ for cineole was observed to be high (K_d 0.7 μM; Table 12.1) and further supported the conclusion that cineole is the substrate for $P450_{cin}$. The gene encoding $P450_{cin}$ (*cinA*) was then isolated from *C. braakii* and following cloning into the expression vector pCW, the hemeprotein was heterologously expressed in *Escherichia coli* in the presence of cineole. As anticipated, the endogenous *E. coli*

P450cin	232	FTILLLGGI**D**NTARFLSSVFWRLA	255
P450cam	242	CGLLLVGGL**DT**VVNFLSFSMEFLA	265
CYP19A1	300	ILEMLIAAP**DT**MSVSLFFMLFLIA	323
CYP17A1	296	IGDIFGAGV**ET**TTSVVKWTLAFLL	319
CYP1A2	309	VNDIFGAGF**ET**VTTAIFWSILLLV	332
		**	

NCBI Accession numbers: P450$_{cin}$ -AAL57614; P450$_{cam}$ -P00183; CYP19A1 -P11511; CYP1A2 -P04799; CYP17A1 -P05093.

Fig. 12.3 Multiple sequence alignment of selected P450s highlighting conserved threonine and aspartic/glutamic acid pair (**) [48]

flavodoxin (Fld)/flavodoxin reductase (FdR) proteins acted as the requisite redox partners for P450$_{cin}$, although poorly, and delivered the electrons necessary for catalysis. Gas chromatography/mass spectrometry (GC/MS) analysis of the organic extract of growth media from *E. coli* expressing P450$_{cin}$ revealed small amounts of 6-hydroxycineole **2** (Fig. 12.2). At this stage, the stereochemistry of **2** was unknown.

Comparative sequence analysis with known P450s indicated that *cinA* exhibited a number of typical P450 features such as the conserved cysteinyl sulphur (Cys347) for heme ligation and other amino acids surrounding this cysteine (e.g. Gly344 and Gly349) [10]. However, P450$_{cin}$ was surprisingly found to lack the conserved threonine that is believed to play a key role in proton delivery for oxygen activation during the P450 catalytic cycle (Fig. 12.3). The equivalent position in P450$_{cin}$ is occupied by an asparagine (Asn242) rather than a threonine. It was initially proposed that this asparagine could possibly be a functional replacement of the threonine. Indeed, a P450$_{cam}$ mutant in which its catalytically-important threonine was replaced with an asparagine (T252N) was found to be functional [11]. However, subsequent site-directed mutagenesis studies with P450$_{cin}$ have revealed that this residue has other functions in this enzyme (*vide infra*) [12, 13].

A number of other genes were also discovered alongside *cinA* on complete analysis of the CIN operon (Fig. 12.4) [10]. In subsequent studies, these genes were found to encode the redox partners cindoxin (Cdx; *cinC*) and the putative cindoxin reductase (CdR; *cinB*) [14, 15] as well as (1*R*)-6β-hydroxycineole dehydrogenase (*cinD*) [16]. The proteins encoded by these genes will be discussed in detail later.

12.3 Structure

The X-ray crystal structure of P450$_{cin}$ was solved in 2004 [17], revealing that this protein was folded in a similar way to most other P450s, with four conserved helical bundles making up the core of the protein. Most P450s differ significantly from each other in the architecture of the substrate-binding pocket above the heme. P450$_{cin}$ is no exception and has converted the B′ helix, typically found in other P450s to be important in substrate binding, to an ordered loop. This loop does not directly interact with the substrate but Tyr81, which is located within it, is utilised via a water molecule to indirectly

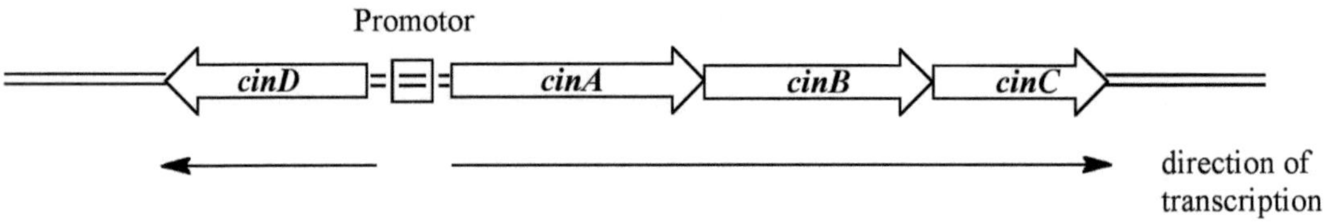

Fig. 12.4 Map of the CIN operon region that contains four open reading frames, *cinA, cinB, cinC* and *cinD* [16]

stabilise Asn242, which in turn directly interacts with the substrate. Asn242 is the only amino acid within the protein that directly binds to the substrate cineole. A hydrogen bond is formed from the terminal amide of Asn242 located in the I-helix to the ethereal oxygen atom of cineole. A number of non-polar interactions of the substrate with the protein, including those with Thr77, Leu88, Leu237, Val386, and Ala285, are also important and these residues form a hydrophobic pocket for cineole.

It has been observed that many P450s have both an open and closed form depending on whether the substrate is absent (open) or present (closed). Initially, it was thought that $P450_{cin}$ only existed in a closed conformation [12] but evidence from crystal structures has recently revealed the presence of an open conformation of the enzyme [18]. The open conformation contains a channel by which cineole can access the active-site heme. The F, G and I helices and the B-C loop shift considerably in the presence of cineole in order to create the substrate and oxygen binding pockets. In the "open" form of the enzyme, Tyr81 that is crucial in the hydrogen-bonding network of the substrate-bound $P450_{cin}$ no longer lies within the active site.

Nitric oxide complexed to substrate-bound $P450_{cin}$ was used as a mimic for oxy-$P450_{cin}$ due to the difficulty in obtaining a crystal structure of dioxygen-bound $P450_{cin}$ [18]. In the open conformation of $P450_{cin}$, a hydrogen bond is present between the carbonyl oxygen atom of Gly238 and the water molecule coordinated to the heme iron. The $P450_{cin}$–NO substrate-bound structure revealed that although Gly238 shifts slightly, it could still interact with nitric oxide. Based on these observations, it has been proposed that Gly238 may be important in assisting the protonation of the distal oxygen atom during the catalytic cycle. Unlike other P450s, $P450_{cin}$ does not have the conserved threonine that is normally responsible for stabilising the ferric hydroperoxy intermediate (Fig. 12.1**F**) and thus controlling dioxygen activation. Other amino acids such as Asn242 have been investigated to determine whether they are involved in oxygen activation but no direct influence has been observed (*vide infra*: Sect. 12.5) [12, 13]. Gly238 may therefore be the residue required for controlling protonation of the bound dioxygen during its activation during $P450_{cin}$–catalysed hydroxylation of cineole.

An additional structurally-conserved feature of P450s is a phenylalanine that resides next to the proximal heme-bound cysteine. In $P450_{cin}$, this is replaced with a leucine (Leu340) and this residue directly interacts with the heme [17]. In $P450_{BM3}$, it has been demonstrated that the phenylalanine is crucial in thermodynamic control of the heme iron redox potential and enables the P450 to efficiently carry out substrate oxidation [19]. It is proposed that Leu340 plays a similar thermodynamic role in $P450_{cin}$ although this has not yet been substantiated [17].

12.4 Electron Transport System

Historically, the auxiliary redox proteins responsible for delivering electrons to P450s in order for oxidation to occur were divided into two classes [20]. The first (Class I: three component system) is usually found in bacterial and mitochondrial systems and consists of two proteins, an NADH-containing reductase and a ferridoxin that together passage electrons from NADH to the P450. The electron transport system utilised by $P450_{cam}$ falls into this class with both the ferridoxin, putidaredoxin (Pdx), and the corresponding reductase, putidaredoxin reductase (PdR), being required for catalysis.

Substitution with other redox partners results in dramatically- lowered activity [5, 21]. Pdx/PdR have also been utilised with mixed success to support catalysis by other P450s for which cognate redox partners were unavailable [22–24]. Conversely, eukaryotic P450s generally have just one redox partner (CPR) that funnels electrons from NADPH via FMN and FAD domains of the reductase before transferring them to the P450 enzyme (Class II: two component system). Typically, this reductase operates with a number of different P450s. In humans, there is a single CPR flavoprotein responsible for electron delivery to the majority of P450s rather than a specialised system for each P450 enzyme [20]. However, over the last 10 years the variety of P450 electron transport systems that have been observed have multiplied significantly, with at least ten different types currently recognised [25].

In *C. braakii*, it is believed that electrons are delivered to $P450_{cin}$ via two proteins. A flavodoxin reductase, CdR (*cinB*), was identified by sequence homology from analysis of the CIN operon and appears to be a 49 kDa protein that is similar to bovine adrenodoxin reductase (31 % identity, 47 % similarity) [10]. Bovine adrenodoxin reductase is an FAD-containing ferridoxin reductase that is responsible for electron delivery to adrenodoxin that in turn transfers electrons to mitochondrial P450s [26]. Unfortunately, despite considerable effort, heterologous expression of CdR in *E. coli* has not been successful [15] impeding any further characterisation of this enzyme.

The second of the two redox partners responsible for providing the electrons required for oxidation by $P450_{cin}$ in *C. braakii* is a flavodoxin, Cdx. Analysis of the CIN operon revealed that Cdx (*cinC*) was a 16 kDa protein that was highly similar to the FMN-domain of human CPR (37 % identity, 56 % similarity) [10]. This was the first indication that $P450_{cin}$ utilised an FMN-containing reductase rather than a ferridoxin typically used in the Class I systems of other bacterial P450s such as $P450_{cam}$ (Table 12.1). To fully characterise Cdx, it was cloned into the expression vector pCW, heterologously expressed in *E. coli* and purified [14, 15]. A combination of mass spectrometry and HPLC confirmed that Cdx did indeed contain an FMN cofactor [15].

As CdR itself was not available, a substitute redox partner was required in order to both characterise the role of Cdx and produce a catalytically-active oxidation system with $P450_{cin}$. It has previously been observed that the redox pair of Fld and FdR found in *E. coli* could be utilised to provide electrons to a number of heterologously expressed P450s [27]. Additionally, expression of $P450_{cin}$ in *E. coli* in the presence of its substrate cineole **1** had produced small amounts of a cineole oxidation product, 6-hydroxycineole **2** (Fig. 12.2) [10]. It was believed that in the absence of its native redox partners, $P450_{cin}$ had recruited both *E. coli* Fld and FdR to provide the necessary electrons for oxygen activation and substrate oxidation. *E. coli* FdR was therefore explored as a potential redox partner and was found to efficiently reduce Cdx, thereby producing a redox system that could be utilised with $P450_{cin}$ for catalytic oxidation (*vide infra*).

Analysis of the crystal structure of $P450_{cin}$ revealed that Cdx is likely to bind to the proximal face of $P450_{cin}$ where it has direct access to the heme [17]. Although different cofactors are utilised by Pdx ($P450_{cam}$) and Cdx, it was predicted that the interaction between these redoxins and their respective P450s may be similar. Like most P450s, $P450_{cin}$ was found to have an electropositive patch over the heme where the electronegative redoxin is likely to bind. In addition, Arg112 in $P450_{cam}$ has been implicated in both the interaction with Pdx and to play a role in electron transfer [28–30]. Arg102 is found in a similar location in the three-dimensional structure of $P450_{cin}$ as compared to $P450_{cam}$ and was postulated to be the functionally-equivalent amino acid in this system [17]. During the preparation of this document, the crystal structure of Cdx has been published [31], confirming both the importance of the Arg102 residue through both mutagenesis and docking studies and highlighting a second essential residue, Arg346. It is believed that both of these residues, Arg102 and Arg346, are involved in the binding of $P450_{cin}$ to Cdx. Arg346 in particular may also play a role in electron transfer. In Cdx, Tyr96 is a unique residue among flavodoxins and extends into solution. Tyr96 is believed to interact with Arg102 in $P450_{cin}$ and even though Arg102 is positioned between the FMN and heme cofactors of the two proteins in models, kinetic and mutagenesis studies have implicated Arg102 in binding and not in electron transfer. Also highlighted was the fact that while Pdx has a significant effector role in $P450_{cam}$ catalysis, Cdx does not play a role in inducing functionally-significant structural changes in $P450_{cin}$. This was attributed to differences in active-site architecture correlated with the mechanism of proton delivery to the bound dioxygen.

The chemistry of electron transfer between Cdx and $P450_{cin}$ also appears to be somewhat atypical for P450s. The semiquinone/hydroquinone (E_2) couple of Cdx is unusually high (E_2 –218 mV at pH 7.5) when compared to other P450 redox partners [15]. Therefore, possibly both redox couples of Cdx, the quinone/semiquinone (E_1) and E_2 are catalytically relevant rather than just one couple as is usually observed with other P450 reductase systems [14, 15]. The rate of electron transfer from Cdx to $P450_{cin}$ is comparable to other P450 systems (15–32 s^{-1}) [14, 15] and is significantly affected by an increase in buffer salt, which is thought to alter the electrostatic interaction between the two proteins [14]. Electrochemical studies indicated that the first electron transfer was coupled to a proton transfer, whereas the second electron transfer from Cdx did not involve proton transfer [15].

The redox potential ($Fe^{III/II}$) for $P450_{cin}$ in the presence of its substrate cineole ($E_{m7.4}$ –202 mV [14]) was observed to be essentially identical to potentials previously reported for other P450s calculated potentiometrically (Table 12.1; $P450_{cam}$: $E_{m7.0}$ –173 mV) [32]. However, contradictory results have been reported for the redox potentials of $P450_{cin}$ in the absence of its substrate. Notably, the redox potential of $P450_{cin}$ shifts +128 mV upon binding of cineole as measured potentiometrically (Substrate-free $P450_{cin}$ $E_{m7.4}$ –330 mV) [14]. In contrast, utilising both potentiometric and cyclic voltametric methods, it was separately observed that the difference in the redox potential of $P450_{cin}$ in the presence or absence of its substrate did not significantly change [33]. If this is the case and the redox potential does not change upon substrate binding, $P450_{cin}$ is unique among bacterial P450s so far studied. It also implies that the change in redox potential is not important in controlling the oxidation state of $P450_{cin}$.

Cdx is also able to both deliver electrons to P450s other than $P450_{cin}$ and in some cases support catalytic turnover. One of these P450s is the bacterial $P450_{BioI}$, which failed to produce a catalytically-active system with the redox partners of $P450_{cam}$ (Pdx/PdR) although it did accept an electron to produce a ferrous heme species [34]. However, the Cdx/FdR couple was observed both to successfully transfer electrons from NADPH to $P450_{BioI}$ and to support catalytic turnover [34]. Additionally, two mammalian P450s (CYP2C10 and CYP2E1) have been observed to accept electrons from Cdx/FdR via NADPH [35]. This was monitored by analysing the reduction of these P450s in the presence of carbon monoxide, thus producing a characteristic peak at 450 nm in the UV-Visible spectrum. However, attempts to generate a catalytically-active system with $P450_{cam}$ and Cdx/FdR were unsuccessful and no product formation was observed [36]. Alternatively, it was shown that Pdx/PdR at high concentrations relative to $P450_{cin}$ ($P450_{cin}$:PdR:Pdx 1:80:20) can deliver electrons via NADH to $P450_{cin}$, albeit at a significantly reduced efficiency (18 % of the reducing equivalents directed toward product formation) [36].

12.5 The Oxidation of Cineole

$P450_{cin}$ was shown to efficiently and enantiospecifically catalyse the hydroxylation of 1,8-cineole **1** to produce (1*R*)-6β-hydroxycineole **2a** (Table 12.1; Fig. 12.2) when reconstituted with Cdx, FdR and a source of electrons in the form of NADPH [15]. Structural data obtained via X-ray crystallography indicates that the pro-*R* methylene of **1** is located 5 Å above the heme iron, in the ideal position to produce the (1*R*)-

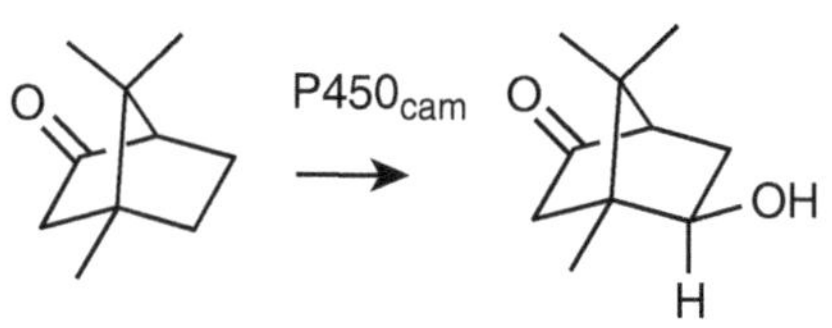

3 **4**

Fig. 12.5 P450$_{cam}$ catalysed oxidation of (1*R*)-camphor **3** to 5-*exo*-hydroxycamphor **4** [5]

6β-hydroxycineole **2a** that is observed experimentally [12]. The production of (1*R*)-6-β-hydroxycineole **2a** by P450$_{cin}$ was found to be highly coupled with 80 % of the reducing equivalents (derived from NADPH) being consumed to form the product **2a** [15]. Similar results have been observed with other P450s such as P450$_{cam}$ where consumed NADH is 95 % coupled to the production of 5-*exo*-hydroxycamphor **4** from (1*R*)-camphor **3** (Table 12.1; Fig. 12.5). With P450$_{cin}$, 20 % of the reducing equivalents were not utilized productively in substrate oxidation and were used either for the production of hydrogen peroxide or water [15]. Under ideal conditions, the maximum rate of the reaction reached approximately 1,500 μM min^{-1} per μM of P450$_{cin}$ [15]. It should be noted that these rate and coupling data were determined using *E. coli* FdR in a reconstituted system rather than the unavailable, proposed natural redox partner, CdR.

12.5.1 Asn242: Importance of the Hydrogen Bond

Initial sequence alignments of P450$_{cin}$ with other P450s revealed that it lacked the conserved threonine that has been implicated in directing protonation of the distal oxygen atom of the bound dioxygen during the catalytic cycle of other P450s (Fig. 12.1**E** to **F**) [10]. In P450$_{cam}$, this threonine (T252) is essential for efficient functioning of the enzyme, with the T252A mutant producing only 5 % of the expected product **4** (Fig. 12.5) as compared to wild-type P450$_{cam}$ [37]. Despite this, the rate of the reaction (NADH consumption) was maintained, with the majority of the reducing equivalents consumed by the P450$_{cam}$ T252A mutant being used for the formation of hydrogen peroxide [37]. Like P450$_{cin}$, P450$_{EryF}$ (CYP107A1) is another P450 that does not have this conserved threonine and instead the native sequence has an alanine at the homologous position. In this case, P450$_{EryF}$ utilises a hydroxyl moiety on the substrate itself in place of the conserved threonine to direct oxygen activation [38, 39]. Removal of the hydroxyl group from the substrate results in significant uncoupling during oxidation, and substitution of the alanine with threonine yields a P450 capable of oxidising a range of substrates. In native P450$_{cin}$, the conserved threonine is absent and an asparagine (Asn242) is present in the homologous position. It was initially postulated that Asn242 could be a functional replacement for the threonine, forming the requisite hydrogen bond to water via its terminal amide. This hypothesis was supported when subsequently it was observed that the T252N mutant of P450$_{cam}$ was catalytically active, with the asparagine being a functional replacement for the threonine [11]. Interestingly, similar experiments with P450$_{BM3}$ did not support this hypothesis [40]. Thus, the P450$_{cin}$ N242A mutant was constructed and characterised. It was found that although the coupling of reducing equivalents to the production of hydroxycineole had been somewhat lessened (to approximately 50 %), the decrease was not as significant as that seen for P450$_{cam}$ [12]. This suggested that the asparagine in P450$_{cin}$ is not a functional replacement for the threonine found in other P450s and that it is not essential for oxygen activation.

The Asn242 residue was, however, found to be vital for the control of the regio- and stereoselectivity of cineole oxidation catalysed by P450$_{cin}$ [13]. X-ray crystallography revealed that the only direct hydrogen bond between the enzyme and substrate was from the terminal amide of Asn242 to the ethereal oxygen atom of cineole **1** [17]. Interestingly, the N242A mutant, which would be unable to maintain this hydrogen bond, had essentially the same binding affinity for cineole as the wild-type enzyme as measured by the dissociation constant (K_d 0.7 μM P450$_{cin}$ vs 0.3 μM N242A; Table 12.2). This mirrors similar observations in P450$_{cam}$ and

Table 12.2 Comparison of data available for $P450_{cin}$ and its mutants

	Spin state change (%)	K_d (μM)	NADPH consumption (%)[a]	Coupling[b]	Uncoupling (H_2O_2)	Uncoupling (H_2O)	Products
Wild-type [10]	100	0.7	100	80	12	8	**2a** (100 %)
N242A [12]	30	0.3	25	49	52	0	**2a** (5 %); **2b** (5 %); **2c** (90 %)
N242T [13]	11	1.5	17	54	17	29	**2a** (47 %); **2b** (22 %); **2c** (31 %)
T243A [13]	73	1.3	134	72	23	5	**2a** (100 %)
N242T/T243A [13]	0	NM[c]	14	40	31	29	**2a** (43 %); **2b** (18 %); **2c** (39 %)
D241N [48]	87	0.8	2	7 (31)[d]	91	ND[e]	**2a** (100 %)

[a]Percentage of the rate of NADPH consumption standardised against cineole with $P450_{cin}$

[b]Ratio of the amount of NADPH consumed compared to the amount of product formed

[c]Not measurable

[d]Background NADPH consumption subtracted

[e]Not determined

its Tyr96 mutant [41]. Tyr96 is known to directly hydrogen-bond to the carbonyl of (1*R*)-camphor **3** (Fig. 12.5), the natural substrate of this enzyme. When Tyr96 is replaced with phenylalanine, the $P450_{cam}$ Y96F mutant still binds (1*R*)-camphor reasonably well (K_d 3.3 μM vs 1.6 μM for $P450_{cam}$ with (1*R*)-camphor) [41, 42]. The $P450_{cin}$ N242A mutant did, however, reveal that the hydrogen bond provided by the asparagine was essential in preserving the stereoselectivity of cineole hydroxylation. In the absence of this hydrogen bond, the production of (1*R*)-6-β-hydroxycineole **2a** fell to only 5 % of the total products formed, with the N242A mutant predominantly producing **2c** (Fig. 12.6; Table 12.2). The production of primarily **2c** is consistent with what was predicted on the basis of the X-ray crystal structure of cineole-bound N242A mutant $P450_{cin}$ [12]. In this structure, two different orientations of cineole within the active site were observed, with one placing the methylene oxidised in the production of **2c** close to the heme iron. This work clearly supports the role of the asparagine in maintaining substrate orientation and allowing enantiospecific oxidation to occur.

Characterisation of an additional mutant in which the asparagine was replaced with a threonine (N242T mutant; Table 12.2) further supports the suggestion that Asn242 is important in controlling the regio- and stereoselectivity of the oxidation rather than having a direct role in oxygen activation. In the N242T mutant, the coupling of NADPH consumption to cineole oxidation was again lower but comparable to that of $P450_{cin}$ (Table 12.2). Although the product profile of the N242T mutant was considerably different to that observed for both the wild-type $P450_{cin}$ and the N242A mutant (Table 12.2), this is presumably due to the altered architecture of the active site with the change in the amino acid residue.

The crucial nature of this hydrogen bond in the stereochemical outcomes of the $P450_{cin}$–catalysed oxidation rather than oxygen activation was further investigated by utilising a number of alternative substrates. Oxidation of a synthetic C_{10} analog of cineole **1**, cinane **5** (Fig. 12.7), in which the ethereal oxygen of cineole is absent,

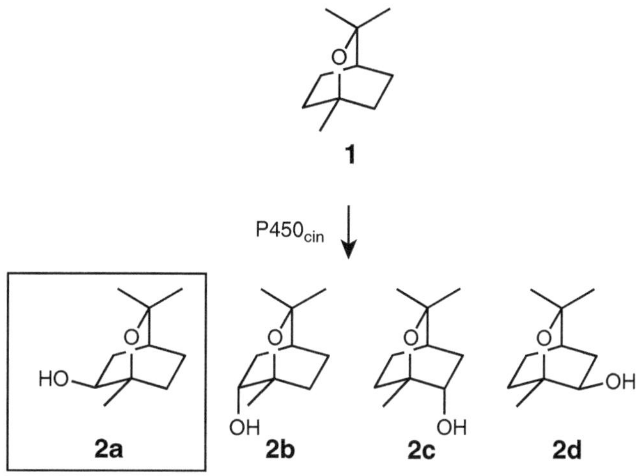

Fig. 12.6 The four possible isomers of 6-hydroxycineole **2** are shown. Wild-type P450$_{cin}$ converts cineole **1** specifically to (1*R*)-6β-hydroxycineole **2a** (*boxed*) [13]

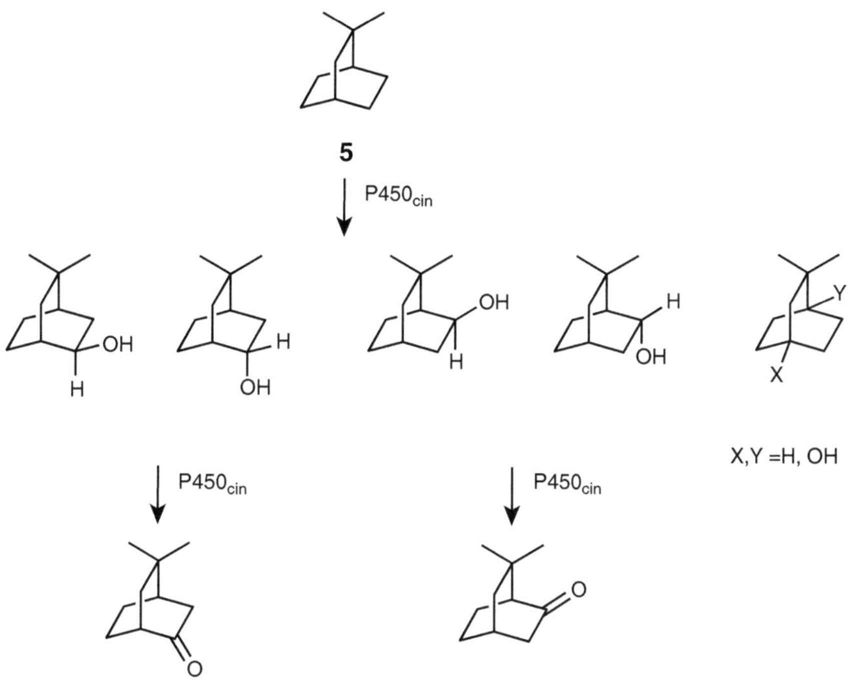

Fig. 12.7 Products formed from the catalytic oxidation of cinane **5** by P450$_{cin}$. Structures are representative of relative stereochemistry only; no absolute configuration or enantiomeric excess is implied [13]

by P450$_{cin}$ was observed to maintain both the rate of the reaction and its coupling. P450$_{cin}$ is therefore unlike P450$_{EryF}$ where the substrate is directly involved in directing protonation of the distal oxygen atom during oxidation [38, 39]. The oxidation of cinane **5** by P450$_{cin}$ resulted in a mixture of products, presumably due to the lack of interaction of **5** with the protein active site because of the loss of the hydrogen bond (Fig. 12.7). A second compound, camphane **6**, in which the carbonyl oxygen atom has been removed from camphor **3**, also yielded a mixture of products when oxidised by P450$_{cin}$ (Fig. 12.8) [13]. Overall, these results suggest that neither

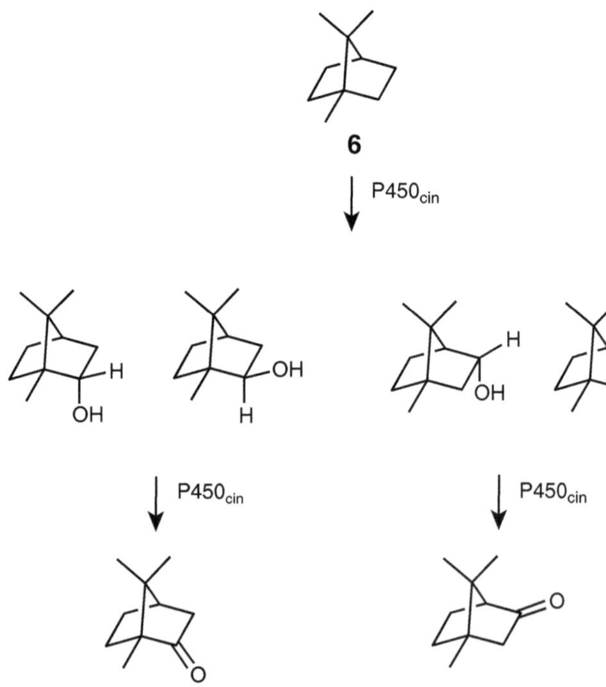

Fig. 12.8 Products from the catalytic oxidation of camphane **6** by $P450_{cin}$. Structures are representative of relative stereochemistry only [13]

the Asn242 nor the substrate itself plays the role of controlling protonation usually performed by the conserved threonine in P450s and that there must be an alternative mechanism by which $P450_{cin}$ directs and controls protonation during oxygen activation (see possible role of Gly238: Sect. 12.3).

Camphor **3** was also utilised as an alternative substrate to cineole to explore the relevance of the hydrogen bond formed between the Asn242 of $P450_{cin}$ and its substrate. It has been established that $P450_{cam}$ uses Tyr96 to form a critical hydrogen bond with the ketone of its substrate (1*R*)-camphor [43]. Just as Asn242 in $P450_{cin}$ interacts with cineole and controls its oxidation, the Tyr96-camphor hydrogen bond is important for the regio and stereoselective control of the oxidation of (1*R*)-camphor by $P450_{cam}$. Hence, the selectivity of $P450_{cin}$ was investigated with camphor, in anticipation that the Asn242 of $P450_{cin}$ may form a similar bond with the carbonyl oxygen atom of this substrate and direct its specific hydroxylation. Interestingly, although camphor could access the active site, there was very little selectivity in its oxidation and surprisingly no differentiation between the product profiles of (1*R*)- or (1*S*)-camphor (Fig. 12.9) [44]. This implies that the hydrogen bond, if it is formed between $P450_{cin}$ and camphor, is not dictating the outcome of substrate oxidation, but rather the inherent reactivity of different positions in camphor itself is driving the selectivity of the oxidation.

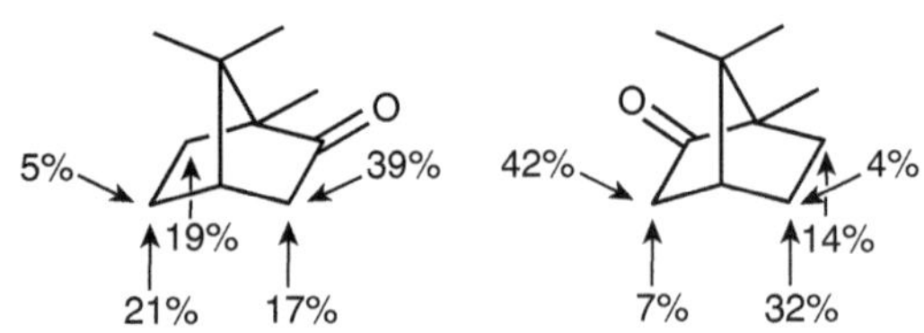

(1*S*)-**3** (1*R*)-**3**

Fig. 12.9 Relative oxidation (% of total product) at the various positions of both (1*S*)- and (1*R*)-camphor **3** by $P450_{cin}$ [44]

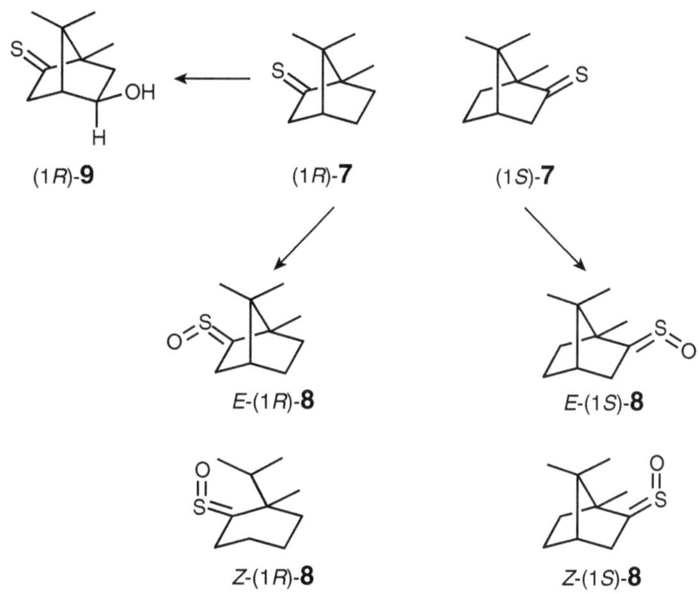

Fig. 12.10 P450$_{cin}$ catalyses the oxidation of (1*R*)- and (1*S*)-thiocamphor **7** to the thiocamphor-*S*-oxides **8** [45]. P450$_{cam}$ also produces a small amount of **9**

Thiocamphor **7** has previously been used to investigate the role of hydrogen bonding in P450$_{cam}$ [41]. Replacing the carbonyl oxygen atom with sulphur results in a much weaker, if any, hydrogen bond between the substrate and the protein and a consequent reduction in selectivity of oxidation. When thiocamphor **7** oxidation by P450$_{cin}$ was investigated concomitantly with our camphor **3** oxidation studies, the reaction was surprisingly specific, with only two products obtained. These were identified as the corresponding *E*- and *Z*-thiocamphor *S*-oxides **8** (Fig. 12.10) that are formed via oxidation of the sulphur of the thiocarbonyl. This product profile effectively masked any information that could be attained about the effect that a weak hydrogen bond has on hydroxylation specificity [45]. The oxidation of either (1*R*) or (1*S*)-thiocamphor by P450$_{cin}$ resulted in the production of only the two thiocamphor *S*-oxide isomers in both cases [45]. Interestingly, re-examination of the P450$_{cam}$ oxidation of thiocamphor revealed that P450$_{cam}$ also predominately produces these same thiocamphor *S*-oxide isomers in addition to a very small amount of 5-*exo*-hydroxythiocamphor **9** (6 %). This is in contrast to the original work in which only a number of hydroxythiocamphors were reported, although these structures had been assigned without the aid of synthetic standards [41]. Together, these results indicate that in the absence of a controlling hydrogen bond, the electronic effects of the easily-oxidised sulphur dominate the oxidation of thiocamphor **7** in both P450$_{cin}$ and P450$_{cam}$. However, some small effect from the interaction of the substrate carbon skeleton with the enzyme apparently remains in P450$_{cam}$, leading to the small amount of 5-*exo*-hydroxythiocamphor **9** observed.

12.5.2 Asp241: Role of the Conserved Acid

Although the conserved threonine is absent in P450$_{cin}$, the enzyme does retain the neighbouring acid (Asp241) that is also preserved in most P450s. This acid residue in P450$_{cam}$ (Asp251) was found to be critical in maintaining the rate of oxidation (P450$_{cam}$ D251N: 1 % of the wild-type) but did not affect the specificity of substrate oxidation [5, 46, 47]. It has been proposed that in partnership with the threonine, this conserved acid moiety is indirectly involved in the protonation of the distal oxygen atom of the bound dioxygen (Fig. 12.1**E** to **F**) [46, 47]. It is believed that either the acid moiety itself or the backbone carbonyl of the acid residue stabilises a water molecule for efficient transfer of protons to the heme-bound dioxygen [6]. When an analogous mutant was analysed in $P450_{cin}$ (D241N), it displayed very similar properties to the D251N $P450_{cam}$ mutant [48]. The rate of the reaction was dramatically slowed (2 % of wild-type), coupling was relatively well maintained (31 %) and the specificity of the substrate oxidation was preserved. Hence, even in the absence of the conserved threonine, the conserved acid (Asp242) in $P450_{cin}$ still remains essential in controlling the efficient activation of oxygen.

12.6 $P450_{cin}$ as a Model P450 Enzyme

12.6.1 Alternative Oxidising Species: Ferric-Peroxo Species

In general, P450s introduce an oxygen atom into a molecule by utilising a highly reactive iron-oxo species (Fig. 12.1**G**). However, not all the chemistry observed as a result of P450-catalysed oxidation can be explained solely using the capabilities of an electrophilic iron-oxo species. It has been proposed that a second oxidising species (ferric-peroxo species; Fig. 12.1**E**) may be responsible for some of the alternative chemistry seen [49–51]. The ferric-peroxo species has been implicated in a number of P450 mechanisms in which a nucleophilic oxidant is hypothesised to be present. For example, aromatase (CYP19A1) is thought to use the ferric-peroxo species as a nucleophile in the generation of estrogen [52–55]. Introduction of the D251N mutation in $P450_{cam}$ was shown to produce an enzyme in which the delivery of the first proton to the P450 bound dioxygen was rate-limiting (Fig. 12.1**E** to **F**) [46, 47]. It was thought that by altering the normal catalytic cycle in this way, the lifetime of the ferric-peroxo species was increased. This was later supported by EPR and ENDOR studies [49–51]. It was anticipated that the equivalent mutation in $P450_{cin}$ (D241N) would produce a protein in which the ferric-peroxo species would also be persistent, and that alternative chemistry that involved nucleophilic oxidants, such as Baeyer-Villiger chemistry, might be observed using this mutant. Unfortunately, even though D241N $P450_{cin}$ appeared to have the same catalytic properties as D251N $P450_{cam}$ (*vide supra*), no chemistry consistent with the ferric-peroxo species was observed with either wild-type or D241N $P450_{cin}$ [48].

(1*R*)-6-Ketocineole **10** (Fig. 12.11) appeared to be the ideal substrate with which to analyse potential Baeyer-Villiger reactions catalysed by the ferric-peroxo intermediate in $P450_{cin}$. It exhibits the same basic skeleton as cineole **1**, the natural $P450_{cin}$ substrate and a carbonyl at the position usually oxidised by $P450_{cin}$ in **1**, ready for nucleophilic Baeyer-Villiger oxidation mediated by the ferric-peroxo species to occur. However, no lactone, the product expected from involvement of the ferric-peroxo species in a nucleophilic oxidation, was observed following incubation of the D241N mutant or wild-type $P450_{cin}$ with (1*R*)-6-ketocineole [48]. The only evidence of oxidation was production of the two epimers of (1*R*)-7-hydroxy-6-ketocineole [48].

An alternate strategy to enhance the potential for observing oxidation involving the ferric-peroxo species was to utilise more reactive molecules as substrates for $P450_{cin}$ itself using a variety of cineole derivatives. No products were observed following the incubation of diketocineole **11** (Fig. 12.11) with $P450_{cin}$, although a significant amount of uncoupling was seen (55 % uncoupled to hydrogen peroxide). Two different ketocinane compounds **12**, **13** (Fig. 12.11) were also investigated. Just as with cinane **5** itself, the hydrogen bond that usually forms between the ethereal oxygen atom of the substrate and the Asn242 of the enzyme was absent (see Sect. 12.5.1). It was hoped that this may allow for movement of the substrate within the active site, leading to ferric-peroxo based oxidation. Unfortunately, only a mixture of

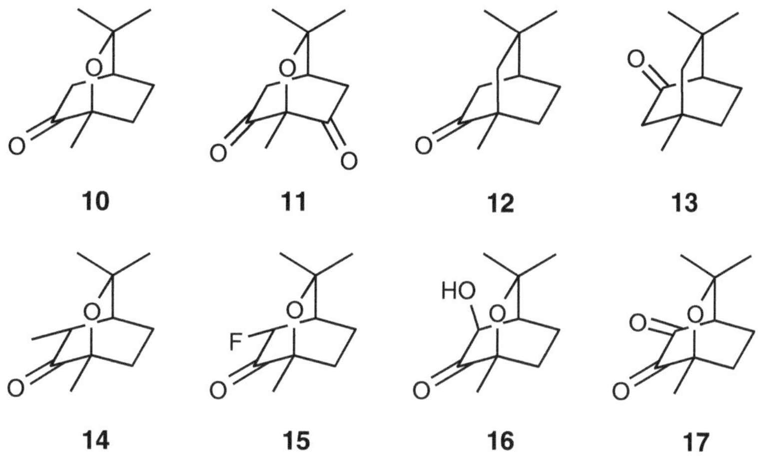

Fig. 12.11 Substrates incubated with a catalytically-active P450$_{cin}$ system to determine whether the ferric-peroxo species is involved in P450$_{cin}$ catalyzed oxidations

alcohols were detected and no lactone was observed that would suggest ferric-peroxo mediated oxidation. Finally, (1*R*)-5β-methyl-6-ketocineole **14**, (1*R*)-5α fluoro-6-ketocineole **15**, (1*R*)-5β-hydroxy-6-ketocineole **16** and (1*R*)-5,6-diketocineole **17** were all incubated with P450$_{cin}$ and all failed to produce a lactone that would be consistent with Baeyer-Villiger type chemistry that relies upon the ferric-peroxo species for oxidation.

A number of other compounds were also tested as potential substrates for both P450$_{cin}$ and the D241N mutant that did not have the core cineole structure but did have functional groups reactive toward nucleophilic oxidants, e.g. α,β-unsaturated ketones. Even though many were substrates and produced oxidised products, none of them showed any sign of chemistry that could be attributable to the ferric-peroxo species [48]. It is not known why the D241N mutant in which the ferric-peroxo species is thought to be persistent failed to produce any evidence of this type of chemistry. It was proposed that the ferric-peroxo species may be electronically or sterically sequestered, thus preventing the substrate from being oxidised [48]. Additionally, the ferric-peroxo species may be stabilised in the D241N mutant, which both slows its protonation and prevents its involvement in substrate oxidation. EPR studies are currently underway that can assist our understanding of these results.

12.6.2 Carbon-Carbon Bond Cleavage

The oxidative cleavage of carbon-carbon bonds by P450s is interesting mechanistically as it often involves multiple oxidative transformations [56]. It is believed that in some instances, a substrate undergoes two consecutive hydroxylation reactions to form a vicinal diol, which is subsequently cleaved during a final oxidative transformation. P450$_{scc}$ (CYP11A1) is an example of a P450 enzyme that catalyses this class of reactions, and the exact mechanism of the final oxidation is still not known although there are number of proposals [56]. A second type of mechanistically distinct, carbon-carbon bond cleaving P450 enzyme is one that utilises a ketoalcohol as a substrate/intermediate. For example, CYP17A1 catalyses the formation and subsequent cleavage of a ketoalcohol in the conversion of pregnenolone to progesterone. The latter step is thought to utilise the ferric-peroxo species to direct carbon-carbon bond cleavage [56]. Unfortunately, it is very difficult to explore the mechanism of these reactions as many of the substrates involved are large and difficult to synthesise, preventing the use of structural analogs as mechanistic probes. In addition, some systems which act on simple substrates, e.g. P450$_{BioI}$ which cleaves an acyl carrier protein-bound fatty acid to pimelic acid, have extremely low rates of catalytic turnover that makes them problematic to investigate [34]. We

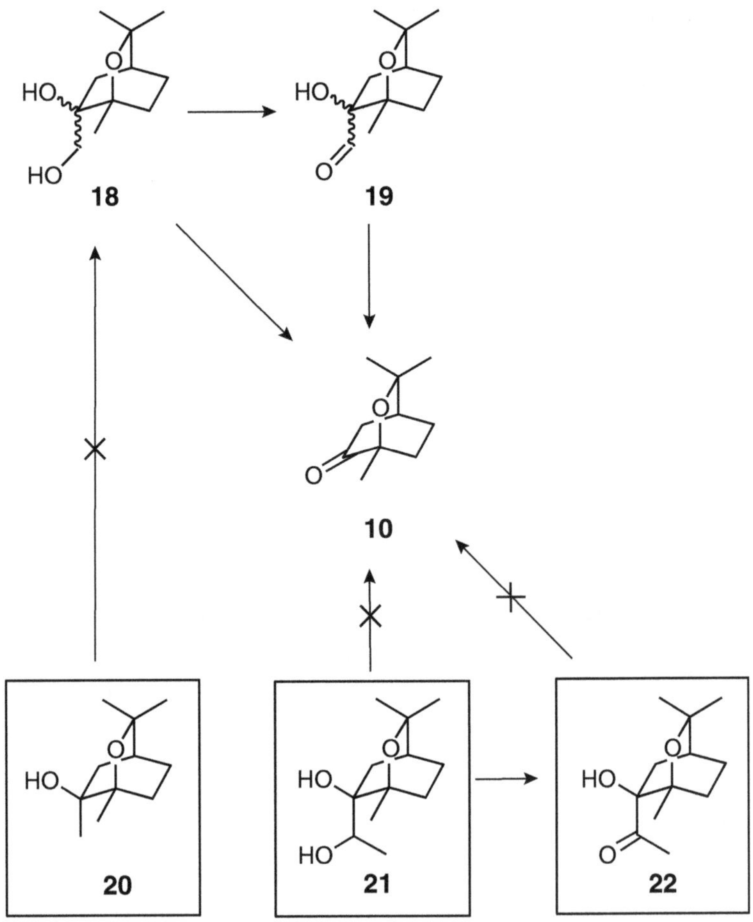

Fig. 12.12 $P450_{cin}$ catalysed oxidative cleavage of either **18** and **19** to (1*R*)-6-ketocineole **10**. The diol **18** was not detected following incubation with a catalytically-active $P450_{cin}$ system and alcohol **20**. $P450_{cin}$ also failed to convert **21** and **22** to (1*R*)-6-ketocineole **10**

postulated that $P450_{cin}$ may provide a useful system in which to study carbon-carbon bond cleavage by utilising the cineole skeleton to facilitate binding of reactive diol or ketoalcohol moieties at its active site.

$P450_{cin}$ was incubated with compounds in which the reactive functionality (either diol or ketoalcohol) was located in essentially the same position on the cineole **1** skeleton as is typically oxidised by the enzyme [36]. Both the diol **18** and the formylalcohol **19** were observed to produce ketocineole **10** (Fig. 12.12), the product expected from carbon-carbon bond cleavage. Diol **18** was also found to be oxidised to the corresponding formylalcohol **19**. It is possible that direct oxidative cleavage of the diol **18** occurs or that **18** is first oxidised to the formylalcohol **19** before cleavage. Studies are ongoing in order to elucidate the exact mechanistic details of this carbon-carbon bond cleavage reaction catalysed by $P450_{cin}$.

No hydroxylation of the alcohol **20** to produce the diol **18** was seen following its incubation with $P450_{cin}$. This suggests that $P450_{cin}$ is unable to mimic completely the initial stages of carbon-carbon bond cleavage as is carried out by P450s that introducc a vicinal diol before carbon-carbon bond cleavage occurs, e.g. $P450_{scc}$ or $P450_{BioI}$. Further substitution via the introduction of a methyl group to produce the methyl ketone **22** or the secondary alcohol **21** was also found to prevent oxidative cleavage. Instead, these compounds were observed to undergo hydroxylation at a number of other positions on the molecule, suggesting a requirement for the electronically and sterically more reactive formyl carbonyl. It is hoped that ongoing investigations with this system will enable us to define the

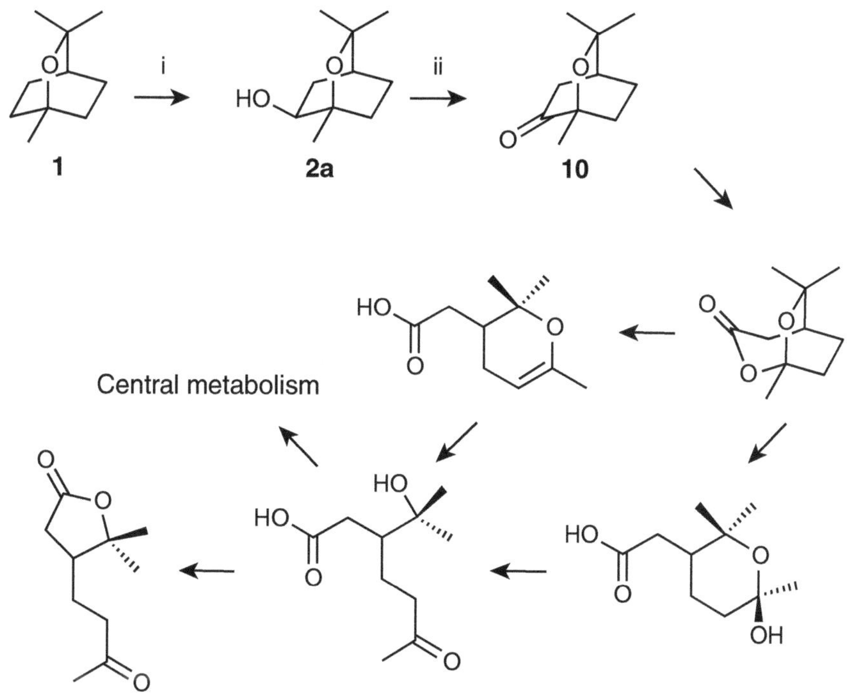

Fig. 12.13 Proposed pathway of 1,8-cineole **1** degradation [69]. (*i*) The first step is catalysed by P450$_{cin}$ in *Citrobacter braakii*. (*ii*) The second step can be catalyzed by P450$_{cin}$ but is likely to be catalyzed by (1*R*)-6β-hydroxycineole dehydrogenase in vivo

mechanistic details of carbon-carbon bond cleavage in this system.

12.7 Cineole Biodegradation

Cineole **1** is an extremely common chemical in the Australian bush, where it is the most abundant component of essential oil from eucalypt trees. It is believed that the Australian eucalypt population produces approximately 50,000 tonnes of cineole annually [57]. Before the isolation of P450$_{cin}$ from *C. braakii*, there was limited information available about the biodegradation of cineole in the environment. This information came predominantly from two studies investigating the degradation of cineole by *Pseudomonas flava* and *Rhodococcus* sp. [58, 59]. It was proposed that the first step in the biodegradation of cineole was the oxidation of 1,8-cineole **1** to 6β-hydroxycineole **2**, which was then converted to 6-ketocineole **10** (Fig. 12.13). 6-Ketocineole **10** could then be further transformed to allow the microorganism to survive on cineole as its sole source of carbon and energy (Fig. 12.13). Interestingly, although the degradation of the cineole appeared to be chemically identical in both species, the two differed in the enantiospecificity of hydroxylation, with either the pro-*S* (*P. flava*) or pro-*R* carbon (*Rhodococcus* species) of the *meso* cineole being oxidised to enantiomeric products [58, 59].

P450$_{cin}$ is the first isolated enzyme known to be involved in cineole biodegradation in *C. braakii* or any other organism. In vitro, together with Cdx and FdR, P450$_{cin}$ catalyses the hydroxylation of 1,8-cineole **1** to (1*R*)-6-β-hydroxycineole **2a**, the first step in the biodegradation of cineole [15]. Further analysis of the CIN operon revealed another gene (Fig. 12.4;

cinD) that was divergently transcribed in relation to *cinA-C*. Following cloning, expression and purification, it was determined that cinD was an alcohol dehydrogenase that specifically converts the (1*R*)-6β-hydroxycineole **2a** to (1*R*)-6-ketocineole **10** using NAD^+ specifically. CinD did not catalyse the oxidation of any of the other isomers of 6β-hydroxycineole **2** [16].

12.8 Biocatalytic Opportunities

P450s are particularly attractive candidates for exploitation as biocatalysts due to the wide array of interesting oxidative chemistry that they catalyse. These reactions include hydroxylation, epoxidation, heteroatom oxidation and carbon-carbon bond cleavage [56]. Not only are some of these reactions difficult to reproduce synthetically, but P450s often carry out these oxidations regio-, stereo-, and entantiospecifically. Unfortunately, there are a number of reasons why the potential of these enzymes can be difficult to translate into useful biocatalysts for either industrial applications or mechanistic studies. Substrate specificity, reaction rates, requirement for auxiliary redox proteins, and expensive cofactors (NAD(P)H) when using purified proteins all limit the potential utilisation of P450s as biocatalysts.

It has previously been demonstrated that bicistronic or tricistronic in vivo systems containing both the P450 enzyme and its redox partners can eliminate the need for the expensive cofactors and, in addition, any requirement for purification of the P450 enzyme and its reductase partners by utilising whole cells to carry out oxidations [60–66]. Based on the success of these in vivo systems, a bicistronic vector containing the genes that encode both $P450_{cin}$ and Cdx was constructed. The proteins were expressed in *E. coli* and the system successfully catalysed the in vivo oxidation of 1,8-cineole **1** to (1*R*)-6β-hydroxycineole **2a**, presumably utilising endogenous *E. coli* FdR to reduce Cdx [44]. This in vivo $P450_{cin}$ system produced preparative yields of (1*R*)-6β-hydroxycineole **2a** at approximately 1 g/L of bacterial culture.

The in vivo $P450_{cin}$ system developed was also shown to be able to provide adequate amounts of compound for structure elucidation after catalysing the oxidation of unnatural substrates. For example, the metabolites of camphor discussed in Sect. 12.5.1 were identified in this way [44]. Modification of $P450_{cin}$ via site directed mutagenesis coupled to the in vivo catalytic system can also provide access to significant quantities of alternative products. Previous reports indicated that the $P450_{cin}$ N242A mutant in vitro could produce (1*S*)-6α-hydroxycineole **2c** as approximately 90 % of the total products formed [13]. Following the construction of a bicistronic vector that included the N242A $P450_{cin}$ mutant and Cdx, expression of this modified $P450_{cin}$ in vivo system was shown to catalyse the oxidation of 1,8-cineole **1** to (1*S*)-6α-hydroxycineole **2c**. Studies to determine the isolatable quantities produced are underway. Manipulation of both $P450_{cin}$ and the substrate may provide more efficient routes to other important oxidised terpenes, as $P450_{cin}$ has been shown to catalyse the oxidation of a number of terpenes (Table 12.3). These may be useful for industries that produce antimicrobial or bactericidal agents in addition to fragrance and flavour compounds.

Table 12.3 Oxidation of alternative substrates by $P450_{cin}$

Substrate	Oxidation	Ref.
Pulegone	Hydroxylation	[48]
Carvone	Hydroxylation and epoxidation	[48]
Borneol	Hydroxylation and ketone formation	[36]
Menthol	Hydroxylation	[36]
Methone	Hydroxylation	[36]
Perillyl alcohol	Hydroxylation	[36]

12.9 Conclusions

The study of $P450_{cin}$ has expanded our overall understanding of P450s since its initial isolation from *C. braakii* [10]. $P450_{cin}$ has revealed that, despite the absence of a number of universal P450 characteristics such as the conserved threonine, well-coupled, stereo- and regiospecific oxidation of its substrate cineole can be achieved. Currently, however, it is still not fully understood how molecular oxygen activation is controlled in this enzyme. $P450_{cin}$ is also a rare example of a bacterial P450 enzyme that receives the electrons required for oxygen activation from an FMN-containing redoxin, (Cdx) rather than from the more common ferridoxin (Pdx with $P450_{cam}$). In this chapter, the potential of $P450_{cin}$ as a model P450 enzyme has been illustrated by the investigation of the ferric-peroxo species in addition to the initial exploration of the mechanism of oxidative carbon-carbon bond cleavage. It is highly likely that $P450_{cin}$ will continue to be used as a model P450 enzyme to further explore other P450 characteristics, as well as be utilised as a terpene biocatalyst.

Acknowledgements The authors would like to acknowledge that this work was supported in part by ARC Grants DP110104455 and DP140103229.

References

1. Guengerich FP (2005) Human cytochrome P450 enzymes. In: Ortiz de Montellano PR (ed) Cytochrome P450: structure, mechanism, and biochemistry, 3rd edn. Kluwer Academic/Plenum Publishers, New York, pp 377–530
2. Ortiz de Montellano PR (ed) (2005) Cytochrome P450: structure, mechanism, and biochemistry. Kluwer Academic/Plenum Publishers, New York
3. Poulos TL, Finzel BC, Gunsalus IC, Wagner GC, Kraut J (1985) The 2.6-Å crystal-structure of *Pseudomonas putida* cytochrome P450. J Biol Chem 260:6122–6130
4. Poulos TL, Finzel BC, Howard AJ (1987) High-resolution crystal structure of cytochrome $P450_{cam}$. J Mol Biol 195:687–700
5. Mueller EJ, Loida PJ, Sligar SG (1995) Twenty-five years of $P450_{cam}$ research. In: Ortiz de Montellano PR (ed) Cytochrome P450: structure, mechanism, and biochemistry, 2nd edn. Plenum Press, New York, pp 83–124
6. Schlichting I, Berendzen J, Chu K, Stock AM, Maves SA, Benson DE, Sweet BM, Ringe D, Petsko GA, Sligar SG (2000) The catalytic pathway of cytochrome $P450_{cam}$ at atomic resolution. Science 287:1615–1622
7. Poulos T (2005) Structural biology of heme monooxygenases. Biochem Biophys Res Commun 338:337–345
8. Whitehouse CJC, Bell SG, Wong LL (2012) P450BM3 (CYP102A1): connecting the dots. Chem Soc Rev 41:1218–1260
9. Deprez E, Di Primo C, Hui Bon Hoa G, Douzou P (1994) Effects of monovalent cations on cytochrome P-450 camphor evidence for preferential binding of potassium. FEBS Lett 347:207–210
10. Hawkes DB, Adams GW, Burlingame AL, Ortiz de Montellano PR, De Voss JJ (2002) Cytochrome $P450_{cin}$ (CYP176A), isolation, expression, and characterization. J Biol Chem 277:27725–27732
11. Kim D, Heo Y-S, Ortiz de Montellano PR (2008) Efficient catalytic turnover of cytochrome P450cam is supported by a T252N mutation. Arch Biochem Biophys 474:150–156
12. Meharenna YT, Slessor KE, Cavaignac SM, Poulos TL, De Voss JJ (2008) The critical role of substrate-protein hydrogen bonding in the control of regioselective hydroxylation in $P450_{cin}$. J Biol Chem 283:10804–10812
13. Slessor KE, Farlow AJ, Cavaignac SM, Stok JE, De Voss JJ (2011) Oxygen activation by $P450_{cin}$: protein and substrate mutagenesis. Arch Biochem Biophys 507:154–162
14. Kimmich N, Das A, Sevrioukova I, Meharenna Y, Sligar SG, Poulos TL (2007) Electron transfer between cytochrome $P450_{cin}$ and its FMN-containing redox partner, cindoxin. J Biol Chem 282:27006–27011
15. Hawkes DB, Slessor KE, Bernhardt PV, De Voss JJ (2010) Cloning, expression and purification of cindoxin, an unusual FMN-containing cytochrome P450 redox partner. ChemBioChem 11:1107–1114
16. Slessor KE, Stok JE, Cavaignac SM, Hawkes DB, Ghasemi Y, De Voss JJ (2010) Cineole biodegradation: molecular cloning, expression and characterisation of (1*R*)-6 β-hydroxycineole dehydrogenase from *Citrobacter braakii*. Bioorg Chem 38:81–86
17. Meharenna YT, Li H, Hawkes DB, Pearson AG, De Voss J, Poulos TL (2004) Crystal structure of $P450_{cin}$ in a complex with its substrate, 1,8-cineole, a close structural homologue to D-camphor, the substrate for $P450_{cam}$. Biochemistry 43:9487–9494
18. Madrona Y, Tripathi S, Li H, Poulos TL (2012) Crystal structures of substrate-free and nitrosyl cytochrome $P450_{cin}$: implications for O_2 activation. Biochemistry 51:6623–6631

19. Ost T, Miles C, Munro A, Murdoch J, Reid G, Chapman S (2001) Phenylalanine 393 exerts thermodynamic control over the heme of flavocytochrome P450 BM3. Biochemistry 40:13421–13429
20. Paine MJI, Scrutton NS, Munro AW, Gutierrez A, Roberts GCK, Wolf CR (2005) Electron transfer partners of cytochrome P450. In: Ortiz de Montellano PR (ed) Cytochrome P450: structure, mechanism, and biochemistry, 3rd edn. Kluwer Academic/Plenum Publishers, New York, pp 115–148
21. Lipscomb JD, Sligar SG, Namtvedt MJ, Gunsalus IC (1976) Autooxidation and hydroxylation reactions of oxygenated cytochrome P-450_{cam}. J Biol Chem 251:1116–1124
22. Bernhardt R, Gunsalus IC (1992) Reconstitution of cytochrome P4502B4 (LM2) activity with camphor and linalool monooxygenase electron donors. Biochem Biophys Res Commun 187:310–317
23. Peterson JA, Graham-Lorence SE (1995) Bacterial P450s. In: Ortiz de Montellano PR (ed) Cytochrome P450: structure, mechanism, and biochemistry, 2nd edn. Plenum Press, New York, pp 151–180
24. Ullah AJ, Murray RI, Bhattacharyya PK, Wagner GC, Gunsalus IC (1990) Protein components of a cytochrome P-450 linalool 8-methyl hydroxylase. J Biol Chem 265:1345–1351
25. Hannemann F, Bichet A, Ewen K, Bernhardt R (2007) Cytochrome P450 systems—biological variations of electron transport chains. Biochim Biophys Acta 1770:330–344
26. Ewen KM, Kleser M, Bernhardt R (2011) Adrenodoxin: the archetype of vertebrate-type [2Fe-2S] cluster ferredoxins. Biochim Biophys Acta 1814:111–125
27. Waterman M, Jenkins C, Pikuleva I (1995) Genetically engineered bacterial cells and applications. Toxicol Lett 82:807–813
28. Holden M, Mayhew M, Bunk D, Roitberg A, Vilker V (1997) Probing the interactions of putidaredoxin with redox partners in camphor P450 5-monooxygenase by mutagenesis of surface residues. J Biol Chem 272:21720–21725
29. Pochapsky T, Ye X, Ratnaswamy G, Lyons T (1994) An NMR-derived model for the solution structure of oxidized putidaredoxin, a 2-Fe, 2-S ferredoxin from *Pseudomonas*. Biochemistry 33:6424–6432
30. Pochapsky TC, Lyons TA, Kazanis S, Arakaki T, Ratnaswamy G (1996) A structure-based model for cytochrome $P450_{cam}$-putidaredoxin interactions. Biochimie 78:723–733
31. Madrona Y, Hollingsworth SA, Tripathi S, Fields JB, Rwigema J-CN, Tobias DJ, Poulos TL (2014) Crystal structure of cindoxin, the $P450_{cin}$ redox partner. Biochemistry 53:1435–1446
32. Sligar S, Gunsalus I (1976) A thermodynamic model of regulation: modulation of redox equilibria in camphor monoxygenase. Proc Natl Acad Sci U S A 73:1078–1082
33. Aguey-Zinsou K-F, Bernhardt PV, De Voss JJ, Slessor KE (2003) Electrochemistry of $P450_{cin}$: new insights into P450 electron transfer. Chem Commun 418–419
34. Stok JE, De Voss JJ (2000) Expression, purification, and characterization of BioI: a carbon-carbon bond cleaving cytochrome P450 involved in biotin biosynthesis in *Bacillus subtilis*. Arch Biochem Biophys 384:351–360
35. Hawkes DB (2003) Cytochrome P450cin, Ph.D. thesis, The University of Queensland, Brisbane
36. Slessor KE (2007) Cytochrome $P450_{cin}$: chemistry and biochemistry, Ph.D. thesis, The University of Queensland, Brisbane
37. Martinis SA, Atkins WM, Stayton PS, Sligar SG (1989) A conserved residue of cytochrome P450 is involved in heme-oxygen stability and activation. J Am Chem Soc 111:9252–9253
38. Andersen JF, Tatsuta K, Gunji H, Ishiyama T, Hutchinson CR (1993) Substrate specificity of 6-deoxyerythronolide B hydroxylase, a bacterial cytochrome P450 of erythromycin A biosynthesis. Biochemistry 32:1905–1913
39. Xiang H, Tschirret-Guth RA, Ortiz de Montellano PR (2000) An A245T mutation conveys on cytochrome P450EryF the ability to oxidize alternative substrates. J Biol Chem 275:35999–36006
40. Clark JP, Miles CS, Mowat CG, Walkinshaw MD, Reid GA, Simon NDA, Chapman SK (2006) The role of Thr268 and Phe393 in cytochrome P450BM3. J Inorg Biochem 100:1075–1090
41. Atkins WM, Sligar SG (1988) The roles of active-site hydrogen-bonding in cytochrome $P450_{cam}$ as revealed by site-directed mutagenesis. J Biol Chem 263:18842–18849
42. Deprez E, Gill E, Helms V, Wade RC, Hui Bon Hoa G (2002) Specific and non-specific effects of potassium cations on substrate-protein interactions in cytochromes $P450_{cam}$ and $P450_{lin}$. J Inorg Biochem 91:597–606
43. Loida PJ, Sligar SG, Paulsen MD, Arnold GE, Ornstein RL (1995) Stereoselective hydroxylation of norcamphor by cytochrome $P450_{cam}$—experimental verification of molecular-dynamics simulations. J Biol Chem 270:5326–5330
44. Slessor KE, Hawkes DB, Farlow A, Pearson AG, Stok JE, De Voss JJ (2012) An *in vivo* cytochrome $P450_{cin}$ (CYP176A1) catalytic system for metabolite production. J Mol Catal B: Enzym 79:15–20
45. Slessor KE, Stok JE, Chow S, De Voss JJ, Unpublished results
46. Gerber NC, Sligar SG (1992) Catalytic mechanism of cytochrome-P450—evidence for a distal charge relay. J Am Chem Soc 114:8742–8743
47. Gerber NC, Sligar SG (1994) A role for Asp-251 in cytochrome $P450_{cam}$ oxygen activation. J Biol Chem 269:4260–4266
48. Stok JE, Yamada S, Farlow AJ, Slessor KE, De Voss JJ (2013) Cytochrome $P450_{cin}$ (CYP176A1) D241N: investigating the role of the conserved acid in the active site of cytochrome P450s. Biochim Biophys Acta 1834:688–696

49. Davydov R, Macdonald IDG, Makris TM, Sligar SG, Hoffman BM (1999) EPR and ENDOR of catalytic intermediates in cryoreduced native and mutant oxy-cytochromes P450cam: mutation-induced changes in the proton delivery system. J Am Chem Soc 121:10654–10655
50. Davydov R, Makris TM, Kofman V, Werst DE, Sligar SG, Hoffman BM (2001) Hydroxylation of camphor by-reduced oxy-cytochrome $P450_{cam}$: mechanistic implications of EPR and ENDOR studies of catalytic intermediates in native and mutant enzymes. J Am Chem Soc 123:1403–1415
51. Benson DE, Suslick KS, Sligar SG (1997) Reduced oxy intermediate observed in D251N cytochrome $P450_{cam}$. Biochemistry 36:5104–5107
52. Akhtar M, Corina D, Pratt J, Smith T (1976) Studies on removal of C-19 in estrogen biosynthesis using $^{18}O_2$. J Chem Soc Chem Commun 854–856
53. Akhtar M, Njar VCO, Wright JN (1993) Mechanistic studies on aromatase and related C-C bond cleaving P450 enzymes. J Steroid Biochem 44:375–387
54. Gantt SL, Denisov IG, Grinkova YV, Sligar SG (2009) The critical iron-oxygen intermediate in human aromatase. Biochem Biophys Res Commun 387:169–173
55. Stevenson DE, Wright JN, Akhtar M (1988) Mechanistic consideration of P450 dependent enzymic reactions–studies on estriol biosynthesis. J Chem Soc Perkin Trans 1:2043–2052
56. Ortiz de Montellano PR, De Voss JJ (2005) Substrate oxidation by cytochrome P450 enzymes. In: Ortiz de Montellano PR (ed) Cytochrome P450: structure, mechanism, and biochemistry, 3rd edn. Kluwer Academic/Plenum Publishers, New York, pp 183–245
57. Carman RM, Fletcher MT (1983) Halogenated terpernoids. XX. The seven monochlorocineoles. Aust J Chem 36:1483–1493
58. Macrae IC, Alberts V, Carman RM, Shaw IM (1979) Products of 1,8-cineole oxidation by a pseudomonad. Aust J Chem 32:917–922
59. Williams DR, Trudgill PW, Taylor DG (1989) Metabolism of 1,8-cineole by a *Rhodococcus species* – ring cleavage reactions. J Gen Microbiol 135:1957–1967
60. Bell SG, Harford-Cross CF, Wong LL (2001) Engineering the CYP101 system for *in vivo* oxidation of unnatural substrates. Protein Eng 14:797–802
61. Blake JAR, Pritchard M, Ding SH, Smith GCM, Burchell B, Wolf CR, Friedberg T (1996) Coexpression of a human P450 (CYP3A4) and P450 reductase generates a highly functional monooxygenase system in *Escherichia coli*. FEBS Lett 397:210–214
62. Gillam EMJ, Wunsch RM, Ueng YF, Shimada T, Reilly PEB, Kamataki T, Guengerich FP (1997) Expression of cytochrome P450 3A7 in *Escherichia coli*: effects of 5′ modification and catalytic characterization of recombinant enzyme expressed in bicistronic format with NADPH-cytochrome P450 reductase. Arch Biochem Biophys 346:81–90
63. Kim D, Ortiz de Montellano PR (2009) Tricistronic overexpression of cytochrome P450cam, putidaredoxin, and putidaredoxin reductase provides a useful cell-based catalytic system. Biotechnol Lett 31:1427–1431
64. Parikh A, Gillam EMJ, Guengerich FP (1997) Drug metabolism by *Escherichia coli* expressing human cytochromes P450. Nat Biotechnol 15:784–788
65. Schneider S, Wubbolts MG, Sanglard D, Witholt B (1998) Biocatalyst engineering by assembly of fatty acid transport and oxidation activities for *in vivo* application of cytochrome P-450BM-3 monooxygenase. Appl Environ Microbiol 64:3784–3790
66. Schneider S, Wubbolts MG, Sanglard D, Witholt B (1998) Production of chiral hydroxy long chain fatty acids by whole cell biocatalysis of pentadecanoic acid with an *E. coli* recombinant containing cytochrome P450BM-3 monooxygenase. Tetrahedron Asymmetry 9:2833–2844
67. Peterson JA (1971) Camphor binding by *Pseudomonas putida* cytochrome P450. Arch Biochem Biophys 144:678–693
68. Atkins W, Sligar S (1989) Molecular recognition in cytochrome P450 – Alteration of regioselective alkane hydroxylation via protein engineering. J Am Chem Soc 111:2715–2717
69. Trudgill PW (1990) Microbial metabolism of monoterpenes – recent developments. Biodegradation 1:93–105

Fungal Unspecific Peroxygenases: Heme-Thiolate Proteins That Combine Peroxidase and Cytochrome P450 Properties

13

Martin Hofrichter, Harald Kellner, Marek J. Pecyna, and René Ullrich

Abstract

Eleven years ago, a secreted heme-thiolate peroxidase with promiscuity for oxygen transfer reactions was discovered in the basidiomycetous fungus, *Agrocybe aegerita*. The enzyme turned out to be a functional mono-peroxygenase that transferred an oxygen atom from hydrogen peroxide to diverse organic substrates (aromatics, heterocycles, linear and cyclic alkanes/alkenes, fatty acids, etc.). Later similar enzymes were found in other mushroom genera such as *Coprinellus* and *Marasmius*. Approximately one thousand putative peroxygenase sequences that form two large clusters can be found in genetic databases and fungal genomes, indicating the widespread occurrence of such enzymes in the whole fungal kingdom including all phyla of true fungi (Eumycota) and certain fungus-like heterokonts (Oomycota). This new enzyme type was classified as unspecific peroxygenase (UPO, EC 1.11.2.1) and placed in a separate peroxidase subclass. Furthermore, UPOs and related heme-thiolate peroxidases such as well-studied chloroperoxidase (CPO) represent a separate superfamily of heme proteins on the phylogenetic level. The reactions catalyzed by UPOs include hydroxylation, epoxidation, *O*- and *N*-dealkylation, aromatization, sulfoxidation, *N*-oxygenation, dechlorination and halide oxidation. In many cases, the product patterns of UPOs resemble those of human cytochrome P450 (P450) monooxygenases and, in fact, combine the catalytic cycle of heme peroxidases with the "peroxide shunt" of P450s. Here, an overview on UPOs is provided with focus on their molecular and catalytic properties.

Keywords

Peroxidase • P450 monooxygenase • Heme-thiolate • Compound I • Hydroxylation • Epoxidation • Dealkylation

M. Hofrichter (✉) • H. Kellner • M.J. Pecyna • R. Ullrich
Department of Bio- and Environmental Sciences,
International Institute Zittau, Technische Universität
Dresden, Markt 23, Zittau 02763, Germany
e-mail: hofrichter@ihi-zittau.de

E.G. Hrycay and S.M. Bandiera (eds.), *Monooxygenase, Peroxidase and Peroxygenase Properties and Mechanisms of Cytochrome P450*, Advances in Experimental Medicine and Biology 851,
DOI 10.1007/978-3-319-16009-2_13,

13.1 Introduction

Peroxygenase activities refer to the transfer of a peroxide-borne oxygen atom to substrates. Biocatalysts that preferably catalyze such reactions are classified in a separate sub-subclass, EC 1.11.2,[1] in the enzyme nomenclature system [www.chem.qmul.ac.uk/iubmb/enzyme/EC1/11/2/] (Fig. 13.1). The sub-subclass was approved in February 2011 and currently comprises four members, among which the unspecific peroxygenase (UPO, EC 1.11.2.1) is the most prominent because of its frequency in fungal organisms and promiscuity for oxygen transfer reactions.

The trivial name "peroxygenase" first appeared in the literature in 1977 in an article of Ishimarua and Yamazaki describing a new type of heme enzyme that catalyzes the hydroperoxide-dependent hydroxylation of several aromatic substrates (A) including indole, phenol and aniline in microsomes of pea seeds (*Pisum sativum*) [1]. The peroxygenase reaction can be illustrated in simplified form as shown in equation (eqn) 13.1.

$$\text{A-H} + \text{ROOH} \rightarrow \text{A-OH} + \text{R-OH} \quad (13.1)$$

where AH is the substrate, ROOH represents the hydroperoxide, R signifies an organic substituent or hydrogen atom, AOH designates the hydroxylated product and ROH depicts the reduced hydroperoxide or H_2O. Nowadays, this enzyme that contains histidine-ligated heme and a caleosin-type calcium binding motif is classified under EC 1.11.2.3 as plant seed peroxygenase [2] that, among others, is thought to be involved in the synthesis of cutin [3].

In the P450 context, the term peroxygenase has been in use since the end of the 1980s [4–7] and is usually related to peroxide-driven substrate oxidation, a side activity that is also known as the "peroxide shunt" pathway [8–10]. Peroxygenase side activities have also been reported for a few dioxygenases [11, 12] as well as for tyrosinase [13]. Interestingly, in deviation from the typical monooxygenase cycle that works with reduced dinucleotides (NAD(P)H), there is one P450 type that prefers H_2O_2 over NAD(P)H. This "true P450-peroxygenase" (CYP152A1, $P450_{BS}$, $P450_{SP\alpha}$, EC 1.11.2.4) is an intracellular enzyme found in bacteria such as *Sphingomonas paucimobilis* and *Bacillus subtilis* [14–16]. It preferably hydroxylates fatty acids (e.g. myristic acid) in the 2- and/or 3-position, as shown in eqns. 13.2 and 13.3, and was therefore designated as fatty acid peroxygenase (EC 1.11.2.4).[2]

$$CH_3\text{-}(CH_2)_{5-10}\text{-}CH_2\text{-}CH_2\text{-}COOH + H_2O_2 \rightarrow CH_3\text{-}(CH_2)_{5-10}\text{-}CHOH\text{-}CH_2\text{-}COOH + H_2O \quad (13.2)$$

$$CH_3\text{-}(CH_2)_{5-10}\text{-}CH_2\text{-}CH_2\text{-}COOH + H_2O_2 \rightarrow CH_3\text{-}(CH_2)_{5-10}\text{-}CH_2\text{-}CHOH\text{-}COOH + H_2O \quad (13.3)$$

The fatty acid substrate can act as a decoy molecule, which widens the substrate spectrum of these peroxygenases. Thus, $P450_{BS\beta}$ and $P450_{SP\alpha}$ were shown to peroxygenate 1-methoxynaphthalene and styrene, respectively, in a carboxylic acid-dependent reaction [17, 18]. Interestingly, the decoy-molecule concept was later also successfully applied to classic P450s such as $P450_{BM3}$ [19].

Mammalian myeloperoxidase (EC 1.11.2.2, formerly 1.11.1.7) is an additional member of the peroxygenase subclass and preferably oxidizes halides into hypohalites (eqn. 13.4), which in turn act as bactericidal agents in phagosomes [20].

[1] EC 1.11.2 With H_2O_2 as acceptor, one oxygen atom is incorporated into the product.

[2] www.chem.qmul.ac.uk/iubmb/enzyme/EC1/11/2/4.html

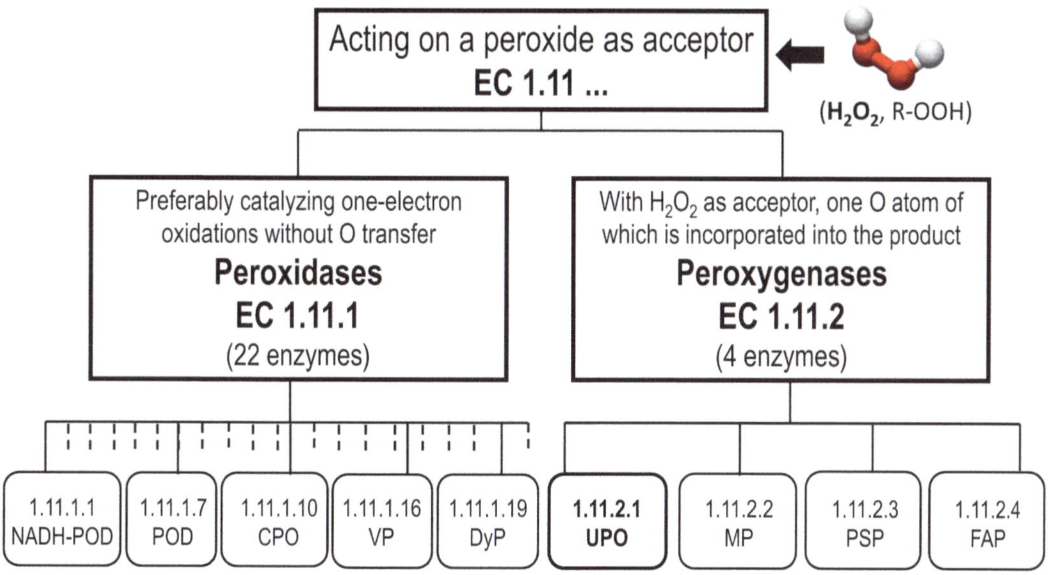

Fig. 13.1 Classification of enzymes using peroxide as the electron acceptor (EC 1.11; peroxidases and peroxygenases) according to the enzyme nomenclature system. *Abbreviations*: *NADH-POD* NADH peroxidase, *POD* peroxidase (phenol oxidizing), *CPO* chloroperoxidase, *VP* versatile peroxidase, *DyP* dye decolorizing peroxidase, *UPO* unspecific peroxygenase, *MP* myeloperoxidase, *PSP* plant seed peroxygenase, *FAP* fatty acid peroxygenase

$$X^- + H_2O_2 \rightarrow OX^- + H_2O,\ X^- = Cl^-,\ Br^-,\ SCN^- \qquad (13.4)$$

Myeloperoxidase differs from catalytically similar fungal CPO (EC 1.11.1.10) in its preference for the formation of hypochlorite (HClO) over the chlorination of organic substrates under physiological conditions (pH 5–8).[3] In addition to halide oxidation, both myeloperoxidase and CPO have strong peroxidase (phenol oxidation) and moderate peroxygenase activities and, as an example, were reported to epoxidize styrene [21]. Beyond that, CPO epoxidizes linear alkenes [22], hydroxylates benzylic carbons to some extent [23], catalyzes sulfoxidations [24, 25] and converts indole to oxindole [26]. However, CPO is not capable of peroxygenating aromatic substrates or stronger C-H bonds as found in alkanes [27]. Nevertheless, from the phylogenetic point of view, CPO can be regarded as an ascomycetous peroxygenase specialized in halide oxidation (compare Fig. 13.2). The following sections will deal exclusively with fungal UPOs, focusing on their catalytic and molecular properties.

[3] www.chem.qmul.ac.uk/iubmb/enzyme/EC1/11/2/2.html

13.2 History and Occurrence of Unspecific Peroxygenases

The first enzyme of this type was described in 2004 as *Agrocybe aegerita* haloperoxidase for the respective fungus (syn. *Agrocybe cylindracea*, Cyclocybe aegerita) that belongs to the Basidiomycota (family Strophariaceae) and is commonly known as the Black poplar mushroom [28, 29]. The fungus grows preferably on wood of poplars (*Populus* spp.) and other broad-leaved trees and causes a moderate white rot. It is found in Europe, North America and Asia and prefers warm and mild climates. *A. aegerita* is a popular edible mushroom in Mediterranean countries, especially in Italy (*ital.* Pioppino or Piopparello), where it is also commercially cultured [30]. The first article had still not used the term peroxygenase and focused on the ability of the enzyme to oxidize halides and aryl alcohols

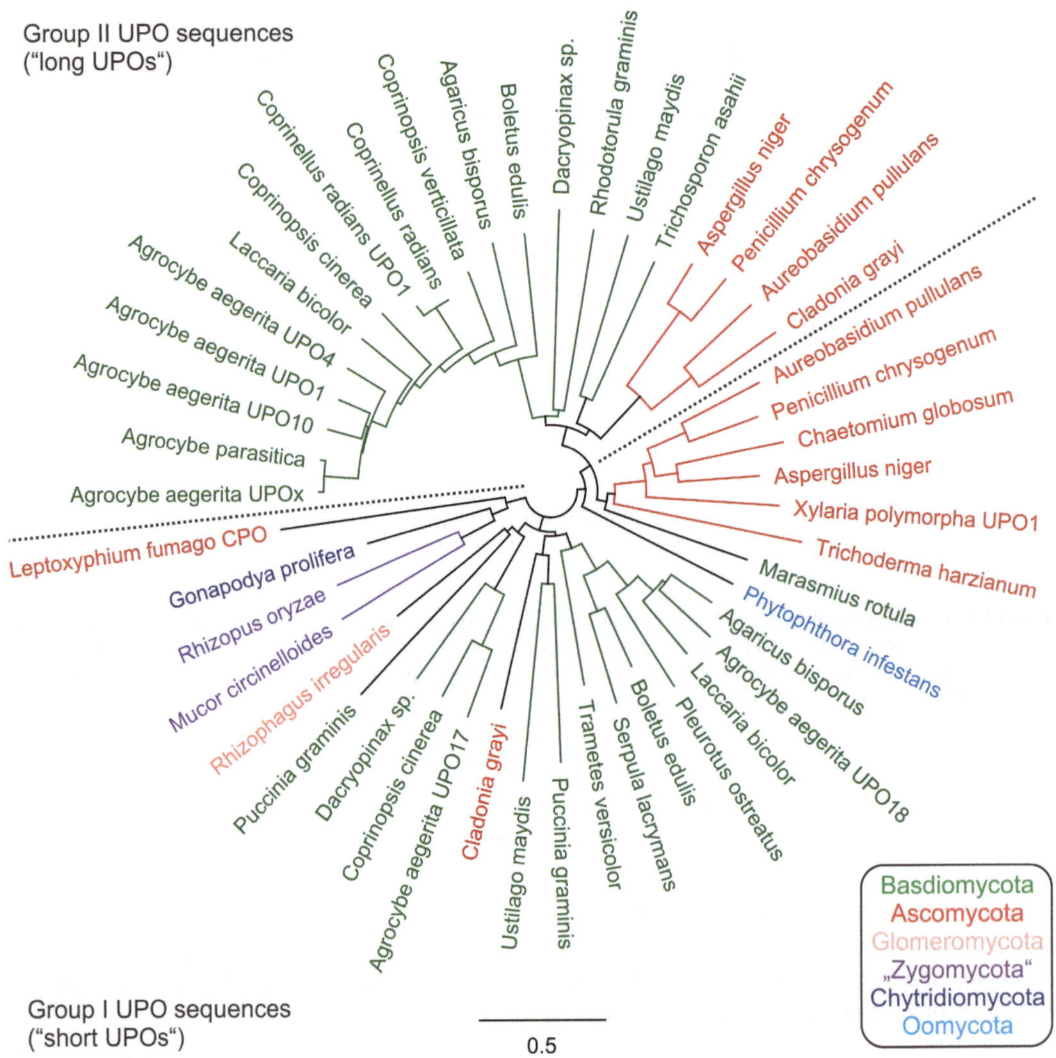

Fig. 13.2 Neighbor-joining phylogentic tree of UPO/HTP-sequences using Jukes-Cantor genetic distances. *Green* – Basidiomycota, *red* – Ascomycota, *blue* – Oomycota, *purple* – Zygomycota, *dark blue* – Chytridiomycota and *rose* – Glomeromycota. The *dotted lines* separate UPO sequences of groups I and II

[31]. Its unique oxygen atom transfer potential was recognized one year later by the hydroxylation of naphthalene [32], a reaction that later turned out to proceed via an initially-formed epoxide intermediate [33, 34]. Over the next few years, additional aromatic, heterocyclic and aliphatic substrates were found to be subjects of peroxygenation [27, 35, 36] (see also Sect. 13.4.3 below) and the name of the enzyme changed from haloperoxidase [31] via haloperoxidase-peroxygenase [33] to *Agrocybe aegerita* aromatic peroxygenase [37] and eventually to unspecific peroxygenase (UPO[4]) [38]. Furthermore, UPOs are also referred to as heme-thiolate peroxidases (HTP), taking into account their characteristic heme ligation by a cysteinate and their relation to CPO [39–41].

[4] Because of the discovery of many more unspecific peroxygenases, they should be systematically abbreviated by the capital letter of the genus plus the first and second letter of the epitheton and the acronym UPO: for example, *Aae*UPO = unspecific peroxygenase of *Agrocybe aegerita*.

The second UPO known as *Cra*UPO was described for the Ink-cap *Coprinellus (Coprinus) radians*, a wood- and mulch-dwelling fungus that belongs to the family Psathyrellaceae closely related to the Strophariaceae [42]. As *Aae*UPO, *Cra*UPO also oxidized naphthalene, aryl alcohols and bromide. Some differences between the two enzymes were observed with respect to the oxidation of aromatic rings vs. alkyl side chains or heteroatoms, as well as in the respective specific activities and kinetic data [43–45]. The third well-studied UPO known as *Mro*UPO was that of the boreo-subtropical Pinwheel mushroom, *Marasmius rotula*, preferably colonizing twigs and belonging to the diverse basidiomycete family of Marasmiaceae. However, *Mro*UPO was unable to oxidize halides. *Mro*UPO exhibits a less pronounced aromatic ring-oxygenating activity [46] but instead oxidizes bulkier substrates than do the other UPOs [47, 48]. Besides these three UPO producers, there are several other mushroom species secreting UPOs (e.g. *A. parasitica*, *A. chaxingu*, *A. alnetorum*, *Agaricus bisporus*, *Coprinus* sp. DSM 14545, *Coprinopsis verticillata*, *Auricularia auricula-judae*, *Mycena galopus*). However, the purification or characterization of these enzymes has not been performed yet and the results have not been published. A recombinant UPO known as r*Cci*UPO from the genome-sequenced model fungus, *Coprinopsis cinerea*, has recently been expressed at laboratory scale in *Aspergillus oryzae* [49].

More information on the occurrence of UPO-like enzymes can be gained from genetic databases where approximately 2,000 sequences of putative UPO enzymes are found. Figure 13.2 illustrates this diversity using a phylogenetic tree of UPOs (HTPs, respectively) covering 30 representative fungal species of different taxonomic and ecophysiological groups. Most sequences were retrieved from databases and genome projects but the tree also comprises 11 full sequences of *A. aegerita*, *A. parasitica*, *C. radians*, *C. verticillata*, *M. rotula* and *Xylaria polymorpha* generated in our laboratory, as well as the sequence of CPO from *Leptoxyphium (Caldariomyces) fumago* [40, 50, 51]. The majority of these sequences belongs to the Dikarya, i.e. Basidiomycota (353 sq.) and Ascomycota (580 sq.). However, other fungal phyla are also represented, such as the Mucoromycotina ("Zygomycetes") by such common genera as *Rhizopus* and *Mucor*, the Chytridiomycota by a genome-sequenced *Spizellomyces*, the Glomeromycota by a *Rhizophagus* (syn. *Glomus*), and the Oomycota (fungus-like heterokonts/stramenopiles, water molds) by several species of the genus *Phytophthora*. The latter finding supports the hypothesis of certain mycologists that an extensive horizontal gene transfer had taken place between phytopathogenic ascomycetes and oomycetes early in evolution [52]. Interestingly, true yeasts such as *Saccharomyces* or fission yeasts such as *Schizosaccharomyces* do not have UPO genes. Also, plants including green algae (Viridiplantae) and animals (Metazoa) are obviously lacking such genes [40, 50].

A more detailed recent analysis of UPO-sequence data has revealed that there are two large groups of these enzymes: the "short and the long peroxygenases" (Fig. 13.2) [53]. The short UPO sequences of group I with an average molecular mass of the putative protein of 29 kDa are found in all fungal phyla, while the long sequences of group II with an average mass of 44 kDa and one internal disulfide bridge are present only in basidiomycetes and ascomycetes. Characterized *Mro*UPO and CPO belong to group I and *Aae*UPO and *Cra*UPO belong to group II. This also means that well-studied CPO, which has been an "orphan" among heme peroxidases for decades [27], is now one out of hundreds of fungal UPO/HTP enzymes. Differences between UPOs of groups I and II exist also in the active sites. In group I, a conserved histidine acts as the charge stabilizer whereas in the long enzymes (group II) of the *Aae*UPO type, an arginine occupies this position. There are conserved amino acids present in UPOs of groups I and II: -PCP-EHD-E- and -PCP-EGD-R-E-, respectively. Deviations may occur from the latter sequence pattern in long UPOs that do not belong to the *Aae*UPO subgroup (Figs. 13.2 and 13.3).

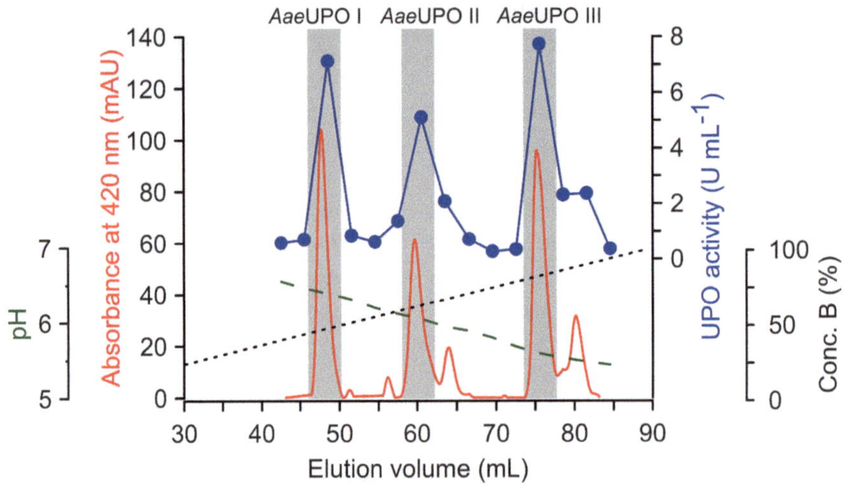

Fig. 13.3 Separation of different *Aae*UPO forms by chromatofocusing on a mixed anion exchanger (Mono P). Major *Aae*UPO forms are highlighted in gray. *Red line*, absorbance at 420 nm; *blue circles*, UPO activity assayed with veratryl alcohol at pH 7.0; *dotted line*, concentration of eluting buffer (%); *dashed green line*, pH gradient (Modified according to [55])

UPOs are seemingly organized in gene clusters (multigene families) in fungal organisms. Thus, transcriptome studies on *A. aegerita* DSM 22459 have indicated (13 long and 3 short UPOs) the presence of at least 16 UPO sequences probably including several gene variants (Pecyna et al. 2013, unpublished results). In the genome of the common White button mushroom (*Agaricus bisporus*), even as much as 24 putative UPO sequences were identified [54]. Notably, among the *A. aegerita* sequences are both group II and a few group I UPOs. Albeit, the major *Aae*UPO forms expressed in soybean medium all belong to the closely-related group II enzymes that hardly differ in their catalytic properties [55]. At the moment, it is still impossible to say how many of these UPOs, and under what conditions, are actually translated and secreted and how this process is regulated on the molecular level.

Despite all the progress in understanding the catalytic mechanisms of UPOs and collecting their molecular data, the natural function of these enzymes in fungal organisms is not clear yet. Of course, the surpassing catalytic versatility may suggest an involvement in all kinds of detoxification reactions (i.e. detoxification of plant phytoalexins, microbial toxins, xenobiotic compounds) but other functions cannot be ruled out (e.g. involvement in lignin and humus degradation/modification or in biosynthetic pathways). The *O*-demethylation and cleavage of non-phenolic lignin model compounds (e.g. adlerol) by *Aae*UPO is at least an indication for UPOs participation in the oxidation of smaller lignin fragments emerging after the action of different enzymes, such as manganese peroxidase (EC 1.11.1.15), on the lignin polymer [56, 57].

13.3 Production, Purification and Properties

UPOs are typically produced with fungal wild-type strains in complex plant-based media rich in carbon and nitrogen. The growth medium must be optimized for each particular species and will vary considerably with respect to the concentration of individual ingredients. However, always soybean (or other legume) components have to be present in order to obtain sufficient amounts of UPOs. Thus, *A. aegerita* prefers slurries containing soybean meal (1–6 % w/w) and bactopeptone (0–2 %) [31], *C. radians* prefers mixtures of glucose (1–4 %) and soybean meal

(1–3 %) [43], and *M. rotula* prefers soluble soybean peptone (4–5 %), yeast extract (4–5 %) and glucose (4 %) [46]. Though the structure of the soybean components triggering UPO production is not known, there are indications that seed storage proteins of the β-conglycinin and glycinin type, or their peptide fragments, are involved in the induction process (Pecyna, unpublished results).

Fungal fermentation can be carried out either simply in agitated or static culture flasks or in stirred-tank bioreactors under constant aeration [31, 43, 46]. As in the case of medium composition, culture parameters must be optimized for each fungal species/strain. UPO production in liquid cultures usually begins 5–12 days after inoculation with gently homogenized mycelium and reaches its maximum in the second to fourth week of cultivation (i.e. during secondary metabolism) [31, 43, 46]. There can be considerable differences in the overall production of UPO by individual fungal strains of the same species. Thus, among five strains of *A. aegerita*, only one strain (TM A1 = DSM 22459) secreted more than 1,500 units[5] per Liter (corresponding to 17 mg L^{-1} UPO protein), while all other strains produced only between 5 and 300 units [31]. At present, the highest amounts of UPO can be obtained with *M. rotula* that produces up to 445 mg L^{-1} UPO protein (corresponding to 41,000 U L^{-1}), which is so far one of the highest levels of a secreted hemeprotein reported for a wild-type basidiomycete [46]. UPO activities (up to 120 U kg^{-1}) are also detectable in solid-state cultures (e.g. beech-wood microcosms) of *A. aegerita* and *C. radians* but they have been too low to establish a functioning purification protocol [58].

UPOs are extracellular enzymes and can be concentrated by ultrafiltration of the culture liquid using appropriate membrane filters (e.g. 10-kDa cut-off). Purification of concentrated crude preparations is achieved by multistep fast protein liquid chromatography (FPLC) using different anion, cation and mixed-ion exchangers (e.g. Mono Q, S, P) and size exclusion chromatography (SEC) columns, depending on the particular UPO [31, 43, 46]. Alternative purification approaches on the basis of preparative isoelectric focusing turned out to be hardly suitable to obtain homogenous UPO fractions [55].

Usually, several UPO forms can be separated from fungal culture liquids. Figure 12.3 exemplarily shows the separation of the three major UPO forms (*Aae*UPO I–III) from *A. aegerita* grown in soybean slurry. The three forms have different isoelectric points (6.1, 5.6 and 5.2, respectively) but showed almost no differences in their catalytic properties, which suggests that these UPOs were considerably different glycosylated forms of the same protein and/or closely related gene products (e.g. of allelic sequences) rather than true UPO isoenzymes with different properties and functions [55].

Molecular masses and isoelectric points of characterized UPOs vary between 32 and 46 kDa as well as 3.8–6.1, respectively, and 16–42 % of mature UPOs are sugars of the high mannose type bound to up to six possible glycosylation sites [50, 59]. Table 13.1 summarizes some physical characteristics of UPOs from

Table 13.1 Physicochemical characteristics of selected unspecific peroxygenases (UPOs) compared to P450 peroxygenase of *Sphingomonas paucimobilis* ($P450_{SP\alpha}$) and lignin peroxidase of *Phanerochaete chrysosporium* (LiP_{Pc}) ApaUPO - UPO of Agrocybe parasitica, CveUPO - UPO of Coprinopsis verticillata LfuCPO - chloroperoxidase of Leptoxyphium fumago

Enzyme[a]	M_W (kDa)	p*I* (pH)	Glycosyl. (%)	Isoforms	Refs.
*Aae*UPO	45–46	4.9–6.1	20	3–6	[31, 50]
*Apa*UPO	37–47	4.5–8.6	N.D.	~7	–
r*Cci*UPO	~44[b]	N.D.	14–44	1	[114]
*Cra*UPO	43–45	3.8–4.2	37	4	[43, 60]
*Cve*UPO	40	4.5–5.2	42	2	[60]
*Mro*UPO	32	5.0–5.3	16	1–6	[46]
*Lfu*CPO	42	4.0	25–30	1	[115]
$P450_{SP\alpha}$	43	N.D.	–	1	[14]
LiP_{Pc}	38–43	3.3–4.7	3–14	6	[116]

[a]*N.D.* no data available. Other abbreviations are listed in Table 13.2

[b]Broad range of molecular weights due to heterogeneous glycosylation

[5] veratryl alcohol units (compare Sect. 13.4.3)

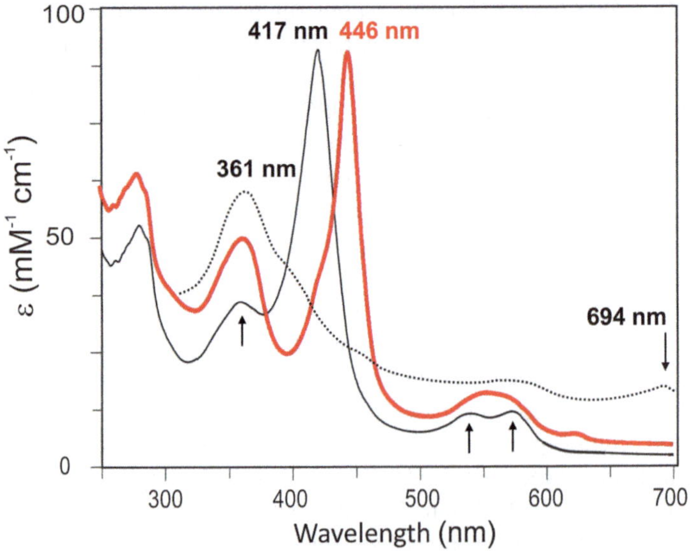

Fig. 13.4 UV-vis spectra of *Aae*UPO states. *Black line* – resting state enzyme, *red line* – CO complex of dithionite-reduced enzyme, *dotted line* – compound I. Spectra were recorded in 10 mM phosphate buffer, pH 7.0 (Based on [31, 66])

different fungi based on six original articles and unpublished results [31, 41, 43, 46, 55, 60]. UV-vis spectra of resting state UPOs are very similar to respective P450 spectra with Soret bands between 415 and 420 nm, which gives the purified UPOs a copper red color and sets them apart from CPO and other heme peroxidases [27]. The dithionite-reduced complex (ferrous UPO) shows a characteristic shift of the Soret band towards 450 nm when it comes into contact with CO, which is typical for heme-thiolate proteins [31, 43, 46, 61] (Fig. 13.4). UV-vis-spectral data of several UPOs are listed in Table 13.2, together with reference data for CPO, a P450 peroxygenase ($P450_{SP\alpha}$) and lignin peroxidase.

The successful crystallization of *Aae*UPO II (corresponding to the gene *apo1* [50]) and solving of the crystal structure at 2.2 Å has recently been reported (Fig. 13.5) [59, 62]. The UPO protein contains ten α-helices and five very short β-sheets, a cysteinate-ligated heme as the prosthetic group and one disulfide bridge between Cys_{278} and Cys_{319} that stabilizes the C-terminal region after the last α-helix. One magnesium ion (probably structure-stabilizing) is located near the heme propionate and additionally coordinated by a glutamate (Glu_{122}) and a serine (Ser_{126}). The latter amino acids (E-XXX-S), as well as the PCP motif exposing a cysteinate as the proximal ligand to the heme (Fig. 13.6), are highly conserved in most UPOs along with another glutamic acid residue (Glu_{196}) involved in acid-base catalysis and peroxide cleavage [50].

Figure 13.7 depicts the amino acid residues at the active sites of groups II and I UPOs by the example of *Aae*UPO and CPO. In both cases, a deprotonated glutamic acid residue (Glu_{196} in *Aae*UPO, Glu_{183} in CPO) near the heme abstracts a proton from the iron-bound peroxide to form compound 0 (Fig. 13.8), but the charge stabilizer of the glutamate is different. While an arginine (Arg_{189}) functions as the charge stabilizer in *Aae*UPO, a histidine (His_{105}) bears this role in CPO (compare also Sect. 13.4.3) [59]. Differences in the formation of compounds 0 and I, as well as in the behavior toward hydrogen peroxide, may be ascribed to these different residues. The heme channel of *Aae*UPO is ~7 Å in diameter and

Table 13.2 Spectroscopic properties of different unspecific peroxygenases (UPOs) compared to chloroperoxidase (*Lfu*CPO) and two P450 enzymes

		Soret band (nm)				Additional maxima of resting enzyme (nm)			
Organism	Enzyme	Resting	Reduced	CO-complex	Cpd I[a]	α	β	δ	Refs.
Agrocybe aegerita	*Aae*UPO	420	409	445	361 (694)	572	540	359	[31, 66]
Agrocybe parasitica	*Apa*UPO	420	N.D.	N.D.	N.D.	573	543	364	–
Coprinellus radians	*Cra*UPO	422	426	446	N.D.	571	542	359	[43]
Coprinopsis cinerea	r*Cci*UPO	416	410	443	N.D.	568	536	359	[114]
Coprinopsis verticillata	*Cve*UPO	417	407	443	N.D.	572	542	359	[60]
Marasmius rotula	*Mro*UPO	418	416	443	N.D.	570	536	353	[46]
Leptoxyphium fumago	*Lfu*CPO	403	409	443	367	542	515	–	[115, 117, 118]
Sphingom. paucimobilis	rP450$_{SP\alpha}$	418	N.D.	N.D.	N.D.	568	536	363	[119]
Sulfolobus acidocaldarius	rCYP119A1	415	N.D.	450	~367 (~650)	N.D.	N.D.	N.D.	[120, 121]

[a]*Cpd I* compound I, *r* recombinant protein, *N.D.* no data available

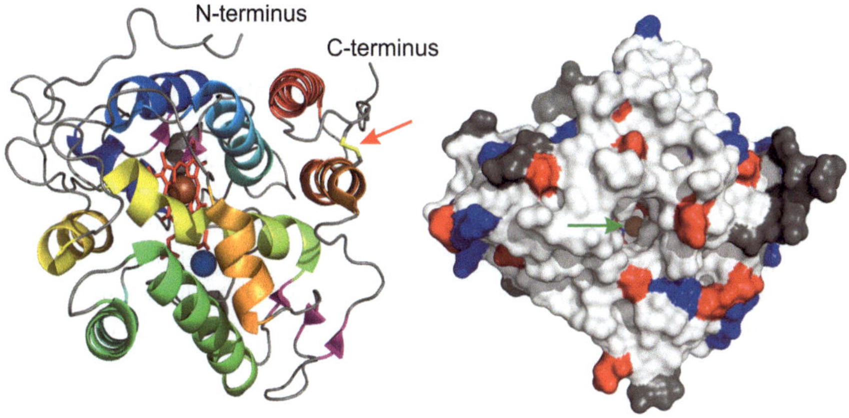

Fig. 13.5 *Left*: Ribbon diagram of the molecular structure of *Aae*UPO; 10 α-helices (*rainbow colored*), 5 short β-sheets (*purple*). The *red arrow* marks the position of a disulfide bond (*yellow*); heme iron (*brown ball*), magnesium (*blue ball*). *Right*: solvent access surface of *Aae*UPO (colors represent electrostatic potentials: *blue* – positive, *red* – negative, *dark grey* – hydrophobic). The picture shows in the middle the channel that provides access to the heme (diameter of the entrance, ~7 Å); the *arrow* points at the iron (*green*) at the end of the heme channel (Based on [50, 59])

contains eight phenylalanines and one tyrosine residue, which make it rather hydrophobic and affined to aromatics. Current studies suggest that the molecular architecture of UPO heme channels is quite variable and that the hydrophobic character can also be brought about by aliphatic amino

Fig. 13.6 Spatial arrangement of the PCP motif found in all characterized UPOs and CPO. Two proline residues (Pro_{35} and Pro_{37} in *Aae*UPO) expose the cysteine (Cys_{36}) in a way that it can perfectly ligate the heme iron (Based on [40, 50, 59, 114])

acids such as valine, leucine and isoleucine (Piontek, unpublished results).

13.4 Catalyzed Reactions and Reaction Mechanisms

13.4.1 Overview

The systematic name UPO according to the EC system is *substrate: hydrogen peroxide oxidoreductase (RH-hydroxylating or -epoxidizing)*. Both the accepted name (unspecific peroxygenase) and the systematic name were chosen in analogy to P450 enzymes that are subsumed under EC 1.14.14.1[6] (unspecific monooxygenase, UMO) and act on a wide range of substrates including diverse xenobiotics, pharmaceuticals, alkanes and fatty acids [63, 64]. Many of these substrates are oxidized by UPOs in a similar manner, although the reaction requirements are different. While microsomal UMOs need molecular oxygen as a cosubstrate that has to be activated by two electrons delivered to the P450 monooxygenase from NAD(P)H via flavoproteins [65], extracellular UPOs need merely peroxide for efficient functioning [27], as shown in eqns. 13.5 and 13.6:

$$\text{P450-UMO}: \text{R-H} + \text{FADH}_2 + \text{O-O} \rightarrow \text{R-OH} + \text{FAD} + \text{H}_2\text{O} \quad (13.5)$$

$$\text{UPO}: \text{R-H} + \text{H-O-O-H} \rightarrow \text{R-OH} + \text{H}_2\text{O} \quad (13.6)$$

A summarizing overview of UPO-catalyzed reactions is shown in Fig. 13.9. The reaction portfolio, among others, includes alkane and alkyl hydroxylation, epoxidation of alkenes and aromatics, heteroatom oxygenation, *O*- and *N*-dealkylation and one-electron oxidation [36, 40]. Roughly calculated, over 300 compounds have been positively tested as UPO substrates and it is expected that even more substrates will be discovered. More details on particular reactions follow in Sect. 13.4.4.

13.4.2 Reaction Cycle

The proposed reaction cycle of UPOs depicted in Fig. 13.10 is, in the first place, based on experimental data obtained with *Aae*UPO [66–68] and considers the comprehensive existing knowledge on P450s, CPO and heme-imidazole peroxidases such as horseradish peroxidase (HRP) [69–73]. It is assumed that this dual catalytic cycle applies for other UPOs/HTPs as well as CPO, with variation in the spectrum of oxidizable substrates (R-H vs. A-OH). It combines elements of the catalytic cycles of P450s and heme peroxidases [35, 36], in which compounds I and II are the key reactive intermediates that catalyze either the two-electron oxidation of a substrate molecule along with oxygen atom incorporation, or two one-electron oxidations resulting in the formation of two free diffusible substrate radicals (A-O•). In other words, UPOs oxygenate carbons in a similar manner as P450s (mono-peroxygenase pathway) and oxidize phenolics (peroxidase route) in a similar manner as prototypical heme peroxidases of the HRP type [74–77].

[6] http://www.chem.qmul.ac.uk/iubmb/enzyme/EC1/14/14/1.html

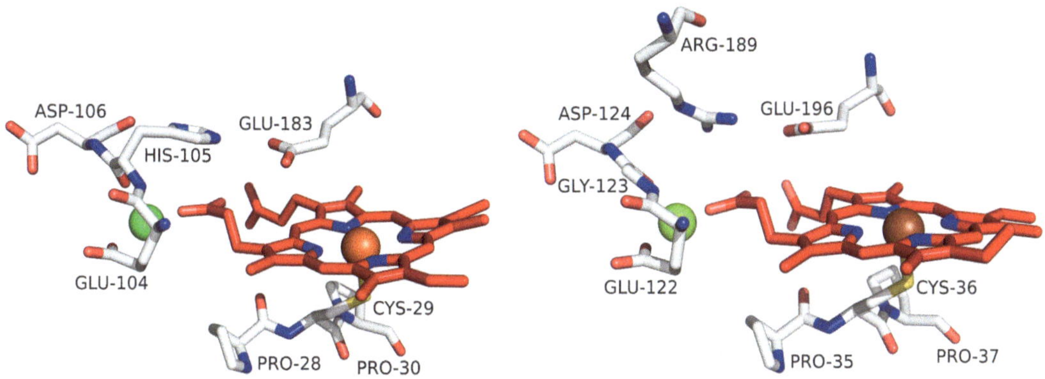

Fig. 13.7 Conserved amino acid residues at the active sites of CPO (*left*) and *Aae*UPO (*right*) (Based on [50, 59, 114])

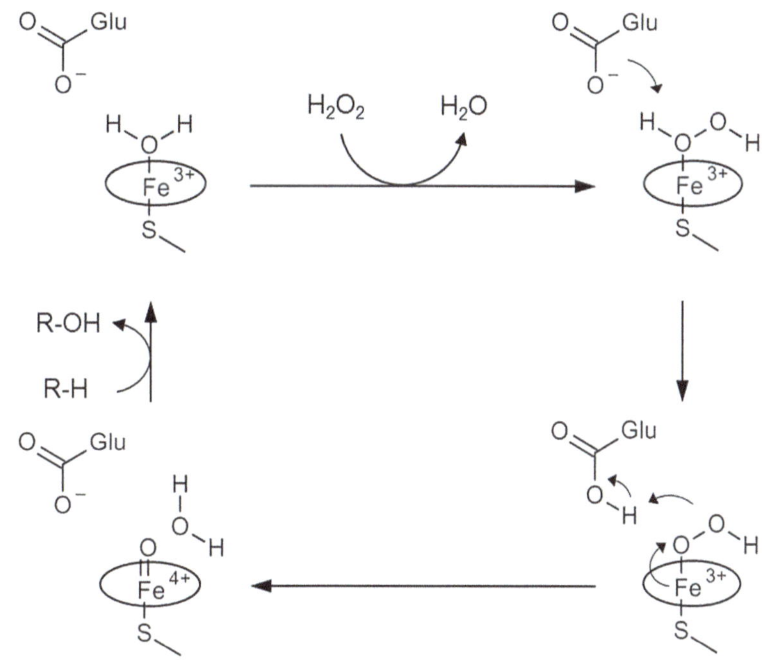

Fig. 13.8 Formation of *Aae*UPO compounds 0 and I and role of deprotonated Glu_{196} that facilitates heterolytic peroxide cleavage and compound 0 formation (Modified after [66, 115])

Now, we examine the catalytic properties of UPOs in more detail. Resting state UPO (*i*) contains a ferric heme that has a water molecule as the 6th (distal) ligand. There are indications that the substrate (R-H) can bind initially to the enzyme, because characteristic type I difference binding spectra [78] have been observed with different substrates such as veratryl alcohol, phenol and kaempferol [79], as well as with the pharmaceuticals dextromethorphan, diclofenac and propranolol [47]. However, this does not rule out that the substrate may also bind in later steps of the catalytic cycle, for example, after formation of compound I, as shown for (iii_b) and (iii_c) in Fig. 13.10 [66, 80]. UPO or the UPO-substrate complex then react with peroxide to form compound 0 (*ii*), a peroxo-complex that is heterolytically cleaved under electron re-arrangement to give the key compound I intermediate (oxo-ferryl cation radical complex) (*iii*) (see also Fig. 13.8). Compound I of *Aae*UPO and its kinetics of formation and decomposition have recently been studied using stopped-flow spectroscopy [66, 67]. In the mono-peroxygenation

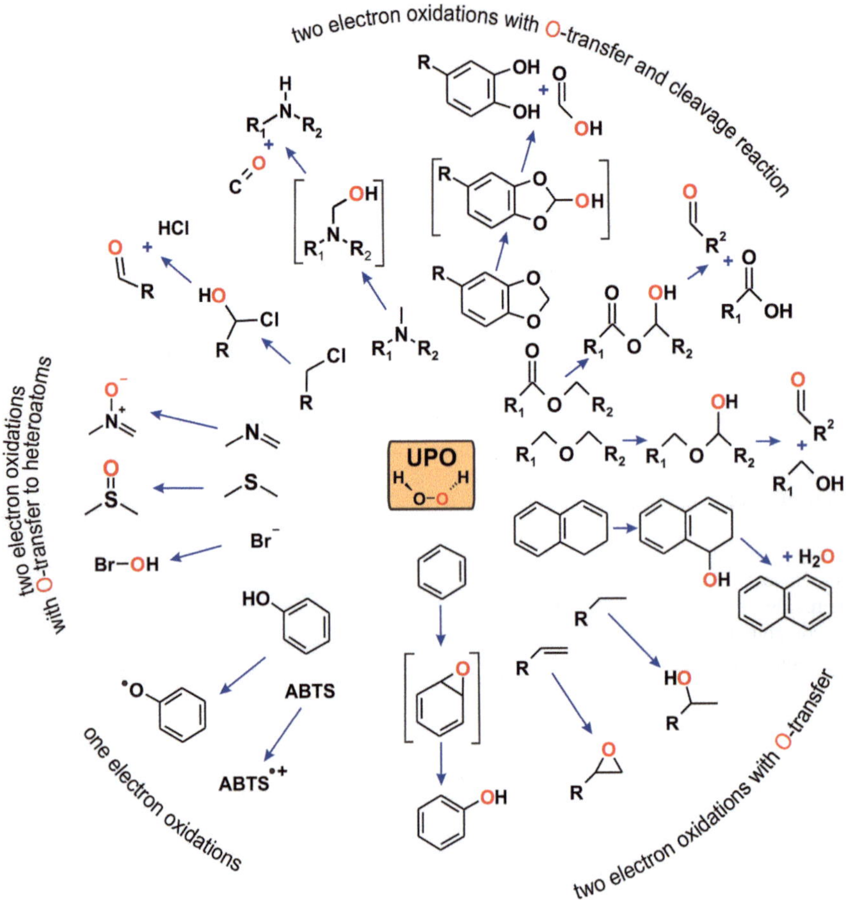

Fig. 13.9 Summarizing overview of UPO-catalyzed reactions

pathway, UPO compound I abstracts an electron and a proton (H-abstraction) from the substrate (R-H, e.g. an alkane) yielding protonated compound II (ferryl hydroxide complex) (*iv*) [69, 72][7] and a substrate radical (R•) (located near the active site), which rapidly recombine to form a hydroxylated product (R-OH, e.g. an alcohol)–ferric enzyme complex. The product complex then dissociates with the release of the hydroxylated product, and a water molecule coordinates to the heme iron (i_b) to begin the catalytic cycle again. The cycle of the monoperoxygenase pathway varies to some extent when epoxidation is the reaction under study. Based on P450 data and our own observations made during alkene oxidation by *Aae*UPO (see also Sect. 13.4.4.2), a modified compound II has to be proposed that binds the substrate (e.g. $CH_2{=}CH\text{-}R$) as a radical via the ferryl oxygen [38, 81–83], i.e. [Heme]-Fe^{IV}-O-CH_2-$C^{\bullet}$H-R (ferryl alkoxy radical complex). In this case, no H-abstraction would take place.

In the peroxidase route, both compounds I and II abstract one electron each from two substrate molecules (A-OH, e.g. a phenol), which are released as free radicals (A-O•, e.g. phenoxyl radicals) and can undergo coupling and/or disproportionation reactions [76]. It can be assumed that there are separate binding sites for

[7] Note that in many, especially older publications on heme peroxidases, compound II is described as a (deprotonated) oxo-ferryl complex with a double bond between iron and oxygen ($Fe^{IV}{=}O$) corresponding to *(v)* in Fig. 13.10. In reality, both ferryl species ($Fe^{IV}{=}O$ and Fe^{IV}-OH) of UPO compound II may be present as shown for CPO.

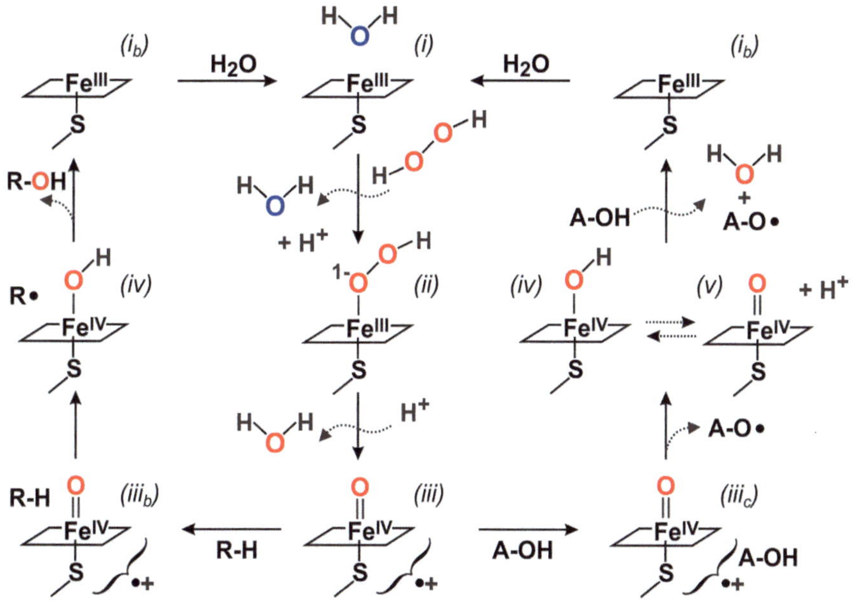

Fig. 13.10 Proposed reaction cycle of UPOs with two routes: mono-peroxygenase pathway (*left*) and peroxidase route (*right*). Details are described in the text under Sect. 13.4.2

one-electron oxidation substrates (A-OH) outside the heme channel, e.g. on the protein surface or at the heme channel entrance, as observed for ligninolytic peroxidases [59, 84, 85]. The route that is followed depends on the particular UPO enzyme and substrate, their redox potentials, the localization of the substrate binding site(s), the size of the heme channel and on the reaction pH [40, 67, 77]. In fact, there are differences between the characterized UPOs as well as UPOs and CPO regarding the substrates that can be oxygenated, as well as the extent to which the enzymes follow the mono-peroxygenase pathway or the peroxidase route. Some examples are presented in the following subsections.

13.4.3 UPO Assays

There are several spectrophotometric assays available to measure UPO activities (Fig. 13.11). They are based on the enzymes' ability to oxidize alcohols to aldehydes, cleave ethers or to oxygenate aromatic rings. For routine measurements, the oxidation of veratryl alcohol to veratraldehyde is monitored at neutral pH. The reaction proceeds via initial hydroxylation of the benzylic carbon to give veratryl *gem*-diol (aldehyde hydrate) that is in equilibrium with veratraldehyde specifically absorbing at 310 nm [31]. Veratraldehyde is also formed in a second assay that uses the cleavage of methyl veratryl ether as a UPO-specific reaction (*O*-demethylation), leading to an unstable hemiacetal intermediate that spontaneously breaks down to veratraldehyde and methanol [86]. Demethylenation is a special case of *O*-dealkylation carried out with 5-nitro-1,3-benzodioxole as a substrate. Oxidation by UPO results in the formation of formic acid and 4-nitrocatechol. The latter product has the advantage that it specifically absorbs in the visible range at 425 nm (yellow color), which facilitates activity measurements in liquids with high background absorption in the UV range [87]. Aromatic ring oxygenation via initial epoxidation and subsequent spontaneous re-aromatization (phenol formation) can be monitored with naphthalene as a substrate at 303 nm [33, 34]. One-electron oxidations catalyzed by UPOs are assayed with classical peroxidase substrates such as ABTS or 2,6-dimethoxyphenol [31, 43, 46]. In addition to spectrophotometric measurements, it is also possible to determine UPO activities and kinetic data

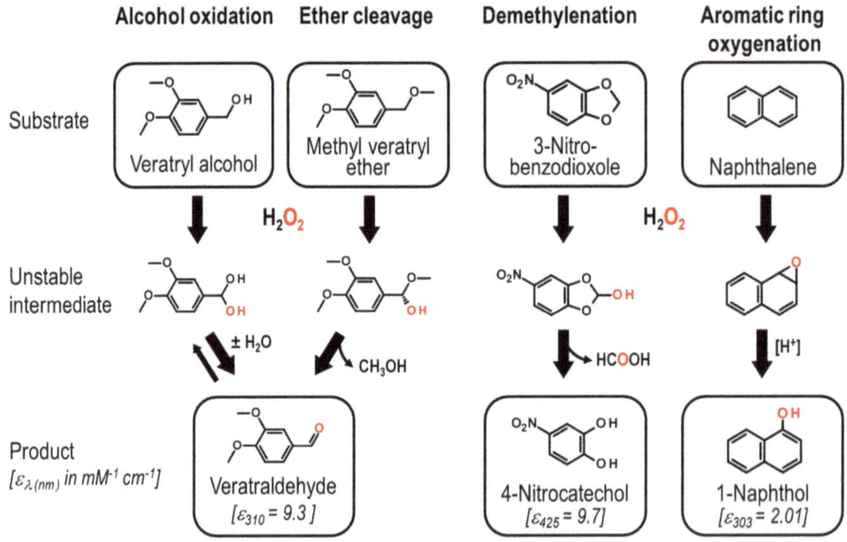

Fig. 13.11 Spectrophotometric assays for the detection of UPO activities [31, 33, 86, 87]; $\varepsilon_{(nm)}$, extinction coefficient of the product at the wavelength indicated in mM^{-1} cm^{-1}

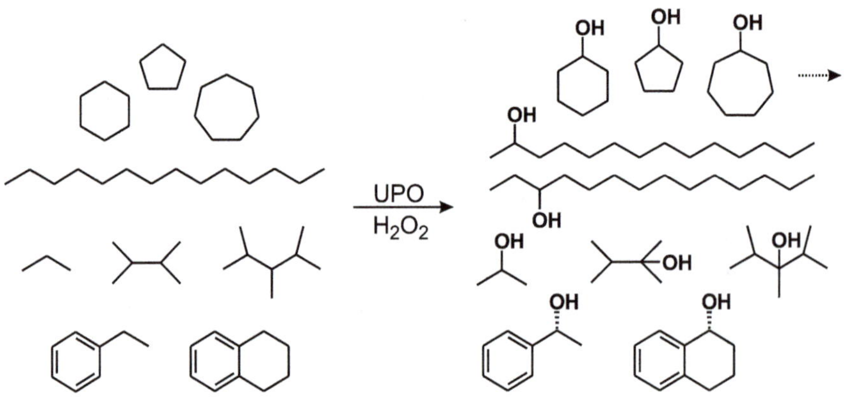

Fig. 13.12 UPO-catalyzed hydroxylation of alkanes and alkyls. Details are described in the text under Sect. 13.4.4.1

with HPLC or GC as was demonstrated, for example, for the oxidation of pyridine, ethylbenzene, benzene, cyclohexane and methylbutene [37, 38, 68, 88, 89].

13.4.4 Exemplary Reactions

13.4.4.1 Alkanes and Alkyl Groups

UPOs catalyze the hydroxylation of various linear, branched and cyclic alkanes as well as of alkyl groups (e.g. attached to aromatic rings) (Fig. 13.12). Most investigations were performed with *Aae*UPO [68], but two recent studies using peroxygenases from different fungi have shown that other UPOs can also efficiently hydroxylate alkanes, sometimes even with higher efficiency [49, 90]. Due to the low solubility of alkane substrates, reactions are usually performed in the presence of a co-solvent (e.g. acetone 4–60 % vol/vol).

The size of linear alkane molecules that are oxidized by *Aae*UPO ranges from gaseous propane (C_3) to viscous *n*-hexadecane (C_{16})

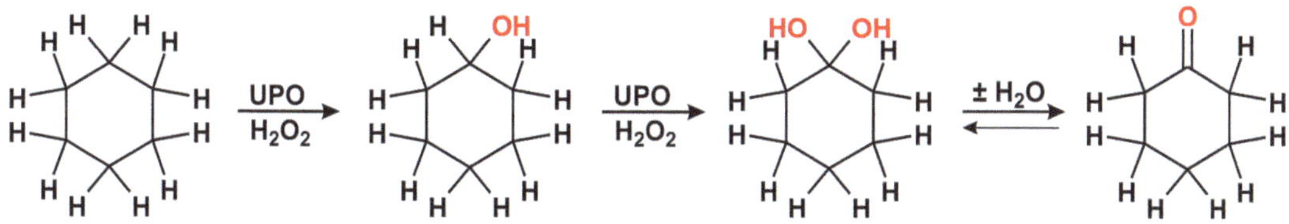

Fig. 13.13 Oxidation of cyclohexane via cyclohexanol and a hypothetical *gem*-diol to cyclohexanone (Modified according to [90])

[68]. The better-soluble fatty acids were even oxidized up to a chain length of C_{20} (arachidic acid) [41]. Alkanols hydroxylated in the 2- and 3-positions and hydroxy fatty acids with hydroxyl groups at (ω − 1) and (ω − 2) were the major products identified. The ratio between 2- and 3-alkanols depended on the chain length and amounted, for example, to 1:2 and 1.5:1 for the hydroxylation of *n*-pentane and *n*-heptane, respectively. In the latter case, an ee of 99.9 % was detected for the (*R*)-enantiomer [(*R*)-3-heptanol] [68]. In addition to monohydroxylated products, the corresponding alkanones were formed as minor over-oxidation products. With *n*-dodecane, *n*-tetradecane and *n*-hexadecane, hydroxylation was observed from both sides yielding small amounts of diols and their oxidation products (hydroxy-keto compounds and diketones) [41]. Traces of ω-hydroxylation products were only observed during the oxidation of fatty acids.

Branched alkanes up to a certain degree of branching are hydroxylated by *Aae*UPO as well, and often the tertiary carbons are preferably attacked. Thus, 2,3-dimethylbutane and isobutane were oxidized to the single products, 2,3-dimethylbutan-2-ol and 2-methylpropan-2-ol, respectively. The hydroxylation of 2,3,4-trimethylpentane yielded two products, 2,3,4-trimethylpentane-3-ol and 2,3,4-trimethylpentane-2-ol (Fig. 13.12) [68, 81]. Regarding the degree of branching, *Aae*UPO reaches its limit with 2,2,3,3-tetramethylbutene that is not subject to peroxygenation.

Cyclic alkanes from cyclopentane to cyclooctane are preferentially oxidized to form monohydroxylated products. Over-oxidation to the corresponding cycloalkanones is possible and depends on the reaction conditions and the UPO used. Recently, the optimization of cyclohexane oxidation via cyclohexanol to cyclohexanone has been reported using different UPOs, among which *Mro*UPO was the most effective [90]. Over-oxidation of cyclohexane proceeds via a *gem*-diol intermediate (cyclohexane-1,1-diol) that spontaneously eliminates water (Fig. 13.13) [90]. In general, the oxidation of primary and secondary alcohols to carbonyls is a typical activity of all UPOs and leads to the formation of aldehydes and ketones (see also veratryl alcohol assay above) [36]. The aldehydes formed can be subjected to further oxidation, generating carboxylic acids (see also toluene oxidation below) [41].

The two-ring system of norcarane (bicyclo [4.1.0]heptane) represents a special case of cycloalkane oxidation, because it is a radical clock substrate that can be converted into a number of different products, whose ratios give information on the oxidation mechanism and the formation of an intermediate substrate radical (R•, compare *(iv)* Fig. 13.10) [66, 68]. With *Aae*UPO, the experiment yielded *exo*-2-norcarenol as a major product and five other products in smaller amounts including the rearrangement product 4-(hydroxymethyl) cyclohexane (Fig. 13.14). All these products have previously also been described for

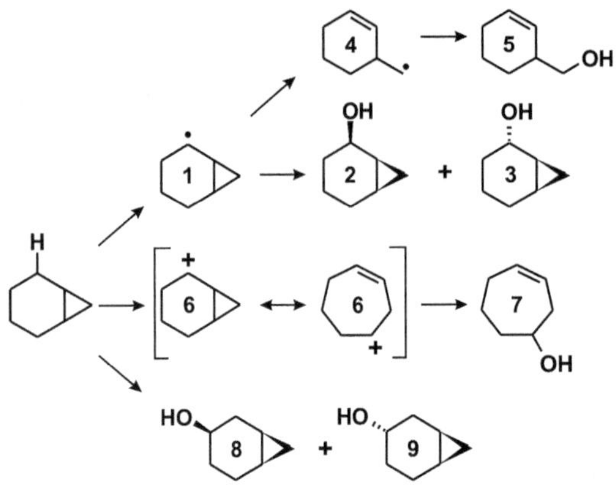

Fig. 13.14 Norcarane oxidation by *Aae*UPO. (*1*) norcarane radical, (*2*) *endo*-2-norcaranol, (*3*) *exo*-2-norcaranol (major product), (*4*) methylcyclohexene radical, (*5*) cyclohexenyl methanol, (*6*) mesomeric forms of norcarane cation, (*7*) 3-cycloheptene-1-ol, (*8*) *endo*-3-norcaranol, (*9*) *exo*-3-norcaranol (Modified after [68])

norcarane oxidation by P450s, which clearly points to an H-abstraction/oxygen rebound mechanism of oxygenation [91]. On the basis of these results, calculations revealed a lifetime of 9.4 ps for the substrate radical and an oxygen rebound rate of $2 \times 10^{11}\ s^{-1}$ for the rebound reaction, which indicates a ~6-fold faster rebound reaction compared to similar functional P450s [68].

Oxidation of the methyl group of toluene was one of the first UPO reactions studied in detail [32]. The molecule can be hydroxylated at both the methyl group and the aromatic ring, resulting in the formation of mixtures of benzyl alcohol, benzaldehyde and benzoic acid as well as *p*- and *o*-cresol, and methylhydroquinone. When 4-nitrotoluene was used as a substrate, the methyl group was oxidized in a similar way but ring hydroxylation was negligible [92]. In the case of toluene, the ratio of alkyl hydroxylation vs. aromatic oxygenation was 2:1 for *Aae*UPO and 26:1 for *Mro*UPO [46]. Interestingly, the aromatic ring is no longer attacked by *Aae*UPO when alkyl benzenes with longer side chains are used as substrates [89, 93]. Thus, ethyl- and propylbenzene were hydroxylated exclusively at the benzylic carbon (C_α) to form (*R*)-1-phenylethanol and (*R*)-1-phenylpropanol, respectively. The reactions were highly enantioselective with an enantiomeric excess of >99 % for the (*R*)-isomers. With increasing alkyl-chain length (C_4–C_6), turnovers and ee values decrease along with an increase in the number and amount of by-products (e.g. ketones). The enzymatic preparation of (*R*)-1-phenylethanol was optimized using a fed-batch reaction design and resulted in a maximum TTN (total turnover number) of 43,000 and a space-time yield of ~60 g per Liter and day. Tetralin (cyclohexylbenzene) can be perceived as a benzene with a cyclic alkyl group and was in fact hydroxylated in a similar manner as ethyl-/propylbenzene with an ee of >99 % for tetralin-(*R*)-1-ol, which was much better compared to its aliphatic counterpart butylbenzene [89].

A set of ten model compounds, including alkylated benzoic acids, cycloaliphatic acids and a branched fatty acid, with ascending C-H bonding dissociation energies (BDEs; 83–100 kcal mol^{-1}), was tested regarding oxidation by *Aae*UPO [66]. The study used a stopped-flow technique to generate *Aae*UPO compound I, and its activity was in turn studied kinetically. The plot of second-order rate constants for C-H hydroxylation by *Aae*UPO compound I vs. the BDE of the model compounds revealed a very distinct, non-linear correlation and a calculated upper limit of "hydroxylizability" of about 102 kcal mol^{-1}, which corresponds to the C-H BDE of ethane. In fact, results of recent experiments have indicated that ethane is barely hydroxylated by *Aae*UPO whereas methane with a BDE of 107 kcal mol^{-1} is definitely not a substrate under normal conditions (Wang and Peter, unpublished results). Whether methane can be hydroxylated by UPO at elevated pressure is currently under investigation.

13.4.4.2 Alkenes and Aromatics

*Aae*UPO oxidizes various alkenes and alkenyls, in which both epoxidation and hydroxylation of the double bond's adjacent carbons (allylic hydroxylation) can occur. In a recent study, 20 alkenes, among them propene and linear 1-alkenes up to C_8, branched alkenes such as 2,3-dimethyl-2-butene, cyclohexene, butadiene and the two enantiomers of limonene, were oxidized by *Aae*UPO in that manner [38]. Considerable differences in conversion rates and product patterns were observed, depending on the size of the molecule and position of the double bond. Surprisingly, branched and cyclic alkenes were much better substrates than linear alkenes. Propene, branched butenes, buta-1,3-diene and *cis*- and *trans*-butene were epoxidized exclusively, while 1-alkenes (C_4–C_8) and cyclohexene were both hydroxylated and epoxidized, i.e. mixtures of 1-alken-3-ols and 1-alkenes epoxides (=1-alkyloxiranes = 1,2-epoxyalkanes) and 2-cyclohexen-1-ol and cyclohexene epoxide, respectively, were formed. However, no products were formed with both oxyfunctionalizations.

The oxidation of *cis*-2-butene and *trans*-2-butene yielded differing amounts of epoxidation products. When *cis*-2-butene was epoxidized to *cis*-2-butene epoxide, more than twice as much product was formed than during the oxidation of *trans*-2-butene to *trans*-2-butene epoxide under otherwise identical conditions. Better conversion of the *cis*-form than the *trans*-form of an alkenyl was also observed for the *Aae*UPO-catalyzed oxidation of styrene derivatives [89]. Thus, *trans*-β-Methylstyrene was oxidized to some extent but only at the terminal carbon. In contrast, *cis*-β-methylstyrene was almost completely oxidized to (1*R*,2*S*)-*cis*-β-methylstyrene epoxide (>99 % ee) as the sole product. Considering these results, it can be concluded that *cis-trans* isomerism strongly influences positioning of alkenes in the active site of UPOs and hence their oxidizibility.

Complex product patterns were observed as a result of the oxidation of the monoterpene limonene (1-isopropenyl-4-methyl-cyclohexane). Both enantiomers, (*R*)-(+)- and (*S*)-(−)-limonene, were rapidly oxidized by *Aae*UPO, which led to the formation of mixtures of alcohol (carveol) and epoxide products (1,2- and 8,9-limonene epoxides) with different ratios of enantiomers and diastereomers (Fig. 13.15) [38, 81].

Aromatic oxygenation was initially studied with naphthalene and toluene as substrates (see also Sect. 13.4.4.1) [32]. Naphthalene is regioselectively epoxidized by different UPOs to naphthalene 1,2-oxide that hydrolyzes in the presence of protons (pH <7.5) to 1- and 2-naphthol. The ratio of both naphthols varied, which depended on the pH, on the manner of H_2O_2 supply and surprisingly also, on the UPO used, indicating the possibility that the active sites somehow affect epoxide hydrolysis [33, 34, 43, 46, 55]. Other polycyclic aromatic hydrocarbons (PAHs) such as methylnaphthalenes, fluorene, anthracene, phenanthrene, pyrene and dibenzofuran were also subject to UPO-catalyzed oxygenation leading to mixtures of mono- and polyhydroxylated products [45]. Differences were observed in the efficiency of oxidation of aromatic vs. non-aromatic carbons. While *Aae*UPO clearly favors the attack on aromatic rings, *Cra*UPO and *Mro*UPO oxidize preferably alkyl side chains or methylene groups in non-aromatic rings (e.g. C9 in fluorene). In the case of *Aae*UPO, the upper limit of molecule size is reached with benzo[*a*]pyrene that is oxidized to a minor degree and only in the presence of high amounts of co-solvents (e.g. acetonitrile).

Eventually, benzene was also oxidized by *Aae*UPO, despite experimental difficulties due to its high volatility and low reactivity. The reaction proceeds via an initial epoxide intermediate that re-aromatizes in aqueous solution to form phenol. Identity of this intermediate as benzene epoxide (that is in equilibrium with oxepine) was proven by a freshly prepared authentic standard [88]. A second and third oxygenation was also observed and resulted in the formation of hydroquinone, catechol and 1,2,4-trihydroxybenzene.

Phenolic products formed during aromatic peroxygenations can be substrates of a subsequent peroxidative activity of UPOs (one-electron oxidations). The phenoxyl radicals formed tend to couple and polymerize. This can be prevented by adding radical scavengers such as ascorbic acid to the reaction mixture. Ascorbic acid reacts with phenoxyl radicals to yield ascorbyl radicals that in turn can disproportionate to dehydroascorbic acid and ascorbic acid. In other words, the phenoxyl radical abstracts one electron from ascorbic acid followed by rapid

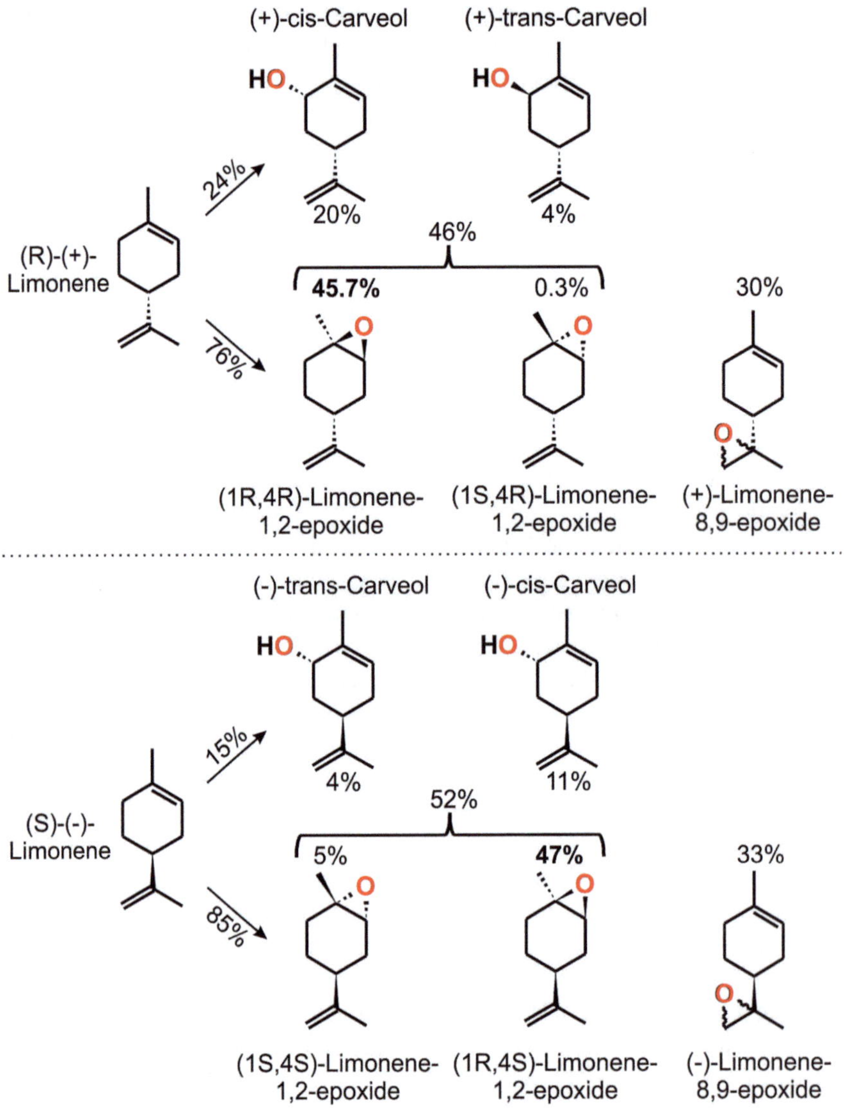

Fig. 13.15 Oxidation of (*R*)-(+)-limonene and (*S*)-(−)-limonene by *Aae*UPO yielding (+)-*cis*-limonene epoxide and (−)-*trans*-1,2-limonene epoxide as major products, respectively (Modified according to [38, 81])

proton rebound that again produces the phenol (Fig. 13.16).

The re-reduction of phenoxyl radicals is of particular relevance when polyphenolic substrates such as flavonoids are oxygenated. Thus, in the presence of ascorbic acid, different flavones, flavonols, flavanones and isoflavones can be hydroxylated by *Aae*UPO, preferably at the C6 position. As an example, Fig. 13.17 shows the hydroxylation of quercetin, a pentahydroxy-flavonol widely distributed in plants [94]. The reaction can proceed via a very unstable epoxide intermediate (7-oxybicyclo[4.1.0]hepta-2,4-diene-2,6,-diol) (Fig. 13.18) and yields quercetagetin (6-hydroxyquercetin) as the sole product. Initial epoxide formation could be demonstrated during the oxidation of unsubstituted flavone to 6-hydroxyflavone [94].

In contrast' to propylbenzene that is hydroxylated at the benzylic carbon (see above and [89]), 2-phenoxypropionic acid is not attacked in the side chain but exclusively in the

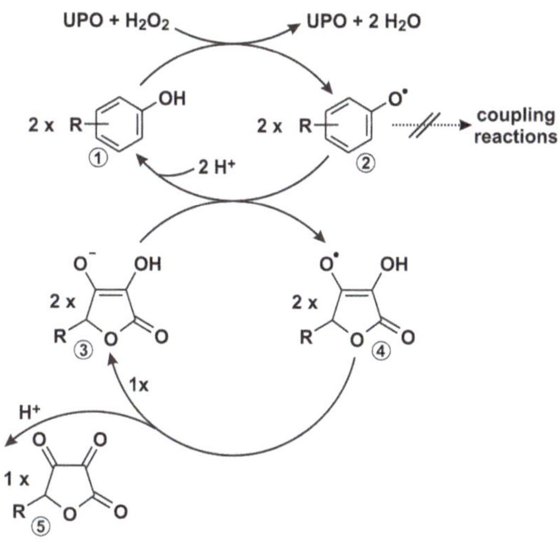

Fig. 13.16 Re-reduction of phenoxyl radicals by ascorbic acid during UPO-catalyzed oxygenations yielding phenolic products. (*1*) phenolic substrate, (*2*) phenoxyl radical, (*3*) ascorbic acid (at pH 7), (*4*) ascorbyl radical, (*5*) dehydroascorbic acid

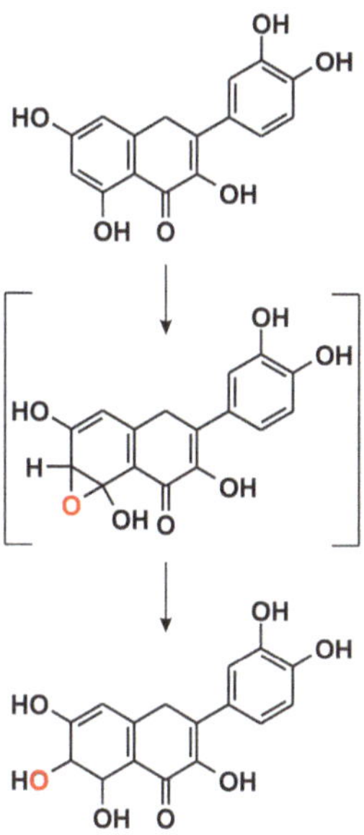

Fig. 13.17 Regioselective oxidation of quercitin by *Aae*UPO in the presence of ascorbic acid via a hypothetical epoxide intermediate into 6-hydroxyquercitin (Based on [94])

para-position on the aromatic ring to yield 2-(4-hydroxyphenoxy)propionic acid. The latter compound is a herbicide precursor and is only formed in appreciable amounts in the presence of ascorbic acid. Chiral analyses after *Aae*UPO-catalyzed oxidation of racemic 2-phenoxypropionic acid revealed that both enantiomers were hydroxylated, but that the (*R*)-enantiomer was clearly the preferred substrate [95]. This interesting finding shows that the spatial orientation of polar side chains can influence the regioselectivity of UPOs and the extent of aromatic ring oxidation.

13.4.4.3 Dealkylation

UPOs catalyze *O*- and *N*-dealkylations of diverse ethers and secondary/tertiary amines, respectively. The mechanism involves, in both cases, initial hydroxylation of one of the heteroatoms' adjacent carbons (e.g. methyl or methylene groups) giving rise to unstable intermediates (hemiacetals, hemiaminals: –HCOH–O–CH_2– or –HCOH–NH-CH_2–, respectively), which spontaneously cleave under release of water. Thus, hemiacetals yield alcohols/phenols and aldehydes, and hemiaminals generate primary or secondary amines and aldehydes. In both cases, the aldehydes indicative for this mechanism can be detected by their corresponding 2,4-dinitrohydrazone adducts [86, 96].

Ether cleavage occurred between aromatic and aliphatic molecules in alkyl aryl ethers (e.g. 1,4-dimethoxybenzene, 1,4-dipropoxybenzene) and in alicyclic and aliphatic ethers (e.g. tertrahydrofuran, dioxane, diisopropyl ether, methyl *t*-butyl ether) [86]. The incorporation of peroxide-borne oxygen into the carbonyl fission product was demonstrated using methyl *p*-nitrobenzyl ether [97, 98][8] and $H_2{}^{18}O_2$ as substrate and cosubstrate, respectively (Fig. 13.18) [86].

As in the case of symmetrically deuterated *n*-hexane (hydroxylation of *n*-hexane-1,1,1,2,2,3,3-D_7 to 3-hexanol-D_7 and 3-hexanol-D_6) [68], a strong intramolecular isotope effect [$(k_H/k_D)_{obs}$ >10] was observed during the *O*-demethylation

[8] Usually, oxygen in aldehyde functionalities rapidly exchanges in water via the corresponding aldehyde hydrates, which prevents the verification of oxygen insertion, but aromatic nitro groups as in *p*-nitrobenzaldehyde slow down the exchange.

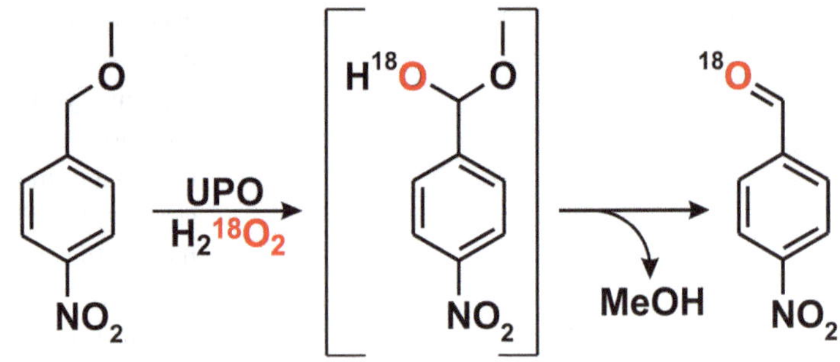

Fig 13.18 Cleavage of methyl *p*-nitrobenzyl ether by *Aae*UPO in the presence of $H_2{}^{18}O_2$ resulting in the formation of ^{18}O-labeled *p*-nitrobenzaldehyde (Based on [86])

of 1-methoxy-4-trideuteromethoxybenzene, indicating in both cases an H-abstraction/oxygen rebound mechanism for oxygen insertion (compare also Fig. 13.10) [86].

Substantial *N*-dealkylation (~60 %) was observed during *N*-methylaniline oxidation by *Aae*UPO along with ring hydroxylation, yielding phenolic products [79]. Other examples of *N*-dealkylated substrates are found among pharmaceuticals such as lidocaine, tamoxifen, methamphetamine and sildenafil [48]. With the two latter drugs ("Crystal meth" and Viagra), the *N*-demethylated metabolites, amphetamine and *N*-desmethyl sildenafil, respectively, formed in the human body by hepatic P450s (CYP2D6), are the actual effective ingredients [99, 100].

13.4.4.4 Additional Reactions and Scope of UPO Oxidations

UPOs are also capable of transferring oxygen to organic heteroatoms such as sulfur and nitrogen. For example, the heterocycle dibenzothiophene is oxidized at the sulfur atom to form the corresponding sulfoxide and sulfone [44]. Differences were observed in the product pattern between *Aae*UPO and *Cra*UPO. While the former enzyme preferably hydroxylated the benzene rings of dibenzothiophene, the latter preferred the heterocyclic sulfur substrate [45]. In a similar reaction, UPO enantioselectively oxidized the side chain of thioanisole into the corresponding (*R*)-sulfoxide with high efficiency [101]. Pyridine and halo-, nitro- and cyanopyridines are oxidized by *Aae*UPO exclusively at the nitrogen atom to form the respective pyridine *N*-oxides. In contrast, methylated pyridines were oxygenated both at the methyl group and at the ring nitrogen [37].

In addition to epoxidation, the formation of naphthalene hydrates (i.e. 1- and 2-hydroxy-1,2-dihydronaphthalene) displays a side activity in the enzymatic transformation of 1,2-dihydronaphthalene by UPOs and accounts for up to 20 % of overall turnover. These arene hydrates decay into naphthalene via spontaneous aromatization. This reaction sequence represents a simple pathway for the selective synthesis of aromatic hydrocarbons via arene hydrates of conjugated cyclic dienes or cycloalkenyl benzenes [102].

*Aae*UPO shows strong bromide oxidation but, in contrast to CPO, only a very low chloride oxidation, even though (according to studies of compound I) its redox potential is higher than that of CPO [67]. The oxidation of halides (X^-) is actually also an oxygen transfer reaction yielding reactive hypohalites (OX^-) that in turn can halogenate organic substrates such as phenols [32]. In contrast to *Aae*UPO, *Mro*UPO has almost no bromide oxidizing activity, indicating that not all peroxygenases have specific halide binding sites [46].

Halogens bound to a carbon atom undergoing hydroxylation are released as the corresponding halides, because the first intermediate (geminal halohydrin) is unstable. An example is the oxidation of chloromethylbenzene (benzylchloride) by *Aae*UPO that yields benzaldehyde and chloride [79]. The analogous reaction has been observed for benzylfluoride and studied with respect to cryptic stereoselectivity. It emerged from this study that *Aae*UPO displays a modest stereospecificity for benzylic *pro*-(*R*) C-H abstraction, although the fluorine atom displays a much-reduced steric influence relative to the methyl group when ethylbenzene is

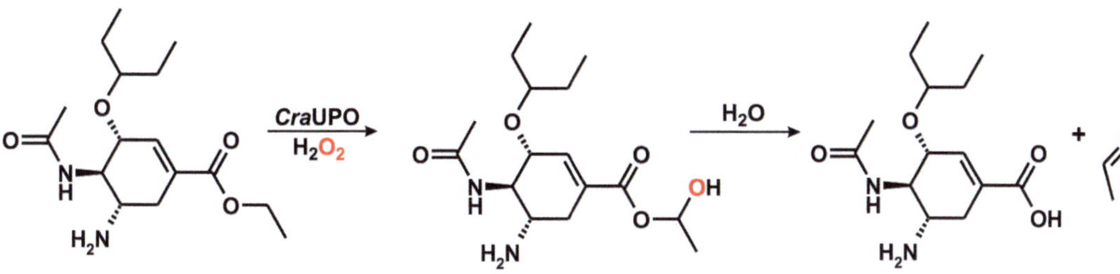

Fig. 13.19 Cleavage of the antiviral drug osaltamivir to the respective carboxylate and acetaldehyde by *Cra*UPO

used as a substrate (Keddie and Kluge 2013, unpublished results).

The promiscuity of peroxygenases in oxygen atom transfer reactions becomes evident when the oxidation of pharmaceuticals and drugs is examined. All reactions mentioned previously can occur and, therefore, the enzymes catalyze, aromatic and aliphatic peroxygenations, *O*- and *N*-dealkylations and even cleavage of ester bonds, depending on the drug used. Altogether, more than 60 different pharmaceuticals and a number of illicit drugs have been shown to undergo oxidative modification by UPOs. Examples for the former agents are the painkillers diclofenac (phenyl hydroxylation) and ibuprofen (isopropyl hydroxylation), the antitussive dextromethorphan (*O*-demethylation), the β-blocker propranolol (naphthyl hydroxylation), the K^+-channel blocker tolbutamide (benzylic hydroxylation), the anti-inflammatory aminophenazone (*N,N*-desmethylation) and the antiviral drug osaltamivir [47, 48]. Osaltamivir is a particularly interesting example, because this ethyl ester is exclusively cleaved by *Cra*UPO [48]. The reaction is a special case of *O*-dealkylation and leads to the formation of acetaldehyde and osaltamivir carboxylate (Fig. 13.19). Among the drugs (of abuse) that are oxidized by UPOs are MDMA ("Ecstasy", demethylenation), LSD (aromatic hydroxylation), THC (methylcyclohexenyl hydroxylation) and cocaine and codeine (*N*-desmethylation). UPOs have also successfully been used to prepare specifically labeled human drug metabolites and drug-drug interaction probes by using deuterated substrates as starting materials [96].

A very recent study on the UPO-catalyzed transformation of 13 steroids revealed considerable differences between the three model UPOs. Whereas *Aae*UPO and *Cra*UPO did not attack any steroid structure, *Mro*UPO oxidized ten of the steroids by 50–100 %. In addition to hydroxylation products, there are mass-spectral indications that, in some cases (e.g. cortisone), the side chain can be removed by C-C bond cleavage ([47] and unpublished results). Whether similar complex reaction sequences are responsible for this cleavage, as in the case of P450s (CYP17A1) [103], is still under investigation.

As already indicated above, there are limitations in the performance of the currently known UPOs. In summary, the following structural characteristics prevent or impede an attack by UPOs: (1) molecular size (e.g. perylene, polyethylene glycol $\geq$ PEG_7); (2) polarity of the substrate (e.g. rutin); (3) abstractability of hydrogen (e.g. biphenyl ether); and specific (still not understood) characteristics of the substrate, as in the case of coumarin.

13.4.4.5 Kinetic Data and Catalytic Performance

A summary of kinetic data of *Aae*UPO for a representative number of substrates and reaction types is given in Table 13.3. More information can be retrieved in the Handbook of Enzymes [104]. Most of the values are apparent, i.e. they were obtained by varying the concentration of one substrate while keeping the concentration of the second substrate (in most cases 1–2 mM H_2O_2) constant [37]. More precise bisubstrate kinetics, which facilitate steady-state conditions and circumvent interfering catalase activity by

Table 13.3 Apparent kinetic data of three model UPOs for different substrates and reaction types compared to chloroperoxidase of *Leptoxyphium fumago* (*Lfu*CPO) and selected P450 enzymes

Substrate	Major product	Enzyme	k_{cat} (s^{-1})	K_m (μM)	k_{cat}/K_m (M^{-1} s^{-1})	pH	Refs.
ABTS	ABTS radical	*Aae*UPO	283	37	7.7×10^6	4.5	[31]
		*Cra*UPO	123	49	2.5×10^6	4.5	[43]
		*Mro*UPO	25	71	3.5×10^5	4.5	[46]
		*Lfu*CPO	N.D.	N.D.	~2×10^{5a}	3.0	[71]
2,6-DMP	Coerulignone	*Mro*UPO	70	133	5.3×10^5	5.5	[46]
		*Aae*UPO	108	298	3.6×10^5	7.0	[31]
		*Cra*UPO	2	342	5.9×10^3	4.5	[43]
Cyclohexanol	Cyclohexanone	*Mro*UPO	31	1,844	1.7×10^4	7.0	[90]
		*Aae*UPO	5	4,977	9.7×10^2	7.0	[90]
		r*Cci*UPO	3	6,571	3.9×10^2	7.0	[90]
Benzyl alcohol	Benzaldehyde	*Mro*UPO	62	118	5.3×10^5	5.5	[46]
		*Cra*UPO	176	635	2.8×10^5	7.0	[43]
		*Aae*UPO	269	1,001	2.7×10^5	7.0	[31]
		*Lfu*CPO	17	1,300	1.3×10^4	6.0	[122]
		P450 2E1[b]	0.06	450	1.3×10^2	7.4	[123]
		P450 2B4[b]	0.06	7,280	7.7×10^0	7.4	[123]
Cyclohexane	Cyclohexanol	*Aae*UPO	72	994	7.2×10^4	7.0	[90]
		*Mro*UPO	43	2,242	4.3×10^4	7.0	[90]
		r*Cci*UPO	13	397	3.2×10^4	7.0	[90]
Ethylbenzene	*R*-1-Phenylethanol	*Aae*UPO	410	694	5.9×10^5	7.0	[89]
Propylbenzene	*R*-1-Phenylpropanol	*Aae*UPO	194	480	4.1×10^5	7.0	[89]
Naphthalene	1-Naphthol	*Aae*UPO	166	320	5.2×10^5	7.0	[33]
		P450 2A13[c]	2.4	36	6.6×10^4	7.4	[124]
		*Mro*UPO	33	791	4.3×10^4	5.5	[46]
		P450 2A6[c]	0.72	23	3.1×10^4	7.4	[124]
		*Cra*UPO	15	584	2.6×10^4	7.0	[43]
		P450 1A1[c]	0.28	244	1.2×10^3	7.4	[124]

[a]k_{cat}/pseudo-K_M

[b]P450s 2B4 and 2E1 originated from rabbit liver and can be reduced by NADPH via CPR

[c]These P450 enzymes metabolize aromatic environmental chemicals in the human liver and respiratory tract, can be expressed in *E. coli*, and need a cooperating reductase such as CPR and a NADPH generating system for maximum activity

varying the concentration of both substrate and co-substrate (H_2O_2), have been reported for the cleavage of methyl 3,4-dimethoxybenzyl ether and the demethylenation of 5-nitro-1,3-benzodioxole (compare Fig. 13.11). The results obtained for these assay substrates are consistent with a ping-pong mechanism that is also characteristic for one-electron oxidations catalyzed by heme peroxidases [47, 69, 86, 105].

Catalytic efficiencies (k_{cat} /K_m), Michaelis-Menten constants (K_m) and turnover numbers (k_{cat}) of UPOs vary in a broad range (10^3–10^6 M^{-1} s^{-1}, 10^1–10^4 μM, and 10^{-1}–10^3 s^{-1}, respectively). Most substrates, however, are oxidized with a catalytic efficiency of ~10^4–10^5 M^{-1} s^{-1}, which again fits classical peroxidases rather than P450 monooxygenases. Compared to the latter, the turnover numbers of UPOs are generally higher (~10- to 1,000-times) whereas the substrate affinities are several times lower (Table 13.3).

The true peroxygenase nature of UPOs has experimentally been verified in various experiments by the incorporation of ^{18}O from

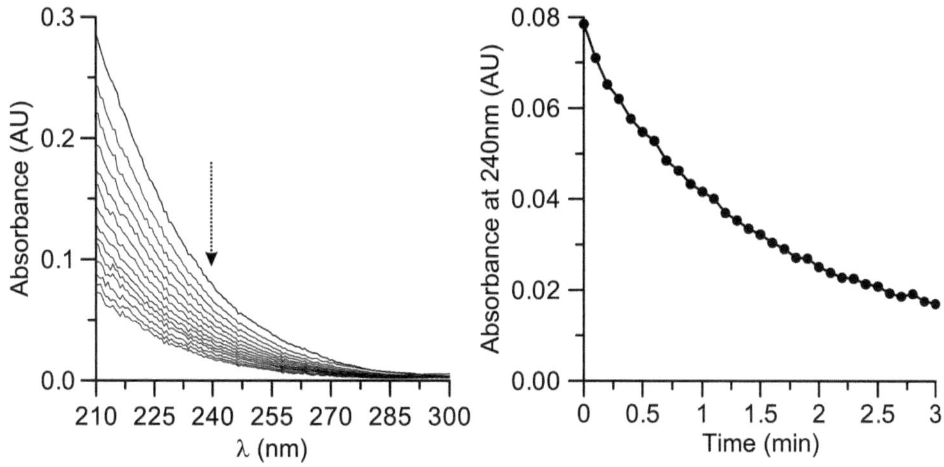

Fig. 13.20 Catalase activity of *Aae*UPO. The reaction solution contained phosphate buffer (10 mM, pH 7), *Aae*UPO (0.2 μM) and H_2O_2 (2 mM). The graphs show the decrease in absorbance over time caused by H_2O_2 decay. Spectral time scans (every 6 s; *left*), decrease at 240 nm (*right*)

$H_2{}^{18}O_2$ into chemically diverse substrate molecules. The following reactions are just a few mentioned in this context: (1) toluene to benzyl alcohol, (2) benzaldehyde to benzoic acid [92], (3) naphthalene to naphthalene oxide and naphthols [34], (4) benzene to phenol [88], (5) pyridine to pyridine *N*-oxide [37] and (6) cyclohexane to cyclohexanol [90].

Ideally, the ratio between oxygenated product and peroxide consumed should be 1:1. In fact, there are examples where this ratio has been reached (e.g. tetrahydrofuran and methyl 3,4-dimethoxybenzyl ether cleavage) [86]. On the other hand, the ratio can be altered to the disadvantage of the substrate to be oxidized and much more peroxide is consumed than is actually necessary for peroxygenation. Then, the catalase activity of *Aae*UPO takes effect and consumes a substantial part of the peroxide without "productive" oxygen atom transfer (Fig. 13.20) [31, 106]. The reason for this activity may be attributed to improper binding of the substrate in the active site, thermodynamic/kinetic obstacles or by competing one-electron oxidations.

In summary, regarding their key oxygenating activities (oxygen transfer potential), the three characterized model UPOs and CPO can be grouped as follows [43, 46]: (1) aromatic oxygenation (*Aae*UPO > *Cra*UPO > *Mro*UPO; CPO does not oxygenate aromatics); (2) alkane/alkyl hydroxylation (*Mro*UPO > *Aae*UPO ~ *Cra*UPO >> CPO); (3) alkene/alkenyl epoxidation (*Aae*UPO > *Cra*UPO ~ *Mro*UPO > CPO); and (4) halide oxidation (CPO > *Aae*UPO > *Cra*UPO ~ *Mro*UPO).

13.5 Conclusions

M. J. Coon has called P450 enzymes "*nature's most versatile biological catalysts*" in his excellent review from 2005 [107]. We do not wish to call this statement into question, as a number of P450 reactions such as the oxidation of coumarin [108], terminal alkane hydroxylation [109] or the aromatase reaction [110] have not yet been shown to be catalyzed by UPOs. However, fungal peroxygenases can at least approach the catalytic versatility of P450s and suitably supplement them in the field of biotechnology in the near future. Some applications of UPOs that are currently under development are biosensors for aromatic compounds [111, 112] as well as new procedures for the synthesis of pesticide precursors [113], drug metabolites [48, 95, 96], chiral alcohols [89] and even bulk chemicals [90].

Acknowledgments We would like to thank K. Barkova, M. G. Kluge, S. Peter, C. Dolge and M. Poraj-Kobielska (TU Dresden-IHI Zittau, Germany) for still unpublished results on the catalytic properties of unspecific peroxygenases and our project partners L. Kalum and H. Lund (Novozymes A/S, Denmark) for enzyme samples as well as useful discussions. We acknowledge fruitful cooperations with the following colleagues: K. Piontek and D. Plattner (University of Freiburg, Germany) in the field of protein crystallography, X. Wang and J. T. Groves (Princeton University, USA) regarding stopped-flow techniques, and J. Atzrodt and W. Holla (Sanofi Frankfurt, Germany) in the field of drug metabolites. UPO work has been financially supported by the European Union (integrated projects BIORENEW, PEROXICATS and INDOX), the Deutsche Bundestiftung Umwelt (DBU; projects AZ 1327 and AZ 13225) and the Bundesministerium für Forschung (BMBF, projects 0313433 and 0315877).

References

1. Ishimaru A, Yamazaki I (1977) Hydroperoxide-dependent hydroxylation involving "H_2O_2-reducible hemoprotein" in microsomes of pea seeds. A new type enzyme acting on hydroperoxide and a physiological role of seed lipoxygenase. J Biol Chem 252:6118–6124
2. Hanano A, Burcklen M, Flenet M, Ivancich A, Louwagie M, Garin J, Blee E (2006) Plant seed peroxygenase is an original heme-oxygenase with an EF-hand calcium binding motif. J Biol Chem 281:33140–33151
3. Lequeu J, Fauconnier ML, Chammai A, Bronner R, Blee E (2003) Formation of plant cuticle: evidence for the occurrence of the peroxygenase pathway. Plant J 36:155–164
4. Salazar O, Cirino PC, Arnold FH (2003) Thermostabilization of a cytochrome P450 peroxygenase. ChemBioChem 4:891–893
5. Estabrook RW, Martin-Wixtrom C, Saeki Y, Renneberg R, Hildebrandt A, Werringloer J (1984) The peroxidatic function of liver microsomal cytochrome P-450: comparison of hydrogen peroxide and NADPH-catalysed *N*-demethylation reactions. Xenobiotica 14:87–104
6. McCallum GP, Weedon AC, Krug P, Bend JR (1996) Microsomal cytochrome P450 peroxygenase metabolism of arachidonic acid in guinea pig liver. J Pharm Exp Ther 278:1188–1194
7. Coon MJ, Vaz AD, Bestervelt LL (1996) Cytochrome P450 2: peroxidative reactions of diversozymes. FASEB J 10:428–434
8. Prasad S, Mitra S (2004) Substrate modulates compound I formation in peroxide shunt pathway of *Pseudomonas putida* cytochrome P450(cam). Biochem Biophys Res Commun 314:610–614
9. Nordblom GD, White RE, Coon MJ (1976) Studies on hydroperoxide-dependent substrate hydroxylation by purified liver microsomal cytochrome P-450. Arch Biochem Biophys 175:524–533
10. Sakaki T (2012) Practical application of cytochrome P450. Biol Pharm Bull 35:844–849
11. Zenser TV, Lakshmi VM, Hsu FF, Davis BB (1999) Peroxygenase metabolism of *N*-acetylbenzidine by prostaglandin H synthase. Formation of an *N*-hydroxylamine. J Biol Chem 274:14850–14856
12. Kuo HH, Mauk AG (2012) Indole peroxygenase activity of indoleamine 2,3-dioxygenase. Proc Natl Acad Sci U S A 109:13966–13971
13. Yamazaki S, Morioka C, Itoh S (2004) Kinetic evaluation of catalase and peroxygenase activities of tyrosinase. Biochemistry 43:11546–11553
14. Matsunaga I, Yamada M, Kusunose E, Miki T, Ichihara K (1998) Further characterization of hydrogen peroxide-dependent fatty acid α-hydroxylase from *Sphingomonas paucimobilis*. J Biochem 124:105–110
15. Lee DS, Yamada A, Sugimoto H, Matsunaga I, Ogura H, Ichihara K, Adachi S, Park SY, Shiro Y (2003) Substrate recognition and molecular mechanism of fatty acid hydroxylation by cytochrome P450 from *Bacillus subtilis*. Crystallographic, spectroscopic, and mutational studies. J Biol Chem 278:9761–9767
16. Matsunaga I, Sumimoto T, Ueda A, Kusunose E, Ichihara K (2000) Fatty acid-specific, regiospecific, and stereospecific hydroxylation by cytochrome P450 (CYP152B1) from *Sphingomonas paucimobilis*: substrate structure required for α-hydroxylation. Lipids 35:365–371
17. Shoji O, Wiese C, Fujishiro T, Shirataki C, Wunsch B, Watanabe Y (2010) Aromatic C-H bond hydroxylation by P450 peroxygenases: a facile colorimetric assay for monooxygenation activities of enzymes based on Russig's blue formation. J Biol Inorg Chem 15:1109–1115
18. Fujishiro T, Shoji O, Kawakami N, Watanabe T, Sugimoto H, Shiro Y, Watanabe Y (2012) Chiral-substrate-assisted stereoselective epoxidation catalyzed by H_2O_2-dependent cytochrome $P450_{SP\alpha}$. Chem Asian J 7:2286–2293
19. Shoji O, Kunimatsu T, Kawakami N, Watanabe Y (2013) Highly selective hydroxylation of benzene to phenol by wild-type cytochrome P450BM3 assisted by decoy molecules. Angew Chem Int Ed 52:6606–6610
20. Gaut JP, Yeh GC, Tran HD, Byun J, Henderson JP, Richter GM, Brennan ML, Lusis AJ, Belaaouaj A, Hotchkiss RS, Heinecke JW (2001) Neutrophils employ the myeloperoxidase system to generate antimicrobial brominating and chlorinating oxidants during sepsis. Proc Natl Acad Sci U S A 98:11961–11966
21. Tuynman A, Spelberg JL, Kooter IM, Schoemaker HE, Wever R (2000) Enantioselective epoxidation

and carbon-carbon bond cleavage catalyzed by *Coprinus cinereus* peroxidase and myeloperoxidase. J Biol Chem 275:3025–3030
22. Geigert J, Lee TD, Dalietos DJ, Hirano DS, Neidleman SL (1986) Epoxidation of alkenes by chloroperoxidase catalysis. Biochem Biophys Res Commun 136:778–782
23. Miller VP, Tschirretguth RA, Ortiz de Montellano P (1995) Chloroperoxidase-catalyzed benzylic hydroxylation. Arch Biochem Biophys 319:333–340
24. Colonna S, Gaggero N, Casella L, Carrea G, Pasta P (1992) Chloroperoxidase and hydrogen peroxide: an efficient system for enzymatic enantioselective sulfoxidations. Tetrahedron Asymmetry 3:95–106
25. Manoj KM, Hager LP (2001) Utilization of peroxide and its relevance in oxygen insertion reactions catalyzed by chloroperoxidase. Biochim Biophys Acta 1547:408–417
26. Zhang R, He Q, Chatfield D, Wang X (2013) Paramagnetic nuclear magnetic resonance relaxation and molecular mechanics studies of the chloroperoxidase-indole complex: insights into the mechanism of chloroperoxidase-catalyzed regioselective oxidation of indole. Biochemistry 52:3688–3701
27. Hofrichter M, Ullrich R (2006) Heme-thiolate haloperoxidases: versatile biocatalysts with biotechnological and environmental significance. Appl Microbiol Biotechnol 71:276–288
28. Stamets P, Chilton JS (1983) The mushroom cultivator: a practical guide to growing mushrooms at home. Agarikon Press, Olympia
29. IndexFungorum (2013) www.indexfungorum.org
30. Manzi P, Marconi S, Aguzzi A, Pizzoferrato L (2004) Commercial mushrooms: nutritional quality and effect of cooking. Food Chem 84:201–206
31. Ullrich R, Nüske J, Scheibner K, Spantzel J, Hofrichter M (2004) Novel haloperoxidase from the agaric basidiomycete *Agrocybe aegerita* oxidizes aryl alcohols and aldehydes. Appl Environ Microbiol 70:4575–4581
32. Ullrich R, Hofrichter M (2005) The haloperoxidase of the agaric fungus *Agrocybe aegerita* hydroxylates toluene and naphthalene. FEBS Lett 579:6247–6250
33. Kluge MG, Ullrich R, Scheibner K, Hofrichter M (2007) Spectrophotometric assay for detection of aromatic hydroxylation catalyzed by fungal haloperoxidase-peroxygenase. Appl Microbiol Biotechnol 75:1473–1478
34. Kluge M, Ullrich R, Dolge C, Scheibner K, Hofrichter M (2009) Hydroxylation of naphthalene by aromatic peroxygenase from *Agrocybe aegerita* proceeds via oxygen transfer from H_2O_2 and intermediary epoxidation. Appl Microbiol Biotechnol 81:1071–1076
35. Ullrich R, Hofrichter M (2007) Enzymatic hydroxylation of aromatic compounds. Cell Mol Life Sci 64:271–293
36. Hofrichter M, Ullrich R (2010) New trends in fungal biooxidation. In: Hofrichter M (ed) Industrial applications, 2nd edn. Springer-Verlag, Berlin, pp 425–449
37. Ullrich R, Dolge C, Kluge M, Hofrichter M (2008) Pyridine as novel substrate for regioselective oxygenation with aromatic peroxygenase from *Agrocybe aegerita*. FEBS Lett 582:4100–4106
38. Peter S, Kinne M, Ullrich R, Kayser G, Hofrichter M (2013) Epoxidation of linear, branched and cyclic alkenes catalyzed by unspecific peroxygenase. Enzyme Microb Technol 52:370–376
39. Ruiz-Dueñas FJ, Martínez AT (2010) Structural and functional features of peroxidases with a potential as industrial biocatalysts. In: Torres E, Ayala M (eds) Biocatalysis based on heme peroxidases as potential industrial biocatalysts, 1st edn. Springer-Verlag, Berlin, pp 37–59
40. Hofrichter M, Ullrich R, Pecyna MJ, Liers C, Lundell T (2010) New and classic families of secreted fungal heme peroxidases. Appl Microbiol Biotechnol 87:871–897
41. Gutiérrez A, Babot ED, Ullrich R, Hofrichter M, Martínez AT, del Río JC (2011) Regioselective oxygenation of fatty acids, fatty alcohols and other aliphatic compounds by a basidiomycete heme-thiolate peroxidase. Arch Biochem Biophys 514:33–43
42. Moncalvo J-M, Vilgalys R, Redhead SA, Johnson JE, James TY, Catherine Aime M, Hofstetter V, Verduin SJW, Larsson E, Baroni TJ, Greg Thorn R et al (2002) One hundred and seventeen clades of euagarics. Mol Phyl Evol 23:357–400
43. Anh DH, Ullrich R, Benndorf D, Svatos A, Muck A, Hofrichter M (2007) The coprophilous mushroom *Coprinus radians* secretes a haloperoxidase that catalyzes aromatic peroxygenation. Appl Environ Microbiol 73:5477–5485
44. Aranda E, Kinne M, Kluge M, Ullrich R, Hofrichter M (2009) Conversion of dibenzothiophene by the mushrooms *Agrocybe aegerita* and *Coprinellus radians* and their extracellular peroxygenases. Appl Microbiol Biotechnol 82:1057–1066
45. Aranda E, Ullrich R, Hofrichter M (2009) Conversion of polycyclic aromatic hydrocarbons, methyl naphthalenes and dibenzofuran by two fungal peroxygenases. Biodegradation 21:267–281
46. Gröbe G, Ullrich R, Pecyna MJ, Kapturska D, Friedrich S, Hofrichter M, Scheibner K (2011) High-yield production of aromatic peroxygenase by the agaric fungus *Marasmius rotula*. AMB Express 1:31
47. Poraj-Kobielska M (2013) Conversion of pharmaceuticals and other drugs by fungal peroxygenases. Ph.D. Thesis, TU Dresden; http://www.ihi-zittau.de/de/dnl/diss_poraj-kobielska_qucosa.4268.pdf
48. Poraj-Kobielska M, Kinne M, Ullrich R, Scheibner K, Kayser G, Hammel KE, Hofrichter M (2011) Preparation of human drug metabolites using

fungal peroxygenases. Biochem Pharmacol 82:789–796
49. Babot ED, del Río JC, Kalum L, Martínez AT, Gutiérrez A (2013) Oxyfunctionalization of aliphatic compounds by a recombinant peroxygenase from *Coprinopsis cinerea*. Biotechnol Bioeng 110:2323–2332
50. Pecyna MJ, Ullrich R, Bittner B, Clemens A, Scheibner K, Schubert R, Hofrichter M (2009) Molecular characterization of aromatic peroxygenase from *Agrocybe aegerita*. Appl Microbiol Biotechnol 84:885–897
51. Nuell MJ, Fang GH, Axley MJ, Kenigsberg P, Hager LP (1988) Isolation and nucleotide sequence of the chloroperoxidase gene from *Caldariomyces fumago*. J Bacteriol 170:1007–1011
52. Richards TA, Soanes DM, Jones MD, Vasieva O, Leonard G, Paszkiewicz K, Foster PG, Hall N, Talbot NJ (2011) Horizontal gene transfer facilitated the evolution of plant parasitic mechanisms in the oomycetes. Proc Natl Acad Sci U S A 108:15258–15263
53. Kellner H, Luis P, Pecyna MJ, Barbi F, Kapturska D, Krüger D, Zak DR, Marmeisse R, Vandenbol M, Hofrichter M (2014) Widespread occurrence of expressed fungal secretory peroxidases in forest soils. PLoS One 9(4):e95557. doi:10.1371/journal.pone.0095557
54. Morin E, Kohler A, Baker AR, Foulongne-Oriol M, Lombard V, Nagy LG, Ohm RA, Patyshakuliyeva A, Brun A, Aerts AL, Bailey AM et al (2012) Genome sequence of the button mushroom *Agaricus bisporus* reveals mechanisms governing adaptation to a humic-rich ecological niche. Proc Natl Acad Sci U S A 109:17501–17506
55. Ullrich R, Liers C, Schimpke S, Hofrichter M (2009) Purification of homogeneous forms of fungal peroxygenase. Biotechnol J 4:1619–1626
56. Kinne M, Poraj-Kobielska M, Ullrich R, Nousiainen P, Sipilä J, Scheibner K, Hammel KE, Hofrichter M (2011) Oxidative cleavage of non-phenolic *β*-O-4 lignin model dimers by an extracellular aromatic peroxygenase. Holzforschung 65:673–679
57. Hatakka A, Lundell T, Hofrichter M, Maijala P (2003) Manganese peroxidase and its role in the degradation of wood lignin. In: Mansfield SD, Saddler JN (eds) Applications of enzymes to lignocellulosics, ACS Symposium series, 855th edn. American Chemical Society, Washington DC, pp 230–243
58. Liers C, Arnstadt T, Ullrich R, Hofrichter M (2011) Patterns of lignin degradation and oxidative enzyme secretion by different wood- and litter-colonizing basidiomycetes and ascomycetes grown on beech-wood. FEMS Microbiol Ecol 78:91–102
59. Piontek K, Strittmatter E, Ullrich R, Gröbe G, Pecyna MJ, Kluge M, Scheibner K, Hofrichter M, Plattner DA (2013) Structural basis of substrate conversion in a new aromatic peroxygenase: cytochrome P450 functionality with benefits. J Biol Chem 288:34767–34776
60. Anh DH (2008) Novel extracellular haloperoxidase-peroxygenases from the coprophilous fungi *Coprinus radians* and *Coprinus verticillatus*: production, purification and biochemical characterization. Ph.D. Thesis, International Graduate School of Zittau
61. Omura T (2005) Heme-thiolate proteins. Biochem Biophys Res Commun 338:404–409
62. Piontek K, Ullrich R, Liers C, Diederichs K, Plattner DA, Hofrichter M (2010) Crystallization of a 45 kDa peroxygenase/peroxidase from the mushroom *Agrocybe aegerita* and structure determination by SAD utilizing only the haem iron. Acta Crystallogr Sect F: Struct Biol Cryst Commun 66:693–698
63. Ullrich V, Kremers P (1977) Multiple forms of cytochrome P450. Arch Toxicol 39:41–50
64. Guengerich FP (1993) Metabolic reactions: types of reactions of cytochrome P450 enzymes. In: Schenkman J, Greim H (eds) Cytochrome P450, 1st edn. Springer, Berlin, pp 89–103
65. Meunier B, de Visser SP, Shaik S (2004) Mechanism of oxidation reactions catalyzed by cytochrome P450 enzymes. Chem Rev 104:3947–3980
66. Wang X, Peter S, Kinne M, Hofrichter M, Groves JT (2012) Detection and kinetic characterization of a highly reactive heme-thiolate peroxygenase compound I. J Am Chem Soc 134:12897–12900
67. Wang X, Peter S, Ullrich R, Hofrichter M, Groves JT (2013) Driving force for oxygen-atom transfer by heme-thiolate enzymes. Angew Chem Int Ed 52:9238–9241
68. Peter S, Kinne M, Wang X, Ullrich R, Kayser G, Groves JT, Hofrichter M (2011) Selective hydroxylation of alkanes by an extracellular fungal peroxygenase. FEBS J 278:3667–3675
69. Dunford HB (1999) Heme peroxidases. Wiley, New York
70. Ortiz de Montellano PR, De Voss JJ (2005) Substrate oxidation by cytochrome P450 enzymes. In: Ortiz De Montellano PR (ed) Cytochrome P450: structure, mechanism, and biochemistry, 3rd edn. Kluwer Academic/Plenum Publishers, New York, pp 183–245
71. Manoj KM, Hager LP (2008) Chloroperoxidase, a janus enzyme. Biochemistry 47:2997–3003
72. Stone KL, Hoffart LM, Behan RK, Krebs C, Green MT (2006) Evidence for two ferryl species in chloroperoxidase compound II. J Am Chem Soc 128:6147–6153
73. Kühnel K, Derat E, Terner J, Shaik S, Schlichting I (2007) Structure and quantum chemical characterization of chloroperoxidase compound 0, a common reaction intermediate of diverse heme enzymes. Proc Natl Acad Sci U S A 104:99–104
74. Hersleth HP, Ryde U, Rydberg P, Görbitz CH, Andersson KK (2006) Structures of the high-valent metal-ion haem-oxygen intermediates in

peroxidases, oxygenases and catalases. J Inorg Biochem 100:460–476

75. Guengerich FP (2007) Mechanisms of cytochrome P450 substrate oxidation: MiniReview. J Biochem Mol Toxicol 21:163–168
76. Ortiz de Montellano PR (2010) Catalytic mechanisms of heme peroxidases. In: Torres E, Ayala M (eds) Biocatalysis based on heme peroxidases—peroxidases as potential industrial biocatalysts, 1st edn. Springer-Verlag, Berlin, pp 80–107
77. Krest CM, Onderko EL, Yosca TH, Calixto JC, Karp RF, Livada J, Rittle J, Green MT (2013) Reactive intermediates in cytochrome P450 catalysis. J Biol Chem 288:17074–17081
78. Lewis DFV (2001) Guide to cytochromes P450: structure and function, 2nd edn. Informa Healthcare, London
79. Kinne M (2010) The extracellular peroxygenase of the agaric fungus *Agrocybe aegerita*: catalytic properties and physiological background with particular emphasis on ether cleavage. Ph.D. Thesis, International Graduate School of Zittau; http://www.qucosa.de/fileadmin/data/qucosa/documents/6207/Diss_Kinne_final.pdf
80. Isin EM, Guengerich FP (2008) Substrate binding to cytochromes P450. Anal Bioanal Chem 392:1019–1030
81. Peter S (2013) Oxyfunctionalization of alkanes, alkenes and alkynes by unspecific peroxygenase (EC 1.11.2.1). Ph.D. Thesis, TU Dresden; http://www.ihi-zittau.de/de/dnl/dissertation_sebastian_peter.4267.pdf
82. de Visser SP, Ogliaro F, Harris N, Shaik S (2001) Multi-state epoxidation of ethene by cytochrome P450: a quantum chemical study. J Am Chem Soc 123:3037–3047
83. de Visser SP, Ogliaro F, Shaik S (2001) Stereospecific oxidation by compound I of cytochrome P450 does not proceed in a concerted synchronous manner. Chem Commun 22:2322–2323
84. Piontek K, Smith AT, Blodig W (2001) Lignin peroxidase structure and function. Biochem Soc Trans 29:111–116
85. Ruiz-Dueñas FJ, Morales M, Garcia E, Miki Y, Martínez MJ, Martínez AT (2009) Substrate oxidation sites in versatile peroxidase and other basidiomycete peroxidases. J Exp Bot 60:441–452
86. Kinne M, Poraj-Kobielska M, Ralph SA, Ullrich R, Hofrichter M, Hammel KE (2009) Oxidative cleavage of diverse ethers by an extracellular fungal peroxygenase. J Biol Chem 284:29343–29349
87. Poraj-Kobielska M, Kinne M, Ullrich R, Scheibner K, Hofrichter M (2012) A spectrophotometric assay for the detection of fungal peroxygenases. Anal Biochem 421:327–329
88. Karich A, Kluge M, Ullrich R, Hofrichter M (2013) Benzene oxygenation and oxidation by the peroxygenase of *Agrocybe aegerita*. AMB Express 3:5
89. Kluge M, Ullrich R, Scheibner K, Hofrichter M (2012) Stereoselective benzylic hydroxylation of alkylbenzenes and epoxidation of styrene derivatives catalyzed by the peroxygenase of *Agrocybe aegerita*. Green Chem 14:440–446
90. Peter S, Karich A, Ullrich R, Gröbe G, Scheibner K, Hofrichter M (2013) Enzymatic one-pot conversion of cyclohexane into cyclohexanone: comparison of four fungal peroxygenases. J Mol Catal B Enz. doi: 10.1016/j.molcatb.2013.1009.1016
91. Auclair K, Hu Z, Little DM, Ortiz De Montellano PR, Groves JT (2002) Revisiting the mechanism of P450 enzymes with the radical clocks norcarane and spiro[2,5]octane. J Am Chem Soc 124:6020–6027
92. Kinne M, Zeisig C, Ullrich R, Kayser G, Hammel KE, Hofrichter M (2010) Stepwise oxygenations of toluene and 4-nitrotoluene by a fungal peroxygenase. Biochem Biophys Res Commun 397:18–21
93. Churakova E, Kluge M, Ullrich R, Arends I, Hofrichter M, Hollmann F (2011) Specific photobiocatalytic oxyfunctionalization reactions. Angew Chem Int Ed 50:10716–10719
94. Barková K, Kinne M, Ullrich R, Hennig L, Fuchs A, Hofrichter M (2011) Regioselective hydroxylation of diverse flavonoids by an aromatic peroxygenase. Tetrahedron 67:4874–4878
95. Kinne M, Ullrich R, Hammel KE, Scheibner K, Hofrichter M (2008) Regioselective preparation of (*R*)-2-(4-hydroxyphenoxy)propionic acid with a fungal peroxygenase. Tetrahedron Lett 49:5950–5953
96. Poraj-Kobielska M, Atzrodt J, Holla W, Sandvoss M, Gröbe G, Scheibner K, Hofrichter M (2013) Preparation of labeled human drug metabolites and drug-drug interaction-probes with fungal peroxygenases. J Label Comp Radiopharm 56:513–519
97. Tien M, Kirk TK (1984) Lignin-degrading enzyme from *Phanerochaete chrysosporium*: purification, characterization, and catalytic properties of a unique H_2O_2-requiring oxygenase. Proc Natl Acad Sci U S A 81:2280–2284
98. Samuel D, Silver BL (1965) Oxygen isotope exchange reactions of organic compounds. In: Gold V (ed) Advances in physical organic chemistry. Academic Press, New York, pp 123–186
99. Maurer HH, Kraemer T, Springer D, Staack RF (2004) Chemistry, pharmacology, toxicology, and hepatic metabolism of designer drugs of the amphetamine (ecstasy), piperazine, and pyrrolidinophenone types: a synopsis. Ther Drug Monit 26:127–131
100. Langtry HD, Markham A (1999) Sildenafil. Drugs 57:967–989
101. Horn A (2009) The use of a novel peroxidase from the basidiomycete *Agrocybe aegerita* as an example of enatioselective sulfoxidation *(Der Einsatz einer neuartigen Peroxidase des Basidiomyceten Agrocybe aegerita am Beispiel der enatioselektiven Sulfoxidation)*. Ph.D. Thesis, University of Rostock, Germany

102. Kluge M, Ullrich R, Scheibner K, Hofrichter M (2013) Formation of naphthalene hydrates in the enzymatic conversion of 1,2-dihydronaphthalene by two fungal peroxygenases and subsequent naphthalene formation. J Mol Catal B Enz. doi: 10.1016/j.molcatb.2013.1008.1017
103. Akhtar M, Wright JN, Lee-Robichaud P (2011) A review of mechanistic studies on aromatase (CYP19) and 17α-hydroxylase-17,20-lyase (CYP17). J Steroid Biochem Mol Biol 125:2–12
104. Schomburg D, Schomburg I (2013) Class 1, oxidoreductases, EC 1, vol S8, 2nd edn, Springer handbook of enzymes. Springer, Berlin/New York, pp 504–516
105. Segel IH (1993) Enzyme kinetics: behavior and analysis of rapid equilibrium and steady-state enzyme systems. Wiley, New York
106. Ullrich R (2008) Some special reactions of *Agrocybe aegerita* peroxygenase (AaP). Conference proceedings, Tampere (Finland), p 20
107. Coon MJ (2005) Cytochrome P450: nature's most versatile biological catalyst. Annu Rev Pharmacol Toxicol 45:1–25
108. Pelkonen O, Sotaniemi EA, Ahokas JT (1985) Coumarin 7-hydroxylase activity in human liver microsomes. Properties of the enzyme and interspecies comparisons. Br J Clin Pharmacol 19:59–66
109. Johnston JB, Ouellet H, Podust LM, Ortiz de Montellano PR (2011) Structural control of cytochrome P450-catalyzed ω-hydroxylation. Arch Biochem Biophys 507:86–94
110. Yoshio O, Tadayoshi H, Fronckowiak M, Nobutaka Y, Yarborough C (1987) Aromatase. J Steroid Biochem 27:781–789
111. Yarman A, Peng L, Wu Y, Bandodkar A, Gajovic-Eichelmann N, Wollenberger U, Hofrichter M, Ullrich R, Scheibner K, Scheller F (2011) Can peroxygenase and microperoxidase substitute cytochrome P450 in biosensors. Bioanal Rev 3:67–94
112. Peng L, Wollenberger U, Kinne M, Hofrichter M, Ullrich R, Scheibner K, Fischer A, Scheller FW (2010) Peroxygenase based sensor for aromatic compounds. Biosens Bioelectron 26:1432–1436
113. Kinne M, Poraj-Kobielska M, Aranda E, Ullrich R, Hammel KE, Scheibner K, Hofrichter M (2009) Regioselective preparation of 5-hydroxypropranolol and 4′-hydroxydiclofenac with a fungal peroxygenase. Bioorg Med Chem Lett 19:3085–3087
114. Sundaramoorthy M, Terner J, Poulos TL (1995) The crystal structure of chloroperoxidase: a heme peroxidase-cytochrome P450 functional hybrid. Structure 3:1367–1377
115. Strittmatter E, Liers C, Ullrich R, Wachter S, Hofrichter M, Plattner DA, Piontek K (2013) First crystal structure of a fungal high-redox potential dye-decolorizing peroxidase: substrate interaction sites and long-range electron transfer. J Biol Chem 288:4095–4102
116. Dolge C, Sass A, Kayser G, Ullrich R, Hofrichter M (2011) Exploration of r*Cci*APO1 from *Coprinopsis cinerea*: first recombinant aromatic peroxygenase. Conference proceedings, Springer: Karlsruhe, p 132
117. Morris DR, Hager LP (1966) Chloroperoxidase. I. Isolation and properties of the crystalline glycoprotein. J Biol Chem 241:1763–1768
118. Farrell RL, Murtagh KE, Tien M, Mozuch MD, Kirk TK (1989) Physical and enzymatic properties of lignin peroxidase isoenzymes from *Phanerochaete chrysosporium*. Enzyme Microb Technol 11:322–328
119. Hollenberg PF, Hager LP (1973) The P-450 nature of the carbon monoxide complex of ferrous chloroperoxidase. J Biol Chem 248:2630–2633
120. Palcic MM, Rutter R, Araiso T, Hager LP, Dunford HB (1980) Spectrum of chloroperoxidase compound I. Biochem Biophys Res Commun 94:1123–1127
121. Imai Y, Matsunaga I, Kusunose E, Ichihara K (2000) Unique heme environment at the putative distal region of hydrogen peroxide-dependent fatty acid α-hydroxylase from *Sphingomonas paucimobilis* (Peroxygenase $P450_{SP\alpha}$). J Biochem 128:189–194
122. Koo LS, Tschirret-Guth RA, Straub WE, Moënne-Loccoz P, Loehr TM, Ortiz de Montellano PR (2000) The active site of the thermophilic CYP119 from *Sulfolobus solfataricus*. J Biol Chem 275:14112–14123
123. Sheng X, Horner JH, Newcomb M (2008) Spectra and kinetic studies of the compound I derivative of cytochrome P450 119. J Am Chem Soc 130:13310–13320
124. Baciocchi E, Manduchi L, Lanzalunga O (1999) Prochiral selectivity and deuterium kinetic isotope effect in the oxidation of benzyl alcohol catalyzed by chloroperoxidase. Chem Commun 17:1715–1716

MIX
Papier aus verantwortungsvollen Quellen
Paper from responsible sources
FSC® C105338

If you have any concerns about our products,
you can contact us on
ProductSafety@springernature.com

In case Publisher is established outside the EU,
the EU authorized representative is:
Springer Nature Customer Service Center GmbH
Europaplatz 3, 69115 Heidelberg, Germany

Printed by Libri Plureos GmbH
in Hamburg, Germany